ESTATES GAZETTE LAW REPORTS

Estates Gazette Law Reports

1987

Volume 1

**Edited by
J Muir Watt O.B.E., M.A.,**
of the Inner Temple, barrister

Ernest Speller
Publishers' editor

THE ESTATES GAZETTE LIMITED
151 WARDOUR STREET, LONDON W1V 4BN

First Published 1987

ISBN for complete set of 2 volumes: 0 7282 0110 0
ISBN for this volume: 0 7282 0111 9

ISSN 0951-9289

©The Estates Gazette Limited, 1987

Typesetting by Digital Graphics, 147-149 Wardour Street, London
Printed and bound at The Bath Press, Avon

CONTENTS

	Page
Preface	vii
Table of Cases	ix
Index of Subject-matter	xi
Agriculture	1
Arbitration	9
Compulsory Purchase and Compensation	12
Estate Agents	15
Housing Acts	26
Landlord and Tenant — General	30
— Business Tenancies	71
— Leasehold Reform	93
— Rent Acts	99
— Rent Review	112
Negligence	155
Rating	164
Real Property and Conveyancing	190
Town and Country Planning	198
Lands Tribunal — Rating	201

SUPPLEMENT
Cases not reported in "Estates Gazette"

Landlord and Tenant — General	209
— Rent Acts	224
Negligence	231
Rating	248

CONTENTS

	Page
Preface	
Table of Cases	
Index to Subject-matter	
Agriculture	1
Arbitration	9
Compulsory Purchase and Compensation	12
Estate Agents	15
Housing Acts	26
Landlord and Tenant — General	30
— Business Tenancies	71
— Leasehold Reform	95
— Rent Acts	99
— Rent Review	112
Negligence	155
Rating	164
Real Property and Conveyancing	190
Town and Country Planning	198
Sales, Tribunal — Rating	201

SUPPLEMENT
(Cases not reported in "Estates Gazette")

Landlord and Tenant — General	209
— Rent Acts	224
Negligence	228
Rating	240

PREFACE

This is the fifth volume of bound Estates Gazette Law Reports to be published since the series was launched 18 months ago. These half-yearly, easy-to-handle volumes are now a firmly established part of the law reporting scene and are increasingly evident in the hands of lawyers and surveyors alike.

The reports, which, it is emphasised, contain the complete judgments delivered, are already being widely referred to and have indeed been cited in a number of decided cases in recent months.

A notable feature of the last year has been an increase in the number of lengthy judgments, and it has been difficult to fit some of them into even the generous space allocated to law reports in the weekly **Estates Gazette** without delaying publication of a significant number of shorter decisions of more immediate importance or wider interest. However, the inclusion in each volume of a supplement of cases not reported in **Estates Gazette** (now commonly being referred to as the "green section") enables decisions of well above average length to be reported within a reasonable time and without delaying the publication of a range of other cases. The "green section" in the present volume has been substantially increased to accommodate several such long reports.

The tide of decisions on rent review clauses has continued to flow strongly with the result that more space is devoted to this subject than to any other in this volume — 16 reports in all.

TABLE OF CASES

Titles of cases shown in bold type
Names of parties reversed shown in ordinary type

Addis Ltd v Clement (VO) (CA), 168
Agricultural Dwelling-House Advisory Committee for Bedfordshire, Cambridgeshire and Northamptonshire, *ex parte* Brough, R v, 106
Allgood, Dixon v, 93
Alton House Holdings Ltd (No 2), Celsteel Ltd v, 48
Aquarius Properties Ltd, Post Office v, 40
Ashton v Sobelman (Ch), 33
Ayers, Brooker Settled Estates Ltd v, 50
Bar v Pathwood Investments Ltd (CA), 90
Barone, Chestertons v, 15
Barpress Ltd, Straudley Investments Ltd v, 69
Bass Holdings Ltd v Morton Music Ltd (CA), 214
Bissett v Marwin Securities Ltd (Ch), 115
Bodfield Ltd, Equity & Law Life Assurance Society plc v, 124
Bostock v Tacher de la Pagerie (CA), 104
Bowen, Cumshaw Ltd v, 30
Bracknell District Council, Westlake v, 161
Braddon Towers Ltd v International Stores Ltd (Ch), 209
Brian Cooper & Co v Fairview Estates (Investments) Ltd (CA), 18
Bristol Rent Assessment Committee, *ex parte* Dunworth, R v, 102
British Crafts Centre, James v, 139
British Rail Pension Trustee Co Ltd, v Cardshops Ltd (Ch), 127
Brooker Settled Estates Ltd v Ayers (CA), 50
Bruce (Robert) & Partners v Winyard Developments (QB), 20
Burton v Timmis (CA), 1
Bush (Eric S), Smith v, 157
Bush Transport Ltd v Nelson (CA), 71
Cabras Ltd, William Hill (Southern) Ltd v, 37
Cardgrange Ltd, Cornwall Coast Country Club v, 146
Cardshops Ltd, British Rail Pension Trustee Co Ltd v, 127
Celsteel Ltd v Alton House Holdings Ltd (No 2) (CA), 48
Charlwood Alliance Properties Ltd, F W Woolworth plc v, 53
Chestertons v Barone (CA), 15
Chilton v Telford Development Corporation (CA), 12
Chin, Dellneed Ltd v, 75
Clayhope Properties Ltd, Evans v, 67
Clement (VO), Addis Ltd v, 168
Collinson, Dresden Estates Ltd v, 45
Cooper (Brian) & Co v Fairview Estates (Investments) Ltd (CA), 18
Cornwall Coast Country Club v Cardgrange Ltd (Ch), 146
Cumshaw Ltd v Bowen (Ch), 30
Dalgety plc, Young v, 116
Debenhams plc v Westminster City Council (HL), 248
Dellneed Ltd v Chin (Ch), 75
Dennis & Robinson Ltd v Kiossos Establishment (CA), 133
Department of the Environment v Royal Insurance plc (Ch), 83
Dixon v Allgood (CA), 93
Dorita Properties Ltd, Oriani v, 88
Dresden Estates Ltd v Collinson (CA), 45
Ebdon (VO), Imperial College of Science and Technology v, 164
Electronic Data Processing Co plc, General Accident Fire & Life Assurance Corporation plc v, 112
Elliott, Swanbrae Ltd v, 99
Enfield London Borough, Wolff v, 119
Equity & Law Life Assurance Society plc v Bodfield Ltd (CA), 124
Eric S Bush, Smith v, 157
Evans v Clayhope Properties Ltd (Ch), 67
Expert Clothing Service & Sales Ltd, Hillgate House Ltd v, 65
Factory Holdings Group Ltd v Leboff International Ltd (Ch), 135
Fairview Estates (Investments) Ltd, Brian Cooper & Co v, 18

Fowler v Minchin (CA), 108
General Accident Fire & Life Assurance Corporation plc v Electronic Data Processing Co plc (Ch), 112
Greggs plc, Warrington and Runcorn Development Corporation v, 9
Griffin, Hill v, 81
H H Property Co Ltd v Rahim (CA), 52
Halil, Rignall Developments Ltd v, 193
Hansel Properties Ltd, Ketteman v, 237
Harris v Wyre Forest District Council (QB), 231
Hemens (VO) v Whitsbury Farm & Stud Ltd (CA), 172
Hill v Griffin (CA), 81
Hillgate House Ltd v Expert Clothing Service & Sales Ltd (Ch), 65
Imperial College of Science and Technology v Ebdon (VO) (CA), 164
International Stores Ltd, Braddon Towers Ltd v, 209
Islington London Borough, Trendworthy Two Ltd v, 184
James v British Crafts Centre (CA), 139
Keith Pople Ltd, Panther Shop Investments Ltd v, 131
Ketteman v Hansel Properties Ltd (HL), 237
Kiossos Establishment, Dennis & Robinson Ltd v, 133
Lambeth Borough Council, *ex parte* Clayhope Properties Ltd, R v, 26
Leboff International Ltd, Factory Holdings Group Ltd v, 135
Macilwraith-Christie, Roberts v, 224
Malbern Construction Ltd, Phipps-Faire Ltd v, 129
Marwin Securities Ltd, Bissett v, 115
Minchin, Fowler v, 108
Monmouth District Council, Stent v, 59
Morris v Patel (CA), 75
Morton Music Ltd, Bass Holdings Ltd v, 214
Nelson, Bush Transport Ltd v, 71
Newcastle upon Tyne City Council, North Eastern Co-operative Society Ltd v, 142
North Eastern Co-operative Society Ltd v Newcastle upon Tyne City Council (Ch), 142
Oriani v Dorita Properties Ltd (CA), 88
Panther Shop Investments Ltd v Keith Pople Ltd (Ch), 131
Patel, Morris v, 75
Pathwood Investments Ltd, Bar v, 90
Phipps-Faire Ltd v Malbern Construction Ltd (Ch), 129
Pople (Keith) Ltd, Panther Shop Investments Ltd v, 131
Post Office v Aquarius Properties Ltd (CA), 40
Power Securities (Manchester) Ltd v Prudential Assurance Co Ltd (Ch), 121
Prudential Assurance Co Ltd, Power Securities (Manchester) Ltd v, 121
R v Agricultural Dwelling-House Advisory Committee for Bedfordshire, Cambridgeshire and Northamptonshire, *ex parte* **Brough (QB),** 106
R v Bristol Rent Assessment Committee, *ex parte* **Dunworth (QB),** 102
R v Lambeth Borough Council, *ex parte* **Clayhope Properties Ltd (QB),** 26
R v Secretary of State for the Environment, *ex parte* **Bournemouth Borough Council (QB),** 198
R v Tower Hamlets London Borough Council, *ex parte* **Chetnik Developments Ltd (CA),** 180
Rahim, H H Property Co Ltd v, 52
Rendall v Duke of Westminster (CA), 96
Rignall Developments Ltd v Halil (Ch), 193

Robert Bruce & Partners v Winyard Developments (QB), 20
Roberts v Macilwraith-Christie (CA), 224
Royal Insurance plc, Department of the Environment v, 83
Saunders (VO), Trustee Savings Bank England and Wales v, 201
Sayer, Sutcliffe v, 155
Secretary of State for the Environment, *ex parte* Bournemouth Borough Council, R v, 198
Smith v Eric S Bush (CA), 157
Sobelman, Ashton v, 33
Stent v Monmouth District Council (CA), 59
Straudley Investments Ltd v Barpress Ltd (CA), 69
Stroud v Weir Associates Ltd (CA), 190
Sun Alliance & London Assurance Co Ltd, United Co-operatives Ltd v, 126
Sutcliffe v Sayer (CA), 155
Swanbrae Ltd v Elliott (CA), 99
Tacher de la Pagerie, Bostock v, 104
Telford Development Corporation, Chilton v, 12
Timmis, Burton v, 1
Tower Hamlets London Borough Council, *ex parte* Chetnik Developments Ltd, R v, 180

Trendworthy Two Ltd v Islington London Borough (CA), 184
Trustee Savings Bank England and Wales v Saunders (VO) (LT), 201
United Co-operatives Ltd v Sun Alliance & London Assurance Co Ltd (Ch), 126
Warrington and Runcorn Development Corporation v Greggs plc (Ch), 9
Watts v Yeend (CA), 4
Weir Associates Ltd, Stroud v, 190
Westlake v Bracknell District Council (QB), 161
Westminster (Duke of), Rendall v, 96
Westminster City Council, Debenhams plc v, 248
Whitsbury Farm & Stud Ltd, Hemens (VO) v, 172
William Hill (Southern) Ltd v Cabras Ltd (Ch), 37
Winyard Developments, Robert Bruce & Partners v, 20
Wolff v Enfield London Borough (CA), 119
Woolworth (F W) plc v Charlwood Alliance Properties Ltd (Ch), 53
Wyre Forest District Council, Harris v, 231
Yeend, Watts v, 4
Young v Dalgety plc (CA), 116

INDEX OF SUBJECT-MATTER

ADVERTISING SIGN — See **Landlord and tenant** (Covenant)

AGRICULTURAL DWELLING-HOUSE — See **Landlord and tenant**

AGRICULTURAL HOLDINGS
Arbitration
Variation of award
Error of law on face of award — Arbitrator had held notice to quit bad as founded on notice to pay rent stating rent due inaccurately — Arbitrator had decided that sum agreed was subject to condition precedent that certain works promised by landlord should first be completed — Held that obligations were independent, not interdependent, that sum claimed for rent was correct and notice to quit good.
Burton v Timmis (CA), **1**

Tenancy within Agricultural Holdings Act
Grazing licence or tenancy
Oral arrangements — Whether arrangements constituted grazing licence or protected agricultural tenancy — Challenge to line of authority establishing that a seasonal grazing licence, not referring to specific dates but to grazing or mowing periods, did not create a tenancy — Questions of burden of proof and weight of evidence — Held that arrangements constituted grazing licence within proviso to s 2(1) of Agricultural Holdings Act 1948
Watts v Yeend (CA), **4**

ARBITRATION
Agricultural holdings — See **Agricultural holdings** (Arbitration)
Rent review — See **Landlord and tenant** (Rent review clause)

ASSIGNMENT — See **Landlord and tenant**

BETTING OFFICE — See **Landlord and tenant** (Covenant — Restriction on signs)

CARAVAN — See **Mobile homes**

COMMISSION — See **Estate agents** (Commission)

COMPENSATION FOR ACQUISITION OF LAND
Date for valuation
Possession taken at different dates
New towns legislation — Eight separate parcels — Whether date of entry into owner-occupied farm was when first parcel taken or whether there were separate dates of entry, and consequently for valuation, according to dates of taking possession of each parcel — Held, disagreeing with Lands Tribunal [1985] 1 EGLR 195 that date of entry, and consequently date for assessment of compensation, in respect of entire farm was date of entry to first parcel.
Chilton v Telford Development Corporation (CA), **12**

DILAPIDATIONS — See **Landlord and tenant** (Repairs)

ENTERPRISE ZONE — See **Rates and rating** (Assessment)

ESTATE AGENTS
Commission
Introduction of purchaser
Purchase by United Arab Emirates of property, renovated by defendants, for over £1m — Introduction antecedent to agreement fixing commission — Subagents not at first identified — Whether intervention by agent acting for purchasers constituted independent introduction prior to that of plaintiffs and so excluded claim — Held that on facts plaintiffs' entitlement to commission established.
Robert Bruce & Partners v Winyard Developments (QB), **20**

Introduction of tenant
Whether commission payable although agents claiming were not effective cause of letting — Terms were that lessors offered full-scale letting fee if agents introduced tenant by whom agents were unable to be retained, with whom lessors had not been in previous communication, and who subsequently completed a lease — Claimants fulfilled all these terms but, after period during which lessors' interest temporarily ceased, another firm of agents made introduction which was effective cause of letting — Held, affirming decision at first instance [1986] 1 EGLR 34, that claimants were entitled.
Brian Cooper & Co v Fairview Estates (Investments) Ltd (CA), **18**

Undisclosed principal
Election — Instructions to sell property came from respondent, a solicitor, but it later appeared he had been acting for an undisclosed principal, a foreign company — Agents' account for commission questioned — Held there was no evidence that agents had made unequivocal election to look to foreign company and to abandon rights against respondent.
Chestertons v Barone (CA), **15**

Negligence — See **Negligence** (Estate agent — Valuation or survey)
EXPERT, APPOINTMENT OF — See **Landlord and tenant** (Rent review clause — Surveyor as expert)
FEES — See **Estate agents** (Commission)
FIXTURES — See **Rates and rating** (Unoccupied property — Exemption) and **Landlord and tenant** (Rent review clause — Fixtures)

HOUSING
Repairs
Local authority notice
Repair grants — Application by freehold owners of block of flats, seeking order to compel local authority to make mandatory repair grants — Issue of such grants depended on validity of notices to do repairs served on leaseholders and, in respect of protected tenants, on freeholders — Held that, as a flat was not a "house", the notices were invalid — Consequently they could not form a base for claims for mandatory grants — Housing Act 1974, ss 71 and 71A, Housing Act 1957, s 9(1A).
R v Lambeth Borough Council, ex parte Clayhope Properties Ltd (QB), **26**

LAND AGENT — See **Agency**

LANDLORD AND TENANT

Agricultural dwelling-house

Advisory committee

Cottage claimed by applicant for forestry worker to be employed by him — Irregularities in committee proceedings — Each party heard in absence of other and allegations made against applicant of bad faith — Administrative law question as to whether certiorari could go to quash report of committee adverse to applicant, although committee's function was advisory only and not determinative — Held that certiorari should go as housing authority likely to be strongly influenced by report — Report quashed — Rent (Agriculture) Act 1976.

R v Agricultural Dwelling-House Advisory Committee for Bedfordshire, Cambridgeshire and Northamptonshire, ex parte Brough (QB), **106**

Assignment

Consent, refusal of

Counterclaim by landlords for specific performance of covenant to keep demised premises as a department store — Proposed assignees had no intention to keep premises as such a store — Tenants argued that landlords could enforce covenant as to user against assignees — Held, distinguishing *Killick* v *Second Covent Garden Property Co,* that landlords justified in refusing consent — Specific performance of user covenant refused but inquiry as to damages ordered.

F W Woolworth plc v Charlwood Alliance Properties Ltd (Ch), **53**

Covenant

Restriction on signs

Signs advertising tenants' first-floor betting office — Dispute as to rights to maintain signs on part of wall belonging to landlords at street level — Decision of Goulding J in favour of tenants [1986] 2 EGLR 62 upheld but on different grounds — Held easement created in favour of tenants by demise of premises "and their appurtenances" — Kerr LJ's view that landlords estopped by convention from disputing tenants' claim.

William Hill (Southern) Ltd v Cabras Ltd (CA), **37**

Forfeiture

Relief

Business subtenant — Trial judge had jurisdiction to grant monthly tenancy by way of relief, but refused to do so because subtenant would not accept repairing liabilities of forfeited tenancy — Held there were no grounds for interfering with judge's discretion — Relief on subtenant's terms would have given landlords less extensive rights in regard to repairs than under forfeited tenancy.

Hill v Griffin (CA), **81**

Peaceable re-entry — Freeholders, with agreement of occupying subtenant, entered premises, a lock-up shop and flat, changed door locks and requested subtenant in future to pay rent to them instead of to leaseholders — Held that re-entry not effected — Arrangement assumed continuing validity of subtenancy, which was inconsistent with forfeiture claim — *Bayliss* v *Le Gros* contemplated that subtenant remained in occupation as direct tenant of superior landlord — Lease not forfeited.

Ashton v Sobelman (Ch), **33**

Wrongful entry

Order for forfeiture of lease reversed by Court of Appeal — Tenants then brought action against landlords for wrongful entry in breach of covenants for quiet enjoyment and non-derogation from grant — Held that acts done in pursuance of court order which is valid until reversed cannot be wrongful — As matter of public policy, people must be entitled to act in pursuance of court order without being at risk of acting unlawfully — Otherwise great confusion would result.

Hillgate House Ltd v Expert Clothing Service & Sales Ltd (Ch), **65**

Landlord and Tenant Act 1954, Part II

Compensation

Amount payable under s 37 — Whether six or three times rateable value — Whether premises occupied for full 14 years before termination of tenancy so as to qualify for higher scale — Occupation did not in fact begin until August 25 1971 and tenancy terminated on August 23 1985 — Various submissions by tenants that short gap should not deprive them of higher amount rejected — Occupation must be literally for full 14 years — Compensation based on three times rateable value.

Department of the Environment v Royal Insurance plc (Ch), **83**

New tenancy application

Amendment — Originating application made within time-limit laid down by s 29(3), but in respect of part of holding only — Amendment to identify whole holding made after expiry of time-limit — Held there was power under County Court Rules to amend originating application and that court would exercise discretion to allow amendment.

Bar v Pathwood Investments Ltd (CA), **90**

New tenancy grant

Rent determination for café-restaurant — Zone A equivalent method agreed — Criticism of calculations by judge, who added together a number of zone A equivalent figures and took average — Complaint that she had determined "the fair and reasonable rent" instead of market rent — Criticism also as to weight given to certain comparables — All criticisms rejected on appeal except that judge had not adjusted comparables for increases in market rents to bring them up to date.

Oriani v Dorita Properties Ltd (CA), **88**

Notice to terminate tenancy

Validity — Use by landlords of out-of-date form — County court judge had held 1969 Form 7 to be substantially to same effect as 1983 Form 1, had rejected tenant's claim that it was invalid and had refused leave to appeal — Held that judge correct in holding form valid — Application for leave to appeal refused.

Morris v Patel (CA), **75**

Lease or licence

Business premises

Chinese restaurant — Whether operator had been granted tenancy by leaseholder — Leaseholder claimed arrangement was merely a licence constituting a "management agreement" — Held that agreement was deliberately misleading to prevent superior landlord from taking action for breach of covenant and that, on principles of *Street* v *Mountford,* a tenancy had been created, there being exclusive possession for a term at a rent.

Dellneed Ltd v Chin (Ch), **75**

Workshop and store — Refinement of *Street* v *Mountford* principles — Document containing agreement showed conflict between provisions pointing to licence and provisions pointing to tenancy — Express limited permission for owners to enter to carry out works indicated tenancy, but provision enabling owners to require occupier to move to other premises during agreement was wholly inconsistent with tenancy — Held that, taken as a whole, the agreement indicated a licence.

Dresden Estates Ltd v Collinson (CA), **45**

Possession claim

Exclusive possession — Occupier had double bed-sitting-room in flat where other bed-sitting-rooms were occupied by other women — Agreement, described as a "licence", asserted that no one had any exclusive possession and that each occupier had right to occupy whole flat — County court judge decided in effect: "This occupant was not a lodger, as defined by Lord Templeman in *Street* v *Mountford.* Therefore she must have had exclusive possession." — Held judge in error — Exclusive possession was to be determined by facts in each case.

Brooker Settled Estates Ltd v Ayers (CA), **50**

Leasehold enfranchisement
"Low rent"
Two cottages reconstructed into one family home and five garages — Whether tenancy a long tenancy at a low rent — Held that county court judge in error in including garages in calculation of rateable value on appropriate day — If rateable value of garages was excluded, as s 4 (1) (a) of the 1967 Act required, rent was more than two-thirds of rateable value.
Dixon v Allgood (CA), **93**

Rateable value limits
Whether rateable value on appropriate day not more than £1,500 — Rateable value had been £1,597, but had been reduced as a result of certified improvements by £88 — This still left value too high and house could only qualify if a reduction from £1,597 to £1,547 in 1985 could be back-dated to April 1 1973 when, together with the £88, the value would be below £1,500 — Held that 1985 amendment could not be back-dated earlier than April 1 1984.
Rendall v Duke of Westminster (CA), **96**

Option
New lease
Conditions precedent to exercise — Effect of past breaches of covenant, spent at date for exercising option — Appeal from decision of Scott J [1986] 2 EGLR 50, who held tenants disqualified from exercising option by reason of breach in past of negative covenant not to apply for planning permission without landlords' consent — Held that general rule, on the authorities, was that conditions precedent of this kind applied only to subsisting breaches, not spent breaches, of covenant, and that this rule applied equally to negative and to positive covenants — Tenants' appeal allowed.
Bass Holdings Ltd v Morton Music Ltd (CA), **214**

Possession
Oral agreement
Vehicle repair shop — Proceedings arose out of long and confusing correspondence between tenant's solicitors and landlords and turned on effect of oral agreement — Held, applying *Jenkin R Lewis & Son Ltd v Kerman,* that oral agreement took effect as agreement to surrender existing term of years and to create new tenancy — Submission that tenant had repudiated oral agreement rejected.
Bush Transport Ltd v Nelson (CA), **71**

See also **Landlord and tenant** (Lease or licence) (Rent Acts)

Premises demised
Extent
Complaint by lessees under 99-year lease that respondents had trespassed on their property by erecting fire escape and ventilator vent — Respondents contended that upper surface of roof, airspace above and external surface of wall were not demised to appellants — Held that in light of parcels clause and repairing obligations, respondents had no arguable case — Injunctions and other relief under Ord 14 granted.
Straudley Investments Ltd v Barpress Ltd (CA), **69**

Quiet enjoyment
Interruption
Whether by persons "claiming under" landlord — Interruption was in fact by tenants of flats and garages whose tenancies had been granted by lessors' predecessors in title — These tenants obtained an injunction restraining a car wash proposed by lessee, who sought to rely on covenant — Held lessee not entitled to rely on covenant — Tenants who obtained injunction did not "claim under" present freeholder but derived their title from freeholders' predecessors.
Celsteel Ltd v Alton House Holdings Ltd (No 2) (CA), **48**

Rent Acts
Agricultural dwelling-house — See **Landlord and tenant** (Agricultural dwelling-house)

Death of tenant
Whether daughter of deceased tenant had been "residing with" her mother — Claimant, who had her own home about two miles away, had for more than 6 months before the death, spent three or four nights a week at her mother's house in order to care for her, but retained tenancy of her own home where her son, aged 21, continued to live — Held claimant had not established she was "residing" with her mother — She had moved in for limited purpose and had not "made her home" there — Rent Act 1977, Sched 1, Part 1, para 3.
Swanbrae Ltd v Elliott (CA), **99**

Possession
Case 1 in Sched 15 to Rent Act 1977 — Suitability of alternative accommodation — Possession sought of basement of very large house, valued with vacant possession at up to £1m, the basement being occupied by statutory tenant and subtenant — County court judge held that Case 1 applied (breach of covenant against subletting) and that alternative accommodation suitable — Court of Appeal rejected criticisms of decision on first ground but held judgment did not deal clearly with general issue of reasonableness and that an opinion, not based on evidence, as to effect on selling value of sitting tenant in basement, amounted to error of law — Nevertheless decision upheld, court themselves deciding issue of reasonableness.
Roberts v Macilwraith-Christie (CA), **224**

Case 9 in Sched 15 to Rent Act 1977 — Landlord's title and application of *McIntyre v Hardcastle* — Landlord was legal owner of flat but was trustee for himself and daughter as joint tenants in equity in equal shares, the flat to be transferred into their joint names when she attained 18 — Held, rejecting suggested gloss on wording of Case 9, that father was sole landlord, that flat was reasonably required for occupation by daughter and that balance of hardship in her favour.
Bostock v Tacher de la Pagerie (CA), **104**

Case 16 in Sched 15 to Rent Act 1977 — Cottage formerly occupied by landlord's cowman and now required for person to be employed by landlord in agriculture — Issue as to condition (b) in Case 16 (notice that possession might be recovered under Case) — Unexpected evidence presented in county court that there had been written agreement by which tenant had undertaken to vacate cottage within 28 days if landlord required it for farm worker — County court judge accepted this as compliance with condition (b) — Held that even if this agreement existed it did not satisfy condition (b) requirements.
Fowler v Minchin (CA), **108**

Rent assessment committee
Committee's decision increasing substantially rent determined by rent officer — Tenant had written ambiguous letters from which it was not clear whether she wished to withdraw objection or merely state her intention not to attend hearing — One letter which did make her intention to withdraw clear never reached committee as friend failed to post it — Committee heard case in her absence — Whether committee had taken course which no reasonable committee in the circumstances could have taken — Held committee had not done so.
R v Bristol Rent Assessment Committee, ex parte Dunworth (QB), **102**

LANDLORD AND TENANT — continued

Rent arrears
Interim order for payment
Failure by tenant to comply with county court judge's interim order — A subsequent county court judge ordered that, unless arrears were paid, defendant's defence would be struck out — Held judge not entitled under inherent jurisdiction to order defence to be struck out in these circumstances — Right course was for landlords to use recognised procedure to enforce payments under interim order — County Courts Act 1984, s 50, CCR Ord 13, r12 and RSC Ord 45.

H H Property Co Ltd v Rahim (CA), 52

Rent review clause
Arbitration
Application for leave to appeal under Arbitration Act 1979 — Complaint that arbitrator appointed under rent review clause disregarded premium paid in connection with lease of "comparable" — Issue as to test to be adopted by judge in deciding whether to grant leave to appeal — Held that test was "Am I left in doubt as to whether the arbitrator was right in law?" — *Lucas Industries* case followed — Leave given but leave to appeal against grant also given.

Warrington and Runcorn Development Corporation v Greggs plc (Ch), 9

Casino premises in Mayfair — Questions referred under Arbitration Act 1979, s 2 — Large gap between parties to sublease as to open market rental value — Questions as to hypotheses in hypothetical letting — Complications due to fact that gaming licence held by Crockford's, who were licensees of sublessee — Judge's warning against dangers of deducing consequences from hypotheses — Eight specific questions answered by judge — Appeal also before judge from arbitrator's refusal to order specific discovery of casino's trading accounts — Arbitrator's refusal based on wrong reasons, but application for discovery not remitted for rehearing.

Cornwall Coast Country Club v Cardgrange Ltd (Ch), 146

Arbitrator or expert
Ambiguity in clause — Rack-rental value to be determined by independent surveyor agreed by parties or, in default of such agreement, by arbitrator nominated by president of RICS — Whether such surveyor to act as arbitrator — Held that independent surveyor was to act as expert if appointed by agreement (as he was) but as arbitrator if appointed by president — No ruling given as to whether surveyor amenable to action for negligence.

North Eastern Co-operative Society Ltd v Newcastle upon Tyne City Council (Ch), 142

Construction
Clause provided for yearly rent at which property might reasonably be expected to be let in open market, but did not refer expressly to willing lessee or willing lessor — Tenants argued there should be no assumption as to willing lessee, as it might be a question as to whether anyone would wish to take lease on terms offered — Held, allowing appeal from deputy judge [1986] 2 EGLR 120, that a willing lessor and lessee must be assumed, but it was a matter for expertise of valuer to determine strength of market.

Dennis & Robinson Ltd v Kiossos Establishment (CA), 133

Named lessees — Problems as to terms of hypothetical lease to be assumed in arriving at "commercial yearly rent" — Specific reference to lessees by name in user clause and restriction on assignment clause — Court of Appeal upheld narrow construction given to hypothetical lease in relation to user clause by Scott J [1986] 1 EGLR, 117, who held that it was intended to grant special personal privilege to actual present lessees — Doubts expressed about wider construction given by Scott J to assignment clause, but no cross-appeal by tenants on this issue.

James v British Crafts Centre (CA), 139

Rental value defined in alternative ways, as full market value, as related to addition based on index of prices or costs, and as previous rent with a compound interest addition "from the date of the immediately preceding rent review" — Landlords, in trigger notice on first review, specified rent based on compound addition, but, on objection from tenant, specified rent based on indices formula — Held that compound interest formula could not apply to first review, as there was no "preceding" review, unless words "or the commencement of the term" could be added, which as a matter of construction they could not.

Bissett v Marwin Securities Ltd (Ch), 115

Disregards
Improvements — Clause provided for disregard of any effect on rent of improvements carried out by tenants, or any person deriving title under them, otherwise than under obligation to landlords — Two improvements carried out, not under present lease but under previous lease between same parties — By time present lease executed these structures had become landlords' fixtures — Held, following *Brett v Brett Essex Golf Club Ltd*, that structures in question were not improvements to, but part of, demised premises, and should not be disregarded.

Panther Shop Investments Ltd v Keith Pople Ltd (Ch), 131

Exclusion of clause
Arbitrator's decision assumed hypothetical lease without rent reviews, there being no express reference in review clause as to assumption of rent review in that lease — Held there was no justification for overriding express terms of lease even if absence of assumption of rent reviews resulted in rent 15% higher than market rent — "Commercial common sense" approach in *British Gas Corporation v Universities Superannuation Scheme Ltd* rejected in favour of "the classic and correct method of approach" in *Philpots (Woking) Ltd v Surrey Conveyancers Ltd*.

General Accident Fire & Life Assurance Corporation plc v Electronic Data Processing Co plc (Ch), 112

"Net rental value" to be assessed on assumption that hypothetical lease would be on terms of actual lease "other than as to duration and rent" — Held, affirming decision of Gibson J [1985] 2 EGLR 144, that hypothetical lease should contain no rent review provisions — Court's observations on unhelpfulness of decisions on construction of other leases — Approach should be to construe lease under dispute without presumptions.

Equity & Law Life Assurance Society plc v Bodfield Ltd (CA), 124

Fixtures
Whether floor covering and light fittings, installed under obligations upon tenants in agreement preceding lease, were landlord's fixtures which should be taken into account in determining rent under rent review clause or tenant's fixtures, which should not — Held, following decision in *Mowats Ltd v Hudson Bros Ltd*, that the items should be treated as tenants' fixtures despite tenants' contractual obligation to install them.

Young v Dalgety plc (CA), 116

Notice and counternotice
Whether tenants' letter headed "subject to contract" was valid counternotice — Provisions of kind not previously before courts — Case distinguished from *Shirlcar Properties Ltd v Heinitz* and *Sheridan v Blaircourt Investments Ltd* — In present case counternotice intended merely to start period during which parties were to negotiate in good faith, resulting in reference to expert only if they failed to agree — Held in context of present case that counternotice was valid — Heading "subject to contract" capable of being explained.

British Rail Pension Trustee Co Ltd v Cardshops Ltd (Ch), 127

Surveyor as expert
   Appointments by president of RICS — Landlords proposed to apply to president to appoint surveyor as expert — Tenants objected on ground that they were contemplating court proceedings on construction or rectification of lease and that in any case appointment would be premature — They sought injunction to restrain president from appointing — Held, dismissing motion, that president under no duty to defer making appointment — There were other remedies to which he would not be party — He should not be a pawn in tactical moves by parties.
   United Co-operatives Ltd v Sun Alliance & London Assurance Co Ltd (Ch), 126

Time-limits
   Arbitration — Clause provided for "trigger" notice by landlords, machinery for reaching agreement on rent and provisions, failing agreement, for appointment of arbitrator — Tenants' letter giving notice that landlords were required within 28 days to refer rent to arbitration — Landlords did not comply with time-limit — Held that, as tenants could have applied for appointment of arbitrator themselves, they were not entitled to make time of the essence for this step.
   Factory Holdings Group Ltd v Leboff International Ltd (Ch), 135

   Valuer's appointment — Whether time of the essence — Paragraph enabling lessees to serve notice on lessors proposing figure which "shall be the revised rent" unless lessors applied within three months to president of RICS to appoint valuer — Lessors failed to apply within time — Lessees proposed rent which was in fact the old rent, contending that time was of the essence — They put forward "contra-indications" to displace *United Scientific Holdings* presumption that time was not of the essence — Held suggested "contra-indications" were not sufficiently compelling and presumption prevailed.
   Phipps-Faire Ltd v Malbern Construction Ltd (Ch), 129

   Whether time of the essence in regard to part of clause relating to agreement as to total income from demised premises, a Manchester shopping centre — Landlords contended that time was of the essence and that time-limit had passed without such agreement — Held that presumption against time being of the essence was not displaced — The fact that there was express provision in case of default of agreement was not conclusive and in circumstances was not sufficient contra-indication.
   Power Securities (Manchester) Ltd v Prudential Assurance Co Ltd (Ch), 121

Use
   Lease provided for use for any purpose within Use Classes Order, Class III or other use permitted by planning authority — Planning permission given before grant of lease for use as non-teaching service unit for polytechnic — Held, affirming decision of Whitford J [1985] 1 EGLR 75, that that use was a composite use and not shorthand for large number of uses — As a result, there was nothing in rent review clause to permit rent to be assessed by reference to any user authorised by a use class except for light industrial uses within Class III.
   Wolff v Enfield London Borough (CA), 119

Repairs
Covenant
   Ingress of rain water through or under door of local authority dwelling — Attempts to cure defect unsuccessful over 30 years, including repairs to and replacement of door — Damage to tenant's carpets — Eventually door replaced for second time, on this occasion by aluminium self-sealing unit, which was successful — Held that as door itself had been damaged and was out of repair, its replacement by a purpose-designed weather-proof door was sensible repair, although correcting design defect, and within covenant to repair.
   Stent v Monmouth District Council (CA), 59

   Tenants' liability to repair under full repairing lease — Basement of office building flooded owing to defects of construction and perhaps design — Despite long periods of ankle-deep flooding, there was no evidence of any actual damage or deterioration — Hoffmann J [1985] 2 EGLR 105 decided that any scheme to cure defects involved substantial structural alteration and improvement and went beyond "repair" — Court of Appeal agreed with decision in tenants' favour but held there was no liability on tenants because there was no disrepair — Disrepair connotes deterioration from previous physical condition.
   Post Office v Aquarius Properties Ltd (CA), 40

Receiver appointed — See **Receiver**

Service charges
Determination
   Retail Price Index changes — Lease in 1960 provided that service charge should increase by reference to rises in index — If, however, index ceased to be published or available increases in service charge to be determined by different method — Effect of changes in index since 1960, both in items and weightings and successive returns to new base of 100 — Held index remained the same despite changes — New bases could be converted to old by simple mathematical formula — Landlords' submission that alternative method should apply rejected.
   Cumshaw Ltd v Bowen (Ch), 30

User covenant
Keep premises open
   Department store in shopping centre — Covenant to keep demised premises as department store — Proposed assignees had no intention to keep premises as such a store — Landlords' counterclaim for specific performance of user covenant — Tenants argued that landlords could enforce covenant against assignees — Held that landlords justified in refusing consent, but, following *Braddon Towers v International Stores*, specific performance "reluctantly" refused — Inquiry as to damages ordered.
   F W Woolworth plc v Charlwood Alliance Properties Ltd (Ch), 53

   Supermarket — Landlords' action to prevent closure of supermarket contrary to covenants in lease — Whether court could grant injunction requiring tenants to keep supermarket open — Weight of authority that court would not require a person to carry on a business — Judge sympathetic to plaintiff landlords on merits but authorities too well established to permit of injunction being granted, at least on interlocutory motion — Little prospect that trial judge would be persuaded to depart from settled practice — Injunction "reluctantly" refused.
   Braddon Towers Ltd v International Stores Ltd (Ch), 209

LEASEHOLD ENFRANCHISEMENT — See **Landlord and tenant** (Leasehold enfranchisement)

LICENCE — See **Landlord and tenant** (Lease or licence)

LIMITATION OF ACTION — See **Negligence** (Mortgage loan report) (Structural damage liability)

LOCAL LAND CHARGE — See **Vendor and purchaser** (Defective title)

MOBILE HOMES
Pitch fee
Determination
   Owner of mobile home dissatisfied with substantial increase in pitch fee — County court judge determined increase equal only to increase in Retail Price Index — Judge correct in rejecting evidence as to pitch fees at other sites and a submission that a "fair market rent" should be determined — He was justified in deciding that the only scope for increase was application of Retail Price Index.
   Stroud v Weir Associates Ltd (CA), 190

NEGLIGENCE
Estate agent
Valuation or survey

Claims against estate agent for alleged negligence in report — Report, described as a valuation, drew attention briefly to settlement and certain other defects, but recommended purchase — A few years later purchasers wished to move but found difficulties in reselling — They had purchased at £8,745 and were asking on resale £19,000 reducing to £17,000 — Held that estate agent had been instructed as a valuer, not a surveyor for a full survey — Report was not misleading, was accurate as a valuation, and there was no duty to warn as to resale difficulties.

Sutcliffe v Sayer (CA), **155**

Mortgage loan report
Damages

Valuer employed by local authority recommended 90% loan, which was granted, on house suffering from extensive settlement — Valuer noticed tie bar and settlement but considered movement was thing of the past — Subsequently would-be purchasers from plaintiff house-owners were put off by state of house, which had to be taken off market — Plaintiffs still in house — Held valuer and authority both negligent, the authority vicariously and primarily.

Harris v Wyre Forest District Council (QB), **231**

Limitation Act

Surveyor employed by local authority in report on house purchased by plaintiffs stated that structural and decorative condition was good, valued at £11,750 and recommended loan of £10,900 — Gap between skirting and floor soon noticed, but surveyor, on second visit, said it was purely settlement and nothing to worry about — House became unsaleable — Held (1) authority liable for negligence, (2) cause of action not statute-barred despite eight years' interval, as there had been "deliberate concealment" within meaning of s 32(1)(b) of Limitation Act 1980, (3) authority liable for special damages and also general damages for distress.

Westlake v Bracknell District Council (QB), **161**

Reasonableness

Surveyors, instructed by building society, gave reassuring report, relied on by house-buyer — Although aware that chimney breasts had been cut away, they failed to check whether chimneys had adequate support — They could have checked by looking through trap door in roof space — Flues collapsed causing damage — Surveyors relied on disclaimers in mortgage application and report and contended they satisfied reasonableness requirement in s 11(3) of Unfair Contract Terms Act 1977 — Held they could not rely on disclaimers as they had not satisfied reasonableness requirement.

Smith v Eric S Bush (a firm) (CA), **157**

Structural damage liability
Limitation Act

Houses with defective foundations — Claim by house-owners against builders, local authority and architects — Liability — Whether statute-barred — Appeal by architects — Unsatisfactory termination of litigation in a division in the House on procedural matter — Majority of House refused to interfere with discretion of Court of Appeal in disallowing amendment of pleadings which would have enabled architects to plead successfully that claim against them was statute-barred — Minority would have allowed architects' appeal.

Ketteman v Hansel Properties Ltd (HL), **237**

PRESIDENT OF RICS (Appointment of arbitrators or experts) —
See **Landlord and tenant** (Rent review clause)

QUIET ENJOYMENT — See **Landlord and tenant** (Forfeiture — Wrongful entry) (Quiet enjoyment)

RATES AND RATING
Assessment
Enterprise zone, effect of

Proposals for reductions in rates of hereditaments situated just outside Lower Swansea Valley Enterprise Zone — Court of Appeal accepted in general valuation officer's submission that s 20(1)(b) of 1967 Act was concerned with physical factors or at least with factors which affect hereditament's physical use and enjoyment — Court, however, indicated that if physical changes could be shown to be consequential on zone's existence, eg if it affected the prosperity of an area in a manner which was manifest, this could be taken into account — General Rate Act 1967, s 20(1)(b).

Addis Ltd v Clement (VO) (CA), **168**

Shops

Units in Killingworth shopping centre — In terms of s 19 of General Rate Act 1967 shops in nearby Longbenton more valuable — Ratepayers contend similar distinction should be made under s 20 — Parties rely on agreed assessments of two larger units in centre — Held that s 20 requires valuation to be made having regard to occupation and use of premises within locality — Criticism of rigid application of zoning system to two essentially different properties — Assessments reduced.

Trustee Savings Bank England and Wales v Saunders (VO) (LT), **201**

Contractor's basis
University college

Valuation criticised on ground that contractor's basis, although appropriate method, had been wrongly applied — Main complaint was as to decapitalisation rate used to arrive at annual rental — Lands Tribunal discounted inflation entirely in fixing rent for a year certain and arriving at real interest rate of between 1.4% and 2.4% — After adding for borrower's premium and depreciation and repairs, tribunal determined $3\frac{1}{2}$% — Held criticisms of tribunal's decision disclosed no error of law.

Imperial College of Science and Technology v Ebdon (VO) (CA), **164**

Rateability
Stud premises

Whether stud premises used for breeding thoroughbred racing stock exempt from rating in same way as agricultural land used for breeding cattle and sheep — Appeal from Lands Tribunal's rejection of claim for exemption [1985] 1 EGLR 227 — Held that grazing such stock was not an "agricultural operation" and that the stock in question were not "livestock" within the Rating Acts — General Rate Act 1967, s 26(3) and (4), and Rating Act 1971, ss 1 and 2.

Hemens (VO) v Whitsbury Farm & Stud Ltd (CA), **172**

Refund of rates
Payment under mistake of law

Rates paid in respect of unoccupied property for period during which occupation would have been unlawful — Whether authority's discretion under s 9 of General Rate Act 1967 was unfettered or whether there had been breach of *Wednesbury* principles — Held that object of s 9 was to remedy injustice where person had paid sums which he was not liable to pay — Orders of certiorari and mandamus granted against rating authority.

R v Tower Hamlets London Borough Council, ex parte Chetnik Developments Ltd (CA), **180**

Unoccupied property
Entry in valuation list

Hereditaments forming part of Angel Centre, Islington — Whether ratepayers liable for unoccupied property rates although addition of hereditament to valuation list had only reached proposal stage and it had not yet been entered in list — Appeal from decision of Mervyn Davies J [1986] 1 EGLR 187 in favour of ratepayers dismissed by a majority (Dillon LJ dissenting) — Cross-appeal by ratepayers as to liability to pay before disposal of appeal against completion notice dismissed by all three judges — General Rate Act 1967, s 17 and Sched 1.

Trendworthy Two Ltd v Islington London Borough (CA), **184**

RATES AND RATING — *continued*
Unoccupied property — *continued*

Exemption
Listed buildings — Position of old Hamleys' toy shop, with main building connected at rear by tunnel and footbridge to building on further side of adjoining street — Whether whole hereditament listed and so exempt, the part being a structure fixed to main building or within latter's curtilage — Held, reversing decision of Court of Appeal [1986] 1 EGLR 189, that the structure must be ancillary, not an independent building — As part only of hereditament listed, exemption did not apply — Town and Country Planning Act 1971, s 54(9).
Debenhams plc v *Westminster City Council (HL)*, **248**

RECEIVER
Appointment — See **Landlord and tenant** (Repairs — Receiver appointed)
Remuneration and expenses
Recovery of sums
Appointment to collect rents and manage block of mansion flats — *Hart* v *Emelkirk* basis — Flats in poor repair and income inadequate — Application by receiver for directions as to recovery of sums for remuneration and expenditure in excess of assets in his hands — Submission that he should be able by interlocutory application to obtain recovery from landlords — Application refused — *Boehm* v *Goodall* cited — Receiver not an agent of or trustee for parties — Court had no jurisdiction to indemnify receiver save to extent of assets in his possession under court order — Warning to receivers as to risks of taking office unless satisfied as to sufficiency of assets.
Evans v *Clayhope Properties Ltd (Ch)*, **67**

RENT ACTS — See **Landlord and tenant** (Rent Acts)

RENT (AGRICULTURE) ACT 1976 — See **Landlord and tenant** (Agricultural dwelling-house)

REPAIRING COVENANT — See **Landlord and tenant** (Repairs)

RETAIL PRICE INDEX — See **Landlord and tenant** (Service charges); **Mobile homes**

"SUBJECT TO CONTRACT" — See **Landlord and tenant** (Rent review clause — Notice and counternotice)

SURVEYORS — See **Landlord and tenant** (Rent review clause); **Negligence**

TOWN AND COUNTRY PLANNING
Blight notice
"Appropriate authority"
Whether borough council or county council was "appropriate authority" for purpose of blight proceedings — Both authorities liable to acquire blighted land but under different sets of circumstances — Secretary of State held that he did not have power to decide which was the appropriate authority — Held that Secretary of State was right — Consequently blight notices could be served on both authorities — Town and Country Planning Act 1971, s 205.
R v *Secretary of State for the Environment, ex parte Bournemouth BC (QB)*, **198**

Listed buildings — See **Rates and rating** (Unoccupied property — Exemption)

VENDOR AND PURCHASER
Defective title
Land subject to local land charge
Incumbrance in form of improvement grant registered in register of local land charges — Knowledge of vendor's solicitor — Effect of conditions of sale deeming purchasers to have made local searches (which they had not in fact done) — Applicability of Eve J's alternative ground in *Re Forsey and Hollebone's Contract* — Held that if there is an incumbrance of which vendor is aware he cannot rely on conditions of sale as in present case without full and frank disclosure of defect — Eve J's doctrine, apart from doubts as to its soundness, not applicable to facts of present case — Declaration that vendor had not shown good title.
Rignall Developments Ltd v *Halil (Ch)*, **193**

AGRICULTURE

Court of Appeal
November 27 1986
(Before Lord Justice KERR and Mr Justice SWINTON THOMAS)

BURTON v TIMMIS AND ANOTHER

Estates Gazette February 21 1987

281 EG 795-798

Agricultural Holdings Act 1948 — Appeal from decision of county court judge holding that arbitrator's award should be varied in accordance with para 25A(2) of Schedule 6 to the Act on the ground of error of law on the face of the award — The issue arose out of an agreement, recorded in two documents, between the landlord and tenant of an agricultural holding, fixing the amount of rent at the end of a three-year interval and avoiding the need to continue with a rental arbitration — The agreement included undertakings by the landlord to execute within a stated time certain works such as the provision of a damp-proof course for the farmhouse, repairs to the granary and outside painting — When the new rent agreed was not paid in response to a notice to pay rent due the landlord served on the tenant a notice to quit in pursuance of Case D in section 2(3) of the Agricultural Holdings (Notices to Quit) Act 1977 — The tenant required arbitration on the notice to quit — The arbitrator decided that the notice to quit was invalid because founded on a notice to pay rent which incorrectly stated the amount of rent due — The arbitrator's ground, which was evident on the face of his award, was that the fulfilment of the landlord's obligation to carry out the works mentioned in the agreement was a condition precedent to the tenant's liability to pay the new rent; in other words that the obligations were mutually interdependent, not independent — As the works had not been carried out by the completion date the new rent had not become payable and the notice to pay rent had misstated the amount due — The landlord applied to the county court to set the award aside or vary it on the ground of error of law on its face — There was a subsidiary allegation of technical misconduct in not resolving a conflict of evidence, but in the end it proved unnecessary to deal with this point — The county court judge held that there was an error of law on the face of the award — The arbitrator had been wrong in his construction of the agreement, the tenant's agreement to pay the increased rent being an independent obligation, not interdependent with the landlord's obligation to carry out the works — Held by the Court of Appeal that the judge had reached the correct conclusion as a matter of the construction of the agreement, a conclusion supported by the serious practical problems to which the opposite conclusion would give rise — The arbitrator's award was rightly varied by the judge with the result that the notice to quit was good — Appeal dismissed — Per Kerr LJ, "I am not at the moment convinced that failure to resolve an issue is misconduct"

The following cases are referred to in this report.

Antaios Compania Naviera SA v *Salen Rederierna AB* [1985] AC 191; [1984] 3 WLR 592; [1984] 3 All ER 229, HL

Graves v *Legg* (1854) 9 Exch 709
Pioneer Shipping v *BTP Tioxide (The "Nema")* [1982] AC 724; [1981] 3 WLR 292; [1981] 2 All ER 1030, HL
Yorkbrook Investments Ltd v *Batten* [1985] 2 EGLR 100; (1985) 276 EG 545, CA

This was an appeal by David Timmis, tenant of the Preston Boats Farm in Upper Magna, Shropshire, from a decision of Judge Peter Northcote at Shrewsbury County Court in favour of the landlord, Robert Lingen Burton, varying an award of the arbitrator, Raymond George Taylor. The arbitrator had determined that a notice to quit served by the landlord was invalid because the notice to pay rent on which it was based stated the rent due inaccurately. On the landlord's application to set aside or remit the award, on the grounds of error of law on the face of the award and misconduct, the judge accepted the submission on the former ground and varied the award by substituting a finding that the notice to quit was good.

Martin Thomas QC and Michael P Farmer (instructed by Scott Lister & Co, of Shrewsbury) appeared on behalf of the appellant; Jonathan Gaunt (instructed by Sharpe Pritchard & Co, agents for Sprott Stokes & Turnbull, of Shrewsbury) represented the respondent landlord; the arbitrator, although named as a respondent, was not represented and took no part in the proceedings.

Giving judgment, KERR LJ said: This is an appeal by the tenant, Mr Timmis, from a judgment of His Honour Judge Peter Northcote given in the Shrewsbury County Court on March 26 1986. He had before him an application to set aside, or remit, the award of an arbitrator made under the Agricultural Holdings Act 1948. For present purposes the relevant provisions are paras 25(2) and 25A(2) of Schedule 6 to that Act, which are in the following terms:

25(2) Where the arbitrator has misconducted himself, or an arbitration award has been improperly procured or there is an error of law on the face of the award, the county court may set the award aside.

25A(2) In any case where it appears to the county court that there is an error of law on the face of the award the court may, instead of exercising its power of remission under the foregoing sub-paragraph, vary the award by substituting for so much of it as is affected by the error such award as the court considers that it would have been proper for the arbitrator to make in the circumstances; and the award shall thereupon have effect as so varied.

For present purposes the main issue is whether or not this award contained an error of law on its face and whether the judge was accordingly entitled to vary it as he did.

I should say in passing that the Arbitration Act 1979 has of course not affected that jurisdiction, although it is now no longer open for the High Court to set aside an award for an error of law on its face. But this statutory arbitration procedure, with an appeal to the county court, is unaffected by that.

The matter arises out of a tenancy agreement made between the plaintiff, Mr Burton, and the defendant, Mr Timmis, dated January 16 1964 relating to a property known as Preston Boats Farm in Upper Magna, Shropshire. The extent of the farm originally let was some 45½ acres, but there have been additions and it is now about 56 acres.

The rent originally agreed was £364 10s per annum, payable half-yearly in arrear on the March and September quarter days; it was increased to £940 with effect from March 25 1975; and it was again increased, with effect from March 25 1978, to £1,250. The matter with which we are concerned arises out of a proposed increase in 1981, which the tenant required to be referred to arbitration. In June 1981 an arbitrator was appointed for the purpose of determining the appropriate increase, if any.

However, before the arbitration took place, the increase and certain other terms were agreed at a meeting between Mr Timmis, Mr Morgan and a Mr Witt (Mr Witt being the agent of the landlord) at the farm on April 6 1981. The terms of that agreement were set out in two documents, to which I shall turn in a moment, but so far as the rent was concerned it was agreed that it should be increased to £2,800 per annum. Mr Thomas QC, who has said everything possible on behalf of the tenant, has drawn attention to the fact that this was a very substantial increase indeed.

I should have mentioned — though it hardly requires mention — that under the terms of the lease the landlord had power to re-enter in default of payment of rent.

The agreement reached at the meeting was incorporated into a letter of April 9 which enclosed a memorandum. The fact that the terms of the oral agreement were incorporated into this letter, together with the memorandum, was found by the arbitrator in his award arising from the subsequent dispute, to which I shall come in a moment; I first read the letter. This is from Mr Witt to Mr Timmis, headed "Preston Boats Farm", and it reads as follows:

> I refer to previous meetings and in particular the meeting held on April 6 when we reached agreement in respect of the new rent to be paid at Preston Boats Farm from March 25 1981.
> This new rent is to be subject to the agreed terms as follows:
> 1 The Landlord will install a new damp proof course to the farmhouse.
> 2 The Landlord will carry out all necessary repairs to the first floor of the granary. This to be the renewal, where necessary, of damaged floor boards and the treatment of timber work with wood preservatives.
> 3 The Landlord will carry out the replacement, where necessary, of the ridge tiles to the granary roof.
> 4 The Landlord will carry out the external painting of the farmhouse and buildings and will recover 50% of the cost from the Tenant.
> All these items to be carried out prior to September 1 1981.
> 5 The Landlord will grant permission for the Tenant to sheet the gable end of the Dutch barn subject to agreed Tenant right terms and formal written approval.
> 6 The Landlord will grant the Tenant permission to take down the sandstone wall running at right angles to the road and to re-erect this beside the road to form a new entrance to the farmyard. This permission is subject to formal approval of the proposals and plans and the Tenant carrying out the work at his own expense with no compensation at the end of the tenancy.

Those were the six terms, and the letter concluded as follows:

> Enclosed with this letter are two copies of the Memorandum showing the agreed new rent based on £50 per acre of 56.049 acres, ie £2,800. I would be obliged if you could kindly sign both copies and attach one to your own Agreement and return the other to this office by Tuesday April 14 so that the Arbitrator can be informed that his services will no longer be required prior to the time limit for submission of Statements of Case being reached.

The enclosed memorandum was in the following terms and was subsequently signed by both parties and dated April 12 1981:

> In consideration of the Landlord of the within-mentioned holding undertaking not to refer to arbitration under section 8 of the Agricultural Holdings Act 1948, the question of the rent to be payable on the holding in respect of any period prior to the 25th day of March 1984 the tenant of the said holding agrees that:—
> 1 The rent payable in respect of the said holding (including all existing increases in respect of improvements or otherwise) shall as from the 25th day of March 1981 be £2,800 (Two Thousand Eight Hundred Pounds) . . . which shall be payable in the same way as the rent of the said holding hereunder.
> 2 The Proviso for re-entry contained in the within-written agreement shall be exercisable in respect of non-payment of the said increased rent or any part thereof.
> 3 In consideration of the premises all the terms and conditions of the within-written Agreement varied as aforesaid shall remain in full force and effect.

The next event was a further meeting at the farm on August 18 1981 — that is to say, fairly shortly before the date of September 1 referred to in the letter — between Mr Timmis and Mr Witt. Mr Witt made a memorandum, or attendance note, of that meeting, which the arbitrator had before him, but Mr Timmis never saw it at the time. According to that memorandum, Mr Timmis agreed that some of the work specified in the letter of April 9 should not be done, and there were various other matters discussed which were relevant to whether or not parts of the work would be done, and if so when and in what manner.

The next quarter day arrived and the estate rent audit was on October 8 1981. The rent not having been paid, a notice to pay within two months was served on the tenant on November 2 1981, requiring him to pay £1,429.08, being £1,400 as the half-yearly rent agreed in the letter which I have read and the balance relating to grazing rights.

The tenant did not pay upon that notice, but on December 17 1981 the tenant's solicitors sent a cheque for £654.07 post-dated to December 31, which was subsequently dishonoured. Consequently the landlord gave notice to quit under Case D of section 2(3) of the Agricultural Holdings (Notices to Quit) Act 1977 on January 8 1982. That notice was due to expire on March 25 1983, in accordance with the Agricultural Holdings Act. On January 27 1982 the tenant, as he was entitled, thereupon required arbitration on the notice to quit. The instalments of rent at the new rent of £2,800 per annum were then duly paid on March 25 and September 29 1982 while the notice period was running.

On November 5 1982 an arbitrator was appointed. He had before him statements of case lodged by the landlord and tenant in the usual way. The hearing took place on January 5 1983 and the arbitrator made his award on January 25 1983. The main issue is whether or not that award contains an error of law on its face, as the county court judge concluded, and I therefore turn to the award.

It was made by Mr Raymond George Taylor, who recites the facts. He recites that it was not disputed that the notice to pay the rent and the notice to quit were in proper form and were effectively served. He rightly concluded that the tenant's explanation as to why the cheque for part of the amount demanded was dishonoured was irrelevant, and he correctly identified the issue for the purposes of the validity of the notice to quit, which was whether or not the rent specified in the notice to pay served on November 2 1981 was correctly stated. That in itself depended upon whether or not the landlord's obligation to carry out the repairs specified in paras 1 to 4 of the letter of April 9 by September 1 1981 was or was not a condition precedent to the tenant's liability to pay the new rent.

I must read certain passages from the award, and comment on them. The arbitrator said that he found that the letter expressed a binding contractual commitment on behalf of the landlord to do certain work, namely the work referred to, prior to December 1 1981. That is undoubtedly correct. Then he said this:

> The question which then arises is whether the Landlord's obligation to carry out the said works and the Tenant's obligation to pay the new rent was discrete and disjunctive or whether the liability for payment of rent is made conditional upon the Landlord carrying out those works prior to September 1 1981.

That, if I may say so with respect, is a perfectly correct way of stating the issue which arises upon the documents to which the arbitrator refers — that is to say, the letter of April 9 and the memorandum of April 12 1981.

The arbitrator then reviews the respective contentions of the parties; he deals first with their conflicting evidence as to what was agreed between Mr Timmis and Mr Witt at the meeting on August 18 1981, of which Mr Witt made an attendance note. The arbitrator records that there was a conflict of evidence as to what was said on that occasion; he then went on as follows:

> As the evidence in relation to the meeting on August 18 1981 is in conflict and the terms of any alleged agreement reached on that date are not recorded in writing I am bound to rely on the contractual documents before me in the form of the letter of April 9 1981 and the subsequent memorandum of April 12 1981.

He goes on as follows:

> At this stage I comment that I have considerable sympathy for the position of the Landlord and his agent. At the meeting on April 6 1981 the Tenant had asked that the damp proofing work should be carried out after the bed and breakfast holiday trade had ceased and yet before September 1 1981 thereby imposing on the Landlord a narrow and rather uncertain time band in which that portion of the agreed works should be carried out. At the meeting on August 18 1981 the Tenant's proposals must have caused confusion in the mind of the Landlord and his agent. Notwithstanding this I am bound to have regard to the fact that it was open to the Landlord to record any variation of the agreement as to the works in writing for acceptance by the Tenant and in the absence of such acceptance to proceed with the works in accordance with the terms agreed before September 1 1981.

Then he says that he reads the two documents of April 9 and 12 together, and he concludes this passage by saying this:

> In deciding upon the accuracy of the Notice to Pay and thereby the validity of the Notice to Quit I am therefore faced solely with the construction of the contractual documents in the form of the said letter and the said memorandum.

Pausing at that point, as I read that award the arbitrator is saying that because there was a conflict in the evidence before him as to what was said or agreed, if anything, on August 18 1981, and because the matter was not reduced to writing, it was not open to him to have any regard to what had occurred at that meeting or, as I understand it, to resolve the conflict of evidence before him, and that he was therefore thrown back solely on the true construction of the documents, to which he then turns. If it were necessary to decide this point, I think I would conclude that that was an error of law on the part of the arbitrator, since he appears to have taken the view that, there being nothing in writing and there being a conflict of evidence as to what was said or possibly agreed orally, he could have no regard to the events of August 18 1981. I think that must clearly be wrong. However, we have not heard Mr Gaunt on behalf of the landlord, and I therefore prefer to express no final opinion about it. I only mention it because the judge appears to have taken the view that the arbitrator's failure to deal with this conflict of evidence constituted misconduct, for which I think he would have remitted the matter to the arbitrator as requested by the landlord. I am not at the moment convinced that failure to resolve an issue is misconduct. However, it is unnecessary to express any final view on that matter either. The reason is that the arbitrator then addressed himself to the true construction of the two documents and that the case can be decided on that issue alone, without any need for a decision on these other points.

Before I come to that, I should make it clear that Mr Thomas repeatedly pointed out that these two documents are not in the form of a deed or a formal agreement in writing but came into existence following the conclusion of an oral agreement, or following negotiations in which agreement was reached, and which were then, in the ordinary course, reduced to writing. He therefore submits that the issue for the arbitrator, and the issue decided by the arbitrator, and therefore the issue of law before the county court and now on appeal to this court, is not concerned with the true construction of these documents, but that the arbitrator found as a fact that the parties had made an oral agreement which accords with his final conclusion, namely — as mentioned below — that the landlord was obliged to carry out the repairs in accordance with paras 1 to 4 by September 1 1981 as a condition precedent to the new rent coming into force.

I cannot for one moment accept this submission. This was a perfectly normal situation of terms being negotiated and agreed orally and then being reduced to writing, as the arbitrator himself finds; and the letter and memorandum in fact contained the terms which he finds were the terms agreed. Accordingly, in order to determine the effect of the terms which were agreed, one must construe the letter and memorandum and, as the arbitrator rightly said, one must construe them together. Whether a document is correctly construed or not is, of course, an issue of law, as has been said in many cases, including recently, as I recollect, by Lord Diplock in delivering the opinion of the House of Lords in *"The Nema"* [*Pioneer Shipping* v *BTP Tioxide*] [1982] AC 724]. So what it comes to is that on this issue one has to consider whether or not the arbitrator's construction of these documents was correct, and that is a question of law.

The arbitrator reviewed the contentions of the parties on this issue. He pointed out, and obviously had considerable sympathy with, the landlord's submission, which is self-evident, that if the agreement bore the meaning for which the tenant contended, all sorts of difficulties would arise. For one thing, it would not be known until September 1 1981, the terminal date for doing the repairs, what the new rent was; whether there had been a new rent as from March 25 1981; or whether the increased rent was in force. Second, it would mean that if the increased rent was not in force, then on one view the prior rent would continue until the works had been completed and the necessary steps taken again, under the tenancy agreement and the 1948 Act, to bring about a further rent revision. There is also the difficulty that it would perhaps be unclear, if this were a condition precedent, whether the landlord's failure to carry out the repairs by September 1 had the effect that the new rent would never come into force or whether it had the effect that it would not come into force unless and until the repairs were done.

The arbitrator refers to some of these points and recognises that they produce problems. But as I read his award, he does not appear to decide that the landlord's construction is incorrect; as I see it, he appears to consider whether it was open to the parties to make an agreement which had the effect for which the tenant contended. I say that because, having dealt with these matters, he said this:

The final question is whether it is good law in the light of the authorities and principles quoted to me to find that the parties were free to contract that the payment of rent by the Tenant and the performance by the Landlord of the Landlord's obligations were inextricably linked such that, the Landlord having failed to carry out the works, the correct rate of rent to specify in the Notice to Pay was the rent at the rate previously obtaining prior to March 25 1981 and not the new rent agreed to run as from that date.

It is to be noted that he asked himself the question whether or not the parties were free to contract in the terms for which the tenant contended and that, as it seems to me, only admits of one answer, that they obviously were free so to contract. But the real question is whether they did so on the true construction of the documents.

However, having posed the question in that way, the arbitrator went on as follows:

On behalf of the Landlord it is argued that the situation is on all fours with the authorities cited by the Landlord. On behalf of the Tenant it is argued that the facts are *sui generis* and that this is uncharted territory. I agree with the Tenant's arguments. The letter of April 9 1981 itself propounded by the Landlord's agent is so clear, and express in its terms, that I must interpret that letter such that the new rent was not payable unless the works were completed by September 1 1981.

Accordingly, he held, and awarded, that the notice to quit was invalid because the notice to pay set out incorrectly the rent which was then due.

The landlord then appealed to the county court, as I have already mentioned, and requested that the award be set aside or varied for error of law on its face, apart from the application for remission on the ground of misconduct which I have already mentioned.

In a brief judgment the judge rejected the tenant's construction and concluded that there was an error of law on the face of the award. He referred to two cases: the first was the decision of this court in *Yorkbrook Investments Ltd* v *Batten* (1985) 276 EG 545*; and the well-known statement of principle in *Graves* v *Legg* (1854) 9 Exch 709. Both these cases (and of course many others) were concerned with the question whether mutual covenants were independent or interdependent, in the sense that the performance of the promise, or promises, on one side was a condition precedent to the obligations undertaken in consideration of those promises by the other side. As is well known, the answer to that question is that it depends on the terms of the agreement, on any statutory background if there is any such background, on the surrounding circumstances and, finally, on the consequences which would follow from one construction or the other. All that amounts to is that the agreement has to be construed in order to give effect to the intention of the parties, and the intention of the parties is to be deduced from matters such as those to which I have referred.

In *Graves* v *Legg* Park B said:

In the numerous cases on the subject, in which it has been laid down that the general rule is, to construe covenants and agreements to be dependent or independent according to the intent and meaning of the parties to be collected from the instrument, and of course to the circumstances legally admissible in evidence with reference to which it is to be construed, one particular rule well acknowledged is, that where a covenant or agreement goes to part of the consideration on both sides, and may be compensated in damages, it is an independent covenant or contract, and an action might be brought for the breach of it without averring performance in the declaration, under the old system of pleading.

The judge took the view that that principle was applicable to the true construction and effect of the letter and memorandum. He concluded, without spelling out his reasons, that on the authorities to which I have referred the tenant's agreement to pay the increased rent as from March 1981 was independent from, and not interdependent with, the landlord's carrying out the repairs by December 1 1981. He clearly took the view that any failure to carry out the repairs by that date was something which could be compensated for in damages. In addition, I have already drawn attention to all the problems which arise if this is to be construed as a condition precedent, and I do not agree with the arbitrator, who said that it would be easy to establish whether or not the landlord had performed his obligations thereunder, since any minor complaint could be disregarded as *de minimis*. One can easily foresee very great difficulties which would arise — and I have already mentioned some of them — if this agreement to carry out the repairs by September 1 is to be treated as a condition precedent to rent due from the previous March. As I have indicated, I am by no means sure that the arbitrator took a different view about the true construction of the agreement. But I can only read his language as suggesting that he may have addressed his mind to a different question, namely whether or not there was any reason

---

* Editor's note: See also [1985] 2 EGLR 100.

why the parties should not have made the agreement for the effect of which the tenant contended; otherwise I cannot understand, for instance, his reference to this being a *sui generis* case and uncharted territory. To my mind it is a fairly simple illustration of an arbitrator, or the court, having to decide on which side of the line, within the principle set out in *Graves* v *Legg*, and in other cases, the true construction of these two documents falls. While not abandoning for one moment that the judge's construction was in fact wrong, Mr Thomas did not — I think rightly — submit any reasons why the tenant's construction is to be preferred as a matter of commercial and common sense. His main submission, as I have mentioned, was that the arbitrator's conclusion is to be treated as a finding of fact of what the parties agreed orally, and I have already explained why I cannot accept that.

Second, Mr Thomas submitted that the arbitrator's construction was a possible construction of the words "subject to". I agree that it was a possible construction in the sense that it would not be perverse if an arbitrator so concluded. It may be — but again it is unnecessary to decide — that if the 1979 Act had applied to this award it would have qualified for a right of appeal within the principles of *"The Nema"* and *"The Antaios"* [*Antaios Compania Naviera SA* v *Salen Rederierna AB* [1985] AC 191] cases; or that the proper conclusion would have been that the award was not so clearly wrong that the court should intervene. That, I must emphasise, is not the issue in the present case. The issue in the present case is whether or not the award contained an error of law on its face. I am quite satisfied that it did, whatever may have been in the mind of the arbitrator on the question of construction, and that the learned judge's construction that the landlord's undertaking was not a condition precedent to the new rent coming into force was correct.

Accordingly, I would dismiss this appeal. The arbitrator's award was rightly varied by the judge with the result that the notice to quit was good.

Agreeing, SWINTON THOMAS J said: In my judgment the true construction of the letter of April 9 1981 and of the memorandum of April 12 1981 is clear, as has been set out in the judgment which has just been delivered by Kerr LJ. Accordingly, I agree that this appeal should be dismissed.

The appeal was dismissed with costs, the order for costs not to be enforced without the leave of the court; legal aid taxation of tenant's costs ordered.

Court of Appeal
November 25 1986
(Before Lord Justice KERR and Mr Justice SWINTON THOMAS)

WATTS AND OTHERS v YEEND

Estates Gazette February 28 1987
281 EG 912-916

Agricultural Holdings Act 1948, section 2(1), proviso — Whether oral arrangements constituted a grazing licence only within the proviso or a protected tenancy of an agricultural holding — "Specified period of the year" — Whether a seasonal grazing licence, not referring to specific dates, was within the proviso — Appeal by defendant from decision of county court judge holding that the arrangements in question constituted such a grazing licence and did not create a tenancy — Defendant made a number of criticisms of judge's decision — He submitted that the judge made no express reference to the proviso to section 2(1) and did not have in mind the words "during some specified period of the year" — Defendant's submissions under this head appeared to challenge the established line of authority which decided that a seasonal grazing licence, not identified by reference to specific dates but related to grazing or mowing periods, fell within the proviso — Authorities on this subject reviewed — Defendant also submitted that the judge had erred in regard to the burden of proof, saying at one point that "so far as the onus of proof is concerned each had a liability to establish what he or she contends for" — The judge's finding was also criticised as being contrary to the weight of evidence — Held by the Court of Appeal, rejecting these criticisms, that there was ample evidence to justify the judge's conclusion that the arrangements in this case constituted a grazing licence within the proviso to section 2(1) — There was plenty of authority that seasonal grazing satisfied the requirement of "some specified period of the year" — The judge was aware that the burden of proof as to the application of the proviso was on the landlords and there was no misdirection in this respect — As regards the evidence that the arrangements constituted a seasonal grazing licence, the judge was entitled to take into account, *inter alia,* statutory returns made by the landlord over a period of years indicating the use of land for seasonal grazing — Appeal dismissed

The following cases are referred to in this report.

Butterfield v *Burniston* (1961) 111 LJ 696; 180 EG 597, CC
James v *Lock* [1978] EGD 6; (1977) 246 EG 395, CA
Lampard v *Barker* [1984] EGD 52; (1984) 272 EG 783, CA
Luton v *Tinsey* [1979] EGD 1; (1978) 249 EG 239, CA
Mackenzie v *Laird* 1959 SLT 268; 1959 SC 266
Reid v *Dawson* [1955] 1 QB 214; [1954] 3 WLR 810; [1954] 3 All ER 498; (1954) 53 LGR 24, CA
Scene Estate Ltd v *Amos* [1957] 2 QB 205; [1957] 2 WLR 1017; [1957] 2 All ER 325; (1957) 56 LGR 14, CA

This was an appeal by the defendant, Brian C Yeend, from a decision of Judge Sir Ian Lewis QC at Bristol County Court, granting possession of certain fields at Rockhampton Green, Rockhampton, Berkeley, in the county of Avon, to the owners, Sylvia May Ann Watts, Joan Elizabeth Bendall and Freda Doreen Davis, executrices of the estate of the late Violet Elizabeth Ann, and respondents to this appeal.

Mark West (instructed by Robbins Olivey & Blake Lapthorn, agents for Blakemores, of Tetbury, Glos) appeared on behalf of the appellant; Colin Sara (instructed by Kirby Simcox, of Bristol) represented the respondents.

Giving judgment KERR LJ said: This is an appeal by the defendant from a judgment given by His Honour Judge Sir Ian Lewis QC at Bristol County Court on September 18 1985. The plaintiffs are the personal representatives of the estate of Violet Elizabeth Ann (to whom I shall refer as "Mrs Ann"). The defendant, Mr Brian C Yeend, is a farmer in the same locality.

Mrs Ann died on October 18 1982. She had been the owner of a house and a number of fields at Rockhampton Green, Rockhampton, Berkeley, in the county of Avon. The dispute is the familiar one, as to whether an agreement or arrangement between a landowner and a local farmer in relation to some fields was for a grazing licence or a tenancy protected by the Agricultural Holdings Act.

An agreed plan shows that, apart from her house and a driveway, Mrs Ann owned an adjoining orchard with some derelict farm buildings and four fields, one of which was a large one adjoining the orchard and the house and the other three all separate. The Ordnance Survey number of the orchard was 116 and the numbers of the fields were 117, 126, 184 and 187. It is common ground that Mr Yeend had some use of those fields, the issue being whether he merely had a grazing licence or a tenancy.

The particulars of claim plead as follows in para 3:

At the date of her death

Mrs Ann

had permitted the Defendant to take the grass keep from

the four fields to which I have referred

during a specified part of the year pursuant to the proviso to section 2 of the Agricultural Holdings Act 1948. The Licence so to take the grass keep was terminated upon the death of the deceased, or, alternatively, at the end of the period of the Licence then current, and has not since been renewed.

Asked for particulars of that agreement, the plaintiffs said that the agreement had been made orally, and then I read from the further and better particulars:

The Plaintiffs have no knowledge of the oral words used by either the Deceased or the Defendant nor when such agreement was made. The

Plaintiffs rely on the fact that the Deceased considered there was an agreement for the sale of the grass keep for a specified portion of the year as evidenced by the Deceased's return to the Ministry of Agriculture in 1979 when the Deceased stated, in writing, that 5 hectares of land was let seasonally in 1979 to another person for cropping, haymaking or grazing.

The 5 hectares referred to the area of the four fields, and there was an additional 0.1 hectare referred to in some of these returns which related to the orchard. The acreage of the area is about 12½.

The amended defence and counterclaim pleaded in substance as follows; I refer to para 4:

By an oral agreement made between the Defendant and the deceased in or about

it was originally December 1970, and that was amended to the spring of 1968; nothing turns on the fact that there was a second thought about the date

[the deceased] granted to the defendant the exclusive occupation of approximately 12.5 acres of agricultural land, together with the use of certain buildings and a yard adjacent thereto, all situate at Rockhampton Green and so forth, at an annual rent of

it was originally £100, but that has been amended to £60

payable half-yearly in advance on December 1 and June 1 in each year.

Then particulars are given of the agricultural land, buildings and yard with reference to the Ordnance Survey numbers, including the orchard and the farm buildings in it. The reference to the yard is to the area surrounded by those buildings.

The defence gives lengthy particulars of what the defendant claims to have done on the land to make good his claim that he was a tenant. There was then a counterclaim for one of two alternative declarations. The first was that he had a tenancy of the land and buildings. The alternative declaration claimed was that he had a licence in respect of the land and buildings which took effect as a tenancy from year to year pursuant to the provisions of section 2 of the Act, and a declaration that this was a protected tenancy.

The defendant farmed other land in partnership with his father, who also gave evidence and played a part in the history. He also rented some other land and had a normal, formal written tenancy agreement concerning that other land. But, as can be seen from what I have already mentioned, it is common ground that the agreement with which this appeal is concerned was informal, oral and was never referred to in any document which passed between the parties during the 14 years or so from 1968 to 1982 when it is common ground that it was in existence.

It is also common ground, I think, that the originally required annual payment was £60; it then rose to £100, and in about 1976 (although nothing turns on the date) it went up to £150. It was payable in two instalments, one on July 1, as the defendant says, and one in December. Whatever the agreement between Mrs Ann and Mr Yeend was, it is also clear from some earlier documents which have survived that it followed on from another yearly arrangement which she had, in that case certainly limited to the taking of the grass, with a firm called Sandoe Luce Panes & Johns. That was, on any view, of a different nature. They had the right to take and dispose of the grass each year, I think from 1966 to 1968, but on the basis that if they took it by mowing, as appears to have been the position throughout, they would then advertise and sell it and charge Mrs Ann with the expenses and a commission. Both sides have sought to draw attention to the difference between that arrangement and the one with Mr Yeend, which was certainly concerned with his actually making use of the land. But the judge clearly felt unable to draw any conclusion as to the reasons why different arrangements were made. I am in exactly the same position. It seems to me that no conclusion can be drawn as to whether, in making a change, Mrs Ann wanted to create a tenancy or merely what I have referred to as a grazing licence. It should be mentioned that she and the Yeend family were originally close friends; in particular, she was friendly with Mr Yeend's mother and also with his father. But there was then some falling out because Mrs Ann had not been invited to the christening of a child of the family. That again is not something on which I can base any inference, let alone conclusion; and nor did the judge. That is the background.

I should say at once that the judge, having heard witnesses on both sides for a considerable time, came to the clear conclusion that the right which Mr Yeend had was limited to a grazing licence which, although he does not mention the point expressly, in his view obviously fell within the proviso of section 2(1) of the Agricultural Holdings Act 1948. He reached the conclusion on a number of grounds, to which I shall be referring.

First I must read the section:

Subject to the provisions of this section, where under an agreement made on or after the first day of March, nineteen hundred and forty-eight, any land is let to a person for use as agricultural land for an interest less than a tenancy from year to year, or a person is granted a licence to occupy land for use as agricultural land, and the circumstances are such that if his interest were a tenancy from year to year he would in respect of that land be the tenant of an agricultural holding, then . . . the agreement shall take effect, with the necessary modifications, as if it were an agreement for the letting of the land for a tenancy from year to year.

That states the general position that the grant of an interest less than a tenancy from year to year, or for a licence in relation to agricultural land, will take effect as an agreement for the letting of the land for a tenancy from year to year; and that of course has the effect that it enjoys the protection of the Act and other statutory provisions in a number of respects.

There then follows the important proviso, and it is on this that most of the cases have turned:

Provided that this subsection shall not have effect in relation to an agreement for the letting of land, or the granting of a licence to occupy land, made (whether or not the agreement expressly so provides) in contemplation of the use of the land only for grazing or mowing during some specified period of the year.

Before I come to the evidence, I should deal with two points of law which, as Mr West submitted, show that the judge had approached the matter on an erroneous basis. Unlike Mr Sara who represents the plaintiffs, Mr West did not appear below. In connection with Mr West's submissions it should also be noted that this is a very experienced judge sitting habitually in that part of the country, and that the problem raised by this case is not at all an unfamiliar one in that county court and, no doubt, many county courts in rural areas. In fact we were referred to another decision of this court which had come from the same county court and in which the same firm of solicitors was involved on behalf of the plaintiffs.

The first point of law is based on the fact that the judge makes no express reference to the proviso to section 2(1). Mr West does not go so far as to suggest that the judge did not have the proviso in mind. It was pleaded, and of course he had it in mind. But what Mr West submits is that the judge did not have in mind the words "during some specified period of the year".

The position in that regard is as follows. It is clear from the judgment and the pleadings, and from what Mr Sara has told us, that the contested issue in the court below was, on the one hand, the plaintiffs' contention that this was a grazing licence — that is to say, a grass-keep agreement, or whatever expression one chooses to use; and on the other hand the defendant's positive contention that he had a full tenancy from year to year. That was the issue, and everyone — the judge, the parties and their legal advisers — assumed, and had in mind, that if the agreement was only in the nature of what I have called a grazing licence, then it would fall within the proviso.

Mr West's position is this. He does not dispute — indeed, he cannot dispute — that that was how the case was conducted and contested below. But, as I follow his submission, he says that this involves the conclusion that everyone proceeded on an erroneous basis of law. What he submits is that, having regard to the evidence, which undoubtedly gave Mr Yeend some rights in relation to the land, it was not sufficient to conclude, as the judge did, that the agreement was merely one for a grazing licence — or, as he referred to it, a seasonal grazing licence — because so to look at it ignores the words "during some specified period of the year". On this appeal Mr West accordingly seeks to take a point which I am quite satisfied was never taken below. But, of course, if he is right in law he is entitled to take it. At any rate I proceed on this basis without deciding whether it is possible for parties, by some agreement, implied or tacit, to produce the same effect as the proviso even if its precise terms are not satisfied.

But I am quite clear that the judge did not err in his approach to the proviso, although he makes no reference to it. He referred repeatedly to a grazing licence, or a seasonal grazing licence, and he clearly had in mind that that would be sufficient, if established, to attract the protection of the proviso. For instance, he says in his judgment:

The Defendant contends that he had a protected agricultural tenancy. The Plaintiffs contend that it was a grazing licence. So far as the onus of proof is concerned each had a liability to establish what he or she contends for.

I shall come back to the last sentence. Then he refers to one of the agricultural returns which were put in evidence in this case, to which he clearly attaches considerable importance. He quotes from that by saying:

It specifically states "I include land on this holding let by you for seasonal grazing".

Finally on this aspect, having reviewed some of the evidence and contentions, he says:

All this confirms the grazing agreement. There was further friendly and informal arrangement that it was to continue on indefinitely but the land was not to be used throughout the year.

Pausing there, Mr West sought to submit that this meant that what was not to continue throughout the year was the "further friendly and informal arrangement"; not the grazing agreement. In my view, that is a clear misconstruction of this passage. When he said "All this confirms the grazing agreement" and went on to say that there was "a further friendly and informal arrangement that it was to continue on indefinitely", he was saying that there was a grazing agreement, and there was a further informal arrangement that "it", the grazing agreement, was to continue on indefinitely — that is to say, in successive years — unless somebody changed their mind. Then he went on to say "but the land was not to be used throughout the year". That was clearly his view of what had been agreed. At the end of the judgment he said: "As to the terms of that arrangement, what was agreed here, I am quite certain, was a grazing licence".

The issue on this aspect is therefore whether a seasonal grazing licence, not related to any specified period in the year but merely to the grazing or mowing periods, falls within the proviso to section 2(1). In that regard I am clear that Mr West is not correct in submitting that by approaching the case on that basis, which, as I have said, was the basis on which both sides and the judge approached it, there was any error of law.

We have been referred to a number of cases and I begin by mentioning them all for the sake of completeness. They were: *Reid* v *Dawson* [1955] 1 QB 214; *Scene Estate Ltd* v *Amos* [1957] 2 QB 205: *Butterfield* v *Burniston*, a decision from the Harrogate County Court reported in (1961) 111 LJ 696; *Luton* v *Tinsey*, a decision of this court reported in [1979] EGD 1; *James* v *Lock* (1977) 246 EG 395; *Lampard* v *Barker* (1984) 272 EG 783; and finally, because I noticed a reference to it in *Butterfield* v *Burniston*, the court referred both parties to an important decision of the Court of Session, *Mackenzie* v *Laird* 1959 SLT 268.

It seems to me that on the authorities, particularly *Scene Estate Ltd* v *Amos* and *Mackenzie* v *Laird*, the judge was clearly entitled to conclude that if the parties contemplated a seasonal grazing licence which, being seasonal, would *ipso facto* be for less than a year, then the requirement that the licence must be for some specified period of the year was satisfied.

In *Scene Estate* v *Amos*, at p 211, Denning LJ (as he then was) put the position quite generally as follows in the context of the proviso:

I do not think that the word "contemplation" in the proviso should be given the meaning which Mr Megarry seeks to put upon it. In my opinion the object of the word "contemplation" in the proviso is to protect a landlord who has not expressly inserted a provision that it is for grazing only, or for mowing only, or that it is for a specified part of the year; but, nevertheless, both parties know that that is what is contemplated. Often a landlord may let a field to a man by word of mouth, saying: "You can have the field this year the same as you had it last year". Both sides mean it to be for grazing only and mean it to be only for a few weeks of the spring, but they do not say so expressly. In such circumstances, even though nothing is expressed in the agreement, nevertheless the landlord can still take advantage of the proviso. That seems to me to be the real object of introducing the "contemplation" of the parties.

Denning LJ's reference to "a few weeks of the spring" was clearly given only as an illustration or in the context of the facts of that case, which was concerned with periods of three months in each year. Parker LJ said that he agreed and that the words "specified period of the year" were equivalent to "specified part of the year".

In the *Butterfield* case in the Harrogate County Court a grazing tenancy granted, or understood to have been granted, by reference to the grazing season was treated as a term well understood by farmers and to be within the scope of the proviso. His Honour Judge McKee cited and followed *Mackenzie* v *Laird*, the Scottish case to which I have referred.

In that case there was a written agreement which provided expressly for a seasonal let for grazing purposes. The issue was whether this satisfied the requirement of the proviso, viz an agreement for the use of the land only for grazing or mowing during some specified period of the year. The three members of the Court of Session all concluded that the proviso was satisfied.

The Lord Justice-Clerk, Lord Thomson, said:

The only question at issue is whether in the absence of definite terminal dates it is shown that what was in contemplation was some specified period of the year. In the present case there are no dates fixed as the beginning and end of the period but I do not see that fixed or specific dates for the beginning and end are essential. What the proviso is excepting is the use of the land only for grazing or mowing and these uses are in essence seasonal. None can give in advance in any particular year or indeed in any particular locality a specific date when the land will be ready for the start of either operation nor can it be said in advance when either will finish. What matters to the agriculturalist is the state of the land. What the tenant wants and what the landlord is prepared to give is the season's grass from the appropriate moment which nature directs and to the subsequent moment when nature calls a halt. Accordingly once it is admitted that what we have to do with is grazing and that it is for part of a year only and once one accepts as the arbiter accepted that what is meant by "grazing season" or "seasonal let" is unambiguous and well understood in farming circles, there is no difficulty in saying that this case falls within the proviso as what was in contemplation was "the use of the land only for grazing or mowing during some specified period of the year".

Lord Patrick agreed and said, at the bottom of p 271, referring to the findings of the arbiter:

. . . some such lettings may specify particular dates as termini in the spring or autumn, but it is a common practice not to specify dates; and that farmers know well what is comprehended in a let of seasonal grazing. . . The sole argument against that view is that the contemplated use must be use for some definite period of days, weeks or months in the year. I do not find any such requirement in the language of the proviso of subsection 2(1). The period must be specified, not exactly defined, and, if the period of the year is one which is capable of reasonably clear ascertainment, the language of the proviso is satisfied.

Lord Mackintosh agreed, and said:

I think that the expression "specified period" as used in the . . . proviso is not limited in its meaning to any period fixed by dates but includes any period which is so named or described as to be identifiable by persons versed in agricultural matters.

He said that he agreed with the sheriff in his conclusion that a seasonal let of grazings

means a let of grazings for a specified period, ie for a period so named or described as to be identifiable in agricultural circles, and that accordingly the proviso applied . . .

I unhesitatingly follow that decision. The Act, or at any rate that part of it, applies in Scotland just as it does in England. It would obviously be undesirable for the Court of Appeal to differ from the Court of Session. But the conclusion is really a matter of commonsense. The courts might of course have interpreted the words "for a specified period" very strictly. For good reason they have not done so. On every occasion when this issue arose directly, they have approached it in the same way as the parties and the judge in this case. That deals with the first point raised by Mr West.

I can take the second point much more shortly. Mr West submitted that the judge had misdirected himself about the burden of proof. He submitted, undoubtedly rightly, that in order to attract the protection of the proviso the burden of proof is on the landowner. The question is, therefore, whether this judgment shows that the judge misdirected himself in that regard. In my view it shows no such thing.

I have already referred to the passage in question where the judge said that the defendant contended that he had a protected agricultural tenancy and the plaintiffs contended that it was a grazing licence, and added:

So far as the onus of proof is concerned each had a liability to establish what he or she contends for.

Given that there was a counterclaim claiming the declarations which I have read, that statement is perfectly correct. Moreover, I do not think that the judge overlooked the fact that in order to recover possession, against the background of the undisputed evidence that Mr Yeend had some rights in relation to this land, the effective onus of proof was on the plaintiffs.

Further on in his judgment the judge said:

The important thing is that the inference is that the arrangement is to be treated as a protected tenancy unless there is evidence

and he went on:

and I find that here there is evidence to the contrary.

In relation to the provision for payment, to which I have already referred, the judge said:

I have to bear in mind that it does not appear to be in dispute that the money was to be paid half-yearly. I don't overlook that this points to a tenancy.

Finally, at the end of his judgment, I have already read the final sentence, in which he said:

As to the terms of that arrangement, what was agreed here, I am quite certain, was a grazing licence.

Having regard to those passages I find it impossible to conclude that the judge misdirected himself as to the onus of proof, which he obviously realised lay on the plaintiffs, to bring themselves within the proviso.

Mr West also faintly argued that the judge may have gone wrong when he said that the inference was that the arrangement was to be treated as a protected tenancy "unless there is evidence and I find that here there is evidence to the contrary". He clearly did not merely mean "unless there is *some* evidence to the contrary". As shown by his judgment as a whole, he meant that he had to be satisfied that there was sufficient evidence to satisfy the burden of proof on the plaintiffs to displace the *prima facie* inference that an agreement for the use of land referable to a year, whether it be a tenancy or a licence, is a protected tenancy.

I then come to the third aspect on which Mr West relied. This concerns the bulk of the evidence and the main parts of the judgment concerned with the findings of fact. Mr West submits that the judge's decision is contrary to the weight of the evidence. The position in that regard is as follows: Mrs Ann, as I have already mentioned, was dead. A number of witnesses were called on both sides and the judge had to do his best with the evidence before him. The witnesses on the side of the plaintiffs included some relations of Mrs Ann and a Mr Freeman, who had worked for Mrs Ann and had had the use of her garden for a number of years, and also a Mr Child, who farms an adjacent farm. On the side of the defendant the main witness was of course Mr Brian Yeend himself. In addition a Mr Weston, who had worked for him for a number of years, gave evidence on the defendant's behalf. Finally, Mr Christopher Yeend, Mr Yeend senior, at the ripe age of 81, also gave evidence.

The only witness who could speak about the actual arrangement was of course Mr Yeend, the defendant. In that regard I am bound to say that the judge was quite clear about the impression which he formed of his evidence, after hearing him for a long time, as can be seen from the lengthy notes of evidence. He said:

Let me say at the outset that I have come to the clear conclusion that Mr Yeend is not a witness on whom I can rely.

In the last paragraph of his judgment he said:

I have therefore come to the clear conclusion that the right granted here is that of a grazing licence. I have said that I do not accept Mr Yeend's evidence. He was confused about a whole series of matters and his recollection was at fault regarding the date of the original arrangement and the amounts and dates of variations in payment.

I have already read the last sentence of the judgment. I also notice from the judge's notes of evidence that Mr Yeend said at one point:

It was arranged that I could mow and graze cattle as I wanted — nothing said of the time of the year I could use the building.

So it is quite clear that the judge did not accept from Mr Yeend that the original agreement was for a full agricultural tenancy.

I must now briefly refer to the evidence on which the judge mainly relied for his conclusion that this was a seasonal grazing licence. The first part of his judgment is concerned with certain returns which were made by Mrs Ann for agricultural purposes. We have returns for a few of the 14 years covered by the agreement and I do not propose to go through them all. But for instance, in 1975 there is a reference to land let for seasonal grazing, against which the words "Let to B C Yeend" appear. I should say in that connection that Mr Yeend senior who, as I have said, was a fairly close friend of Mrs Ann, helped her to complete these returns, and a number of the entries appear in his writing. In the return for 1978 a number of boxes had to be completed by ticks where applicable. Box 92 had to state the total number of cattle and calves and the figure six was there inserted by Mr Yeend senior. There followed Box 93, which also contained a tick against the following wording: "Please tick this box if all the cattle entered at 92 above belonged to someone else and you are only providing grazing". Similarly in relation to the return for 1979 there was the following entry. Under the heading "Seasonal use of Land",

Box 41 contained the following wording: "Area of land let seasonally this year to another person for cropping, haymaking or grazing". The words "5 hectares" were inserted, which the judge thought, no doubt correctly, referred to the area of the four fields to the exclusion of the orchard. There was the same type of entry as before in Boxes 92 and 93, with a slight increase in the number of cattle and calves. It is true that at the end of that return, in what was apparently Mrs Ann's handwriting in a box headed "Land given up", she put the words "No Change", and then added "land is let to B C Yeend", giving his address. But that is of no weight against the entries referring to grazing. Mr Yeend senior said, in relation to those entries which clearly and expressly point to seasonal grazing, that they must have been mistaken, although he agreed that he had helped Mrs Ann to complete these returns.

As I have already mentioned, Mr Yeend senior farmed in partnership with his son, the defendant. He was clearly experienced and must have known — as, I think, did everybody in this case — the crucial difference between a seasonal grazing licence on the one hand and an agricultural tenancy on the other. It must also be borne in mind that Mrs Ann was only required to make these returns if she had *not* let the land in question to Mr Yeend. On the other hand it was for Mr Yeend to make returns which included this land if he had had a tenancy of it as he claimed. The fact is that it was Mrs Ann who made the returns, with the knowledge and assistance of Mr Yeend senior, and that the defendant, Mr Yeend junior, made no returns including this land. Those facts speak for themselves. Of course I agree with Mr West when he emphasises that Mr Yeend, the defendant, never saw these returns and had no part in them, and that they could be construed as having been used by Mrs Ann to serve her own purposes. But in the circumstances that is an extremely far-fetched suggestion.

Mr West has also submitted that one should not look at events after the original agreement in order to interpret its effect. But, rightly, he does not seek to go to the full length of that submission, which would often be impracticable in circumstances such as those envisaged by this proviso. The cases show that one has to have regard to what was in fact done on the land. The court is not concerned with construing or interpreting a written agreement and therefore precluded from deciding what the parties meant by the words which they used by looking at their subsequent conduct. The proviso refers to what was in the contemplation of the parties concerning the use of the land. In that context one must have regard to what was happening on the ground, particularly if the arrangement continued over a period of years.

Accordingly, I conclude that the judge was entitled to take these returns into account. For the reasons which I have explained, they clearly point in only one direction.

The remainder of the evidence was concerned with accounts of the various activities which, according to the various witnesses, Mr Yeend carried on or did not carry on in relation to the land; to what extent he used the buildings; to what extent he worked the land as if he were a tenant of it; or whether what he did was consistent with his having only a seasonal grazing licence. In that regard two main aspects were important. First, his use of the buildings and whether or not he used them and the land for the whole of the year or only during the grazing season. Second, whether the work that he did on the land was compatible with his being a tenant, with full responsibilities for the land, or whether Mrs Ann retained these responsibilities on the ground that she had not let it. In that connection there were references to hedging, ditching, gates and matters of that kind.

What the judge said about those parts of the evidence is as follows:

So far as the buildings are concerned I accept Mr Bendall's evidence that he did not see the buildings used in the winter months. So far as the buildings are concerned I accept that the buildings were used at times but not throughout the year. I believe that the buildings were used contemporaneously with the grazing and that if the land was wet it was reasonable that they must have been.

A little later on, dealing with hedging and ditching, his finding is favourable to Mr Yeend to the extent that Mr Yeend did the work or had it done. But the important question was who paid for it. The judge said:

I accept that Mr Yeend did the hedges and ditches and fixed the gates. But I bear in mind that Mrs Ann was asked by him to pay the bill and she agreed. The importance of this was that he asked her for money though she died before she could pay. This would not be the position if it was a tenancy. He also asked her about the draining. If Mr Yeend had a tenancy he would have drained without asking. Also with the draining he asked her for money although she could not in fact afford it. This was done really to improve the grazing for his own benefit and to make better use of the land subject to the grazing licence.

That was the judge's conclusion on a very large part of the evidence through which Mr West had taken us. Some aspects supported either view. But I find it quite impossible to hold that the judge was not entitled to reach the conclusions about the buildings, the hedging and ditching and so forth which I have just read. On the contrary, on the weight of the evidence, I think it supported his findings.

In all the circumstances I am left in no doubt but that this appeal must be dismissed.

Agreeing, SWINTON THOMAS J said: As I see it, the first of the two main issues which arise on this appeal is whether the appellant's occupation was by way of a tenancy or by way of a grazing licence, and whether before the learned judge it was proved by the plaintiff that the occupation was by way of a grazing licence.

On that issue the trial judge found quite clearly that the occupation was by way of a grazing licence. On the last page of his judgment the judge said:

As to the terms of the arrangement, what was agreed here, I am quite certain, was a grazing licence.

That issue was a question of fact; in my judgment there was ample evidence upon which the learned judge was entitled to come to the conclusion to which he in fact came. My lord has reviewed that evidence in the course of his judgment and it is not necessary for me to do so again. The learned judge was entitled to rely, as he did, upon the fact that he found the appellant to be an unsatisfactory witness, upon whose evidence he could not rely.

The second, and in my view crucial, question is whether the grazing licence was made in contemplation of use of the land for grazing during some specified period of the year, as set out in the proviso to section 2(1) of the Act of 1948. As has been pointed out, those words are set out in para 3 of the particulars of claim; in the defence that paragraph is formally denied but, more relevantly for this purpose, there is a counterclaim. In para 12 of the defence and counterclaim the appellant alleges:

On a true construction of the said agreement and in the events which have happened the Defendant at all material times held the said land, buildings and yard from the deceased as a tenant thereof and such tenancy is protected by the provisions of the Agricultural Holdings Act 1948.

He then makes two alternative counterclaims:

(a) A declaration that on a true construction of the said agreement made between the Defendant and the deceased in or about the Spring of 1968 and in the events which have happened, the Defendant holds a tenancy of the said land, buildings and yard and that such tenancy is protected by the Agricultural Holdings Act 1948.

Then alternatively:

(b) A declaration that if the said agreement made between the Defendant and the deceased in or about the Spring of 1968 was a licence to occupy the said land, buildings and yard such licence has taken effect as a tenancy from year to year pursuant to the provisions of section 2 of the said Act and that such tenancy is protected thereby.

In those circumstances, and bearing in mind the experience of the learned judge, I find it quite impossible to accept that he would not have had in the forefront of his mind in this case the terms of the proviso to section 2(1) of the relevant Act.

The learned judge had said, on the penultimate page of his judgment, that there was further friendly and informal arrangement that it — quite clearly that refers to the grazing agreement — was to continue on indefinitely, but the land was not to be used throughout the year.

Clearly, in my view, the judge was there finding that the appellant's occupation was by way of a seasonal grazing licence. "Seasonal grazing" is a term which is well understood in agricultural circles and, as was said *Mackenzie* v *Laird*, seasonal grazing does not have to be for the same period of time during every year, but is none the less in contemplation of the use of the land for grazing during some specified period of the year. In my judgment, that clearly is exactly the position in the instant case.

Accordingly, I am quite satisfied that the plaintiffs were entitled to possession of the land the subject-matter of the claim, and that the learned judge came to the right conclusion, and therefore I, too, would dismiss this appeal.

The appeal was dismissed with costs. Possession was ordered within 28 days of judgment.

ARBITRATION

Chancery Division
December 10 1986
(Before Mr Justice WARNER)

WARRINGTON AND RUNCORN DEVELOPMENT CORPORATION v GREGGS PLC

Estates Gazette March 7 1987

281 EG 1075-1076

Arbitration Act 1979, section 1(3)(b) — Arbitration under rent review clause in lease of shop — Application by lessors to court for leave to appeal on a question of law arising out of award — Test to be applied by court in deciding whether to give leave to appeal — Lease provided for the current market rental value for the purposes of the rent review to be determined, in default of agreement between the parties, by an arbitrator appointed by the president of the RICS — The applicants complained about the way in which the arbitrator had dealt with a particular "comparable" in arriving at his determination — The current market rental value of the subject shop was that of a shop with substantial fixtures and fittings — The "comparable" in question was another shop in the same shopping centre let for a term of 20 years at a rent of £7,500 per annum with a premium of £9,000 paid for the landlord's fittings — The arbitrator in valuing the subject property had, in agreement with the submissions of the lessees' surveyors, but contrary to those of the lessors' surveyors, disregarded the premium of £9,000 paid in respect of the "comparable" — The point of law raised by the applicants was that there was no justification for this attitude — It was argued on behalf of the lessees that the point taken by the lessors was not a point of law at all but one of valuation practice — However, the question whether the arbitrator was entitled as a matter of law to deal with the premium in the way that he did was the very question which, if leave to appeal were given, would have to be decided by the judge hearing the appeal — The present issue before Warner J was what test should be applied in deciding whether to give leave to appeal — Held, applying the test adopted by the Vice-Chancellor in *Lucas Industries plc* v *Welsh Development Agency*, and distinguishing the cases of *The Nema* and *The Antaios*, that the question was whether he was left in real doubt as to whether the arbitrator was right in law — The answer to that was "yes" — Following further the Vice-Chancellor's guidance, Warner J did not state his reasons for this conclusion — He was satisfied, however, as required by section 1(4) of the 1979 Act, that the determination of the question of law concerned could substantially affect the rights of one of the parties, namely, the lessors, if only because there were pending rent reviews in relation to 11 other shops in the centre — Finally, Warner J gave leave to appeal to the Court of Appeal against his decision, because a question of principle was involved, that is, whether in a rent review case, where the dispute was not as to the interpretation of the review clause itself, as it was in the *Lucas Industries* case, it was nevertheless correct to apply the Vice-Chancellor's test rather than those laid down in *The Antaios* and *The Nema*

The following cases are referred to in this report.

Antaios Compania Naviera SA v *Salen Rederierna AB* [1985] AC 191; [1984] 3 WLR 592; [1984] 3 All ER 229, HL
Lucas Industries plc v *Welsh Development Agency* [1986] Ch 500; [1986] 3 WLR 80; [1986] 2 All ER 858; [1986] 1 EGLR 147; (1986) 278 EG 878
Pioneer Shipping v *BTP Tioxide ("The Nema")* [1982] AC 724; [1981] 3 WLR 292; [1981] 2 All ER 1030, HL

This was an application by Warrington and Runcorn Development Corporation, lessors of a shop at 31 Dewhurst Road, in Birchwood Shopping Centre, Warrington, for leave to appeal under section 1(3)(b) of the Arbitration Act 1979 on a question of law arising out of the award of Mr J G Fifield FRICS under a rent review clause in the lease of the property to the respondents, Greggs plc.

John M Male (instructed by Speechly Bircham) appeared on behalf of the applicants; John V Martin (instructed by Maughan & Hall, of Newcastle upon Tyne) represented the respondents.

Giving judgment, WARNER J said: This is an application under section 1(3)(b) of the Arbitration Act 1979 for leave to appeal on a question of law arising out of an award made on June 10 1986 by Mr J G Fifield, a Fellow of the Royal Institution of Chartered Surveyors, as arbitrator under a rent review clause contained in a lease granted by the applicant, the Warrington and Runcorn Development Corporation (which I will call "the corporation") to the respondent, Greggs plc (which I will call "Greggs"). The lease, which was dated June 2 1981, was for a term of 25 years from January 27 1981. The premises thereby demised were a shop, 31 Dewhurst Road, in a modern shopping centre built by the corporation in Warrington which is called the Birchwood Shopping Centre. The rent review clause provides for rent reviews in the fifth and each subsequent fifth year of the term and that for the purposes of each review the current market rental value of the demised premises shall, in default of agreement between the parties, be determined by an arbitrator appointed by the president of the Royal Institution of Chartered Surveyors. Mr Fifield was, it appears, actually appointed by a vice-president of that institution, but nothing turns on that.

Mr Fifield determined that the rent payable for 31 Dewhurst Road with effect from January 27 1986 should be £7,300 pa, which was rather more than had been contended for on behalf of Greggs (which was £6,000) but substantially less than had been contended for on behalf of the corporation (which was £11,000). Mr Fifield reached his decision on the basis of two rounds of written submissions by the parties' respective surveyors, and of inspections that he himself made of the demised premises and of the premises cited to him by the parties as "comparables". There was no cross-examination of the parties' surveyors. The suggested comparables were five other shops in the same shopping centre. The basis of the corporation's application is the way in which Mr Fifield dealt with one of those comparables, 19 Dewhurst Road, of which the corporation is also the owner. It is common ground, so I understand, that no 19 was relet in the open market by the corporation itself on March 4 1985 as a fitted-out shop to a tenant called Bianca Needlecraft for the retail sale of wool and related items and that the previous tenant of that shop had been in the same trade.

In considering the submissions of the parties, and the decision of Mr Fifield relating to 19 Dewhurst Road, one has to bear in mind that, as a result of the combined effect of a provision in the rent review clause itself and of an agreement between the parties which preceded the lease, the relevant current market rental value of no 31 was not that of the shop as a shell but that of the shop with substantial fixtures and fittings which were listed in a schedule to that agreement and which were put in by Greggs.

No 19 Dewhurst Road was cited as a comparable by the corporation's surveyor (who was in fact its chief estates officer) in his initial written submissions. He mentioned, among other things, that those premises had been let to Bianca Needlecraft by a lease for a term of 20 years from March 4 1985 at a rent of £7,500 and that a premium of £9,000 had been paid by the lessee for landlord's fittings on the signing of the lease. In working out the rental value of no 19 for the purposes of comparison he took the premium into account.

Greggs' surveyor had not relied on no 19 as a comparable in his initial submissions. In his subsequent submissions commenting on the initial submissions of the corporation's surveyor, he expressed the opinion that there was no benefit to be derived from consideration of no 19 as a comparable because the information given by the corporation's surveyor about it was, and I quote, "wholly inadequate". He pointed out that, and I quote again, "premiums are paid for a variety of reasons, not always relating to value". While acknowledging, after having spoken to the tenant, that "undoubtedly, in this case the premium related to cost and, specifically, the cost of fitting out the unit", he added: "However, this unit was formerly occupied by an identical trader and part of the premium must relate to goodwill."

I must now, I fear, use some more surveyors' jargon. One basis on which surveyors estimate the rental value of a shop is known, so it appears, as "devaluing to zone A", which means, so far as I can understand and so far as is relevant, ascertaining the rental value of the shop in pounds per sq ft by reference to the value of the front part of it. Mr Fifield, at the beginning of the statement of his reasons for his award, announced that he would adopt that method.

The corporation's surveyor in his submissions had, taking into account the premium, reached a rental value for no 19 "in terms of zone A" of £14.95 per sq ft. Ignoring the premium it would have been £10.44 per sq ft.

In his statement of reasons Mr Fifield gave brief details of the five "comparables" that had been cited to him and indicated why he proposed to discard one of them. He then dealt in successive paragraphs with a number of specific points. One of those paragraphs is in these terms:

The Landlords have amortised the premium of £9,000 in respect of 19 Dewhurst Road. The Tenant's Surveyors have stated that this should be totally disregarded, as it was a payment for fixtures, fittings and, possibly, goodwill, as the premises were previously occupied by a similar trader. I agree with the Tenant's Surveyors and I have not made any allowance in my valuation.

Mr Fifield then commented on each of the four "comparables" that he had retained. He took one of them, a letting of a shell unit at a rental value in terms of zone A of £8.32, "as the minimum", by which I understand him to have meant the bottom of the bracket by reference to which he would assess the rent for no 31. On no 19 he commented as follows:

Devaluing to Zone A £10.44 plus a premium of £9,000 some ten months before this review. I have already stated earlier that I agree with the Tenant's Surveyors that the premium should be ignored. Both parties agree that this shop is better located than the subject premises and I have, therefore, taken this rent, making an allowance for "age", as the maximum.

Mr Fifield concluded that the appropriate rate for no 31 was "zone A £10" plus an allowance of $7\frac{1}{2}\%$ for the fact that no 31 has a return frontage.

The point of law that the corporation wishes to take is that there was, in the circumstances, no justification for Mr Fifield's ignoring the premium paid in respect of no 19, or at all events no justification for his ignoring it completely. Mr Male, who appears for the corporation, submits that, moreover, in so doing Mr Fifield went outside the evidence because Greggs' surveyor had not said that the premium should be ignored but that, because of the uncertainties concerning the premium, no 19 should not be used as a "comparable" at all. Mr Male concedes that that point of law is not accurately formulated in the notice of motion that is before me and that if I grant leave to appeal he will need leave to amend it.

It is material that there are at present pending rent reviews concerning 11 other shops owned by the corporation in the Birchwood Shopping Centre. In one of those cases Mr Fifield has been appointed arbitrator. In the other 10 no arbitrator has yet been appointed, but it is common ground that Mr Fifield may well be appointed in each of them, because he is likely to be regarded by the president of the Royal Institution of Chartered Surveyors (or by its vice-president) as particularly qualified to deal with them. Even if Mr Fifield is not appointed, his decision in the present case, if it stands, is likely to be used as evidence in those other cases — see *Bernstein & Reynolds' Handbook of Rent Review*, p 815, para 8-5. So this is by no means a "one-off" case from the point of view of the corporation.

Mr Martin, who appears for Greggs, submits in the first place that the point taken by the corporation is not a point of law at all, but one of valuation practice. In support of that submission Mr Martin referred me to the paragraph headed "Premiums" at p 816 of *Bernstein & Reynolds' Handbook of Rent Review*. I will not take up time reading that paragraph now (though I have read it over and over again) because it seems to me that the question whether Mr Fifield was entitled, as a matter of law, to deal with the premium here in the way that he did is the very question that, if I give leave to appeal, the judge who hears the appeal will have to determine. So I turn to the question what is the test that I should apply in deciding whether to give leave to appeal.

As to that, Mr Male submits that the correct test is that adopted by the Vice-Chancellor in *Lucas Industries plc* v *Welsh Development Agency* [1986] 3 WLR 80*, ie whether I am left in real doubt whether the arbitrator was right in law. Mr Martin submits that that case is distinguishable and that I should apply the test laid down by the House of Lords in *The Nema* [1982] AC 724 and *The Antaios* [1985] AC 191, ie whether I am satisfied that there is a strong *prima facie* case that the arbitrator was wrong in law. Mr Martin does not suggest that I should apply the even more stringent test laid down by the House of Lords in *The Nema* and *The Antaios* for the "one-off" type of case, ie whether I am satisfied that the arbitrator was obviously wrong.

Mr Martin points out, quite rightly, that none of the three considerations mentioned by the Vice-Chancellor at p 83 B-E of the report of the *Lucas Industries* case is material here, because we are not here concerned with the interpretation of the rent review clause itself. In particular, the third consideration, which the Vice-Chancellor regarded as the most important, does not apply. The decision in this case will not be material on future reviews between the same parties under the same lease, because by the time the next review is due under that lease the letting of no 19 on March 4 1985 will be far too remote in time to be acceptable as a "comparable". Nor can Mr Fifield's decision give rise to any material issue estoppel.

The choice before me is between holding, as Mr Martin urges me to do, that the test adopted by the Vice-Chancellor in the *Lucas Industries* case is applicable only in the type of case with which he was dealing, that is a case turning on the interpretation of the rent review clause itself, and holding, as Mr Male urges me to do, that that test is applicable to any arbitration arising from a rent review clause, with the possible exception of a true "one-off" case.

I have come to the conclusion that the right course for me to adopt is that urged upon me by Mr Male. The considerations that have led me to that conclusion are these.

First, I respectfully agree with the Vice-Chancellor that neither in *The Nema* nor in *The Antaios* was the House of Lords considering problems raised by rent review clauses. A close examination of the speech of Lord Diplock in *The Nema* evinces that he, and one must presume the other members of the House of Lords who agreed with him, had in mind primarily arbitrations arising from contracts made in the course of what may be described broadly as international commerce. That is apparent from the whole tenor of his speech, from the first paragraph onwards. The actual principle laid down by that decision was stated by Lord Diplock at p 739 F-H of the report in these terms:

The judicial discretion conferred by subsection (3)(b) to refuse leave to appeal from an arbitrator's award in the face of an objection by any of the parties to the reference is in terms unfettered; but it must be exercised judicially; and this, in the case of a dispute that parties have agreed to submit to arbitration, involves deciding between the rival merits of assured finality on the one hand and upon the other the resolution of doubts as to the accuracy of the legal reasoning followed by the arbitrator in the course of arriving at his award, having regard in that assessment to the nature and circumstances of the particular dispute.

It seems to me that the tests, or guidelines, laid down later in Lord Diplock's speech (at pp 742-743) are simply those appropriate in the kind of dispute with which the House was there primarily concerned

---

* Editor's note: Also reported at [1986] 1 EGLR 147; (1986) 278 EG 878.

As the Vice-Chancellor pointed out, it is right also to have regard to the passage in Lord Diplock's speech in *The Antaios* where (at p 200 of the report) he said, of the guidelines laid down in *The Nema*:

Like all guidelines as to how judicial discretion should be exercised they are not intended to be all-embracing or immutable, but subject to adaptation to match changes in practices when these occur or to refinement to meet problems of kinds that were not foreseen, and are not covered by, what was said by this House in *The Nema*.

Second, the point made by the Vice-Chancellor that the need to avoid delay and expense is possibly not as great in rent review arbitrations as it is in "commercial" arbitrations is valid for all rent review arbitrations, not only for those where the point at issue turns on the interpretation of the rent review clause.

Third, it seems to me that if I were to draw the distinction suggested by Mr Martin I should be laying the foundations for copious and time-consuming arguments in future cases arising from rent review clauses as to whether they fell into one category or the other. It is not difficult to imagine the sort of fine distinctions that might come to be made as a result. It seems to me far better that it should be beyond argument that in all rent review cases (except possibly true "one-off" cases, as to which I say nothing) the discretion conferred on the court by section 1(3)(b) of the Arbitration Act 1979 will be exercised in accordance with the principle laid down by the Vice-Chancellor in the *Lucas Industries* case.

I accordingly ask myself whether, in this case, I am left in real doubt whether the arbitrator was right in law. The answer is "yes". Following further the guidance of the Vice-Chancellor in the *Lucas Industries* case, I refrain from stating my reasons for saying that. I am also satisfied (as required by section 1(4) of the Act) that the determination of the question of law concerned could substantially affect the rights of one of the parties, namely the corporation, if only because of those 11 other cases. I therefore grant the leave to appeal sought.

After considering submissions as to leave to appeal against his decision to the Court of Appeal, WARNER J said: I think this is obviously a case where I should grant leave to appeal to the Court of Appeal against my decision. Since I am required, it appears, to give my reasons for doing that, I shall do so briefly. They are the reasons that Mr Martin has put forward in asking for that leave, namely that my decision does involve a question of principle, that is to say whether, in a case arising under a rent review clause, where the dispute is not as to the interpretation of the rent review clause itself, as it was in the *Lucas Industries* case, it is none the less right to apply the test laid down by the Vice-Chancellor in that case rather than the tests laid down in *The Antaios* and *The Nema*.

Having regard to the vast number of rent review clauses and of disputes arising out of them that there are in this country today, that is obviously a question of fairly wide importance.

COMPULSORY PURCHASE AND COMPENSATION

Court of Appeal
December 8 1986
(Before Lord Justice PURCHAS, Lord Justice NEILL and Lord Justice BALCOMBE)

CHILTON v TELFORD DEVELOPMENT CORPORATION

Estates Gazette March 28 1987

281 EG 1443-1446

Compulsory purchase — Date for assessment of compensation — New Towns legislation — Possession of land taken in several parcels over a period — Preliminary point of law — Whether date of entry was date when first parcel was taken or whether there were a number of separate dates for the purpose of valuation — Owner-occupied farm — Notice to treat and notice of entry relating to whole farm given on May 3 1978 — Possession of a part first taken on June 5 1978 — Remaining acts of taking possession spread over a lengthy period — The issue before the court, on a case stated by the Lands Tribunal at the request of the claimant, was as to the effective date or dates of entry for the purpose of compensation — The Lands Tribunal accepted the submission on behalf of the acquiring authority that there were eight separate valuation dates, being the specific dates on which the authority entered into possession of the several parts of the land — The claimant contended that the tribunal was in error and that the material date was the date when possession was taken of the first parcel, at which date the authority should be treated as having entered upon the entire farm — The Court of Appeal, agreeing with the claimant's submission, discussed the purpose of the notice of entry provisions in Schedule 6 to the New Towns Act 1965 (the statute operative at the time) — The purpose of requiring the authority to give not less than 14 days' notice of their intention to enter and take possession could not be to cover "a contingent intention, not formally adopted, and not carrying with it the intention to act upon it within a reasonable time" — The court adopted the construction which was favourable to the owner-occupier because the provisions in question, although incidentally dealing with compensation and interest, were primarily enacted for the protection of such persons — The authority in the present case must be treated as taking possession of the whole farm at the date of the first entry, June 5 1978, that being the material date for the assessment of compensation — Appeal allowed — Comments by Purchas LJ on "a provision which has on its face an open-ended power granted to the authority to act or not to act"

No cases are referred to in this report.

This was an appeal by case stated at the request of the claimant, Mr A R H Chilton, challenging the conclusion of the Lands Tribunal (V G Wellings QC) that there were eight dates at which entry was made and possession taken of the several parts of the claimant's farm by the Telford Development Corporation, these being the relevant dates for the assessment of compensation. The farm, of which the claimant was the owner-occupier, was Trench Lodge Farm, Trench, Telford, Shropshire.

The Lands Tribunal's decision was reported at [1985] 1 EGLR 195; (1985) 274 EG 1037.

A Anderson QC and R Fookes (instructed by Treasures & Rivers Wyatt, of Gloucester) appeared on behalf of the claimant; R J A Carnwath QC and Miss A Robinson (instructed by J C H Bowdler & Sons, of Shrewsbury) represented the acquiring authority, Telford Development Corporation.

Giving judgment, PURCHAS LJ said: This is a case stated by the Lands Tribunal at the request of Mr A R H Chilton, the claimant, under section 3(4) of the Lands Tribunal Act 1949. The single member of the tribunal had been asked to determine a preliminary point of law, and he did so on April 1 1985.

The point arises out of the compulsory purchase and compensation provisions contained in the New Towns Act 1965 and the Compulsory Purchase Act 1965. Although the former Act has been repealed by the New Towns Act 1981, we are told by learned counsel that the corresponding provisions in the later Act are the same in most relevant particulars, and that the point is one of some interest generally, beyond the interests of the parties particularly involved in this reference.

The point raises a consideration of the relevant date, or dates, of entry and taking possession of land by the acquiring authority for the purpose of compensation. The acquiring authority involved is the Telford Development Corporation.

The question posed was summarised in the decision of the single member in these words:

Possession of the land acquired was taken by the acquiring authority in several parcels and the preliminary issue is concerned with the question whether there is a single date, namely, the date on which possession of the first parcel was taken, or several dates namely, the individual dates on which possession of the several parcels was taken, which is or are material for the purposes of valuation.

The parties had agreed the facts so far as they were relevant; they are as follows: the land of which the claimant at all material times was owner-occupier, and which he farmed, consisted of 67.87 acres or thereabouts of agricultural land at Trench Lodge Farm, Trench, Telford, Shropshire.

On July 10 1973 the acquiring authority made the Telford Development Corporation (Hadley Park No 1) Compulsory Purchase Order 1973 under Section 7 of the New Towns Act 1965. The order was confirmed by the Secretary of State for the Environment on March 10 1978. The order related to the 67.87 acres and other lands

required by the development corporation for their activities under the New Towns Act.

A notice to treat in accordance with the statutory provisions was served on, and dated, May 3 1978; it related to the whole of the 67.87 acres. On the same date the acquiring authority served on the claimant one notice of entry relating to the whole of the 67.87 acres.

As the history evolved, the acquiring authority went into physical occupation of individual parcels of the whole area over a period of 28 months. The first date of entry, June 5 1978, involved 4.62 acres; thereafter there were three more areas involved and entries made, until December 6 1978, which I mention specifically because in respect of that area, 4.9 acres, the acquiring authority purported to serve a "revised notice". No point has been taken on this, and for the purposes of this appeal it can be ignored. It is common ground that the "revised notice" was of no formal effect.

The remaining acts of taking possession and entering occurred, as to 3.95 acres, on January 1 1980; as to 1.60 acres on January 1 1980, and then an area which is over half the whole of the area involved, 35.501 acres, on October 4 1980. After June 5 1978 in fact the claimant remained in occupation of the parcels not previously entered until such time as the acquiring authority in fact entered into physical occupation. Saving only the area of 4.9 acres entered on December 6 1978, there is no evidence or information as to any further sort of notice, but it must be assumed that there was some kind of communication between the parties. That is how things progressed.

The dispute, as is clear from the question posed, relates to the effective date of entry and taking possession for the purposes of compensation. But the particular statutory provisions which must be considered relate mainly to inhibit or restrict the exercise of the powers to acquire compulsorily by way of granting some limited protection to the owner or occupiers involved.

In very short summary, and by way of introduction, the statutory procedure involves two main steps: first of all, the obtaining of the compulsory purchase powers; there are provisions in that process whereby the landowner or occupier can oppose the confirming of the order; and, second, those powers having been confirmed, how they are to be exercised. They are exercised subject to these restrictions: first of all, there must be a notice to treat served under section 5 of the Compulsory Purchase Act 1965; then, that having been done, there are further restrictions on the acquiring authority preventing that authority from entering upon the land without giving at least 14 days' notice.

It is now convenient to turn to the statutory provisions with which we are concerned. I start by reading section 12(1) of the New Towns Act 1965:

Part I of the Compulsory Purchase Act 1965 shall apply in relation to the acquisition of land under this Act subject to any necessary adaptations and to the provisions of Part I of Schedule 6 to this Act.

Section 11(1) of the Compulsory Purchase Act is replaced by para 4 of Schedule 6, the relevant provisions of which are:

4(1) If the acquiring authority have, in respect of any of the land, served notice to treat on every owner of that land, they may at any time thereafter serve a notice —
(a) on every occupier of any of that land, and
(b) on every person (other than such an occupier)
...
describing the land to which the notice relates and stating their intention to enter on and take possession thereof at the expiration of such period (not being less than fourteen days) as may be specified in the notice.

I pause to emphasise the words "describing the land to which the notice relates"; that is the notice of intention to enter. The period is one of not less than 14 days, and therefore envisages periods of greater notice if appropriate.

Subpara (2) of para 4 of Schedule 6 reads as follows:

At the expiration of the period specified in such a notice (or, where two or more such notices are required, and the periods specified in the several notices do not expire at the same time, of the last of those periods to expire), or at any time thereafter, the acquiring authority may enter on and take possession of the land to which the notice or notices relate without previous consent of or compliance with section 11 of the Compulsory Purchase Act 1965, but subject to payment of the like compensation for the land of which possession is taken, and interest on the compensation agreed or awarded, as they would have been required to pay if those provisions had been complied with.

Subpara (2) of para 4 of the Schedule refers first of all to the cases where there may be more than one occupier or owner in respect of whom notices of intention to enter must be served, but gives power "at any time thereafter [to] enter on and take possession of the land to which the notice or notices relate without previous consent" and then "or compliance with section 11 of the Compulsory Purchase Act 1965 but subject to" the provisions that interest will run from the date of entry and taking possession.

The member considered the arguments of counsel which, generally speaking, were in line with the arguments they have put before us. Mr Anderson argued to the effect that the notice of entry of May 3 related to the whole of the land and for that reason, when the acquiring authority on June 5 1978 first took possession of a part of the land, it must be treated as having done so in the name of the whole; he developed his argument along those lines, saying also that any other interpretation of the statute would lead to doubt, confusion and hardship, and in particular relating to identifying not only the area involved if part of the land in respect of which the notice of entry was served was in fact entered, and the unfairness that would be involved, as could be said in this case, by a long delay as to a substantial part of the land under threat. I do pause to comment, however, that counsel have been scrupulous to point out that it is not alleged that there has been either unfairness or difficulty in this case; the matter comes before us essentially as one of construction of the statute and, as Mr Carnwath submitted, merely construction of the statute in the accepted and established facts of what indeed happened in this case.

Mr Carnwath submitted before the single member that there was no basis in law for the proposition that the original entry into possession of part of the land was to be treated as entry in the name of the whole. There was no factual basis on the agreed statement of facts for inferring the grant of a licence by the authority in respect of the parts which, after June 5, continued in the occupation of the claimant. I pause to emphasise that in fact the claimant did remain farming the land which had not been occupied by the acquiring authority. Upon one of the two alternative assessments of compensation which had been agreed between the parties, credit is given for what is described, probably inaccurately, as mesne profits to relate to the benefit received by the claimant from his continued occupation of the land, if the contentions of Mr Anderson are right, namely that the claimant is entitled to compensation, and interest thereon, to run from June 5 1978 in respect of the whole of the land involved in the notice of entry.

The single member accepted Mr Carnwath's argument and said this:

In my judgment, on the agreed facts there is no reason to infer that when the authority took possession of part of the land on June 5 1978 they did so in the name of the whole. Equally, notwithstanding the use of the expression "mesne profits" (an expression more appropriate to a tenancy than to a licence) there is no reason to infer that the authority purported to grant a licence to the claimant to remain in possession of the remainder of the land. In my opinion, in the present case, there are eight valuation dates, namely, the specific dates on which the authority entered into possession of the several parts of the land as set out in agreed facts No (6). I answer the preliminary issue accordingly.

The questions stated for this court are threefold:

1 Whether I was correct in law in determining as the Respondent contended that for the purposes of calculating compensation for disturbance and interest on the purchase price, the acquiring authority should be treated as having taken possession of each individual parcel of land on the date on which the acquiring authority took actual physical possession of the same . . .
2 Whether I erred in law in not adopting the Appellant's contention that for the purposes of calculating compensation for disturbance and interest on the purchase price, the acquiring authority should be treated as having taken possession of the whole 67.87 acres on June 5 1978, being the date of the first entry onto that land following the notice of entry dated May 3 1978 relating to the whole property the subject of the Notice to Treat on May 3 1978.
3 Whether I was correct in law to find on the agreed evidence no reason to infer that the authority either purported to or did grant a licence to the Appellant to remain in possession of part of the land.

The statutory provisions which I have already set out do not give a simple definitive answer to the two contending interpretations. It is perfectly right to say that nowhere in Schedule 6 does one find anything other than the expression "intention to enter on and take possession of the land". Para 4(1) does refer to "intention to enter on and take possession of the land", and para 4(2) refers to "the land of which possession is taken shall carry interest at the rate". So the statute refers to "taking possession" (para 4(2) of Schedule 6) or intention to enter on and take possession thereof (para 4(1) of Schedule 6). In addition to that there is no provision providing a period during which a notice of intention to enter should lapse. Subject to reasonable exercise of power, which has not been argued before us, if the acquiring authority took no step in relation to any part of the land, albeit that they had served a notice of intention to enter, then the provisions of para 4(2) of Schedule 6 remain as a threat to the enjoyment of the land, which can be determined at any time without notice.

It must of course be remembered that before the stage of serving a notice of intention to enter and take possession the acquiring authority has already been obliged to serve a notice to treat, which must incorporate the whole of the land subject to the notice of intention to enter; but of course it may, and very frequently does, extend to other lands in respect of which a notice of intention to enter will not have been served. But this does afford a remedy to the owner of the land who can, in the absence of agreement as to compensation, refer to the Lands Tribunal, but his remedies are not immediately or easily available.

So one is left here with a provision which has on its face an open-ended power granted to the acquiring authority to act or not to act. It may at least be questioned whether that was entirely the intention of Parliament in enacting what are essentially provisions for the protection of an owner or occupier who is subjected to the necessary, but nevertheless draconian, powers of the acquiring authority to dispossess him of his title and occupation of his land.

For my part, in approaching the construction of these statutory provisions I have two concepts in mind: first, to give to this part of the legislation, if it is open to do so, a purposive construction, bearing in mind the general considerations and need for these provisions. Second, to bear in mind that it is a statute which is depriving the citizen of his rights in property and his title, and the right to enjoy occupation of his lands or perhaps somebody else's lands.

So, looking at section 12 of the New Towns Act, and Schedule 6, I pose the question: what is the purpose of requiring the acquiring authority to give notice of not less than 14 days of an intention to enter on and take possession of the land? What it cannot cover, in my judgment, is a contingent intention, not formally adopted, and not in fact carrying with it the intention to act upon it within a reasonable time. Otherwise, the statute provides that the acquiring authority does not have to serve a notice of intention to enter in respect of the whole of the land, subject to the notice to treat, or of the compulsory purchase order itself as relates to the particular owner or occupier; and that they may serve, as and when necessary — which I would construe as meaning as and when they form the intention necessary to bring into effect the provisions of para 4(1) — the intention to enter on and take physical possession thereof.

I am conscious of the argument to the contrary, that there is also no express reference in this statute to give effect to the concept that entry upon part of the land, subject to the notice of intention to enter, should be deemed to constitute entry of the whole, and it would not have been beyond the wit of man to have put that in the drafting of the statute. But, choosing between the two alternative approaches to this legislation, and I hope fairly recognising the lacuna which exists for practical purposes in the day-to-day carrying out of these exercises, I adopt the construction which is favourable to the owner and occupier of the land, because these sections, although incidentally dealing with calculation of compensation and interest, were primarily enacted for the protection of such a person.

I would therefore construe the statute in that light, and that would bring me to the conclusion that the answers to the questions posed by the single member for our consideration are that with regard to the first question the answer must be "no"; the answer to the second question must be "yes", and that leaves the answer to the third question still to be considered. For my purposes the substance of this reference is dealt with by answering the first two questions. Whatever the status may be, whether as licensee or some other formally acquired status as between the parties involved, will depend from case to case, from attitude to attitude adopted by the parties concerned — that is, the landowner or occupier and the acquiring authority — and I do not consider that any formal general answer to question 3 will further the matter in any useful way at all. If I had to choose on the facts of this case, I would have thought that Mr Chilton was present there as a bare licensee with few or no rights, bearing in mind that under the statute he is not entitled to receive any notice of change of intention or of the termination of his licence; nor is he entitled to any description of the land which the acquiring authority will choose to possess and enter on no notice. That is a further argument for supporting the conclusion I have reached on the interpretation of the statute, namely, as I emphasised when reading para 4 of Schedule 6, it does specifically require the acquiring authority to describe the land which it intends to enter and of which it intends to take possession. By adopting the construction for which Mr Carnwath in his admirable submissions contends, it would merely defeat the object of that precaution which has been put in the Schedule for the protection of the landowner and occupier.

I would therefore answer questions 1 and 2 in the way that I have described, and I would not give any formal answer to question 3.

Agreeing, NEILL LJ said: I add a few words of my own only because we are differing from the construction given to the statutory provisions by the member of the Lands Tribunal.

It was argued on behalf of the acquiring authority that, for the purpose of para 4(2) of Schedule 6 to the New Towns Act 1965, they entered on and took possession of the parcels of land set out in para 6 of the agreed facts on the several dates therein set out. It was further argued that for the purpose of calculating the interest to be paid on the compensation for the land, these several dates constituted the time of entry on the individual parcels.

However, it seems to me, with respect, that these arguments do not pay due regard to the provisions of para 4 of Schedule 6. It is clear from this paragraph (a) that a notice of entry may relate to part only of the land comprised in the relevant notice to treat; (b) that any notice of entry must describe the land to which it relates; (c) that the notice of entry must state the intention of the acquiring authority to enter on and take possession of the described land at the expiration of a specified period; and (d) that at the expiration of the period specified in the notice or at any time thereafter, the acquiring authority may enter on and take possession of that land.

It follows, therefore, that an acquiring authority may, if so minded, serve a series of notices of entry and can enter on to the land described in each of such notices in accordance with a programme which suits its requirements. But where, as here, a single notice of entry is given, it seems to me that, save perhaps where the *de minimis* rule applies, a subsequent entry on to the land, or any part of the land, is an entry made in accordance with the permissive power given in para 4(2) of Schedule 6, and is an entry on the land described in the notice to enter. The first entry on to the described land constitutes the entry on to the described land foreshadowed by the statutory notice; the date of that entry is the time of entry for the purpose of the interest provisions contained in para 4(2).

For these reasons, as well as for those given by Purchas LJ, I, too, would answer the questions in the manner in which my lord has proposed.

BALCOMBE LJ also agreed, for the reasons given in the previous judgments, that the appeal should be allowed and he did not add anything further.

Appeal allowed and questions in case stated answered as proposed in judgment of Purchas LJ. Costs awarded in favour of claimant. Leave to appeal to House of Lords refused.

ESTATE AGENTS

Court of Appeal
February 11 1987
(Before Lord Justice MAY, Lord Justice CROOM-JOHNSON and Lord Justice NOURSE)

CHESTERTONS (a firm) v BARONE

Estates Gazette April 4 1987

282 EG 87-91

Estate agents' commission — Successful appeal by agents from county court decision rejecting claim — Undisclosed principal — Election — Whether abandonment by estate agents of rights against vendor's agent and unequivocal election to look to principal, a foreign company incorporated in Panama or Liberia and managed from Channel Islands — Estate agents' instructions to sell came from the respondent, a solicitor, and initially there was no mention of the existence of any principal — After estate agents had obtained an acceptable offer they were informed that the respondent was acting for a foreign company — The estate agents acknowledged the company as a client and matters proceeded to an exchange of contracts, but when the estate agents submitted their account for commission the respondent alleged that they had not earned it in full — This led to proceedings in the county court against the respondent, in which the assistant recorder held that the estate agents had elected to look to the principal, the foreign company, alone and that the respondent was under no liability — Alternatively, the assistant recorder held that there had been a novation of the contract whereby the company had been substituted for the respondent as the other party — Held that, although the assistant recorder had stated the law correctly in general, he had erred in its application — There was no evidence that the appellant estate agents had made an unequivocal election to look to the foreign company as the principal alone and to abandon their rights against the respondent — Letters on which the assistant recorder relied were consistent with the estate agents still looking to both the company and the respondent rather than indicating that they had decided to look to the company alone — As to the suggested novation, here again there was no evidence of a giving-up by the appellants of their rights against the respondent in exchange for a consensual right against the company — Appeal by estate agents allowed — *Per* May LJ, "The clearest evidence of an election is at least the commencement of proceedings against one or other of the two relevant parties"

The following case is referred to in this report.

Clarkson Booker Ltd v *Andjel* [1964] 2 QB 775; [1964] 3 WLR 466; [1964] 3 All ER 260, CA

This was an appeal by Chestertons from a decision of Mr Assistant Recorder A W E Wheeler at Amersham County Court dismissing their claim for commission against the defendant, the present respondent, Arturo Barone, in respect of the sale of a leasehold property at 24 Daska House, Chelsea, London SW3.

P W Birts (instructed by Roche Hardcastles) appeared on behalf of the appellants; F A Philpott (instructed by Barone, of Amersham, Bucks) represented the respondent.

Giving judgment, MAY LJ said: This cautionary tale involves an appeal from the dismissal of a plaintiff estate agents' claim for commission by a judgment of Mr Assistant Recorder A W E Wheeler in the Amersham County Court on June 20 1986. The appellants now seek to have that dismissal of their claim set aside and judgment entered for them for a total sum of £2,070, about which there is no dispute, comprising what would have been the commission payable in respect of the transaction together with VAT.

The facts of the case are uncomplicated. The plaintiffs are a well-known firm of estate agents and surveyors carrying on business with a substantial number of offices in London and elsewhere. The defendant is a solicitor who had as a client, for the purposes of this litigation, a foreign company incorporated either in Panama or Liberia, but managed from the Channel Islands. An officer of that company owned the leasehold property at 24 Daska House, Chelsea, London SW3, which he or she wanted to sell. They instructed the defendant to set the necessary proceedings in motion and on September 14 1984, having received oral instructions from Mr Barone, the plaintiff appellants wrote back to him confirming his instructions to offer the flat for sale on the stated terms and in the antepenultimate para in the letter wrote:

We do assure you of our best endeavours to find a purchaser and would confirm that in the event of a sale resulting through our introduction, we shall look to you for payment of our commission being 3% of the agreed sale price including any contents, plus VAT. This commission will include normal promotion costs.

Thereafter, the estate agents set about obtaining an offer which would be acceptable to the vendor of the property and having obtained one of £60,000 communicated it to the defendant, or to somebody in his office, when they were told that Mr Barone was in truth acting for a foreign limited company known as Rosemary Inc, the existence of which and his authority to act on behalf of which had not theretofor been disclosed to the estate agents.

It seems that the offer of £60,000 for the property was acceptable to the vendors and consequently on November 2 1984 the appellants wrote to Rosemary Inc c/o Mr Barone for the attention of Mrs Jeffries in his office as follows:

In accordance with your instructions, we are pleased to confirm that we have accepted on your behalf and subject to contract, the offer made by our applicant, Mr S Allabert, to purchase the above . . .

We take this opportunity to confirm that upon completion we shall look to you for payment of our commission being 3% plus VAT of the agreed sale price in accordance with our letter of September 14 1984.

The parties' solicitors were in due course put in touch with each other in the usual way and shortly after the new year contracts were exchanged for the sale and purchase. On January 3 1985 the appellants wrote to Mr Barone, again for the attention of Mrs Jeffries:

We were pleased to learn that the contracts for the sale of the above have been exchanged and we have pleasure in enclosing our commission account. We request that you kindly seek our mutual client's authority to settle this immediately upon completion.

Enclosed with that letter was the appellants' commission account addressed to "Rosemary Inc Per [Mr Barone's firm]" for the sum of £1,800, being 3% on the £60,000 sale price, together with £270 VAT, making a total of £2,070.

Thereafter Mr Barone, either acting on behalf of the foreign corporation but without their specific instructions or alternatively on their express instructions, contended that the estate agents had not earned in full the commission for which they had invoiced Rosemary Inc. After the matter had been put into the hands of the estate agents' solicitors and there had been an exchange of correspondence into which it is unnecessary to go further, these proceedings were issued by the estate agent appellants against Mr Barone personally.

The learned assistant recorder, in a judgment which has been of the greatest assistance, clearly and succinctly set out the facts and issues between the parties. He stated the basic principle of law applicable in a quotation from *Halsbury's Laws of England,* vol 1, 4th ed, para 853:

Where a person makes a contract in his own name without disclosing either the name or the existence of a principal, he is personally liable on the contract to the other contracting party, though he may be in fact acting on a principal's behalf. He will continue to be liable even after discovery of the agency by the other party, unless and until there has been an unequivocal election by the other contracting party to look to the principal alone.

That was a defence which Mr Barone in this litigation raised to the appellants' claim for commission. Albeit that the original instructions had been given to the appellants by his acting on behalf of an undisclosed principal, so that *prima facie* he was personally liable with the principal for the commission, his contention was, as was accepted by the assistant recorder, that there had been a sufficient election as referred to in that passage from *Halsbury* and that accordingly there was no liability. Further or alternatively, the defendant contended that there had in any event been a novation of the contract between him and the appellants, whereby his principal, Rosemary Inc, had been substituted for him as the other party to the appellants in the contract under which they were instructed to find a purchaser for the relevant flat. The learned judge found in favour of the defendant on that contention also.

The submission on behalf of the appellants is that the learned judge erred on both those points, that there was insufficient material upon which he could conclude that there had been a sufficient election within the relevant principles of law and similarly, because there is very little dispute that the two points go together, that there was insufficient material upon which the learned judge could have found a novation.

We have been referred to one or two authorities. I think it is sufficient merely to refer to *Clarkson Booker Ltd* v *Andjel* [1964] 2 QB 775. That was a case concerning the supply of goods and services by the plaintiffs to the defendant, who at the material time had been acting (as had Mr Barone in the instant case) as an agent for an undisclosed principal.

This type of case, as appears from the textbooks and other authorities, has to be dealt with on its own particular facts and accordingly I think it unnecessary to go into detail to the facts of the *Clarkson Booker* case. I merely go to the judgment of Russell LJ at p 794 and a short passage at p 795 where the precise scope of the doctrine of election in this particular context is set out. At p 794 the learned lord justice said:

The defendant having contracted as agent for an undisclosed principal, the plaintiffs were entitled to enforce the contract either against the defendant on the footing that he was contracting and liable as principal, or against the principal on the footing that the defendant was not liable, being merely an agent. The plaintiffs could not enforce the contract against both. Their right against the defendant and their right against the principal were inconsistent rights. At some stage the plaintiffs had to elect to avail themselves of one of those inconsistent rights and abandon the other. The question is whether the correct conclusion from the facts of this case is that, prior to the issue of their writ against the defendant, the plaintiffs had so elected.

Then at p 795:

The position is that in every case the external acts of the plaintiff must lead to the conclusion, as a matter of fact, that the plaintiff has settled to a choice involving abandonment of his option to enforce his right against one party.

The essential elements in the principle of election, in so far as applicable to the instant case, are made clear not only in those two passages from the learned lord justice's judgment but also in the quotation from *Halsbury* read by the assistant recorder, which I have already mentioned, that there must be an unequivocal election to look to the principal and to look to the principal alone. Where the plaintiff is entitled to look to either principal or agent and he indicates his intention of enforcing his rights, it is only when it can be demonstrated that there has been such an unequivocal election and that the plaintiff has abandoned his rights against one or other of the two parties that the doctrine of election bites and the plaintiff is no longer entitled to seek to recover against both principal and agent.

It is, with respect, in that regard that I think that the learned assistant recorder, in the course of his reasoning twice took too large a step. I say that without in any way seeking to derogate from the comments that I have already made about the excellence of his judgment and the assistance that the court has obtained from it.

On the facts of the instant case, looking at the evidence to which our attention has been drawn, and the correspondence, the overwhelming probability seems to me to be that, at least until objection was taken to the estate agents' invoice for commission, neither party applied their minds to the strict legal position between the estate agents on the one hand and Mr Barone and his client, Rosemary Inc, on the other. For my part I do not think either of them, although it might be said that both of them were remiss not to have done so, really considered the legal principles involved in the liability of agents acting for undisclosed principals or the question of election at all. That they did not apply their minds necessarily leads to the conclusion (upon which I shall elaborate in a moment) that there was not the necessary election and abandonment on the part of the estate agents sufficient to preclude them from recovering successfully against Mr Barone.

I say that because of the comment which the learned assistant recorder made in his judgment in relation to the letter of November 2 which they wrote to Rosemary Inc after they had heard that the corporation was in truth the intended vendor rather than Mr Barone. In contrasting the two letters of September 14 and November 2, which I have quoted, the assistant recorder said of the second:

It shows that the plaintiffs were looking back at the letter of September 14 and indicating that they were no longer looking to Mr Barone the defendant as being under an obligation to pay the commission, but were clearly looking forward to Rosemary Inc as being responsible for that sum.

That is the first place in which the assistant recorder has, in my view, taken too large a jump in the course of his reasoning. The two letters of September 14 and November 2 are, I think, equally consistent with the estate agent appellants still looking to both the principal and the agent, rather than indicating that they were no longer looking to Mr Barone but only to Rosemary Inc. Similarly, when the learned assistant recorder moved on to the letter of January 3 1985, enclosing the commission invoice, he quoted the relevant passage from the letter and continued:

That account is addressed to Rosemary Incorporated per the defendant and indicates the amount of the commission. The words "We request that you kindly seek our mutual client's authority to settle this immediately upon completion" seem to me to categorically indicate that, especially when read with the earlier letter of November 2 1984, the defendant was no longer being treated as a party to whom the plaintiffs would look for payment. They were looking for payment to Rosemary Inc and to them alone. The defendant was no longer regarded by them as a client: their client was Rosemary Inc.

The assistant recorder was there correctly setting out the legal principles to be applied, but on the facts of the case it is his finding that the estate agents were looking to Rosemary Inc and to them alone that in my judgment does not logically follow from either that which had gone before or the facts of the case. There is nothing in any of the three letters to which I have referred, or the facts, to indicate that there had been an *abandonment* of the estate agents' rights against the agent, Mr Barone. In my opinion, there was certainly the suggestion that in the circumstances Rosemary Inc were, it might be said, the principal party liable for the commission, but I can see nothing in the facts or letters to indicate an abandonment; a giving up of the simultaneous right, if it arose, to go against the agent, Mr Barone, at the same time.

Again, when one turns over the page of the judgment, the learned judge said:

There is not at this stage or since September 14 1984 any indication anywhere, orally or in the documentary evidence, that the defendant would in addition be looked to for payment.

I think that is putting it the wrong way round. The doctrine of election, in order to excuse the agent acting on behalf of the undisclosed principal, requires that agent in those circumstances to show that the third party has abandoned or will no longer look to him for payment; it is not sufficient that he would be looked to for payment merely in addition to his principal. He, the agent for the undisclosed principal, remains liable at the same time as the principal, unless and until he can demonstrate that the other party to the contract has abandoned the legal right that he has against him.

That the learned assistant recorder so erred is in my judgment also shown at the bottom of that same page of his judgment:

> At this point in time the defendant was entitled to conclude from the correspondence, from the attitude of the plaintiffs, that he had dropped out as a principal, and that payment was no longer being sought from him. The plaintiffs were now looking to Rosemary Inc . . .

So far so good. But then the assistant recorder adds:

> and no one else.

There is in the correspondence an indication that the estate agents were looking to Rosemary Inc because Rosemary Inc were the vendors of the relevant flat and they would be the party who, in all the circumstances, would pay the commission out of the sale price which they received from the purchaser of their flat. But there is nothing that I can see in the correspondence or the facts to support the conclusion that the estate agents were looking to Rosemary Inc *alone*.

In those circumstances, I am driven to the conclusion that in the present case the learned judge was wrong in the conclusion to which he came on the facts and material before him that there had been an election.

I would wish to add this. Whether or not there has been an election is, I think, largely a question of fact and we have in this case a finding of fact by the learned judge below that there had been by the appellants an election — an abandonment of their rights against Mr Barone. However, as Willmer LJ said in *Clarkson Booker Ltd* at p 792:

> Since the relevant evidence is all contained in the correspondence, we have been invited to review his findings and to draw our own inferences from the correspondence. In a case such as the present

as the learned lord justice thought was the *Clarkson Booker* case and as I think is the instant case

> [this court] is entitled to take this course, for we are in as good a position to draw inferences

from the correspondence and the other circumstances of the case

as was the judge.

Turning briefly from the question of election to the question of novation, Mr Philpott, on behalf of the respondent to this appeal, made it clear that he was relying not upon any express contract of novation but on an inferred contract of novation to be inferred from all the circumstances of the case. That being so, I think that the point which falls for decision on the novation question is precisely the same point as fell for decision on the election question. Was there an express abandonment? Was there a clear giving-up by the estate agents of their rights against the agent, Mr Barone, in exchange for a consensual right against the vendor principal and against the vendor principal alone? I think that it necessarily follows from what I have said that in my view the learned assistant recorder erred in concluding that there was any novation in this case.

Before leaving this appeal, I would add this. We were shown in the course of the argument references in *Bowstead on Agency*, 15th ed, and a like passage in the second volume of *Chitty on Contract*, each of which it appears were edited by the same editor, which indicate that there is little, if any, authority in which there has been held to have been an election in the general circumstances of the instant appeal short of there being judgment in favour of the plaintiff against one or other of the principal or agent, save in two cases, to which *Bowstead* refers, where the defence could have been based equally well on an estoppel, as it was based on the principle of election.

For my part, I think it is difficult to think of facts where a plaintiff is claiming against both principal and agent, the agent having acted originally on behalf of the principal undisclosed, where an election can be shown without legal proceedings having been started. However, I would not like to be thought to be saying that in exceptional circumstances such a situation could not arise. It seems to me, however, that the clearest evidence of an election is at least the commencement of proceedings by the plaintiff against one or other of the two relevant parties. That, I think, was probably the difficulty which Willmer LJ found in the *Clarkson Booker* case, because, as he said at the end of his judgment, he regarded the case as being very near the borderline.

Be that as it may, for the reasons that I have indicated, I think that the learned judge erred in his reasoning in reaching the conclusion either that there had been an election on the part of the estate agents to sue only the foreign corporation or that there had been any question of a novation by which the corporation was substituted for Mr Barone, the original agent.

In those circumstances I, for my part, would allow this appeal and substitute for the judgment below a judgment for the plaintiffs for the sum of £2,070, together with any appropriate interest there may be and about which we can no doubt be told by counsel in due course.

Agreeing, CROOM-JOHNSON LJ said: There was a dispute in the course of the trial as to whether the defendant had been an agent for the undisclosed principal and had in fact contracted with the plaintiffs at the same time disclosing that there was a principal in existence and that therefore he was contracting only as an agent. The learned assistant recorder, after having heard the evidence about that, came to a clear conclusion. He accepted entirely the evidence of Mr Pallot, the witness called on behalf of the plaintiffs, and indeed several other witnesses who were called for them, that there was initially no mention of the existence of any principal standing behind Mr Barone. Accordingly, the whole question which arose was this, that in law Mr Barone was therefore taken to have initially contracted as a principal. When the plaintiffs discovered, as they did discover at about the beginning of November, the existence of Rosemary Inc and agreed to treat Rosemary Inc as a client, did that involve an abandonment by them of their rights against Mr Barone?

It was suggested by counsel for the respondent that this matter had only been raised because after Mr Barone had written to the plaintiffs on January 21 1985 indicating that in his view and in the view of the principal the plaintiffs had not earned their fee (it being suggested that during protracted negotiations on the sale they had formed the view that the plaintiffs were acting as agents for the purchaser and not the vendor), it was thereafter the appearance on the scene of solicitors acting for the plaintiffs which had raised for the first time the suggestion that Chestertons were entitled to sue Mr Barone as primarily liable himself for having contracted as he did. That must involve that only the circumstances as at that date would have to be looked at in order to see whether or not there had been an abandonment by the plaintiffs of their rights against Mr Barone.

It was said in *Clarkson Booker* v *Andjel* at the bottom of p 792 and the top of p 793 by Willmer LJ that there would have to be something which was a truly unequivocal act so as to preclude the plaintiffs in that case from subsequently suing the defendant. He went on to say:

> This, I think, involves looking closely at the context in which the decision was taken, for any conclusion must be based on a review of all the relevant circumstances.

The reliance which was placed upon the letter of November 2 by Mr Barone in the present case was that, when Chestertons wrote, as they then did, to Rosemary Inc c/o Messrs Arturo Barone at his address, saying that they were "pleased to confirm that we have accepted on your behalf and subject to contract, the offer made by our applicant . . . to purchase the [premises]", they were in those circumstances switching their expectations of payment from Mr Barone to Rosemary Inc. It was suggested that para three of that letter, "We take this opportunity to confirm that upon completion we shall look to you for payment of our commission being 3% plus VAT of the agreed sale price in accordance with our letter of 14th September 1984," indicated that. On the other hand, it was quite clear that that letter came to be written because it was only immediately before then that Chestertons had been told for the first time that Mr Barone was acting as an agent for Rosemary Inc. The evidence apparently had been that Mr Shingles, who was one of the negotiators acting on behalf of Chestertons, was happy to accept Rosemary Inc as clients from then on, because that is what he had been told and he had been asked to send the letter to them in those terms.

What clearly did not happen, either in a letter or at all, was that there was an abandonment by Chestertons of their rights against Mr Barone. They were quite content to add Rosemary Inc as another client to Mr Barone. But there was never any clear intention, shown in writing or in any other way, that they were electing to go only against Rosemary Inc. The letter was quite sufficient to meet the test laid down in *Clarkson Booker* v *Andjel* and in my view that was not sufficient to allow Mr Barone to escape his liability. Moreover, when the account was ultimately sent to Rosemary Inc, addressed to them with the request that Mr Barone would make the usual arrangements for the commission due to the estate agents to be retained out of the purchase price, that was done as a matter of routine and certainly not as a way of indicating that Mr Barone was no longer to be regarded as responsible, as he was hitherto in law.

A   For the reasons which my lord has given, I, too, agree that this appeal ought to be allowed, both on the question of election and on the question of novation.

NOURSE LJ agreed with both judgments and did not add anything of his own.
The appeal was allowed with costs.

Court of Appeal
March 13 1987
B (Before Sir John DONALDSON MR, Lord Justice WOOLF and Lord Justice RUSSELL)

BRIAN COOPER & CO v FAIRVIEW ESTATES (INVESTMENTS) LTD

Estates Gazette May 30 1987
282 EG 1131-1134

Estate agents' commission — Entitlement to commission for letting a property — Whether commission was payable under the terms of the contract although the agents claiming
C **commission were not the effective cause of the letting — The terms accepted were that "we confirm that we are pleased to offer a full scale letting fee to your company should you introduce a tenant by whom you are unable to be retained and with whom we have not been in previous communication and who subsequently completes a lease" — Appeal by defendant lessors from decision of Judge Tibber in favour of the plaintiff agents' claim for commission — The facts were that the agents (present respondents) were the first to introduce the company which eventually completed the lease, but there was a period during which the company in question ceased for various**
D **reasons to be looking for a property — During this period contact between the respondents and the company was interrupted — When the company's interest in finding a property revived, an introduction was made to the property by a different firm of agents and through a different officer of the company — The trial judge found as a fact that the effective cause of the letting was the introduction by the second firm of agents — This finding, however, did not, in the judge's opinion, disentitle the respondent agents to commission — The judge held that the question of the effective cause was in this case irrelevant owing to the express terms of the instructions — The words "should you introduce a tenant . . .**
E **who subsequently completes a lease" were the very negation of causation and were indeed evidently framed precisely to obviate arguments as to who was the effective cause — On this view the respondent agents satisfied all the terms of the instructions — They introduced a tenant by whom they were not retained, with whom the appellants had not been in previous communication and who subsequently completed the lease — On appeal it was submitted by the appellant lessors that the authorities on estate agents' commission established that entitlement was always subject to an implied term that the agents must be at least *an* effective cause, if not *the* effective**
F **cause, of the transaction — Held by the Court of Appeal, affirming the trial judge's decision, that the express language of the instructions was inconsistent with the implication of a term imposing an additional requirement that the estate agent must be at least an effective cause of the lease being granted — Some doubt expressed by the court as to whether it was true as a general proposition that the implied term, when relevant, referred to "an" rather than "the" effective cause — A suggestion made in the course of argument that the court would be less ready to imply such a term in the case of a vendor or lessor who was a commercial developer (such as the present appellants) than in the case of a private individual was mentioned by the court but does not appear to be part of the** *ratio decidendi* **— Appeal dismissed**

The following cases are referred to in this report.

Lordsgate Properties Ltd v *Balcombe* [1985] 1 EGLR 20; (1985) 274 EG 493
Luxor (Eastbourne) Ltd v *Cooper* [1941] AC 108; [1941] 1 All ER 33, HL
Millar Son & Co v *Radford* (1903) 19 TLR 575

This was an appeal by Fairview Estates (Investments) Ltd, the defendants, from a decision of Judge Tibber, sitting as a judge of the High Court, in favour of the plaintiffs, present respondents, Brian Cooper & Co, a firm of estate agents. The respondents had claimed to have earned commission by the introduction of a company, Metier Management Systems Ltd, as the lessees of an office building, Fairview House, at Station Road, Hayes, Middlesex. The decision of Judge Tibber was reported at [1986] 1 EGLR 34; (1986) 278 EG 1094.

John Chadwick QC and Miss Elizabeth Gloster (instructed by Lovell, White & King) appeared on behalf of the appellants; Tom Morison QC and Stephen Suttle (instructed by H Davis & Co) represented the respondents.

Giving the first judgment at the invitation of Sir John Donaldson MR, WOOLF LJ said: This appeal arises out of yet another dispute over the entitlement of an estate agent to commission. The dispute arises because although the respondent estate agent was responsible for introducing the tenant who eventually entered into a lease, the estate agent was not an effective cause of the letting.

On March 21 1986 judgment was given by His Honour Judge Tibber, sitting as a deputy High Court judge, in favour of the estate agent in the sum of £73,831.64, of which £58,190 was commission and the balance was interest.

Neither the appellant landlord nor the respondent seeks to go behind the judge's findings of fact and in order to resolve the issue raised by the appeal it is possible to deal with the facts very shortly.

In the spring of 1982 the defendant company ("Fairview") was in the process of completing the development of a new office building, Fairview House, Station Road, Hayes, Middlesex. On March 22 1982 Mr McCulloch, Fairview's senior development surveyor, wrote to the plaintiff estate agents ("Cooper"). He enclosed brochures of Fairview House, indicating that the property was scheduled for completion in approximately nine months' time and went on to say:

We confirm that we are pleased to offer a full scale letting fee to your company should you introduce a tenant by whom you are unable to be retained, and with whom we have not been in previous communication and who subsequently completes a lease.

At that time Mr Spencer, one of the partners of Cooper, had been approached by a company, Metier Management Systems Ltd ("Metier"), to assist them in finding accommodation and, as Mr Spencer thought that Fairview House would meet Metier's requirements, on April 5 1982 he sent a copy of the brochure to Mr Ross of Metier. On the same day Mr Spencer had a series of conversations with Mr McCulloch and then wrote a letter dated April 15 1982 in which he confirmed his introduction to Fairview of Metier, stated that that company hoped to arrange to inspect the building, confirmed that he was not retained by Metier and went on to say:

. . . in the event that the company do complete a lease on the building I would wish to take advantage of your offer of full scale commission.

In April 1982 Fairview House was inspected by two employees of Metier, Mr Clark and Mr Ross, who were not very senior in the hierarchy of Metier. Later Mr Spencer discussed with a more senior employee of Metier, Mr Hood, a number of properties including Fairview House.

On December 2 1982 Mr McCulloch telephoned Mr Spencer and suggested that he should get in touch with Metier again and see whether they were interested in the property and offered him double scale commission. As a result of that offer, on the same day Mr Spencer wrote to Mr Hood of Metier about Fairview House and also wrote to Mr McCulloch confirming his "generous offer of double scale commission in the event that the company do proceed to a lease".

However, by that time Metier were no longer interested in taking new premises and so, on December 21 1982, Mr Hood replied to Mr Spencer making this clear and telling him that as and when the situation changed he would contact him again. There were subsequent conversations between Metier and Mr Spencer, but for practical purposes thereafter Mr Spencer's contact with Metier was at an end.

On March 2 1983, after Fairview House had been completed, Mr McCulloch again wrote to Mr Spencer and sent new letting brochures in respect of the building, informing Mr Spencer of the new terms for the letting and concluding by saying:

> I confirm that, should your company introduce a tenant by whom you are unable to be retained, and with whom we have not been in previous communication, then, should your applicants enter into a lease, we are pleased to pay a double scale letting commission.

In August 1983, as a result of a visit by a senior executive of Metier from the States, Metier once more became interested in finding new accommodation. The search for accommodation was conducted by members of Metier's staff who were wholly unaware of the previous role of Mr Spencer. Metier engaged their own estate agents, Phillips Roth, to act on their behalf. The result of the new initiative was that in October 1983, without Mr Spencer or anyone else from Cooper being involved in any way, Metier entered into a lease of Fairview House. Mr McCulloch then, as a "matter of courtesy", contacted Mr Spencer and told him of the letting. Mr Spencer was pleased to learn the news and responded by sending in his account for his commission at the double scale rate in accordance with the December 1982 and March 1983 letters. Fairview denied liability and proceedings followed.

Before this court and before the learned judge there was no dispute that if Fairview are liable to Cooper, they are liable for double commission and interest on the amount claimed. The judge found that Mr Spencer's introduction had by the time they entered into the lease been forgotten by Metier and "that Mr Spencer's introduction did not operate on its mind in coming to sign a lease". He held: "if effective cause of the letting is a relevant consideration in this case I have come to the conclusion that that effective cause was not Mr Spencer but was probably Mr Ross" of Phillips Roth. However, on the basis that it was not necessary for Cooper to establish that they were an effective cause of the letting the judge went on to find in favour of Cooper.

Although the language of the relevant term in the letters of March 22 1982 and March 2 1983 is not identical, the differences are of no significance. On the literal interpretation of this language, Mr Chadwick accepts that the judge's decision was right. He, however, submits that the terms as to commission must be subject to an implied term that Cooper were only entitled to the commission if they were at least an effective cause, if not the effective cause, of the letting. He submits that the combined effect of a series of cases starting in 1903 with the case of *Millar Son & Co* v *Radford* (1903) 19 TLR 575 and concluding with the case of *Lordsgate Properties Ltd* v *Balcombe* (1985) 274 EG 493* makes it clear that, in the absence of clear language to the contrary, the courts will always imply a term that commission is earned only if the estate agent is an effective cause of the letting or sale which is the subject of the claim.

In support of this submission he relied upon Article 59 in *Bowstead on Agency,* 15th ed, p 229, which states:

> Subject to any special terms in the contract of agency, where remuneration of an agent is a commission on a transaction to be brought about, he is not entitled to such commission unless his services were the effective cause of the transaction being brought about.

He also relies upon a similar statement of principle which appears in *Chitty on Contracts,* 25th ed, at para 2312. Mr Chadwick submits that when the letter of March 22 1982 is examined, there is no special or express term excluding the implied term. He submits, and I agree with this construction, that when the relevant paragraph is read it is requiring the estate agent to "introduce a tenant" and then setting out three qualifications of that tenant which had to be fulfilled before commission would be payable.

The first qualification is that the tenant should not have retained Cooper; the intention being that Fairview should not have to pay commission if Cooper was the tenant's agent, since in that situation the tenants should be responsible for Cooper's remuneration.

The second qualification is that the tenant must be someone with whom Fairview "have not been in previous communication"; the intention being to exclude the obligation to pay commission where the person introduced by the agent has already been the subject of a direct approach by Fairview or has made a direct approach to Fairview or has already been introduced to Fairview by another agent.

The third qualification is that the tenant must subsequently complete a lease and this requirement is obviously inserted so that Fairview would be responsible for commission only if a letting takes place.

Mr Chadwick submits that when the paragraph is properly construed there is nothing inconsistent with the implied term. So, as there was nothing in the circumstances in which the offer to pay commission was made which was inconsistent with the implied term, it should be a condition of Cooper's entitlement to commission that they were an effective cause of the subsequent letting.

When the cases to which I have already referred and the other cases upon which Mr Chadwick relies are examined, and six of the decisions are decisions of this court, it is clear that the court very readily infers an implied term either that the agent is required to be an, or the, effective cause of the subsequent purchase. This is not surprising when it is remembered that in the ordinary way, and in particular in the case of agents retained by private individuals to sell their homes, what the agent is being employed to do is to find a prospective purchaser or a prospective tenant who actually purchases or takes a lease. From the viewpoint of the vendor in such a case, the estate agent has not fulfilled his engagement unless he is an effective cause of the sale or the tenancy. As Viscount Simon LC in the leading case on estate agents' commission, *Luxor (Eastbourne) Ltd* v *Cooper* [1941] AC 108, at p 117, says of the role of an estate agent:

> He is commonly described as "employed": but he is not "employed" in the sense in which a man is employed to paint a picture or build a house, with the liability to pay damages for delay or want of skill. The owner is offering to the agent a reward if the agent's activity helps to bring about an actual sale . . .

However, Mr Morison, while not quarrelling in general with Mr Chadwick's construction, submits that in the case of a developer the court should be substantially less ready to infer such an implied term. He points out a developer such as Fairview has its own sales staff and what they require is not so much the agent's assistance to conclude a sale but the agent's assistance in obtaining an introduction, and it is for the introduction that the commercial developer will be prepared to pay the commission. Mr Morison is unable to point to any authority which precisely supports his submission but instead relies upon the general approach laid down by the House of Lords in the *Luxor* case as to the interpretation of clauses of this nature.

In particular he refers to another passage in the speech of Viscount Simon at p 119 where he says:

> There is, I think, considerable difficulty, and no little danger, in trying to formulate general propositions on such a subject, for contracts with commission agents do not follow a single pattern and the primary necessity in each instance is to ascertain with precision what are the express terms of the particular contract under discussion, and then to consider whether these express terms necessitate the addition, by implication, of other terms . . . in contracts made with commission agents there is no justification for introducing an implied term unless it is necessary to do so for the purpose of giving to the contract the business effect which both parties to it intended it should have.

He also refers to passages in the speech of Lord Russell at p 124 and in Lord Wright's speech in particular at p 130.

Adopting the approach laid down in these speeches in the House of Lords, but having, as I must confess, changed my mind more than once in the course of the admirable arguments which were presented on both sides in this court, I have ultimately come firmly to the conclusion that Mr Morison's submissions and the decision of the learned judge are correct. I can see no necessity in this case to imply a term. On the contrary, I regard the relevant language as being inconsistent with implication of a term imposing an additional implied requirement that the estate agent must be at least an effective cause of the lease being granted.

In a case where there are no express qualifications to be fulfilled other than that a purchaser should be introduced by the estate agent, then the need to imply a term as to effective cause can be readily appreciated, since otherwise if the vendor engages more than one agent there will be no way in which he can avoid being faced with an obligation to meet the claims for commission of more than one agent who each introduced the tenant. However, in this case there is virtually no danger of this happening because of the words "with whom we have not been in previous communication". Furthermore, there is no danger of a requirement to pay commission in the absence of a lease being completed. It is only in a very rare case such as the one

---

*Editor's note: Also reported at [1985] 1 EGLR 20.

under consideration where an agent is responsible for the first introduction and then disappears from the scene altogether that any problem will arise. In such circumstances, the insertion of the implied term contended for by the landlord will change the problem but not necessarily overcome it. If the other estate agent, Phillips Roth, had not been retained by Metier, although they were an effective cause of the letting, they would not be entitled to a commission because undoubtedly Metier would have been in previous communication with Fairview.

Difficulties of a different sort could also occur if the time-lag between the introduction and the completion of the lease were longer than that which occurred in this case. In such a situation it might be necessary to consider whether there is a different implied term, namely that the lease has to be entered into within a reasonable time of the introduction. However, no such implied term was contended for here.

It seems to me that the present clause works perfectly satisfactorily from a developer/landlord point of view. He obtains an introduction to somebody with whom he has not communicated previously, the person enters into a lease and it is then and only then that he becomes liable to pay a commission. An additional requirement could be imposed (and it is not without interest to note, as the Master of the Rolls pointed out in argument, Fairview now impose such a requirement) that the agent should be an effective cause of the letting, but, if this is required, this should be stated expressly.

It is only necessary for me to add that Mr Chadwick submitted that, nowadays, when a term is to be implied the appropriate term to imply is not that set out in the passage from *Bowstead* which I have quoted but an implied term that the agent is "an", not "the", effective cause of the letting. Mr Chadwick may be right as to this in the case of some commission agreements, but I am not satisfied that he is right as to all. It could also create problems where there are two or more effective causes, each of which could be the subject of a claim for commission.

I would dismiss this appeal.

THE MASTER OF THE ROLLS and RUSSELL LJ agreed and did not add anything.

The appeal was dismissed with costs. A stay of execution for one month was granted on condition that the sum awarded was brought into court. Leave to appeal to the House of Lords was refused.

Queen's Bench Division

March 11 1987
(Before Mr R M STEWART QC, sitting as a deputy judge of the division)

ROBERT BRUCE & PARTNERS v WINYARD DEVELOPMENTS

Estates Gazette June 6 1987

282 EG 1255-1261

Estate agents' commission — Claim for commission on introduction of purchasers upheld — Renovated and refurbished property purchased by United Arab Emirates for over £1m — Facts somewhat complicated and some conflict of evidence — The rules of law applicable to the facts as found by the judge were the following — An agreement to pay commission on business introduced does not include an introduction antecedent to the agreement in the absence of an express provision to that effect — An estate agent does not have any implied authority to appoint a subagent — An "introduction" means directing the attention of a person who hitherto has not applied his mind in that direction to the fact that a property is for sale — The judge held, applying the law to the facts as found, that the plaintiffs were entitled to the commission agreed (to be shared on a 50-50 basis with the subagents) — There had been an express acceptance that the commission agreement covered the antecedent introduction — It had also been accepted that subagents, although not identified until later, were involved, so that their participation was authorised — A submission by the defendant vendors that an intervention by the agent acting for the purchasers, before a critical meeting, constituted an effective independent prior introduction, thus excluding the plaintiffs' claim, was rejected by the judge — The plaintiffs had satisfied the conditions required for entitlement to commission under their agreement with the vendors — Judgment for plaintiffs in the sum of £36,225

The following cases are referred to in this report.

McCann (John) & Co v Pow [1974] 1 WLR 1643; [1975] 1 All ER 129; [1974] EGD 184; (1974) 232 EG 827, CA
Samuel & Co v Sanders Brothers (1886) 3 TLR 145
Wyld v Sparg [1977] 2 SLAR 75

In this action the plaintiffs, Robert Bruce & Partners, a firm of estate agents, of St James House, 13 Kensington Square, London W8, sued the defendants, Winyard Developments, a firm of property developers, for commission claimed to be due on the sale to the United Arab Emirates of a property which the defendants had developed at 16 Young Street, Kensington, London W8.

G W Bishop (instructed by Joynson Hicks & Co) appeared on behalf of the plaintiffs; T M Wormington (instructed by Bartlett & Son, of Liverpool) represented the defendants.

Giving judgment, MR R M STEWART QC said: I start my judgment by expressing my gratitude to counsel for their assistance in a case which, though simple in the nature of the claim, has proved to be complex and complicated to hear and to resolve.

The background
The plaintiffs are a firm of estate agents in Kensington. They deal, I accept, in commercial and residential property. They are a small firm by some standards, but size is not a measure of probity.

The defendants are a firm — a partnership — of property developers. I accept what Mr Michael Squire, one of their partners, said in his evidence that prior to the development in question they had only had one development, so that in terms of property development they were comparative newcomers. Having said that, it is clear that they are men of the world.

In late 1983 the defendants found a property for development, 16 Young Street, Kensington. The plaintiffs found it for them, the intermediary being one Johnny Robertson of Mistral. Completion was in about mid-January 1984. The price was £440,000. In addition, the defendants negotiated the purchase from Barkers, or House of Fraser, of a plot of land for a lavatory block for £25,000. That was the work of Mr Michael Squire. It made an enormous impact on the development, changing the whole potential of its profitability. I accept what Mr Squire and others have said about that. In the event, the plaintiffs' commission on the acquisition was limited to the basic price of £440,000, as the documents show.

Building, renovation and refurbishment commenced in about February 1984. The builders were Hatton Builders Ltd. Mr Edward Charles Markes was the relevant contracts manager and in charge of the works and daily on site while they were being carried on. Though well advanced at the times of the matters relevant to this claim in late July/early August 1984, I accept that the works were not completed: they were not expected to be completed until late September of 1984, and the defendants had planned that no marketing of the property should commence until completion, when the pristine refurbished state should make a strikingly attractive office building to any would-be buyers or lessees.

As it happens, the building was sold earlier to the Military Section of the United Arab Emirates. They came on the scene in July 1984 and they exchanged contracts in early August 1984 and duly completed. The plaintiffs claim that they introduced that sale and that, pursuant to a contract with the defendants, they are entitled to commission of 3% on the sale price, plus VAT.

Fortunately, I am not called upon to try, as on a *quantum meruit*, what rate of commission would be reasonable. Quantum is agreed, subject to liability. But that is hotly disputed.

The evidence called and the essence of the dispute
I had the benefit of a number of witnesses. Indeed, the evidence has lasted several days. For the plaintiffs I have had, first, Mr Robert Doxford, a partner in the plaintiffs since 1966; second, Miss Dorothy

Fay Jerrom, "Miss Dee", secretary and director of Old Leadenhall Estates Ltd; third, Mr John Ferrand, then working for the plaintiffs on a commission basis but now a director of a wholesale fish marketing company; fourth, Mr Robin Langton, a director of Aylesfords, estate agents, mainly residential, in Kensington, and now a director of a property company.

Essentially, the plaintiffs' case is this. They had originally been agents for the disposal of 16 Young Street. That appointment was terminated when their Charles Barrow left. So far, common ground. Dee Jerrom had clients, the Military Section of the UAE, who wanted to acquire a freehold property building in Kensington of around 4,000 sq ft and were prepared to pay in excess of £1m. She contacted, *inter alia*, Robin Langton of Aylesfords in her search. Langton knew of the plaintiffs, knew that they dealt in commercial property in Kensington, knew that they had been involved in the acquisition of 16 Young Street, thought they might still be involved as agents, and contacted, therefore, Robert Doxford of the plaintiffs.

The plaintiffs were not then retained by the defendants. Doxford told Langton that the property might be available for over £1m. At that level he hoped to be able to negotiate commission at 3%, to be split 50-50. The lay clients should look at the outside of the building to say if they were interested. They did, and said they were. The chain therefore on the plaintiffs' case was Doxford to Langton to Jerrom to the UAE and vice versa.

On July 31 1984 Doxford met the defendants Bartlett and Squire, disclosed that he had an applicant, said he had told the applicant that the asking price was £1.25m and expected a sale in excess of £1m, said other agents were involved (in fact, Aylesfords) — Mr Doxford claimed that he had named them, but I think he is wrong in that — and said that he required a commission of 3%, and that the defendants agreed.

Doxford organised a site visit for August 1, and thereafter channelled a bid of £1m, which was rejected, and a second bid of £1.1m, which was accepted. In a subsequent meeting on August 6, the clients face-to-face negotiating — the plaintiffs and Aylesfords were not there — the price was reduced to £1.05m, at which price the deal went through. The fact that the offers at £1m rejected and £1.1m accepted and reduced to £1.05m were made and dealt with in that way is common ground. How or by whom they were made is not.

So much for the plaintiffs' case in outline. The defendants called before me Mr Anthony Bickford Bartlett, a partner in the defendants, who was called to the Bar many years ago by the Honourable Society of Gray's Inn; Mr Edward Charles Markes, contracts manager for Hatton Builders Ltd and, as it happens, the brother-in-law of Mr Bartlett; and, third, Mr Michael James Squire, an architect and partner in the defendants.

The defendants contend that they are not liable to the plaintiffs. There are a number of grounds, but they can fairly be brought down to this: first, because the introduction to the property was by Aylesfords and not by the plaintiffs; second, the introduction was antecedent to the agreement of July 31 1984 and was covered neither by the contract nor by the pleadings; third, the plaintiffs anyway were obliged by the contract as a precondition of their entitlement to commission to perform some further act of introduction of the buyers to the defendants after the meeting of July 31. A further act of introduction would be bringing the parties into contact with each other. The defendants claim that the plaintiffs did nothing for that. All was done by Miss Jerrom direct and thus outwith the plaintiffs' contract entitling them to commission. Mr Bartlett put in his evidence that he thought that he was being conned by the plaintiffs and their claim; in other words, that the claim was phoney. As I indicated earlier in the course of argument, strong though that allegation is, the defendants are perfectly entitled to make it and pursue it, and it is covered by their pleadings.

Quality of the evidence

It is appropriate to make some comments generally on the quality of the evidence before I give my primary findings of fact and, from those, my conclusions. As Mr Squire himself said, it is very difficult for a witness to think back the best part of three years and to be precise and completely accurate as to the specific events that happened or, I add, as to the precise dates and times at which they happened. Some witnesses were less impressive than others. I start with the lesser actors before I come to the principal protagonists. Though I say "lesser actors", it does not mean that they are necessarily less impressive.

Mr Markes I accept as entirely honest. I do not think he was very accurate when he said, in relation to the meeting on August 1, that Ferrand of the plaintiffs turned up three-quarters of an hour after the meeting had started. I think Ferrand did turn up after the others, but really at the beginning of the meeting. In making that finding against Mr Markes' evidence, I should make it clear that I am satisfied that he was trying to be accurate. I think his recollection of the particular timing was wrong.

Miss Jerrom — a witness somewhat flamboyant. I approach her evidence with some caution because I am satisfied that in some material respects she was wrong, and materially wrong, and her reasoning justifying her original stance does not stand up. I take one specific example. She said that she spoke to Mr Bartlett but not before about 8.30 pm on the night of Friday, August 3. She said that she was then saying there was a need for a quick deal and the clients wanted to meet. She gave as her justification for that memory and that timing of it: "I felt it was in order to talk direct to the vendors, once the bid had been put to and accepted by Aylesfords." I am satisfied that she is wrong, not just in her recollection of the date or time she first spoke to Mr Bartlett, but wrong in her reasoning. That is important. Is it that she has forgotten the truth? Or is it that she does not want to let it be known that she was in contact with Mr Bartlett earlier than the time when her offer was accepted, and really quite a lot? Either way, her evidence must be approached with care, and I do so. Not only was she in contact before the offer of £1.1m was accepted — and that is the clear inference to be drawn from the documents — but I am satisfied that she did in fact have a conversation with Mr Bartlett, much as he described, as having taken place on July 31. Whether he is right in all its material detail or as to the date will emerge hereafter, but I am satisfied that she had at least three telephone conversations with Mr Bartlett before the offer of £1.1m was accepted.

Mr Langton — essentially honest, but my overall impression is that he was not as impressive or careful a witness as some. I comment on two specific matters. I accept his evidence that his diary entries at the disputed pp 29, 30 and 35 were made contemporaneously. Second, I reject any suggestion that he received or ever asked for or ever thought himself entitled to commission from Miss Jerrom.

Mr Ferrand — similar comments apply in a way as they do to Mr Langton. I think he was essentially honest, but the impression that he gave was of a man who was not exactly dedicated to the details of his work for the plaintiffs, and therefore of a person not likely necessarily to recollect the nicest details and minutiae after nearly three years.

The principal protagonists

Mr Squire — I am satisfied that he is a basically honest and decent man who was trying to be as accurate and frank as possible. He himself accepted the difficulty to recollect back nearly three years and to be precise and accurate on all details after that time. If criticism lies against him or his evidence on that, it arises not by way of any reflection against him but because of a factor of time; indeed, he is frank that it is difficult to recollect all the details after time. I am satisfied in particular that he had a conversation with Mr Bartlett after Miss Jerrom had telephoned Bartlett and that he, Squire, realised the significance of the plans to which Miss Jerrom had referred to Mr Bartlett, and that they could only have emanated from the plaintiffs. He thereby appreciated that Miss Jerrom was, in fact, the agent of the applicants spoken of by Mr Doxford.

Mr Doxford and Mr Bartlett — I have carefully considered their manner and demeanour, as I have indeed of all the witnesses. I have considered all the points raised in cross-examination and argument, particularly the point made for the defendants that the whole aura of activity by Mr Doxford smacked of a lack of professional probity. I do not think it did. True, his agency had been terminated by the defendants on June 18 1984. He had asked to be reinstated and had not been. No doubt he was upset. But he knew the figures advised by Gooch & Wagstaff that:

£950,000 is perhaps higher than we may anticipate achieving.

He knew that the defendants were entrusting matters to Gooch & Wagstaff, and it was reasonable for him therefore to assume that they would abide by Gooch & Wagstaff's advice on price. I think he rightly suspected that an offer in excess of £1m would delight the defendants. He was acting as an entrepreneur, not unprofessionally, and, indeed, he was acting in the interests of the defendants in the end, knowing where their sights were set and reckoning he could better their hopes.

I have considered carefully the criticisms, and they are valid, that no documents or memoranda emanate from the plaintiffs to support their case. Equally, it can be said that no documents were adduced to support Mr Bartlett's being inaccessibly en route to or in Ireland throughout Friday, August 3, though I accept that Mr Bartlett wrote in Mr Squire's diary Dee Jerrom's telephone number — a diary only very belatedly produced. In fact, the entry does not prove the date on which it was written, only that it appears on the space for Friday, August 3. I say in parenthesis that I accept that it was written not later than then.

In the end, where there is a clash between Mr Doxford and Mr Bartlett on the essential areas of dispute, I prefer the evidence of Mr Doxford. It is not a decision that I have reached without very careful consideration. I am satisfied that Mr Doxford was essentially accurate on most issues in dispute, specifically as to who arranged the meeting of August 1. I am also satisfied that he transmitted the first offer of £1m and the second offer of £1.1m to the defendants. I return later to what role Miss Jerrom was playing at the same time. I comment that I am confirmed in my view of these matters, particularly as to who set up the meeting of August 1, by the fact that Mr Ferrand attended it. It is common ground that he attended it. He said he attended it because he was asked to by Mr Doxford, and Mr Doxford essentially confirmed that. I am satisfied that he was asked to attend because Mr Doxford had arranged the meeting. If, as was suggested by the defendants, Miss Jerrom had arranged the meeting, as it were, as a frolic of her own, there would have been no way that Mr Doxford would have known that it was on unless he had seen activity on the defendants' premises. But he would not have known or had any means of knowing that these were the applicants that he was getting through Mr Langton of Aylesfords. If he had suspected that they were, and that something was being done behind his back, I do not think that he would have sent Mr Ferrand; he would have gone himself. I am satisfied that the reason why Ferrand went to the meeting was that Doxford had in fact arranged it.

Primary findings of fact

I make the following primary findings of fact:

1. The UAE delegates were expected to visit the property on July 23 1984. There is no direct evidence that they did so, but the diary entries of Miss Jerrom and Mr Langton record the expected visit. If pushed to it, but it is not a necessary finding for this judgment, I would infer that they probably did visit the outside then or thereabouts.

2. Miss Jerrom had been informed of 16 Young Street as a property on the market prior to July 23 1984 by Mr Langton of Aylesfords, probably some time between July 17 and 23. He, in turn, had got his information from Mr Doxford of the plaintiffs.

3. Mr Doxford told Langton that he was not instructed, but that he knew the property and believed that at an offer of £1m plus he could probably get a commission agreement and would go for 3% and share it 50-50.

4. Miss Jerrom did not know until Ferrand turned up at the meeting on August 1 that Aylesfords were not the vendors' agents. Until August 1 she thought they were, which no doubt is why she expressed the view, as I find she did, that negotiations must go through Aylesfords.

5. The UAE delegates made I think a second visit, but whether it was a *second* visit matters not; they certainly made a visit on July 30 1984. Mr Khaldi was recorded by Mr Markes as taking "his clients" round, and they went inside. They met Mr Markes inside. He told them that they could not visit inside without an appointment. He gave them Mr Bartlett's telephone number as the representative of the owners. His minute is, I find, accurate, as is his evidence on this part. He sent the minute to Mr Bartlett and I am satisfied that Mr Markes spoke to Mr Bartlett about that particular visit prior to either of the afternoon meetings on July 31 1984.

6. In making that visit inside the premises, the UAE delegates went outside their brief from Miss Jerrom and outside the brief that Mr Doxford had given to Langton or Langton had given to Jerrom in consequence. Certainly, neither the plaintiffs nor Aylesfords knew of the visit inside on July 31 then or for some time thereafter. What Miss Jerrom knew as to when it took place is not clear, but she did know that a visit had taken place. She was given Bartlett's telephone number by her clients on July 30 or 31.

7. On the morning of July 31 1984, Mr Langton contacted Mr Doxford to say that the UAE were definitely interested and wanted a formal inspection visit. Whether Mr Langton had been contacted by Miss Jerrom on July 30 or 31 I am not clear, and I do not think that question has to be answered. As to the line of communication — Jerrom to Langton to Doxford — I am clear. Doxford then wished to set up a meeting with the defendants to agree a commission basis and, if that could be agreed, then to arrange for the inspection. He instructed Ferrand to contact the defendants. Ferrand did not do so or was unable to do so. Eventually, in the afternoon, Mr Doxford saw activity of the defendants out of his window, which looked on to 16 Young Street, so Mr Ferrand was sent to speak to them. He spoke to Johnny Robertson of Mistral, who was with Messrs Squire and Bartlett, and so a meeting was arranged.

8. At about 4.30 p.m on July 31 1984 there was a meeting at Mr Doxford's offices. Present were Mr Doxford of the plaintiffs, Mr Bartlett and Mr Squire of the defendants. The question arises as to what was said and agreed. I am satisfied:

(a) Doxford said he had clients interested in the building. He had told them of the property. They had seen the outside.

(b) Doxford said he had another agent involved. I do not think that he mentioned Aylesfords by name. I do not think the defendants realised it was Aylesfords until a day or so later. He simply said that another agent was involved with him.

(c) Doxford said that the deal was on in excess of £1m. He had, he said, quoted a price of £1.25m.

(d) Doxford said if he was to process the deal and introduce those clients to the defendants, a commission was required of 3% — at that level because he was having to share it with his subagents.

(e) The defendants, principally by Mr Bartlett, were delighted with the prospect of the price, and agreed to this specific deal and specifically to the 3% commission. It was on the basis, first, that the introduction was to take place to the defendants, and I construe that as meaning the bringing of the clients into contact the one with the other so that each party was in contact with the other; second, the price was to exceed £1m before any entitlement to commission would arise; third — and this came from Mr Squire's evidence rather than Mr Bartlett's — solicitors were to be in contact within a week. That is not a pleaded term but it matters not.

(f) Mr Doxford then, after the agreement on commission was reached, said he wanted to arrange a formal site visit for the next day. I am satisfied that this was so because the whole purpose of the meeting of July 31 was to get an agency agreement and, if successful, to arrange the site visit. It would have been ludicrous not to raise the topic of the site visit. I am satisfied that Mr Doxford suggested the site visit. It was accepted, and it was in this context that Doxford said he was unable to disclose then the precise identity of his clients, though there was some mention of Arabs. And Mr Bartlett said no matter, they would learn that the next day.

9. Those conditions by the defendants were met in this sense:

(a) The meeting took place as proposed and agreed on site on August 1. The clients were then brought into contact with the defendants.

(b) The sale took place at a price in excess of £1m.

(c) The solicitors were in contact within the requisite time.

10. At the meeting on July 31 1984, Mr Bartlett asked whether the clients had visited the premises. I am satisfied that Mr Doxford had said that they had seen the outside. Mr Bartlett asked whether they were connected with a Miss Dee. Mr Doxford said that he had never heard of her. Mr Bartlett asked whether they were Arabs, because that was the nationality of the people that Miss Dee had sent, and Mr Doxford said that there might be a connection. It was clear to all parties that the applicants in respect of whom the agreement for commission was reached had had their attention drawn to the fact that the building was for sale, had visited at least the outside of the building, and had therefore previously been introduced to the property.

11. Following the meeting on July 31 1984, Doxford got on to Langton, who got on to Jerrom to set up the meeting for August 1. By the same chain in reverse came back the confirmation that 10.30 am for August 1 1984 was convenient. I am satisified that Mr Doxford telephoned Mr Bartlett to confirm that. I specifically reject the suggestion that Dee Jerrom set up this meeting herself by some direct action.

12. I am satisfied that Dee Jerrom in fact telephoned Mr Bartlett on

the evening of July 31. That was after the arrangements had been made and confirmed for the meeting of August 1 but before the meeting took place. She wished to know more details about the building, its size and what was likely to be an acceptable price. There have been a lot of submissions as to the date, and I should say briefly why I am satisfied that it took place on July 31. It is not just the evidence of Mr Bartlett that it took place then.

First of all, Dee Jerrom had plans for the property. She referred to them in that telephone conversation and did not appreciate that the plans were wrong and out of date and did not include the lavatory block. That would be odd indeed if she had visited the premises already and spent a long time being taken around by Mr Markes, as I am satisfied it happened on the morning of Wednesday, August 1. Second, she told Mr Bartlett in terms that her information as to the property came from Aylesfords and that negotiations should go through Aylesfords if the sale proceeded. Mr Bartlett said that he had never heard of Aylesfords, and Dee Jerrom was unable to help him further. At that stage on July 31, she did not know that Aylesfords had got their information through Robert Bruce and were acting as subagents for Robert Bruce. She did learn that when Ferrand turned up on the morning of August 1 at the site meeting. I find it quite incredible that this telephone conversation could have taken place after the site meeting of August 1 and Miss Jerrom not to explain "Well, Aylesfords got it from Robert Bruce, and you know all about them." I am quite sure that the conversation took place on the evening of July 31 at or about the time that Mr Bartlett said that it did.

I accept, therefore, Mr Bartlett's evidence that the telephone conversation took place on the date and at the approximate time that it took place. I reject only his evidence that in that conversation Miss Jerrom was asking for a meeting the next day. The meeting for August 1 had already been asked for by Mr Doxford. I think there were a number of telephone calls by Miss Jerrom to Mr Bartlett and I think that what he has done is in no sense dishonest; I think he has got muddled in which conversation Miss Jerrom asked for a meeting. What does seem clear is that there was a meeting set up for August 2 — that would be the Thursday — at which the clients did not turn up. August 2, Mr. Markes thought, or it could have been the 3rd; but subsequent at any rate to August 1. I think that Miss Jerrom asked in a subsequent telephone call for a meeting. I am quite satisified that she did not try on July 31 to set up a meeting for August 1 because that to her knowledge had already been done.

I further accept Mr Squire's evidence that Mr Bartlett telephoned him later that evening to tell him of the call. Mr Bartlett's attitude was one of pleasure that there were now two lots of people potentially interested — the applicants of the plaintiffs, and Dee Jerrom and her clients. But it was when Squire heard that Jerrom had old plans without the lavatory block that he realised that they could only have come from the plaintiffs and that these were the plaintiffs' applicants. I am satisfied that he gave that reasoning to Mr Bartlett.

Finally, I accept and find that in the telephone call on July 31 Dee Jerrom gave to Mr Bartlett not the identity of her clients, save that they were Arabs, but her name, Dee Jerrom — not just "Miss Dee" — and telephone number, as the agent acting for her Arab clients in the purchase.

13. On August 1 1984 a meeting took place on site. Dee Jerrom turned up with Mr Khaldi and the UAE delegates. Mr Markes hosted or acted for the defendants. Mr Ferrand of the plaintiffs was there for a time. The meeting was a long and thorough meeting and inspection; I accept the evidence that it lasted probably an hour and a half or thereabouts. As I have said, I do not think that Mr Ferrand turned up three-quarters of an hour after the meeting had started. I think Mr Markes is wrong in that recollection. Equally, I think it likely and I so find, that Mr Ferrand was the last of all the parties to turn up, but not by long. The meeting was scheduled for 10.30. He was there at about that time. It may be that Miss Jerrom and her clients were there a little early.

At the meeting Ferrand introduced himself to Miss Jerrom. She asked him who he was and what he was doing, and he said he was from Bruce & Partners. That was when she realised that Aylesfords were not directly the agents for the owners but came through Robert Bruce. I think that is the way the introduction went, and it is a further pointer to the meeting being set up by Mr Doxford, because if Mr Doxford had even sent Mr Ferrand to find out what other people were doing, Mr Ferrand would have opened the batting "I'm Ferrand from Robert Bruce. What on earth are you doing here?" He did not open that way. I am satisfied that the opening came from Miss Jerrom, establishing who he was and why he was there.

Ferrand did not speak to Mr Markes or introduce himself to Mr Markes. He spoke to Miss Jerrom, nodded to the UAE delegates but did not speak to Mr Markes. That was not because the plaintiffs should not be represented, but because I think Mr Ferrand was doing the bare minimum to comply with some rather boring instructions given to him by Mr Doxford. As a result, Mr Markes did not then know who Mr Ferrand was.

14. Following the meeting, probably on Thursday, August 2, Miss Jerrom visited Langton's office and put an offer of £1m to Langton. He put it to Doxford, who put it to the defendants. It matters not at this stage whether Doxford was right that it was to Squire or whether it was to Bartlett. I am satisfied that it was put to the defendants and it may indeed have been put by way of a message for one of them rather than direct. It was rejected on Doxford's advice that more could be negotiated. The answer went back down the same chain, quite likely with an indication that an increase would have to be made but not all that large.

15. On August 3 Miss Jerrom left a message for Langton that her clients would offer £1.1m. He passed it to Doxford, who passed it to the defendants.

16. Meanwhile, and before an answer to that offer was given, Jerrom contacted Bartlett direct. Her evidence — I have already referred to it — was that she did not contact him until after the offer was accepted. I am satisfied that she is wholly wrong on that. It is inconsistent with the letter to her from Mr Langton at p 37 on which she was cross-examined. She did not dissent from the proposition set out by Mr Langton in that letter. And it is inconsistent with her letter at p 53. As I have said, I reject her account that she did not contact Bartlett until the evening of Friday, August 3. I am satisfied that this second offer was in fact communicated by Doxford to the defendants. Dee Jerrom had left a message that the increased offer was being made of £1.1m for Langton in his office. It was transmitted by Aylesfords — and it may well be that Mr Langton was out at the time — to the plaintiffs, and by Doxford of the plaintiffs to the defendants. If Bartlett was then in Ireland or en route to Ireland, it must have been to Squire. I am bound to say that I find the evidence as to when Bartlett went to Ireland very unsatisfactory. I am far from persuaded that he was incommunicado and unreachable, at any rate on the morning of August 3. I think he was still in communication on that morning.

I am satisfied that before Dee Jerrom had an answer to that increased offer of £1.1m, she had talked to Bartlett about it. It may be that she also put it directly to Bartlett as well as setting up the chain via Langton for transmission by Doxford. It matters not whether the approach or the conversation between Jerrom and Bartlett was on August 2 or early on August 3. At that stage — and this is either late on August 2 or first thing on August 3 — Jerrom and Doxford were both involved in negotiations. Mr Bartlett and Mr Squire decided that the situation was becoming chaotic and that everything should be transmitted through Gooch & Wagstaff. The letter from Gooch & Wagstaff, dated August 3, is significant in this context. I think it was written on August 3 and written because on that day or, at the very earliest, at the end of the preceding day they had received instructions that they were to channel all negotiations. I do not think those instructions were given before, at the very earliest, the end of the working day of August 2.

17. I also accept, as I have indicated, that there were several telephone calls from Dee Jerrom to Mr Bartlett between July 31 and August 3, and at least one other than the two I have already dealt with. Probably on Thursday, August 2, there was a call in which Mr Bartlett got hot under the collar. He had previously said on July 31 that an offer of the order of £1.05m would be acceptable. On this occasion he got hot under the collar because of the complications of the long telephone calls from Dee Jerrom and he upped his minimum price to £1.1m from the previously indicated figure of £1.05m. I am satisfied that there was such a conversation as Mr Bartlett speaks of, but, equally, in so far as Mr Bartlett says that in this conversation Dee Jerrom said she had never heard of Robert Bruce & Partners and there was no need to pay Aylesfords, I reject it. I am satisfied that Dee Jerrom said she had never heard of Robert Bruce & Partners on July 31. She had heard of them on August 1 and I do not believe for a minute that she said she had not heard of them afterwards. I am satisfied that at all times Dee Jerrom said that Aylesfords were not her baby for commission but that she expected negotiations to go

through Aylesfords and she expected Aylesfords' commission to be met by the defendants. This conversation on Thursday, August 2, was the second in time of the three telephone calls. The conversation referred to in finding 16 took place subsequent to this one, after a sufficient interval of time for Miss Jerrom to get her clients' instructions on the £1.1m.

18. The plans meanwhile had been conveyed by Doxford to Langton to Jerrom to the UAE. They were the old plans, not showing the lavatory block. I am satisfied that Mr Bartlett realised that they were the old plans when he was talking to Miss Jerrom on July 31. Thus it follows that the plans had been transmitted at the latest on July 31.

19. It follows that in so far as Mr Bartlett differs on these facts to which I have referred, I reject Mr Bartlett and accept Mr Doxford. I add that I find it significant that Mr Squire considered, after the final deal was struck and the contracts had been exchanged and, indeed, completion had taken place, that the plaintiffs should get their commission. I do not think that he then thought that the plaintiffs had done nothing in setting up the arrangements for the meetings or transmitting offers and conducting negotiations. On the contrary, I am satisfied that the plaintiffs had done those things to the knowledge of Mr Bartlett and Mr Squire.

20. On August 3 two things happened. Gooch & Wagstaff were directly involved as agents for the defendants, though Miss Jerrom was not particularly happy about it. Second, Dee Jerrom went to the forefront of negotiations direct, with Squire, who succeeded Bartlett when the latter was in Ireland (and I am satisfied that he went to Ireland some time around about then, though not necessarily quite as early as he says).

21. On August 6 a meeting took place. The UAE delegates conducted negotiations with Mr Squire. Dee Jerrom was taking a back seat. Neither the plaintiffs nor Aylesfords were represented. I accept that the price was reduced to £1.05m. Dee Jerrom said that the meeting was to take place because her clients wanted a meeting face to face because they were not happy with the price of £1.1m. I accept Mr Squire's evidence that the UAE were claiming that they had not agreed to more than £1m; that he started to walk out; and that he would have done so if they had not come up with £1.05m. What they had authorised Miss Jerrom to offer on their behalf I simply cannot decide, But I think the way the UAE were conducting themselves was probably part of their negotiating or bargaining habit and no more.

22. The sale at £1.05m duly went through.

The law

I have been referred to a number of authorities, in particular a Court of Appeal decision in the case of *Samuel & Co* v *Sanders Brothers* (1886) 3 TLR 145, which is authority for the proposition that an agreement to pay commission on business introduced does not include an antecedent introduction of business in the absence of express provision to that effect.

The second authority is *John McCann & Co (a firm)* v *Pow* [1975] 1 All ER 129, again a decision of the Court of Appeal, to the effect that an estate agent does not have implied authority to appoint a subagent. If therefore he appoints a subagent without express authority, an introduction by that subagent is not one within the agent's contract with his principal and he is not entitled to commission on it.

Third, a South African case of *Wyld* v *Sparg* [1977] 2 SALR 75, where a helpful definition is given of introduction:

directing the attention of a person who hitherto has not applied his mind in that direction to the fact that a property is for sale

and provides authority for the proposition that the introduction by the agent must be the cause of the transaction. That, of course, is within the concept of law that an event may have several causes. What is relevant is that there is a causal link. That proposition is confirmed by the passage in *Chitty on Contract*, vol 2, para 2312, which I summarise as this, that the agent must be the effective cause of the transaction. I do not think I need refer to any other matters of law that have been urged before me.

Applying those principles of law to the present case and to my primary findings of fact, I find:

1. The original introduction of the UAE to the property was as a result of Aylesfords introducing it to Dee Jerrom, the UAE's agent. Aylesfords in turn got it from the plaintiffs. At that time Aylesfords acted by agreement with the plaintiffs as their subagents, but the defendants had not then appointed the plaintiffs as their agents and, *a fortiori*, they had not then authorised the appointment of Aylesfords as subagents.

2. At the meeting on July 31 1984 in the plaintiffs' offices, both parties realised that the plaintiffs' applicants, albeit the defendants did not know the name, had been introduced to the property within the definition from the South African case to which I have referred. The agreement to pay commission to the plaintiffs was in respect of those specific, unnamed applicants already introduced. Therefore there was express agreement that the commission and contract covered the antecedent introduction to the property.

3. At the meeting on July 31 both parties knew that the plaintiffs had other agents involved in getting the UAE. I say "UAE"; the identity was not then known. Aylesfords were not then named, only later. In my judgment, it matters not that they were not named. The commission was assented to on the basis that the plaintiffs already had other agents involved. Therefore, there was express agreement to and authorisation of the antecedent appointment by the plaintiffs of Aylesfords as their subagents in this deal.

4. The agreement between the plaintiffs and the defendants envisaged a further act of introduction by the plaintiffs as a precondition of entitlement to commission; that is to say, in my judgment, introduction by bringing together of the parties or bringing the parties into contact with each other. That is what the introduction meant. I have already found that the meeting of August 1 was arranged by the plaintiffs. It was clear on July 31 that the applicants were Arabs, whether or not UAE was mentioned (and I do not think any more was mentioned than that they were Arabs from some Arab government). The meeting of August 1 was arranged and took place so that thereby (a) the parties were brought together, causing (b) negotiations and a sale thereafter to take place.

5. The issue is complicated, and materially complicated, by the action of Dee Jerrom on July 31, before the meeting of August 1 had taken place but after it was arranged, contacting the defendants and identifying herself as the purchasing agent for would-be Arab buyers but giving no further clue as to the identity of her clients. The defendants submit, and submit cogently, that that act constituted the effective introduction in the sense of bringing the parties into contact with each other, because from the act of Miss Jerrom contacting Bartlett and giving her telephone number as the agent for the buyers the parties had the means of carrying out negotiations between each other and certainly of contacting each other. The defendants further submit that that act took place because Mr Markes, on July 30 and before the agreement between the plaintiffs and the defendants, gave Bartlett's telephone number.

Therefore they submit that, the effective introduction or coming together having occurred before the meeting of August 1, the effective introduction was outside the contract. For the plaintiffs to succeed, they submit it must be shown that the plaintiffs were the first to bring the parties together.

6. That submission has great attraction at first sight, but I have concluded that it is wrong because

(a) The act of Miss Jerrom in contacting the defendants was, I am satisfied, not an act independent of the acts of the plaintiffs and Aylesfords. She contacted the defendants after the site meeting was arranged and because of the arrangement, and did so to sort out the particular size of the property, which itself would give a guide to its market price and the likely acceptable price.

(b) In fact, she did not identify her clients then. That came only physically at the site meeting arranged by the plaintiffs on August 1 and in name thereafter, as a result of the plaintiffs' meeting and the transmission of the initial offers by the plaintiffs to the defendants.

(c) She said she had details from Aylesfords and that all negotiations must go through them. Aylesfords were in fact the plaintiffs' subagents, and though the defendants did not know the name, they had authorised the plaintiffs to proceed with subagents.

I am therefore satisfied that this was not an introduction of the applicants, the bringing together of the parties outside the plaintiffs' contract. I am far from satisfied that what Miss Jerrom did in fact constituted the bringing together of the parties, but if I am wrong in that, then, in my judgment, she was not doing it outside and independent of the chain of the plaintiffs and Aylesfords. Rather, she was expressly bringing herself within the chain from Aylesfords.

It follows, therefore, that, in my judgment, the bringing into contact of the UAE buyers was in fact done by the plaintiffs. That was the introduction for which they were contractually responsible as a precondition of entitlement to commission. The sale then proceeding to £1.05m (the precondition being that it should exceed £1m) it follows that, in my judgment, the effective cause of the ultimate sale was that act of the plaintiffs in setting up the meeting and, indeed, in carrying out such negotiations as they did thereafter. Therefore they are entitled to the commission claimed, which I understand to be £36,225.

HOUSING ACTS

Queen's Bench Division
October 8 1986
(Before Mr Justice HODGSON)

R v LAMBETH BOROUGH COUNCIL, EX PARTE CLAYHOPE PROPERTIES LTD

Estates Gazette February 14 1987

281 EG 688-694

Housing Act 1974, sections 71 and 71A, and Housing Act 1957, section 9(1A) — Application by freehold owners of a mansion block of flats, occupied by a mixture of long leaseholders and protected tenants, for judicial review, seeking an order of mandamus to require the local authority to make mandatory repair grants under sections 71 and 71A of the Housing Act 1974 — Mansion block in a bad state of repair; substantial works, particularly to common parts, required — The mandatory repair grants depended on the execution of works required by notice under section 9(1A) of the 1957 Act — Notices under section 9(1A) were served on the leaseholders in respect of their flats and on the freehold owners in respect of the flats occupied by the protected tenants — The question for the court was whether these notices were valid — The applicants' argument was that a flat was part of a building within section 18 of the 1957 Act and the authority could take the same action in regard to it as to a house, so that it was a house within the meaning of the Act; also, the common parts could be regarded as its "appurtenances" within section 189 — This argument was rejected by Hodgson J, who held that a flat in a block was not a "house" and, in any case, common parts, such as the roof, could not be regarded as "appurtenances" — It followed that the notices served under section 9(1A) on the leaseholders in respect of their flats and on the freeholders in respect of the flats occupied by the protected tenants were alike invalid — An argument that a local authority was estopped from challenging the validity of its own notices in the circumstances of this case was rejected — The notices could not therefore form the basis of a claim for grant under the 1974 Act — The effect of this decision together with that of the House of Lords in the case of *Pollway Nominees Ltd* v *London Borough of Croydon* might make it impossible for a local authority to exercise control under Part II of the Housing Act 1957 over the external and common parts of a block of self-contained flats — If this was an unintended lacuna it was a matter for the legislature to consider — Application for judicial review dismissed

The following cases are referred to in this report.

Critchell v *Lambeth London Borough* [1957] 2 QB 535; [1957] 3 WLR 108; [1957] 2 All ER 417, CA

Lever Finance Ltd v *Westminster (City) London Borough Council* [1971] 1 QB 222; [1970] 3 WLR 732; [1970] 2 All ER 496; (1970) 21 P&CR 778, CA

Pollway Nominees Ltd v *Croydon London Borough* [1986] 2 EGLR 27; (1986) 280 EG 87, HL

R v *Inland Revenue Commissioners, ex p Preston* [1985] AC 835; [1985] 2 WLR 836; [1985] 2 All ER 327, HL

Western Fish Products Ltd v *Penwith District Council* [1981] 2 All ER 204; (1978) 38 P&CR 7; 77 LGR 185, CA

In this case Clayhope Properties Ltd, the freehold owners of the block of flats known as Dover Mansions, Canterbury Crescent, London SW9, applied for judicial review, seeking an order of mandamus directed to Lambeth Borough Council, requiring the council to make mandatory grants for repairs to the block under the Housing Act 1974 as amended.

John Colyer QC and Roger Cooke (instructed by Bernstein & Co) appeared on behalf of the applicants; Andrew Arden (instructed by R G Broomfield, chief solicitor, Lambeth Borough Council) represented the respondent council.

Giving judgment, HODGSON J said: In this case the applicant company seeks judicial review by way of mandamus directed to the respondent council requiring the council to make mandatory repair grants pursuant to sections 71 and 71A of the Housing Act 1974.

The case concerns a mansion block of 20 flats known as Dover Mansions. The applicant is the freehold owner of the block. Fourteen of the flats are let on 99-year leases with small ground rents; the remaining six are let on short-term protected tenancies. The block has two entrances, one to Flats 1-10, another to Flats 11-20. The block has one common roof and there are, of course, common parts including passages and staircases enjoyed by the occupants of all the flats. I refer to the two classes of flat as leasehold flats and tenanted flats.

So far as the leasehold flats are concerned, I have been shown a copy of the lease of no 1, which I assume to be in common form with the other 13. The lease provides for the demise of:

All that ground floor flat being on the ground floor of the Mansion including as appropriate one half in depth of the joists between the ceiling of the flat and the floor of the flat above it and the internal and external walls between the flat and any adjoining flats the situation whereof is shown on the plan annexed hereto . . .

The lessor covenants:

That (subject to contribution and payments as hereinbefore provided) the Lessors will maintain repair and renew: (a) the main structure and in particular the roof foundations chimney stacks gutters and rainwater pipes of the Mansion (b) the gas and water pipes sewers drains and electric cables and wires in under and upon the Mansion and enjoyed or used by the Lessees in common with the owners and lessees of the other flats (c) the forecourt main entrance passages and right of way of the Mansion so enjoyed or used by the Lessees in common as aforesaid and (d) the boundary walls and fences of the Mansion.

Towards the cost of these works each long leaseholder undertakes to contribute one twentieth, but, of course, from the protected tenants the landlord can recover nothing.

The block is in disrepair, and substantial work of repair, particularly to the common parts, is urgently needed.

Under the provisions of section 71 and 71A of the Housing Act 1974 (71A being added by the Housing Act 1980) it is provided: "In so far as an application for a repairs grant relates to the execution of works required by a notice under section 9 of the Housing Act 1957" the award of a grant is mandatory. Section 9(1A) was added to the 1957 Act by section 72 of the Housing Act 1969. It provides, so far as relevant:

Where a local authority . . . are satisfied that a house is in such a state of disrepair that, although it is not unfit for human habitation, substantial repairs are required to bring it up to a reasonable standard, having regard to its age, character and locality, they may serve upon the person having control of the house a notice requiring him, within such reasonable time, not being less than twenty-one days, as may be specified in the notice, to execute the works specified in the notice, not being works of internal decorative repair

In 1981 the council served a notice purportedly under section 9(1A). It was served on the applicant and required works to be done to Dover Mansions as a whole, including work to individual flats. As Mr George [solicitor of Lambeth Borough Council] put it in his affidavit, this "in substance represented the view that Dover Mansions could be treated as 'a house' for the purposes of Part II of the 1957 Act". The 1957 Act gives a right of appeal against notices to the county court. Against the 1981 notice the applicant exercised this right, but the appeal was never heard and has been overtaken by events.

In 1982, in the course of negotiations, the applicant says that the council indicated that it would give grants for the work comprised in the notices, and an application for a grant was made in 1983, but that also has been overtaken by events.

Also in 1983 the leasehold tenants began an action against the applicant for specific performance of the landlord's repairing covenants and for damages. In that action a receiver and manager of Dover Mansions has been appointed to receive the rents and profits and to manage the block and receive grants. The action itself stands adjourned.

The immediate events out of which this litigation arises began in the spring of 1984. By then the legal advice given to the council had changed and they were advised that Dover Mansions as a whole could not be treated as "a house". Also, deterioration since 1981 had occurred, which meant that the original schedule of works would no longer be sufficient. A further factor was that the tariff rate for mandatory grants was to be lowered in respect of applications submitted after March 31 1984. (It is one of the peculiarities of this case that the parties have a common aim in that all are anxious that as much money as possible should be made available; this is a less generous attitude on the part of the council than appears on the surface as the grants are funded wholly or largely from central government funds.)

The council were now advised that the way in which notices should be served was this. In respect of the leasehold flats the leaseholder should be served; in respect of the tenanted flats the applicant.

Accordingly, by a letter dated March 20 1984 addressed to one of the leaseholders and copied to the others, the council set out in great detail the steps it proposed to take and the reasons for doing so. It is unnecessary to cite in detail the contents of that letter; it made clear that the council were going to proceed on the basis that the mansion block as a whole could not be classed as a house as it "comprises a number of self-contained purpose-built dwellings" but that each flat was itself "classed as 'a house' within the meaning of" the (1957) Act, and that in respect of each such "house" a notice would be served on the "person having control". The council further intimated in that letter that, as a "house" by section 189 of the 1957 Act includes "appurtenances" the "main structure, common parts and roof can therefore be deemed to be appurtenances".

Acting on behalf of the residents' association, solicitors replied to this letter on March 22 1984 contending in strong terms that:

To contend that service (of the notices) must be effected on the lessees themselves, is with the greatest of respect, nonsense. The lessees have absolutely no authority or power to do works to the external/common parts as you suggest.

However, acting on the legal advice they then had, the council served 20 notices on March 23 1984. In respect of the leasehold flats the notice in respect of Flat 1 is typical. It reads, so far as relevant:

To (the name of the tenant) . . . being the person having control of the house known as Flat 1, Dover Mansions, Canterbury Crescent, SW9. TAKE NOTICE that — (1) the LONDON BOROUGH OF LAMBETH are satisfied that the above-mentioned house is in such state of disrepair that, although it is not unfit for human habitation, substantial repairs are required to bring it up to a reasonable standard, having regard to its age, character and locality; and (2) in pursuance of section 9(1A) of the Housing Act 1957, the Council require you within a period of Forty two (42) days ending on the 4th (FOURTH) day of May, 1984, to execute the following works, not being works of internal decorative repair, namely SEE ATTACHED SPECIFICATION.

The notice goes on:

The attached schedule is divided into two parts as follows: Part A — Internal works of repair in connection with the specified flat. Part B — External works of repair to the whole block and internal works to the shared common parts.

Each recipient of this notice is responsible for all the works specified in Part A and one twentieth of the cost of the works in Part B of the schedule.

In respect of the phrase "person having control" the recipient was referred to section 39(2) of the Act.

So far as Part B was concerned, each notice specified the whole of the work required so that each recipient was required, for example, "to examine all chimney pots and remove broken, cracked, rusted or otherwise defective pots".

The notices in respect of the tenanted flats were served on the applicant. Each was in the same terms as the 14 notices to leaseholders and each required the applicant to be "responsible" for one twentieth of the cost of the external and common part works.

Despite the solicitors' response of March 24 fresh applications for grants were already being prepared. The applicant, the receiver and all the tenants co-operated in making the grant applications; each was made in the name of the tenant, countersigned by the applicant. They in fact predate the notices: they were sent by the receiver to the council on March 30 1984 together with other documents and information about tenders which had been received for doing the work. In a letter dated April 16 1984 the council appeared to be accepting the validity of the grant applications.

As well as making the grant applications (or rather countersigning them) the applicant gave notice of appeal to the county court against them. The grounds of appeal are set out in the affidavit of Mr Allen [of Bernstein & Co, applicant's solicitors]: grounds (b) and (c) have since been abandoned and, if the relief sought in these proceedings is granted, the appeals will be withdrawn.

But the respondents, perhaps because of a circular from the Department of the Environment, which I have not been shown, took fresh legal advice. As a result of that advice they have made no determination on the grant applications and no grants have been paid.

On October 17 1984 the council wrote to the receiver:

Dear Sir Re: Dover Mansions, Canterbury Crescent
I write further to our recent telephone conversation regarding grants.

As explained, current legal opinion is that we are not able to grant aid works to common parts and areas of shared responsibility.

The whole issue of dealing with Mansion Blocks highlights a number of deficiencies in the current housing law relating to purpose-built flats and maisonettes.

Our legal department is currently seeking a further counsel's opinion but until this has been received I have been advised by the Housing Directorate that they will not be approving any further grants where there is an element of shared works.

Clearly this means everything has ground to a halt at the moment until the grant situation has been resolved. I shall keep you informed of any developments at the first opportunity.

On December 10 1985 the council wrote again to the receiver and to all the tenants:

I must inform you that the above Notices served on you . . . are hereby withdrawn in line with the legal advice received by this Council.

The withdrawal of the Notice(s) is without prejudice to the fact that the Notice(s) shall stand if the Court so rules in the action for judicial review between Clayhope Properties and the Council.

As these Notice(s) have been withdrawn (subject to the Court's decision as stated above) the grant application(s) received cannot be treated as mandatory and must be treated as application(s) for discretionary grants. The Council regrets that there is no funding presently available for the payment of discretionary grants and the grant applications as such extend beyond admissible works.

Should there be any further developments I shall be contacting you further.

The respondent now accepts that this letter can be ignored. The notices stand until and if they are declared to be invalid.

The question which I have to decide (subject to an argument based on estoppel) is whether the notices purportedly served under section 9(1A) of the 1957 Act were valid notices.

During the course of argument I was told that the case of *Pollway Nominees Ltd* v *London Borough of Croydon* was about to be argued in the House of Lords. As that case concerned the application of section 9(1A) to a purpose-built block of flats (although a solution of the problem was differently attempted in that case) I thought it right to defer giving judgment in this case until I had had an opportunity of considering the speeches in *Pollway*. They were made on July 17 1986 and I have now had the opportunity of seeing them.[1]

What early becomes apparent when one considers the statutory provisions in Part II of the Housing Act 1957 is that they were plainly not framed to cope with the situation where there is a purpose-built block of flats.

The compulsory powers of local authorities contained in Part II of the 1957 Act as amended have their origin in Part II of the Housing

---

[1] Editor's note: See [1986] 2 EGLR 27; (1986) 280 EG 87 for House of Lords report.

Act 1930. In the 1930 Act the heading to Part II read "Provision for securing the repair, maintenance and sanitary condition of houses", and the Act required the person having control to effect repairs to a house which was unfit for human habitation and which was "occupied or of a type suitable for occupation by persons of the working classes". It was also provided by section 20 that the "like proceedings" could be taken "in relation to any part of a building which is let for human habitation as a separate tenement" save that in relation to part of a building a demolition order could not be made, only a closing order. The heading to section 20 was "Power of local authority to deal with part of a building".

In the 1957 Act the heading to the fasciculus of sections in Part II remained the same, but the reference to the working classes was omitted and the powers widened to include "any house". And, in 1969, the power was extended to cases where, although fit for human habitation, substantial repairs are required to bring it up to a reasonable standard, having regard to its age, character and locality.

The 1957 Act in section 18 makes similar provisions as to part of a building as were contained in section 20 of the 1930 Act, but some indication of the draftsman's lack of concern with purpose-built flats is perhaps shown by the substitution of a new heading "Power to make a closing order as to part of a building". As Lord Bridge said in *Pollway*: "The truth, I suspect, is that generations of parliamentary draftsmen have been content to use the time-honoured formula without ever contemplating its application to the circumstances presently under consideration."[2]

The argument put forward by the applicant in this case is basically that which the respondent was advancing in its long letter of March 20 1984. I hope I summarise it accurately as follows. Section 9(1A) empowers a local authority, in the circumstances set out and already recited, to "serve upon the person having control of the house a notice" requiring him to execute works. By section 18 a local authority may take the like proceedings in relation to "any part of a building which is used . . . as a dwelling" as they are empowered to take in relation to a house. A flat in a block is plainly part of a building used as a dwelling; therefore a local authority can take proceedings under section 9(1A) in relation to it.

So far, despite the doubt raised by the restrictive heading to section 18, I do not think that the respondent would be disposed to quarrel with the argument.

But the argument has to proceed two steps further if the notices are to be valid notices in respect of the outside and common parts. First, it is contended that, because of section 18, a flat is a house within the meaning of the Act so that the interpretation section 189 applies to it. That section provides that "house" includes "(a) any yard, garden, outhouses, and appurtenances belonging thereto or usually enjoyed therewith . . . ".

I have then to be satisfied that all the outside and common parts are appurtenances to all the flats.

The respondents contend that these submissions fail *in limine* because a flat in a block is not a "house"; it is only a part of a building. I agree. I can find nothing in section 18 which would justify me in holding that a flat in a block is a house for the purposes of Part II of the Act. If that had been intended it could have been specifically provided. And, in this connection, it is, in my judgment, importantly indicative of statutory intention that specific provision is made under (b) relating to "house" in section 189 that the word includes "part of a building" for the purposes of provisions in the Act relating to the provision of housing accommodation; these provisions one finds in Part V of the Act. See also *Critchell* v *London Borough of Lambeth* [1957] 2 All ER 417 at p 420A per Lord Evershed MR: "In any case, in my judgment, s9, s10 and s11 of the Act of 1936" — sections 9, 10 and 16 of the 1957 Act — "use the word 'house', in their context, as meaning what is commonly called a house — that is, a separate structure".

If I am wrong and a flat is a house and includes its appurtenances, are the common parts, the roof etc all appurtenances of each flat? I have heard lengthy argument in support of this proposition, but I am bound to say that it would offend my sense of reality to find that the chimney pots above Flats 1-10 were appurtenances of Flats 11-20.

As Mr Arden points out, the whole basis of Part II of the Act is that the notice is served on the "person having control" as defined in section 39(2) of the Act. As Lord Bridge pointed out in *Pollway*, the formula there used has a long statutory history:

It has been used in a wide variety of legislative contexts most commonly for the purpose of identifying the person entitled to that interest in property upon whom Parliament thought it appropriate to impose some obligation to undertake work on or in connection with the property required in the public interest or to meet a proportionate share of the cost of public works from which the property would derive the benefit.[3]

I find it impossible to think that Parliament intended that the long leaseholder of a flat should be an appropriate person on whom to impose an obligation in respect of the roof of the block in the public interest.

In respect of the notices served on the long leaseholders three other issues arise. The first is whether these notices are severable so that they apply in so far as Part A is concerned but not Part B. I do not think that they can be so severed for a number of reasons. Notices have to tell the recipients what works they have to carry out. The whole scheme depends upon this. But these notices require the recipients in the main to carry out works on parts of the premises which they do not "control" either in the wide sense of having the power to do work or the narrow sense defined in section 39. In my judgment, a notice which required a recipient to do work which is the main part of the requirement and which the local authority have no right in law to require him to do is as invalid and void as is a notice served on the wrong recipient: see *Pollway*.

Even if this court in theory has the power to sever these notices, there is no evidence before the court as to whether or not the works required in Part A would achieve the result required by section 9(1A) if the Part B work is not done. In the light of this judgment it will be possible for the local authority, if it wishes, to serve fresh notices in respect of work required to the flats themselves which can then, if necessary, be appealed to the county court.

Second, it is submitted that, because there is given to the recipient a right of appeal to the county court, that is the route which should be taken. I think there is nothing in that point. Here it is the local authority who is taking the point that the notices were invalid and void, and it would be absurd to suggest that it should do so by appealing to the county court in respect of its own notices.

Last, is the local authority in some way estopped from challenging the validity of its own notices? I do not think such an argument could possibly succeed. *Lever Finance Ltd* v *Westminster (City) London Borough Council* [1971] 1 QB 222 is relied on by the applicant, but that was a case where the planning authority had the power to do that which the court held it was not entitled to deny it had done. Here, in my judgment, the local authority had no power under section 9(1A) to serve the notices at all. See *Western Fish Products Ltd* v *Penwith District Council* [1981] 2 All ER 204. As *de Smith* puts it (4th ed, p 104) "the general principle remains that a public authority may not vary the scope of its statutory powers and duties as a result of its own errors or the conduct of others".[4]

I am aware that in his speech in *R* v *Inland Revenue Commissioners, ex p Preston* [1985] AC 835 Lord Scarman said that "judicial review should in principle be available where the conduct of the commissioners in initiating such action would have been equivalent, had they not been a public authority, to a breach of contract or a breach of a representation giving rise to an estoppel". But, in this case, it does not seem to me that there are any grounds for holding that, if not a public authority, the respondents could be estopped. The applicant has in no way relied upon the notices; all it has done is appeal against them and make an application for a grant. If works had actually been commenced prior to the approval of the grant, the situation might have been different for the reasons adumbrated in argument by Mr Arden.

In respect, therefore, of the notices served upon the long leaseholders I find that they were invalid and void and that they cannot therefore form the basis of a claim for a grant under the 1974 legislation.

Is the position any different in respect of the six notices served upon the applicant itself as being the legal person falling in respect of those flats within section 39(2)? I think not. The validity of the requirement that work be done on the external and common parts still depends, in respect of each flat, upon the arguments which I have held to be inadmissible in respect of the long leases.

I am aware that the effect of this judgment together with the decision of the House of Lords in *Pollway* may make it impossible for a local authority to exercise Part II control over the external and

---

[2] Editor's note: See [1986] 2 EGLR 27 at p 29L; (1986) 280 EG 87 at p 91.

[3] Editor's note: See [1986] 2 EGLR 27 at p 28H; (1986) 280 EG 87 at p 88.
[4] Editor's note: *Judicial Control of Administrative Action*, by S A de Smith.

common parts of a block of self-contained flats. However, I do not find the fact that there is this lacuna in the local authority's statutory powers particularly surprising. The whole purpose of this and similar legislation is to fix the responsibility for doing works upon the person who in fairness ought to bear the burden. To achieve this over the years the legislature has used the formula set out in section 39(2) of the 1957 Act. In *Pollway* Lord Bridge showed how this formula has been repeated time and time again with but minor variations of language. But in the context of a block of flats the formula does not work in respect of, for example, the roof. In common parlance the person who has "control" of the roof is the applicant. It has the power, and indeed the contractual obligation, to do the necessary repairs to the roof, but the applicant, in respect of the roof, does not fall within the formula.

I do not myself find this lacuna (if it be one) particularly surprising. In *Pollway* Lord Bridge said this:[5]

I appreciate that this conclusion may cause inconvenience for local authorities. But I imagine that normally the contractual rights of the owners of long leasehold interests in flats to enforce repairing obligations against their lessors will provide an adequate solution of the problem. This may be the explanation of the fact that, though the formula found in the definition has been in common use in statutes since at least 1847, it was not until 1982 that its application to buildings divided into units let on long leases had to be considered by the courts. The truth, I suspect, is that generations of parliamentary draftsmen have been content to use the time-honoured formula without ever contemplating its application to the circumstances presently under consideration. That must surely be true of section 39(2) of the Act of 1957, which simply re-enacts the formula first used in its present context in section 17(4) of the Housing Act 1930. That Act introduced the compulsory procedure which we now see in expanded and amended form in Part II of the Act of 1957 requiring "the person having control of the house" to effect repairs to a house which was unfit for human habitation and which was "occupied or of a type suitable for occupation by persons of the working classes." The draftsman in 1930 can hardly be blamed if it did not occur to him to make suitable provision for dealing with problems arising from flats let on long leases at low rents.

On the facts of this case the contractual rights enjoyed by the tenants are capable of providing, and will no doubt provide, an adequate solution to the problem at which Part II of the 1957 Act is aimed.[6] The misfortune is that, by limiting the grant legislation to work performed under a section 9 notice, the legislature have failed to provide for the making of grants in respect of the "outside" work. Whether that comes about by oversight or intention matters not. If it is an unintentional lacuna it is for the legislature, if it wishes, to fill it; if it is an intentional restriction of grant aid (and this cannot be excluded) then, of course, while one may or may not think it fair, the court is in no way concerned. The application therefore fails.

The application was dismissed with costs.

---

[5]Editor's note: See [1986] 2 EGLR 27 at p 29K; (1986) 280 EG 87 at p 91.

[6]Editor's note: The proceedings in the present case were governed by the pre-1985 Act legislation. The Housing Act 1985 repealed the whole of the 1957 Act.

LANDLORD AND TENANT
GENERAL

Chancery Division
July 11 1986
(Before Mr Justice SCOTT)

CUMSHAW LTD v BOWEN AND OTHERS

Estates Gazette January 10 1987
281 EG 68-75

Landlord and tenant — Point on construction of lease — Retail Price Index — Lease for a term of 61 years from June 24 1960 provided that the service charge should be £100 per annum plus £1 per annum for each point or part of a point by which the Retail Price Index rose above the figure of 110.4 — If, however, the Retail Price Index should cease to be published or to be available to the public, increases in the service charge were to be based on a fair proportion of increases in the landlords' costs, to be determined by an independent surveyor — The question, to be decided as a preliminary point in proceedings by the plaintiff landlords was whether, as the result of certain changes, the Retail Price Index had ceased to be published or available — The history of the index showed that since the date of the subject lease not only had there been changes in the commodities and services represented in the index, and in their weightings, but also in the bases, which had been brought back to a base of 100 on two occasions since 1960 — This meant *inter alia* that the direction in the lease about the addition of £1 per annum for each rise of a point in the index could not be literally applied — The plaintiffs accordingly submitted that the alternative machinery in the lease for ascertaining the service charge should be used — Held, rejecting this submission, that the revised index remained the index to which the lease referred and that it had neither ceased to be published nor ceased to be available — Changes in the basket of items and in their weightings merely reflected changes in habits and tastes and were made in the interests of greater accuracy — Changes in the bases posed an immediate literal difficulty, but by a simple mathematical formula even the less numerate could convert the new figures to increases on the old base — Declaration accordingly

No cases are referred to in this report.

This was a preliminary point heard by virtue of a consent order made by Sir Nicolas Browne-Wilkinson V-C in proceedings by the plaintiff landlords, Cumshaw Ltd, by originating summons seeking a determination of the court on a dispute as to the amount of a service charge payable by the defendant tenants under leases of flats in property at 15 Chelsea Embankment, London SW3. The defendants were lessees of the flats, the leases of which were for all material purposes in the same terms.

Jonathan Brock (instructed by A Kramer & Co) appeared on behalf of the plaintiffs; David Neuberger (instructed by Victor Mishcon & Co) represented the first, second and third defendants; M Rosen (instructed by Tarlo Lyons Randall Rose) represented the fourth defendant.

Giving judgment, SCOTT LJ said: I have before me a short point of construction of a lease. The point turns, as do all points of construction, on the language used in the lease, and also on certain underlying facts, to which I must refer. The demised property is 15 Chelsea Embankment. It seems that it is divided into 10 flats. The defendants in the case are the lessees of three of the flats. The first defendant is the lessee of flat 9, the second and third defendants are the lessees of flat 5, and they all appear by Mr Neuberger; the fourth defendant is the lessee of flat 3, and she appears by Mr Rosen.

The plaintiff, Cumshaw Ltd, acquired the reversion to the property in 1975. The plaintiff appears by Mr Brock.

The leases held by the defendants were granted on different dates in 1960. They are for all material purposes in identical terms. Each contains a demise of the flat comprised therein for a term of 61 years odd from June 24 1960. There is provision for payment of a fairly small ground rent, a premium having been charged, and for payment of a service charge. The provision for payment of a service charge to the landlord is natural enough, because in the body of each lease is to be found a covenant by the lessor to be responsible for maintaining, repairing, decorating and renewing the structure, the roof, the foundations, the exterior, party walls, chimney stacks, gutters and rainwater pipes of 15 Chelsea Embankment.

I turn to the service charge provision. I should read it in full. Following the reservation of the ground rent the lease continues:

And second, paying by way of additional or further rent (hereinafter called the service charge) the sum of £100 per annum plus £1 per annum for each point or part of a point Her Majesty's Government Index of Retail Prices rises above the figure of 110.4

that figure is not constant in all the leases, but the variation therein is not material for any present purposes

being the figure at which the Index stood on the 16th day of August 1960

that date is not constant, but again, the difference is not material —

the service charge to be paid quarterly on the usual Quarter Days, provided always (a) that the figure of the said Index to be taken for this purpose shall be the figure stated in such Index at the commencement of the relevant Quarter; (b) if at any time during the term hereby created the said Index shall cease to be published by Her Majesty's Government, or for any reason is no longer available to the public, the service charge shall be £100 per annum plus a fair proportion of any increase from time to time in the cost incurred by the lessors in providing the services and complying with the lessor's obligations mentioned in clause 3(b) and of the fourth schedule hereto over and above the cost incurred by the lessors in respect of the same as at the date hereof, such proportion to be determined by an independent surveyor for the time being by reference to the percentage increase in the cost of services at such time as the said Index table shall cease to be published or made available to the public; (c) the service charge shall not in any circumstances be less than £100 per annum whether or not the said Index table falls below the said figure of 110.4 and (d) in the case of any dispute the matter shall be referred to arbitration under the Arbitration Act 1950.

A dispute has arisen between the plaintiff and the defendants as to the amount of service charge that the respective lessees ought to be paying.

On May 29 1985 the plaintiff issued the originating summons seeking, in effect, the determination by the court of the issue between the parties as to the service charge. The plaintiff contends that the Index of Retail Prices mentioned in the leases has either ceased to be published by Her Majesty's Government or is no longer available to the public for the purposes of proviso (b) of the service charge provisions. If that is so, it is common ground that the quantum of the service charge falls to be determined under proviso (b), subject, however, to an estoppel, which in that event the defendants seek to raise against the plaintiff. If the plaintiff is wrong in contending that one or other of those events has happened, then, subject to some adjustments which it is common ground would then have to be made, the amount of the service charge falls to be determined under the main part of the service charge provision and under proviso (a) of the proviso. In that event the plaintiff contends that an estoppel can be raised against the defendants.

The originating summons came before the Vice-Chancellor, Sir Nicolas Browne-Wilkinson, on May 16, when the case was opened and the question was raised of an adjournment in order to enable the plaintiff to answer certain evidence filed on behalf of the defendants relating to the estoppels. The parties, after discussion, decided that it would be sensible to obtain a preliminary ruling on the question whether one or other of the events specified in proviso (b) had in fact happened, that is to say, whether the Index of Retail Prices referred to in the leases had ceased to be published by Her Majesty's Government, or was no longer available to the public. The Vice-Chancellor accordingly on May 16 made a consent order for the hearing as a preliminary point of that question. That is the question which is now before me, and any answer to it must, given the manner in which the point comes before me, be without prejudice to the rights of the loser of the argument before me to seek to cure the consequences of the loss by reliance on the alleged estoppel. The question is whether within the meaning of the language used in the leases the Index of Retail Prices there referred to has ceased to be published or is no longer available to the public. First, I must endeavour to describe the nature of the Index of Retail Prices to which the parties were in the leases referring, a little of its background, and some of the changes that have occurred in it since the date of the leases.

The Index of Retail Prices is an index published by the government for the purpose of recording changes in the level of retail prices. The purpose of the index was usefully described in a paragraph in a report published in 1967. The report was entitled "Method of Construction and Calculation of the Index of Retail Prices." Under para 5 the report says this:

It is important to understand that the Index is an index of price changes and not a cost of living index. It does not measure changes in the kinds and amounts of goods and services people buy, or in the total amount spent in order to live, nor does it measure differences in living costs between different localities. However, one of the most important factors determining changes in the cost of living is the extent to which retail prices of goods and services change from month to month. It is this particular aspect of the cost of living which is measured by the Index of Retail Prices.

As appears from that paragraph, publication of new and up-to-date figures is made monthly. The leases in adopting the Index of Retail Prices as the basis for calculating rises in the service charge were, plainly enough, linking the cost to the landlord of discharging its repairing and other obligations to the rises from time to time in retail prices. Whether this link would provide an accurate measure of the increase in the landlord's costs is, to my mind, an irrelevance. Whether there was a more accurate formula available is also an irrelevance. The fact of the matter is that the parties chose to use the changes in the Index of Retail Prices for the purpose of determining the size of the service charge to be paid by the lessees.

The purpose of the Index of Retail Prices was, and is, to identify changes in retail prices over the country as a whole. The manner in which that is done is relevant. I can best describe the manner in which the index is prepared by reading a paragraph which seems to have been included in each of the monthly editions of the *Ministry of Labour Gazette* in which the monthly retail prices figures are set out. The paragraph is in these terms:

The Index of Retail Prices measures the change from month to month in the average level of prices of the commodities and services purchased by the great majority of households in the United Kingdom, including practically all wage earners and most small and medium salary earners. The Index is not calculated in terms of money, but in percentage form, the average level of prices at the base date being represented by 100. Some goods and services are relatively much more important than others, and the percentage changes in the price levels of the various items since the base date are combined by the use of weights. The index figures for each month are first calculated as index numbers with prices at January 16 1962 taken as 100, and the weights used have been computed from information provided by the Family Expenditure Surveys made in 1958 to 1961 adjusted to correspond with the level of prices ruling in January 1962. Lists of these weights are given on page 88 of this *Gazette*.

So, a basket of commodities and services is taken, a relative weight is attached to each commodity or service, depending upon the extent to which that commodity or service is the object of expenditure in those households, the survey of which produced the statistics on which the index was based. When in 1960 these leases were executed there had been in existence since 1956 an index known as the Index of Retail Prices. It had had a predecessor known as the Interim Index of Retail Prices. That Interim Index had come into existence in June 1947. According to a note to a report on the revision of the Index of Retail Prices made by a committee in March 1962 the Interim Index "was regarded as a temporary measure to cover a period of abnormal conditions of spending after the war". The note goes on to describe the basis on which the 1956 Index then came into existence. It reads:

In the Interim Report dated June 26 1951 the committee recommended that a new large-scale budget inquiry should be held as soon as possible to provide up-to-date information about the pattern of household expenditure to serve as a basis for a new Index of Retail Prices. The committee also recommended that there should in future be smaller-scale inquiries at frequent intervals in order to keep a check on the weighting basis of the Index. The Interim Index was re-weighted in January 1952 on a temporary basis, and a large-scale inquiry, the Household Expenditure Inquiry, was held during 1953. A new Index, the present Index of Retail Prices, was introduced in January 1956 with weights based on the expenditure patterns derived from this inquiry. A continuous inquiry on a smaller scale, the Family Expenditure Survey, was started in January 1957. The main results obtained in this survey for 1957, 1958 and 1959 were published in October 1961, and some results for 1960 were published in December 1961.

Accordingly, in 1960 when the leases were executed, there was in existence the Index of Retail Prices which had started in 1956 with a base of 100. It will be recalled that in the service charge that I read, the figure of 110.4 was mentioned. That figure was the figure at which the index stood on August 16 1960. The committee whose report I have just mentioned was appointed to consider a revision of the Index of Retail Prices. It undertook a re-examination of the basket of commodities and services on which the index was based, and of the relative weighting given to the various items comprised in that basket. It produced its report in March 1962. Its report led to a revision of the Index of Retail Prices. Alterations were made in three respects. There were changes made to the commodities and services in the basket. These changes were relatively few. There were some items added, but the categories under which the various items were listed remained the same. Second, the weightings were revised. These revisions were of course important, because it was on the selections of the items in the basket and on their relative weightings that the accuracy of the index depended. The intention of the revision was plainly to produce an index which would more accurately reflect changes in retail prices than the previously organised index had done. None the less, the extent of the changes in the items in the basket and in their weighting, though important and though changes of substance, did not, in my view, change the character of the index. It was still an index which endeavoured to reflect changes in retail prices in a broad band of commonly used household commodities or services. It is rightly referred to in my view as a *revision* of the Index of Retail Prices.

The third change was that the base reference on which subsequent changes in retail prices would be based was brought back to the figure of 100. The reasons for this are set out in the 1962 report. They are various in number. One reason was that since changes were being made in the commodities and services included in the basket and in their relative weighting, post-revision and pre-revision comparisons would not be comparing like with like. Another reason for bringing the base back to 100 as part of the process of revision was, it seems, a somewhat cosmetic one. It was thought that unless that were done it would not be properly recognised by the public at large that the index had been brought up-to-date and rendered accurate by the process of revision to which I have referred. These reasons seem to have been behind the decision to have a new base reference of 100.

For a period of a year the Index of Retail Prices, adjusted by the changes in the basket and the weighting that I have mentioned, but based upon the original 1956 base reference of 100, continued to be published monthly. The index thus based was described for the year 1962 as the Official Index of Retail Prices. Side by side with it for each month that year was published a figure based upon the new 1962 base reference of 100. The monthly publications containing both these figures contained, for the assistance of those whose numerate skills might not have been up to the task, a formula for converting the new retail prices figure based upon the new 100 base into a figure which would be applicable to the old 1956 100 base. The formula involved taking the figure of 117.5, which was the retail price figure current at the beginning of 1962 and based upon the 1956 100 figure, and multiplying it by a fraction of which the numerator was the new retail price figure based upon the 1962 reference of 100, and the denominator was 100. The figures were published in that manner for the year of 1962. As I have said, the official figure was that which was based upon the 1956 100 base. From the beginning of 1963, however, a figure based upon the old 1956 base of 100 was no longer published. But the monthly publications continued to include a small paragraph to enable those who needed the assistance to calculate what the retail price figure would have been if still based on the 1956 base of 100.

In 1974 or 1975 (I am not clear what the exact date is, and it does not matter), there having been a considerable rise in the Retail Price Index since 1962, and the current figure based upon the 1962 base of 100 having risen to what might have seemed to some to be an unacceptably high level, it was decided to start again with a new 100 base figure on which future increases would be based. So from 1974 or 1975 the Index of Retail Prices was brought back to 100 and subsequently shown as an increase on that base figure. There was no change in the commodities or services in the basket, or in their relative weighting. The adjustment was simply mathematical. One of the recommendations made by the committee which had reported in 1962 was that for the future there should be annual adjustments to the basket and to weighting. There was to be a continuous process of inquiry in order to enable annual adjustments to be made. The recommendation was adopted and annual adjustments were thereafter made. For that reason a particular revision of the basket or of weighting in 1974 or 1975 was not to be expected.

One of the consequences of the annual adjustments made since 1962 has been that over the years fairly considerable changes in national spending habits have been reflected in fairly considerable alterations in weighting. In particular, the weighting given to transport costs and transport services has risen considerably, so that by the 1970s the weighting attributed to those matters was not far below twice the level at which those matters had been weighted in the period 1956 to 1962. That increase in weighting was reflected by corresponding reductions in weighting of other items; in particular, the weighting in respect of food seems to have come down. Those are the facts. The question then is how in the light of those facts the leases are to be construed. Mr Brock has made a simple submission. He has submitted that the reference in the lease to the Index of Retail Prices must be taken to be a reference to that index as it was organised and as it was based on the dates in 1960 when the leases were executed. If he be right, it would follow that any change in the base figure, whether or not accompanied by a change in the commodities and services in the basket, or in the weighting of those commodities and services, would require the conclusion that the index referred to in the leases had ceased to be published, or was no longer available to the public. The index referred to, he submitted, was the 1960 Index, with the 1956 base of 100, and any alteration, even if simply mathematical, meant that the index referred to in the leases had gone. Second, he submitted that the changes which there had been to the commodities and services in the basket, made first by the 1962 revision and later by the annual revisions, had produced an index which could no longer, as a matter of substance, be described as the same index as that referred to in the leases.

I will deal with the second point first. In my view the question whether changes in the commodities and services in the basket, or changes in the relative weightings given to the items in the basket, justify the conclusion that the index referred to in the leases has gone is a question of degree. It is easy to see that if the nature of the commodities and services were to suffer a sea-change and were no longer intended to reflect the general household commodities and services that people spend their money on, the conclusion might very well be justified that the index referred to in the leases had gone. But the sort of changes in the basket that have taken place are, in my judgment, a mile away from that. The changes in the basket are minor, reflecting merely somewhat of a change in styles and tastes. One of the changes that I observed was that jeans came into the basket as an item of clothing used by the young. Other changes are perhaps less notable but are not of dissimilar character. As to the changes in weighting, the purpose of those was and is to make the index more accurate. As spending habits change, weighting based on past spending habits will detract from the accuracy of the index. A continuous inquiry, keeping under review spending habits and, in consequence, the attribution to particular commodities and services of particular weighting, is necessary in order to retain for the index the maximum accuracy possible. The fact that changes have been made to the basket and to the weighting in order that the index should remain accurate cannot possibly, in my judgment, justify the conclusion that the revised index is no longer the index to which the leases referred. I would therefore reject Mr Brock's submission based upon the changes since 1960 in the commodities and services in the basket and in the relative weighting given to the items in the basket. There remains his submission relating to the mathematical change to the base by which changes in retail prices are to be measured – a change to 100 in 1962 from the 1956 base of 100 and a change again, for that matter, to 100 in 1974 or 1975.

Mr Brock has relied heavily, and is entitled to rely heavily, on the references in each of the leases to a particular figure of the index. One lease specifies 110.4 as the figure at which the index stood on August 16 1960. The lease provides for the service charge to increase by £1 per annum for each point or part of a point above that figure that the index rises. Mr Brock's point was that once the base figure had been brought back to 100 in 1962, and again in 1974, the terms of the lease providing for increases in service charge could no longer be applied. No one could suppose when the figure came down to 100 in 1962, and then rose, say, to 109 subsequently, that that rise would produce no rise in the service charge. That 109 could, by the calculation to which I have referred, be expressed as, say, 125 based on the 1956 base of 100 and, so expressed, would justify a rise in the service charge accordingly. But if the literal meaning of the language used in the lease is adhered to, a rise from 100 in 1962 to 109 subsequently would not be a rise above the figure of 110.4 and so, the argument went, would not justify any rise in the service charge. So Mr Brock submitted that, once the base figure had been brought back to 100, was no longer the index referred to in the leases, that the terms of the leases could no longer be properly applied to the index, and that the alternative machinery set out in the proviso (b) had to be used.

In my judgment Mr Brock had adopted the wrong approach to the problem. The question is whether the condition precedent to the coming into effect of the proviso (b) machinery has been or has not been fulfilled. There are two limbs to that condition precedent. One is that the index shall cease to be published; the other is that the index shall no longer be available to the public. I must ask myself if either of those events has happened. If they have happened, then the para (b) machinery comes into effect. If they have not happened, then it does not. In my judgment neither of those events has happened. The Index of Retail Prices is in my view the same Index of Retail Prices to which the leases referred, notwithstanding that the mathematical base has been altered. I think it obvious that the parties to the leases did not contemplate the subsequent alteration of the mathematical base on which the increase in retail prices was to be calculated. But it does not follow that an alteration of the mathematical base has produced a different index to that to which the leases referred. In my view the present index is the same index. If it is the same index, then, as a matter of construction of the terms of the leases, I have no difficulty in concluding that the formula for the purpose of calculating rises in service charge and in the events which have happened, the Retail Price Index figure current from time to time must, by the simple calculation necessary, be expressed as a figure based on the 1956 base of 100. That, as I have said, requires merely a simple calculation, and, as a matter of construction of the lease in order to reflect the parties' evident intention, it is what, in my judgment, should be done. I would answer the question before me by declaring that neither of the events referred to in para (b) of the proviso has happened, subject of course to any right of the plaintiff to hold the defendants estopped from denying that they have happened. As to that, I have not heard the evidence and I say nothing at all.

Chancery Division

April 29 1986
(Before Mr John CHADWICK QC, sitting as a deputy judge of the division)

ASHTON AND OTHERS v SOBELMAN

Estates Gazette January 24 1987

281 EG 303-309

Landlord and tenant — Whether lease had been forfeited by peaceable re-entry and, if so, whether relief against forfeiture should be granted to tenants — The lease, for 80 years from 1932 at a yearly rent of £50, was, subject to the question of forfeiture, vested in the plaintiffs — Defendant was the present freeholder — There was a subtenancy, for 10 years from 1976, of the premises, a lock-up shop and flat — Defendant's predecessors in title as freeholders devised a scheme to secure the forfeiture of the lease by peaceable re-entry, the steps being to enter the premises with the consent of the occupying subtenant, to change the locks on the door and to request the subtenant in future to pay rent direct to them — The subtenant agreed to co-operate on the understanding that his own security would not be adversely affected — It was submitted on behalf of the defendant freeholder that these arrangements resulted in a forfeiture of the leasehold interest by peaceable re-entry — *London & County (A & D) Ltd* v *Wilfred Sportsman Ltd* and *Bayliss* v *Le Gros* considered — Held that the arrangements made in the present case did not constitute a re-entry — Here both the freeholders and the subtenant were proceeding on the footing of the continuing validity of the subtenant's underlease — This was wholly inconsistent with the determination by forfeiture of the superior lease — *Bayliss* v *Le Gros* was at the most an authority that a re-entry might be effected by an arrangement whereby the subtenant remained in occupation as the direct tenant of the landlord on the terms of a new tenancy — In view of this decision the question of relief against forfeiture did not arise, but the judge expressed the opinion that it would have been a proper case for relief — The defendant, who had purchased the freehold at auction, had sufficient knowledge of the facts to put him on notice that the leaseholders were likely to have an unanswerable claim for relief — Declaration that the lease had not been forfeited

The following cases are referred to in this report.

Bayliss v *Le Gros* (1858) 4 CB(NS) 537
Howard v *Fanshawe* [1895] 2 Ch 581
London & County (A&D) Ltd v *Wilfred Sportsman* [1971] Ch 764; [1970] 3 WLR 418; [1970] 2 All ER 600; (1970) 21 P&CR 788, CA
Lovelock v *Margo* [1963] 2 QB 786; [1963] 2 WLR 794; [1963] 2 All ER 13, CA

The plaintiffs, who had become beneficially entitled under a will to the leasehold interest in a lock-up shop and flat at 195 Burnt Oak Broadway, Edgware, Middlesex, sought a declaration to the effect that in the events which had happened their leasehold interest had not been forfeited. The defendant, Samuel Sobelman, was the present freeholder.

D Burton (instructed by Gamlens, agents for Charlsley Harrison, of Windsor) appeared on behalf of the plaintiffs; Miss C Hutton (instructed by Duke-Cohan & Co) represented the defendant.

Giving judgment, MR JOHN CHADWICK QC said: The property known as 195 Burnt Oak Broadway, Edgware, comprises a lock-up shop with a flat above. The freehold title to the property is registered at Her Majesty's Land Registry under title number NGL 27938. Between March 17 1982 and January 17 1985 the registered proprietor of the property was Twogates Properties Ltd. On January 17 1985, the defendant, Samuel Sobelman, was registered as proprietor in the place of Twogates.

At all material times since the first registration of the property, entry no 3 in the charges register has shown the property to be subject to a lease dated October 24 1932 for a term of 80 years from September 29 1932 at the yearly rent of £50. The leasehold interest under the lease of 1932 is itself registered at the Land Registry, formerly under title no P115020, but now under title NGL540312. At all times between July 2 1938 and October 22 1985 the registered proprietor of the leasehold interest was Norman Frederick Stockbridge.

Norman Frederick Stockbridge died on September 23 1950. On October 22 1985 the plaintiffs in this action, who are the three children of Norman Frederick Stockbridge and the persons who, in the events which have happened, are now entitled beneficially to the real property devised by his will, were registered as the proprietors of the leasehold interest. The questions raised in this action are whether the 1932 lease was forfeited by a peaceable re-entry made on behalf of the then freeholders, Twogates Properties Ltd, on October 19 1984; and, if so, whether the plaintiffs should be relieved from such a forfeiture.

The 1932 lease contains a proviso for re-entry in the following terms, so far as material:

provided always and these presents are upon this condition that if the said yearly rent hereby reserved or any part thereof shall at any time be in arrear and unpaid for 21 days after the same shall have become due and whether legally demanded or not. . . then and in any case it shall be lawful for the lessor or any person or persons duly authorised by him in that behalf into or upon the hereby demised premises or any part thereof in the name of the whole to re-enter and the said premises peaceably to hold and enjoy thenceforth as if this demise had not been made.

The yearly rent of £50 payable under the 1932 lease was to be paid by equal quarterly payments on the usual quarter days in each year. It is admitted by the plaintiffs that rent was not paid on September 29 1983 or on any of the four quarter days thereafter. Accordingly, by October 19 1984, the arrears of rent owing by the plaintiffs to Twogates Properties Ltd amounted to not less than £62.50.

At the relevant time in 1984 the directors of Twogates Properties Ltd included Mr Michael Moss, his brother Mr Sydney Moss and Mr Bryan David Lipson, a partner in the firm of solicitors, Cowan, Lipson & Rumney. Mr Michael Moss and Mr Sydney Moss were chartered surveyors who carried on business under the name Quennell Moss & Co. It is clear that, by February 1984, Mr Sydney Moss and his brother were giving consideration to the remedies which might be open to Twogates Properties Ltd as landlords in relation to the non-payment of rent under the 1932 lease; and, in particular, to the possibility of forfeiture of the lease. In a memorandum dated February 27 1984, Mr Lipson advised Mr Sydney Moss as to the legal position: after setting out what he conceived to be the relevant legal principles, Mr Lipson continued:

On these basic principles if you wish to effect a forfeiture by peaceable re-entry then in my opinion you must take the following steps on the day on which you peaceably re-enter the premises:-
1 Peaceably enter the premises (which will obviously have to be done with the consent of the occupation lessee who is in any event unlikely to refuse consent)
2 Take physical possession of the premises by changing the locks on the door
3 Accept the present subtenant as your tenant by giving that subtenant a set of keys to the new lock
4 Inform the subtenant in writing of the steps you have taken with a request that he should pay all future rent to Twogates Properties Ltd.

As appears from that memorandum it was appreciated by the landlords at the time that the property was then in the occupation of a subtenant. The subtenancy was upon the terms of a lease dated April 29 1976 made between Ada Lillian Stockbridge, the widow of Norman Frederick Stockbridge, and Gerald Clayton for a term of 10 years from March 25 1976. Following a rent review in 1983, the rent payable by Mr Clayton under the 1976 underlease was £3,500 per annum. Mr Clayton carried on the business of a retail furniture shop at the property. He used the flat premises above the shop for storage purposes in connection with that business. Mr Clayton also carried on the business of a retail do-it-yourself shop under the name "Home Handyman" at 188 Burnt Oak Broadway, which was on the opposite side of the street from the property.

It is clear that by the middle of September 1984 Twogates Properties Ltd had decided to take steps to forfeit the 1932 lease. The property was entered for sale at an auction which was to be held by Harman Healy & Co on November 20 1984. On September 13 1984, Cowan, Lipson & Rumney sent to Mr Barnett at Harman Healy & Co draft special conditions of sale in respect of the freehold interest, and a copy land certificate of the freehold title. Although the land certificate showed the property to be subject to the 1932 lease, it is clear from the draft special conditions that it was intended to be sold free from that lease, and subject only to the occupation lease; although this was, at that stage, wrongly described as a 15-year lease granted in 1971 at a current rent of £2,600 per annum.

The third quarter's rent was due on September 29 1984. I will assume, but without deciding, that a demand for that rent was sent by Quennell Moss & Co, on behalf of Twogates Properties Ltd, to Mrs Verena Evans on behalf of the plaintiffs. Some three weeks went past and no rent was received. It was decided by Twogates Properties Ltd and Mr Lipson that they would effect a forfeiture on October 19 1984. In preparation, Mr Lipson dictated two letters. I set out the first of these in full. The letter is addressed to Mr Clayton at 195 Burnt Oak Broadway, and reads:

Re: 195 Burnt Oak Broadway, Edgware
We act for Twogates Properties Ltd, the freeholders of the above property which they own subject to the following leases:-
1 A lease dated October 24 1932 made between Robert Wilson Black and The Oak Property Co Ltd for a term of 80 years from September 29 1932 at a rent of £50 per annum payable quarterly in arrears on the usual quarter days in each year.
2 An underlease made in or about 1971 for a term of 15 years at a current rent of £2,600 per annum payable quarterly in advance on the four usual quarter days in each year. We understand that you are in occupation of the above premises under the provisions of the underlease referred to in 2 above, but we do not know whether that underlease was granted to you direct by the head lessees or whether you subsequently acquired it by way of assignment.

Our clients acquired the freehold on December 17 1981 and from that date until in or about June 1983, they sent rent demands to Stockbridge Holdings Ltd, care of Mrs V Evans, the Old Crispin, Windsor Forest, Berkshire and the rent was always paid. Since September 28 1983, despite sending demands, the rent owing to our clients has not been paid and, as at today's date, our clients are owed rent for the period from September 29 1983 to September 29 1984 (five quarters at £50 per annum amounting to £62.50).

We have advised our clients that in these circumstances, they are entitled to forfeit the lease dated October 24 1932 on the grounds of non-payment of rent by peaceably re-entering the premises and re-taking possession. Our clients propose to exercise their legal rights today by taking the following steps:-
(1) Peaceably re-entering the premises.
(2) Changing the locks on the front door.
(3) Instructing you to pay all future rent payable by you under the underlease referred to in 2 above to our clients or as they may direct.

Our representative, Mr L M Bloch, a partner in this firm (who will probably be accompanied by another member of our staff, Miss A Prizeman) will call at your premises today together with our clients' locksmith and they will hand you this letter and will supervise the changing of the locks, and when that work had been completed they will hand you the following:-
(1) A complete set of keys to the new lock.
(2) A letter of authority relating to the payment of future rent.

We would like to make it perfectly clear that our clients are not in any way challenging your right to remain in occupation of the premises under the provisions of the underlease under which you are the present lessee. Their sole concern is to enforce their legal rights of forfeiture against their tenant and the effect of the steps which our clients propose to take will be that after the forfeiture has been effected, you will become our clients' direct lessee.

As our clients would like to know the exact terms upon which you occupy the premises, we would be grateful if you would kindly hand Mr Bloch a copy of your underlease. If you do not have a copy available, then we will be quite prepared at our clients' expense to make a photocopy of the original in your presence at a nearby photocopying agency so that the original document never leaves your possession.

We hope that we have made the position clear, but if there is any further information you require, Mr Bloch will be able to provide it and the writer of this letter (Mr B D Lipson) will also be available to answer any questions you may have on the telephone.

Mr Bloch gave evidence before me. He told me that he had been asked by Mr Lipson on October 18 1984 to attend at the property, 195 Burnt Oak Broadway, on the following day for the purpose of making a peaceable re-entry. Mr Bloch told me that he had no experience in effecting a peaceable re-entry, and that he had no detailed knowledge of the relevant law other than what was stated in the letter which I have set out above and which was handed to him by Mr Lipson. Mr Bloch's evidence, which was confirmed by Miss Prizeman (then an articled clerk with Cowan, Lipson & Rumney) and which was not seriously at variance with that given by Mr Clayton, may be summarised as follows.

At 10am on October 19 1984, Mr Bloch and Miss Prizeman went to 195 Burnt Oak Broadway. They found Mrs Clayton there, and she directed them to the do-it-yourself shop at 188 Burnt Oak Broadway on the other side of the road. Mr Clayton was busy in his shop and Mr Bloch joined the queue and waited for customers to be served. When his turn came Mr Bloch told Mr Clayton that he was a solicitor, and that there was a matter which he wanted to discuss privately. He said that he acted for the landlords, but he assured Mr Clayton that nothing that he wanted to discuss would be prejudicial to Mr Clayton personally. Mr Clayton told Mr Bloch that he was busy and asked him to return at 11 o'clock.

When Mr Bloch and Miss Prizeman returned to 188 Burnt Oak Broadway at 11 o'clock the shop was less busy. Mr Bloch again told Mr Clayton that he acted for the landlords, and on this occasion he showed him the letter of October 19 1984, which I have set out above. Mr Clayton read the letter and, as it appeared to Mr Bloch, understood what was proposed. Mr Clayton expressed his concern as to his own position, and Mr Bloch assured him that he would not be prejudiced. Mr Bloch told me that he did not think that he had applied his mind to the question whether forfeiture of the head lease would involve a forfeiture of the underlease when he told Mr Clayton that he would remain in possession under his underlease. He accepted, under cross-examination, that it would be consistent both with the terms of the letter dated October 19 1984 and with his, Mr Bloch's probable understanding at that time, that Mr Clayton's underlease was not to be affected by what was being done.

Mr Clayton was content with these assurances. Understandably, he did not wish to incur unnecessary legal costs; but he telephoned his solicitors and asked them to let Mr Bloch have a copy of the underlease. Mr Bloch recalls that Mr Clayton told his solicitors in the course of that telephone call that what Mr Bloch was proposing to do would not prejudice him. I am satisfied, first, that Mr Bloch did not seek to influence Mr Clayton against consulting his solicitors; but, second, that Mr Clayton was content not to do so because of the assurances that he was being given. Mr Clayton then telephoned his wife, at 195 Burnt Oak Broadway, to tell her that Mr Bloch would be coming to change the locks with a locksmith and that he was to be allowed to do this.

Mr Bloch and Miss Prizeman then went to 195 Burnt Oak Broadway, with the locksmith, and instructed him to start changing the locks. Miss Prizeman and the locksmith remained there with Mrs Clayton. Mr Bloch went to Mr Clayton's solicitors to collect a copy of the underlease. When he returned to the property he found that there had been some difficulty with the locks, but that the locksmith had finally succeeded in fitting a new lock. Mr Bloch told me that he wanted to be sure that the new lock did in fact work and that he satisfied himself of this. He then gave all the keys to the new lock to Mrs Clayton, and he returned to 188 Burnt Oak Broadway, where he handed to Mr Clayton the second of the two letters which had been dictated by Mr Lipson on the previous day. The second letter was also addressed to Mr Clayton and was in these terms:

Re: 195 Burnt Oak Broadway, Edgware
We act for Twogates Properties Ltd, the freeholders of the above property, and write to give you formal notice that our clients have today forfeited the lease of the above premises dated October 24 1932 and made between Robert Wilson Black (1) and the Oak Property Co Ltd (2) by peaceably re-entering the premises and re-taking possession.
As a result of our clients' having taken physical possession of the premises, you are now our clients' direct lessee. Will you please accept this letter as our clients' formal request and authority to you to pay all future rent payable by you under the provisions of the underlease under which you hold the premises to this firm on behalf of Twogates Properties Ltd.

After handing over that letter, Mr Bloch thought that the act of forfeiture by re-entry had been completed. He returned to the offices of Cowan, Lipson & Rumney. In due course, he made a statutory declaration, dated October 29 1984, setting out the events of October 19.

As I have said, Mr Clayton's evidence did not differ substantially from that of Mr Bloch. He confirmed that Mr Bloch had assured him that he had nothing to worry about and that his lease was safe. Mr Clayton thought that he had suggested to Mr Bloch that he, Mr Clayton, should telephone Mrs Evans. I accept that Mr Clayton considered this possibility; but I do not think that it was the subject of discussion between him and Mr Bloch, nor that Mr Bloch made any attempt to persuade him against such a course. In my judgment, the true position was that Mr Clayton took the view, upon reading the first of the two letters of October 19 1984, that Mr Bloch was entitled to do what he was proposing to do; that it was inconceivable that Mrs Evans could be unaware of what was to be done; and that having been given assurances that the course proposed was not going to affect his own underlease, he was not concerned to object.

After Mr Bloch had left, Mr Clayton found that the new lock which had been installed at 195 Burnt Oak Broadway was unsatisfactory.

He told me that although the locksmith had changed the lock, he had not changed the locking plate; with the consequence that the tongue of the new lock would not enter into the style of the door. He telephoned Cowan, Lipson & Rumney during the afternoon of that day to complain about this. His complaint was met by a third letter dated October 19 1984 from Cowan, Lipson & Rumney. The second paragraph of that letter is of some significance:

We confirm that our clients have asked us to apologise for any inconvenience that may have been caused and we further confirm that we told you on the telephone this afternoon to arrange for a new lock to be fitted and we stated that as long as the cost was reasonable our clients would pay the bill. We invite you to send us an invoice for the installation of the new lock and as long as the figure is reasonable we have instructions to pay the cost on our clients' behalf direct to your locksmith.

In the event, Mr Clayton replaced the new lock with the original lock that had been left behind. The matter progressed towards the auction. On October 29 1984 Cowan, Lipson & Rumney sent to Mr Barnett at Harman Healy & Co amendments to the special conditions of sale relating to the property at 195 Burnt Oak Broadway, and a copy of the statutory declaration made by Mr Bloch and the exhibits thereto. The amendment to the special conditions was in these terms:

2 ... The purchaser shall assume (as is the case) that the lease dated October 24 1932 and made between the Oak Property Co Ltd (1) and Robert Wilson Black (2) referred to in entry number 3 in the charges register to the said title was forfeited on the grounds of non-payment of rent by a peaceable re-entry on October 19 1984 on which date the vendor retook physical possession of the property and authorised and instructed Gerald Clayton, the lessee under the provisions of the underlease referred to in special condition 3 hereof, to pay all future rent due from him to the vendor. On completion, the vendor's solicitors shall hand the purchaser a statutory declaration approving the facts which amount to a lawful re-entry on October 19 1984. A copy of the said statutory declaration is available for inspection at the offices of the auctioneers and the vendor's solicitors, and the purchaser having had an opportunity to inspect it prior to the date hereof shall be deemed to purchase with full knowledge of the contents thereof and shall raise no requisition or objection in relation thereto. The purchaser shall not be entitled to require the vendor to procure the deletion of entry number 3 from the charges register of the said title prior to completion and shall not be entitled to refuse or delay completion by reason of the fact that the charges register of the said title still retains an entry of the said lease dated October 24 1932.

On November 12 1984 Cowan, Lipson & Rumney sent a further amendment to the special conditions to Harman Healy & Co. That was in the following terms:

3. The property is sold subject to and with the benefit of a lease dated April 29 1976.

By reason of the matters referred to in special condition 2 hereof the vendor does not have in its possession the original counterpart of the said lease and can only provide a copy thereof. The purchaser shall not be entitled to refuse or delay completion by reason of the fact that the vendor does not hand over either the original counterpart lease or a certified copy thereof and only has in its possession a copy thereof referred to in paragraph 4 of the said statutory declaration referred to in special condition 2 hereof. The purchaser shall not be entitled to base any claim against the vendor whether for damages for breach of contract, compensation for loss of bargain, interest or otherwise by reason of any of the matters referred to in special conditions 2 and 3 hereof.

Some time prior to the auction, the defendant, Mr Sobelman, obtained copies of the auction particulars and the special conditions in their unamended form. It is, I think, clear that there is nothing in the particulars or the conditions (as printed, and before amendment) which would put a prospective purchaser on notice that the property was subject to the 1932 lease. Mr Sobelman attended the auction on November 20 1984. The auction was conducted on behalf of Harman Healy & Co by Mr Barnett. Mr Barnett is a fellow of the Royal Institution of Chartered Surveyors, an experienced auctioneer and has been a partner in Harman Healy & Co for some 21 years.

Mr Barnett told me that he was well aware, and had expressed the view to Cowan, Lipson & Rumney, that it had to be made abundantly clear to any purchaser that the property at 195 Burnt Oak Broadway had been the subject of a ground lease which had been forfeited. In particular, it had to be made clear that the landlords would not have the original counterpart of the 1976 underlease to produce to a purchaser.

There were over 90 lots to be sold on November 20 1984; and I accept that it would not be unusual in a sale of this nature for there to be a number of last minute addenda to the particulars and conditions as published in the printed catalogue. To meet this point Harman Healy & Co produced a series of pink roneoed sheets drawing attention to the additions. On one of those sheets there is found the entry:

Lot 35 — 195 Burnt Oak Broadway, Edgware, Middlesex.
Extra special conditions together with revised plan available at the rostrum.

Mr Barnett told me, and I accept, that some 500 copies of these sheets were produced and were placed on every chair in the room at the auction hall.

The property, which was lot 35, was to be sold at the beginning of the afternoon session. I have had the advantage of hearing in the course of the trial a tape-recording of Mr Barnett's introduction to the afternoon session. That recording was, I understand, taken as a matter of normal procedure. It is clear from that recording, and Mr Barnett confirmed in his oral evidence before me, that he drew specific attention to the addendum and to the entry relating to lot 35. I read from a transcript of the relevant parts of that recording. At the beginning of the afternoon session Mr Barnett said this:

On lot 35 there is a two page extra special condition together with a revised plan. I'll just mention that one in particular now. If there is anyone contemplating bidding on lot 35, which is Burnt Oak Broadway, I would like you to be aware of what is in the additional special conditions. Don't ask me in the middle of the bidding, so please, anyone for lot 35 who hasn't seen the additional three page special conditions and the plan, please make sure you see it before I get to that lot because it is only in two lots' time.

And, immediately before offering lot 35 for sale, he said:

Lot 35 is 195 Burnt Oak Broadway, Edgware, Middlesex. It is a lock-up shop with a self-contained flat above. A separate lock-up shop at the side plus a rear building. It's freehold. It is let on the tenancy as set out. They have paid £3,500 a year, it is worth quite a lot more, you will see the reference to the comparable, and I would just again tell you that there are these special conditions that we are selling subject to, so don't anyone come in at the very last minute not knowing what is in these special conditions. The plan will show the additional land at the back that wasn't shown on the catalogue. I advise you that you shouldn't bid unless you have read these special conditions. That is all.

Mr Sobelman did not recall seeing the pink addenda sheets in the auction hall. But he did recall Mr Barnett's speech; and, in particular, he recalled that Mr Barnett had said that anyone interested in lot 35 should come up and take the special conditions. Mr Sobelman went to the rostrum and asked for the special conditions for lot 35. He told me that he was handed one white sheet of paper; and it is clear that this was the first of the two additional special conditions. He read the document sitting in his seat in the auction hall and appreciated that there was an additional lease referred to in the special condition about which he had not previously been aware. He told me that he thought it was an old lease which had expired before Mr Clayton had taken possession, and that his knowledge of it had no effect on his bidding. It is clear from his evidence that he was aware that some step had been taken to obtain peaceable re-entry within the previous months, but I accept that he had little time to consider the matter — and no time on which to take legal advice — and that he would not have appreciated the full significance of what was stated in the special condition.

The property was offered for auction on the afternoon of November 20 1984, and was knocked down in the course of the auction to Mr Sobelman at the price of £54,500. It is common ground that this price would reflect the value of the property at that time on the basis that it was subject only to the 1976 underlease. The value of the freehold subject to the 1932 lease would be very substantially less. At the conclusion of the auction, Mr Sobelman signed a memorandum evidencing his purchase. In due course, after the title had been investigated by his solicitors, Mr Sobelman took a transfer of the freehold interest dated December 8 1984 at a purchase consideration of £54,500. He charged the property to his bank to secure an advance of some £30,000 which he applied towards the purchase price. I should, perhaps, mention, that the bank has taken no part in these proceedings, although I have been informed that it is aware of them.

On January 2 1985 Cowan, Lipson & Rumney sent to Brecher & Co, the solicitors acting for Mr Sobelman in the purchase, the executed transfer, the land certificate, the counterpart lease of 1932, the original statutory declaration of Mr Bloch and the exhibits thereto, and a letter of authority addressed to Mr Clayton in respect of all rent falling due on or after December 24 1984.

On January 4 1985 Brecher & Co sent that letter of authority to Mr Clayton and requested him to forward to them a cheque for the current quarter's rent due under the 1976 underlease — an amount of £875. On January 11 1985, Mr Clayton replied to the effect that, although he had been about to send the rent to Brecher & Co in response to their letter, he had now had a demand from Stockbridge

Trust, who were astounded to be told of the events of the past two months. Mr Clayton asked Brecher & Co to "get together with Stockbridge Trust and resolve the matter".

At about the same time, the plaintiffs, through their solicitors Charlsley Harrison, by a letter dated January 10 1985 tendered the arrears of rent due under the 1932 lease. That tender was not accepted, and shortly thereafter, on February 13 1985, the present proceedings were commenced by the plaintiffs.

The first question which I have to consider is, of course, whether there was a re-entry by the landlords on October 19 1984. If there was no re-entry, then there was no forfeiture. There appears to be little authority on the nature of the acts required to effect a re-entry against an intermediate tenant in a case where the premises are in the occupation of a subtenant. I was referred to a passage in the current edition of *Woodfall on Landlord and Tenant,* at para 1-1899:

Peaceable re-entry may be effected by the forfeiting landlord accepting as tenant a subtenant who is already in occupation, or by letting into occupation some third party and maintaining him there as tenant. But, it is clear that some unequivocal act or words are necessary to constitute a peaceable re-entry.

The authority cited in support of the first sentence of that passage is *London & County (A&D) Ltd* v *Wilfred Sportsman* [1971] Ch 764. But that was a case in which the defendants had gone into possession as trespassers against the tenant. The position was explained by Russell LJ at p 785:

Finally, the plaintiffs argued that there was in fact here no re-entry. There could be none by the third party until the reversion was vested in it on August 31 1965. Prior to that date the situation was that no 5 was being occupied by the defendants at the instance of the third party in trespass against Miah as the lessee. Mere continuation of that situation (it was argued) could not operate as a re-entry. It would have been necessary for the third party and the defendants in some way to withdraw from the premises and then return physically in order to achieve a re-entry. As it was put, the trespass must be discontinued before there can be a forfeiture by re-entry. I am not able to accept that argument. I see no point in the law requiring what was described in *Bayliss* v *Le Gros* (1858) 4 CBNS 537 as an "idle ceremony" — a case in which there was sufficient re-entry by acceptance of a subtenant already in occupation as tenant of the forfeiting landlord. Before August 31 1965 the third party was supporting the defendants in occupation in trespass against Miah. Thereafter the third party was reversioner, entitled to forfeit, and by asserting the right in law to keep the defendant in occupation as tenant of the third party could only be understood in law to be asserting the determination of Miah's lease by reason of a forfeiture.

The right in law which the third party, Greenwoods, was asserting in that case was the right to maintain the defendants, Sportsman, in occupation of the premises known as 5 Upper High Street, Bargoed, Glamorgan, as their tenants under a new tenancy upon the terms of a draft lease which had been negotiated and agreed between Greenwoods and Sportsman. That was not a case in which Sportsman had ever been the subtenants, in respect of 5 Upper High Street, of the tenant, Miah, whose lease Greenwoods were held to have determined by re-entry. In my judgment *London & County (A&D) Ltd* v *Wilfred Sportsman Ltd* is no authority for the proposition that a landlord may effect a re-entry of premises against his tenant by an arrangement made with an existing subtenant under which the subtenant is to remain in occupation of the premises as the tenant of the landlord for the residue and otherwise upon the terms of his existing sublease. But, although it appears to me clear that the *Wilfred Sportsman* case does not, on its own facts, afford the defendant any assistance in the present case, the dictum of Russell LJ to which I have referred does suggest that *Bayliss* v *Le Gros* may be authority for that proposition.

In *Bayliss* v *Le Gros* (1858) 4 CB(NS) 537, the facts were these. Factory premises at Tottenham were let under a lease dated September 28 1850 by one Barnewell to the plaintiff's father for a term of 21 years at the yearly rent of £165.15s, payable quarterly. The lease contained a proviso for re-entry in the event of non-payment of rent or breach of covenant. The plaintiff's father died on October 23, 1850, having by his will bequeathed the leasehold premises to his executors, namely his wife, one Joseph Fletcher and the plaintiff, on his attaining age 21. The widow alone proved the will; and she continued to carry on business at the factory premises.

On June 1 1851 the widow deposited the lease with Fletcher as security for a debt of £2,500, and also secured the same debt by a chattel mortgage over the plant and machinery of the premises. On June 22 1852 the plaintiff came of age, but he did not then come in to prove his father's will. On November 1 1852 Fletcher died, and his will was proved by his executors. In February 1853 the widow died intestate.

At some time after the widow's death, the defendants, Messrs Le Gros, went into possession of the premises. On October 24 1853, the landlord Barnewell, having inspected the premises and found them out of repair, allowed the defendants to remain in possession as his tenants under an oral agreement until Lady Day 1854 at a yearly rent of £150 and upon condition that they should purchase the plant and machinery from Fletcher's executors. This they did by an assignment dated November 5 1853.

On January 9 1854 the defendants gave Barnewell notice of their intention to quit the premises at Lady Day 1854. Nevertheless, they remained in possession, and a tenancy for a further year was agreed at the same rent. On December 21 1854, the plaintiff took out letters of administration to his mother's estate; and on February 24 1855 he took probate of his father's will. By a writ dated February 12 1855 he brought an action of ejectment to recover possession of the premises from the defendants.

The question for the court was whether the lease dated September 28 1850 had been determined by a re-entry by Barnewell. The court were unanimous in holding that there had been a re-entry. Cockburn CJ said:

Finding the premises in a dilapidated state the landlord comes upon them and enters into an agreement with a man he finds in possession, to become his tenant — intending thereby to act upon the forfeiture and to oust the lessee. I think that was quite sufficient to constitute an entry by the landlord so as to put an end to the lease.

Williams J was of the same view. He said:

As to the other point, if Barnewell had entered and desired the person he found on the premises to go out, and then desired him to resume possession as his tenant, the case would have been clear beyond all doubt. They did not go through that idle ceremony: but the facts set out in the special case show a re-entry by the landlord, and something more.

It was urged upon me on behalf of Mr Sobelman that *Bayliss* v *Le Gros* had been decided on the basis that Messrs Le Gros had been let into possession by the plaintiff. Some support for this contention can be found in an answer given by counsel for the plaintiff to a question put by the court in the course of the argument in that case — see the last few lines at the bottom of p 553 in the report. But it appears to me that, upon a true analysis of the facts, there was no evidence to explain how the defendants in that case came to be occupying the premises before October 24 1853; and, in particular, no evidence that they had ever been subtenants of, or paid rent to, the plaintiff before that date. It must be kept in mind that the plaintiff had no estate in 1853 out of which he could grant a sublease. But, whether Messrs Le Gros were subtenants, licensees or trespassers in relation to the plaintiff immediately before October 24 1853, it is clear that the arrangement which they made with Barnewell on that date was not that they should remain in possession under any existing tenancy, but rather that they should be allowed to remain only on the basis that they became the tenants of Barnewell under a new tenancy, and for a different term.

It follows that, notwithstanding the dictum of Russell LJ in *London & County (A & D) Ltd* v *Wilfred Sportsman Ltd,* to which I have referred, I am unable to regard *Bayliss* v *Le Gros* as an authority for the proposition which I have set out above. In my judgment, the most that can be derived from *Bayliss* v *Le Gros* is that a landlord may effect a re-entry against his tenant by an arrangement with an existing subtenant under which the subtenant is to remain in occupation as the tenant of the landlord upon the terms of a new tenancy. So understood, I do not think that that authority assists the defendant in the present case.

In the present case, it appears to me apt to describe the changing of the lock at 195 Burnt Oak Broadway as "an idle ceremony". This is, I think, exemplified by the fact that, when told by Mr Clayton that the new lock was unsatisfactory, Twogates Properties Ltd, through their solicitors, were content that the lock should be replaced by Mr Clayton himself. There was never any intention on the part of the landlords to exclude the subtenant from possession. There is, to my mind, no doubt that Mr Bloch and his locksmith would not have been permitted by Mr Clayton to interfere with the existing lock if Mr Clayton had been told that that act was intended in any way to interfere with his rights under his existing sublease. Mr Clayton was not told this. On the contrary he was assured, by Mr Bloch in person and by the first of the letters dated October 19 1984, that Twogates Properties Ltd were not in any way challenging his right to remain in occupation of the premises at 195 Burnt Oak Broadway under the provisions of his existing underlease. If there was a re-entry in the present case, it was not effected by the changing of the lock.

The real question on this part of the case, as it appears to me, is whether the landlords effected a re-entry, constructively, by obtaining Mr Clayton's consent to their actions upon the terms of the first letter of October 1984. In my judgment, even if it could be said that Mr Clayton attorned tenant to Twogates Properties Ltd by tacitly accepting the terms of that letter, such an attornment would not be evidence of an unequivocal intention on the part of the landlords to re-enter under the provisions of the 1932 lease. It is clear that both Twogates Properties Ltd and Mr Clayton were acting on the basis that the 1976 underlease would continue. If Mr Clayton was making an attornment at all he was doing so as tenant under that underlease.

It is equally clear, although perhaps not appreciated at the time, that the continuation of the 1976 underlease was wholly inconsistent with the determination, by forfeiture, of the 1932 lease. In these circumstances, it is, in my judgment, impossible to regard the arrangements which Twogates Properties Ltd made on October 19 1984 as amounting to a re-entry under the 1932 lease.

On the view which I take of the matter, it is unnecessary to consider whether an attornment by Mr Clayton upon the terms of the 1976 underlease would be made void by the provisions of section 151(2) of the Law of Property Act 1925. It is, I think, a question of some difficulty whether a landlord who is asserting that notwithstanding that the head lease has been forfeited nevertheless the underlease survives is "a person claiming to be entitled to the interest of the land of the lessor" (meaning the intermediate lessor) for the purposes of that subsection. It seems to me that that difficulty arises because the subsection is not framed to meet a situation which the draftsman would have recognised as being incapable of existing at law.

It is, strictly, unnecessary also for me to go on to consider the second question raised in the action; namely whether, if the 1932 lease had been determined by forfeiture, this would be a proper case for the court to grant relief from forfeiture.

Nevertheless, in case this matter should go further, it may be of assistance if I express the view that this would have been a proper case for relief. The principles upon which relief from forfeiture for non-payment of rent is granted are well known, and they may be found in the line of authority of which *Howard* v *Fanshawe* [1895] 2 Ch 581 and *Lovelock* v *Margo* [1963] 2 QB 786 are leading examples. It was urged upon me, on behalf of the defendant, that relief ought not to be granted where to do so would prejudice the rights of third parties. Mr Sobelman, so it was argued, had purchased the freehold at auction on the basis that it was not encumbered by the 1932 lease. I do not think that it would be right to give weight to this consideration. Before Mr Sobelman made his bid at the auction on November 20 1984, he had in his possession the first of the additional special conditions, which he had obtained from the auctioneers. This disclosed that the property had been subject to the 1932 lease; and that, if that lease had determined, it was by reason of a peaceable re-entry for non-payment of rent which had been effected within the past few weeks. Knowing these facts, Mr Sobelman was on notice, before he purchased the property, that the tenant under the 1932 lease was likely to have an unanswerable claim for relief from forfeiture. Whether or not Mr Sobelman was actually aware that such a claim could or would be made, I do not know of any grounds upon which (if they had need to do so) the plaintiffs should have been unable to pursue that claim against Mr Sobelman to as full an extent as they could have done against his predecessors, Twogates Properties.

Accordingly, I declare that the lease dated October 24 1932 has not been forfeited by the events which took place on October 19 1984 and, further, that the plaintiffs are entitled to give a good receipt to Gerald Clayton for the sum of £875, being the rent due under the underlease on December 25 1984 and for all subsequent rent payable thereunder.

The plaintiffs were awarded costs.

Court of Appeal

October 21 1986

(Before Lord Justice KERR, Lord Justice NOURSE and Lord Justice STOCKER)

WILLIAM HILL (SOUTHERN) LTD v CABRAS LTD

Estates Gazette January 24 1987

281 EG 309-312

Landlord and tenant — Signs advertising tenants' licensed betting office — Dispute as to tenants' right to maintain signs on landlords' property — The signs in question, which were illuminated and were regarded as essential to the success of the tenants' business, were placed on the wall of the building just above the entrance at street level — Tenants' betting office was on the first floor and was approached from the street by an external staircase not included in the demise — The signs had been affixed with the consent of, or without objection from, the previous landlords — After the reversion changed hands, however, difficulties arose — New landlords contended that tenants had a revocable licence only to maintain the signs and requested their removal — Tenants sought a declaration as to their rights in regard to the signs and Goulding J decided in their favour — He did so, not on the ground that the rights were included in a demise of the first-floor offices and their "appurtenances", nor that they passed under section 62 of the Law of Property Act 1925, but by implication from a negative covenant forbidding the exhibition of signs in, on or to the premises without the landlords' consent — Held by the Court of Appeal that Goulding J had come to the correct conclusion but that the particular ground on which he based it could not be supported — The view that the grant of an easement could be spelt out of a prohibition against placing signs without the landlords' consent, even where consent had been given before the grant of the lease, was too novel and unorthodox to be accepted — The easement in the present case arose from the demise of "appurtenances", which included the signs in question — Kerr LJ, while agreeing with this ratio, would if necessary have decided the case in favour of the tenants on the ground that communications between the parties' solicitors before the lease was executed constituted a "convention" which estopped the landlords from disputing the tenants' claim — Appeal by landlords dismissed

The following cases are referred to in this report.

Cuthbert v *Robinson* (1882) 51 LJ (NS) 238
Francis v *Hayward* (1882) 22 Ch D 177
Moody v *Steggles* (1879) 12 Ch D 251
Wheeldon v *Burrows* (1879) 12 Ch D 31

This was an appeal by the landlords, Cabras Ltd, from a decision of Goulding J granting a declaration in favour of the rights of the tenants, William Hill (Southern) Ltd, to maintain certain illuminated signs on a part of the landlords' wall at 27 Park Lane, Mayfair, on the first floor of which the tenants carried on their business as a licensed betting office and turf accountants.

Gerald Godfrey QC and S Bickford-Smith (instructed by Wallis & Co, of Bromley) appeared on behalf of the appellants; Charles Sparrow QC and K F Farrow (instructed by Titmuss Sainer & Webb) represented the respondents. (The reversion had been transferred to new freeholders since Goulding J's judgment and they had been added as appellants for the purposes of the appeal). Goulding J's judgment is reported at [1985] 2 EGLR 62; (1985) 275 EG 149.

Giving the first judgment at the invitation of Kerr LJ, NOURSE LJ said: On June 20 1978 the appellants' predecessors in title granted to the respondents a 20-year lease of a licensed betting office on the first floor of 27 Park Lane, Mayfair, London W1, together with a right to use the staircase leading from the entrance at street level. The respondents were already in possession under an earlier lease of the property and they had previously, either with the consent of, or without objection from, their then landlords, affixed two illuminated name signs to the landlords' exterior wall of the building just above the entrance at street level.

In 1979, this time with the express consent of their then landlords, the respondents removed the lower of the two signs and replaced it with an illuminated quadrant canopy bearing the name "William Hill" on each of its three exterior faces.

Between 1978 and 1983 they continued to maintain and illumine the two signs. By the terms of the lease the respondents are prohibited from using the premises otherwise than as offices in connection with their business as turf accountants and a licensed betting office. Goulding J found that at all material times the success of the business which the respondents carry on from the premises has depended upon their ability conspicuously to advertise the existence of the office to passers-by and that, both by reason of the physical situation of the premises and by virtue of section 10(5) of the Betting, Gaming and Lotteries Act 1963, such advertisement can only be by a sign or signs at street level.

To anyone uninitiated in the mysteries of English property law that bare statement of the essential facts of this case might suggest that the respondents were entitled as of right to keep their two existing signs in position during the remainder of the 20-year term. However, in 1983 the ownership of the building of which the premises form part and of the reversion to the premises was acquired by the appellants, who have ever since contended that there is no such entitlement. They say that the respondents have no more than a licence to maintain the signs, which is revocable by the appellants at any time on reasonable notice. They say that they want the signs removed, because they would greatly interfere with, if not actually inhibit, their proposed redevelopment or redesign of the front of the building, although it must be said that if the appellants' contention is correct it is not one which they have to justify.

Goulding J was, I think, disposed to reject the appellants' contention if he properly could, but in so doing he accepted only one of the respondents' five arguments to the contrary.

The appellants now appeal to this court against his decision. We have heard this morning that they have now assigned the reversion in the premises — and, I take it, in the whole building — to another company. However, that company, which is now a party to the proceedings, through Mr Godfrey, who appears for the appellants, adopts the same stance as they do.

The material terms of the lease are as follows. It starts with nearly two pages of particulars, which are, by clause 1(a), expressed to form part of the lease. Paras 5 and 12 of the particulars define "demised premises" and "user" respectively as follows:

ALL THAT suite of offices situated on the first floor of 27 PARK LANE MAYFAIR W1 in the CITY OF WESTMINSTER WITH TOILET ACCOMMODATION AT THE REAR THEREOF up to but excluding the joists of the floor above down to and including the floors thereof (together with joists) (as the same is defined on the plan and thereon edged red together with the Landlords fixtures and fittings therein.

Then "user" is defined as:

Use as offices in connection with the Lessee's business as Turf Accountants and Licensed Betting office and for purposes ancillary thereto or associated with such activity or for such other purposes as may first be approved in writing by the Lessor such approval not to be unreasonably withheld.

By clause 1A, "the Premises" is defined to mean:

The "Demised Premises" described in the Particulars and each and every part thereof together with the appurtenances thereto and any building now or hereafter on any part thereof including all additions alterations and improvements thereto and all Landlords fixtures and fittings and plant machinery and equipment now or hereafter in or about the same.

The first part of clause 2 is as follows:

The Lessor HEREBY DEMISES unto the Lessee the premises together with the rights expressed in the First Schedule hereto but EXCEPTING AND RESERVING unto the Lessor the rights and easements specified in the Second Schedule hereto.

The rights expressed in the First Schedule are, first, the right to use the staircase from Park Lane and, second, the right to the free passage and running of water, soil, gas and electricity serving the demised premises.

Clause 3 is in the following terms:

The demise hereby made shall not be deemed to include or confer and shall not operate to convey or demise unto the Lessee any right of light or air liberties privileges easements or advantages (except such as are specifically granted by this Lease) in through over and upon any land or premises adjoining or near to the demised premises.

There are, as is usual in a lease of this kind, a large number of tenant's covenants, no (xv) which is in the following terms:

That no figure letter pole flag signboard advertisement inscription bill placard or sign whatsoever shall be attached to or exhibited in on or to the Premises or the windows thereof so as to be seen from the exterior without the previous consent of the Lessor which shall not be unreasonably withheld in respect of a sign stating the Lessee's name and business or profession (such sign if the Lessor so requires to be removed and any damage caused thereby made good by the Lessee at the end or sooner determination of the said term) and in the case of the premises demised this covenant shall not apply to any reasonable and usual display of trade or business notice inside the shop window.

Although Mr Sparrow, for the respondents, also advanced arguments based on tenant's covenants nos (xiii) and (xiv), on the view which I take of the case it is unnecessary for me to refer to them.

The decisive passage in Goulding J's judgment starts at p 7 E of the transcript* where, having stated that Mr Sparrow had put forward five alternative grounds for the respondents' alleged right to maintain the signs during the whole term of the lease he said this:

The most obvious, and in my opinion the most satisfactory, approach to the matter is this. At the time of the lease the exhibition on the lessor's retained property of sign no 1 and the finger sign was an advantage actually enjoyed with the betting office. Therefore, if no intention was expressed to the contrary in the lease, it was granted thereby as an easement to be enjoyed with the betting office, whether or not it had been the subject of any such right under the previous regime. That conclusion follows from the demise of the office "together with the appurtenances thereto" or alternatively from the incorporation of the general words set out in section 62 of the Law of Property Act 1925: see *Watts* v *Kelson* (1870) LR 6 Ch App 166; *Bayley* v *Great Western Railway Co* (1881) 26 Ch D 434; *Henry Ltd* v *M'Glade* [1926] NI 144. However, the lease expresses an intention to limit the wide effect of such words, and even of the demise itself considered apart from general words. That is the purpose of clause 3 . . .

The judge then read clause 3, and continued at p8C:

Rights over adjoining premises are therefore limited to those "specifically granted" by the lease. One looks first for such rights in the first schedule, but the rights there specified do not help the plaintiff in the present controversy. I turn next to tenant's covenant no xv.

Then he read that covenant continued at p8F:

That covenant, in my judgment, specifically grants a right to exhibit a sign stating the lessee's name and business or profession and fit for approval by a reasonable lessor. If, on its true construction, it can extend to signs not on the demised premises but on the ground-floor property of the lessor, there is accordingly nothing in clause 3 to cut it down. But does the covenant go so far? In construing it, I find two points of particular relevance, one in the grammar of the subclause, the other in the surrounding circumstances. In its language the scope of the covenant embraces not only signs attached to the premises, signs exhibited *in* the premises and signs exhibited *on* the premises, but also signs exhibited *to* the premises. Second, it was the fact, and was within the knowledge of both parties, that the success of the business for which the premises were let depended upon a sign at ground-floor level. The right granted to exhibit a proper sign therefore extends in my view to a sign on the lessor's retained property indicating the way to the betting shop, such as the signs known to both parties as erected before, and subsisting at, the time of the lease.

The learned judge then said that there remained two points regarding consent under covenant (xv), both of which he resolved in favour of the respondents. It was on that ground that he gave judgment for the respondents. He made a declaration appropriate to reflect the construction of the lease which they had urged upon him and he gave them liberty to apply for an injunction to restrain the appellants from interfering with the signs.

In the passage which I have quoted from his judgment Goulding J started by considering two possible ways in which there might have been the grant of an easement to maintain the signs. I should state here that it is not in dispute that such an easement is capable of existence. If authority is needed for that, it will be found in *Moody* v *Steggles* (1879) 12 Ch D 251.

The first way in which the judge thought that an easement might have been granted was by virtue of the demise of the suite of offices "together with the appurtenances thereto"; second, by virtue of section 62 of the Law of Property Act 1925. He then went on expressly to reject the second possibility and impliedly to reject the first as well. He rejected the second possibility by reason of the terms of clause 3 of the lease. In my view he was correct in so doing. The language of that clause clearly demonstrates that it was intended to exclude the effect of section 62, a state of affairs for which express

---

Editor's note: See [1985] 2 EGLR 62 at p 64.

provision is made by subsection (4) of that section. I think that it is clear that the judge must also have relied on clause 3, to reject the first possibility, although that is a point on which, as will appear, I respectfully disagree with him.

Having rejected both those possibilities, the judge then went on to find a specific grant of the right to exhibit a sign in tenant's covenant no (xv). There are, as it seems to me, two difficulties about that. First, the judge's construction of the covenant, with the weight which it places on the word "to", is in my view a very strained one and I do not find myself able to accept it. Second and conclusively, the proposition that a tenant's covenant which prohibits him from placing signs on the landlord's other land without consent can operate as a grant of an easement over that land, even in a case where consent has been given before the grant of the lease, is novel and unorthodox and I have no hesitation in rejecting it. It seems to me that that proposition, if it were to be accepted, might be of far-reaching effect in the law of landlord and tenant and it would certainly cause consternation among conveyancers. It is right to say that Mr Sparrow did not rely on that point before the learned judge, and that he has not relied on it in this court either.

His case was based on the construction of covenant no (xv) which was favoured by the learned judge, and also on an anterior consent granted for the erection of the signs and for the maintenance of them during the term of the lease. Since I am unable to agree with the learned judge on the question of construction, it becomes unnecessary for me to consider the question of consent.

I return then to the demise of the suite of offices "together with the appurtenances thereto". What effect did those last words have?

It is to be noted that clause 2 of the lease demised "the premises", being something which had previously been given a very studied definition, ie the "demised premises" (the suite of offices situated on the first floor) together with the landlord's fixtures and fittings therein and "together with the appurtenances thereto". In the circumstances I do not think that those last words can be treated as being mere surplusage. I would accept that in certain contexts they can be so treated, but it would not seem to me that that would be a permissible course in the context of the present lease.

Moreover, since clause 2 contains an express demise of the rights expressed in the First Schedule, it would appear that the "appurtenances" were intended to include either some other rights appertaining to the demised premises or some other parcel or parcels of land. The word "appurtenances" in its strict sense means the former and not the latter. Bearing in mind the permitted user of the premises and the judge's finding as to the practical and commercial considerations for keeping the two signs in their present positions if that user is to be continued, I do not find it very difficult to suppose that the appurtenances include the right to maintain the two signs.

Then this question arises. Is that view of the position invalidated by clause 3 of the lease, on which Mr Godfrey so strongly relies? I do not think that it is. For one thing, that clause refers to a number of different rights, or quasi rights, but not to appurtenances as such. For another, and perhaps more significantly, it makes an express exception for rights and quasi rights which are "specifically" granted by the lease. They would obviously include the two rights expressed in the First Schedule, but they must also, I think, include the appurtenances demised by clause 2. There was some debate as to whether it could be said that the appurtenances were specifically granted, on the ground that the word is an entirely general one. However, I do not think that there is much in that point. The appurtenances were clearly granted expressly, and I think that that is enough, particularly when the general principle to which I now come is borne in mind.

Mr Sparrow relied on the well-known judgment of Thesiger LJ in *Wheeldon* v *Burrows* (1879) 12 Ch D 31, at p 49, where, having said that there were two general rules covering cases of this kind, he continued as follows:

The first of these rules is, that on the grant by the owner of a tenement of part of that tenement as it is then used and enjoyed, there will pass to the grantee all those continuous and apparent easements (by which, of course, I mean *quasi* easements), or, in other words, all those easements which are necessary to the reasonable enjoyment of the property granted, and which have been and are at the time of the grant used by the owners of the entirety for the benefit of the part granted.

That is the rule which would, in the absence of clause 3, apply to this case. The learned lord justice then went on to state the second rule, namely that rights intended to be reserved over the tenement granted must be expressly reserved, and he dealt with certain exceptions to the second rule, in particular the case of ways of necessity. Then comes the sentence on which Mr Sparrow particularly relies:

Both of the general rules which I have mentioned are founded upon a maxim which is as well established by authority as it is consonant to reason and commonsense, viz that a grantor shall not derogate from his grant.

Mr Sparrow submits that that shows that the court will as a general rule construe provisions such as clause 3 of the lease in the present case so as not to derogate from the grant made by the lease, that is to say, in this case, the grant made for the purpose of using the premises as a licensed betting office and for that purpose only. In other words, Mr Sparrow submits that the court will not construe a general provision in a lease, particularly an exception and most of all an exception couched in very general terms such as those in clause 3, so as to take away with the other hand that which has already been granted by the one hand in the dispositive provisions of the lease. Although *Wheeldon* v *Burrows* was a case on implied rights, I accept Mr Sparrow's proposition in regard to the construction of express rights, it being, as Thesiger LJ said, consonant to reason and commonsense and also, I would add, to the commercial realities of a case such as this. I think that there is enough latitude in the language of clause 3 to enable the proposition to be applied in the present case. I would therefore accede to this argument of Mr Sparrow, and hold that the right to maintain the signs was granted to the respondents by the demise of the appurtenances to the demised premises. If that view prevailed it would be enough to dispose of this appeal in favour of the respondents. However, I wish to mention more briefly two further arguments which have been advanced by Mr Sparrow.

The first of these arguments seeks to treat the signs as corporeal hereditaments and to interpret "appurtenances" accordingly. Whatever its strict meaning may be, there is undoubtedly authority for the view that the word "appurtenances" can mean corporeal hereditaments. An example to which Mr Sparrow referred us is *Cuthbert* v *Robinson* (1882) 51 LJ (NS) 238. The foundation of his argument in this respect is the decision of this court in *Francis* v *Hayward* (1882) 22 Ch D 177. Undoubtedly the facts of that case do bear some similarity to those of the present. There it was held, both by Kay J and by this court, that a fascia formed of cement or stucco plastered on the brickwork of other premises belonging to the landlord was part of the premises demised to the tenant.

However, Goulding J, in rejecting Mr Sparrow's argument, said this at p 11B of the transcript:

The industry of counsel has not discovered any judicial reference to the case of *Francis* v *Hayward* during the century that has now passed since it was decided. I for my part regard the decision not as laying down any principle but as applying to the very special facts of this case. Certainly I find it quite unreal to regard the easily detachable signs now under consideration as part of the premises demised and I do not think that *Francis* v *Hayward* could legitimately be used to circumvent the operation of clause 3 of the lease.

Although I was at one time greatly attracted by Mr Sparrow's argument, I have in the end come to the conclusion that the learned judge's decision on this point was correct. The question depends to a great extent on whether the signs should be regarded as being tenant's fixtures on the ground that they were intended to be fixed to the premises during the period of the lease, so as to become part of the realty, or whether, as Mr Godfrey contends, they were intended to be enjoyed as chattels, in which event there would not have been the necessary purpose of annexation in order to make them fixtures. The point is an interesting one. I dare say that it would repay a closer study of some of the authorities on fixtures in a case in which a decision on it was necessary. That is not this case.

The second of Mr Sparrow's further arguments was that by virtue of two communications which took place between the parties' solicitors before the lease was executed, to which the judge referred as "the preliminary exchange", the former landlords' solicitors had made a statement which constituted a convention, or an underlying assumption, as to the applicability and effect of tenant's covenant no (xv) to the signs which are in issue in these proceedings. Mr Sparrow contended that after the respondents had apparently incurred expenditure on the property on the faith of their continued right to maintain the signs with the consent of their former landlords they and the appellants were estopped from claiming that the respondents were not entitled to do so. That contention is undoubtedly improved by the fact that on September 14 1979 the former landlords' solicitors wrote to the respondents' solicitors giving an express consent to the substitution of the canopy to which I have already referred. That is another point of some interest, but it is one which I do not feel it necessary to decide. I therefore express no view on it, beyond saying that I see great force in the argument which Mr Sparrow has advanced.

For the reasons which I have given, although on a different ground from that favoured by the learned judge, I have come to the conclusion that his decision was correct. I would therefore dismiss this appeal.

Agreeing, STOCKER LJ said: I think it is possible that the decision can be supported on some of the other grounds which have been argued before us, and in particular that consents given within the terms of tenant's covenant (xv) are irrevocable during the continuance of the lease. But for my part I am content to rest my judgment on the grounds which have been stated by my lord.

I also would dismiss the appeal.

Also agreeing, KERR LJ said: I share the view expressed by Nourse LJ that the judge was not correct in the conclusion, which he evidently formed unaided by argument, that a specific grant can be extracted from the negative covenant in clause (xv), albeit subject to the qualification of consent not being unreasonably withheld.

But I am not convinced that the judge's construction of that covenant is also necessarily incorrect, ie the meaning which he gave to the word "to" in the phrase ". . . in on or to the premises". It seems to me that the word "to" means "relating to" the premises. Having regard to the pre-existing facts found by the judge concerning the location of these signs, the word "to" may be sufficient for the purposes of covenant (xv), even though the signs were situated on the landlords' property.

But assuming that this goes too far, I am certainly of the view that the appeal should be dismissed for the reasons stated by Nourse LJ and also on the last ground to which he briefly referred. The exchange between the solicitors before the lease was executed, which the judge quotes at p 3 of his judgment, was as follows. The lessee's solicitors inquired:

Please confirm that the Lessee's present signs are approved.

The reply was clearly intended to refer to covenant (xv):

There should not be any problem about this, and in any event the particular covenant dealing with this is open to the extent that the landlords' consent is not to be unreasonably withheld.

It seems to me that once that reply had been given and the lessees thereupon remained content to rely and act upon it, as they clearly did, it is not open to the landlords subsequently to contend that because of some alleged restrictive meaning of the word "to", covenant (xv) cannot be applied to these signs on the ground that they were situated on the landlords' retained property. Both parties were in my view in agreement to proceed on the basis that the then existing signs, and the tenants' right to maintain them in their present location, fall within the scope of covenant (xv). In these circumstances I do not think that the landlords can go back on their solicitors' assurance.

If that is correct, as I think it must be as a matter of commercial sense and justice, then it becomes relevant to refer briefly to what the judge concluded on the issue of consent, beginning at p 9C of the judgment. He said:

There remain two points regarding consent. Did Python Properties Ltd, on the grant of the lease, consent to the existing signs for the purposes of covenant no xv, and, if so, was such consent revocable?
I incline to the view that, as the existing signs were known and obvious to both parties and some sign on the ground floor was patently necessary, the grant of the lease without any

he is reported as saying "observation", but I think it must be "reservation"

on the subject by the lessor would have authorised the continuance of the signs as approved signs within covenant no xv. That it in fact did so after the point had been brought into the open by the preliminary exchange, I feel no doubt. The plaintiff's inquiry was made and the vague but reassuring reply thereto was given, expressly in relation to the covenant, and the matter was not raised again on behalf of Python Properties Ltd before the plaintiff accepted the lease. Thus if my construction of the covenant is correct, sign no 1 and the finger sign were authorised under its terms.

Pausing at that point, when the judge referred to his construction of the covenant, it seems to me that in the present connection he was only referring to the meaning of the words "to the premises" and not to his view that covenant (xv) also contained an implied specific grant.

He then went on as follows:

No authority precisely covering the second point has been cited. In my judgment a landlord's consent under such a clause as that now in question, once acted upon by a tenant, cannot be withdrawn during the term unless given with an express reservation for that purpose; and here the tenant at once acted on the consent by continuing to maintain and illuminate the signs as it had been doing under the previous lease.

If it had been necessary to decide this appeal on this basis, I would have done so and dismissed it on that ground. However, my primary reason is the same as that stated by Nourse LJ concerning the demise of "appurtenances", which in my view remains unaffected by the generality of clause 3.

I should add that the reversion has been sold once again, the present freeholders having evidently acquired their title since the judgment and before this appeal, as was discovered only today. In that regard it has been agreed that they should be added as appellants for the purposes of this appeal and that they will be bound by its result.

The appeal was dismissed with costs. Leave to appeal to the House of Lords was refused.

Court of Appeal

December 18 1986

(Before Lord Justice SLADE, Lord Justice RALPH GIBSON and Sir Roger ORMROD)

POST OFFICE v AQUARIUS PROPERTIES LTD

Estates Gazette February 21 1987

281 EG 798-804

Landlord and tenant — Tenants' liability to repair under full repairing lease — Appeal from decision of Hoffmann J in favour of tenants — Basement of office building consisting of basement, ground floor and six upper storeys flooded owing to rise of water-table combined with defects of construction and possibly of design — Basement ankle deep in water for most of the time between 1979 and 1984, but owing to lowering of water-table since 1984 basement had been dry — Although the water had been in the basement for years, there was no evidence of any actual damage to the building — Although the defects existed, and had existed, since the original construction, no part of the building had suffered deterioration — It appeared that the defects were due to a failure of the "kicker" joint between floor and walls which had resulted in weak areas of concrete of a relatively porous texture — Experts differed as to the measures required to remedy the defects, the tenants' expert recommending an asphalt tanking of the basement while the landlords' expert recommended either an asphalt tanking scheme, but of a lesser height, or a cheaper, waterproof-rendering method — The issue was as to the tenants' liability in these circumstances under their covenant in the lease, the critical words of which were to "keep in good and substantial repair the demised premises and every part thereof" — Hoffmann J decided the issue in favour of the tenants on the ground that any of the schemes put forward to cure the defects entailed substantial structural alterations to the basement which did not come within the scope of "repair" — This ratio assumed that the premises were in fact in disrepair — The Court of Appeal agreed with the decision in favour of the tenants but reached it by a different route — Tenants were not liable because there was at present no disrepair — Disrepair connotes a deterioration from a previous physical condition — Here, although water had been present for long periods, there was no evidence that it had caused deterioration — Court reserved their opinion on what the position might be in the future if these original defects combined with further flooding were to cause damage, for example, to plaster-work or electric fittings

— **Landlords' appeal dismissed — A number of authorities considered**

The following cases are referred to in this report.

Anstruther-Gough-Calthorpe v *McOscar* [1924] 1 KB 716
Brew Bros Ltd v *Snax (Ross) Ltd* [1970] 1 QB 612; [1969] 3 WLR 657; [1970] 1 All ER 587; (1969) 20 P&CR 829; [1969] EGD 1012; 212 EG 281, CA
Elmcroft Developments Ltd v *Tankersley-Sawyer* [1984] EGD 348; (1984) 270 EG 140, CA
Halliard Property Co Ltd v *Nicholas Clarke Investments Ltd* [1984] EGD 341; (1983) 269 EG 1257
Pembery v *Lamdin* [1940] 2 All ER 434
Proudfoot v *Hart* (1890) 25 QBD 42
Quick v *Taff-Ely Borough Council* [1986] QB 809; [1985] 3 WLR 981; [1985] 3 All ER 321; (1985) 84 LGR 498; 276 EG 452, CA
Ravenseft Properties Ltd v *Davstone (Holdings) Ltd* [1980] QB 12; [1979] 2 WLR 897; [1979] 1 All ER 929; (1978) 37 P&CR 502; [1979] EGD 316; 249 EG 51, DC
Wright v *Lawson* (1903) 19 TLR 510; 68 JP 34; affirming 19 TLR 203

This was an appeal by the landlords, Aquarius Properties Ltd, from a decision of Hoffmann J, reported at [1985] 2 EGLR 105; (1985) 276 EG 923, in favour of the tenants, the Post Office, in regard to repairing liability under the repairing covenants of their lease of Abbey House, 74-76 St John Street in the City of London.

Michael Spence QC and A D Dinkin (instructed by Grangewoods) appeared on behalf of the appellants; Paul Morgan (instructed by B A Holland, Solicitors' Department, The Post Office) represented the respondents.

Giving the first judgment at the invitation of Slade LJ, RALPH GIBSON LJ said: This is an appeal by landlords from the decision of Hoffmann J given on July 26 1985 by which he made a declaration at the suit of the Post Office, who are the tenants and the plaintiffs in the action, that on the true construction of an underlease dated June 25 1969 the Post Office is not liable to carry out certain work to the basement floor and walls at Abbey House, 74-76 St John Street, London EC1. The appeal raises questions of principle upon the proper construction in law of repairing covenants in common form. The facts which have given rise to these questions, however, appear to be highly unusual. In short, there is shown to be a defect in the structure of the basement of an office building which was present from the time of the construction of the building. During a period of time when the water-table rose at the place where the building stands, the defect permitted ground water to enter the basement so that water stood ankle deep on the floor for some years. The defect has not grown worse but is in the same condition as when the building was built. Apart from permitting water to enter, which disappeared when the water-table dropped, leaving the basement dry for the last two years, no damage to any part of the building is shown to have been caused by the defect.

The first question, accordingly, is whether it has been proved that the building was out of repair so as to give rise to an obligation under the covenant to put it into repair. The second question — which appears to have been treated as the main, if not the only, question at the trial — is whether, assuming the building to be in a state of disrepair by reason of the existence of the defect, any of the schemes of treatment put forward for curing the defect were capable of being regarded as work of repair as opposed to being structural alterations and improvements.

On this appeal the findings of fact of the learned judge have not been questioned. I set out his primary findings in his words:

Abbey House is an office building in the City of London constructed in the mid-sixties. Since 1969 it has been let to the Post Office on a full repairing lease by Aquarius Properties Ltd. For most of the time between 1979 and 1984 the basement was ankle deep in water. This appears to have been the result of a rise in the level of the local water tables combined with defects in the construction and possibly the design of the building. In 1984 the water table subsided again and since then the basement has been dry.

The tenant's lease expires in 1991 but the building has a life expectancy of many years and it is therefore agreed by landlord and tenant that remedial work is necessary to make the basement waterproof in case the water table should rise again. . . .

The building consists of front and rear sections. The basement runs under both. It has 12-in thick reinforced concrete walls. Under the front section of the building the basement floor is a reinforced concrete raft 3 ft thick which, together with the walls, supports the ground and six upper storeys. At the rear there are only two upper storeys and the basement slab is of lighter construction. For the most part it is 8 in thick . . . The floor has been constructed integrally with the walls by forming a 5-in upstand around the edge of the floor and then using that as the base for the walls. This upstand is called the "kicker".

Concrete has to be cast in sections with each new section of wet concrete being poured alongside or above a section which has already dried. There is also a tendency for concrete to shrink as it dries, partly from chemical reaction and partly on account of evaporation. The result is to produce a construction joint between sections of concrete which, unless suitably bridged, may admit water. In the design of Abbey House basement PVC water bars were specified for insertion at the construction joints beneath the concrete floor and on the outside of the walls up to a height of 6 ft from a datum line corresponding to the surface of the basement concrete floor. These water bars are, in effect, strips of PVC which are keyed into the concrete and overlay and construction joints by some inches on each side. . . .

The contract documents record that in November 1964 trial holes had shown the water table to lie $1\frac{3}{4}$ ft below the datum line to which I have referred, that is about a foot below the 8 in section of the rear basement floor. In May 1965 the engineer engaged for the construction of the building reported the water table 3 in higher. In 1979 it appears to have risen to at least 6 in above the basement floor and the flooding took place. There is no evidence that ground water entered at any higher level. . . .

I have had the benefit of the views of two eminent structural engineers on the causes of the flooding and the remedies which should be adopted. Both agreed that there had been a failure of the kicker joint between the floor and the walls. This was caused by poor workmanship which had produced weak areas of concrete or what Mr Reith, the plaintiffs' expert, described as honeycombing of the concrete. This means that the concrete consists in places almost entirely of aggregate and is deficient in sand and cement. It is therefore relatively porous. Unless care is taken this phenomenon tends to occur at the bottom of the section being cast and in this case affected the concrete immediately above the kicker. There was also evidence that some construction joints had not been formed with care and that the water bars had therefore been inadequate to prevent the ingress of water.

The learned judge then considered whether, in addition to the porous concrete in way of the kicker joint and the inadequate water bars, other causes of the entry of ground water had been proved. He considered shrinkage of concrete as a possible cause of the detachment of water bars, which may have allowed water to enter between them and the concrete, and possible cracks in the 8-in slab, but no such defects were positively shown to have existed.

Next the learned judge considered the remedial measures proposed by the two experts. He approached the matter in that way because he had defined the issues in the case as whether the remedial work necessary to make the basement waterproof was "repair" within the meaning of the tenants' covenants. Mr Reith, the tenants' expert, recommended the more elaborate and expensive asphalt scheme, costing, at £175,000, rather more than twice the less elaborate rendering scheme, at £86,000, recommended by Mr Deverill, the landlords' expert. Reliance by the tenants upon evidence showing more expensive work to be necessary and by the landlords upon evidence showing that much less expensive work would suffice resulted no doubt from the concern of the parties over whether this work could properly be regarded in law as work of repair.

The scheme proposed by Mr Reith at £175,000 was, as described by the learned judge, as follows:

Mr Reith recommends the tanking of the basement with a layer of asphalt covering the floor and the walls up to a height of 5 ft. Since asphalt does not bond very well with concrete it would have to be held in place by an additional concrete slab of 12-in thickness laid over the floor and an inner concrete skin 6 in thick within the walls. In order to add additional weight to the floor, he also recommends the extension of the inner concrete walls to the ground floor slab so as to enable its weight to be transmitted to the basement floor and the construction of heavy internal partitions for the same purpose.

Mr Deverill proposed two alternative schemes. The learned judge described them as follows:

One is an asphalt tanking scheme similar to that of Mr Reith. The differences are that Mr Deverill considers that it would be excessive to allow for the possibility of a 5 ft head of water. In his opinion 3 ft would be enough . . . He agrees that it would be advisable to cover the asphalt on the floor with an additional 12-in concrete slab. Mr Deverill's second and preferred scheme involves the use of waterproof rendering in place of asphalt. This would be strongly bonded to the floor and walls and would, therefore, not need an inner concrete skin or covering. On the floor, however, Mr Deverill still thinks it advisable to add another 9 in of concrete slab firmly bonded to the existing slab and reinforced on the upper side. The waterproof rendering would then be applied to the upper surface of the new slab. Mr Deverill proposed the rendering scheme as a cheaper and more elegant solution than the traditional asphalt scheme. He agreed that the application of rendering was technically more demanding than asphalt and that unless the workmanship was good there was a higher risk of failure. Mr Reith said that his experience taught that the rendering scheme was too risky and that asphalt was worth the additional expense.

The learned judge observed that the choice between the schemes involved an evaluation of the additional risk in the cheaper scheme, and the likely damage and inconvenience if the risk eventuated, against the additional cost. The question before him was whether the work required to waterproof the basement amounted to repair within the meaning of the tenants' covenants. The answer to that question might have depended upon the scheme which he thought appropriate, but in this case, in his view, the answer to that question was the same whichever of the three schemes is adopted. In the result the learned judge expressed no view between the schemes proposed.

Finally, the learned judge found certain facts which he later considered as relevant to the question whether in all the circumstances any of the schemes of remedial work were shown to be work of repair. The landlords held the property under a long lease for 125 years from September 29 1966 granted in consideration of the rents and covenants and the erection of the building. The Post Office had an underlease dated June 25 1969 for a term from March 25 1969 until September 19 1991, or about 22½ years. The rent was £11,250 until a rent review at March 24 1983. Thereafter the rent was to be the market rent at the rent review date on the assumption that the Post Office had complied with its covenants. Since the parties had been unable to agree upon whether that meant that the Post Office must be assumed to have waterproofed the basement, the determination of the new rent awaited the outcome of these proceedings. Upon the assumption that the basement was waterproof, the annual rent would be in the region of £50,000 to £60,000. The capital value of the building was about £687,000, but the cost of rebuilding it today would be a good deal more.

It was said at the beginning of this judgment that this appeal raised questions of principle upon the proper construction in law of repairing covenants in common form. The learned judge referred to the repairing covenant in this case in clause 2(3) of the Post Office's underlease as in fairly standard form. He set out only what he called the critical words:

keep in good and substantial repair the demised premises and every part thereof . . .

In the course of argument Slade LJ pointed to the fact that the covenant includes also the following:

. . . during the term well and substantially to repair . . . amend . . . renew and keep in good and substantial repair and condition . . .

No reliance has been placed in this case upon any contention that the words "amend" or "renew" or "condition" add anything to the words referred to by the learned judge as the critical words.

Finally, the learned judge approached the question for decision thus:

I have found most assistance in the judgment of Sachs LJ in *Brew Brothers Ltd* v *Snax (Ross) Ltd* [1970] 1 QB 612. This says, in effect, that the whole law on the subject may be summed up in the proposition that "repair" is an ordinary English word. It also contains a timely warning against attempting to impose the crudities of judicial exegesis upon the subtle and often intuitive discriminations of ordinary speech. All words take meaning from context and it is, of course, necessary to have regard to the language of the particular covenant and the lease as a whole, the commercial relationship between the parties, the state of the premises at the time of the demise and any other surrounding circumstances which may colour the way in which the word is used. In the end, however, the question is whether the ordinary speaker of English would consider that the word "repair" as used in the covenant was appropriate to describe the work which has to be done. The cases do no more than illustrate specific contexts in which judges, as ordinary speakers of English, have thought that it was or was not appropriate to do so.

After considering the submissions of counsel and noting that it was agreed by counsel that the question was one of degree and, to a large extent, one of impression on which different people could reasonably give different answers, the learned judge continued:

I think one is entitled to take into account, first, as part of the context, the commercial relationship between the parties at the time of the demise. This was that the landlords had a head lease of 125 years and the tenants had been given an underlease of about 22 years. Secondly, in considering whether the work was improvement rather than repair, one must have regard to its substantiality. In this case both experts advise that whatever scheme of waterproofing is adopted, there should be a very substantial structural addition to the basement . . .
Thirdly, I think I am entitled to take into account the probable cost of the work . . . This

at the lower end of the range

is twice the likely annual market rent for the whole building with waterproof basement and over 15% of its total capital value. . . .
Taking these matters into consideration and deploying my ordinary understanding of language, I do not think it would be appropriate to describe any of the three schemes of treatment as work of repair. In my judgment, they involve structural alterations and improvements to the basement. Consequently, they do not fall within the tenants' obligations under the lease and I shall so declare.

By their notice of appeal the landlords asked, *inter alia*, that the declaration made by the learned judge be set aside and that this court should declare that on a true construction of the underlease the Post Office is liable to carry out one of the schemes of works considered in evidence. The main contentions in the notice of appeal were that, in reaching his conclusion, the learned judge had failed to apply the correct legal tests. In particular, it was said that (i) he failed properly to apply the test as to whether the work would involve giving back to the landlord a wholly different thing from that which was demised, and (ii) he failed to hold that works can constitute repair notwithstanding that they may also constitute a substantial addition or alteration or improvement if those works are the only reasonable way in which the defect can be remedied.

Upon opening the appeal in this court, Mr Spence, who did not appear below, contended primarily that the work required to be done would be work of repair to cure the failure of the kicker joint between the floor and the walls caused by poor workmanship and would not give the landlords a wholly different thing from that which they demised; and that as such it would be work to a portion, a subsidiary portion of the demised property, and not to the whole or substantially the whole. Reliance was placed upon the test stated by Forbes J in *Ravenseft Properties Ltd* v *Davstone (Holdings) Ltd* [1980] QB 12. There, in holding that the cost of certain work to a building was recoverable under the tenants' covenant to repair, Forbes J (at p 21C) said:

The true test is, as the cases show, that it is always a question of degree whether that which the tenant is being asked to do can properly be described as repair, or whether on the contrary it would involve giving back to the landlord a wholly different thing from that which he demised.

In that case under the terms of an underlease the tenants covenanted to be liable for the repairs of the demised building. The building had been constructed in concrete with an external cladding of stone. No expansion joints had been included when the building was being constructed because it had not been realised that the different coefficients in expansion of stone and concrete made it necessary to include such joints. The stones had not been tied in properly to the building, so, instead of cracking as a result of pressure as the building expanded, they bowed away from the concrete frame and there was danger of stones falling. The landlords required the tenants to carry out the necessary work, but the tenants denied that they were liable under the covenant to repair the damage caused by the inherent defect in the building of the exclusion of expansion joints. The landlords carried out the necessary work of taking down the cladding stones, retying the stones and inserting expansion joints. Forbes J gave judgment for the landlords. He rejected (p 21B-C) the contention that, if the cause of the want of reparation was an inherent defect, the tenant was thereby given a complete defence. It is to be noted that the inherent defect, ie the absence of an expansion joint, had caused damage and deterioration to the building and the making good of the inherent defect was a necessary part of the making good of the damage and deterioration suffered by the building. This authority had, of course, been relied upon by Mr Dinkin for the landlords before the learned judge.

On July 29 1985, some three days after the decision of Hoffmann J in this case, the decision of Forbes J in the *Ravenseft* case was considered by this court (Lawton, Dillon and Neill LJJ) in *Quick* v *Taff-Ely Borough Council* [1986] QB 809. In that case the plaintiff was the tenant of a house owned by the defendant council. As a result of severe condensation throughout the house, decorations, woodwork, furnishings, bedding and clothes rotted, and living conditions were appalling. The condensation was caused by lack of insulation of window lintels, single-glazed metal-frame windows and inadequate heating. The plaintiff brought proceedings in the county court, alleging that the council was in breach of its covenant, implied in the tenancy agreement by section 32(1) of the Housing Act 1961 "to keep in repair the structure and exterior" of the house and seeking an order for specific performance of the covenant. The judge held that the council was in breach of the repairing covenant in respect of, *inter alia*, the condensation and made an order requiring the council to insulate the lintels and to replace the metal-frame windows. This court allowed the appeal of the council. Liability under the covenant did not arise because of lack of amenity or inefficiency, but only when there existed a physical condition which called for repair to the structure or exterior of the dwelling-house; and, as there was no evidence to indicate any physical damage to, or want of repair in, the windows or lintels themselves or any other part of the structure and exterior, the council could not be required to carry out work to alleviate the condensation. The decision of Forbes J in *Ravenseft Properties* v *Davstone* was approved. Dillon LJ described the reasoning in the careful judgment of the judge as follows:

(1) recent authorities such as *Ravenseft Properties Ltd* v *Davstone (Holdings) Ltd* [1980] QB 12 and *Elmcroft Developments Ltd* v *Tankersley-Sawyer* (1984) 270 EG 140 show that works of repair under a repairing covenant, whether by a landlord or a tenant, may require the remedying of an inherent defect in a building; (2) the authorities also show that it is a question of degree whether works which remedy an inherent defect in a building may not be so extensive as to amount to an improvement or renewal of the whole which is beyond the concept of repair; (3) in the present case the replacement of windows and the provision of insulation for the lintels does not amount to such an improvement or renewal of the whole; (4) therefore, the replacement of the windows and provision of the insulation to alleviate an inherent defect is a repair which the council is bound to carry out under the repairing covenant.

Dillon LJ continued:

But . . . this reasoning begs the important question.
It assumes that any work to eradicate an inherent defect in a building must be a work of repair, which the relevant party is bound to carry out if, as a matter of degree, it does not amount to a renewal or improvement of the building.

Later in his judgment Dillon LJ said (p 818D):

In my judgment, the key factor in the present case is that disrepair is related to the physical condition of whatever has to be repaired, and not to questions of lack of amenity or inefficiency. I find helpful the observations of Atkin LJ in *Anstruther-Gough-Calthorpe* v *McOscar* [1924] 1 KB 716, 734 that repair "connotes the idea of making good damage so as to leave the subject so far as possible as though it had not been damaged". Where decorative repair is in question one must look for damage to the decorations but where, as here, the obligation is merely to keep the structure and exterior of the house in repair, the covenant will only come into operation where there has been damage to the structure and exterior which requires to be made good.
If there is such damage caused by an unsuspected inherent defect, then it may be necessary to cure the defect, and thus to some extent improve without wholly renewing the property as the only practicable way of making good the damage to the subject matter of the repairing covenant. That, as I read the case, was the basis of the decision in *Ravenseft*.

Lawton LJ (p 822E) said:

It follows that, on the evidence in this case, the trial judge should first have identified the parts of the exterior and structure of the house which were out of repair and then have gone on to decide whether, in order to remedy the defects, it was reasonably necessary to replace the concrete lintels over the windows, which caused "cold bridging", and the single glazed metal windows, both of which were among the causes, probably the major causes, of excessive condensation in the house.

Later (p 823B) Lawton LJ said:

There must be disrepair before any question arises as to whether it would be reasonable to remedy a design fault when doing the repair. In this case, as the trial judge found, there was no evidence that the single glazed metal windows were in any different state at the date of the trial from what they had been in when the plaintiff first became a tenant. The same could have been said of the lintels.

When asked early in the course of argument by Slade LJ whether the first question in this case was not more accurately to be stated as whether the premises were shown to have been out of repair, Mr Spence accepted that that was the correct approach. Mr Morgan told the court that the point, which is not expressly discussed in the judgment of Hoffmann J, had been argued for the Post Office before the learned judge, namely that there was no proof of disrepair or of damage or deterioration resulting from any inherent defect, and that reliance had been placed upon the dictum of Atkin LJ in *Anstruther-Gough-Calthorpe* v *McOscar* cited by Dillon LJ in *Quick's* case.

Addressing the question whether disrepair had been proved, Mr Spence relied upon the findings by the judge as to the failure of the kicker joint owing to porosity of the concrete and as to the construction joints not having been formed with care. At one stage Mr Spence suggested that the porosity of the concrete owing to honeycombing had not necessarily existed since the date of the underlease in 1969 or from any earlier date, but had "come on over the years". Upon consideration of the words used by the judge in his findings of fact, which indicate that the defects had existed since the building was constructed, and in the absence of any finding of deterioration or worsening of the condition caused by the presence of the water when the water-table rose, or from the passing of the water through the porous concrete, Mr Spence was unable to pursue the contention. It is clear to me that the defects which were found by the judge to have caused the flooding were present in the condition in which they now are from the time that the building was constructed.

Mr Spence submitted that it was not relevant to inquire whether the original defects in the concrete or construction joints had grown worse or had caused any other damage or deterioration to the demised premises. In summary form his submission was as follows: (i) at the date of the lease the premises were defective in that by reason of the porosity of the concrete and the defective construction joints ground water could enter the basement and did enter when the water-table rose; (ii) the defects existed by reason of bad workmanship and not from a decision as a matter of design not to incorporate some amenity or advantage: the defects, accordingly, constituted disrepair; (iii) to hold that such a state of a building did not constitute disrepair — and such a holding would be applicable to the terms of a similar covenant whether by landlord or tenant — would be a departure from, or an undesirable limitation of, the important principle established in *Proudfoot* v *Hart* (1890) 25 QBD 42 and never since doubted, namely that

under a contract to keep the premises in tenantable repair and leave them in tenantable repair, the obligation of the tenant, if the premises are not in tenantable repair when the tenancy begins, is to put them into, keep them in, and deliver them up in tenantable repair

per Lord Esher MR at p 50; (iv) the reference in *Quick's* case to deterioration or damage, such as the statement by Lawton LJ that "that which requires repair is in a condition worse than it was at some earlier time", or that of Dillon LJ that a covenant to repair the structure or exterior "will only come into operation where there has been damage to the structure and exterior which requires to be made good" are not to be taken as applicable to a case of this nature, and their lordships in *Quick's* case did not have such a case as this in mind. In particular, they were not dealing with a case like this where the defective part of the premises is such that it has interfered, and may again interfere, with the ordinary use and occupation of the premises contemplated by the demise and, having been caused by defective work, was "worse" than it was required to be if that part of the premises was to be regarded as in good repair.

For my part I am unable to accept the submission made for the appellant landlords by Mr Spence. The facts of this case seem to me to be, as I have said, highly unusual. I found it at first to be a startling proposition that, when an almost new office building lets ground-water into the basement so that the water is ankle deep for some years, that state of affairs is consistent with there being no condition of disrepair under a repairing covenant in standard form whether given by landlord or tenant. Nevertheless, as was pointed out in the course of argument, the landlord of such a building gives no implied warranty of fitness merely by reason of letting it; and neither a landlord nor a tenant who enters into a covenant to repair in ordinary form thereby undertakes to do work to improve the demised premises in any way. I see no escape from the conclusion that, if on the evidence the premises demised are and at all times have been in the same physical condition (so far as concerns the matters in issue) as they were when constructed, no want of repair has been proved for which the tenants could be liable under the covenant.

When the water entered by reason of the original defects damage might have been done to the premises, whether to plaster on walls, or to the flooring, or to electrical or other installations. But no such damage was proved. If such damage is done, the authorities show that the resulting state is a condition of disrepair: see *Ravenseft Properties Ltd* v *Davstone (Holdings) Ltd* [1980] QB 12 and *Elmcroft Developments Ltd* v *Tankersley-Sawyer* (1984) 270 EG 140. As to the submission that the court in *Quick's* case was not considering a defect which had been caused by defective work, I accept that such were the facts in that case — the house was built in accordance with the regulations in force and standards accepted at the time (see [1986] 1 QB 816B-C). In my judgment, however, the reasoning of the court in *Quick's* case is equally applicable whether the original defect resulted from error in design, or in workmanship, or from deliberate parsimony or any other cause. If on the letting of premises it were desired by the parties to impose on landlord or tenant an obligation to put the premises into a particular state or condition so as to be at all times fit for some stated purpose, even if it means making the premises better than they were when constructed, there would be no difficulty in finding words apt for that purpose.

Neither Mr Spence for the landlords nor Mr Morgan for the Post Office could refer the court to any reported case in which a defect, whether of design or workmanship, and present unaltered since construction of the premises, and which had caused or permitted entry of water into the premises, had nevertheless caused no damage to the premises demised. It seems to me that the unusual facts of this case are covered by the plain meaning of the word "repair" and by the decision of the court in *Quick's* case. It is not possible to hold that the wetting of the basement floor or the presence of the water upon the floor, coupled with the inconvenience caused thereby to the tenant, constitutes damage to the premises demised. There is, accordingly, no disrepair proved in this case and therefore no liability under the tenant's covenant to repair has arisen.

So to hold is not, in my judgment, to depart from, or to cast doubt upon, the principles established by the decision of the Court of Appeal in *Proudfoot* v *Hart*. Upon examination of the judgment of Lord Esher in that case, it is apparent that he was only directing his observations to premises of which the condition has deteriorated from a former better condition. Mr Spence accepted that that was so. It might be said that there is no difference in principle between, on the one hand, imposing a liability on a tenant to put premises into a better condition than they were at the date of the demise by reference to an earlier state before deterioration had occurred, and, on the other hand, imposing on a tenant, as Mr Spence asks the court to do, liability to put premises into a better condition than ever they were in by reference to a state in which they ought to have been if they are to be in some state of fitness or suitability. That, however, in my judgment is not the point. This court in this case is not laying down rules to govern the relationships and mutual responsibilities of landlords and tenants of office buildings. Our task is to construe a repairing covenant in a particular underlease. The decision may be of some general importance, despite the unusual set of facts, because the covenant so far as concerns the words upon which reliance is placed is in common form. But landlords and tenants of such premises are free to modify the repairing covenants as they think appropriate and can agree. There is no basis to be found in the decision in *Proudfoot* v *Hart* for holding that the tenant can be held liable under an ordinary repairing covenant to carry out work merely to improve premises so as to remove a defect present since construction of the building.

That conclusion, if right, is sufficient to dispose of this appeal. The questions considered by the learned judge as to whether, as a matter of degree, any of the schemes of work qualified as "repair", as contrasted with works of improvement or alteration, do not arise. These issues were, however, argued before us and I think it is appropriate to make some comments upon them. The main criticism advanced by Mr Spence was that the learned judge had failed to have due regard to, or to apply correctly, the test enunciated by Forbes J in *Ravenseft Properties Ltd* v *Davstone (Holdings) Ltd*, namely that it was a question of degree whether work carried out on a building was repair or work that so changed the character of the building as to involve giving back to the landlord "a wholly different thing" from that which he demised. It was said that if the judge had given sufficient force to the fact that the necessary work in this case was directed to a subordinate part only of the building he must have reached a different conclusion. For my part I do not accept that the learned judge is shown to have misapprehended or misapplied in this way the appropriate test. The relationship of the part upon which work is required to the whole of the subject-matter of the demise is also, in my judgment and as Slade LJ suggested in argument, a question of proportion and degree. At one end of the scale there has never been any doubt that a party liable under a covenant to repair may have to renew in effect a whole part such as a floor, or a door, or a window. An example of a substantial part is the stone cladding of the building in the *Ravenseft Property* case. I do not accept, however, that it is only open to the court to hold that work involves giving back a wholly different thing if it is possible to say that the whole subject-matter of the demise, or a whole building within the subject-matter of the demise, will by the work be made different.

Mr Morgan in a comprehensive review of the authorities, which lost nothing by its brevity of reference to them, drew attention to cases in which, applying the substance of the test propounded by Forbes J, courts have held that liability for work could not be imposed upon a party bound by the covenant although the work was required only to a part of the relevant premises: eg the bay window in *Wright* v *Lawson* (1903) 19 TLR 203; the basement walls in *Pembery* v *Lamdin* [1940] 2 All ER 434; and the roofing of the back addition in *Halliard Property Co Ltd* v *Nicholas Clarke Investments Ltd* (1983) 269 EG 1257.

The notice of appeal in this case included as a ground of appeal complaint that the learned judge had wrongly taken into account, in considering as a question of degree whether the proposed works were works of repair or of alteration and improvement, the respective lengths of the landlords' and tenants' leasehold interests and the cost of the scheme of work proposed. Upon the hearing of this appeal Mr Spence did not argue that ground. He conceded that those matters were relevant but argued that they could be of little weight. In my judgment, Mr Spence was right to concede the relevance of those matters and I accept also that in a case of this nature, assuming disrepair to have been proved, the weight that could properly be given to those factors would not be large.

As to whether upon the evidence before the judge he was right to reach the conclusion which he did upon the assumption that the premises were in a state of disrepair, I have seen in draft the judgment of Slade LJ and I, too, prefer to express no view upon that matter. For these reasons I would dismiss this appeal.

Agreeing, SLADE LJ said: I think that there is quite a short answer to the landlords' claims under present circumstances.

The only provision of the tenants' covenants contained in the Post Office's underlease on which reliance has been placed by the landlords in argument is that which obliges the tenants to "keep in good and substantial repair . . . the demised premises and every part thereof". The Post Office cannot yet be under any obligation to do any work pursuant to this covenant unless the demised premises are at present out of repair. However, a state of disrepair, in my judgment, connotes a deterioration from some previous physical condition. I would have reached this conclusion even in the absence of authority, but its correctness is shown by the decision of this court (later in time than that of Hoffmann J in the present case) in *Quick* v *Taff-Ely Borough Council* [1986] QB 809. As Lawton LJ there observed (at p 821):

As a matter of the ordinary usage of English that which requires repair is in a condition worse than it was at some earlier time.

Dillon LJ, adopting a similar train of thought, said (at p 818E):

Where decorative repair is in question one must look for damage to the decorations but where, as here, the obligation is merely to keep the structure and exterior of the house in repair, the covenant will only come into operation where there has been damage to the structure and exterior which requires to be made good.

In the present case, there was evidence at the trial that, in the original construction of the building, poor workmanship (or perhaps, in the case of the construction joints, poor design) had produced weak areas in the concrete or the kicker joint and also in some construction joints, so as to allow the ingress of flood water. However, despite the prolonged flooding in past years, there was, as Ralph Gibson LJ has pointed out, no finding (and we have been referred to no evidence) that any part of this defective building had suffered deterioration since its original construction. Thus, there was no finding at the trial that any of the building was out of repair.

In these circumstances, there would in my opinion have been no grounds upon which the learned judge could properly have held that the words of the repairing covenant quoted above imposed any present obligation on the Post Office to do work to the premises. Mr Spence on behalf of the landlords suggested that any such conclusion would conflict with the principle established by *Proudfoot* v *Hart* (1890) 25 QBD 42 that a tenant's covenant to keep premises in good repair obliges the tenant, if the premises are not in good repair when the tenancy begins, to put them into that state. However, as he accepted in the course of argument, the relevant statements of the law in that case were only directed to the case where the condition of premises has deteriorated from an earlier better condition. They were not directed, and in my judgment have no application, to a case such as the present where the structural defect complained of by the landlords has existed from the time when the premises were originally built. Though Mr Spence sought to draw a distinction in this context between structural defects due to errors in design and those due to faulty workmanship, I can see no grounds on principle or authority for drawing any such distinction.

For these reasons, albeit different from those of the learned judge, I think that he was right to conclude that the Post Office is under no present liability to the landlords by virtue of the repairing covenants and to dismiss the landlords' counterclaim. I do not reach this conclusion with regret on the very unusual facts of this case. It seems to me that, if, as in the present case, landlords let to tenants a newly built premises of which parts are defectively constructed, clear words are needed to impose a contractual obligation on the tenant to remedy the defects in the original construction, at least at a time before these have caused any damage. This is not an obligation which tenants under a commercial lease might reasonably be expected readily to undertake.

It is, however, possible that, before the underlease expires in 1991, the original defects in construction could result in actual damage to the demised premises. For example, if there was another rise in the water-table, there might be further flooding and this might cause damage to the plaster work or electrical fittings of the building. I would like to leave entirely open any questions as to the liability of the Post Office in that event.

Forbes J in *Ravenseft Properties Ltd* v *Davstone (Holdings) Ltd* [1980] QB 12 (at p 21) specifically rejected a contention that, if it can be shown that any want of reparation has been caused by an inherent defect in a building, then that want of reparation is necessarily not within the ambit of a covenant to repair. He considered that, whether or not the lack of repair has been caused by an inherent defect, "it is always a question of degree whether that which the tenant is being asked to do can properly be described as repair, or whether on the contrary it would involve giving back to the landlord a wholly different thing from that which he demised". The *Ravenseft* decision was referred to by this court in *Elmcroft Developments Ltd* v *Tankersley-Sawyer* (1984) 270 EG 140 without criticism and in *Quick* v *Taff-Ely Borough Council (supra)* with approval.

In the future contingency now under discussion, the extent of the liability of the Post Office would, in my opinion, depend on precisely what works it was being asked to do by way of so-called repair and in what circumstances. In that contingency, those works would not by any means necessarily be the same as those envisaged by any of the schemes of work now proposed. On the further evidence then available, the questions of degree involved might be different from those considered by Hoffmann J on the evidence before him. For these reasons I would prefer to express no opinion on any of these particular schemes beyond saying that *under present circumstances* none of them can properly be described as repair and the Post Office is *at present* under no obligation to carry out any of them.

I agree that this appeal should be dismissed.

SIR ROGER ORMROD agreed with both judgments and did not add anything.

The appeal was dismissed with costs.

Court of Appeal

January 30 1987

(Before Lord Justice LLOYD and Lord Justice GLIDEWELL)

DRESDEN ESTATES LTD v COLLINSON

Estates Gazette March 21 1987

281 EG 1321-1327

Landlord and tenant — Whether occupier of workshop and store a tenant or licensee — Refinement of principles formulated in *Street* v *Mountford* — Successful appeal by owners from decision of county court judge that occupier was a tenant — The relevant document throughout described itself as a licence, emphasised that exclusive occupation was not conferred and even provided that the licensees could be moved from time to time to other premises in the owners' adjoining property — There were, however, a number of provisions normally found in tenancies and the express grant of a limited right for the owners to enter for the purpose of carrying out work, which pointed to a tenancy — *Addiscombe Garden Estates Ltd* v *Crabbe*, *Shell-Mex & BP Ltd* v *Manchester Garages Ltd* and *Street* v *Mountford* considered and the passage in *Halsbury's Laws of England* approved which stated that the decisive consideration was the intention of the parties — *Street* v *Mountford* was itself a decision in regard to residential premises and the suggested restriction of the inquiry to whether the occupier is a tenant or a lodger has no application to business premises, as there is no such person as a lodger in relation to the latter — There was a conflict in the present case between provisions suggesting a tenancy and provisions suggesting a licence, but the judge had come down on the wrong side — Looking at the agreement as a whole, the indications were in favour of a licence — In particular, the provision by which the occupier could be required to move to other premises was wholly inconsistent with a right to exclusive possession during the continuance of the agreement, and consequently wholly inconsistent with a tenancy — Owners' appeal allowed — Warning against reading this decision as a way round *Street* v *Mountford*

The following cases are referred to in this report.

Addiscombe Garden Estates Ltd v *Crabbe* [1958] 1 QB 513; [1957] 3 WLR 980; [1957] 3 All ER 563, CA
Allan v *Liverpool Overseers* (1874) LR 9 QB 180
Errington v *Errington and Woods* [1952] 1 KB 290; [1952] 1 All ER 149, CA
Shell-Mex & BP Ltd v *Manchester Garages Ltd* [1971] 1 WLR 612; [1971] 1 All ER 841, CA
Street v *Mountford* [1985] AC 809; [1985] 2 WLR 877; [1985] 2 All ER 289; [1985] 1 EGLR 128; (1985) 274 EG 821, HL

This was an appeal by the plaintiffs, Dresden Estates Ltd, from a decision of Judge Kenneth Taylor, at Stoke-on-Trent County Court, holding that the defendant (the present respondent), Mr A Collinson, held premises at Sneyd Hall Works, Burslem, Stoke-on-Trent, under a tenancy and refusing the appellants an order for possession.

C J Coveney (instructed by Breton Deacon & Co, of Stoke-on-Trent) appeared on behalf of the appellants; Peter Rank (instructed by Leslie Moon & Co) represented the respondent.

Giving the first judgment at the invitation of Lloyd LJ, GLIDEWELL LJ said: This appeal from the judgment of His Honour Judge Kenneth Taylor, sitting at Stoke-on-Trent County Court on June 2 1986, raises the familiar but difficult question: does the occupier of premises occupy as a tenant or as a licensee? I am tempted to say that answering the question in some ways has been made easier and in other ways more difficult by the admirable way in which it has been argued on both sides. For myself, my mind swung back and to during the course of the argument.

The appellants, Dresden Estates Ltd, own property at Sneyd Hall Works, Sneyd Hill, Burslem, Stoke-on-Trent. We are told it was formerly a pottery. The respondent, Mr Collinson, is a builder and scaffolder. In 1985 he required accommodation in which to store his plant and equipment. He therefore entered into a written agreement dated September 6 1985 with Dresden Estates Ltd to which I must refer in some detail.

It commences with the words "Licence dated 6th September 1985" and throughout the agreement is referred to as a "licence". It begins by defining a number of terms. It identifies the parties, Dresden Estates Ltd and Mr Collinson. It describes the premises as:

Ground Floor Unit at Sneyd Hall Works, Sneyd Hill, Burslem, Stoke-on-Trent, Staffs, as shown on attached Plan and subject to a right of way to First Floor

It defines the use to which the property was to be put as a workshop and store, and the commencement date of the licence. Then it defines the licence fee as:

£200 per calendar month payable monthly in advance. The first payment of the Licence Fee to be the sum of £154 in respect of the period from the Commencement Date to 30th September, 1985 and thereafter on the first day of each succeeding month by Bankers Order.

It defines "the Required Notice", and this is a matter upon which there has been some discussion, as:

Not less than 3 months notice in writing to be served before the first day of any month.

There is then a reference to a deposit. It then provides:

The Licensors hereby grant to the Licensee a licence to use and occupy the Premises from the Commencement Date upon the terms and conditions set out in the Terms and Conditions.

Those terms and conditions are upon the next sheet of the document, and they start with an agreement by Mr Collinson — who is throughout described as "the Licensee" and Dresden Estates are described throughout as "the Licensors" — ". . . to pay the Licence Fee on the dates and in manner specified".

Clause 2 then contains, if this is a tenancy, what are effectively covenants by the tenant, but are described as "agreements" by the licensee, to pay the general and water rates, to pay charges in respect of gas and electricity, and

(c) To permit the Licensors with necessary workmen and contractors and equipment to enter on the Premises to carry out any work deemed necessary by the Licensors to the premises or to adjoining premises or services.

And there follow another six subclauses of a kind familiarly found in tenancy agreements relating to not carrying on noisy, offensive or dangerous trades, not making alterations to the premises, not doing anything that would affect the owners' insurance policy, an agreement "to keep the interior of the Premises and the decorations and fittings of the Premises in good repair and condition . . . ", and two others of a somewhat similar type.

Clause 3 grants to Mr Collinson the right to use shared facilities such as delivery areas and parking areas. Clause 4 then contains important provisions, to most of which I must refer:

It is agreed between the Licensors and the Licensees as follows:
(a) this Licence is personal to the Licensees and the licensees shall not transfer this interest in the same in any manner whatsoever.
(b) this Licence confers no exclusive right for the Licensees to use and occupy the Premises and the Licensors shall be entitled from time to time on giving the Required Notice to require the Licensees to transfer this occupation to other premises within the Licensor's adjoining property.
(c) this Licence does not constitute any tenancy or lease of the Premises.

(d) concerns the deposit and I do not need to read it.

(e) this Licence may be determined by either party giving to the other the Required Notice to terminate on a stated date and may be determined forthwith by the Licensors if the Licensees are in breach of any of its obligations set out in this Licence.
(f) the Licensors may by giving the Required Notice to the Licensees increase the Licence fee to such amount as may be specified in such notice.

I do not need to read (g).

Mr Collinson having entered into occupation of the premises under that agreement within a very short time, namely on December 20 1985, Dresden Estates Ltd purported to give him notice to terminate the agreement under clause 4(e), and that notice was expressed to terminate on March 31 1986. The notice itself, I should say, is not among the documents put before us or presumably before the learned judge, but those facts are apparently accepted on both sides.

Mr Collinson claims that the agreement created a tenancy and that that tenancy is subject to the provisions of the Landlord and Tenant Act 1954. If, as the learned judge has found, that contention is correct, he has the protection of that Act. The appellants, Dresden Estates Ltd, contend that the agreement created a mere licence terminable in accordance with its terms, which, therefore, came to an end on March 31 1986. That is the issue which has to be determined on this appeal.

As a matter of fact, I understand that Mr Collinson is still in occupation of the premises, or certainly was at the time of the hearing before the learned judge, and that he has paid the rent or licence fee, whichever it is, up to date.

We were referred to three authorities which are relevant to this issue among the very considerable number of authorities which bear upon this or similar questions.

In *Addiscombe Garden Estates Ltd* v *Crabbe* [1958] 1 QB 513 this court was concerned with a tennis club which occupied tennis courts and a pavilion under an agreement described as "A Licence". The agreement was for a term certain of two years. It contained what, if it were a tenancy, was a covenant for quiet enjoyment land a repairing covenant. This court held that it was a tenancy. The passage upon which Mr Rank particularly relies in that authority is in the judgment of Jenkins LJ at p 522 at the very top of the page:

As to the first question — whether the so-called licence of April 12 1954 in fact amounted to a tenancy agreement under which the premises were let to the trustees — the principles applicable in resolving a question of this sort are, I apprehend, these. It does not necessarily follow that a document described as a licence is, merely on that account, to be regarded as amounting only to a licence in law. The whole of the document must be looked at; and if, after it has been examined, the right conclusion appears to be that, whatever label may have been attached to it, it in fact conferred and imposed on the grantee in substance the rights and obligations of a tenant, and on the grantor in substance the rights and obligations of a landlord, then it must be given the appropriate effect, that is to say, it must be treated as a tenancy agreement as distinct from a mere licence.

I note that Jenkins LJ did not say that the description of the document in the document itself as "a licence" is irrelevant: nor did he say that no account was to be taken of it. What he did say, in effect, was that it was not conclusive.

The next authority in point of time was also a decision of this court, in *Shell-Mex & BP Ltd* v *Manchester Garages Ltd* [1971] 1 WLR 612. The petrol company made an agreement described as "a licence" with Manchester Garages Ltd under which the latter company occupied a petrol filling station. The agreement contained a specific agreement by Manchester Garages Ltd not to impede Shell-Mex & BP Ltd in their right of possession and control. It was held that the agreement did what it said it did, that is to say, it constituted "a licence". At p 615 Lord Denning MR, giving a judgment with which Sachs and Buckley LJJ agreed, said, at D:

I turn, therefore, to the point: was this transaction a licence or a tenancy? This does not depend on the label which is put on it. It depends on the nature of the transaction itself: see *Addiscombe Garden Estates Ltd* v *Crabbe* [1958] 1 QB 513. Broadly speaking, we have to see whether it is a personal privilege given to a person (in which case it is a licence), or whether it grants an interest in land (in which case it is a tenancy). At one time it used to be thought that exclusive possession was a decisive factor. But that is not so. It depends on broader considerations altogether. Primarily on whether it is personal in its nature or not: see *Errington* v *Errington and Woods* [1952] 1 KB 290.

Then he turned to the facts of that particular case. Over the page at p 616 at F he said:

It seems to me that when the parties are making arrangements for a filling station, they can agree either on a licence or a tenancy. If they agree on a licence, it is easy enough for their agreement to be put into writing, in which case the licensee has no protection under the Landlord and Tenant Act 1954. But, if they agree upon a tenancy and so express it, he is protected. I realise that this means that the parties can, by agreement on a licence, get out of the Act. But so be it. It may be no bad thing. Especially as I see that the parties can now, with the authority of the court, contract out of the Act, even in regard to tenancies: see section 5 of the Law of Property Act 1969.

Finally, the recent decision of the House of Lords in *Street* v *Mountford* [1985] AC 809 was much cited to us. That was a case concerning the occupation of residential as opposed to business premises. Mrs Mountford occupied two furnished rooms as her residence under an agreement, and it was held, all the members of the committee of the House agreeing, that the agreement constituted a tenancy, although it was described as a licence. The leading speech was that of Lord Templeman. He said, at p 817H:

In the case of residential accommodation there is no difficulty in deciding whether the grant confers exclusive possession. An occupier of residential accommodation at a rent for a term is either a lodger or a tenant. The occupier is a lodger if the landlord provides attendance or services which require the landlord or his servants to exercise unrestricted access to and use of the premises. A lodger is entitled to live in the premises but cannot call a place his own.

Then he quoted a passage from the judgment of Blackburn J in *Allan* v *Liverpool Overseers* (1874) LR 9 QB 180. At p 824B he came to consider *Shell-Mex & BP Ltd* v *Manchester Garages Ltd*. Having referred to that case and quoted the second of the passages from the judgment of Lord Denning MR which I have quoted, Lord Templeman said this, at p 824E:

In my opinion, the agreement was only "personal in its nature" and created "a personal privilege" if the agreement did not confer the right to exclusive possession of the filling station. No other test for distinguishing between a contractual tenancy and a contractual licence appears to be understandable or workable.

I make two comments about that last authority. First, both counsel are agreed, and for my part I agree, that in the circumstances of this case, although one has to look at all the circumstances, certainly the most important factor is that of exclusive possession. Did the agreement give Mr Collinson an exclusive right to the possession of the premises which he occupied? Second, *Street* v *Mountford*, as I have said, was concerned with residential premises. Mr Coven-

conceded that there was no material difference, at least for present purposes, between the law applicable to residential premises and the law applicable to business premises. As a broad, general proposition that may be right, but I am not sure that his concession may not have gone too far in this respect, that the attributes of residential premises and business premises are often quite different.

The passage that I have already quoted from the speech of Lord Templeman, where he says in effect that all you have to decide in relation to residential premises is whether the occupier is a tenant or a lodger, is, of course, of itself not applicable to business premises because there is no such person as a lodger in relation to business premises. For myself, I think that the indicia, which may make it more apparent in the case of a residential tenant or a residential occupier that he is indeed a tenant, may be less applicable or be less likely to have that effect in the case of some business tenancies.

To my mind, the law generally is accurately summarised in *Halsbury's Laws of England,* vol 27 (4th ed) at para 6, which reads:

In determining whether an agreement creates between the parties the relationship of landlord and tenant or merely that of licensor and licensee the decisive consideration is the intention of the parties. The parties to an agreement cannot, however, turn a lease into a licence merely by stating that the document is to be deemed a licence or describing it as such. The parties' relationship is determined by law on a consideration of all relevant provisions of the agreement, and an agreement labelled by the parties to it as a "licence" will still be held to create a tenancy if the substance of the agreement conflicts with that label. Similarly, the use of operative words ("let", "lessor" etc) which are appropriate to a lease will not prevent the agreement from conferring only a licence if from the whole document it appears that it was intended merely to confer a licence. Primarily the court is concerned to see whether the parties to the agreement intend to create an arrangement personal to its nature or not, so that the assignability of the grantee's interest, the nature of the land and the grantor's capacity to grant a lease will all be relevant considerations in assessing what is the nature of the interest created by the transaction.

The first sentence of the next paragraph is:

The fact that the agreement grants a right of exclusive possession is not in itself conclusive evidence of the existence of a tenancy, but it is a consideration of the first importance, although of lesser significance than the intention of the parties.

I return to apply the law so stated to the terms of this agreement. To my mind, clause 2 of the agreement, which I have already said, if this is a tenancy, really contains tenant's covenants, contains a number of provisions which are wholly appropriate to and some of which are certainly indicative of this agreement creating a tenancy. In particular, as counsel pointed out, the agreement by which Mr Collinson permitted Dresden Estates Ltd to enter with workmen and contractors to carry out any work on the premises or adjoining premises is one which would not be necessary if this agreement constituted a mere licence not reserving exclusive possession to Mr Collinson. But the provisions of that clause seem to me to be in conflict with many of the provisions of clause 4, and the real difficulty in this case, and the difficulty that confronted the learned judge, is to resolve that conflict. To my mind, the opening words of clause 4:

This licence is personal to the Licensees and the Licensees shall not transfer this interest in the same in any manner whatsoever

cannot be disregarded and are of importance.

What is even more important is to decide what clause 4(b) and also clause 4(f) mean. Clause 4(b) is the clause that starts by saying in terms:

This Licence confers no exclusive right for the Licensees to use and occupy the Premises.

It then goes on to give Dresden Estates Ltd the right

... from time to time on giving the Required Notice to require Mr Collinson

to transfer this occupation to other premises within

Dresden's

adjoining property.

Clause 4(f) entitles Dresden Estates Ltd by giving the required notice to increase the licence fee to such amount as the notice may specify. Both those clauses, if they have their apparent meaning, are inconsistent with there being a tenancy. You cannot have a tenancy granting exclusive possession of particular premises subject to a provision that the landlord can require the tenant to move to somewhere else. The landlord can do that only by terminating the tenancy and creating a new one in other premises. So, too, with regard to the rent and licence fee. It is axiomatic that unless there is a rent review clause a landlord cannot for the duration of the tenancy alter the rent unilaterally. All he can do is to terminate the tenancy and then enter into a new agreement for the letting of the same premises at a new rent. Of course, the whole thing can be done by agreement. The tenancy agreement itself cannot give a landlord the power to alter a rent unilaterally.

Mr Rank says that this agreement, properly read, does not. All that it really does is to say that the landlord (Dresden Estates Ltd) can, if it wishes to require Mr Collinson to move to some other part of the total premises or if it wishes to raise rent, give notice to terminate. He claims that the phrase the "Required Notice" in the definition clause of the agreement means "notice to terminate".

Mr Coveney argues, I think correctly, that that is not so. This agreement carefully defines the phrase "the Required Notice" simply in relation to the length of notice —

Not less than 3 months' notice in writing to be served before the first day of any month.

It then uses the phrase in relation to three different concepts: giving notice to terminate the agreement; giving notice to move out of the particular premises and move to other accommodation; and giving notice to increase the rent. Mr Coveney says (and I conclude that he is right in this) that this agreement entitles Dresden Estates Ltd to take either of those other steps unilaterally, the remedy which Mr Collinson had, if he did not like it, being himself then to serve three months' notice to terminate, as the agreement provided that he could.

If, as I believe, that is right, then clauses 4(b) and 4(f) militate very strongly against this agreement creating a tenancy. I should say that there is some difficulty in Mr Coveney's way, because it is difficult to see how clause 4(b) could allow Dresden Estates Ltd to require Mr Collinson to move to some other part of the wider premises which was wholly unsuitable for him. Mr Coveney says, and I think he is right, that clause 4(b) must be read as relating to alternative premises that were reasonably comparable and reasonably suitable. But that comment apart, those two clauses do seem to me to permit what Dresden Estates Ltd wish to achieve to take place within the context of the agreement and without terminating. If that is so, Mr Coveney argues that it means that there is strong evidence that Mr Collinson did not have exclusive possession of these premises during the period of the agreement, because he could be required to go out of these premises while the agreement subsisted and go into other premises.

That argument, I think, is right; and though it does conflict with clause 2, I think on balance the considerations set out in clause 4, added to the express terms of the agreement which refer to it time and again as "a licence", outweigh the considerations based upon clause 2.

In the note of his judgment that we have, and I accept that it is an abbreviated note, the learned judge is recorded as saying:

As to clause 4(b), one must look at what it says. It refers to the right of the Plaintiff to move the Defendant to alternative accommodation in the same building if the "Required Notice" is given. The Required Notice is defined in the agreement as 3 months.

Clause 4(b) has no meaning which is inconsistent with the grant of a tenancy.

That last sentence, with the greatest respect to the learned judge, can only be right if clause 4(b) is interpreted as doing no more than giving the right to Dresden Estates Ltd to terminate by three months' notice. If clause 4(b) is interpreted in the sense in which I interpret it, then clause 4(b) does have a meaning which is inconsistent with the right of grant of a tenancy.

For these reasons, I would allow the appeal and hold that the agreement was effective to constitute a licence and, thus, that Mr Collinson does not have and did not have the benefit of the Landlord and Tenant Act 1954.

I make one last observation. In my experience, this is an unusual form of agreement. That may be because these are somewhat unusual premises. If it be right that the owners are seeking to put to good use an old pottery, which is no longer needed for its former use, there are no doubt practical difficulties in achieving that objective and it may be (I know not) that the agreement was tailored to achieve that. But I want to make it clear that, for my part, my decision is based upon the particular facts of this agreement; it is not intended to be read and should not be read as laying down any guidelines for the future going outside agreements containing these unusual provisions. Subject to that, as I say, I would allow the appeal.

Agreeing, LLOYD LJ said: Clause 2 of the agreement looks like an agreement for a tenancy, despite the numerous references to "Licence", "Licensor" and "Licensee". In particular, the grant of a limited right to the licensor to enter for the purpose of carrying out work is consistent only with a tenancy. The grant of an *unlimited* right of entry would be consistent with a licence, even superfluous. But the grant of a *limited* right would seem to have no place at all in a licence.

But when one comes to clause 4 the agreement wears a different complexion. Clause 4(b) confers on the licensor the right to require the licensee to transfer to other premises. Mr Coveney argues that that is a right which the licensor can exercise during the continuance of the licence. It is a right which is wholly inconsistent with a right to exclusive possession during the continuance of the agreement and is therefore wholly inconsistent with a tenancy.

Mr Rank meets that argument in this way. He submits that a notice under clause 4(b) must, in reality and of necessity, bring the old agreement to an end. There is no way in which the existing agreement can be made to apply to the new premises. The parties must, therefore, have contemplated that, on the exercise of the right under 4(b), they would enter into a new agreement. In support of that argument, Mr Rank says it is significant that the period of notice required under clause 4(b), namely three months, is the same as the period of notice to terminate the agreement under clause 4(e).

If that argument of Mr Rank's be correct, then it would undermine the basis of Mr Coveney's argument that the licensor can require the licensee to move to other premises during the continuance of the agreement.

Mr Coveney replies by drawing attention to the words of clause 4(b), "this occupation". He suggests that that means that the agreement was indeed to be capable of continuing after the transfer to new premises. No doubt the parties could have reached the same result by giving notice to terminate and then entering into a new agreement. But that is not what clause 4(b) contemplates.

I was nearly persuaded by Mr Rank's submission, and, like Glidewell LJ, I would pay tribute to his excellent argument, as indeed to the arguments on both sides. But Mr Rank's construction of the agreement does not, in my judgment, do full justice to clause 4(b). That means that clause 2 points in one direction and clause 4 points in another. Looking at the agreement as a whole, I agree with Glidewell LJ that clause 4 must prevail. If that be right, then the appeal must be allowed.

I would only add, like Glidewell LJ, that our decision today should not be regarded as providing a way round the decision of the House of Lords in *Street* v *Mountford* [1985] AC 809. It will be in only a limited class of case that a provision such as is found in clause 4(b) would be appropriate. If it is included in an agreement where it is not appropriate, then it will not carry the day.

The appeal was allowed, possession being ordered within 28 days.

Court of Appeal

December 18 1986

(Before Lord Justice FOX, Lord Justice STEPHEN BROWN and Lord Justice PARKER)

CELSTEEL LTD AND OTHERS v ALTON HOUSE HOLDINGS LTD AND ANOTHER (NO 2)

Estates Gazette March 28 1987

281 EG 1446-1448

Landlord and tenant — Covenant for quiet enjoyment — Express covenant of qualified kind giving protection against "any interruption by the landlord or any person lawfully claiming through under or in trust for the landlord" — Interpretation of claiming "under" — Predecessors in title of present freeholders granted a number of tenancies of flats and garages with rights of way over driveways and parking areas — Present freeholders, after acquiring the reversion, granted a long lease of part of the land to an oil company which established a petrol filling station and proposed to set up a car-wash — The effect was that the demise to the oil company included land over which the tenants of the flats and garages enjoyed and were currently exercising rights of way, which would be interfered with by the car-wash — These tenants in previous proceedings before Scott J obtained an injunction restraining the construction of the car-wash — The oil company then contended that the injunction constituted a breach of the covenant for quiet enjoyment contained in the lease to the company by the freeholders — In the proceedings giving rise to the present appeal Scott J rejected this contention, holding that the tenants of the flats and garages did not "claim under" the freeholders — On appeal by the oil company the Court of Appeal held that Scott J had reached the right conclusion — Although the tenants might well be properly described as "holding under" the freeholders, they did not "claim under" them in the context of the covenant — "Claim" in this context meant claiming a lawful right to interrupt the oil company's occupation of the demised property — The tenants did not claim this right under the freeholders; the right was derived from the freeholders' predecessors in title — It was created by a title paramount to that of the freeholders — Textbooks cited and Law Commission's recommendations noted — Appeal dismissed

The following case is referred to in this report.

Griffiths v *Riggs* (1917) 61 SJ 268

This was the second instalment of litigation concerning Cavendish House, 21 Wellington Road, London NW8. The first instalment ended in a judgment given by Scott J in November 1984, reported in [1985] 1 WLR 204. The judgment from which the present appeal was brought was reported at [1986] 1 WLR 666. The appellants were Mobil Oil Ltd, the second defendants in the action *Celsteel Ltd and Others* v *Alton House Holdings Ltd and Another*, and the respondents were Alton House Holdings Ltd, the first defendants in the action. Third parties, a firm of solicitors, against whom Alton House Holdings Ltd had commenced third party proceedings, were represented, although the issue in such proceedings was not before the judge in the present case.

Michael Barnes QC and Edward Davidson (instructed by Metson Cross & Co) appeared on behalf of the appellants; Dennis Levy QC and George Laurence (instructed by Crellins, of Walton-on-Thames, Surrey) represented the respondents; A G Steinfeld (instructed by Reynolds Porter Chamberlain) represented the third parties.

Giving judgment, FOX LJ said: This is an appeal from a decision of Scott J, and it raises a question of construction on a covenant for quiet enjoyment in a lease.

The essential facts are as follows. Calflane Ltd was the owner of the freehold of a property called Cavendish House in St John's Wood. The site was acquired by Calflane in the 1970s. By 1981 the site was developed as a block of flats and, in that year, Calflane leased the flats, together with garages, to various persons. The leases contained grants of rights of way to the tenants over "the drives" and certain other areas of the property.

On March 1 1982 Calflane conveyed the freehold of Cavendish House to Alton House Holdings Ltd subject to the leases — all of which Scott J found had then been granted.

On October 27 1982 Alton House granted to Mobil Oil Company Ltd, for the term of 99 years, a lease of part of the property. The effect of this lease was that Alton House demised to Mobil part of the driveway over which the tenants of the flats and garages enjoyed, and were then currently exercising, rights of way.

Mobil proposed to build a car-wash on the land demised to it. The tenants of the flats and garages then commenced this action to restrain Mobil from building the car-wash which, they said, would interfere with their rights of way.

That issue was tried before Scott J and, on June 27 1984, he gave judgment for the tenants, and granted injunctions against Mobil and Alton House restraining the construction of the car-wash. That judgment (which was not appealed from) is reported at [1985] 1 WLR 204. Mobil contended that the injunction constituted a breach of the covenant for quiet enjoyment contained in the lease by Alton House to Mobil. Consequently, Mobil claimed an indemnity against the costs of the action, and damages for the breach of covenant. It was upon that claim that Scott J gave the judgment which is now appealed from. The judgment contains a more detailed account of the background facts than I have given.

I should mention that, upon the agreement for the lease to Mobil, the material parts of the leases to the tenants were disclosed, but it does not seem to have occurred to either party that the rights of the tenants would give rise to the problems which arose.

I come now to the covenant for quiet enjoyment in the lease to Mobil. It was in these terms:

The Landlord hereby covenants with the Tenant that . . . the Tenant shall peaceably hold the demised premises for the term hereby granted without any interruption by the Landlord or any person lawfully claiming through under or in trust for the Landlord.

Mobil contend that the exercise by the flat and garage tenants of their lawful rights of way over part of the premises leased to Mobil so as to prevent Mobil from building the car-wash constitutes an interference with Mobil's enjoyment of the premises demised to it. It is Mobil's case that the tenants are persons claiming "under" Alton House, and that Alton House are accordingly liable on the covenant.

It is accepted by Mobil:
(a) that the express covenant for quiet enjoyment excludes any implied covenant for quiet enjoyment, and
(b) that the tenants do not claim "through" or "in trust for" Alton House.

The case, therefore, turns on the ambit of the words "claiming under".

Scott J rejected Mobil's claim and held that the tenants of the flats and garages did not claim "under" Alton House. Mobil, appealing against that decision, accepts that covenants for title are commonly qualified to exclude liability for acts of predecessors in title and that the covenant in this case is qualified. Mobil contends, however, that when an assignor of a reversion upon a term warrants to an assignee freedom of interruption by the assignor's lessees, he is not taking on liability for acts of his predecessors in title, but merely of his own immediate tenants whose rights will be known to him, and that this is so whether the leases were granted by the assignor himself or by a predecessor.

In my opinion Scott J came to the right conclusion. It may well be that the tenants can properly be described as "holding" under Alton House, but I do not think that they "claim" under Alton House in the context of this covenant. The covenant warrants that Mobil shall hold the premises ". . . without interruption by . . . any person lawfully claiming . . . under . . . (Alton House)". This language, it seems to me, raises the question: "claiming what?"

The answer to that, I think, is: claiming a lawful right to interrupt the lessee's occupation of the demised property. That indeed is what the tenants of the flats and garages did claim — and successfully so, as was determined at the 1984 hearing.

The next question is: did the tenants claim that right "under" Alton House? In my judgment they did not. In asserting and establishing the right, the tenants did not need to refer to Alton House, or any act or disposition of Alton House, at all. They derived their right from Calflane, and it would be from and through Calflane alone that they would deduce the title which gave them the right to claim the injunctions. Alton House formed no part of that title and it never conferred upon the tenants any rights in respect of the premises at all. The rights were created by virtue of a title paramount to Alton House which, at all times, had effect in priority to Alton House's title and was superior to it. In those circumstances, I do not think it would be a proper use of English to say that the tenants of the flats and garages claimed "under" Alton House. Their relevant rights were wholly independent of anything ever done by Alton House.

We were referred to certain paragraphs in the leading textbooks, which were as follows:

Woodfall, 28th ed, 1978, vol 1, para 1-1289:

The wrongful acts of a tenant of the lessor, under a previous lease, who does things not authorised by such lease, do not amount to a breach of the usual qualified covenant for quiet enjoyment, and, where a subsidence was caused by the lessees under a lease of mineral rights, a lessor on a subsequent lease of the surface, who was not a party to the mineral lease, will not be liable under the covenant for quiet enjoyment in the surface lease, as the damage was not caused by a person "claiming from or under" him.

To that statement there is a footnote citing *Re Griffiths, Griffiths v Riggs* (1917) 61 SJ 268, as authority. I will deal with that case later.

Hill and Redman, 17th ed, 1982, vol 1, p 196:

The restricted form

which corresponds to that in this case

is commonly adopted, and under it the lessor, whether or not his own title is defective, is not liable for acts of persons claiming by title paramount even though those acts are the consequence of his own default.

Hill and Redman further states, at p 197, that

in view of the limited effort of the restricted covenant, it must be considered unsatisfactory from the tenant's point of view and where it is desired to give him the fullest protection it should be extended so as to apply to the acts of persons rightfully claiming by title paramount.

Halsbury's Laws, 4th ed, 1981, vol 27, p 253, para 323:

Under the heading *"Effect of Qualified Covenant"*, it is stated:

. . . the covenant only protects against the acts of persons claiming under the landlord so far as they are successors in title of the landlord, or actually have authority from him to do the acts

These statements are, at any rate, not inconsistent with the view which I have expressed above as to the effect of the words "claiming under" the landlord.

Foa's General Law of Landlord and Tenant, 8th ed, 1957, pp 299 to 300:

Lessor and persons claiming under the lessor —

The covenant is usually expressed to extend to the acts of the lessor and persons claiming "by, from or under" him. A person claiming under a settlement made by the lessor, . . . is within such words; and so is a prior lessee of the premises demised or a lessee of adjacent premises holding under the same lessor. Where a lease granted by two trustees of a will contained a covenant, as to their own respective acts only, that there should be no lawful interruption by any person rightfully claiming "from or under" them, it was held that disturbing acts of a company, to whom a previous lease of the underlying minerals had been granted by earlier trustees of the same will (only one of whom was a party to the later lease), were an interruption at the instance of a person rightfully claiming, by virtue of the devolution of the reversion "under" them . . . and that the trustee common to both leases, and he alone, was liable.

Re Griffiths, Griffiths v Riggs, to which I have already referred, is cited as authority for the example about the disturbance by the mineral lessees.

Mobil place some reliance upon the *Griffiths* case, and I should refer to it in more detail. It was a summons to determine whether the plaintiff trustees were liable for damage caused by colliery workings. Griffiths and Martin, who were the trustees of the will of one Webb, demised to the defendant for a term of 99 years from 1904 a piece of land on which a house was being built; the defendant contracted to complete the house. Minerals were excepted from the lease. There was, in the lease, a covenant that the lessee should quietly enjoy the premises "without any lawful interruption from or by the lessors or any person rightfully claiming from or under them".

The mines underlying the premises had been demised in 1896 by the then trustees, who were Griffiths and Shenton, to a colliery company for a 60-year term. The house being completed, it suffered damage by subsidence from the mineral workings. The question was whether the colliery company (the lessee under the mineral lease) claimed "under" the plaintiffs. The report is very short.

Younger J said he would have been glad to read the covenant for quiet enjoyment as protecting the lessee by anybody claiming or holding under the present or any previous trustee. That is understandable, since the trust owned the freehold at all times and the changes in the legal title to the reversions were simply consequent on changes of trustees.

The judge's actual decision, however, was that only Griffiths was liable. Griffiths was a party to the mineral lease as well as the surface lease. Therefore, as the judge said, the lawful interruption was due to Griffiths' own act as a joint grantor of the mineral lease. That seems correct. But if Mobil's case on this appeal is correct, it is not clear why Martin was not liable as well. The judge's reason for making Griffiths liable seems to have been simply that he was a party to the grant of the mineral lease. In view, however, of the brevity of the report, I do not feel that much help can be obtained from the case, at any rate in relation to the question of the liability of Martin.

In support of its case, Mobil puts the following example as demonstrating the practical objections to Alton House's contentions:

(1) L demises parcel B to B and parcel C to C.
(2) In the demise of parcel B there is a grant of rights to B over a part of parcel C.
(3) The exercise of B's rights would constitute a breach of covenant for quiet enjoyment in the present form in the lease

to C. It is agreed that C could sue A for that breach.

(4) Suppose, however, that A assigns the reversion in parcels B and C to X and that C then assigns his lease of parcel C to Y. What then would be the remedy of Y if his enjoyment of parcel C were interrupted by B's exercise of his rights over parcel C?

Mr Barnes, for Mobil, says that if Alton House is right, Y could not sue X; B, though holding under X, would not be "claiming under" X for the purposes of the covenant. Nor could Y sue A, since there was neither privity of estate nor privity of contract between them. Thus Y is without remedy.

This suggested anomaly was considered by Scott J on pp 9 to 11 of his judgment, and he was not satisfied that it represented the true state of the law. In particular, he was of opinion that section 142(1) of the Law of Property Act 1925 was wide enough to enable an action to have been brought by Y against A, and if that were so, he was not satisfied the cause of action could not be asserted against X, the successor in title to A. I take the same view for the same reasons as Scott J. It is not, however, necessary to decide the matter since, even if the anomaly exists, it does not alter my view of the construction of what is undoubtedly a limited covenant in the present case.

As regards policy, the Law Commission's Report (No 67) of June 1975 on *Obligations of Landlords and Tenants* states, in para 37:

The obligation

under the common form of covenant for quiet enjoyment or the implied covenant which is similarly qualified

is also limited to the lawful acts of people claiming through or under the landlord; it does not extend to lawful acts of anyone with a title better than the landlord's own title. It was originally uncertain whether the implied covenant extended also to interruption or disturbance of the tenant by people having title paramount to that of the landlord . . . but it is now settled that the obligation is limited to the acts of people claiming under the landlord.

The commission (in para 43) was of opinion that the present effect of covenants for quiet enjoyment, whether implied or in the common-form express covenant, gives inadequate protection to the tenant.

In para 51 (b) the commission recommends:

A landlord's responsibility ought not to be limited to the acts of people who derive their rights from him; it should extend to the lawful acts of anyone, whether the justification for the disturbance depends on a title superior to the landlord's or on a title created out of the landlord's title.

Limitations upon the landlord's liability were, however, proposed. In para 54 it is stated:

If he

a landlord

does disclose everything that he can of his own title and everything that he knows of which might interrupt or disturb the tenant's occupation, we do not think that he should be liable to the tenant for risks he cannot know about. The obligation arises out of a covenant for quiet enjoyment and we are not proposing an absolute covenant for title. This qualification of the landlord's ignorance would result in a clear distinction between a case in which the person disturbing the tenant derived his right from the landlord himself and one in which the right was superior to the landlord's own title. This defence could only exist in the latter case; if the right were derived from the landlord, the landlord could not claim that he did not know of it.

Clause 5 (2) of the draft Bill in the report deals with the matter.

In the present case the judge, after hearing oral evidence, found that "everything Alton House actually knew had been disclosed to Mobil". The parties evidently misapprehended the effect of the documents in relation to the state of the land. It seems, therefore, that even under the Law Commission's recommendation, Mobil would probably be without a remedy against Alton House. I mention these matters only for completeness. They do not affect my view of the law as it stands in relation to the covenant in the present case.

For the reasons which I have given, I take the view that Scott J was right in his conclusion. I would dismiss the appeal.

PARKER and STEPHEN BROWN LJJ agreed and did not add anything.

The appeal was dismissed with costs.

Court of Appeal

February 12 1987

(Before Lord Justice O'CONNOR and Lord Justice NICHOLLS)

BROOKER SETTLED ESTATES LTD v AYERS

Estates Gazette April 18 1987

282 EG 325-326

Landlord and tenant — Whether occupier a tenant or licensee — Whether occupier had exclusive possession — Further analysis of Lord Templeman's speech in *Street* v *Mountford* — Appeal by owners of flat from decision of county court judge who held that occupier was a tenant — She occupied a double bedsitting-room in the flat, which consisted of this room, two single bedsitting-rooms, a kitchen-dining room and a bathroom with lavatory — The two other rooms were each occupied by another woman — She entered under an oral agreement, the exact terms of which were not clearly found by the county court judge, and subsequently signed a written agreement described as a licence, there being a question as to whether the terms of this agreement were imported into the oral agreement — The written agreement purported to license each occupier of a room to occupy the entire flat and repeatedly stated that no one had exclusive possession of anything — It also reserved a right to put another person in the respondent occupier's room — In deciding that the occupier was a tenant the county court judge purported to follow the principles explained by Lord Templeman — In particular the judge cited the passage from Lord Templeman's speech which said "An occupier of residential accommodation at a rent for a term is either a lodger or a tenant. The occupier is a lodger if the landlord provides attendance or services which require the landlord or his servants to exercise unrestricted access to and use of the premises . . . If on the other hand residential accommodation is granted for a term at a rent with exclusive possession, the landlord providing neither attendance nor services, the grant is a tenancy" — The judge held that in the present case there was no evidence that attendance or services were provided and that by the above definition the occupier must have had exclusive use of the room — "She was not a lodger, *ergo* she was a tenant" — Held by the Court of Appeal that the judge had fallen into error, understandably, "because of the succinct way in which Lord Templeman in his speech has attempted to try to simplify this situation" — In *Street* v *Mountford* it was conceded that Mrs Mountford was entitled to exclusive possession and was not a lodger, but here that crucial matter was not conceded — It is a question based on the facts in each case whether the occupier has exclusive possession — The judge's error in this case was to say "She is not a lodger. Therefore she must have exclusive possession" — Unfortunately there was not enough material for the Court of Appeal itself to come to a conclusion on the question of exclusive possession — Appeal allowed and new trial ordered

The following cases are referred to in this report.

Radaich v *Smith* [1959] 101 CLR 209
Street v *Mountford* [1985] AC 809; [1985] 2 WLR 877; [1985] 2 All ER 289; [1985] 1 EGLR 128; (1985) 274 EG 821, HL

This was an appeal by Brooker Settled Estates Ltd, owners of a furnished flat at 18 Agnes Road, Acton, London W3, from a decision of Judge Birks in Brentford County Court in favour of Miss Bridget Louise Ayers, the occupier of double bedsitting-room in the flat, in proceedings brought by the appellants as plaintiffs for possession. The judge upheld a counterclaim by Miss Ayers, the defendant, for a declaration that she was in occupation under an oral agreement as a tenant protected by the Rent Act 1977.

Robert Thoresby (instructed by Fairchild Greig & Co, of Acton, London W3) appeared on behalf of the appellants; the respondent did not appear and was not represented, having decided to leave the flat and not to take any part in the appeal.

Giving judgment, O'CONNOR LJ said: The appellant company is the landlord of a furnished flat on the first floor of 18 Agnes Road, Acton. The flat has its own entrance from the street and consists of one double bedsitting-room, and two singles: a kitchen/dining-room and a bathroom with a lavatory.

The respondent occupied the double bedsitting-room in the flat. The other two rooms were each occupied by another woman. The appellant claimed that the respondent was a licensee in occupation under the terms of a written agreement, dated April 25 1985, that she had broken the terms of the licence, and as a result they were entitled to possession. The respondent pleaded that she was in occupation under an oral agreement made in July 1982, and was a tenant at £20.00 a week, protected by the Rent Acts. She claimed a declaration to that effect and repayment of sums above that rate paid in 1984 and 1985. The learned judge found in favour of the respondent, made the declaration prayed, and gave judgment for £510 on her counterclaim. The appellant appeals to this court.

The respondent has not appeared. She gave instructions to her solicitors that she did not wish them to oppose the appeal, the reason being that she wants to leave the flat and get other accommodation. No doubt she may be helped in that search if she is not in protected accommodation. At all events, the solicitors informed the court of their instructions, and as they were operating under a legal aid certificate they very properly inquired of the court as to whether it was necessary for them to appear. The court told them that it was unnecessary, but, of course, it does not follow that the court for that reason can allow the appeal. Mr Thoresby, who appeared for the appellant, and to whose submissions we are very much indebted, recognised that he had to establish that there were proper grounds for interfering with the learned judge's judgment.

In view of the fact that there has been no contested argument in this case, and as I have come to the conclusion that Mr Thoresby is right in saying that the learned judge has fallen into error, and that the case must go back for a re-hearing, it is desirable that I should say as little as possible about the case; and what I do say must not be taken as being authoritative for any other case because, as I have said, there has not been full argument on the various topics.

The appellant runs this flat, and, I think, two others in the same house, in order to provide accommodation for single women who are working in London. The appellant has a standard form of agreement which it requires those who get accommodation in these flats to enter into. That agreement is described as a "licence", and it purports to license each individual occupier of a room to occupy the whole flat. It repeatedly asserts that nobody has exclusive occupation of anything, and it contains a number of standard terms that one would expect, including one which requires each of the occupants to behave reasonably towards the others. I am paraphrasing clause 3 of the licence agreement.

I would like to get one matter out of the way. There is a term that no child shall be in the premises. What happened here was that the respondent, having gone into occupation in July 1982, under an oral agreement, as the judge found — and that is no longer disputed — in fact used to sign annual written agreements from April 1983 onwards. In April 1985, she told Mr Brooker, when it came to signing a new agreement for a year, that she was pregnant and that she could not properly sign the agreement because she would be in breach of the child clause. He pointed out that there was to be no child for six months and there was no reason why she should not sign, and she did. Therefore, the relationship appears to have been of an amicable kind.

The learned judge, having found that the respondent had gone into occupation under an oral agreement made in July 1982, properly directed himself that she was not in occupation on the terms of the written agreement, save that a submission was made that in the oral agreement the terms of the written agreement were imported into it. The learned judge appears to have found that that was not so. However, with great respect to him, that finding itself is suspect, and I will say why in a moment.

In July 1982, the double bedsitting-room, which at one stage had two beds in it, had only one bed in it, and it was vacant. An advertisement was put up in All Souls in Langham Place stating:

Flat to share: Acton W3. Flat to share with two other Christian girls. Facilities include: Own room — approx 15′ × 15′, Gas Fire Heating. Kitchen/diner. Bathroom. Roof Garden. Gas/Electricity/Telephone to be shared. Easy access to Acton and Shepherds Bush Stations. Situated close to main bus route. Quiet Road. Fully furnished. £20 per week (Contact: 01-524 7137 for appointment).

The respondent applied to Mr Brooker. They met. She was shown the room. She agreed to take it and he explained that the terms were the same as for other occupants and that it was a licence and not a tenancy. Having made the bargain, Mr Brooker wrote a letter to her, dated July 27, in these terms:

I am writing to confirm my agreement with you to occupy the master bed/sitting room at 18 Agnes Road, Acton W3, as licensee; and to share the other common facilities with the existing occupants and with those who from time to time we may introduce. The rent is to be £20 per week paid monthly in advance. We look forward to you moving in on Friday, August 6 or before.

She accepted that, and, as I have said, written agreements were entered into in April 1983, 1984 and 1985.

Both the appellant and the respondent appeared by counsel before the learned judge and the case was decided by an examination of the facts against the judgment of the House of Lords in *Street* v *Mountford* [1985] AC 809. In that case Lord Templeman, delivering the only speech of their lordships, reviewed the authorities dealing with this tortured question as to whether an occupant is in occupation as a licensee or a tenant — on which topic there are a large number of cases — and sought to introduce some order into the law for the better administration of the law and guidance of the learned judges, particularly in the county court, who have to deal with this problem.

Unfortunately, the learned judge has used some parts of Lord Templeman's speech to assert a proposition which the speech does not support. I can illustrate that by showing what happened. The appellant's case is that the respondent was not given exclusive possession of her room for two reasons. The first is that the appellant reserved the right to put another person into the room — and remember that it was a double room — and, second, that all the occupants of the flat had a right of access to all parts of the flat. The appellant submitted that in those circumstances the respondent was not a tenant because she did not have exclusive possession of any part of the premises and that that was an essential requirement for her to be held to be a tenant.

The learned judge had the letter, which I have read, the advertisement, which I have read, and he had the evidence of Mr Brooker, who gave evidence and was cross-examined. There is a note of what he said. The respondent gave no evidence at all; she did not go into the witness box. The learned judge, in his judgment, dealt with this in this way, and I must read, in explanation to him as to why I have come to the conclusion that he has fallen into error, what he said:

Miss Ayers moved in. She occupied the master bedroom. She shared the facilities with other girls. The following year, she did sign a licence agreement. I am not concerned to construe it. It goes as far as is possible to say that there is no exclusive possession. It is not in dispute that it has no bearing on this case. I have to construe the agreement that in fact was entered into.

The leading case is *Street* v *Mountford* [1985] 2 All ER 289. Mrs Mountford entered into a "licence agreement" to take two rooms in Boscombe. There were various terms, and the owner or his agent had a right to enter the room for various purposes. The sort of terms one expects. In substance, there is not much to distinguish this from the present case. Miss Ayers occupied her room on a periodical term, making a payment. Reading from the headnote in *Street* v *Mountford* a person who: i) has exclusive possession; ii) for a fixed or periodic term; iii) at a rent; is a tenant. The question I have to decide is whether she had exclusive possession. It is argued on behalf of the plaintiff that there was no lock and there was nothing to show that she had the entitlement to keep anybody else out. The test is laid down in *Street* v *Mountford* at p 817. An occupier is either a lodger or a tenant.

Then the learned judge cites the following passage from the foot of p 817:

An occupier of residential accommodation at a rent for a term is either a lodger or a tenant. The occupier is a lodger if the landlord provides attendance or services which require the landlord or his servants to exercise unrestricted access to and use of the premises . . . If on the other hand residential accommodation is granted for a term at a rent with exclusive possession, the landlord providing neither attendance nor services, the grant is a tenancy.

That is the end of the passage which the learned judge quoted, and he continued:

In this case there is no evidence that Brooker & Co, provided any attendance or services to Miss Ayers. By that definition, she must have had the exclusive use of the room. She was not a lodger, *ergo* she was a tenant.

In my judgment, the learned judge has fallen into error, because it will be seen that what he has done is to say: "This occupant was not a lodger, as defined by Lord Templeman. Therefore she must have had exclusive possession." That does not follow. I think it is understandable that the learned judge fell into this error because of the succinct way in which Lord Templeman in his speech has attempted to try to simplify this situation.

I turn back to *Street* v *Mountford*. The passage which the learned judge cited goes on to say at p 818:

Any express reservation to the landlord of limited rights to enter and view the state of the premises and to repair and maintain the premises only serves to emphasise the fact that the grantee is entitled to exclusive possession and is a tenant. In the present case it is conceded that Mrs Mountford is entitled to exclusive possession and is not a lodger.

There was a sharp distinction between the present case and *Street* v *Mountford* because that crucial matter was not conceded and was indeed the only live factual issue in the case.

Lord Templeman went on to say at p 818E:

There can be no tenancy unless the occupier enjoys exclusive possession; but an occupier who enjoys exclusive possession is not necessarily a tenant.

He then gives examples of an owner in fee simple and others.

It is quite unnecessary to analyse that part of the speech, which may have to be done in another case because, as Mr Thoresby pointed out, there appear to be some inconsistencies as to the meaning which Lord Templeman was attaching to the phrase "exclusive possession", but it does not arise in the present case.

That Lord Templeman intended exactly what he said I have no doubt. Part of the problem has arisen because of the introduction at the foot of p 817 of a reference to residential accommodation. What he said was: "In the case of residential accommodation there is no difficulty in deciding whether the grant confers exclusive possession." In the context of *Street* v *Mountford* of course that was right because it was conceded.

Later in his speech at p 826, where he was dealing with the cases concerned with the intention of the parties, he said:

My Lords, the only intention which is relevant is the intention demonstrated by the agreement to grant exclusive possession for a term at a rent. Sometimes it may be difficult to discover whether, on the true construction of an agreement, exclusive possession is conferred. Sometimes it may appear from the surrounding circumstances that there was no intention to create legal relationships. Sometimes it may appear from the surrounding circumstances that the right to exclusive possession is referable to a legal relationship other than a tenancy.

He went on to give some examples.

He was recognising that each case had to be looked at on its own facts, and his expression of opinion that where residential accommodation was concerned there was no difficulty in deciding whether exclusive possession had been granted, may have been induced by the facts of *Street* v *Mountford* or may have been a hope that that would usually be so. But it is the latter, I think, that appears from the end of his speech where he said, having dealt with the Australian case of *Radaich* v *Smith* (1959) 101 CLR 209:

My Lords, I gratefully adopt the logic and the language of Windeyer J. Henceforth the courts which deal with these problems will, save in exceptional circumstances, only be concerned to inquire whether as a result of an agreement relating to residential accommodation the occupier is a lodger or a tenant.

That is because it will usually be clear that the occupier starts with "exclusive possession".

In the present case, that is not so. The learned judge simply, having asked himself the right question in the passage which I have read, namely "The question I have to decide is whether she had 'exclusive possession'", has purported to answer that by saying: "She is not a lodger. Therefore she must have exclusive possession." He did not deal with either of the submissions made by the appellant.

I have to consider in the circumstances, from the facts of the case and the note of the evidence which we have, whether we could come to a conclusion. Unfortunately, I find that impossible to do. The note of the evidence from Mr Brooker shows that he said in chief that he agreed with her that she would have her own room; that she would not be expected to share with anyone else; and that there were no locks on the doors of any of the rooms. In cross-examination he said:

[Her] own room meant that she would not have to share. We reserve the right to introduce people to share rooms. I don't recall the exact words. I said we had usual licence agreement.

That called for a decision by the learned judge as to what evidence he accepted on the facts of the case, and for a decision on it. I think it right to say that, of course, one has to bear in mind what Lord Templeman said about sham devices at p 825 of his speech. In the present case where there was only the evidence of Mr Brooker it was for the judge to decide as to whether he was a credible and honest witness and as to what the nature of the possession given to the respondent really was when the agreement was made, and that, for the reasons which I have given, very unfortunately, perhaps understandably, he has not done.

It follows that, in my judgment, this case must go back for a re-hearing. I would allow the appeal and order a new trial.

NICHOLLS LJ agreed and did not add anything.

Appeal allowed and new trial ordered; costs of appeal reserved to county court judge.

Court of Appeal
October 24 1986
(Before Lord Justice LAWTON and Lord Justice STEPHEN BROWN)

H H PROPERTY CO LTD v RAHIM AND OTHERS

Estates Gazette April 25 1987

282 EG 455

Landlord and tenant — Incorrect procedure adopted following interim order for payment of arrears of rent in proceedings by landlords claiming possession and other relief against the tenant and other defendants — The interim order had not been fully complied with and the county court judge whose decision gave rise to the present appeal ordered that, unless the sum due, amounting to £13,960, was paid, the defendants' defence should be struck out — Defendants appealed, claiming to have this order set aside under the provisions of section 50 of the County Courts Act 1984, Ord 13, r 12 of the County Court Rules and under RSC Ord 45 — Held "with considerable regret" that the order must be quashed — The judge was not entitled in the circumstances to make an order for the defence to be struck out under the inherent jurisdiction of the court — The correct procedure for the plaintiffs was not to seek to have the defence struck out but to use the recognised procedure to enforce the interim order for payments — RSC Ord 45, which was applicable, through CCR Ord 13, r 12 and RSC Ord 29, r 10, provided for the way in which interim orders were to be enforced — Appeal allowed

No cases are referred to in this report.

This was an appeal by defendants Abdul Aziz Abdul Rahim and others against an order made by Judge Deborah Rowlands, conditionally striking out the defendants' defence in proceedings begun by the landlords, H H Property Co Ltd, before the late Judge Curtis-Raleigh, who had made an interim order for payment of £13,960 to the landlords on account of arrears.

G Zelin (instructed by G Lebor & Co) appeared on behalf of the appellants; A R Connerty (instructed by Radcliffes & Co) represented the respondents.

Giving judgment, LAWTON LJ said: This case comes before the court, first as an application by the first and second defendants for leave to appeal against an order made by Her Honour Judge Deborah Rowlands on February 4 1986 whereby she ordered that unless the sum of £13,960, due under an interim order made earlier by the late Judge Curtis-Raleigh, was paid, the defendants' defence should be struck out. We have granted leave to appeal and, having granted leave, have gone on to consider the merits. (I put the word "merits" mentally in inverted commas.)

The plaintiffs are the owners of property at Porchester Court, Porchester Gardens, London W2. They claim that they are entitled to possession of premises known as Flat 3 and 3A at that address. They say that by a lease made on July 13 1976 between the plaintiffs as lessors and the first defendant as lessee, the premises were demised to the first defendant for a term of 10 years from December 25 1975 at a yearly rent of £2,000. By the lease the first defendant covenanted, among other things, to pay the lessor on demand each year of the said term by way of additional rent an annual service charge (called in the lease "the service charge"). The lease contained a proviso for re-entry and forfeiture in the event of the lessee failing to observe and perform any of the lessee's covenants.

The first defendant is in arrears with the said rent. By the time the summons was issued, the arrears are alleged to amount to no less than £6,179. It is also said by the plaintiffs that the first defendant does not occupy the premises and that he has allowed the second defendant to do so.

The defence puts in issue the various allegations made by the plaintiffs and in particular alleges that there have been breaches of covenant by the plaintiffs, breaches which are not uncommonly alleged in this class of case. There is also a statutory defence under the provisions of section 91A of the Housing Finance Act 1972, the allegation being that some of the service charges, which are to be treated as if they were rent, were not previously discussed with the tenant; and, inevitably, there is a counterclaim for damages for breach of covenant.

At a preliminary hearing held in October 1984 His Honour Judge Curtis-Raleigh made an order as follows: "That the defendants do pay to the plaintiffs an interim payment of £14,000 plus £80 per week from October 31 1984."

The defendants have not complied with that order. We have been told today by Mr Zelin that something like £5,000 has been paid, but there is clearly outstanding a considerable amount of money. It is understandable in those circumstances that, when the matter came before Her Honour Judge Rowlands, she was concerned about the fact that there had clearly been a defiance of the court's order. No explanation appears to have been given to her and none has been given to this court as to why the defendants are in breach of the order.

It is relevant to point out that there has been no attempt whatsoever to appeal the first order. In those circumstances it is with some courage that the defendants have appeared in this court, asking that the court intervene on their behalf, when in fact they are in clear breach of an order made as long ago as October 1984.

The defendants claim to be entitled to have this order set aside under the provisions of section 50 of the County Court Act 1984, the terms of Ord 13, r 12 of the County Court Rules, and RSC Ord 45 r 1. Section 50 of the County Courts Act 1984 envisages that the county court may make orders for interim payments. In my judgment, subsection (2) of that section also envisages that a defence put forward by a defendant will be in existence when the matter is finally decided by the court. Under subsection (2) a defendant who has overpaid pursuant to an interim payment may recover when the matter is finally tried. Ord 13, r 12 of the County Court Rules provides by sub-rule (2) that RSC Ord 29, r 10 shall apply to the county court. RSC Ord 29, r 10 is the order which provides for the making of interim payments. Ord 45 provides for the way orders for interim payments are to be enforced and is in these terms:

Subject to the provisions of these rules, a judgment or order for the payment of money, not being a judgment or order for the payment of money into court, may be enforced by one or more of the following means, that is to say —

There are then set out a number of ways which amount to how orders of courts are to be executed.

Mr Zelin says (correctly in my judgment) that in the circumstances of this case the plaintiffs' remedy was not to come to the court and ask for the defence to be struck out but to use the machinery of the law to enforce the interim payments. It seems to me impossible to accept Mr Connerty's argument that the judge was entitled to make the order she did under the inherent jurisdiction of the court. It was not an order for conditional leave to defend; it was an order for the payment of rent, which was due. Had the judge given judgment for the plaintiffs, then the only way in which the plaintiffs could have enforced it was by one of the ways set out in Ord 45, r 1. In my judgment, having regard to the provisions of Ord 45, there is no room for the inherent jurisdiction of the court to be called in aid. Striking out the defence runs counter to the intention of section 50 (2) of the County Courts Act.

It is therefore with considerable regret that I have come to the conclusion that the order of the learned judge must be quashed and the appeal allowed.

STEPHEN BROWN LJ agreed and did not add anything.

The appeal was allowed: no order was made as to costs.

Chancery Division

October 9 1986

(Before His Honour Judge FINLAY QC, sitting as a judge of the High Court)

F W WOOLWORTH PLC v CHARLWOOD ALLIANCE PROPERTIES LTD

Estates Gazette May 2 1987

282 EG 585-594

Landlord and tenant — Consent to assignment of underlease — Claim by tenants that consent to a proposed assignment had been unreasonably withheld by landlords — Counterclaim by landlords for specific performance of covenant to keep open demised premises as a departmental store — Premises consisted of part of second floor of Arndale Centre, Middleton, north of Manchester — Underlease included, in addition to covenants restricting assignment, various covenants designed to ensure that the demised premises were kept open as, and were not used otherwise than as, a retail departmental store or shop — Tenants operated a "Woolco" store for non-stop shopping on the premises — Present litigation arose as a result of information which led landlords to believe that tenants were not intending to keep the premises open as a department store — Tenants subsequently sought consent to an assignment — They refused to give any assurance that the premises would be kept open as a store after assignment — Proceedings by landlords to restrain assignment without consent were commenced, but later stayed, and the present originating summons by the tenants seeking a declaration as to their entitlement to assign was issued — Tenants relied mainly on *Killick* **v** *Second Covent Garden Property Co Ltd,* **submitting that, unless the proposed assignment necessarily involved a breach of covenant, it was not reasonable to refuse consent; the landlords after assignment would be able to enforce the user covenant against the assignees — After reviewing a number of other authorities, including the principles set out by Balcombe LJ in** *International Drilling Fluids Ltd* **v** *Louisville Investments (Uxbridge) Ltd,* **Judge Finlay distinguished the** *Killick* **case from the present — The user covenant there was positive, and thus capable of enforcement by prohibitory injunction, while here the covenant was negative; here the breach had already taken place and the proposed assignees had no intention of running a department store; consent to the assignment would inevitably involve a continuation of the breach — On these grounds the landlords were not acting unreasonably in refusing consent — They were also entitled to consider the probable adverse effects of the assignment on their ability to let satisfactorily other premises in the centre — The balance of detriment was on their side — Tenants' declaration refused — Landlords' counterclaim for specific performance of the user covenant "reluctantly" refused, having regard to the state of the authorities on this point, but inquiry as to damages ordered**

The following cases are referred to in this report.

Attorney-General v *Colchester Corporation* [1955] 2 QB 207; [1955] 2 WLR 913; [1955] 2 All ER 124; (1955) 53 LGR 415
Braddon Towers Ltd v *International Stores Ltd,* March 28 1979, unreported
Dowty Boulton Paul Ltd v *Wolverhampton Corporation* [1971] 1 WLR 204; [1971] 2 All ER 277; (1971) 69 LGR 192
Fuller's Theatre & Vaudeville Co v *Rofe* [1923] AC 435; 128 LT 774; 39 TLR 236, PC
Giles (CH) & Co Ltd v *Morris* [1972] 1 WLR 307; [1972] 1 All ER 960
Gravesham Borough Council v *British Railways Board* [1978] Ch 379; [1978] 3 WLR 494; [1978] 3 All ER 853; (1977) 76 LGR 202
Greene v *West Cheshire Railway Co* (1871) LR 13 Eq 44
Hooper v *Brodrick* (1840) 11 Sim 47

International Drilling Fluids Ltd v *Louisville Investments (Uxbridge) Ltd* [1986] Ch 513; [1986] 2 WLR 581; [1986] 1 All ER 321; (1985) 51 P&CR 187; [1986] 1 EGLR 39; (1985) 277 EG 62, CA
Killick v *Second Covent Garden Property Co Ltd* [1973] 1 WLR 658; [1973] 2 All ER 337; (1973) 25 P&CR 332; [1973] EGD 377; 227 EG 1849, CA
Posner v *Scott-Lewis* [1986] 3 WLR 531; [1986] 3 All ER 513; [1986] 1 EGLR 56; (1985) 277 EG 859
Powell Duffryn Steam Coal Co v *Taff Vale Railway Co* (1874) LR 9 Ch 331
Premier Confectionery (London) Co Ltd v *London Commercial Sale Rooms Ltd* [1933] Ch 904
Ryan v *Mutual Tontine Westminster Chambers Association* [1893] 1 Ch 116
Shanly v *Ward* (1913) 29 TLR 714
Shiloh Spinners Ltd v *Harding* [1973] AC 691; [1973] 2 WLR 28; [1973] 1 All ER 90; (1973) 25 P&CR 48, HL
Tito v *Waddell (No 2)* [1977] Ch 106; [1977] 2 WLR 496; [1977] 3 All ER 129
Town Investments Ltd Underlease, Re [1954] Ch 301; [1954] 2 WLR 355; [1954] 1 All ER 585

This was an originating summons by the plaintiff tenants, F W Woolworth plc, seeking a declaration that consent to the assignment of their underlease of a substantial part of the second floor of the Arndale Centre at Middleton, some 6 miles north of the centre of Manchester, had been unreasonably refused by the landlords, Charlwood Alliance Properties Ltd. The defendant landlords counterclaimed for specific performance of certain user covenants in the underlease, the object of which was to ensure that the demised premises were kept open as a department store.

Jonathan Gaunt (instructed by Clifford-Turner) appeared on behalf of the plaintiffs; Malcolm Spence QC and Peter W Smith (instructed by Gorna & Co, of Manchester) represented the defendants.

Giving judgment, JUDGE FINLAY said: This is an originating summons in which the plaintiff tenants seek a declaration that consent to an assignment has been unreasonably withheld by the defendant landlords, and by way of counterclaim the landlords are seeking in these same proceedings specific performance of certain covenants in the relevant underlease, being covenants, putting it briefly, to keep open the demised premises as a departmental store or shop at the times specified in the underlease. The underlease was dated April 27 1976 and it demised the premises in question for 99 years from March 25 1971. It was indeed from about that date in 1971 that the premises were occupied by the plaintiffs, and the reason for the delay in the execution and dating of the lease has not emerged in the course of the evidence.

Although the underlease contains many provisions, I need refer only to a few of them. In clause 4, subclause 23(a), the tenants covenanted not to use the demised unit or any part thereof or permit the same to be used otherwise than as a retail departmental store or shop or shops, the tenants having the right, if they so desired, to use part of the demised unit for the service of motor vehicles and as a restaurant and café with or without an excise licence and for no other purpose whatsoever. The demised unit was, perhaps it is obvious, the premises demised by the underlease and they comprise the substantial part of the second floor of a building known as the Arndale Centre and situated in Middleton some 6 miles to the north of the centre of Manchester. The Arndale Centre comprises, as well as those premises on the second floor, a large number of other shop premises of varying sizes, some of which are large enough to be occupied by concerns carrying on a supermarket business.

In clause 4(24) of the underlease the tenants covenanted:

to keep the shop portion of the demised unit open for retail trade at least during the normal trading hours of the shopping centre, provided that at the date hereof the normal trading hours of the shopping centre in each week should be as follows:

and then there is set out: Sunday closed; Monday 9 am to 5 pm; Tuesday to Friday inclusive 10 am to 8 pm; Saturday 9 am to 6 pm; and the clause continued:

And any change in these times shall be agreed between the landlord and the tenant provided further that the commencement of trading by the tenant not later than 10.00 am on any day shall not be deemed to be a breach of this covenant.

And in subclause 25 the tenants further covenanted:

To keep the display windows, if any, of the demised unit dressed and illuminated in a suitable manner in keeping with a good class shopping centre unless prevented by matters or circumstances beyond the tenant's reasonable control and to keep the windows of those parts of the demised unit which are used for storage purposes obscured to the satisfaction of the landlord.

A portion of the demised premises is used for storage and is screened off, but the major part of the premises comprised in the underlease was used as a sales area for the purposes of the tenants' business.

The tenants, F W Woolworth & Co, in or about 1971 commenced to operate in a number of locations throughout England, and I think Wales and Scotland also, an operation which was called a Woolco store and it was for the purposes of a Woolco store that these premises at the Arndale Centre at Middleton, demised to the plaintiffs, were in fact used and continued to be used until a date earlier in this year. A Woolco store was, as I understand it, a departmental store in which the customer could carry out what is called one-stop shopping; that is to say, go into the store and emerge having bought all the kinds of produce or goods which a shopper might reasonably in the ordinary way be expected to wish to buy.

The only other clause which I think I need refer to in the lease, and that which indeed occasions the application of the plaintiffs, is clause 4(34), where the tenant covenants not to assign, underlet, or part with, or share the possession of the whole of the demised unit without the previous consent in writing of the landlord, which shall not be unreasonably withheld, and of any superior landlord; and on the grant of any such permitted underlease, either under this subclause or under subclause 33(b) of this clause, to obtain — and I should say that clause 33(b) dealt with underletting, with the consent of the landlord, part of the premises — (a) an unqualified covenant on the part of the underlessee not to assign, underlet, part with, or share the possession of the part only of the premises thereby demised and (b) a covenant on the part of the underlessee that the underlessee would not assign, underlet, or part with the possession of the whole of the premises thereby demised without obtaining the previous written consent of the superior landlord under the landlord's lease and of the landlord, and continued with other provisions which I do not think are material.

In 1984 there was some discussion between the plaintiffs and the defendants about the possibility of refurbishment of the Arndale Centre and in the course of those discussions it was suggested on behalf of the defendants that the plaintiffs might consider remodelling their operation in the Woolco store so that two other subsidiaries of the plaintiff company, namely B & Q, which is a do-it-yourself retail operation, and Comet, a concern which in various outlets throughout the country sells such articles as refrigerators, cookers, and so on (I think what are called white goods) might occupy the premises; but that suggestion, although not turned down, was never in fact taken up by the plaintiffs or by the subsidiary Woolco, and towards the end of 1985 the Woolco subsidiary of the plaintiffs was considering whether or not to carry on the Woolco operation both at Middleton and at the 10 other Woolco stores throughout the United Kingdom.

In February 1986 negotiations with the Dee Corporation were commenced by Woolco and these in the event led to a contract whereunder the Dee Corporation agreed to buy from — I have not seen the contract, it may be from the Woolco subsidiary or from the plaintiffs — the 11 Woolco stores, including Middleton, at a price of £26 m, and provision was made in that contract, as I understand it, that if in the case of any Woolco store the consent of the landlord to an assignment to the purchaser should be refused, then there was provision under which the store in question would be in effect withdrawn from the contract and the contract price reduced by an amount which differed in the case of each of the 11 stores and which in the case of the store at Middleton was the sum of £570,000.

On April 14 1986, Mr Wilcox-Wood, the defendants' manager of the Arndale Centre at Middleton, received some information which led him to call on the second-floor premises to find out what was happening there and he was told that they were conducting some staff-training operation, but later in the day, at about 6 pm, he was informed by the manager of the Woolco store that that earlier information had in fact not been correct; that the employees at the store, including the manager himself, had earlier on that day been informed of the sale to the Dee Corporation, and furthermore he informed Mr Wilcox-Wood that it was planned that the Woolco store should be closed down, with the last trading day on July 19 1986.

Early in June, Mr Ford, the chairman of the Arndale company, went to meet a Mr Harford, a director of one of the subsidiaries of the Dee Corporation called Carrefour, or Carrefour Ltd, at Milton Keynes. There had been some confusion apparently over Mr Harford's arrangements; Mr Ford did not see him but saw instead some other, some colleague of his, a Mr Whitehead, and on June 6 — I think the meeting was in fact on May 30 — but on June 6 Mr Ford wrote to Mr Whitehead mentioning their meeting on the 30th and recording what he had been told by Mr Whitehead at that time in these terms:

> We were very pleased to receive your assurance that the Dee Corporation and/or its subsidiary company will continue to trade from this unit,

the reference being to the Woolco unit at the Arndale Centre, Middleton.

It subsequently emerged that that assurance given by Mr Whitehead was ill-founded, and indeed it was heard in evidence that he had no authority to give any such assurance, but the defendants continued for some time in the belief that it was at least a possibility that the Dee Corporation was intending to carry on business in the way of a retail trading operation in the premises; but although they were not disabused of that idea they began to have doubts about it and those doubts became relevant and credible when on June 10 1986 an application was made on behalf of the plaintiffs for consent to an assignment of the premises. The letter requesting such consent read:

> Please accept this letter as an application on behalf of our client for consent to an assignment of the lease to Eldergate Property Co CV Ltd. This is a new company, but it is proposed that the Dee Corporation should act as surety by accepting the previous guarantee. A copy of the 1985 report and accounts of the Dee Corporation is enclosed.

I should say at this juncture that no point is taken either in or outside these proceedings as to the reliability of the Dee Corporation as the guarantor of the rent and other obligations due from the tenant under the lease; the landlord's objections to the assignment are of a different character.

On June 20, the solicitors acting for the defendants wrote to the plaintiffs' solicitors stating that their clients were giving consideration to the request for a licence to assign and, in the meantime, asking for an undertaking for their costs and referring also to:

> The assurance of your clients and the proposed assignee and the proposed surety that, after a reasonable period which may be required for fitting out works, the store will be kept open in compliance with clause 4(24) of the underlease dated April 27 1976,

and they ask to hear from them on that matter. But such an assurance was not in fact forthcoming and on June 24 the defendants' solicitors again wrote to the plaintiffs' solicitors saying, *inter alia*, this:

> Mr Peter Ford of our client's company yesterday spoke to Mr Harford, the property director of the Dee Corporation, to seek an assurance that if the lease were to be assigned in the manner envisaged in the letter of June 10, the store would thereafter be kept open in compliance with clause 4(24) of the underlease. Mr Harford would not give an unequivocal assurance to that effect, indeed our client has described Mr Harford's attitude as being guarded. In the circumstances our clients are extremely concerned about the situation. They are liable to sustain very substantial damage if the store closes down in breach of the covenant. It is therefore essential that we receive forthwith the assurance requested in point 2 of our letter to you of June 20.

That is to say, the assurance in the terms which they had already proposed.

There not having been a satisfactory reply to that request, on June 26 the defendants issued a writ, not in these but in other proceedings, seeking an injunction to restrain the plaintiffs in the present proceedings from assigning the property, that is, assigning the underlease, without consent. An *ex parte* injunction was obtained and then an undertaking that the premises would not be assigned without the consent of the landlords or pursuant to some declaration of the court was given, and those proceedings were, I think, on or about July 2 stayed.

There followed correspondence between the respective solicitors for the parties in which the defendants' request for an assurance on the lines of that I have already indicated was continuously sought and as continuously declined. There was no unequivocal undertaking and assurance that the assigned lease would keep open the store in compliance with the covenant.

On August 4 1986, following a communication from the plaintiffs' solicitors stating that the assignees could not give the unequivocal confirmation required on behalf of the defendants, the defendants' solicitors wrote to those acting for the plaintiffs saying:

> We note that you say it is unreasonable to expect any retail concern to give an unequivocal undertaking to keep a retail store open. With respect, it is not a question of reasonableness. Your clients' underlease contains a specific covenant to remain open. Your clients entered into that obligation freely and honourably. So far as the request for a licence to assign is concerned, our clients are currently in a position neither to grant nor to refuse such consent. Our clients are seeking information to enable them to make a decision. Our clients are entitled to take into account whether the prospective assignees are or are not likely to comply with the obligations of the Tenant under the underlease. Furthermore, our clients are entitled to ask to see and approve plans and specifications for any works of alteration. Until our clients have full information in this matter, they will not be able to make a decision. We are instructed that, in breach of clause 4(24) of the underlease, your clients closed the store on about July 18. Our clients require your clients to reopen the store forthwith. If the current breach of clause 4(24) of the underlease is not remedied within the next seven days, our clients will institute proceedings forthwith and without further notice for a mandatory injunction and damages.

The result of that was a response which indicated that the plaintiffs' solicitors did not believe that a mandatory injunction would be obtained by the defendants while the application for the licence to assign was yet to be dealt with. That communication was dated August 7 and on the day before the originating summons in these proceedings was issued. By that summons the plaintiffs claim that on the true construction of the underlease and in the events which have happened, the refusal of the defendants to grant a licence to assign the said underlease to Eldergate Property Co Ltd was unreasonable; and second, a declaration that notwithstanding that refusal the plaintiffs without any licence from the defendants are entitled to assign the said underlease to Eldergate Property Co Ltd.

Mr Gaunt, who appears for the plaintiffs, relies in particular upon a decision of the Court of Appeal in *Killick* v *Second Covent Garden Property Co Ltd* [1973] 1 WLR 658*. That was a case where, by the user covenant in a lease, the lessee covenanted not to use the premises:

> for any other purpose than the trade or business of a printer nor have or permit any sale by auction in or upon the demised premises or any part thereof without the landlord's written consent which shall not be unreasonably withheld.

There was a question as to the construction of that covenant, briefly as to whether the words about consent not being unreasonably withheld governed only the part of the covenant which prohibited sale by auction, or whether it governed the whole of the covenant, that is to say including the covenant not to use the premises for any other purpose than the trade or business of a printer.

The Court of Appeal held that as a matter of construction the proviso governed the whole clause, but the part of the decision which is relied upon by Mr Gaunt is that contained in the judgment of Stamp LJ (with whom both Edmund Davies LJ and Cairns LJ agreed) in which he deals with the matter on the basis that the construction of the covenant is either left in doubt or at any rate is not that which ultimately the court decided was the true construction.

At p 661, Stamp LJ says this:

> Mr Priday, on behalf of the landlords, submitted that a landlord may reasonably refuse to consent to an assignment if the assignment would necessarily involve a breach of covenant, and I will accept that submission as being well founded. But whatever view one takes as to the construction of the user covenant, I cannot accept that, if the landlords did consent to the proposed assignment there would as a necessary consequence be a breach of the user covenant. As a result of the assignments Primaplex

that is the proposed assignee

> would step into the shoes of the lessee and underlessee

I should say that the proposals were to assign both the lease and the underlease under which the assignors would occupy the premises

> and would thereupon become subject to the user covenant. The landlords would be in the same position, neither better nor worse, to enforce the user covenant as would be the case if the present underlessee was itself proposing to seek planning permission for use of the premises as offices and proposed so to use them. On that short ground I would hold that the landlord's withholding of consent is unreasonable.

So there was a case where the proposed assignee intended to use the premises as offices, that is in breach of the covenant not to use them otherwise than for the trade or business of a printer, but because the assignment itself was not regarded as a breach the court held that the withholding of consent was unreasonable. Stamp LJ went on to refer to the authorities — I read this following passage because of the submissions which have been made and with which I propose to deal in due course:

> Two authorities were cited in support of the proposition that the withholding of consent in such a case as this necessarily involved a breach of the user covenant: namely, the decision of this court in *Packaging Centre Ltd* v *Poland*

---

*Editor's note: Reported also at [1973] EGD 377; (1973) 227 EG 1849.

Street Estate Ltd (1961) 178 EG 189 . . . and *Granada TV Network Ltd* v *Great Universal Stores Ltd* (1962) 187 EG 391. Those cases were cases in which the tenant sought the consent of the landlord to an underlease to an underlessee who under the terms of the underlease was to use the premises in breach of the user covenant in the lease. In *Packaging Centre Ltd* v *Poland Street Estate Ltd* the underlease provided that the premises would be used exclusively as office premises, which was a breach of the user covenant in the lease. It followed that, had the landlords consented to that subletting, they could not thereafter successfully object to the user which the lease prohibited and they would therefore be prejudiced by the giving of consent.

I draw attention to the fact that there the consent which was sought was consent to the sublease.

Not dissimilar was the *Granada TV Network* case, in which the underletting was to be in favour of a person who would necessarily be unable to use the premises for the purposes for which they could alone be used under the terms of the head lease. Again,

I interpose this, that that was because on its true construction the user covenant required user by the tenants themselves or some associated company

the giving by the landlord of consent to a subletting on those terms would have precluded him from thereafter relying on the user covenant.

Those cases are, in my judgment, not applicable where, as here, the giving of consent to an assignment does not of itself preclude the landlords from thereafter insisting that the terms of the user covenant be strictly complied with. Of course, a landlord who gives his consent to an assignment knowing that the assignee intends to use the premises in breach of the user covenant may incautiously estop himself from thereafter relying upon the covenant or may waive the right to enforce it. But a landlord who is minded to refuse consent to an assignment on account of the user covenant is not acting incautiously; and nothing could have been easier than for the landlords here, while giving their consent, expressly to reserve their right to enforce the user covenant against the assignee. Here, be it observed, the proposed assignee was content to accept that position, relying, as I understand it, on the view that the part of the user covenant prohibiting use otherwise than for printing was qualified by the last eleven words of the user covenant and that if the landlords refused their consent to use as offices that consent would be unreasonably withheld.

Mr Gaunt also relied upon another decision of the Court of Appeal, *International Drilling Fluids Ltd* v *Louisville Investments (Uxbridge) Ltd* [1986] 2 WLR 581.* The passage in the judgment of Balcombe LJ, with whom Mustill LJ and Fox LJ agreed, upon which reliance is placed, is one in which he, having referred to the numerous cases which had been cited during the course of argument, set out the principles that he derived from them. Balcombe LJ says at p 586:

During the course of argument many cases were cited to us, as they were to the judge. I do not propose to set them out in detail here; many of the older cases were considered in the full judgment of the Court of Appeal in *Pimms Ltd* v *Tallow Chandlers Co* [1964] 2 QB 547. From the authorities I deduce the following propositions of law.

(1) The purpose of a covenant against assignment without the consent of the landlord, such consent not to be unreasonably withheld, is to protect the lessor from having his premises used or occupied in an undesirable way, or by an undesirable tenant or assignee: *per* A L Smith LJ in *Bates* v *Donaldson* [1896] 2 QB 241, 247, approved by all the members of the Court of Appeal in *Houlder Brothers & Co Ltd* v *Gibbs* [1925] Ch 575.

(2) As a corollary to the first proposition, a landlord is not entitled to refuse his consent to an assignment on grounds which have nothing whatever to do with the relationship of landlord and tenant in regard to the subject-matter of the lease: see *Houlder Brothers & Co Ltd* v *Gibbs*, a decision which (despite some criticism) is binding on this court: *Bickel* v *Duke of Westminster* [1977] QB 517. A recent example of a case where the landlord's consent was unreasonably withheld because the refusal was designed to achieve a collateral purpose unconnected with the terms of the lease is *Bromley Park Garden Estates Ltd* v *Moss* [1982] 1 WLR 1019.

(3) The onus of proving that consent has been unreasonably withheld is on the tenant: see *Shanly* v *Ward* (1913) 29 TLR 714 and *Pimms Ltd* v *Tallow Chandlers Co* [1964] 2 QB 547, 564.

(4) It is not necessary for the landlord to prove that the conclusions which led him to refuse consent were justified, if they were conclusions that might be reached by a reasonable man in the circumstances: *Pimms* v *Tallow Chandlers Co*.

(5) It may be reasonable for the landlord to refuse his consent to an assignment on the ground of the purpose for which the proposed assignee intends to use the premises, even though that purpose is not forbidden by the lease: See *Bates* v *Donaldson* [1896] 2 QB 241, 244.

(6) There is a divergence of authority on the question, in considering whether the landlord's refusal of consent is reasonable, whether it is permissible to have regard to the consequences to the tenant if consent to the proposed assignment is withheld. In an early case at first instance, *Sheppard* v *Hongkong & Shanghai Banking Corporation* (1872) 20 WR 459, 460, Malins

*Editor's note: Reported also at [1986] 1 EGLR 39.

V-C said that by their withholding their consent the lessors threw a very heavy burden on the lessees and they therefore ought to show good grounds for refusing it. In *Houlder Brothers & Co Ltd* v *Gibbs* [1925] Ch 575, 584, Warrington LJ said: "An act must be regarded as reasonable or unreasonable in reference to the circumstances under which it is committed, and when the question arises on the construction of a contract the outstanding circumstances to be considered are the nature of the contract to be construed, and the relations between the parties resulting from it."

In a recent decision of this court, *Leeward Securities Ltd* v *Lilyheath Properties Ltd* (1983) 271 EG 279, concerning a subletting which would attract the protection of the Rent Act, both Oliver LJ and O'Connor LJ made it clear in their judgments that they could envisage circumstances in which it might be unreasonable to refuse consent to an underletting, if the result would be that there was no way in which the tenant (the sublandlord) could reasonably exploit the premises except by creating a tenancy to which the Rent Act protection would apply, and which inevitably would affect the value of the landlord's reversion. O'Connor LJ said at p 283: "It must not be thought that, because the introduction of a Rent Act tenant inevitably has an adverse effect upon the value of the reversion, that that is a sufficient ground for the landlords to say that they can withhold consent and that the court will hold that that is reasonable."

To the opposite effect are the dicta, *obiter* but nevertheless weighty, of Viscount Dunedin and Lord Phillimore in *Viscount Tredegar* v *Harwood* [1929] AC 72, 78, 82. There are numerous other dicta to the effect that a landlord need consider only his own interests: see, eg, *West Layton Ltd* v *Ford* [1979] QB 593, 605, and *Bromley Park Garden Estates Ltd* v *Moss* [1982] 1 WLR 1019, 1027. Those dicta must be qualified, since a landlord's interests, collateral to the purposes of the lease, are in any event ineligible for consideration: see proposition (2) above. But in my judgment a proper reconciliation of those two streams of authority can be achieved by saying that while a landlord need usually only consider his own relevant interests, there may be cases where there is such a disproportion between the benefit to the landlord and the detriment to the tenant if the landlord withholds his consent to an assignment that it is unreasonable for the landlord to refuse consent. (7) Subject to the propositions set out above, it is in each case a question of fact, depending upon all the circumstances, whether the landlord's consent to an assignment is being unreasonably withheld: see *Bickel* v *Duke of Westminster* [1977] QB 517, 524 and *West Layton Ltd* v *Ford* [1979] QB 593, 604, 606-607.

In that case (*International Drilling Fluids Ltd* v *Louisville Investments (Uxbridge) Ltd*), there are, as evidenced from the passages I have read from the judgment of Balcombe LJ, references to numerous authorities; but other cases were cited in argument. On one of them reliance is placed before me, that is *Re Town Investments Ltd Underlease* [1954] Ch 301, a decision of Danckwerts J, as he was at that time. That was a case where the underlessee of premises which contained a covenant against assigning without consent proposed to underlet part of the premises at a rent well below the current market rate and in consideration of a substantial fee and the defendant, the immediate lessor, objected and it was held that the defendant had not unreasonably withheld his consent to the proposed underletting.

At p 314, Danckwerts J, having referred to and cited the passage from the judgment of Denning J in *Premier Confectionery (London) Co Ltd* v *London Commercial Sale Rooms Ltd* [1933] Ch 904 (the case of the kiosk associated with a tobacconist's shop) went on to say:

It is, of course, true as pointed out by counsel for the plaintiff that in that case Bennett J found in the evidence that the landlords' apprehension of injury to their interests as owners of the properties was well founded. But the case is authority for the proposition that, in considering whether to give or withhold consent, the landlords were entitled to consider the effect which the transaction might have upon their ability in the future to let satisfactorily the different parts of their property, particularly in case of default on the part of the tenant in performing his obligations.

Having referred to *Shanly* v *Ward* (1913) 29 TLR 714, Danckwerts J said:

Counsel for the defendant relied upon this decision as showing that it was sufficient for his purpose if a reasonable man in the defendant's position might have regarded the proposed transactions as damaging to his property interests, even though some persons might take a different view.

This is a proposition which appears to accord with that which is numbered (4) in the passage from Balcombe LJ's judgment in *International Drilling Fluids Ltd* v *Louisville Investments (Uxbridge) Ltd*. I do not find the fact that there is no reference in the judgment of Balcombe LJ to Danckwerts J's decision in *Re Town Investments Ltd Underlease* as indicating in any way that the Court of Appeal was tacitly expressing disapproval of the decision in that case, but I will return to the impact upon the present case of what is said by Danckwerts J at a later stage.

The main point in the argument of Mr Gaunt for the plaintiff is, I think, founded upon the decision in *Killick* v *Second Covent Garden Property Co Ltd*; that is to say, on the proposition there enunciated by Stamp LJ that unless the assignment necessarily involves a breach of covenant it is not reasonble to withhold consent to the proposal, and that is so on the basis that once the assignment has been effected the landlord will be able to enforce the user covenant against the assignee.

In determining whether that principle is applicable here, in my judgment it is clearly permissible to consider whether there are actual differences between the facts that were there considered and the facts of the present case. Now in *Killick* the covenant was a negative covenant; not to use the premises for any other purposes than the trade or business of a printer. Such a covenant is enforceable by a prohibitory injunction. In the present case the covenant which is in question, that is clause 4(24), is a positive covenant to keep the premises open for retail trade. Such a covenant is enforceable, if at all, only by a mandatory injunction. Second, in *Killick* there had been no breach of covenant when the licence to assign was sought. Here a breach has already taken place and although it had not taken place at the time when the licence to assign was sought, it was at that stage being indicated that the store would be closed and whether that would constitute a breach or merely the temporary closure resulting in the reopening by the assignee was not at that juncture clear.

Third, in *Killick* there would be no breach of the user covenant unless or until the assignee took some action, but in the present case there would be a breach if the assignee did not take action; that is to say, there would be a breach if the assignee did nothing because there would inevitably be a breach of covenant unless the assignee took the action of opening the store for retail trade. The ratio in *Killick* was that it was unreasonable to refuse consent to the assignment because, as the landlords could enforce the covenant, the assignment would not necessarily cause a breach. So, it was held, there was no ground for refusal of consent. That, in my judgment, is no doubt so where the covenant is negative and no breach has taken place. In such a case, assuming a covenant in terms which give rise to no relevant difficulty with construction, an injunction either obtained *ex parte* or *inter partes* would prevent a breach before it happened. Is that the case here?

Here it is said that the proposed assignee neither will nor could comply with the covenant. Eldergate Property Holding Co CV Ltd has no practical capacity and very possibly has no legal capacity to run a retail business in the premises. The evidence makes it clear, however, that there is no intention whatsoever of Eldergate doing so. Even assuming, and I will later have to consider this point, that there is no rule of law or settled practice which would militate against the grant of a mandatory injunction or order for specific performance, can it possibly be said that a landlord who took the view that in such circumstances the prospects of procuring compliance with the covenant by means of a mandatory injunction (if he had prior to such an application consented to an assignment to a tenant in such a situation and having such character as Eldergate) were negligible was taking an unreasonable view of the matter? I think not. It appears to me that in such circumstances a landlord who refused consent to the assignment on the grounds that, were he to grant it with knowledge of the assignee's intentions and, seemingly, inability to comply with the covenant, he would be powerless to prevent a breach, could not be said to be acting unreasonably. I have not ignored what Stamp LJ says about the landlord expressly reserving the right to enforce the user covenant. Such a reservation would no doubt have practical results in the case of a negative covenant, but expressly to reserve rights is of little utility if there are no effective rights.

Mr Gaunt submitted that the landlords could grant consent expressly, reserving the right (assuming they were unable to obtain an injunction) to "enforce" — and I put that word in inverted commas — the covenant by pursuing a remedy in damages. I cannot accept the suggestion that Stamp LJ, when concluding that the landlord could enforce the user covenant against the assignee, had in mind enforcement by any means other than by injunction. The words he uses are quite clear. He says, at p 662:

. . . here the giving of consent to an assignment does not of itself preclude the landlords from thereafter insisting that the terms of the user covenant strictly be complied with.

It is evident that he is thinking of effective insistence, not of other remedies, for example a remedy in damages available if the insistence were ineffective.

I come, therefore, to the clear conclusion that for the reasons I have already stated, *Killick* v *Second Covent Garden Property Co Ltd* is distinguishable on the facts from the present case. The present case is one where consent to an assignment would inevitably involve the continuance of the subsisting breach of covenant in circumstances where the landlords could reasonably predict that they would be unable to prevent that continuance. That is how the matter stands on the facts now known. At the time when the originating summons was issued the plaintiffs through their solicitors were declining to state whether or not the assignee intended to reopen the store, although writing in terms calculated to induce the belief that they would open the store.

I say that because of what is said by the plaintiffs' solicitors in a communication from them to the defendants' solicitors dated August 7 1986, where at para 3 they say:

This is not a case where our clients can simply close the store in breach of covenant to keep it open. Prior to closing they had found a very substantial assignee well able to perform all the covenants of the lease and experienced in trading from premises of the kind in question and sought your client's consent to the assignment. In the circumstances our clients have not preferred to open the shop premises.

The suggestion that the assignees were well able to carry on trading is followed by the indication that the clients were not prepared to open the shop premises; and the clients there referred to are not the assignees but the present plaintiffs. There was, in my judgment, good ground at that time, that is at the beginning of August, for the landlords to entertain the view that the facts were as they now turn out to be; that is that the assignee does not intend to reopen the store but intends to dispose of the underlease. It follows, in my judgment, that the landlords cannot be said to have been acting unreasonably in refusing consent in the circumstances.

There are, however, further grounds which lead to the same conclusion. In *Fuller's Theatres & Vaudeville Co* v *Rofe*, a decision of the Privy Council reported in [1923] AC 435, and dealing with the kind of question that I am concerned with here, Lord Atkins, in giving their lordships' judgment, said at p 440:

The lessor is, in their Lordships' view, entitled to be told what is in substance the true nature of the transaction to which he is asked to assent.

I find that the lessors were not told the true nature of the transaction. They were being led to believe that the assignment was an assignment to an intending occupier of the premises. The reality is that the premises in question are proposed to be assigned to a company which has no intention of occupying but intends merely to dispose of the property and apparently to do that only after making alterations to the premises, the nature of which has not even now been disclosed. In saying that they were led to believe that the assignment was to an intending occupier, I equate for that purpose the holding company Eldergate and the group, the Dee Corporation, which was acquiring the property and one of whose subsidiaries might be the occupier of the premises. It is clear that at one juncture the defendants were led to believe that such a subsidiary, Carrefour Ltd, was the intended occupant.

The lessor has not been told what in substance is the true nature of the transaction. There was ample material in correspondence between the parties' solicitors to indicate with a high degree of probability that what has turned out to be the case was so, and in these circumstances this is a further reason why the plaintiffs failed to show that the landlords were acting unreasonably in not granting consent. Furthermore, in the words which I have already cited in the judgment of Danckwerts J in *Re Town Investments Ltd Underlease* (see p 314):

. . . in considering whether to give or withhold consent the landlords were entitled to consider the effect which the transaction might have upon their ability in the future to let satisfactorily the different parts of their property, particularly in case of default on the part of the tenant in performing his obligations.

The landlords here are, in my judgment, entitled to consider the likely effect upon their ability to let other parts of the property and, indeed, to obtain the appropriate rents for their other property in the centre. At all material times there was a high likelihood, now shown to be a certainty, that the assignee would not keep the store open and the landlords are entitled to consider the effect which that would have upon their ability not only to let the other property in the centre but to obtain satisfactory rents for them. I accept Mr Spence's submission that whatever the case may be for the ground- and first-floor premises, those on the second floor would inevitably be adversely affected by an empty store.

The Arndale Centre contained property, various shops and so on, which are let both on the ground floor and first floor and, in addition to the store in question, there are other shops on the second floor. The evidence to the effect that properties on the second floor would be adversely affected by an empty store on their level is all on one side. It is no answer that Woolco's closure has already caused damage; it is not unreasonable for the landlords to object to being party to a continuance of the damage. As to the ground and first floors, I find that these also have been adversely affected by the closure. The evidence of the shop proprietors of the figures as to car parking and as to entrance to the centre all indicate a decline in custom to the centre and in my view commonsense indicates that whether or not other causes are important one need look no further than the obvious cause, the closure of Woolco, to find what at least is a contributory factor, and a very significant one, to that decline.

Mr Gaunt submits that the balance of detriment in respect of parties should be considered in determining whether or not the landlords should give consent. He relies on the passage in Balcombe LJ's judgment in *International Drilling Fluids Ltd* v *Louisville Investments (Uxbridge) Ltd* where in para six he deals with that matter. What Balcombe LJ says there gives rise to this question:

Is this a case where

and I quote the lord justice's words

there is such a disproportion between the benefit to the landlord and the detriment to the tenant if the landlord withholds his consent to the assignment that it is unreasonable for the landlord to refuse consent?

Mr Gaunt submits that the tenants' detriment is threefold. First, loss of the sum of £570,000; second, loss of the Dee Corporation's indemnity against rent and repairing and other liabilities; and third, the need, in the light of the tenants' alleged inability to reopen the store, to market the premises.

As to the first, Mr Spence submitted that loss of a gain is not the same thing as a detriment; the converse submission being made by Mr Gaunt. The point I think is this. The plaintiffs have contracted to sell all 11 Woolco properties for £26m. If in any case the landlords refuse consent to an assignment to the Dee Corporation of one of the properties, that price then falls to be reduced by the portion of it which under the contract is attributable to the property in question. The Middleton property portion is £570,000. The evidence is that the cost of repairing the demised property so that the residue of the term of the underlease can be sold is more or less equal to the value of the residue of the term once the property is repaired. So it is said that if consent is refused the detriment will be the difference between £570,000 and in effect little or nothing, that is to say the detriment will be approximately £570,000.

The detriment arises, however, from the terms of the contract made with the Dee Corporation and that contract is not one relating to the Middleton property alone. Had there been a contract relating only to the demised premises, there is nothing to indicate that the plaintiffs would have suffered any damage had they had to find a purchaser prepared to trade from the store instead of the one who was not prepared to do so and to whom the landlords objected. The detriment arising from the terms of the Dee Corporation contract is, in my judgment, not a relevant detriment. In considering detriment it is necessary when the landlord considers the detriment to himself that he confines his consideration to relevant detriment; that is, detriment not collateral. The same, in my judgment, applies to detriment suffered by the tenant. The continuing liability to pay rent is detrimental only because the plaintiffs closed the store and are making no profit to meet the rent; it does not arise from the refusal by the landlords to grant consent to the assignment. Furthermore, the need to market the premises once again arises from the closure not from the refusal.

As to the detriment to the landlord, Mr Gaunt submits that it is the plaintiffs' closure that gives rise to the detriment, not the assignment. Certainly the closure is detrimental, but an assignment to an assignee, to an assignee who is not going to trade, aggravates the detriment already caused. In my view this is not a case where the disproportion of relevant detriment makes it unreasonable for the landlord to refuse consent. On the contrary I consider that the balance of detriment tends to the opposite conclusion. When the detriment to the landlord is not counterbalanced by relevant detriment to the tenant, the reasonableness of refusing consent is underlined. I conclude, therefore, that the plaintiffs are not entitled to a declaration that the defendant landlords unreasonably withheld consent to the assignment.

I turn to the counterclaim. Mr Gaunt relied upon a series of cases as showing that the courts would not grant a mandatory injunction to enforce the performance of such a covenant as that in question, namely to keep open a shop or store. The cases upon which he relied were *Hooper* v *Brodrick* (1840) 11 Sim 47, where an injunction was refused, being in the nature of an injunction to run an inn; *Powell Duffryn Steam Coal Co* v *Taff Vale Railway Co* (1874) LR 9 Ch App 331, where no injunction was granted to run a railway; *Attorney-General* v *Colchester Corporation* [1955] 2 QB 207, where an injunction was refused where it was sought to compel the defendant to run a ferry; *Dowty Boulton Paul Ltd* v *Wolverhampton Corporation* [1971] 1 WLR 204, where the injunction sought and refused was that the defendant be compelled to run an airfield; and finally, *Gravesham Borough Council* v *British Railways Board* [1978] Ch 379, where once again an injunction to compel the running of a ferry was refused.

These cases were all considered in an unreported judgment of Slade J given on March 28 1979, *Braddon Towers Ltd* v *International Stores Ltd*. That was a case where the facts were significantly similar to those of the present case. The defendant tenants had covenanted in their lease at all times of the year during the normal business hours of the locality except when the demised premises might be closed for alterations or upon a change of occupier to keep the shop open as a first-class shop and at all times to give the shop a display window suitably and attractively placed and use their utmost endeavours to develop, improve and extend the business and not to do or permit or suffer to be done anything which might injure the goodwill of the said business and/or any other businesses carried on in the Centre. It is to be noted that there is there a covenant to use the utmost endeavours to develop, improve and extend the said business; but I do not think that anything turns on that because the relief sought related to the keeping open of the store and preventing the defendants from ceasing to operate the premises as a supermarket.

The relief sought in the notice of motion, that is in interlocutory proceedings, is for an order restraining the defendants, until judgment or further order, from ceasing to operate the premises as a first-class supermarket and from opening another supermarket in the area; so that no relief was being sought on the lines of compelling the defendants to use their utmost endeavours to develop, improve and extend the business. Accordingly, I approach the judgment of Slade J on the basis that he was dealing with, in substance, a covenant of the same character as that with which I am concerned.

In the course of his judgment Slade J observed:

The plaintiffs might fairly, in the case before me, be said to be fighting a battle not only on their own behalf but on behalf of the tenants of the sweet shops and the many members of the public who wish to see the supermarket kept open.

Then, having dealt with the conduct of the defendants, which he found may have been rather unattractive, and adverted to the fact that they had expressed no kind of apology or regret for their clear breach of covenant, he said:

The principal question for this court, however, is not whether the defendants have behaved in a shabby manner, it is whether Mr Godfrey is correct in submitting that on the present state of the law the court is precluded from granting a mandatory injunction compelling someone to carry on a business.

He went on: "At least so far as this court is concerned this submission is correct." He dealt with the point as to whether the reason for the court's reluctance to make such an order was the difficulty of informing the defendants by the terms of the order of what they were obliged to do, and he said:

In particular for all Mr Godfrey's colourful examples of hypothetically borderline cases, I am not convinced that the defendants in the present case would have any real practical difficulty in complying with an order of the court effectively compelling them to observe the provisions of the clause in the lease.

I would observe that here I think that an injunction ordering the plaintiffs to keep the store open for retail trade at certain hours and on certain days could again be one which would not give rise to any real difficulty in their knowing what it was they were obliged by the order to do. Then Slade J, having referred to a number of the cases dealing with the matter, went on to say:

It appears, however, from recent authority that the court may to some extent be abandoning the distinction between mandatory and prohibitory injunctions. In *Shiloh Spinners Ltd* v *Harding* [1973] AC 691, 724, Lord Wilberforce in a speech with which Viscount Dilhorne and Lord Pearson and

Lord Kilbrandon wholly agreed, said: "Where it is necessary and in my opinion right to move away from some 19th-century authorities is to reject as a reason against granting relief the impossibility for the courts to supervise the doing of work."

He referred then to what was said by Megarry J in *Giles (CH) & Co Ltd* v *Morris* [1972] 1 WLR 307, 318, that the so-called rules that contracts involving the continuous performance of services could not be specifically enforced is plainly not absolute and without exception and that it could not be based on any narrow consideration such as difficulties of constant superintendence by the court.

He quoted this passage from Megarry J's judgment:

Mandatory injunctions are by no means unknown, and there is normally no question of the court having to send its officers to supervise the performance of the order of the court. Prohibitory injunctions are common, and again there is no direct supervision by the court. Performance of each type of injunction is normally secured by the realisation of the person enjoined that he is to be punished for contempt if evidence of his disobedience to the order is put before the court.

And then he referred to a recent decision of his own, *Gravesham Borough Council* v *British Railways Board* [1978] Ch 379, 405, where, after brief reference to authority, he expressed the view:

that it cannot be regarded as an absolute and inflexible rule that the court will never grant an injunction requiring a person to do a series of acts requiring the continuous employment of people over a number of years,

and went on to say that the "jurisdiction is one that will be exercised only in exceptional circumstances".

Having referred to some other authorities and in particular to the decision of Bacon V-C in *Greene* v *West Cheshire Railway Co* (1871) LR 13 Eq 44, where Bacon V-C ordered specific performance of an agreement to maintain a siding alongside a railway line upon land belonging to the plaintiff, he said:

In the *Greene* case, however, Bacon V-C was not dealing with a covenant to carry on the business nor was he faced with at least three reported cases where it was thought in the first instance that clearly the court will never grant a mandatory injunction of the particular nature under consideration.

He then said:

Whether or not this may be properly described as a rule of law, I do not doubt that for many years solicitors have advised their clients that it is a settled and invariable practice of this court never to grant mandatory injunctions requiring persons to carry on business. In my judgment, there is no real prospect of the trial judge in this case being persuaded to depart from the principle acted upon by at least three different judges sitting at first instance, still less likely would it be right for the court at first instance to depart from it on an interlocutory application. The principle that now lies behind the rule of practice may perhaps need rethinking at least in relation to those cases where it is impossible to find with sufficient accuracy, with sufficient certainty, the obligations of the person enjoined to carry on the business. This process, however, is not an appropriate function for this court on this present interlocutory motion.

I think, however, it is clear that the reason for Slade J refusing relief was his view that at the trial the injunction would not be sustained.

There is one other case which I have considered and that was the decision of Mervyn Davies J, *Posner* v *Scott-Lewis,* reported in [1986] 1 EGLR 56, now reported in [1986] 3 WLR 531, where the covenant which the plaintiffs sought to enforce was one whereby the lessor covenanted to employ, so far as the lessor's power allowed, a resident porter for the following purposes and no other purposes: (a) to keep clean the staircases and entrance hall, landing and passages and lift; (b) to be responsible for looking after the central heating, and domestic hot-water boilers; (c) to carry down rubbish from the flats to the dustbins outside the building every day.

There, Mervyn Davies J, having referred to *Ryan* v *Mutual Tontine Westminster Chambers Association* [1893] 1 Ch 116, a case where the landlord had undertaken to employ a porter, but where the porter was to be and act as the servant of the tenant, and the court refused to grant an injunction, and also referred to the decision of Megarry J which I have already mentioned (*Giles* v *Morris*), read a passage from his judgment which ends by saying:

The present case, of course, is *a fortiori*, since the contract of which specific performance has been decreed requires not the performance of personal services or any continuous series of acts, but merely procuring the execution of an agreement which contains a provision for such services or acts.

Having also referred to what was said in *Shiloh Spinners Ltd* v *Harding*, which I have already mentioned, and that something to the like effect was said by the Vice-Chancellor in *Tito* v *Waddell (No 2)* [1977] Ch 106, Mervyn Davies J concluded thus:

In the light of those authorities, it is, I think, open to me to consider the making of an order for specific performance in this case, particularly since the order contemplated is in the *a fortiori* class referred to by Megarry J in the last sentence of the extract from *C H Giles & Co* v *Morris* . . . Damages here could hardly be regarded as an adequate remedy.

That decision in *Posner* v *Scott-Lewis* is, I think, distinguishable from the present case because here, were an order made, the plaintiffs themselves, that is until they find a suitable assignee, would have to carry on business in the store. It is not a case simply of their entering into a contract until somebody else moves there. This case cannot really be distinguished from the *Braddon Towers Ltd* v *International Stores Ltd* case, the decision of Slade J, and for the like reason that Slade J found compelled him to refuse injunctive relief, I come with like reluctance to the conclusion that I cannot grant an injunction by way of an order for specific performance by the plaintiffs. I do not, however, refuse that relief on the grounds that there has been delay on the part of the defendants in the proceedings. When the defendants learned that the store was to be closed on July 19 their understanding was that that was for the purpose of a sale to an assignee who, so it was said in the first place, would trade from the premises. In these circumstances, there was no ground for then applying for relief.

Having regard to the uncertainty in which the plaintiffs kept the defendants about the proposed intentions of the proposed assignee, the reason for the present application only arose in the course of these present proceedings.

Although I find myself reluctantly compelled to refuse to grant the specific performance which the defendants seek, they are, I find, entitled to an inquiry as to damages.

Court of Appeal
January 21 1987
(Before Sir John ARNOLD, President, and Lord Justice STOCKER)

STENT v MONMOUTH DISTRICT COUNCIL

Estates Gazette May 9 1987

282 EG 705-715

Landlord and tenant — Covenant to repair — Appeal by local authority from county court decision in favour of tenant holding authority liable for breach of covenant to repair and maintain the structure and exterior of dwelling-house — The dispute concerned the front door of the dwelling-house, which stood on an exposed site facing the prevailing south-west wind — Tenant complained of the constant ingress of water blown through or under the door which, over a period of 30 years, had been a source of trouble and inconvenience and had *inter alia* caused damage to carpets — A variety of remedial works had been carried out by the local authority landlords over this period without success, including replacement of parts of the door which had rotted and, in 1979, replacement of the whole door — The trouble was eventually cured in 1983 by the installation of a purpose-designed weatherproof door, an aluminium self-sealing door unit — In an action in the county court the tenant was awarded £250 for damage to carpets, £100 as general damages for loss of amenity and £37.50 interest — As the issue was one which affected other houses in the same row, the landlords appealed — Although the tenant's claim mentioned the implied obligations under section 32 of the Housing Act 1961 and also the provisions of the Defective Premises Act 1972, the argument was based on the express covenant to maintain and repair the structure and exterior — The appellants submitted that the water penetration was not due to a condition calling for repair under the covenant but to an inherent design defect outside its ambit; any duty to rectify the underlying defect could arise only if there was an existing want of repair — Respondent contended that the house was not in repair if the door did not keep out the wind, water and

the elements; alternatively, there was evidence of actual damage and defects in this particular door — The court considered *Ravenseft Properties Ltd* v *Davstone (Holdings) Ltd, Quick* v *Taff-Ely Borough Council, Post Office* v *Aquarius Properties Ltd* and *Elmcroft Developments Ltd* v *Tankersley-Sawyer* — Held (1) the fact that the door did not fulfil its function of keeping out the rain was not *ipso facto* a defect for the purpose of the repairing covenant; (2) but in this case the door had itself become damaged, had rotted and became out of repair to the extent that it twice required replacement; (3) the replacement of the wooden door in 1983 by the self-sealing aluminium door was a sensible and practicable repair which should have been carried out much earlier — The damage suffered by the respondent was within the ambit of the repair covenant — Appeal by local authority landlords dismissed

The following cases are referred to in this report.

Brew Bros Ltd v Snax (Ross) Ltd [1970] 1 QB 612; [1969] 3 WLR 657; [1970] 1 All ER 587; (1969) 20 P&CR 829; [1969] EGD 1012; 212 EG 281, CA
Elmcroft Developments Ltd v Tankersley-Sawyer [1984] EGD 348; (1984) 270 EG 140, CA
Foster v Day (1968) 208 EG 495, CA
Lurcott v Wakely and Wheeler [1911] 1 KB 905, CA
Post Office v Aquarius Properties Ltd [1985] 2 EGLR 105; (1985) 276 EG 923 (Hoffmann J); [1987] 1 EGLR 40; (1987) 281 EG 798, CA
Quick v Taff-Ely Borough Council [1986] QB 809; [1985] 3 WLR 981; [1985] 3 All ER 321; (1985) 84 LGR 498; [1985] 2 EGLR 50; 276 EG 452, CA
Ravenseft Properties Ltd v Davstone (Holdings) Ltd [1980] QB 12; [1979] 2 WLR 897; [1979] 1 All ER 929; (1978) 37 P&CR 502; [1979] EGD 316; 249 EG 51, DC
Wates v Rowland [1952] 2 QB 12; [1952] 1 All ER 470, CA

This was an appeal by Monmouth District Council, as landlords, from a decision of Assistant Recorder Eifion Morgans, at Pontypool County Court, in favour of Arthur John Stent, tenant of a dwelling-house at 102 Old Barn Way, Abergavenny, Gwent, in an action by the latter for damages for breach of the landlords' covenant to keep in repair the structure and exterior of the dwelling-house.

Nigel Hague QC and J Milwyn Jarman (instructed by K N A Raynor, solicitors' department, Monmouth District Council) appeared on behalf of the appellants; William Gage QC and Brian Watson (instructed by Gabb & Co, of Abergavenny) represented the respondent.

Giving the first judgment at the invitation of Sir John Arnold P, STOCKER LJ said: This is an appeal from so much of the judgment of Mr Assistant Recorder Eifion Morgans given at Pontypool County Court on November 26 1985 as adjudged that the defendants were in breach of their duty to keep in repair the structure and exterior of a dwelling-house known as 102 Old Barn Way, Abergavenny, in the county of Gwent, by failing to prevent water penetration to the front external door of the said property. The learned judge also found that the water penetration had caused damage to the plaintiff's carpets in the sum of £250, and he awarded £100 general damages in respect of loss of amenity and £37.50 interest. Therefore, he awarded a total sum of £387.50, with costs. No issue arises on the causation or remoteness of the damage or of the amount of the damages so assessed.

The issue, which is in form a simple one, but which in fact has presented numerous problems which have been skilfully argued before this court, is whether such damages admittedly caused by water penetration through or under the front door of the premises arose from the failure on the part of the appellants to carry out their obligations under the repairing covenant.

So far as the factual outline of the case is concerned, the situation is this. The premises in question, no 102, is one of a number of similar houses in Old Barn Way built in 1953. The respondent became a tenant in that year and was the original tenant under a weekly tenancy of those premises. He therefore took over the house as the first person to have that experience. The landlords at that date were the Abergavenny Council and the appellants are their successors in title. It is not entirely clear what were the precise terms of the landlords' obligations at the date of the first letting or the document in which such terms were contained, but that does not matter for the purposes of this case, since it is accepted by the appellants that they were under an obligation imposed by a covenant in the lease to keep the exterior and structure of the property in repair. There are among the documents certain printed terms of conditions of tenancy, condition 2 of the landlords' obligations being:

The Landlord will repair and maintain the structure and exterior of the dwelling-house.

Mr Hague agrees that that represents the contractual obligation of the landlords in this case.

The respondent's house and the others in Old Barn Way stand on an elevated and very exposed site facing into the direction of the prevailing south-west wind. It is not disputed that the respondent's house, and for that matter others in the same row, had almost from the outset of the tenancy in 1953 suffered from the ingress of water through or under the front door. From time to time efforts were made by the appellants to rectify that situation.

The appellants rely, in support of the arguments put before this court, on the detail of the evidence given by the plaintiff, and accordingly I will now refer to that evidence.

The plaintiff having described that his front door was wet within six months after he moved in continued:

Windows, letter box and panelling in the door. I reported to the Town Hall (Rent Offices). Houses had just been built. Clerk of Works on Site — Advice was given to put a bolt on top of the door to prevent it from warping. Reported it again, time and time again, we more or less complained every time it rained. The door would become swollen, water came in via door frame. Wood of the frame and door became rotten. The workmen cut off the rotten parts and put a new one in — it happened a few times.
When the Monmouth District Council took over — Report to the Town Hall. Workmen would come and see — the problems like before. Door replaced roughly 1979. No good at all. Water still came in through the glass panel, sides and underneath. Door stuck, frame wanted doing. If you stood inside you could see outside — there was a gap rain, snow, wood lice, anything would get in.
We kept complaining, very rarely they would do anything about it. Put aluminium step below the door — no effect at all. Water stayed in the groove, rain into hall via frame and damaged the carpet. Summer 1983, new aluminium sealed door unit fitted, no problem since then. Door opens inwards, water back well beyond the first step of the staircase. . . .

Then he describes the damage to his carpets, to which it is not necessary to refer.

When he was cross-examined, he said:

For 30 years the problem remained. New house when I went there. Water started coming in through the edge of the glass and panelling. Most of the water came from under the door. There was a groove put under the door a couple of years later . . .
Aluminium door fixed in Summer, 1983. . . .

He describes that groove in re-examination thus:

Groove a couple of inches, 2" deep. It got full up with water. No threshold or a piece of wood across the bottom of the door.

That, in outline, is the evidence that was given to the court as recorded by the learned recorder.

There was also called in support of the respondent's case a Mrs Morgan, a resident in the same street, who lived at 116 Old Barn Way. She said:

I am friendly with the plaintiff. We have had the same problems ourselves. Seen the carpet wet about 2 years after we moved in. Went [to the respondent's] for coffee. The problem went away when the aluminium door fitted in 1983. New door in 1979 no good. It was saturated half-way back to the hall. Bottom step of stairs began to rot. . . .

So she was describing fairly succinctly a similar state of affairs as that which existed and was described by the respondent.

The appellants disclosed a certain number of their work sheets and documents which are contained in the bundle. They do not go back further than 1979. The first document records a complaint of May 11 1979 which apparently was not dealt with by work being carried out until April 13 1980. The details shown upon that job specification are:

Take off door and rehang on new frame and bed in glass door.
Fix new 6' 6" × 2' 9" door frame.
Fix 2 ADS weather strips . . . and repair doors.
Fix new mortice lock to rear door

A complaint of January 7 1980, dealt with apparently on the next day, January 8, indicates as the reported defect:

New door frame fitted recently, but wasn't completed.

Then apparently the work done was to seal around the front door-frame with mastic. That was referring clearly to the new door fitted in 1979.

There is a further report of a complaint on June 28 1982, the defect reported being:

Front door leaking very badly, water running into hall.

The work carried out was apparently:

Remove glass in front door, re-bed 4 pieces of glass.
Take off defective weather board and fix new board.
Fix a storm guard threshold.

That is perhaps sufficient indication. The records from their nature are not necessarily complete or comprehensive, but it would appear, in my view, that the defects dealt with on those occasions were treated by the respondents as being repairs under the repairing covenant. Of course, that does not conclude the matter, because they may have been acting from paternalistic ideals rather than under the compulsion of legal obligation. But that does seem to be how they then dealt with it.

The plaintiff wrote a letter of complaint in 1983, and proceedings being imminent or having started, a surveyor, a Mr Frecknall, was instructed on his behalf. He gave evidence at the trial and produced his report, which included details of his inspection and his conclusion. This report was not formally admitted by the appellants, but its conclusions and findings were not challenged or seriously controverted by the appellants. Indeed, it is that report which forms a great deal of the substance of the appellants' own case. That report describes instructions being received on April 2 1983. An inspection was carried out on April 3, and its objective is stated to have been that the aim of the survey was to assess the likely cause of the problem being ingress of water. Six houses in Old Barn Way were examined by Mr Frecknall, one of them being the respondent's house. The report describes by way of introduction:

The Old Barn Way houses are built on an elevated highly exposed site. With the exception of number 49

which is not the respondent's house

all the houses seen are directly exposed to the prevailing south to south-westerly wind.
The houses were built nearly 30 years ago, and from my inspection it is evident that maintenance has been a constant problem. Painting, loose tiles, eaves details, metal windows and the front doors and frames appear to have been in need of regular repair and maintenance.

The report includes a number of diagrammatic representations of the door in its original form (though that, of course, would have had to have been to a degree speculation on the part of Mr Frecknall) and details of certain attempted rectifications by the appellants though not necessarily to the respondent's house. By way of observation, the report states:

The top surface of the concrete step is at the same level as the finished floor level inside the entrance hall.
There is no evidence that a planted threshold or weather bar was included in the original design detail.

He points out there are certain electrical risks which may arise through the ingress of water or the state of the carpet, and by way of assessment, so the report says:

Water runs down the face of the door and collects on the step which extends in front of the door and frame. Under normal conditions it would be unrealistic to expect the typical base of door detail to be weatherproof.
The exposed nature of the site in question requires a much more effective design solution for the base of the door to achieve a weatherproof detail.

That, of course, was an inspection prior to the insertion of the aluminium self-sealing door.

Modifications, as shown on [the attached plan] have been carried out in an attempt to solve the problem of water penetration by containing the water and draining it back towards the outside, either through the step or through an aluminium extrusion by means of weep holes.
These details have failed because of the severe wind pressure that undoubtedly builds up under the door leaf.

By way of conclusion he says:

The main reason for water penetration under the doors is that the concrete step holds water which has run down the face of the door.
The exposure of the site determines that no amount of run-off provided by chamfering the top surface of the step would prevent water being blown under the door. Water held on the step by wind pressure would eventually cause the bottom of the door frames to rot. This appears to have happened already, and at least two of the houses visited had had new pieces of door frame spliced into the existing frame.

One adds that in 1979 the respondent's house had a completely new door. The conclusions continue:

The modifications which have been carried out include the use of various flexible sealing strip, increasing the size of weather bars and rebates by adding external timber linings, . . .

And then he continues with the other efforts which have been made. Finally, under the heading of "Conclusions", he says:

Unless this particular design detail is treated from first principles it is unlikely that the tenants of these houses will ever be free from the problem of water penetration and the subsequent recurring damage to carpets and internal fittings and decorations.

He then made three recommendations, the second of which appears to have been one adopted by the appellants, for the second recommendation is:

Purpose-designed weatherproof doors and frames with integral seals are available and could be used. These are sophisticated components and require fitting by specialists.

It is unnecessary to deal with his other two recommendations. The recommendation adopted is in fact the cheapest of the three recommendations. One perhaps should add that it was not suggested in the course of the trial that such a purpose-designed weatherproof door would not have been available at a much earlier date, and certainly in 1979.

As has been observed, it is largely upon that report, together with the respondent's evidence, that the appellants found their case as argued in this court, and indeed before the trial judge, and it is for that reason that I have thought it proper to cite fairly extensively from those parts of the evidence.

Put very shortly at this stage, the appellants contend that the cause of the water ingress either through or under the door was not due to any defective condition of the door calling for repair under the covenant, but was due to an inherent design defect, and that accordingly the damage sustained by the respondent was due to design defects outside the ambit of the repair covenant and, therefore, cannot found a valid claim for breach of covenant in respect of the repairing obligations.

The respondent gave evidence concerning the damage and described it. It is unnecessary to relate that part of his evidence. Nor is it necessary to go in any detail into the contentions in the pleadings. It is perhaps sufficient to say that in the amended particulars of claim by para 3 the covenant to repair is set out, and the provisions of section 32 of the Housing Act 1961 are alleged to apply to the tenancy and the claim was founded in the alternative under the provisions of that implied term under the Housing Act. It is also contended that the provisions of section 4 of the Defective Premises Act 1972 applied and were breached. It is unnecessary to express any view on that, because neither statutory provision seems to be relevant to the real issues in this case and neither was relied upon in argument before us.

So far as the defence is concerned, the only matter of relevance seems to be that contained in subpara (3) of para 4 of the defence, in which it is affirmatively alleged that the defendants (that is the appellants) had responded to complaints received from the plaintiff and carried out the following repairs, being such repairs as were reasonably necessary at the time of the relevant complaint, and they are listed:

Changed the door.
Increased the width of the door stop and inserted additional capillary grooves.
Inserted normal weather stripping.
Inserted high performance weather stripping.
Inserted aluminium storm guards.

The only possible relevance of those pleadings, as it seems to me, is the point mentioned earlier, that the appellants at all times seem to have accepted that their attempts to prevent the ingress of water included those matters which might be said to involve a change in the nature of the structure itself and accordingly attempts to deal with a latent or design defect under the provisions of the covenant.

The judge's findings were fairly shortly expressed. He found that:

. . . water did seep under the door and between the wall and frame and that [the respondent] had a genuine complaint, as indeed one can gather from the evidence of Mr Frecknall, a well qualified architect and arbitrator of fifteen years' standing.
His report was not agreed as such, but he had seen six houses in Old Barn Way, including the plaintiff's. His evidence indicated that there was no threshold or weatherboard bar in the original design and it is true, as Mr Jarman [for the appellants] suggests, that this may be a matter of improvement as opposed to repair, but in my judgment it is a repair, notwithstanding *Quick* v *Taff-Ely*

Borough Council [1985] 3 All ER 321 and *Pembery* v *Lamdin* [1940] 2 All ER 434

which are matters to be considered hereafter

which in my judgment are not on all fours with the present case. Although I am bound by any decision of the Court of Appeal the case of *Pembery* was in relation to a damp proof course, and the case of *Quick* was mostly in relation to condensation, as opposed to a patent defect. I find as a fact here, having heard all the evidence in this case, including Mr Taleman's,

I interpose that he was an official of the appellants

who gave evidence in the best manner he could, that his authority (the defendant) — [appellants] — has no records before 1979. I am not going to go through the subsequent reports here as to the repairs which have been carried out, but they include replacement of a defective weatherboard: P2 p 4. These are repairs not improvements. Even if they were improvements, as Mr Jarman invited me to hold, I would find that they amounted to repairs since they were defects here since the beginning of this (Council) estate.

The judge continued:

Under the circumstances, therefore, I have come to the conclusion rightly or wrongly that it was the defendant's duty to repair the external and structural parts and incur any capital expenditure which was required in the replacement of any doors, windows etc, in accordance with the terms of the tenancy and as laid out in P3, which was given to the plaintiff in about 1980.

Those conclusions, of course, beg specifically the very issue which is challenged in this court.

It is again unnecessary, as it seems to me, to refer to the formal grounds of appeal, since all the relevant matters have been reflected in the cogent argument of Mr Hague before this court. Those arguments, I hope, can be summarised accurately as follows. First, and this is an important matter, that the damage did not result from disrepair, but as a result of the door not being effective in the first place. That is to say that the cause of the ingress of water was original design defect. He argued that, in such a case, a duty to repair arose only if there was an existing defect in fact which itself called for repair, and the only "sensible" way of achieving this repair was to rectify the design defect also. (Many synonyms have, in the authorities cited, been substituted for the word "sensible" used by counsel in his argument.) He submitted that in the instant appeal those facts did not arise, since the door could have been replaced in its original form, and the replacement of the original door would be the limit of the appellants' obligations. He agreed that those comments were subject to a general observation that it was possible that a defect calling for repairs under the covenant might be due to inherent underlying defect and was repairable as such, but that that comment was subject to the criticisms to which I have just referred. He submitted that it was not correct in law that a duty to repair applies simply because the door lets in water or is not wind and watertight. There must be some specific defect or lack of repair within the terms of the covenant that must be proved; and in this case no such specific defect or lack of repair was established. He gave as a contrasting example of that proposition that if slates of a house were loose due to defect, then there would be a liability because the house would be in disrepair in that respect, but if the slates had never been there, then, although that would give rise to the ingress of water there would be no disrepair and no liabilities. He enlarged upon those submissions in the context of various authorities to which reference has to be made. Second, he submitted generally that on the facts of this case such damage or lack of repair to the door as may have been proved over the years did not itself cause the damage which was solely due to the design defect presented by the alignment of the outside step and the inside floor in conjunction with the exposed site upon which this building had been built.

Mr Gage for the respondent submitted, first, a general proposition that the purpose of a door is to provide access and keep out wind, water and the elements, and in the context of a private dwelling-house and of a covenant to keep in repair such a house is not in repair if the door in question does not fulfil those basic purposes. He submitted in the alternative that there was evidence of lack of repair, actual damage and actual defects in this particular door.

I therefore turn to consider the various authorities which have been cited before this court. A convenient starting point, though not chronologically the first of the cases cited, is the case of *Ravenseft Properties Ltd* v *Davstone (Holdings) Ltd* [1980] QB 12. This was a case of a building constructed of concrete with a facing of stone cladding. When constructed there were no expansion joints to deal with the differential expansion of those two materials respectively, and also the facing stones had not been properly tied in and had been erected in a way which was not a workmanlike one. The consequence of those two matters was that there was a bowing of the stonefacing with a consequence of the danger of the stones falling, to the danger of persons below. The lack of the expansion joint was an inherent design factor. The failure to tie in the defective stone was a matter of defective building practice. The learned trial judge, Forbes J, held that both, in the circumstances of the case, fell within the repairing covenant. At the time the building was erected the importance of expansion joints had not been appreciated in the trade, and it was not then current practice to erect a building with such expansion joints. The unsuccessful defendants put forward a proposition which had two limbs, which are set out by the learned judge at p 17 at A of the report. That reads:

Now despite somewhat lengthy cross-examination of the landlords' witnesses, there is here no dispute on fact and the tenants called no evidence. In these circumstances the landlords claim to be reimbursed by the tenants the cost of the work carried out and they put their claim under three headings.

He deals with those headings, and continues:

The tenant's defence is twofold. Mr Colyer says first, there is, in that branch of landlord and tenant law concerned with repairing covenants, a doctrine of inherent defect which is applicable to such covenants to repair. This is where wants of reparation arise which are caused by some inherent defect in the premises demised, the results of the inherent defect can never fall within the ambit of a covenant to repair . . . Secondly,

and this is what has been referred to in the argument as the second limb

he says, if that proposition is wrong the covenantor is still not bound to pay for any works which, in fact, remedy the inherent defect.

On the first limb of that proposition, the learned judge's findings and conclusion are set out at p 21 at C, where he says:

I find myself, therefore, unable to accept Mr Colyer's contention that a doctrine such as he enunciates has any place in the law of landlord and tenant. The true test is, as the cases show, that it is always a question of degree whether that which the tenant is being asked to do can properly be described as repair, or whether on the contrary it would involve giving back to the landlord a wholly different thing from that which he demised.

On the second limb, after considering the case of *Lurcott* v *Wakely and Wheeler* [1911] 1 KB 905, a case in which the court had found recoverable under a repairing covenant the provision of footings in a wall, Forbes J said, on p 22 at D:

. . . By this time it was proper engineering practice to see that such expansion joints were included, and it would have been dangerous not to include them. In no realistic sense, therefore, could it be said that there was any other possible way of reinstating this cladding than by providing the expansion joints which were, in fact, provided. It seems to me to matter not whether that state of affairs is caused by the necessary sanction of statutory notices or by the realistic fact that as a matter of professional expertise no responsible engineer would have allowed a rebuilding which did not include such expansion joints to be carried out. I find myself, therefore, bound to follow the guidance given by Sir Herbert Cozens-Hardy MR in *Lurcott's* case [1911] 1 KB 905, 914-915:

"It seems to me that we should be narrowing in a most dangerous way the limit and extent of these covenants if we did not hold that the defendants were liable under covenants framed as these are to make good the cost of repairing this wall *in the only sense in which it can be repaired*, namely, by rebuilding it according to the requirements of the county council."

The next case to which reference requires to be made is the case of *Quick* v *Taff-Ely Borough Council* [1986] QB 809*. This is a case upon which the appellants strongly relied. The headnote, to give the facts, reads:

The plaintiff was the tenant of a house owned by the defendant council. As a result of very severe condensation throughout the house decorations, woodwork, furnishings, bedding and clothes rotted, and living conditions were appalling. The condensation was caused by lack of insulation of window lintels, single-glazed metal-frame windows and inadequate heating. The plaintiff brought proceedings in the county court, alleging that the council was in breach of its covenant, implied in the tenancy agreement by section 32 (1) of the Housing Act 1961, "to keep in repair the structure and exterior" of the house.

I think that that is, perhaps, a sufficient résumé for present purposes.

At p 817 between C-F, Dillon LJ, giving the first judgment of the court, said this:

The judge delivered a careful reserved judgment in which he reviewed many of the more recent authorities on repairing covenants, starting with *Pembery* v *Lamdin* [1940] 2 All ER 434.

---

* Editor's note: Also reported at [1985] 2 EGLR 50; (1985) 276 EG 452.

Submissions had been made upon which the judge founded his judgment as follows:

(1) recent authorities such as *Ravenseft Properties Ltd* v *Davstone (Holdings) Ltd* [1980] QB 12 and *Elmcroft Developments Ltd* v *Tankersley-Sawyer* (1984) 270 EG 140 show that works of repair under a repairing covenant, whether by a landlord or a tenant, may require the remedying of an inherent defect in a building; (2) the authorities also show that it is a question of degree whether works which remedy an inherent defect in a building may not be so extensive as to amount to an improvement or renewal of the whole which is beyond the concept of repair; (3) in the present case the replacement of windows and the provision of insulation for lintels does not amount to such an improvement or renewal of the whole; (4) therefore, the replacement of the windows and provision of the insulation to alleviate an inherent defect is a repair which the council is bound to carry out under the repairing covenant.

He then stated, by way of comment upon those submissions:

But, with every respect to the judge, this reasoning begs the important question. It assumes that any work to eradicate an inherent defect in a building must be a work of repair, which the relevant party is bound to carry out if, as a matter of degree, it does not amount to a renewal or improvement of the building. In effect, it assumes the broad proposition urged on us by Mr Blom-Cooper for the plaintiff that anything defective or inherently inefficient for living in or ineffective to provide the conditions of ordinary habitation is in disrepair. But that does not follow from the decisions in *Ravenseft's* case [1980] QB 12 and *Elmcroft's* case 270 EG 140 that works of repair *may* require the remedying of an inherent defect.

On p 818 at F there appears this passage:

If there is such damage caused by an unsuspected inherent defect, then it may be necessary to cure the defect, and thus to some extent improve without wholly renewing the property as the only practicable way of making good the damage to the subject-matter of the repairing covenant. That, as I read the case, was the basis of the decision in *Ravenseft* [1980] QB 12. There there was an inherent defect when the building, a relatively new one, was built in that no expansion joints had been included because it had not been realised that the different coefficients of expansion of the stone of the cladding and the concrete of the structure made it necessary to include such joints. There was, however, also physical damage to the subject-matter of the covenant in that, because of the differing coefficients of expansion, the stones of the cladding had become bowed, detached from the structure, loose and in danger of falling. Forbes J in a very valuable judgment rejected the argument that no liability arose under a repairing covenant if it could be shown that the disrepair was due to an inherent defect in the building.

That was limb 1 of the argument put before us.

He allowed in the damages under the repairing covenant the cost of putting in expansion joints, and in that respect improving the building, because, as he put it, at p 22, on the evidence "In no realistic sense . . . could it be said that there was any other possible way of reinstating this cladding than by providing the expansion joints which were, in fact, provided."

That is a passage which has already been cited. Dillon LJ then goes on to consider the *Elmcroft* case, to which reference will be shortly made in this judgment; and he sets out the facts in the *Elmcroft* case.

Dillon LJ, having considered the cases of *Ravenseft* and *Elmcroft*, concluded at p 820 at H:

But the crux of the matter is whether there has been disrepair in relation to the structure and exterior of the building and, for the reasons I have endeavoured to explain, in my judgment, there has not, *quoad* the case put forward by the plaintiff on condensation as opposed to the case on water penetration.

He therefore, as I understand it, rejected the submission on the basis that there was no damage which could fall within a repairing covenant other than condensation which was due to inherent causes.

Lawton LJ, at p 821 at F, said this (and I quote this passage at some length):

It has to be approached in the same way as the letting of any house which is outside the provisions of section 6 of the Housing Act 1957, as amended, in respect of which there is a covenant by the landlord "to keep in repair the structure and exterior . . ." The standard of repair may depend on whether the house is in a South Wales valley or in Grosvenor Square; but, wherever it is, the landlord need not do anything until there exists a condition which calls for repair. As a matter of the ordinary use of English that which requires repair is in a condition worse than it was at some earlier time. This usage of English is, in my judgment, the explanation for the many decisions on the extent of a landlord's or tenant's obligation under covenants to keep houses in repair. Broadly stated, they come to this: a tenant must take the house as he finds it; neither a landlord nor a tenant is bound to provide the other with a better house than there was to start with; but, because almost all repair work requires some degree of renewal, problems of degree arise as to whether after the repair there is a house which is different from that which was let. I do not find it necessary to review the cases which were decided before 1980.

During the last 20 years the way in which houses and other buildings have been constructed has produced new problems. Traditional materials may not have been used: new methods of construction may have been employed. The materials may fail; the methods may prove to have been unsatisfactory, causing damage; the building may have got into a worse condition than it was when the lease was granted. In such a case there is need for repair. The landlord or the tenant may be under an obligation to put right what has gone wrong; and, in putting right what has gone wrong, it may be necessary to abandon the use of the defective materials or to use a different and better method of construction.

When something like this happens, does the landlord or the tenant have a better building? In one sense he does: he gets a building without the design defect which caused the damage; but the repair could only have been done in a sensible way by getting rid of the design defect.

In parentheses, I observe that Mr Hague places importance on the use there of the word "only".

Forbes J had to consider this problem in *Ravenseft Properties Ltd* v *Davstone (Holdings) Ltd* [1980] QB 12. In that case the repair work could not be done satisfactorily without getting rid of a design fault. He adjudged that doing so did not amount to such a change in the character of the building as to take the works out of the ambit of the covenant to repair: see pp 21-22. This court in *Smedley* v *Chumley & Hawke Ltd*, 44 P & CR 50, approached the problem in the same way. The *Ravenseft* case [1980] QB 12 does not seem to have been cited. It was, however, cited to this court in *Elmcroft Developments Ltd* v *Tankersley-Sawyer* (1984) 270 EG 140 and clearly approved. It was not cited to this court in *Wainwright* v *Leeds City Council* (1984) 270 EG 1289. . . . I am satisfied that the approach of Forbes J in the *Ravenseft* case [1980] QB 12 was right.

Again I observe in parentheses that there are a number of cases in this court, including this one, in which the ratio of Forbes J's decision has been specifically approved. Lawton LJ continued:

It follows that, on the evidence in this case, the trial judge should first have identified the parts of the exterior and structure of the house which were out of repair and then have gone on to decide whether, in order to remedy the defects, it was reasonably necessary to replace the concrete lintels over the windows, which caused "cold bridging", and the single glazed metal windows, both of which were among the causes, probably the major causes, of excessive condensation in the house. An argument along the following lines was put before this court: the evidence established that some of the wooden frames into which the single glazed metal windows were inserted had rotted and that nearby plaster had crumbled away. Mr Hague, for the purposes of this case, accepted that the plaster was part of the structure. Repairing the wooden frames and the plaster could only be done sensibly if the single glazed metal windows and the lintels were replaced by ones of better design. The council should have appreciated that this was so. A submission of this kind would have required the trial judge to make findings of the same kind as Forbes J made in *Ravenseft Properties Ltd* v *Davstone (Holdings) Ltd* [1980] QB 12. He made none, almost certainly because he was not asked to do so. He referred to Forbes J's judgment in these terms:

"He held that want of repair due to an inherent defect could fall within the ambit of a repairing covenant, and that it was a question of degree whether work could properly be described as repair or whether it so changed the character of the building as to involve giving back to the landlord a different building from that demised."

He seems to have overlooked the important fact in the *Ravenseft* case that the cladding around the building was in disrepair and could only be repaired in a sensible way if the design fault were put right.

In my judgment, there must be disrepair before any question arises as to whether it would be reasonable to remedy a design fault when doing the repair . . .

It seems to me that in the *Quick* case had there been evidence that there was actual disrepair in some material respect, that is to say some respect material to the rectification, the decision might have been the other way. But, of course, I cite the case only for the dicta that it contains, and not the conclusion that might arise had the findings of fact been different from those that they were.

Neill LJ made this observation at p 823 at E:

The authorities to which we were referred establish that, in some cases, the only realistic way of effecting the relevant repairs is to carry out some additional work which will go somewhat further than putting the property back into its former condition and will indeed result in some improvement. But this case does not fall into that category. The repair work consisting of the replacement of the defective parts of the wooden surrounds and the replacement of the areas of plaster did not require as a realistic way of effecting those repairs the replacement of the metal windows by wooden-framed windows or windows with PVC frames.

Mr Gage on behalf of the respondent distinguishes that case from the instant case on the basis that the actual observable physical damage did involve or could realistically or sensibly involve the repair also by altering the design of what has been described as design defect.

Mr Hague relied on the case of *Foster* v *Day* (1968) 208 EG 495. For my part, despite his argument to the contrary, I do not derive any great assistance from it, since the judgment of Edmund Davies LJ (which is very shortly set out, in no more than eight lines) was a matter which was clearly obiter and, in my view, begs rather than resolves the issues before this court.

I now turn to consider a very recent case relied upon strongly by Mr Hague, *Post Office* v *Aquarius Properties Ltd* (1985) 276 EG 923, the only report of which before this court is the transcript itself of a judgment handed down on December 18 1986*. It was a case which concerned a defect in the structure of a basement which had existed from the time of the construction of the building itself, and on p 15, near the bottom of the page, Ralph Gibson LJ, giving the first judgment of the court, said this:

In my judgment, however, the reasoning of the court in *Quick's* case is equally applicable whether the original defect resulted from error in design, or in workmanship, or from deliberate parsimony or any other cause. If on the letting of premises it were desired by the parties to impose on landlord or tenant an obligation to put the premises into a particular state or condition so as to be at all times for some stated purpose, even if it means making the premises better than they were when constructed, there would be no difficulty in finding words apt for that purpose.

On p 17 he observed:

There is no basis to be found in the decision in *Proudfoot* v *Hart* for holding that the tenant can be held liable under an ordinary repairing covenant to carry out work merely to improve premises so as to remove a defect present since construction of the building.

Slade LJ, at p 21, defined what in his view disrepair meant in these terms:

However, a state of disrepair, in my judgment, connotes a deterioration from some previous physical condition.

Finally, I turn to the case of *Elmcroft Developments Ltd* v *Tankersley-Sawyer* (1984) 270 EG 140. The facts were that the respondents, by their counterclaim, claimed, *inter alia*, damages for breach of covenant of repair. The respondents were tenants of flats in a block of which the appellants were the landlords. The Court of Appeal upheld the judge's finding that the appellants were in breach of their covenants to repair, and at p 140 there is set out the condition of the premises in question. It is said:

He held — and this is not disputed — that there was constructed into the walls what was intended to be a damp-proof course, consisting of slates laid horizontally. These existed in the external and the party walls of the flat, but, owing either to a defect in design or construction or bad workmanship, this layer of slates intended to be a damp-proof course was ineffectual because it was positioned below ground. The result was obvious. It allowed moisture to be drawn up from the ground by the capillary action, with the inevitable consequence that the flats were in a damp condition, rising damp resulting from what was described as the bridging of this damp-proof course, and parts of the interior of the main walls of the flats had been adversely affected up to a height of about 1 to 1½m. The rooms in the flats were damp, and the plaster, decoration and woodwork needed repair or renewal.

There was then cited a graphic description of the condition of two of those flats. The remedial work necessary was referred to on p 141 as being:

The remedial works necessary to eradicate rising dampness in the walls is the installation of a horizontal damp-proof course by silicone injection and the formation by silicone injection of vertical barriers where the front and back external walls meet the dividing walls.

The court then considered and cited from the case of *Lurcott* v *Wakely* [1911] 1 KB 905, and made citations from the judgment of Fletcher Moulton LJ, first the definition of the expression "tenable repair", and then citing from Buckley LJ at p 141 these words:

But if that which I have said is accurate, it follows that the question of repair is in every case one of degree, and the test is whether the act to be done is one which in substance is the renewal or replacement of defective parts or the renewal or replacement of substantially the whole.

The court then considered a passage from the judgment of Lord Evershed MR in *Wates* v *Rowland* [1952] 2 QB 12, quoted by this court in *Brew Bros Ltd* v *Snax (Ross) Ltd* [1970] 1 QB 612, as follows:

Between the two extremes, it seems to me to be largely a matter of degree, which in the ordinary case the county court judge could decide as a matter of fact, applying a common-sense man-of-the-world view; . . .

And in the *Brew Bros* case, a quotation from Sachs LJ at p 640:

It seems to me that the correct approach is to look at the particular building, to look at the state which it is in at the date of the lease, to look at the precise terms of the lease, and then come to a conclusion as to whether, on a fair interpretation of those terms in relation to that state, the requisite work can fairly be termed repair. However large the covenant it must not be looked at *in vacuo*.

The court then cited a passage from the judgment of Forbes J in *Ravenseft Properties Ltd* v *Davstone (Holdings) Ltd* [1980] 1 QB 12, a passage which has already been cited in this judgment and which I, therefore, will not repeat. Ackner LJ at p 142 said this, and it is a relevant passage in my view:

I therefore conclude that the learned judge was wholly right in the decision which he made as to the failure by the appellants to comply with the repairing covenant and their obligation in regard to curing the damp by using the only practical method at this price, namely, injecting silicone into the wall. Mr Whitaker was at one stage prepared to concede that, as the plaster became saturated (which, of course, it was), his clients had the obligation to do the necessary patching — that is removing — the perished plaster and renewing it. I am bound to say that concession made the resistance to inserting the damp-proof course a strange one. The damp-proof course, once inserted, would on the expert evidence cure the damp. The patching work would have to go on and on and on, because, as the plaster absorbed (as it would) the rising damp, it would have to be renewed, and the cost to the appellants in constantly being involved with this sort of work, one would have thought, would have outweighed easily the cost in doing the job properly. I have no hesitation in rejecting the submission that the appellant's obligation was repetitively to carry out futile work instead of doing the job properly once and for all.

One may observe there that that would appear to be an explicit approval of the judgment of Forbes J in *Ravenseft* so far as this second limb of the proposition is concerned and, accordingly, that judgment of Forbes J would, in my view, have clearly received the approbation of this court.

What are the conclusions which should be drawn from those authorities in the light of the evidence which has been accepted by the learned judge and which has been put before us? I find from those authorities, though I must confess with some regret, since in the context of council letting of small houses it conforms in my view with common sense, that Mr Gage's first proposition is stated in terms which are indeed too wide. This is a hypothesis: if the only defect in the door was that it did not perform its primary function of keeping out the rain, and the door was otherwise undamaged and in a condition which it or its predecessors had been at the time of the letting, then it seems to me, on the authorities of *Quick* and *Aquarius*, this cannot amount to a defect for the purpose of a repairing covenant even though, as it seems to me in layman's terms, that a door which does not keep out the rain is a defective door, and one which is in need of some form of repair or modification or replacement. That seems to have been the view indeed of the appellants themselves throughout the year, since they did take steps (though, as it turned out, ineffective steps) to deal with that very problem. However, though common sense would dictate to me that a door which does not keep out the rain is not performing the primary function of a door and is, therefore, defective, and in want of repair it seems to me that in so far as the door was the original one and was wholly undamaged, if that were to be the factual position, then the first proposition of Mr Gage is expressed in terms which are too wide.

In this case, however, the factual position is that the damage undoubtedly did occur. The appellants' own documents illustrate that clearly and graphically. There was damage such as to require the replacement of the door in 1979. The same applies in 1983, and clearly this was also the position on many other occasions both prior to and between those dates. Accordingly, applying the reasoning of this court from the cases cited, and in particular *Ravenseft Properties Ltd* v *Davstone (Holdings) Ltd* and *Elmcroft Developments Ltd* v *Tankersley-Sawyer*, the former having been specifically approved by this court, in my judgment the replacement of the wooden door by a self-sealing aluminium door was a mode of repair which a sensible person would have adopted; and the same reasoning applies if for the word "sensible" there is substituted some such word as "practicable" or "necessary". The argument reflected by Ackner LJ in the passage recently cited from *Elmcroft* seems to me to be precisely in point here. There has been a history of nearly 30 years of difficulty with this door, the difficulty being that it did not keep the rain out, and itself became damaged; and because of the design, from time to time parts of it rotted. It became distorted. It needed accordingly replacement in order to enable it to perform its function at all, quite apart from the question of repairing obvious defects which it had exhibited.

Accordingly, in my view and upon those authorities, in this case

---

*Editor's note: The report of the case in the Court of Appeal appears at p 40 *ante* and at (1987) 281 EG 798. Hoffmann J's judgment will be found at [1985] 2 EGLR 105 as well as at (1985) 276 EG 923.

the repair carried out in 1983 by the installation of a purpose-built, self-sealing aluminium door was one of the methods which could have been adopted much earlier, and which in my view should have been adopted. Of course, it does not follow that the self-sealing door is the only sensible way in which that object could be achieved. There may well have been others, but in my view the obligation under the covenant in this case was one which called upon the appellants to carry out repairs which not only effected the repair of the manifestly damaged parts but also achieved the object of rendering it unnecessary in the future for the continual repair of this door. Accordingly, some such steps as were in fact taken in 1983 should in my view have been carried out at any rate by 1979 and perhaps earlier, there being no suggestion either before the trial judge or before this court that the steps adopted were not ones which were known to the trade in 1979 or which were for any other reasons impracticable at that date. Accordingly, and for these reasons I agree with the conclusion of the trial judge, though perhaps the reasons that I have given are not identical with those expressed in his judgment.

I would dismiss this appeal.

Agreeing, SIR JOHN ARNOLD P said: It is no doubt the case that where there is no repair requiring to be done as in such cases as *Quick* v *Taff-Ely BC* [1985] 3 WLR 981 and *Post Office* v *Aquarius Properties Ltd* (1985) 276 EG 923, cited by my lord, the repairing covenant on its true construction does not require any design defect to be made good. It is, in my judgment, undoubtedly the case on the basis of the authorities relating to cases where there was a repair requiring to be done, such as *Ravenseft Properties Ltd* v *Davstone (Holdings) Ltd* [1980] QB 12 and *Elmcroft Developments Ltd* v *Tankersley-Sawyer* (1984) 270 EG 140 that on the true construction of the covenant to repair there is required to be done, not only the making good of the immediate occasion of disrepair, but also, if this is what a sensible, practical man would do, the elimination of the cause of that disrepair through the making good of an inherent design defect at least where the making good of that defect does not involve a substantial rebuilding of the whole.

Different adjectives have been employed in different cases to describe the degree of necessity which inspires in the particular case the doing of the further remedial works — necessary, sensible, reasonable, satisfactory and the like. But those are only compendious descriptions of the underlying state of affairs in so far as they can be afforded by the use of a single adjective. If one requires an effective exegesis of the conception sought to be embodied in those separate adjectives, one cannot do better, in my judgment, than to look at the descriptive passage in the judgment of Ackner LJ (as he then was) in *Elmcroft Developments Ltd* v *Tankersley-Sawyer* (1984) 270 EG 140 at p 142 in the passage cited by my lord, where he proposes the test in these terms:

The patching work would have to go on and on and on and on, because as the plaster absorbed (as it would) the rising damp, it would have to be renewed, and the cost to the appellants in constantly being involved with this sort of work, one would have thought, would have outweighed easily the cost in doing the job properly. I have no hesitation in rejecting the submission that the appellants' obligation was repetitively to carry out futile work instead of doing the job properly once and for all.

And when he refers to "the appellants' obligation", it is plain in the context that what he means is the obligation upon the true construction of the relevant covenant.

Accordingly, in my judgment, the approach to be adopted is that which my lord has described, and it is plain that if all that was done to the door which stood in need of repair was to patch it or even to renew it and to leave, when so doing, the cause of the damage, which was the absence of any agent to defeat the collection of the rotting water beneath the door, then one was not doing that which the sensible, practical man would have advised as a sensible way of dealing with the problem. Accordingly, on the true construction of this covenant, it seems to me that the right conclusion is that the appellant council had the obligation of making good the design defect which caused the collection of water which occasioned the rotting, and that the failure so to do was in the circumstances a breach of the appellants' covenant for which they were properly required to pay damages by the learned judge.

I agree that the appeal should be dismissed.
The appeal was dismissed with costs.

Chancery Division

July 24 1986
(Before Sir Nicolas BROWNE-WILKINSON V-C)

HILLGATE HOUSE LTD v EXPERT CLOTHING SERVICE & SALES LTD

Estates Gazette May 9 1987
282 EG 715-718

Landlord and tenant — Landlords, having obtained an order for possession on the ground of forfeiture of the lease, went into possession — Judge's order was subsequently reversed by Court of Appeal — Tenants thereupon brought an action against landlords claiming damages of a substantial nature alleging wrongful entry by landlords in breach of express covenant for quiet enjoyment and implied covenant not to derogate from grant — The facts were that the Court of Appeal had reversed the forfeiture decision on the ground that the notice served by the landlords under section 146 of the Law of Property Act 1925 was invalid for failing to specify a reasonable time for remedying the breach which was the subject of complaint — In the present action the plaintiff tenants were claiming that the acts carried out by all parties pursuant to the judge's forfeiture order were unlawful and gave rise to a cause of action — Held, rejecting this claim, that acts done pursuant to an order of the court which is valid until reversed cannot be wrongful — As a matter of public policy people must be entitled to act in pursuance of a court order without being at risk of acting unlawfully — If this were not so there could be great confusion — Support for this view can be found in the decision of the Privy Council in *Isaacs* v *Robertson* and to some extent in Order 59, rule 13(1) — Judgment for defendant landlords

The following cases are referred to in this report.

Isaacs v *Robertson* [1985] AC 97; [1984] 3 WLR 705; [1984] 3 All ER 140, PC
Official Custodian for Charities v *Mackey* [1985] Ch 168; [1984] 3 WLR 915; [1984] 3 All ER 689
Rodger v *Comptoir d'Escompte de Paris* (1871) LR 3 PC 465

This was a preliminary point of law in a second action in which the tenants, Hillgate House Ltd, claimed damages against the landlords, Expert Clothing Service & Sales Ltd, in respect of the taking of possession of Hillgate House, 13 Hillgate Street, London W8, under a forfeiture order made in the first action when the landlords were the plaintiffs and the tenants the defendants.

David Neuberger (instructed by Baker Bell & Baker, of Southwick, West Sussex) appeared on behalf of the plaintiffs; Paul Collins (instructed by Brecher & Co) represented the defendants.

Giving judgment, SIR NICOLAS BROWNE-WILKINSON V-C said: This is a preliminary point of law which comes before me in this way. In an action between Expert Clothing Service & Sales Ltd as plaintiffs, and Hillgate House Ltd as defendants, the plaintiff landlords obtained an order for possession on the grounds of forfeiture of the lease. The plaintiff landlords went into possession pursuant to that order, but the order of the judge was subsequently reversed by the Court of Appeal*. In this second action, the tenants, Hillgate House, are claiming damages arising from that taking of possession by the landlords during the period between the judgment at first instance and the judgment of the Court of Appeal.

Shortly stated, the facts are these. The tenants were under an obligation to reconstruct the premises by September 28 1982. In fact, the tenants failed to do so, and on October 8 1982 the landlords served a section 146 notice which did not require the breach to be remedied, or a reasonable time for it to be remedied, since in the landlords' view the breach was incapable of remedy. On October 26 1982 the landlords issued a writ claiming forfeiture; that writ was served, either on the same day or shortly thereafter.

The action came before His Honour Judge Baker on April 2 1984.

---
* Editor's note: Reported at [1985] 2 EGLR 85; (1985) 275 EG 1011 and 1129.

The judge reached the conclusion that the lease had been forfeited, and declined to give relief against forfeiture. He made an order in these terms:

1. That the First Defendants

that is to say the tenant

do deliver up to the landlords forthwith possession of the property,

and it is then described.

The tenants applied to the judge for a stay but it was refused. That application for a stay was not repeated before the Court of Appeal. Pursuant to the judge's order, on April 2 or very shortly thereafter the landlords re-entered the premises. The landlords remained in possession until April 2 1985 on which date the Court of Appeal gave judgment on the appeal. They reversed the decison of the learned judge, holding that the section 146 notice was invalid by reason of its having failed to specify a reasonable time for remedying the breach, as a result of which there had never been any valid forfeiture of the term.

In this second action, the tenants by their statement of claim allege that wrongly, and in breach of the express covenant for quiet enjoyment and the implied covenant not to derogate from grant, the landlords entered on to the premises on April 2 1984 and excluded the tenants from those premises until April 2 1985. By reason of those facts the tenants say that they have suffered damage of a substantial nature — in the region of £100,000.

The matter came before me first on an application to strike out the claim as disclosing no cause of action. But after the argument had developed for a time it became clear that it raised quite a difficult point of law unsuitable to be dealt with by way of striking out. I therefore directed that the case should go forward on the basis of a formulated preliminary issue, which has been agreed in the following terms:

Can an action for trespass, and/or breach of covenant for quiet enjoyment, and/or derogation from grant, be maintained by a tenant against his landlord who has entered on to the demised premises and excluded the tenant therefrom following the making of an order for possession in forfeiture proceedings by a High Court judge whose decision is subsequently reversed by the Court of Appeal?

Mr Collins for the landlord puts his case broadly in two ways. First he says that during the period during which the learned judge's order was in force there was no lease. The effect of the judge's order, he says, was to put an end to the lease; so that until it revived by the reversal of the judge's decision by the Court of Appeal, there was no right to possession in the tenant, no relevant covenant for quiet enjoyment, no question of derogation from grant. He submits that the order of the judge, so long as it is in force, was decisive as to the existence or non-existence of the term.

He relies in support on the decision of Scott J in *Official Custodian for Charities* v *Mackey* [1985] Ch 168. The facts there were very complicated. I think it is sufficient for my purposes to explain that at the time the matter came before Scott J on an application for an interim injunction, the Court of Appeal had declared that the lease in question had been forfeited and that the tenant (Parway Estates Developments) was not entitled to relief against forfeiture. However, there was pending an application to the House of Lords for leave to appeal. It was therefore possible that the House of Lords would reverse the Court of Appeal judgment and reinstate the decision of the trial judge that there had never been a forfeiture, or alternatively give relief against forfeiture. The question arose because receivers appointed by mortgagees of the tenant were in receipt of the rents and profits of the land and wished to go on receiving them in respect of the period after the Court of Appeal judgment declaring the lease, together necessarily with any mortgage on it, to have been forfeited.

In that context, Scott J says:

On the other hand, if the House of Lords should give Parway leave to appeal, and if Parway's appeal to the House of Lords should succeed, the lease will stand as though it had never been forfeited. This will be the result whether Parway succeeds on the waiver point, in which case there will be shown never to have been a forfeiture, or whether Parway succeeds in claiming relief from forfeiture.

Then I can omit certain words.

The reversal by the House of Lords of the Court of Appeal order and the restoration by the House of Lords of the relief from forfeiture granted to Parway by the deputy judge would, therefore, retrospectively reinstate the lease, the mortgages and the status of the first and second defendants as receivers, and would retrospectively validate the management acts of the receivers done since July 19 1982. It would retrospectively disentitle the plaintiffs to receipt of the rents or monies paid or payable by sub-lessees or occupiers of the property since the Court of Appeal order, and would retrospectively invalidate any arrangements made by the plaintiffs with those sub-lessees or occupiers by way of management of the property.

What may be done by the House of Lords with the forfeiture action and with the Court of Appeal order is obviously a matter of speculation. What is not a matter of speculation is that at present the first and second defendants have no title or right as receivers to do that which they are now doing, that is to say, collecting rents and managing the property. What is not a matter of speculation is that at present the plaintiffs are entitled to the property freed from the lease and, subject to the mortgagees' claims for relief under section 146(4), freed from the mortgages.

On the basis that the receivers under the Court of Appeal order had no right at that time to be collecting rents since the lease had gone, Scott J granted the injunction restraining the receivers from continuing to collect the rents.

Mr Collins says that the latter part of the passage that I have read shows that in the view of Scott J there was not at the time in question any right or title to the lease, notwithstanding the fact that the House of Lords might subsequently reverse the Court of Appeal decision. In one sense that is right. But I think what the learned judge was there directing his mind to, as the earlier part of the passage indicates, was the position at the moment that he was dealing with it, namely, before the House of Lords had heard the matter. I do not think he is saying that if the House of Lords should have reversed the Court of Appeal decision there would, during the time in question, never have been any title or right or lease in existence during the intermediate period between the Court of Appeal decision and the House of Lords decision.

Certainly, in my judgment, the correct analysis is this. The claim in the present case was a claim to forfeit the lease. The order of Judge Baker declared that the tenants had forfeited. When the case went to the Court of Appeal, the Court of Appeal declared that they had not forfeited*. In my judgment, as the Court of Appeal's judgment discloses, the true view all along was that the lease had remained in existence. What was in doubt was what was the true legal effect.

Therefore, in my judgment, Mr Collins is wrong on the first ground. In my judgment, throughout the period between Judge Baker's judgment at first instance and the Court of Appeal judgment the lease was in existence and the obligations under it remained. Pending the decision of the Court of Appeal there was a misunderstanding of what the law was. But once the Court of Appeal had spoken, the true position which had existed throughout was disclosed.

In my judgment, Mr Collins' alternative way of putting his case is, however, right. On analysis what the plaintiffs are claiming in this case is that the acts done by them, the tenants, and by the landlords, directly pursuant to the order of the trial judge, themselves constitute a breach of legal duty which gives rise for the first time to a cause of action. In my judgment, that cannot be right. As the judgment of Scott J indicates, when an order is in force, and so long as it is in force, it is to be obeyed and is in law correct. It is true that it may be subsequently altered on appeal; but unless and until it is altered, it is an order of the court and acts done under it are lawful.

The position is clearly established in relation to contempt of court. For example, the decision of the Privy Council in *Isaacs* v *Robertson* [1985] AC 97 shows that an order even irregularly made has to be obeyed, and failure to comply with it constitutes a contempt, so long as it stands. So here, in the absence of a stay, the order of Judge Baker ordering the tenants to deliver up possession had to be obeyed. In my judgment, an act done pursuant to an order of the court which at the time the act was done is valid cannot constitute a wrongful act by the party doing it. So when one turns to the various causes of action relied on here, in my judgment the taking of possession by the landlords under the order of the court, which possession was given to them by the tenants under the order of the court, cannot have constituted a trespass. Likewise, so long as the order of the court persisted, the tenants had no immediate right to possession, such as is necessary to found a cause of action in trespass, since the order of the court directs that they do deliver it up. So the cause of action in trespass, in my judgment, was not a good cause of action.

Similarly, when one comes to the cause of action based on the covenant for quiet enjoyment, it is a covenant that the tenant should enjoy without any interruption by the landlord; but in my judgment it

---

* Editor's note: Decision reported at [1985] 2 EGLR 85; (1985) 275 EG 1011 and 1129.

is clear that must mean without any unlawful interruption. Since the landlords were acting under an order of the court, any interruption was lawful at the time it took place and cannot retrospectively be made unlawful.

Similarly, on derogation from grant, in my judgment the doctrine of derogation from grant cannot apply to acts done pursuant to a court order.

If the case were otherwise, there would, in my judgment, be very great confusion. People must be entitled to act in pursuance of a court order without being at risk that they are thereby acting unlawfully. Public policy requires it. I am not in any sense casting doubt on, or seeking to cut down, those cases to which I have been referred which indicate that where a judgment is reversed, the objective of the court should be to put back the litigants into the position in which they should all along have been had the law been properly appreciated — cases such as *Rodger v Comptoir d'Escompte de Paris*, reported in (1871) LR 3 PC 465. Those cases are concerned with reimbursing to the parties moneys lost as a result of the execution of the judgment by the payment of money. They are not cases, such as the present, in which it is sought to found a separate cause of action on the carrying out of the court order.

In my judgment, though I find the terms of RSC Ord 59, r 13(1) obscure, it gives me some support in the view that I have reached. The rule reads as follows:

Except so far as the Court below or the Court of Appeal or a single Judge may otherwise direct — (a) an appeal shall not operate as a stay of execution or of proceedings under the decision of the Court below; (b) no intermediate act or proceeding shall be invalidated by an appeal.

As Mr Neuberger in his persuasive argument indicated, and on this I am in agreement with him, it is difficult to find out exactly what rule 13(1)(b) means. He suggests that when it refers to something being invalidated by an appeal, it means being invalidated by the bringing of the appeal, not by the order made on the appeal. There are certain linguistic reasons in support of that view. However, I do not think that can be right because when the rule refers to an "intermediate act or proceeding" it must be referring to something done between two dates, which can only be the date of the decision in the court below and the date of the decision in the Court of Appeal. Likewise, the word "invalidated" is not very easy to apply to the present case. Nobody is suggesting that the acts of the landlords are invalidated. But in my judgment, giving it the best consideration I can, I think it does indicate that in the absence of a stay, it is legitimate for the successful party to proceed under the order of the judge below without being at risk that thereafter the acts he is doing will be rendered unlawful, or invalid, by the subsequent decision of the Court of Appeal.

For those reasons, I answer the particular point of law in the negative. I will give judgment for the defendants.

Chancery Division

November 3 1986
(Before Mr Justice VINELOTT)

EVANS v CLAYHOPE PROPERTIES LTD

Estates Gazette May 16 1987

282 EG 862-868

Receiver — Appointment by court to collect rents and manage a block of flats in poor state of repair — Appointment on *Hart v Emelkirk* basis — Interlocutory application by receiver in action by a tenant (at present adjourned pending conclusion of separate judicial review proceedings concerning local authority's alleged obligation to make repair grants) for directions as to recovery of sums due in respect of expenditure and receiver's remuneration which were in excess of moneys which he was appointed to receive — It was submitted that the receiver was entitled to recover from the landlords at this interlocutory stage expenses properly incurred and remuneration properly claimed — The income which the receiver was appointed to receive was wholly inadequate to meet the expenses of management — Held, rejecting this submission, that the court had no power on an interlocutory application to order the landlords to meet the receiver's expenses or remuneration — It was clear from the decision in *Boehm v Goodall* that the receiver was not an agent of the parties or a trustee for them — The court had no jurisdiction to indemnify a receiver save to the extent of the assets of which he was put in possession by the order of the court — The position was most unfortunate — A receiver should not take office unless he is satisfied that the assets will be adequate to meet his remuneration and necessary expenditure — Here the income receivable from rents payable by long leaseholders and by protected or statutory tenants was small and a service charge due from the leaseholders was not payable until expenditure on repairs had actually been incurred by the landlords — The use of the court's power to enforce a landlord's obligation to repair property was a new development posing novel questions which would have to be answered in time

The following cases are referred to in this report.

Boehm v *Goodall* [1911] 1 Ch 155
Hart v *Emelkirk Ltd* [1983] 1 WLR 1289; [1983] 3 All ER 15; (1982) 267 EG 946

This was an interlocutory application under RSC Ord 30 r 3 by Richard Denis Collins, the receiver and manager appointed by the court in respect of Dover Mansions, Canterbury Crescent, Brixton, London SW9, in regard to his remuneration. The main action was between Rudolph Bayfield Evans, a tenant of one of the flats in Dover Mansions, against the landlords, Clayhope Properties Ltd. This action had been adjourned pending the outcome of other litigation.

A decision of the Court of Appeal on another aspect of this dispute, the registration of a caution in respect of the order appointing the receiver, is reported at [1986] 2 EGLR 34; (1986) 279 EG 855.

Christopher Heath (instructed by Rooks Rider) appeared on behalf of the receiver, Richard Denis Collins; Patrick Ground QC and C Digby (instructed by Zelin Bale) represented Rudolph Bayfield Evans, the plaintiff in the action; Roger Cooke (instructed by Bernstein & Co, of Stoke Newington) represented the defendant landlords, Clayhope Properties Ltd.

Giving judgment, VINELOTT J said: The question raised in this application is whether, where a receiver has been appointed by the court in circumstances such as those which arose in *Hart* v *Emelkirk Ltd* [1983] 1 WLR 1289* and the receiver incurs expenditure or becomes entitled to remuneration in excess of moneys which he was appointed to receive, the court has jurisdiction on an interlocutory application while the litigation is still proceeding to order that the balance of that expenditure or remuneration over the sums available to the receiver should be paid by one of the parties to the action.

The question arises in the following way. The defendant, Clayhope Properties Ltd, which I will call "Clayhope", is the owner of a mansion block in Lambeth called Dover Mansions. The freehold interest was transferred to it in December 1979. Dover Mansions comprises 20 flats. Fourteen of them were let on long leases in the course of 1975, in each case for a term of 99 years from 1975. The leases were granted in each case at a premium. The rents payable are small. The aggregate of the rents payable under all the leases is £420 per annum. Each lease contains a covenant by the landlord to keep the main structure and the common parts in repair and to clean and light the common parts and to decorate the entrance. Each tenant covenanted both to repair his own flat and to pay a service charge equal to one-twentieth of the cost to the landlord of complying with his obligations. However, the service charge is not payable until the expenditure has been incurred by the landlord. Thus the landlord must fund the expenditure before he can reclaim fourteen-twentieths from the tenants holding under long leases. Of the six remaining flats two are unlet. The other four are held on controlled or protected tenancies. Under those tenancies the landlord owes obligations, derived from the original tenancy agreements or implied by law or imposed by statute, for the repair of the structure of the block and the repair and maintenance of the common parts. The extent of these

*Editor's note: Reported also at (1982) 267 EG 946.

obligations is in issue. The landlord has no right in relation to those tenancies to recover any part of the cost by way of a service charge.

It is common ground that Dover Mansions is in a poor state of repair, although the extent of that disrepair and the extent to which Clayhope is liable under any express or implied covenant for repair are in issue. On July 12 1983 a Mr Rudolph Bayfield Evans, who is the statutory tenant of one of the flats in Dover Mansions, issued a writ. Although a statutory tenant, he claims to represent all tenants, including those holding under long leases. In his statement of claim, he claims specific performance of the obligations imposed on Clayhope as landlord to carry out repairs, either by express or implied covenant or by statute, or, in the alternative, damages. On November 14 1983 the Chief Master, Master Heward, made an order appointing one William Arthur Johnson, a salaried partner in a firm of chartered building surveyors, to receive the rents, profits and other moneys payable under the leases of "flats in Dover Mansions" and to manage the same (that is the block of flats) "in accordance with the rights and obligations of" Clayhope. On February 9 1984 another partner in the same firm, a Mr Richard Denis Collins, was appointed receiver in his place and on the same terms. The order appointing Mr Johnson, which was opposed by Clayhope, was modelled upon the order made by Goulding J in *Hart* v *Emelkirk*. In the case at least of the order appointing Mr Collins, it was also ordered that he be at liberty to receive any grant payable in respect of Dover Mansions from any local authority.

In July 1985 Mr Collins applied for an order under Order 30, rule 3, of the Rules of the Supreme Court, allowing proper remuneration. He claimed, on behalf of himself and his predecessor, fees totalling some £26,000, representing, as to approximately £10,000, fees for managing Dover Mansions; as to approximately £14,000, a scale fee for preparing a specification for repairs; and, as to approximately £2,000, a fee for extracting work relating to the common parts and writing a separate specification for it. Those fees were disputed by Clayhope. The disputes as to the amounts claimed were heard by Master Chamberlain. In the course of a number of hearings, Master Chamberlain authorised an interim payment of £5,000. On January 13 1986 he gave a written judgment setting out the principles on which the fees and remuneration ought to be calculated. The final figures have not yet been agreed, but I understand that Clayhope accept that, applying the principles laid down by Master Chamberlain, the total remuneration that would be authorised by the court would be not less than £15,000 and the fee for managing Dover Mansions well in excess of £3,000. I mention that figure because it is in broad terms the amount which has come into the hands of the present receiver and his predecessor. In the course of the hearing the parties also sought from the master directions "as to the manner, if any, in which the receiver can recover the remuneration that it is entitled to charge and generally in relation to the receivership". The master adjourned that paragraph to the judge. The only question that has been argued before me is as to the manner in which the receiver can recover the remuneration to which he is ultimately held to be entitled.

Before turning to that question, I think I should say something about the history of the litigation. In its defence, Clayhope deny that at the date of the writ it was in breach of any express or implied covenant to keep the structure and common parts in good repair or, if it was, that the plaintiff was entitled to an order for specific performance. It also claims that at August 30 1983 it was ready, able and willing to carry out works of repair specified in a schedule annexed to the defence and that it has been prevented from doing so by the order appointing a receiver. The appointment of a receiver operated to divest Clayhope of possession of Dover Mansions and to put into suspense its obligations under the covenants for repair.

On November 20 1984, it was ordered that the action should be set down for trial on or before January 15 1985. However, no step has been taken to bring the action on for trial. The reason is shortly as follows. It is common ground that Dover Mansions requires extensive works of repair, the cost of which would be disproportionate to the rents now payable or which could be derived from the unlet flats. The cost of the work set out in the receiver's specification would be of the order of a quarter of a million pounds. Clayhope, of course, deny liability and also say that they do not have the resources to fund this expenditure. It is common ground that if they were to do the repairs and claim a due proportion from the tenants holding under long leases by way of service charge, many if not all of them would be unable to meet the contribution due from them. In addition, they may assert claims for damages for consequential loss by way of set-off. The receiver similarly cannot fund the expenditure by borrowing in the hope that it could repay the borrowing out of damages awarded against Clayhope in the action or from service charges. However, the tenants and Clayhope claim that the local authority is under a statutory obligation (under the Housing Act 1974) to make grants for the repair of each flat as a separate dwelling. That is denied by the local authority. Clayhope applied to the Divisional Court of the Queen's Bench for an order by way of judicial review directing the local authority to make repair grants. That application was refused by the Divisional Court.* Notice of appeal has been given. Final determination of this litigation, which raises important questions of principle, may be the subject of an appeal to the House of Lords. In July 1985, after application for leave to proceed by way of judicial review had been made, the plaintiff and Clayhope agreed that the hearing of the action be adjourned while Clayhope sought to compel the local authority to make repair grants and agreed that, if the local authority made grants amounting in the aggregate to not less than £98,000, Clayhope would agree to the carrying out of the works in the specification in respect of which the grants were made and any further work that it agreed to be necessary to comply with its repairing obligations.

That is the current position. There is no possibility that any works will be carried out unless the local authority is compelled to pay repairing grants amounting in the aggregate to over £98,000 or, if it is not, until after this action has been heard. In the meantime, the receiver is in office charged with the management of the block. The income which he was appointed to receive is wholly inadequate even to meet the expenses of management. I have already referred to the amount of the rents payable under the long leases. The rents payable by the statutory tenants are also modest and I understand that the receiver has encountered difficulty in recovering them. The unlet flats cannot be sold or let otherwise than on depreciatory terms while the block remains in its present state of disrepair.

It is submitted by Mr Ground, on behalf of the plaintiff, that the court has jurisdiction, even at an interlocutory stage, to order one of the parties to proceedings, in which a receiver has been appointed, to pay expenses properly incurred and remuneration properly claimed by the receiver. That general proposition is, in my judgment, contrary to principle and to authority. A receiver appointed by the court is an officer of the court appointed to get in the property over which he is appointed receiver and to deal with it in accordance with the directions of the court. He is not the agent or the representative of either of the parties to the litigation. In *Boehm* v *Goodall* [1911] 1 Ch 155, a receiver and manager was appointed in an action for the dissolution of a partnership to carry on the business with a view to its sale as a going concern. The usual order for dissolution was made. The receiver, carrying on the business, borrowed moneys and incurred debts in excess of the sum received on the sale of the business. An application by him asking that the parties to the action pay him the deficiency was refused. Warrington J, as he then was, said at p 161:

Such a receiver . . .

That is to say, one appointed by the court,

. . . is not the agent of the parties, he is not a trustee for them, and they cannot control him. He may, as far as they are concerned, incur expenses or liabilities without their having a say in the matter. I think it is of the utmost importance that receivers and managers in this position should know that they must look for their indemnity to the assets which are under the control of the court. The court itself cannot indemnify receivers, but it can, and will, do so out of the assets, so far as they extend, for expenses properly incurred; but it cannot go further. It would be an extreme hardship in most cases to parties to an action if they were to be held personally liable for expenses incurred by receivers and managers over which they have no control.

Mr Ground submitted that the explanation of this case is not that the court lacks jurisdiction but that in a case where a receiver may incur expenses or liabilities without parties to the litigation having a say in the matter, the court will not, as a matter of practice, throw any deficiency on to the parties. He submitted that, in a case where a receiver is appointed with a view to ensuring that repairing obligations in leases of flats in a mansion block are carried out for the common benefit of the landlord and the tenants, the court can and should order the landlord to meet the expenses incurred by the receiver. The landlord, it is said, has control over the expenditure incurred by the receiver in that the landlord can apply to the court to

---
*Editor's note: Reported at p 26 *ante* and (1986) 281 EG 688.

regulate his conduct by prescribing a limit to the expenditure to be incurred. I do not think that that proposition is consistent with the principle stated by Warrington J, who clearly considered that it was outside the powers of the court to indemnify a receiver save to the extent of assets of which he was put in possession by the order of the court. The proposition asserted by Mr Ground is plainly too wide. There is simply no foundation for any general power in the court to impose on a party to an action liability for expenditure by or remuneration due to a receiver who is not his agent.

Mr Ground submitted that, to the extent at least of the part of the remuneration relating to the cost of preparing a specification of repairs, the remuneration was part of the cost of carrying out repairs which Clayhope is obliged to carry out. But the liability of Clayhope to carry out repairs in accordance with the receiver's specification is in issue. No application is made for summary judgment and I can see no ground on which the court could make an interim order imposing on Clayhope part of the cost of such repairs. Mr Heath submitted that the remuneration of the receiver was part of the cost of the action and that the court's inherent jurisdiction to order the party to pay the cost of an action extended to the remuneration of a receiver appointed by the court. But again, even if that proposition is well-founded, I can see no justification for making an interim order for costs. Whether when the action is heard the court will have power to order Clayhope to pay the remuneration of the receiver, either as part of the cost of complying with its repairing obligation or as part of the cost of the action, is a question on which I express no opinion. The use of the court's power to appoint a receiver to enforce a landlord's obligation to repair property is a new development which poses many novel questions. They will have to be answered in time. I think that it is undesirable that I should attempt to do more than answer the precise question raised by the application now before me. Mr Heath also submitted that the court could and should extend the receiver's powers so as to authorise him to charge Dover Mansions to raise money for the purpose of meeting his remuneration. That application is made at a late stage. On the basis of the evidence before me, the prospect that moneys could be raised by way of charge is simply chimerical. The income derived from Dover Mansions is very small and, as matters stand, there is no real prospect that it could be enhanced by letting the unlet flats, whether at rack rents or at a premium, while the block remains in the present state of disrepair. There is, therefore, effectively no security which could be offered to a lender. Whether the court could authorise a receiver appointed in these circumstances to create a charge and meet his own remuneration, so indirectly throwing the burden on the landlord, is again a question which I do not think I should attempt to answer.

The position is a most unfortunate one. It may serve as a reminder of the limitations inherent in the power of the court to appoint a receiver. A receiver should not take office unless he is satisfied that the assets of which he is appointed a receiver will be adequate to meet his remuneration or that he has an enforceable indemnity by a party to the litigation capable of meeting his remuneration. The appointment of a receiver in order, indirectly, to ensure that property is kept in repair in accordance with covenants to repair entered into by a landlord is a valuable extension of the power of the court to appoint a receiver for the preservation of property in cases where urgent repairs are necessary, where the landlord is plainly in breach of a covenant to repair and where the income of the property will enable the receiver to remedy the want of repair (in particular where, as in *Hart* v *Emelkirk*, a service charge capable of being recovered by the landlord and capable of being vested in the receiver can be imposed in advance of the carrying out of works of repair to meet their eventual cost). But the appointment of a receiver in a case where the income, including any service charge, or any other property or money which can be put under the control of the receiver, is patently inadequate to meet the cost of repair is likely to prove ineffective and may even frustrate the carrying out of repairs that the landlord is willing to carry out.

I propose only to declare that the court has no power on an interlocutory application to order Clayhope to meet the receiver's remuneration. He can, of course, have recourse to the funds in his hands to meet remuneration approved by the master for the management of the property.

Court of Appeal
February 17 1987
(Before Lord Justice O'CONNOR and Lord Justice NICHOLLS)

STRAUDLEY INVESTMENTS LTD v BARPRESS LTD

Estates Gazette May 30 1987

282 EG 1124-1125

Landlord and tenant — Extent of property demised by long lease of a building — Appeal from decision of Mervyn Davies J refusing appellants' claim to mandatory injunctions under Ord 14 requiring respondents to remove a fire escape and ventilation vent erected by them on and against the roof of appellants' property and constituting a trespass thereto — Respondents' defence was that the upper surface of the roof, airspace above and external surface of the wall were not part of the premises demised to the appellants — Mervyn Davies J, although accepting that there was a very strong case for a summary judgment in favour of appellants, considered that the respondents should be given leave to defend — He was influenced by dicta in *Cockburn* v *Smith* and *Douglas-Scott* v *Scorgie*, which left a doubt in his mind as to whether the matter should be concluded summarily against the respondents — The appellants' lease was for 99 years from 1936 and the parcels clause demised "all that piece or parcel of ground with the messuages and buildings erected thereon" comprising numbers 67 to 81 (odd) in Mortimer Street, London W1 — The lease was a full repairing lease and the repairing covenant required the lessee to "repair support and uphold the said messuages buildings and premises" and to keep the premises in repair "with all additional erections and improvements" — Held by the Court of Appeal that it was really unarguable that the lease did not demise to the appellants the roof of the buildings and the exterior walls, which were the subjects of the trespass — The cases of *Cockburn* v *Smith* and *Douglas-Scott* v *Scorgie* were distinguishable and did not justify the doubt felt by the judge — Appeal allowed and appropriate relief under Ord 14 granted to the appellants

The following cases are referred to in this report.

Cockburn v *Smith* [1924] 2 KB 119
Douglas-Scott v *Scorgie* [1984] 1 WLR 716; [1984] 1 All ER 1086; (1984) 48 P&CR 109; [1984] EGD 325; (1984) 269 EG 1164, CA

This was an appeal by the plaintiffs, Straudley Investments Ltd, from the decision of Mervyn Davies J, refusing to grant summary relief under Ord 14 against the defendants, the present respondents, Barpress Ltd, requiring them to remove a fire escape and ventilation vent, the erection of which was claimed by the appellants to be a trespass on their property. Paul de la Piquerie (instructed by Binks Stern & Partners) appeared on behalf of the appellants; P J Susman (instructed by Elkan David & Co) represented the respondents.

Giving judgment, O'CONNOR LJ said: This is the plaintiffs' appeal from a refusal of Mervyn Davies J to grant relief under Ord 14 and order mandatory injunctions against the defendants to remove a fire escape and a ventilation vent which they had erected on and against the roof of the plaintiffs' property. The statement of claim was in the writ issued on February 13 and it alleged that in December 1985 the defendants had put up the fire escape across a part of the roof of the plaintiffs' property in Mortimer Street, London W1, and, as I have said, put up a ventilation duct which had impinged in part on to the plaintiffs' property.

By para 5 of their defence the defendants, who had admitted that the fire escape and the extractor duct were erected without any leave or licence of the plaintiffs, said:

> It is denied that the erection of the said fire escape or of the said air extractor duct or the fact of the same remaining in position constituted or constitutes a trespass by the Defendants or either of them on the said premises so demised, in that it is denied that the upper surface of the said roof or the air space above the same or the external surface of the said wall were or are part of the said premises so demised.

The plaintiffs, in their affidavit in support of their summons for mandatory relief, exhibited the lease under which they held the premises. It is a lease made in 1936 for 99 years and is a full repairing lease. The parcels clause, so far as material, demises:

ALL THAT piece or parcel of ground with the messuages and buildings erected thereon situate and being on the South side of and Numbered 67, 69, 71, 73, 75, 77, 79 and 81 in Mortimer Street.

There is a full repairing covenant in clause II(2) requiring the lessee to "repair support and uphold the said messuages buildings and premises with all pavements sinks sewers" etc. Later in that clause it says: "And the said premises so repaired and kept with all additional erections and improvements together with all windows shutters leaden gutters ridges and hips leaden and other pipes" etc and a whole stack of things are to be kept in repair.

In my judgment, it is quite unarguable that that lease does not demise the roof of the buildings and the exterior walls. The learned judge came to a conclusion that it was in some fashion arguable. He said in the note of his judgment:

[There is a very] strong case for Order 14 judgment in this case, but after some hesitation I do not accede to [the] request.

There should be leave to the defendant to defend. I am led to this conclusion by the references to dicta in *Cockburn* v *Smith* [1924] 2 KB 119 and *Douglas-Scott* v *Scorgie* [1984] 1 WLR 716 which suggest that, although it is unlikely, consideration of the physical circumstances of the buildings and of the terms of the demise might lead to the conclusion that the roof never was demised to the plaintiffs. It is unlikely but there is a doubt in my mind planted there by Mr Susman's submission.

I feel uneasy concluding this matter once and for all by a mandatory injunction.

Mr Susman's submission was that in certain circumstances it may be that the roof does not pass with a demise. Of the two cases cited and referred to by the learned judge, *Cockburn* v *Smith* in 1924 was a claim by a tenant of a top-floor flat against the landlords for personal injuries and ill health suffered by her as a result of non-repair of the roof, which had let in water. An argument was put up that the roof had passed to her and it was her roof. The Court of Appeal held that the judge who had come to that conclusion was wrong and stated in terms that it was a question of construction of the individual leases as to whether the roof passed or not and that, on the terms of the lease in that case, it was quite clear that the roof did not pass to the tenant.

The other case, *Douglas-Scott* v *Scorgie*, was a landlord and tenant case which really turned on the true construction of section 32(1)(a) of the Housing Act 1961, which implies a term into certain classes of letting that there shall be implied a covenant by the lessor to keep in repair the structure and exterior of the dwelling-house. That again was a question as to whether the roof of the flat formed part of the structure or exterior of the tenant's dwelling-house. In that case *Cockburn* v *Smith* was cited and it was conceded that the roof was not demised.

For my part I get no help whatsoever from those two cases in construing the lease in the present case. In my judgment its terms are crystal clear. It is quite apparent and certain that the roof did pass under the demise and there is no defence to this action. It has not been suggested by Mr Susman that the second ground given by the learned judge was a proper ground to take into account in giving time, namely that the defendants were negotiating with others to provide an alternative fire escape from their premises, and can have any effect on this case. It is not suggested that that creates any defence to the claim.

In my judgment the learned judge fell into error and I would hold that the plaintiffs are entitled to relief in the terms of paras 1, 2 and 3 of the prayer in their statement of claim and an order that the damages should be assessed.

Agreeing, NICHOLLS LJ said: I add only a few words because we are differing from the learned judge. The short question raised by this appeal is one of the construction of the words governing the extent of the property demised by the lease granted on January 6 1936. The lease is a 99-year full repairing lease. The parcels clause is in these terms:

ALL THAT piece or parcel of ground with the messuages and buildings erected thereon situate and being on the South side of and Numbered 67, 69, 71, 73, 75, 77, 79 and 81 in Mortimer Street in the Parish of St Marylebone in the County of London which said premises with the dimensions and abuttals thereof are more particularly delineated and described in the plan drawn hereon TOGETHER with all yards areas vaults ways lights easements watercourses and appurtenances to the said premises belonging

with an immaterial exception.

The factual position now (it is not suggested that there was any material difference in 1936 when the lease was granted) is that the building or buildings 67 to 81 Mortimer Street form part of a terraced block. On those simple facts, for my part I can see no escape from the conclusion that the demise was of the whole of the building or buildings 67 to 81 Mortimer Street shown on the plan including, as part of that building or those buildings, the roof of the relevant two-storey area and the roof of the rest of the building or buildings and, in the normal way, the air space above those roofs. If that construction were not correct, one of the conclusions which would inevitably follow in this case is that the lessee's repairing obligations would not extend to the roof of the building or buildings. Plainly that could not have been intended in the case of this lease. I add that this lease, being a long lease of a whole building or whole buildings, is quite different from a lease or tenancy of a top-floor flat of a building which has been divided horizontally into flats.

I agree that this appeal should be allowed.

The appeal was allowed with costs and mandatory injunctions granted, not to be enforced for six weeks; damages to be assessed.

For further cases on this subject see p 209

LANDLORD AND TENANT
BUSINESS TENANCIES

Court of Appeal
April 24 1986
(Before Lord Justice SLADE and Mr Justice EASTHAM)

BUSH TRANSPORT LTD v NELSON

Estates Gazette January 17 1987
281 EG 177-183

Landlord and tenant — Appeal by landlords from decision of county court judge rejecting their application for possession of premises used as a repair shop — Premises, originally consisting of an open yard on which a lean-to shed had been erected, had been let to the respondent by a tenancy agreement for a term of 7 years from 1977 at a rent of £2,600 a year, payable by weekly payments of £50 — Appellant landlords subsequently offered to extend the term of the respondent's tenancy until 1991 at the higher rent of £65 per week with a three-year rent review — Respondent was not, however, happy with this arrangement and did not take up the offer — There was a long and confusing correspondence between the appellants and the respondent's solicitors in the course of which there appeared to be various misunderstandings — Eventually the appellants brought proceedings for possession against the respondent in the county court, based on a section 25 notice served under the Landlord and Tenant Act 1954 — The judge found that, some time after the appellants' offer above mentioned, there had been an oral agreement to grant the respondent an extended tenancy until 1991 at the old rent of £50 per week, with one rent review in 1984 — The Court of Appeal held, in the light of _Jenkin R Lewis & Son Ltd_ v _Kerwan_, that this must have taken effect as an agreement to surrender the existing term and to create a new term — They also agreed with the judge in rejecting a submission that the respondent through his solicitors had repudiated the agreement in question — The correspondence on which the appellants relied did not show a repudiation of the oral agreement for an extended tenancy at £50 per week but an intention to decline the offer of a new lease at £65 per week — Appeal dismissed

The following cases are referred to in this report.

Jenkin R Lewis & Son Ltd v _Kerman_ [1971] Ch 477; [1970] 3 WLR 673; [1970] 3 All ER 414; (1970) 21 P&CR 941, CA

Walsh v _Lonsdale_ (1882) 21 ChD 9

This was an appeal by Bush Transport Ltd, the landlords, plaintiffs in an action before Judge Parker at the West London County Court, against the judge's decision in favour of the defendant tenant, the present respondent, Hughie Nelson, dismissing the appellants' application for possession of Bush Garage, 2a Godolphin Road, Shepherds Bush, London W12.

Roger Smith (instructed by Webber & Co) appeared on behalf of the appellants; Jonathan Chrispin (instructed by W R Bennett Emanuel & Co) represented the respondent.

Giving judgment, SLADE LJ said: This appeal involves a long, and somewhat complicated, story, but the point which ultimately falls for decision by this court is a short one and, if I may say so, the case has been argued with admirable economy on both sides.

It is an appeal by the plaintiffs in an action, Bush Transport Ltd, from a judgment of His Honour Judge Parker, given in the West London County Court on December 20 1985. On that occasion he dismissed an application by the plaintiffs for possession of certain premises known as Bush Garage, 2a Godolphin Road, Shepherds Bush, London W12, but ordered that they should recover the sum of £925 from the defendant, Mr Hughie Nelson, for the use and occupation of the premises from November 1 1984 to the date of the hearing. He made no order as to costs.

In his judgment the learned judge described the case as a very troublesome and puzzling one, and I am not surprised. The history of the matter is as follows. The defendant began to occupy the premises in or about November 1977 and has occupied them ever since. When he acquired them they consisted of an open yard upon which a lean-to shed had been erected. He paid £50 a week for his occupation, and he says that he spent considerable capital sums on the premises in 1978 and 1979, and indeed later.

On November 1 1979 a formal agreement was entered into between the plaintiffs and the defendant, by which the plaintiffs let the premises to the defendant for a term of seven years from November 1 1977 at a yearly rent of £2,600 payable by equal weekly payments of £50 in advance. This document, which called itself a lease, was only executed under hand, so I shall call it "the 1979 tenancy agreement". It recorded that the first of the weekly payments had been made on November 1 1977.

The defendant used the premises as a repair shop. The learned judge found that for some time there continued what he described as a "reasonable relationship" between the defendant and the plaintiffs, and in particular with a Mr Little, who had a 50% shareholding in the plaintiff company. The plaintiffs themselves held no more than a leasehold interest in the premises. In the beginning of 1980 they entered into a new agreement with their head landlords which not only increased the rent payable by the plaintiffs in respect of the premises sublet to the defendant, but also provided for a rent review every three years, beginning in 1982. The plaintiffs' lease was to run for twelve years from September 29 1979 until 1991.

Early in 1980 Mr Little, who appears to have been more concerned about the triennial rent review provision than the rent immediately payable, spoke to the defendant and suggested that if he was willing to pay £65 per week, with a three-yearly rent review, the plaintiffs would be willing to extend his lease until 1991, when their own term ended.

The defendant consulted solicitors about this proposal. There followed a long correspondence. Some of the letters contained in that correspondence were marked "Without Prejudice", but privilege has clearly been waived in respect of all of them now before the court. On May 12 1980 the defendant's solicitors wrote to the plaintiffs referring to the rent payable under the 1979 tenancy agreement and saying:

Nevertheless we understand that you have been demanding from him

the defendant

rent of £65 per week which he has paid. You are not entitled, of course, to any increase in rent until the Lease granted in 1979 expires.

However Mr Nelson says that he understands that you have recently had an increase in rent imposed against you by your Landlord in order to get a new Lease of 12 years from 1980 and he is therefore prepared entirely without prejudice if you will grant him a Lease for 12 years from 1980 at a rent of £65 per week until 1 November 1984 with a rent review on that date to carry the rent on to the end of the term then he would be prepared to enter into such a Lease provided further, of course, that you obtain a consent from your

71

Landlord so to do.

We await your early reply to this letter.

That letter was not answered. It is common ground that rent was not paid at the rate of £65 per week.

By a further letter of May 29 1980 the defendant's solicitors wrote to the plaintiffs saying:

> You have not replied to our letter May 12 and the offered rent beyond that reserved by the Lease of November 1 1979 is now withdrawn.

That letter was written without prejudice, but in an open letter of the same date they asserted that, despite the terms of the 1979 tenancy agreement, the plaintiffs had been demanding rent at the rate of £65 per week; they said that the defendant would pay only the rent reserved by the 1979 tenancy agreement, namely £50 per week, and that they understood that he had already overpaid the plaintiffs. The defendant continued to pay £50 per week. The learned judge found that it was not paid regularly but in lump sums on certain occasions, and there were arrears.

On July 3 1980 the defendant visited the plaintiffs' premises. There then ensued a discussion between the defendant and Mr Little; I shall have to revert to the evidence as to this discussion later in this judgment. It is, however, common ground that at the end of the meeting the defendant took away with him a typed document which was addressed to the defendant and signed by Mr Little and Mr Ackland-Snow, another director of the plaintiff company, dated July 3 1980. That document read as follows:

> Dear Sirs, Re:- Occupation of Premises near to Goldhawk Road, W12. This is to certify that while Bush Transport Ltd, of 164 Dalling Road, London W6 is the lessee of the piece of land at the rear of numbers 152-154 Goldhawk Road, Shepherds Bush, London W12 they will allow Hughie Nelson of 32 Elm Way, London NW10 to carry out his business at "Bush Garage" until September 1 1991, at the weekly rent of £65.00: four weekly in advance, with a rent review every three years as from 1st September 1979. Conditional to the following:-
> 1. H Nelson to keep up to date with his payments.
> 2. Keep premises in a reasonable state of repair.
> 3. H Nelson will insure and keep insured the above said premises against loss or damage by fire and such other risks he may from time to time deem desirable.
> Yours faithfully
> pp Bush Transport Limited

There then followed the two signatures to which I have referred. For the time being I shall continue simply to refer to the correspondence.

On January 7 1981 the defendant's solicitors wrote to the plaintiffs an important letter. It was headed with a reference to the premises and read:

> As you know, we act for Mr Hughie Nelson.
> We understand that you have demanded from him rent in excess of that reserved by the Lease which you granted to him on 1st November, 1979 which will not expire until 1986.
> We also understand that our client has paid you sums in excess of the rent reserved.
> Our client will no longer do so and we are getting out an account showing the up to date position, of which a copy will be forwarded to you in due course. Please do not approach our client with accounts for rent which is not due, demand excess rent or offer him a Lease. He already has a Lease and is perfectly satisfied with it.
> If you wish to dispose of your interest in these premises, please let us have details of the price required and our client will consider whether it is worth his while to purchase.
> We shall be writing to you again quite shortly with an account showing the date up to which our client has discharged the rent, which we think is some time in the Autumn of 1981.

On January 14 1981 Mr Little's wife, on behalf of the plaintiffs, wrote to the defendant's solicitors a letter in which she enclosed a statement of account purporting to indicate the rent outstanding from the defendant.

On January 23 1981 the defendant's solicitors wrote another important letter to Mr Little headed in the matter of the premises and reading as follows:

> Further to our letter to you January 7 we have now checked the position with our client and are informed that in fact the payment in the sum of £500 effected on November 17 1980 has not been taken into account in computing the figures set out in the summary sent to us on January 14.
> We are accordingly instructed to write to you in order to inform you yet again not to approach our client for rent which is not due or demand excess rent or offer a Lease.
> We sincerely trust that no further unnecessary correspondence and expense will be entailed by this matter.

On February 18 1981 the defendant's solicitors wrote to Mr Little a letter which dealt solely with accounting matters concerning rent; I need not refer further to it.

On March 16 1981 the plaintiffs' solicitors again wrote to the defendant's solicitors, in effect complaining that rent had not been paid as due under the arrangements between the parties. Nothing turns on the precise terms of that letter.

So far as the documentation is concerned, I next come to a notice served by the plaintiffs' solicitors on the defendant under section 25 of the Landlord and Tenant Act 1954, purporting to give him notice terminating his tenancy on November 1 1984. That notice was dated February 24 1984.

On March 5 1984 the defendant's solicitors wrote to the plaintiffs' solicitors asserting that the plaintiffs could not bring the tenancy to an end on November 1 1984, on the grounds that Mr Little had "overlooked the agreement of July 3 1980 whereby the premises were let to our client for a term expiring September 1 1991". They asserted that the notice was therefore invalid and they returned it.

On March 9 1984 the plaintiffs' solicitors replied, saying that the plaintiffs had no knowledge of any tenancy agreement dated July 3 1980. I pause to say that this was a rather surprising assertion in the light of what subsequently transpired.

On March 13 1984 the defendant's solicitors replied enclosing a copy of the document of July 3 1980.

On March 19 1984 the plaintiffs' solicitors wrote to the defendant's solicitors, saying that they had taken the plaintiffs' instructions, and asserting that the plaintiffs had no knowledge of the document, and could only conclude that the document was a forgery. At first sight this letter was even more surprising than that of March 9, but it now appears that the plaintiffs' solicitors had taken their instructions not so much from Mr Little as from Mrs Little who, it would seem, was probably not aware of the document of July 3 1980.

On March 21 1984 the defendant's solicitors wrote to the plaintiffs' solicitors expressing astonishment at the allegations made in the letter of March 19. On April 18 they wrote saying that, without admitting the validity of the notice served, the defendant was not willing to give up possession.

Against this background, on November 2 1984 the plaintiffs issued the present proceedings against the defendant seeking possession and arrears of rent and mesne profits. In their particulars of claim they pleaded that the defendant was occupying the premises under the 1979 tenancy agreement; that by virtue of the section 25 notice of February 24 1984 they had duly terminated the tenancy on November 1 1984; that the defendant had failed within the prescribed period to apply to the court for a new tenancy; that accordingly the tenancy terminated on November 1 1984 and that he remained in occupation of the premises as a trespasser.

The defendant, by his defence and counterclaim, denied that the tenancy had come to an end and pleaded that

> by an oral agreement of unknown date made between the plaintiffs' servant or agent one Little and the defendant and evidenced in writing by a document dated July 3 1980, the plaintiffs varied the said lease by inter alia granting the defendant a term expiring on September 1 1991.

He denied that the notice of February 24 1984 was a valid notice terminating the tenancy. Further or in the alternative he pleaded that, if it was a valid notice, the plaintiffs were estopped from relying on it by reason of the document of July 3 1980. By his counterclaim he asked that if, contrary to his contention, it should be found that he had become liable to forfeiture, he should be granted relief from forfeiture.

The defence and counterclaim was followed by a lengthy reply and defence to counterclaim, which was subsequently amended. By this amended pleading the plaintiffs admitted the document of July 3 1980 and its terms, but pleaded in effect that there was never any concluded agreement between the parties on the terms of that document. They pleaded in the alternative that if there was a concluded agreement in the terms evidenced by the document, the defendant had repudiated it by refusing to pay the increased weekly rent of £65, and also by the terms of the letter from his solicitors to the plaintiffs of January 7 1981.

At the hearing the learned judge heard evidence from, among other persons, Mr Little and the defendant. They gave conflicting accounts of what had happened at the meeting on July 3 1980. I think the learned judge sufficiently accurately summarised Mr Little's evidence as to this conversation as follows:

Mr Little says that there was a discussion and agreement was reached in accordance with the points set out in the letter of July 3 1980. The letter was typed out by Mr Little. The letter recorded that the rent was to be increased and there were to be rent reviews and was subject to three conditions. There was evidence that up till then the obligation was on the landlords to insure the premises. They had not in fact insured the premises. Mr Little when asked seemed rather muddled about it. Something must have caused Mr Little to put this proviso in the letter. Perhaps Mr Little was more aware of the need for this at the time. Mr Little says the letter was a gesture of good intent and he expected Mr Nelson to take it to his solicitor who would then draw up a formal agreement.

The learned judge went on to say that this last point was consistent (as indeed it was) with the plaintiffs' amended reply.

He then proceeded to summarise the defendant's evidence as to this meeting, and again I think sufficiently accurately:

Mr Nelson says that on July 3 after the discussion he was presented with the document. He took it away and within a short time brought it back and said he was not happy with it because he wanted to spend a considerable sum of money on renovating the premises and wanted Mr Little to contribute towards the costs. Mr Little refused says Mr Nelson but agreed, if Mr Nelson was going to spend substantial sums, instead of Mr Little making a contribution the rent would stay at £50 per week until the expiry of the original agreement and he could stay until 1991 with the rent being subject to review every three years. In other words there was to be a new lease incorporating those conditions and the rent was to stay at £50 per week until then. Later Mr Nelson spent money on improvements and took on responsibility for insurance. It is said that Mr Little had no real reason to want the premises improved and it was no real advantage to him. Mr Little says it was not really discussed.

Thus the defendant's evidence actually given to the learned judge did not exactly accord with his pleading. He was asserting to the trial judge that, following the signature of the document of July 3, an oral agreement was concluded between him and the plaintiffs under which the plaintiffs granted him a tenancy until September 1 1991 at a rent of £50 per week, but otherwise on the terms of the document of July 3 1980. It appears that no case based on estoppel in the strict sense was put forward on behalf of the defendant at the hearing; his case seems to have been argued on the footing that the alleged oral agreement amounted to a variation of the original lease, or a new agreement, the effect of which was, in either case, to give him a term expiring on September 1 1991.

The learned judge held that there was an agreement in the terms alleged by the defendant. As to that he said:

I have come to the conclusion that while Mr Little was not interested in the improvements Mr Nelson wanted to do the improvements and was concerned about how to pay for them if the rent was increased. Mr Little was not concerned at that time, only in the future when the rent was reviewed. I think it more likely than not that Mr Little took the view that £50 per week would be alright if there was a rent review in Nov 1984 and thereafter every three years. I take the view that Mr Little assented to the suggestion that was made. Mr Nelson went away and no formal agreement was drawn up.

Then, at the end of his judgment, the learned judge said:

I have come to the conclusion that an agreement had been reached that Mr Nelson was to have a tenancy until 1991 at a rent of £50 per week with a rent review in November 1984.

He went on to find, implicitly, that there had been performance, or part performance, of the agreement by the defendant. This finding has not been challenged on behalf of the plaintiffs before this court, because there was evidence before the learned judge that during the months, or years, following July 1980, and more particularly, I think, from 1981 onwards, the defendant had spent substantial sums on the property.

The learned judge rejected the plaintiffs' submission that the defendant had repudiated the agreement. He accordingly dismissed the claim for possession but gave judgment for the plaintiffs for £925, which he found due to them in respect of the use and occupation of the premises since November 1984.

In their notice of appeal the plaintiffs attack the learned judge's decision on a number of grounds. Grounds 1 to 5 are essentially narrative. Grounds 6 to 8 in effect attack his finding that there was such an oral agreement as the defendant alleged. Grounds 9 to 11 read as follows:

(9) The learned judge failed to deal at all with the issue raised by the plaintiffs in their reply that the agreement under which they had agreed to grant the defendant a term expiring in 1991 had been repudiated by the defendant by his solicitors' said letter of January 7 1981 though the said letter was equally capable of being repudiation even if the said agreement had been in the terms alleged by the defendant rather than those of the letter of July 3 1980.

(10) Alternatively the learned judge dealt with the said issue by finding that the said letter was written as a result of a misunderstanding between the defendant and his solicitors and overlooked the point that the defendant's solicitors had apparent authority to write the said letter and he was bound by its contents.

(11) The learned judge erred in law by failing to hold that the said letter of January 7 1981 was a repudiation of any prior agreement to grant the defendant a new lease or an extended term.

Mr Smith, on behalf of the plaintiffs, having seen the learned judge's notes of evidence, now accepts that there was material on which he could make the finding of fact attacked in paras (6), (7) and (8) of the notice of appeal; accordingly, these particular grounds have not been pursued before us. Grounds (9) to (11) are the only material ones for the purposes of this appeal.

Since the learned judge found that an oral agreement of the nature alleged by the defendant had been concluded, this involved an implicit finding that an oral agreement had been concluded which amounted either to a variation of the 1979 tenancy agreement or a new agreement, the effect of which was, in either case, to give the defendant a term expiring on September 1 1991.

The judgment of this court in *Jenkin R Lewis & Son Ltd v Kerman* [1970] 3 All ER 414, which is referred to in Mr Smith's skeleton argument, shows that where a landlord and tenant wish that the period of the subsisting term of a tenancy shall be extended, this object can be achieved only either by granting the tenant a reversionary lease to take effect on the expiry of the existing lease or by the surrender of the existing term and the creation of a new term for the extended period: see *per* Russell LJ at p 419 G to J of the report.

In the present case the oral agreement which the learned judge found to have been concluded between the parties within the short time after July 3 1980 was clearly not intended by either of them to take effect merely as the grant of a reversionary lease. In these circumstances I accept Mr Smith's submission that the agreement could only have taken effect, if at all, as an agreement to surrender the existing term and to create a new term. That much, I think, is common ground.

Furthermore, since the agreed new term exceeded three years in length, it could not have been created by parol, in view of the provisions of section 54 of the Law of Property Act. The oral agreement, therefore, could amount to no more than an executory agreement for the lease. As Mr Smith has pointed out, while a tenant in occupation pursuant to an agreement for a lease is to be treated as holding under the same terms in equity as if the lease had been granted, in view of the doctrine of *Walsh v Lonsdale* (1882) 21 ChD 9, he can only rely on that doctrine if he is entitled to specific performance of the relevant agreement. Mr Smith has referred us to a passage in *Snell's Principles of Equity,* 28th ed p 592, which states the principle that a plaintiff who seeks to enforce a contract by way of specific performance must show that he has performed, or has been ready and willing to perform, all terms and conditions to be performed by him, and that he has not acted in contravention of the essential terms of the contract.

The more controversial part of Mr Smith's submissions begins with his proposition that the only possible construction to put on the defendant's solicitors' two letters of January 7 1981 and January 23 1981 is that they evinced a clear intention on the part of the defendant not to be bound by any agreement to give him a new lease, and constituted a repudiation of any previous agreement for the grant of an extended lease which might have previously subsisted.

As to this submission, the learned judge said this:

I bear in mind (and counsel for the plaintiffs stresses) the letter written by the defendant's solicitors on his behalf in January 1981 about demands for excess rent and with a reference to a lease not being required. It is said that it is not consistent with either the letter of July 3 1980 or the version given by Mr Nelson. There was another letter from Mr Nelson's solicitors saying again do not demand further rent or offer him a lease. Said is consistent with repudiation by Mr Nelson and not consistent with his version of the agreement arrived at with Mr Little. In almost any case I would have found these propositions irresistible. But having seen Mr Nelson in the witness box I can well understand particularly as the solicitors had not seen the document of July 3 1980 there may well have been a misunderstanding. The sum of £65 was not asked for by Mrs Little who was doing the accounts but some request for it was made by Mr Little. Mr Nelson's understanding was that the rent was to remain at £50 per week until November 1984. Having seen Mr Nelson I have concluded he does have difficulty in explaining things. It is quite clear that in 1980 and 1981 Mr Nelson did spend substantial sums on improving the premises and I cannot think he would have done that if there had not been an agreement with Mr Little such as he alleges. He also says he increased the insurance because he had improved the premises, in accordance with his liabilities under the agreement. I have come to the conclusion not without hesitation. The witnesses on both sides do not have a clear recollection of what happened which does not make my task any easier. There was an appalling muddle when the defendant referred in 1984 to the document of July 3 1980.

Then the judge, in the final sentence of his judgment (which I have already quoted) reaffirmed his conclusion that an oral agreement of the nature mentioned had been concluded in 1980.

Before this court Mr Smith has submitted that in dealing with this important issue of alleged repudiation, the learned judge erred, first because there was no evidence to support the view that the defendant's solicitors had made any mistake in writing the two letters in question; and second because the solicitors were in any event the defendant's agents, and he was bound by what they said on his behalf. On the basis of the learned judge's now undisputed findings as to the existence of the oral agreement of 1980, and the fact that the defendant's solicitors had apparently never seen the document of July 3 1980 when they wrote the letters in question, I think there was evidence to support the view that at the time when those letters were written there was a misunderstanding between the defendant and his solicitors. However, this seems to me to be irrelevant, because for my part I would entirely accept the submission of Mr Smith that, whatever may have been the extent of their understanding, or the extent of their actual authority, the defendant's solicitors had *ostensible* authority to act on behalf of the defendant in writing the two letters in question, and that he must be bound by what they said on his behalf. The crucial question, therefore, is: What is the true meaning and effect of the two letters?

It has to be borne in mind that the agreement found by the learned judge to have existed and now alleged by the plaintiffs to have been repudiated is an oral agreement concluded shortly after July 3 1980, under which, in effect, there was to be a surrender of the existing tenancy, and the plaintiffs were to grant, and the defendant was to take, a tenancy until 1991 at a rent of £50 per week, with a rent review in November 1984. A statement as to the manner in which repudiation of contractual obligations can take place is to be found in a short passage in *Chitty on Contracts,* 25th ed, at para 1601, which I think Mr Smith was prepared to accept as correctly stating the law:

If there is an absolute refusal to perform, the other party may treat himself as discharged. Short of an express refusal, however, the test is to ascertain whether the action or actions of the party in default are such as to lead a reasonable person to conclude that he no longer intends to be bound by its provisions.

The letters of January 7 and 23 1981, relied on by the plaintiffs, did not contain any explicit renunciation of any oral agreement. The question as I see it, therefore, is whether they amounted to an implicit renunciation of it. The answer to this question, in accordance with the principles I have stated, must depend on whether a reasonable man in the shoes of the plaintiffs, on receipt of one or both of those two letters, would have believed that the defendant had no intention of carrying out his part under the oral agreement.

Mr Smith drew our attention to the fact that Mr Little, in the course of his evidence, said that when the plaintiffs' solicitors received the letter of January 7 1981, he took the view that the defendant did not want the new lease. But I think it is not so much what Mr Little actually thought as what he could reasonably have thought, which is the relevant question for this court. Mr Smith, in a forceful argument, has directed our attention most particularly to the letter of January 7. He reminded us of the second sentence, in which the defendant's solicitors said:

We understand that you have demanded from him rent in excess of that reserved by the Lease which you granted to him on November 1 1979 which will not expire until 1986.

Though the actual date of expiration of the 1979 tenancy agreement was incorrectly stated, Mr Smith submitted, and I would be prepared to accept, that this was on the face of it an affirmation by the defendant's solicitors that he was relying on his rights under the 1979 tenancy agreement.

Then Mr Smith naturally stressed very strongly the sentence in the fifth paragraph of that letter:

Please do not approach our client with accounts for rent which is not due, demand excess rent or offer him a Lease.

This, he submitted, could mean only one thing, on any reasonable construction of that sentence, namely that the defendant was not prepared to take up a new lease of the premises.

However forceful these submissions are, I think that the letter of January 7 1981 and the other letter, of January 23 1981, containing similar phraseology, have to be read against the context in which they were written and received. What is perfectly clear from the terms of both letters is that, quite shortly before the letters were written, there must have been some kind of offer by Mr Little, or by the plaintiffs, to the defendant of a new lease. They cannot merely have been referring to the document of July 3 1980. There must have been further communications of one kind or another, which the defendant had interpreted as an offer of something.

Unfortunately, the evidence as to these communications is scanty in the extreme. However, we do get a little indication of what was happening between the parties from the evidence of Mr Little and from the defendant himself. Mr Little, in the course of his evidence, having referred to the document of July 3, went on to say (I am quoting from the learned judge's notes):

I phoned Nelson and asked him when he was going to start paying the £65.00 a week. Can't remember precisely what I said or he said. Doc 6 Bundle B

that is the letter of January 7 1981

I remember that letter, and when I got it I thought he did not want the new lease.

It thus seems fairly clear, therefore, from Mr Little's own evidence, that following the conclusion of the oral agreement made in or about July 1980, he was making requests, or demands, on the defendant to pay rent at the rate of £65 (not £50) a week.

Then the defendant, in the course of his evidence, deals with this point very briefly. He was asked about the letter of January 7 and he said this:

I think I asked them to write this letter. Forget why I wanted them to write this letter. I suppose they

that must mean the plaintiffs

had asked me to pay at £65.00 a week. Never have paid rent at rate of £65.00 per week.

It is true that, when one reads this evidence, one finds Mr Little saying that, on receipt of the letter of January 7, he thought that the defendant did not want the new lease. However, one important point appears to be common ground both in the material evidence of the defendant and of Mr Little himself. The letter of January 7 1981 had been preceded by a demand, or request, by Mr Little that the defendant should pay rent at a rate, not of £50 per week, as had been provided for by the oral agreement of 1980, but at a rate of £65.00 per week. This was £15 per week more than the rent which the learned judge found to have been agreed under the oral agreement.

In these circumstances, whilst I accept that the matter is not entirely clear, I for my part think that, on the evidence, the reasonable interpretation for the plaintiffs to have placed on the particular passages in the two letters of January 7 and January 23 1981, which are so heavily relied upon by them, is that the defendant did not wish to be bothered with further proposals which would involve him with payment of a rent of more than £50 a week and that he was content to stand on his existing rights. In short, I think that the judge was right to find that there had been no repudiation by the defendant of the oral agreement which he found to have been made.

To put the point in one sentence, in my view a man cannot be said to repudiate a contract to take a lease at a rent of £50 a week by refusing an offer of a lease at £65 a week.

We are not helped, as the learned judge was not helped, by the obscurity of much of the evidence, but I think it is worth adding this comment. The subsequent conduct of the plaintiffs seems to me hardly consistent with the view that there had in January 1981 been any repudiation by the defendant of the oral agreement providing for the surrender of the old lease and the grant of a new one. They did not suggest that any repudiation had taken place. They were content to allow him to continue on the premises, and indeed to incur substantial expenditure on them. It was only several years later that they took the present proceedings to obtain possession of the premises.

In all the circumstances — though I reach this conclusion by a somewhat different route from that taken by the learned judge — I think that he reached the right conclusion, and I would dismiss this appeal.

Agreeing, EASTHAM J said: In view of the fact that my lord has so fully and accurately dealt with the documents in this case, it seems to me that this appeal falls within a very narrow compass. Notwithstanding the terms of the respondent's defence in the court below, he proceeded to give evidence, which the learned judge accepted, to the effect that after the date of the important document of July 3 1980 he reached an oral agreement with Mr Little of the appellant company, under which he was to have a tenancy until 1991 at a rent of £50 per week until the expiration of his current tenancy in 1984, but otherwise subject to the terms of the document of July 3.

In spite of the fact that that evidence was inconsistent with the terms of the defence and some of the letters which the respondent's solicitors wrote prior to the hearing, the learned judge, who had the advantage of seeing both Mr Little, on behalf of the appellants, and the respondent, in the witness box, came to the conclusion that the respondent had established such an agreement.

When the legal advisers to the appellants had an opportunity of seeing the judge's notes of evidence, very rightly, and very frankly, they came to the conclusion, which is reflected in the skeleton argument on the appellant's behalf, that there was material on which the learned judge could make that finding relating to the agreement alleged by the respondent. Accordingly, as appears from para 5 of the skeleton argument, paras 6, 7 and 8 of the notice of appeal, which challenge that agreement, were no longer to be pursued.

That leaves only the possible question of repudiation, and if there was a repudiation, whether that repudiation was accepted. The learned judge dealt with that upon the basis that the letters alleged to constitute the repudiation are those of January 7 1981 and January 23 1981. The learned judge thought that there had been some misunderstanding as between the respondent to this appeal and his solicitor; I agree with Mr Smith that if that is the only ground on which the judge's finding that there was no repudiation is founded, that would not be a good ground. But as I think Mr Smith conceded, whether or not those letters, or either of them, amounted to a repudiation depends on whether a reasonable man in the shoes of the plaintiffs, on receiving those letters, would have believed that the defendant had no intention of carrying out his part of the oral agreement. We have all the material before us; we have the factual matrix, namely, the previous negotiations by Mr Little, and, at any rate in my judgment, when one comes to consider in particular the first letter on which most reliance was placed, it cannot be construed by any reasonable person as being a reference to not wanting, under the terms of the oral agreement, an extended tenancy, but was declining what had been offered by Mr Little, namely, a further lease to 1991 at a rent of £65.00 per week.

In my judgment at any rate there was no repudiation, and it is therefore unnecessary to decide whether it was accepted. If it were necessary to go on to that, I would have had very considerable doubt that the so-called repudiation had been accepted, by reason of the facts mentioned by my lord, Slade LJ, in his judgment.

I agree that this appeal should be dismissed.

The plaintiffs' appeal was dismissed with costs. Legal aid taxation of respondent's costs was ordered.

Court of Appeal

July 15 1986
(Before Lord Justice DILLON)

MORRIS v PATEL AND OTHERS

Estates Gazette January 31 1987

281 EG 419

Landlord and Tenant Act 1954, Part II — Application for leave to appeal and extension of time for appealing by plaintiff tenant from order in county court proceedings — County court judge had dismissed plaintiff's application for declaration that landlords' notice purporting to terminate plaintiff's business tenancy was invalid — Landlords' notice had been in the old Form 7 in Appendix 1 to the Landlord and Tenant (Notices) Regulations 1969 instead of the new Form 1 in Schedule 2 to the Landlord and Tenant Act 1954 Part II (Notices) Regulations 1983 — The new form differed in several ways including a large and prominently printed box warning the recipient of the notice to act quickly and, in case of doubt, to obtain advice immediately — However, the absence of the box had no adverse consequences in the present case, as the tenant did obtain advice and took the appropriate steps according to the timetable — The judge decided that the old form was substantially to the same effect as the new one and refused leave to appeal — The present application for leave was out of time because neither the plaintiff nor his solicitors had appreciated the urgency of such application — Held that there was not enough to merit the granting of leave — Application dismissed

No cases are referred to in this report.

This was an application by Roger William Morris for leave to appeal and an extension of time for appealing from an order made by Judge Macnair at Lambeth County Court. The applicant had as plaintiff sought a declaration that a notice of termination of his business tenancy served by his landlords, Mukesh Kanubhai Patel, Parul Narendra Patel and Arun Jashbahi Patel was invalid as being in the wrong form. Judge Macnair had decided against the applicant and refused him leave to appeal.

Jonathan Gaunt (instructed by Proctor Gillett) appeared on behalf of the applicant; M A P Hopmeier (instructed by J B Wheatley & Co) represented the respondents.

Giving judgment, DILLON LJ said: This is an application for leave to appeal and an extension of time for appealing by a Mr Morris, who is the plaintiff in proceedings in the Lambeth County Court. The order against which leave to appeal is sought was made by His Honour Judge Macnair on April 16 of this year. It was an order whereby he dismissed an application by the plaintiff for a declaration that a notice served on him under Part II of the Landlord and Tenant Act 1954 purporting to terminate his business tenancy of certain premises was invalid. The judge refused leave to appeal. The application for leave to appeal and an extension of time was only issued on May 28. That is out of time, though not so far out of time as to cause serious prejudice. The reason why it is out of time is that neither the plaintiff nor his solicitors appreciated the urgency of applications for leave to appeal.

The substantive matter arises in this way. The notice purporting to terminate the plaintiff's tenancy, which was given on March 21 1985 by the solicitors for the defendants, was in the familiar old Form 7 in Appendix 1 to the Landlord and Tenant (Notices) Regulations 1969. The solicitors for the landlords had apparently not appreciated that that form had been replaced by a new Form 1 in Schedule 2 to the Landlord and Tenant Act 1954, Part II (Notices) Regulations 1983. That new form contains differences. Firstly, there is a large and heavily printed box advising the recipient of the notice to act quickly and, if in any doubt, to take immediate advice from a solicitor, surveyor or Citizens' Advice Bureau. The absence of that box did not have any consequences in the present case since the plaintiff did consult solicitors and did take the appropriate steps according to the timetable required by the Act. The other difference is in the drafting of the notes on the back of the notice; these are more explanatory and useful in the new form of notice.

The substantive question for Judge Macnair to decide was whether the old Form 7 was a form substantially to the like effect as the Form 1 of Schedule 2 to the new regulations. He said that it was. It seems to me he was right and there is not enough in this appeal to merit the granting of leave to appeal. Consequently, this application is dismissed.

The application was dismissed with costs.

Chancery Division

June 25 1986
(Before Mr Justice MILLETT)

DELLNEED LTD AND ANOTHER v CHIN

Estates Gazette February 7 1987

281 EG 531-539

Landlord and tenant — Arrangements for operation of Chinese restaurant — Whether document between leaseholder and restaurant operator granted a tenancy — Document described as a licence creating a "management

agreement" and specifically stating that "nothing in it shall be construed as creating a legal demise" and that "this agreement confers no tenancy" — After a detailed analysis of the clauses of the agreement the judge decided that it did in fact confer on the restaurateur exclusive possession — The document was intended to mislead, the leaseholder being anxious to avoid any indication of subletting or parting with possession which might have led to action by his superior landlord based on breach of covenant — The agreement had the ingredients mentioned in *Street* v *Mountford* as constituting the hallmarks of a tenancy, namely, the grant of exclusive possession for a term at a rent — The judge, although agreeing that the categories of exceptional circumstances referred to by Lord Templeman in *Street* v *Mountford* might not be closed, rejected a suggestion that the circumstances in the present case constituted an exception — Held accordingly that the plaintiffs, the operators of the restaurant, had a tenancy of the premises

The following cases are referred to in this report.

Matchams Park (Holdings) Ltd v *Dommett* (1984) 272 EG 549
Shell-Mex & BP Ltd v *Manchester Garages Ltd* [1971] 1 WLR 612; [1971] 1 All ER 841, CA
Street v *Mountford* [1985] AC 809; [1985] 2 WLR 877; [1985] 2 All ER 289; [1985] 1 EGLR 128; (1985) 274 EG 821, HL

The plaintiffs in this case, Dellneed Ltd and Mr Chung Yow Chan, sought a declaration that they occupied premises at 37-38 Gerrard Street, London W1, which were the site of the Loon Fung Chinese restaurant, as subtenants of the defendant, Mr James Chin. A claim for specific performance of an alleged oral agreement by the plaintiffs to purchase the defendant's leasehold interest in the premises was abandoned during the hearing. There was a counterclaim by the defendant for possession and mesne profits.

Timothy Jennings (instructed by Barnett, Alexander & Chart) appeared on behalf of the plaintiffs; Patrick Ground QC and J W Haines (instructed by Gerard Dunne & Co, of Lowestoft) represented the defendant.

Giving judgment, MILLETT J said: This action concerns leasehold premises at 37-38 Gerrard Street, London W1, which consist of a ground floor and basement only, the upper storeys having been demolished some years ago following the service of a dangerous structures notice.

The premises are held under a lease dated March 23 1964 and the term granted by the lease will expire in March 1991. At all times material to this action the defendant, Mr James Chin, has been the lessee, and the rent payable under the lease has been £37,500 per annum.

The lease contains the usual prohibition against assigning, subletting or parting with possession of the demised premises, or any part thereof, without the landlord's consent.

The premises are the site of a Chinese restaurant called the Loon Fung Restaurant, which was formerly carried on by Mr Chin's company, Loon Fung London Ltd, but which has for the past four years been carried on by the first plaintiff, Dellneed Ltd, pursuant to a so-called management agreement dated December 6 1982 and made between Mr Chin and Dellneed. Dellneed is a company formed or acquired by the second plaintiff, Mr Chung Yow Chan, for the purpose of operating the restaurant, and all its issued shares belong beneficially to him.

By the action Mr Chan originally sought specific performance of an oral agreement alleged to have been made between himself as agent for and on behalf of Dellneed to purchase the lease from Mr Chin for a sum of £180,000, of which it was alleged no less than £157,100 had been paid to Mr Chin, all but £20,000 in weekly sums of cash for which no receipt was sought or given. Further or alternatively Dellneed seeks a declaration that it occupies the premises as a subtenant by virtue of the management agreement as extended by a supplemental agreement dated June 15 1985, such subtenancy not having been determined in accordance with the provisions of Part II of the Landlord and Tenant Act 1954. Mr Chin denied that there was an agreement for the sale and purchase of the lease or that any weekly payment was made as alleged, and contends that the management agreement and the supplemental agreement created a licence which has since been determined and not a tenancy,
and he counterclaims for possession and mesne profits. At the conclusion of the plaintiffs' case Mr Jennings very properly abandoned the claim based on the alleged oral agreement. Mr Chan's story, improbable from the start, disintegrated in cross-examination. It was not, I am afraid, restored by his wife's attempt to corroborate it. It had long since passed from the implausible to the incredible. I did not believe a word of it.

I have already dismissed Mr Chan's claim to enforce the alleged agreement and I must now deal with what remains of the action and with the counterclaim which are concerned with the management agreement. In doing so I prefer the evidence of Mr Chin and his witnesses wherever it differs from that of Mr and Mrs Chan.

Mr Chin is a Chinese businessman of considerable repute. He is life president of the Chinese Chamber of Commerce in England, president of the Chinese Food Traders' Association in the United Kingdom and honorary president of the China Town Chinese Association of London. He has had more than 20 years' experience as a Chinese restaurateur. He has been concerned in the development of Gerrard Street as the centre of modern China Town. In addition to the Loon Fung Restaurant at 37-38 Gerrard Street Mr Chin or his company owns a Chinese takeaway at 39 Gerrard Street and a craft shop at 31 Gerrard Street. His company Loon Fung London Ltd also owns freehold premises at 42, 43 and 44 Gerrard Street, where Mr Chin runs the Loon Fung Supermarket on the ground floor and, since February 1985, a Chinese restaurant upstairs called the New Loon Fung Restaurant. From May 10 1982 the upstairs of these premises was leased by Loon Fung London Ltd to a Mr David Chan by a formal lease for a term of $2\frac{1}{2}$ years at a rent of £31,200 per annum payable at the rate of £600 per week. Mr David Chan used the premises for the purposes of a Chinese restaurant run by him which was then called the Peking Garden.

The Loon Fung Restaurant has had an eventful recent history. In 1978 or 1979, after the upper storeys had been demolished, there was a serious fire on the premises which gutted them. The landlord had not maintained a fire policy and Mr Chin had to reinstate and completely refurbish and re-equip the restaurant out of his own pocket. In 1981 he brought proceedings against the landlord, claiming damages in excess of £50,000 for breach of the landlord's covenants to insure and to repair the structure. Mr Chin withheld rent because of his claim and in 1983 the landlord brought proceedings to forfeit the lease, *inter alia* for non-payment of rent. The landlord has since withdrawn the claim to forfeit the lease, but the proceedings are still continuing as a claim for arrears of rent.

From June 1980 the restaurant was run by a Mr Bill Tang, whose company Lionbear Ltd had a management agreement with Mr Chin for two years from June 29 1980. In January or February 1982 there was a serious gang fight in the restaurant in which one person was killed, and Mr Tang decided to give up the restaurant prematurely. Mr Chan heard of this and approached Mr Chin for a similar management agreement. Mr Chan, who is aged 51 and speaks no English, had no previous experience of running a restaurant. He sold bean sprouts and bean curds to Chinese restaurants and food stores, trading under the name of C Y Chan & Co. The Loon Fung Supermarket was one of his customers and he had known Mr Chin for some nine years. Mr Chin agreed to give Mr Chan a management agreement for three years on financial terms which were virtually identical to those previously enjoyed by Mr Tang. Mr Chan was to form a company and join in the agreement as surety. Although the document was not executed until December 6 1982 the main terms were agreed in March 1982 and the first draft was forwarded by Mr Chin's solicitor, Mr Michael Tang, to Mr Chan's solicitors on March 31 1982. Dellneed was acquired and took possession for the purpose of redecoration on May 1 1982 and the restaurant officially opened under Dellneed's management on May 31 1982. Part of the delay in completing the management agreement was due to a fire in the restaurant in July 1982 in which several lives were lost, and which also burnt down Mr Michael Chan's offices which were next door to the restaurant. By a supplemental agreement dated June 15 1985 and made between the same parties the terms of the management agreement were extended until December 31 1985 and an additional £5,000 deposit was also paid by Dellneed. By a letter dated December 18 1985 Mr Chin, through his solicitors, informed Dellneed that he was not willing to extend the agreement for a further period and would require the premises to be vacated on December 31 1985 and the business and the property to be returned to him on that date. Hence this action and counterclaim.

The management agreement and the agreement with Lionbear which preceded it were both prepared by Mr Michael Tang, who gave evidence before me and told me that they represented a form of agreement which was well known among the Chinese community and was known as a Mai Toi Agreement, literally a table use or table rent agreement. Mr Tang, who was English educated and speaks no Chinese, was admitted a solicitor in 1970. He soon became aware of the nature of a Mai Toi Agreement and how it was intended to operate. It was frequently used by Chinese businessmen who owned property and was informal and flexible as the parties well understood. It would normally be used between relatives or close friends who trusted each other and where the parties had no long-term plans. It was confined to the restaurant business. A young would-be restaurateur, often a recent immigrant, perhaps with little or no knowledge of English, desirous of running his own restaurant but without capital, contacts, ability to borrow, experience or reputation would look to an established restaurant owner to give him a start. The gist of the arrangement was that for a specified period, usually fairly short, the owner would provide the newcomer with a restaurant, fully furnished and equipped and ready for trade, and with an established name and reputation. The newcomer would operate the business on his own account, keeping the profits and bearing the losses. The newcomer would indemnify the owner against all outgoings in respect of the property, and in addition would pay a weekly sum which would reflect the fact that he had the use of a restaurant fully equipped and decorated and with a good name and reputation. He would also pay a deposit to secure his obligation to maintain the restaurant and its equipment in good condition. The newcomer would look to the owner for advice and assistance, and the owner would supervise the business at least in the early stages. The extent of the owner's supervision would depend on how well the newcomer was doing. He would give advice on prices, menu and staff. To the outside world it would appear still to be the owner's restaurant. The owner would still retain the keys and pay the rent and rates for which he would be reimbursed. He would expect to eat in the restaurant without charge and to be able to inspect the kitchen. He would be surprised if he could be excluded from the premises. His family and associates would still regard it as his restaurant. If things went well, by the end of the stipulated period the newcomer would have enough capital and experience to move to other premises and set up on his own. If things did not work out or serious disagreement arose the owner would simply take back the restaurant fairly summarily. I have no doubt that even where the arrangement was between relatives or friends who trusted each other it was intended to create legal relations, but I have a strong impression that the enforcement of the owner's rights, while intended to be effective, was expected to be primarily by extra legal means.

The management agreement in the present case is a formal written document, executed by the parties after its terms had been considered and negotiated by their respective solicitors. The question whether it created a tenancy might therefore be thought to be primarily a question of construction. As will appear, however, the management agreement was designed to obscure rather than reveal the parties' true relationship, and this must therefore be discovered from their conduct, the correspondence between them and the surrounding circumstances.

Lionbear had had the use of the restaurant, together with its fixtures, fittings and equipment, on payment of a deposit of £15,000 and a weekly sum of £400, and an indemnity against the rents, rates and other outgoings of the premises other than insurance. Mr Michael Tang told me that the Lionbear agreement was a Mai Toi Agreement. He clearly recognised the difficulty in drawing up an agreement which reflected the arrangements which the parties wished to make but which did not create a tenancy. He adopted the device, which he admitted was artificial, of splitting the agreement in two. First, there was a licence agreement by which Mr Chin authorised Lionbear "to enter upon and use" the premises for the purpose of a restaurant upon indemnifying Mr Chin against outgoings, but which reserved no licence fee or rent at all because, as Mr Michael Tang told me, "I saw danger in that". And, second, there was a consultancy agreement between Lionbear and Loon Fung London Ltd (then Loon Fung Restaurant Ltd) under which Lionbear agreed to pay £400 a week in return for the use of the equipment and advice. Mr Michael Tang subsequently discovered that the Chinese community preferred their Mai Toi Agreements to be contained in a single document and he abandoned the device of having two documents.

Counsel for Mr Chin invited me to approach the construction of the management agreement with Mr Chan on the footing that it was also a Mai Toi Agreement intended to replace a Mai Toi Agreement on virtually identical terms which had just come to an end. I do so, but I must not be taken to assume that Mai Toi Agreements in general, or the Lionbear agreement in particular, do not create a tenancy in law.

The terms agreed between Mr Chin and Mr Chan were identical to those previously agreed with Lionbear, save that Mr Chan agreed also to indemnify Mr Chin against the insurance of the premises. Mr Michael Tang forwarded a copy of the draft management agreement to Mr Ko, Mr Chan's solicitor, under cover of a letter dated March 31 1982. That letter included the following passage:

Our client wishes to be absolutely certain that the landlord would not effectively be able to take proceedings against him to terminate the lease as a result of this management agreement. Obviously we refer specifically to the user covenant against parting with possession. On counsel's advice we suggest that in the limited company which is to be incorporated our client takes and holds a nominal shareholding in the company which he will then declare will be held by him in trust for your client. The effect therefore will be that your client will be the beneficial owner of the company.

Mr Chin had also had a shareholding in Lionbear which he had held as nominee for Mr Bill Tang. This was Mr Michael Tang's idea. Mr Chin told me: "He thought that this would protect my lease." Mr Chin told me that he knew his landlord wanted to forfeit the lease so he had to be very careful. He had no intention of seeking the landlord's consent and, as will appear, Mr Chan was aware of this.

On April 23 1982 Mr Chan saw his solicitor, Mr Ko, and handed him a copy of the lease of the premises. According to Mr Ko's attendance note Mr Ko took him through the draft management agreement and explained its terms to him. He advised him on the terms and conditions of the lease, particularly with regard to the restrictions on assignment, subletting and parting with possession of the premises. He told him that:

Although the scheme of management agreement had been designed by Mr Chin's solicitor to bypass consent of the landlord there was always a possibility that the landlord might start legal proceedings for breach of the covenant restricting assignment, etc, of the premises without consent. Whether the landlord would succeed in any such move was a separate question, but Mr Chan must be prepared to face such a possibility.

According to Mr Ko's attendance note Mr Chan fully understood his position.

Mr Chin agreed with Mr Chan that on payment of £5,000 towards the deposit of £15,000 provided by the draft management agreement Mr Chan could go into occupation of the premises for the purpose of redecorating them on April 30 1982. Mr Chin had left for Canada and Hongkong shortly before the end of April 1982, but before leaving he wrote out a form of receipt for his cousin, Mr Harold Chin, a director of Loon Fung London Ltd, to sign. It was in the following terms:

April 30 1982, re-occupancy of premises at 37/38 Gerrard Street, W1. On receipt of cheque amounting to £5,000 the keys to the above premises have been surrendered to Mr C Y Chan of 88 Chatfield Road, Kenton, Harrow, Middlesex. He is given full authority to decorate the above premises as he requires. Any dispute with the landlord is solely his responsibility.

That was signed Mr Harold Chin, pp James Chin, director.

The phrase "surrender the keys" is not without significance. Mr Chin explained that the last few words "Any dispute with the landlord is solely his responsibility" were included because Mr Chan had told him that he wanted to enlarge the front entrance of the restaurant. This would require the landlord's consent, and Mr Chin had no intention of asking for consent because, as he told me, he did not think he would get it. He told Mr Chan that if the landlord made trouble as a result Mr Chan would have to reinstate the former entrance.

Mr Chan paid the £5,000 and collected the receipt and the keys on May 1 1982. He went into possession and redecorated the restaurant. He had told Mr Chin what he wanted to do. Mr Chin did not agree with all the changes that Mr Chan proposed. Many of them he thought were unnecessary. But Mr Chan went ahead. According to Dellneed's audited accounts over £22,000 was spent on furniture, fixtures and fittings and over £11,000 on equipment. This included over £4,000 on dim sum trolleys which Mr Chin advised against. But Mr Chan insisted, so Mr Chin took the opportunity to sell some to him.

On May 26 1982 Mr Michael Tang wrote to Mr Michael Ko concerning the draft management agreement. One passage in his letter, explaining the purpose of clause 15(e) of the draft, was in the following terms:

> With regard to clause 15(e) the intention behind this sub-clause is that our client should be able to terminate the lease *[sic]* if it is necessary to protect the lease should the landlord take action against our client for parting with possession and in the unlikely event of this proving successful. In such an event our client would wish to terminate the lease *[sic]* and put an end to the landlord's action.

The restaurant was officially opened for business with a big party on May 31 1972. From then on Dellneed paid a sum of £1,291 weekly in arrear to Mr Chin. Mr Chan described it as rent. Dellneed's audited accounts described it as "rent and rates". In the correspondence with Mr Ko Mr Michael Tang explained the breakdown. £831 was for rent due under the lease, rates, water rates and the insurance premium on the building, and would increase as they increased. And £400 was a "licence fee". The £400 was subsequently attributed to the hire of the equipment and an extra £60 a week VAT was charged in consequence, making a weekly total of £1,291. In his evidence Mr Chin was more realistic. The £400, he said, was for the equipment, his advice and the use of the premises. In fact it was a package. I find that the consideration was not £400 a week but £1,291 inclusive of VAT, capable of increase as I have mentioned, and that it was paid for the use of a fully furnished and equipped restaurant, and no doubt for Mr Chin's advice and assistance.

Dellneed carried on the business on its own account, taking the profits and bearing the losses. It consistently lost money. It lost £29,878 in its first year. After three years' trading it had accumulated losses of £88,989. It had a paid-up share capital of only £100. At all material times it was insolvent.

I find as a fact that Dellneed went into exclusive possession. Mr Chan was given the keys. There is no evidence that Mr Chin retained a set of keys or exercised any rights of possession. Mr Chin used to visit the restaurant and eat there several times a week, but he paid for his meals. He used to inspect the kitchens, but he was proposing to reserve the right to do so in the draft management agreement and Mr Chan did not object. For a short period he continued to store some possessions in an unused part of the basement, but this seems to have been with Mr Chan's permission. Mr Chin gave advice from time to time and no doubt his experience and contacts were useful to Mr Chan. But the decisions were Mr Chan's to take. Mr Harold Chin, who spoke English, assisted as an intermediary with a local environmental health officer and told Mr Chan what that officer required to be done. When asked what was the difference between Dellneed's position and that of Mr David Chan (who it will be recalled was granted a formal lease of the Peking Garden for 2½ years) Mr Chin replied that he was the freehold owner of the Peking Garden and did not have a landlord on top of him. He added that Mr David Chan had some experience of the restaurant business and needed less advice. But I was left with the impression that this was added almost as an afterthought.

On July 18 1982 there was a serious fire on the premises. The premises were reinstated at a cost of £24,290. The work was done to the order and at the expense of Dellneed. The fire policy in respect of the premises was in the name of Loon Fung London Ltd. Despite this Dellneed made a claim and obtained payment of a sum of £20,000. The claim form was completed in the name of Dellneed Ltd trading as Loon Fung London Ltd, which was incorrect and unauthorised. However, there is no reason to think that anyone was misled. In correspondence the insurance company referred to the policy as covering "the premises owned by Loon Fung London Ltd and leased to Dellneed Ltd trading as Loon Fung Restaurant". Mr Harold Chin complained to the assessors that the payment had been made to Dellneed without authority. It is clear, however, that his only concern was that the apportionment between the parties of the global settlement in respect of the various policies should not be prejudiced.

After the fire Mr Chan decided to install a dim sum factory in the basement in order to make dim sum for supply not to the restaurant but to shops and offices in the area. This was a new venture, not part of the restaurant business. Mr Chin removed the goods he had been storing in the basement in order to make room available for this purpose. This episode provides further evidence, if evidence were needed, that Dellneed's occupation of the premises was not merely in order to carry on the previous business of the restaurant.

The management agreement was finally executed on December 6 1982. It was made between Mr Chin, described as the licensor for the first part, Dellneed described as the contractor of the second part and Mr Chan, described as the surety of the third part. I must, I think, read virtually the whole of that agreement.

The first recital recites:

> Whereas the licensor has up to the date hereof been operating a Chinese restaurant at the premises (hereinafter called "the premises") described in the first schedule hereof under the style or name of Loon Fung Restaurant (hereinafter called "the business").

The first schedule is in the following terms:

> All those premises situate on the ground floor and basement of 37/38, Gerrard Street in the City of Westminster.

Recital 2 recites:

> It has been agreed that the contractor shall take over the management of the business on the terms hereinafter appearing.

Clause 1 provides:

> The contractor shall take over and continue throughout the period referred to in this clause the management, subject as herein provided, of the business at the premises from the 15th day of June, 1982 *[sic]* until and including the 15th day of June, 1985.

Clause 2:

> The contractor shall pay to the company *[sic]*

that is an obvious misprint for "licensor"

> the fees and other sums mentioned in the second schedule hereof for the use of chattels referred to in clause 8 hereof payable weekly in advance *[sic]* save as otherwise provided without any deduction whatsoever. The first such weekly payment to be made on the 15th June, 1982 and subsequent weekly payments to be on the Sunday of every week thereafter and the other said sums to be paid within seven days of receiving notification thereof from the licensor. The contractor shall pay to the licensor the sum of £15,000 by way of deposit (hereinafter called "the deposit") returnable to this agreement *[sic]*

again an obvious misprint for "the contractor"

> without interest but subject to deduction by the licensor of any sum payable by the contractor to the licensor by virtue of this agreement or its breach, the said deposit being payable on or before the signing hereof.
>
> (4a) The contractor shall maintain and improve the reputation and goodwill of the business: (b) The contractor and the licensor or their duly authorised representative shall meet on the premises from time to time as either party may reasonably require for discussion on the conduct, management and improvement of the business.

Clause 5:

> As from the 15th day of June, 1982 the contractor shall be entitled to retain all money received in the business over and above the weekly and other payments referred to in clause 2 hereof and shall be responsible for all outgoings including but not limited to rent payable under the lease, general and water rates, insurance, rent and taxes and shall pay for all purchases and services including but not limited to gas, electricity and telephone charges, cleaning, laundry, refrigeration and laundry.

Clause 6:

> The contractor shall not redecorate or make any alterations or additions to the premises without first submitting to the licensor details and plans for the proposed work and obtaining the prior consent in writing of the licensor and the contractor shall hand back to the licensor or as he may direct any existing furniture and fixtures, decorative chattels or other items which are no longer required as a result of such work.

Clause 7:

> The contractor shall employ and shall have power to dismiss as principal and not as agent in the business whomsoever it wishes and on whatever terms provided that in exercising such rights the contractor shall not damage the reputation of the business.

Clause 8:

> The contractor shall have the use of the licensor's furniture and furnishings, fixtures, equipment and chattels at the premises as detailed in the inventory agreed and signed by the parties on the date hereof and shall keep them in the same condition as they are in on the 15th June, 1982 and shall make good repair or restore or at the option of the licensor pay the value of all or any part of the same which may be broken, lost damaged or destroyed, fair wear and tear and damage by accidental fire excepted, and in such condition shall return the same to the licensor at the termination of this agreement howsoever it may be terminated and shall insure the same in the full value thereof comprehensively against loss or damage.

Clause 9 (a):

> The contractor shall keep the premises, including by way of example and no of limitation, the drains, heating, water and sanitary pipes, lift and apparatus, gas pipes and fittings, electricity wiring and fittings and all other fixtures and fittings in the same good condition as they are in on the 15th June 1982, fair wear and tear excepted, and in such condition shall return the same

to the licensor at the termination of this agreement howsoever it may be terminated.

Clause 9 (b):

The contractor shall at his own expense comply with all byelaws, whether relating to cleanliness, hygiene, fire precaution, public safety or otherwise, and shall at his own expense complied [sic] with any requirements of any local or public authority relating to the maintenance and/or improvement of existing facilities. (c) The contractor shall not do anything which would constitute a breach of the lessee's covenants and obligations under the lease. (d) The contractor shall immediately notify the licensor if any writ shall be served upon it or any other proceedings commenced against it and act therein as the licensor shall direct.

Clause 10:

The licensor and the contractor shall use their best endeavours to keep in force and renew when necessary the justices' licence granted to the licensor.

Clause 11:

The contractor shall not change the name of the present business from that of Loon Fung Restaurant.

Clause 12:

The licensor is to be at liberty to enter the premises at any time and [sic] to ensure that the business is operated in any efficient manner and that the covenants on the part of the contractor herein contained are being observed.

Clause 13 (a):

This agreement is personal to the contractor and the benefit of this agreement shall under no circumstances be assignable by the contractor.

Clause 13 (b):

The directors and shareholders of the contractor company shall be

and there were six names inserted of which one is Mr Chan and another is Mr Chin.

And the contractor may not without the prior written consent of the licensor change its directors or shareholders other than to members of their immediate families or between the directors and the shareholders themselves.

Clause 14:

Nothing in this agreement shall be construed as creating a legal demise and it is hereby agreed between the parties that this agreement confers no tenancy upon the contractor and the possession of the premises is retained by the contractor [sic]

an obvious misprint for "licensor"

subject however to the rights created by this agreement.

Clause 15 provides for circumstances in which the agreement may be determined by notice given by the licensor. I need not go through it in detail, but it includes the failure to pay any payment due within seven days after becoming due whether demanded or not, a failure to remedy breaches capable of remedy, and so on. But clause 15 (e) is in the following terms:

"If the licensor will not give consent under clause 13 (b) hereof" — that is to say to the change of directors or shareholders of the contractor company.

Clause 16:

The licensor agrees to pay any land charge that may be registered with the City of Westminster up to the date hereof and to indemnify the contractor against any claims, actions or legal proceedings in respect thereof.

Clause 17 was an ordinary surety covenant.

The second schedule is in the following terms:

From the 15th day of June, 1982 until the 15th day of June, 1985 the fee of £400 per week and all payments due from the licensor to the lessor under the lease and all other payments of an annual or recurring nature payable by the contractor in respect of the business.

I observe first that the document is described as a management agreement, and the grantor as a licensor, and that the draftsman has studiously avoided any words indicative of a tenancy. Indeed, he has gone so far as to avoid even a grant of the right to enter upon and use the premises, a right which had been granted in the Lionbear agreement but which is conspicuously absent from the present agreement. But it must obviously be inferred. How could Dellneed carry on the business of a restaurant upon the premises, which it was not only entitled but bound to do, or enjoy the right to use Mr Chin's furniture and furnishings, *fixtures*, equipment and chattels granted by clause 8 unless it had the right to enter upon and use the premises? To leave the primary right intended to be granted to be inferred in this way excites suspicion that the document is drawn to mislead.

Next I observe that any impression gained from the fact that the document is called a management agreement or from recital 2 or from clause 1 that Dellneed was merely taking over the management of an existing business owned by Mr Chin would be quite false. Dellneed was starting up its own business on its own account on premises and with equipment provided by Mr Chin and with the benefit of the name, reputation and goodwill attaching to those premises. The fact that Dellneed was placed under a positive obligation to carry on the business on the premises is not of course inconsistent with the creation of a tenancy. On any footing, Mr Chin retained a valuable interest in the premises and the goodwill attaching thereto, and needed to impose such an obligation for its protection.

Next it is to be observed that the premises in question are defined as the ground floor and basement of 37-38 Gerrard Street, a self-contained defined area, the whole of the premises now comprised in the lease, and obviously not an unsuitable subject-matter for a demise. It is to be observed that under clause 2 the whole of the payments and not merely the £400 is expressed to be for the use of the chattels referred to in clause 8, but not the furniture, furnishings or fixtures also mentioned in that clause. This is an extravagant notion. None of the witnesses have suggested that anything beyond the £400 per week was paid for the chattels and it will be remembered that VAT was charged on only £400. Even the £400 was paid in part at least for the use of the premises and the furnishings and fixtures thereof. In so far as clause 2 is intended to suggest that the substance of the agreement is the hire of equipment it is misleading. The chattels were not hired separately but as the equipment of the restaurant, and were intended to be used *in situ* and nowhere else. As I have already mentioned, not only the £400 but all the other payments were made weekly in arrear and not as provided by clause 2.

Clause 4 in my view is neutral on the question whether the document creates a tenancy. On any footing Mr Chin retained a valuable interest in the reputation and goodwill of the restaurant. But it is to be observed that the clause does not require Dellneed to follow Mr Chin's instructions on any matter. Clauses 5 and 7 show that the business in fact belongs to Dellneed. Clause 6 is significant. It was thought necessary to prohibit Dellneed not only from redecorating but from making alterations or even additions to the premises without Mr Chin's consent. If Dellneed were a mere licensee or manager it would plainly not dream it had the right to do so. I have already referred to clause 8, which entitles Dellneed to the use of Mr Chin's furniture and furnishings, *fixtures*, equipment and chattels. This shows that Dellneed was to have the use and enjoyment of the premises, though this was not expressly granted.

Clause 9 (a) requires Dellneed to keep not only the drains, heating, water, sanitary pipes, lift and apparatus and so on but the premises themselves in repair. This seems to be an inappropriate obligation for a mere licensee or manager to undertake.

Clause 11, which prohibits Dellneed from changing the name of the restaurant, is not inconsistent with the creation of a tenancy for the reasons I have already mentioned, but I observe that it is inconsistent with a management agreement. A manager would have no right to change the name of the restaurant he was managing and it would be unnecessary to prohibit him from doing so. On the contrary, the owner would almost certainly insist upon reserving an express right to change the name without the manager's consent, which Mr Chin significantly did not do.

Clause 12 is another provision of great significance. Despite Mr Ground's argument to the contrary it is plain as a matter of construction in my judgment that the word "and" is intrusive and should be disregarded. Properly read, the clause reserves to Mr Chin the right to enter the premises at any time for the purpose of ensuring that the business is operated in an efficient manner and that Dellneed's obligations are being performed. This is completely inconsistent with a mere licence. The reservation of an express right to Mr Chin to enter the premises for certain limited purposes only demonstrates in my view that subject thereto Dellneed was being granted exclusive possession.

Clause 13 (a) is an absolute prohibition of assignment. This is some indication that a personal licence may have been intended, but it is not, of course, inconsistent with the creation of a tenancy. Clause 13 (b) is part of the facade that this is merely a management agreement. Its true purpose, which was frankly admitted in evidence, was to protect Mr Chin's lease by reducing the risk of forfeiture. It was obviously designed to mislead the landlord, should he call for the agreement under which Dellneed appeared to be in possession, into thinking that Dellneed was merely a management company in which Mr Chin was interested, so that any change in possession was merely technical. The fact that Mr Chin held his shares as nominee for Mr Chan was not, of course, disclosed by the management agreement.

Clause 14 is an express declaration that the agreement is not to create a tenancy and that possession of the premises is retained by Mr Chin. I have already stated my finding that it was not in fact retained by Mr Chin, so that this is another part of the facade to mislead the landlord.

Clause 15 (e) was varied in correspondence. I have already referred to the letter dated May 26 1982 in which its purpose was explained by Mr Michael Tang. The parties' respective solicitors agreed in correspondence as early as June 1982 that clause 15(e) should in effect be capable of being invoked if, but only if, the landlord should bring proceedings on the ground that Mr Chin had parted with possession of the premises without consent, but that the clause should remain as drawn. The reason for the suppression of the true ambit of clause 15 is too obvious to need to be stated.

It is plain from what I have already said that the management agreement does not represent the true relationship between the parties and that one of its purposes was to mislead the landlord should he make inquiries of the basis upon which Dellneed was or appeared to be in possession. It is plain that what was paraded as a management agreement was in truth nothing of the kind. Whatever the true relationship of the parties created by the agreement, it was not the relationship of owner and manager of a business. The business which Dellneed was "to manage" was its own.

In *Shell-Mex & BP Ltd* v *Manchester Garages Ltd* [1971] 1 WLR 612 Buckley LJ said at p 618:

It is clear on authority that in considering whether a transaction such as we have before us in this case constitutes a licence or a tenancy the court is not to have regard to the label which the parties give to the document or to the formal language of the document but to the substance of the transaction.

The substance of transaction in the present case was abundantly established by the evidence. It was that for a period of three years Dellneed was to have the use of a fully furnished and equipped restaurant, with an established name and reputation on which to carry on its own restaurant business; that Mr Chin was to make his advice and experience available to Dellneed; and that Dellneed was to pay to Mr Chin the outgoings of the premises and an additional £400 a week so that Mr Chin should receive this latter sum clear. I am quite satisfied that this arrangement necessarily involved, did involve and was intended to involve the granting of exclusive possession to Dellneed.

Mr Ground submitted that Mr Chin retained the right to enter the premises at any time and for any purpose, provided only that he did not derogate from the grant that he had made of the right to run a restaurant on the premises. In my judgment not only does that submission fly in the face of the express terms of clause 12 of the management agreement as I construe them but it is a contradiction in terms. The right granted to Dellneed to use the premises and the fixtures, fittings and equipment for the purpose of running a restaurant was so extensive that I find it impossible to conceive of the retention of any significant right of possession by Mr Chin which would be consistent with it.

Mr Ground submitted that even so Dellneed was not granted exclusive possession because, he says, this was contrary to the intention of the parties. Mr Chin had a difficult relationship with his landlord, who was looking for an opportunity to forfeit the lease. The existence of a covenant against subletting or parting with possession without the landlord's consent was known to both parties, and it was common ground that no application for consent would be made. Indeed I would add it must have been obvious to both parties that any such consent would be most unlikely to be granted. Dellneed was a new company with a paid-up share capital of £100. Mr Chan had no experience of the business and had little in the way of financial resources. Naturally, Mr Ground submitted, Mr Chin did not wish to commit a breach of the covenants in his lease and accordingly must be taken to have had no intention of granting a tenancy or parting with possession of the premises; and Mr Chan, who was aware of this, must be taken to have entered into the transaction on this basis. In my judgment, however, Mr Chin's intentions were more complex than those attributed to him by Mr Ground. His primary intention was to grant to Dellneed the use of the premises, furniture and equipment and all the rights necessary to enable it to run the restaurant. If this could be done without committing a breach of covenant in the lease so much the better. If not, he was still willing to go ahead, but in order to protect himself he wanted to make it appear that no breach had occurred, and if that did not work he wanted the right to terminate the arrangements and resume possession.

Mr Ground also relied upon the other unusual circumstances of the case as demonstrating an intention on Mr Chin's part not to grant exclusive possession. He cited the fact that this was a Mai Toi Agreement made between personal friends, both members of the Chinese community, and that Mr Chin's advice and experience were to be made available to Mr Chan. In my judgment none of these circumstances, separately or cumulatively, are sufficient for the purpose. Mr Chin was not required to provide services on the premises as, for example, the owner of a lodging-house or old people's home is, which would require him to retain possession and control of the premises. A management or business consultant does not need to have possession of the premises on which the business is carried on in order to provide the services for which he is employed. In my judgment the circumstances and the conduct of the parties show that Dellneed was to have exclusive possession of the restaurant premises for a term at a rent.

Mr Ground also relied upon the features I have mentioned as special circumstances negativing an intention to create a tenancy despite the grant of exclusive possession. In my judgment, this submission depends upon a misreading of the decision of the House of Lords in *Street* v *Mountford* [1985] AC 809*. It is clear from that case that once there is found to be a grant of exclusive possession of premises for a term at a rent then whether the arrangements create a tenancy or licence does not depend upon the parties' intentions to create the one rather than the other but upon the legal effect of the transaction into which they have entered. The only intention which is relevant is the intention to grant exclusive possession (see p 826 H). Mr Ground relied upon the recent decision of the Court of Appeal in *Matchams Park (Holdings) Ltd* v *Dommett* [1984] 272 EG 549, in which, despite the grant of exclusive possession, the parties' intention to create a licence rather than a tenancy was apparently held to be decisive. In my judgment, that approach is inconsistent and cannot stand with the even more recent decision of the House of Lords in *Street* v *Mountford* (see, for example, the passage at p 819 between E and F).

In *Street* v *Mountford* it was made quite clear that save in exceptional circumstances the grant of exclusive possession of premises for a term at a rent constitutes the grant of a tenancy. Only three examples of such exceptional situations were given. First, where the circumstances showed that there was no intention to create legal relations at all. Second, where the possession was referable to some other legal relationship such as vendor and purchaser, master and servant or, I might add, owner and manager of a restaurant. Third, where the grantor had no estate or interest in the land out of which a tenancy could be created, as in the case of a requisitioning authority.

None of those exceptional circumstances exists in the present case. It was not and could not be suggested that the management agreement was not intended to create legal relations. The transaction was a commercial one and contractually binding from the start. Nor was there any legal relationship between the parties other than that of owner and occupier of the land to which the occupier's possession could be ascribed. Nor, despite Mr Ground's spirited argument to the contrary, was Mr Chin's position comparable to that of a requisitioning authority. Mr Chin had capacity to grant a tenancy, he had an estate out of which a tenancy could be granted and a grant by him would effectively vest an estate in the land in the grantee. He had a motive for not granting a tenancy and a still more powerful motive for not being found out, since it risked a forfeiture of the lease. The Rent Acts and Part II of the Landlord and Tenant Act 1954 provide equally strong inducements in many cases to landowners to avoid granting tenancies of their properties, but if their desire to grant exclusive possession for a term at a rent is acted upon they cannot point to such considerations to enable them to escape the consequences of their own actions.

I am prepared to accept Mr Ground's submission that the categories of special circumstances mentioned by Lord Templeman in *Street* v *Mountford* are illustrative and not exhaustive and that the categories are not closed, but I can find no exceptional circumstances in the present case which would justify me in finding that despite the grant of exclusive possession there was no tenancy created. It was conceded that Part II of the Landlord and Tenant Act 1954 applies to

---

* Editor's note: Also reported at [1985] 1 EGLR 128.

a tenancy granted in breach of a covenant in a superior lease. I conclude, therefore, that Dellneed has a tenancy of the premises for a period extended to December 31 1985 subject to a break clause in the event of the landlord bringing proceedings for possession on the grounds I have mentioned and that the tenancy has not been determined in accordance with Part II of the Landlord and Tenant Act 1954.

The furniture, fittings and equipment were provided to Dellneed for use in the premises, and by clause 8 of the management agreement Dellneed is not bound to redeliver them to Mr Chin until the termination of that agreement. Accordingly, Dellneed continues to be entitled to retain possession of those items.

Court of Appeal
October 20 1986
(Before Lord Justice SLADE and Mr Justice WAITE)

HILL AND ANOTHER v GRIFFIN AND OTHERS

Estates Gazette April 4 1987

282 EG 85-87

Landlord and tenant — Forfeiture of tenancy — Position of business subtenant — Appeal from county court judge's refusal of relief to subtenant — Normal rule that when a lease is forfeited any subtenancy is also destroyed (subject to grant of relief) applies although subtenancy is a business subtenancy protected under Part II of Landlord and Tenant Act 1954 — Real issue in present case was whether relief against the forfeiture should be granted to the subtenant and, if so, on what terms — Provisions of section 146 (4) of the Law of Property Act 1925 — The term of the superior tenancy had already come to an end prior to the hearing before the county court judge by reason of the forfeiture of the tenancy and the tenant's bankruptcy, but there was jurisdiction, as was clear from *Cadogan* v *Dimovic*, to grant a monthly tenancy by way of relief to the business subtenant, the restriction to a monthly tenancy being in accordance with the prohibition at the end of section 146 (4) — The county court judge, however, refused relief to the subtenant as the latter was not prepared to enter into repairing obligations of the extent under the forfeited lease — It was submitted on behalf of the appellant that the judge had erred in the exercise of his discretion and should have been prepared to grant a new tenancy to the subtenant on much less onerous terms — *Creery* v *Summersell* considered — Held by the Court of Appeal that there were no sufficient grounds for interfering with the judge's discretion — Relief on the suggested terms would have imposed on the landlords a person whom they had never accepted as their own tenant and would have given them less extensive rights in regard to repairs than they had against the tenant whose lease was forfeited — Appeal dismissed

The following cases are referred to in this report.

Cadogan v *Dimovic* [1984] 1 WLR 609; [1984] 2 All ER 168; (1984) 48 P&CR 288; [1984] EGD 128; 270 EG 37, CA
Creery v *Summersell and Flowerdew & Co* [1949] Ch 751
Gray v *Bonsall* [1904] 1 KB 601

This was an appeal by Mr R Soni, the fifth defendant in a county court action for possession and other relief before Judge Monier-Williams at West London County Court. The appellant was the subtenant of the basement of a property at 307A North End Road, London W14, holding a business tenancy protected by Part II of the Landlord and Tenant Act 1954 — The present respondents, Nicholas Lorraine Edmund Hill and Richard Alan Kenneth Bacon, had been substituted for the original plaintiff lessors, National Westminster Bank plc and Richard Josiah Ritchie. The first, second, third and fourth defendants in the action took no part in the appeal.

R T Good (instructed by P C D York & Co) appeared on behalf of the appellant; D N Barnard (instructed by Sutton-Mattocks & Co) represented the respondents.

Giving judgment, SLADE LJ said: This is an appeal by Mr R Soni from a judgment given by His Honour Judge Monier-Williams in the West London County Court on June 27 1985. By that judgment he ordered that the then plaintiffs in the action, National Westminster Bank plc and Mr R J Ritchie, should recover against the appellant possession of the basement of a property known as 307A North End Road, London W14, and also the sum of £910 for rent and mesne profits, and the plaintiffs' costs of the action to be taxed on scale 2. He further ordered that the appellant should give possession of the property on August 12 1985 and should pay mesne profits at the rate of £15 per week until possession was given.

I should mention, by way of preliminary, two small points relating to the formal order which was drawn up to embody the judge's judgment. Though the title to the order, at least in my copy, makes specific reference only to the National Westminster Bank plc as plaintiff, it is plain from the notes of the judge's judgment, which he approved, that he intended to make the order in favour of both of the then plaintiffs. Second, though the formal order in its terms provided for recovery of possession of 307A North End Road, the learned judge in his judgment had pointed out that his understanding was that, while no 307A included the whole building, the claim against Mr Soni related only to the basement.

The judgment given by the learned judge against Mr Soni was given *extempore*. However, at the same time as he heard the case against him, the judge heard another claim by the plaintiffs for possession of the building against a Mr John Watson, who was, I think, occupying the second floor. The judge heard evidence relating to both claims at the same hearing, but he considered that the latter claim raised questions of fact and law which were more complex than those raised by the claim against Mr Soni. He therefore reserved judgment in regard to the claim against Mr Watson and in due course delivered a considered judgment by which he made an order for possession in 28 days against Mr Watson. However, the two judgments were delivered in the same proceedings, as I have indicated, and the issues involved were closely interrelated. I have therefore referred to both judgments for the purpose of eliciting the facts relevant to this appeal.

The history of the matter, as I understand it, is briefly as follows. Mr Soni entered into occupation of the basement in 1966. The judge described him as a respectable person and a useful member of the Indian community. Ever since 1966 he has carried on in the basement the business of printing a well-known newspaper for the Indian community. His landlord was a Mr R A Medina, who himself had a lease or tenancy of the property. By a lease of November 1 1969 the bank and Mr Ritchie, as lessors, formally demised the building to Mr Medina (who already had held a tenancy of it) for a term of 14 years running from November 1 1969. Clause 2.4 of this lease imposed upon the tenant a very wide repairing covenant, the precise terms of which I need not read. Clause 2.18 contained a covenant on the part of the tenant not to assign, transfer, underlet or part with the possession of the premises or any part thereof without the written consent of the lessors first obtained, such consent not to be unreasonably withheld.

Clause 3(3) contained the usual proviso for re-entry in the event of a breach of covenant, and also permitted the lessors to re-enter in the event of the lessee becoming insolvent and/or committing an act of bankruptcy.

The judge found that from 1981 onwards the rent payable by Mr Soni to Mr Medina was £71 a month. Mr Medina subsequently became insolvent and was adjudicated bankrupt on December 13 1982. He died shortly after his adjudication. His lease then became vested in his trustee in bankruptcy, Mr Griffin, on Febrary 23 1983. Mr Griffin disclaimed the lease in April 1983, a few months before it was due to expire. At the date of Mr Medina's bankruptcy and afterwards Mr Soni was still in occupation of the basement.

On March 27 1983 a notice under section 146 of the Law of Property Act 1925 was served on Mr Griffin on behalf of the lessors, stating *inter alia* that Mr Medina had become insolvent and/or committed an act of bankruptcy, and further alleging that Mr Medina had broken the covenant against subletting in the lease by subletting various parts of the property to various named persons, including the basement to Mr Soni, without first obtaining the consent of the landlords.

On October 19 1983 the landlords issued proceedings against Mr

Griffin as first defendant, a Mr Raczkowski as second defendant, Mr Duncan Watson as third defendant, Mr John Watson as fourth defendant, and Mr Soni as the fifth defendant, claiming possession of the premises and the several parts thereof and various other subsidiary forms of relief. The fifth defendant, Mr Soni, filed a defence admitting and averring that in 1966 Mr Medina had let the basement to him. He claimed that a business tenancy, within Part II of the Landlord and Tenant Act 1954, was thereby created; and he counterclaimed for a declaration that he had a tenancy of the basement falling within Part II.

On September 12 1984 Master Turner gave summary judgment against the first defendant, Mr Griffin, and the second defendant, Mr Raczkowski, who had formerly occupied the third floor of the premises. The action was remitted to the county court. By that time Mr Duncan Watson had also vacated the premises, so that the remaining claims that fell to be pursued were those against Mr John Watson and Mr Soni, the appellant.

At an early stage in the proceedings, counsel appearing for the plaintiffs specifically accepted that, as at the date of the bankruptcy of Mr Medina, Mr Soni was entitled to a tenancy protected by Part II of the Landlord and Tenant Act 1954 and that his subtenancy was a lawful tenancy. Accordingly, the allegations which had been previously made in relation to Mr Soni that the subtenancy had been granted unlawfully by Mr Medina were not pursued. At the hearing before the learned judge Mr Soni appeared in person and Mr Barnard (who has also appeared before us) appeared on behalf of the plaintiffs.

For reasons which I will explain a little later, it soon became more or less common ground before the learned judge that, in relation to Mr Soni, the only effective issue which fell to be dealt with was whether relief should be granted and, if so, on what terms. As appears from his judgment, the learned judge considered that he would have no jurisdiction to grant relief to Mr Soni by ordering the grant of a tenancy for a period of more than one month. As appears from the judge's notes of evidence, Mr Soni's evidence was to this effect, in the context of relief. He described the state of the premises as "pretty bad". He said that he thought it would cost £10,000 to £15,000 to put the basement into a good state of repair. He asked for the grant of a 15-year lease with the rent to be revised after five years. He said that he was prepared to pay £71 a month up to the time that the repairs were finished. He submitted to the judge that the fair course was that the repairs should be done by the landlords, but that, if he had to do the repairs himself, he would ask for a period of three years within which to do them. He said that if he were merely granted a monthly tenancy and was at the same time expected to do all the necessary repairs, this was really impossible having regard to his financial means. He was not prepared to commit himself to the repairs on the basis of a monthly tenancy. In the face of this evidence from Mr Soni and the indication of his attitude, the learned judge came to the conclusion that this was not an appropriate case for the grant of relief and declined to give it.

Following the learned judge's judgment the reversion has, I understand, been transferred by the former plaintiffs to Mr N L E Hill and Mr R A K Bacon and an order has been made substituting them as plaintiffs in the action in the place of the bank and Mr Ritchie. So they are now the respondents to this appeal.

We have before us a notice of appeal, by which the appellant asks for a retrial and gives as the grounds of his appeal that "[The] finding of the learned judge was incorrect in law and against the weight of the evidence"; and, second, "The judge failed to take account of the fact that the plaintiff had been aware of the 5th defendant's occupation of [the] premises for several years." I infer that this was drafted by the appellant himself. However, he has appeared before us by counsel, Mr Good, whose argument has been primarily directed to the issue of relief. He and Mr Barnard, on behalf of the respondents, have given us considerable assistance.

A certain amount of the legal background is, I think, common ground. As I have already indicated, in the court below the plaintiffs' counsel made it clear that they did not seek possession against Mr Soni on the grounds that the subletting to him was unlawful. They relied on the general rule of law that when a lease is forfeited the subtenancy also is destroyed. This rule is to be found stated in *Halsbury's Laws of England*, 4th ed, vol 27, para 422, where it is said: "The forfeiture of the lease also destroys the rights of underlessees." But of course the situation in the present case was a special one to this extent. As at the date of Mr Medina's bankruptcy when his lease was forfeited, Mr Soni, as is common ground, had a tenancy protected by the Landlord and Tenant Act 1954, so that it is necessary to look at the relevant special provisions relating to such tenancies.

Section 24(1) of the Landlord and Tenant Act 1954 (as amended) provides:

A tenancy to which this Part of this Act applies shall not come to an end unless terminated in accordance with the provisions of this Part of this Act.

The subsection goes on to give the right to the tenant under such a tenancy to apply to the court for a new tenancy.

Section 24(2) so far as material provides:

The last foregoing subsection shall not prevent the coming to an end of a tenancy . . . by surrender or forfeiture or by the forfeiture of a superior tenancy . . .

It is, I think, therefore common ground that, in view of section 24(2), the forfeiture of Mr Medina's superior tenancy had the effect of terminating the appellant's tenancy, subject only to his right to apply for relief.

As to relief, section 146(4) of the Law of Property Act 1925 provides as follows:

Where a lessor is proceeding by action or otherwise to enforce a right of re-entry or forfeiture under any covenant, proviso, or stipulation in lease, or for non-payment of rent, the court may, on application by any person claiming as under-lessee any estate or interest in the property comprised in the lease or any part thereof, either in the lessor's action (if any) or in any action brought by such person for that purpose, make an order vesting, for the whole term of the lease or any less term, the property comprised in the lease or any part thereof in any person entitled as under-lessee to any estate or interest in such property upon such conditions as to execution of any deed or other document, payment of rent, costs, expenses, damages, compensation, giving security, or otherwise, as the court in the circumstances of each case may think fit, but in no case shall any such under-lessee be entitled to require a lease to be granted to him for any longer term than he had under his original sub-lease.

The term of Mr Medina's lease had already come to an end prior to the hearing before Judge Monier-Williams, by virtue of forfeiture, following on his bankruptcy. There was, therefore, no question of the judge being able to make an order for the vesting of any term which had previously been vested in Mr Medina in the appellant, because there was no such term left to vest. This court considered the construction of section 146(4) in the case of *Cadogan* v *Dimovic* [1984] 2 All ER 168 and that decision establishes that section 24(2) of the 1954 Act does not preclude the court, in a situation such as the present, from giving the sublessee under a business tenancy relief by way of the grant of a new lease for an appropriate term. However, it will be seen from its wording that, though section 146(4) applies to the case where an underlease has been forfeited as a result of the forfeiture of the superior lease, it contains a restriction which prevents the grant of a new lease to an underlessee "for any longer term than he had under his original sub-lease". In *Cadogan* v *Dimovic* this court held that this restriction refers to the term which the under-lessee would have had but for the forfeiture, and that in the case of a business tenancy this term is, by virtue of section 24(1) of the 1954 Act, the period which would elapse before the tenancy could be terminated in accordance with Part II of the 1954 Act following the expiry of the term granted by the underlease. In the present case that period was one month.

In the result, the learned judge was, I think, right in concluding that under section 146(4) of the 1925 Act he had jurisdiction to order the grant of a monthly tenancy to the appellant by way of relief but no more than a monthly tenancy. As I have understood his argument in this court, Mr Good on the appellant's behalf has not challenged that conclusion. His argument has centred on the question whether a new monthly tenancy should have been ordered by the learned judge in favour of the appellant and, if so, on what terms relating to repair.

The learned judge in this context was referred to the decision of this court in *Creery* v *Summersell* [1949] Ch 751, where Harman J at p 767 made some important statements of principle as regards relief from forfeiture:

I think this remains a jurisdiction to be exercised sparingly because it thrusts upon the landlord a person whom he has never accepted as tenant and creates *in invitum* a privity of contract between them. It appears to me that I ought only to vest the head term in the under-lessees upon the footing that they enter into covenants in all respects the same, or at least as stringent, as the covenants in the head-lease. . . . Now this the under-lessees are not content to do; it would be useless to them; they wish to substitute for cl 4 of the head-lease a new clause widening the purpose for which they may use the property. I am not prepared to oblige the plaintiff to put up with this, and therefore in this instance also I refuse relief.

In the present case the learned judge, having referred to *Creery* v *Summersell*, commented:

Thus one should only be prepared to entertain such a claim [that is to say, a claim for relief] if the same, or not less onerous, covenants are included as those in the head-lease. Here Mr Soni seeks a new clause widening the purposes for which the property can be used, and Mr Soni is not prepared to commit himself to the repairs on the basis of a monthly tenancy.

In those circumstances, bearing in mind the principles stated by Harman J and Mr Soni's attitude as expressed in evidence, the judge did not think that this was an appropriate case for the grant of relief.

Mr Good on behalf of the appellant, while accepting the correctness of the general principle stated by Harman J in *Creery* v *Summersell* (which, I might add, has been referred to as representing the present law in *Halsbury's Laws*, 4th ed, vol 27, at para 441 — see note 7), nevertheless submitted that the present is a rather special case. He submitted that at the date of the forfeiture of Mr Medina's lease the state of repair of the basement was clearly very bad; and that anybody who had been granted a new tenancy of the basement at that date would have been very unlikely to agree to undertake an obligation to repair in anything like so wide a form as that imposed by clause 2.4 of Mr Medina's lease. They would, he suggested, have asked for the landlords, first, to put the property into a decent state of repair themselves. In his submission, at all material times the landlords knew, or ought to have known full well, that Mr Medina was not complying with the covenants in the lease. If they were to ask that Mr Soni should be subjected to covenants as onerous as those relating to repair in the original lease, they would, as he put it, be "taking out" on him what they ought to have "taken out" on Mr Medina. Thus, in his submission, the court ought to grant relief on terms that any covenant to repair imposed on Mr Soni by a new monthly tenancy should not be such as to place upon him the obligation to restore the property to any better condition than that in which it was when the bankruptcy of Mr Medina occurred. Alternatively, he submitted, before Mr Soni took his new lease, the lessors should be required to bring the premises up to the standard to which they wish it to be maintained. He pointed out that, if the only relief to be afforded to Mr Soni were the grant of a monthly tenancy coupled with an obligation to repair in the terms of Mr Medina's original lease, he would be immediately faced with a very heavy repairs bill but would have no guarantee at all that he would thereafter receive any proper return for his money. For it could well be that within a few months' time the landlords would be entitled to recover possession on one or other of the grounds upon which landlords can recover possession under the 1954 Act.

In all the circumstances, Mr Good submitted that the learned judge erred in the exercise of his discretion, in that he should have been prepared to make an order for the grant of a new tenancy on terms much less onerous than those which he suggested to Mr Soni and which Mr Soni felt unable to accept.

While I have considerable sympathy with Mr Soni's position, I am not able to accept the submissions made on his behalf to this court. First, it has to be borne in mind, as Mr Barnard has pointed out, that section 146(4) of the Law of Property Act 1925 has vested in the judge a wide discretion and that this court will only be entitled to interfere with the exercise of his discretion on the well-established grounds, that is to say, broadly, if he has misapplied the law or taken into account irrelevant matters or failed to take into account relevant matters.

Mr Barnard has submitted that there is no reason why the learned judge should not have applied the principle of *Creery* v *Summersell* in its full force. He has also referred us to what Romer LJ said in the earlier case of *Gray* v *Bonsall* [1904] 1 KB 601. That again was a case where the question of relief to underlessees against forfeiture was considered. In the course of his judgment (at p 608) Romer LJ said:

That being so, I think the argument against the application of s 4 [section 4 of the Conveyancing and Law of Property Act 1892] in this case falls to the ground, and that the proper course, in giving relief to the underlessees, is to proceed under that section, by vesting the premises in them for the term of the underlease, they covenanting with the lessors during that term to pay the rent reserved by the lease, and to perform the covenants therein contained as to the demised premises. In that view it appears unnecessary to bring either the original lessee or the assignee of the lease before the court. It was urged for the lessors that there may have been breaches of the covenants for repair contained in the original lease, and that before any vesting order is made, or any deed is executed by the underlessees under such an order, such breaches ought to be made good by the underlessees. I agree that, if there are any such breaches, they ought to be made good as a condition of granting relief to the underlessees.

For my part, despite Mr Good's forceful submissions to the contrary, I can see no sufficient grounds for departing from this approach in the present case. Immediately before the date of his bankruptcy, it would have been open to the lessors to proceed against Mr Medina under his repairing covenant according to its full force and effect, and, even after the bankruptcy, his obligations or those of his trustee in bankruptcy to make financial compensation for breaches would still have continued. If relief were now to be granted on the terms suggested by Mr Good, the effect would be to impose upon the landlords a person whom they had never accepted as their own tenant and would be to give them far less extensive rights in relation to repair than they had against Mr Medina, from whom Mr Soni derived his title. It does not seem to me that, in accordance with the principle of *Creery* v *Summersell*, this would be right.

Quite apart from this point, there is another pertinent point which Mr Barnard has drawn to our attention. While Mr Good has made much in this court of the allegedly bad state of repair of the basement, it would appear that in the court below no evidence, or at any rate no clear evidence, was given either that the basement was in a bad state of repair or, still less, that the landlords had inspected the basement and found it to be in a bad state of repair. In these circumstances, even if the general approach which Mr Good invites us to adopt were correctly founded, it seems to me that he has not been able to point to the necessary evidence to support its application. For there is no sufficient evidence that the landlords stood by and knowingly allowed Mr Medina to let the premises, to break his repairing covenant and let the basement premises fall into decay.

In all the circumstances, I can see no sufficient grounds for us to interfere with the exercise of the learned judge's discretion. Indeed, it appears to me that in all the circumstances of this case, having regard to Mr Soni's unwillingness or inability to enter into the appropriate obligation as to repair, he had little choice but to refuse the relief sought.

Accordingly, while, as I have said, I have some sympathy with the appellant, who has suffered not through his personal defaults but through the defaults of his own landlord, Mr Medina, I for my part would dismiss this appeal.

Agreeing, WAITE J said: I wish only to add my own expression of sympathy for Mr Soni in the loss of the very long-standing occupancy of his business premises through circumstances largely outside his control; but it would have been inequitably harsh on the head-lessors to allow any other result.

The appeal was dismissed with costs.

Chancery Division

November 12 1986
(Before Mr Justice FALCONER)

DEPARTMENT OF THE ENVIRONMENT v ROYAL INSURANCE PLC

Estates Gazette April 11 1987
282 EG 208-214

Landlord and Tenant Act 1954, section 37 — Court precluded from ordering a new tenancy by reason of the ground specified in section 30(1)(f) of the Act — Entitlement to compensation — Whether amount of compensation at the higher rate of six times the rateable value or at the lower rate of three times — Whether the condition required for the higher rate was satisfied, namely, that during the whole of the 14 years immediately preceding the termination of the tenancy the premises had been occupied for the purpose of a business carried on by the occupier — Department of the Environment were the tenants and the tenancy terminated on August 23 1985 — As a physical fact the department's occupation of the subject premises did not begin until August 25 1971, although the department's lease was for a term of 14 years from August 23 1971 — It was submitted on behalf of the department that

they should be treated as having been in occupation for the full 14 years despite the short gap and thus qualified for the higher scale of payment — It was suggested that the requirement should not be interpreted too literally, that it was a matter of fact and degree, that the intention of the parties was relevant, and, finally, that the *de minimis* rule should be applied to produce a reasonable result — The department should not be deprived of the higher scale because the occupation was one or two days short of 14 years — The cases of *Lee-Verhulst (Investments) Ltd v Harwood Trust* and *Morrisons Holdings Ltd v Manders Property (Wolverhampton) Ltd* were cited in support of these submissions — Held, rejecting the department's construction, that Parliament had made it clear that there must be occupation for the whole period of 14 years immediately preceding the termination — Cases where there had been a break in the period of occupation, eg because of a fire, differed from the present, where there was a gap, although short, before the occupation commenced — The entitlement to compensation was on the lower scale of three times the rateable value

The following cases are referred to in this report.

Caplan (I&H) Ltd v Caplan (No2) [1963] 1 WLR 1247; [1963] 2 All ER 930
Cardshops Ltd v John Lewis Properties Ltd [1983] QB 161; [1982] 3 WLR 803; [1982] 3 All ER 746; (1982) 45 P&CR 197; [1982] EGD 305; 263 EG 791, CA
Edicron Ltd v William Whiteley Ltd [1984] 1 WLR 59; [1984] 1 All ER 219; (1983) 47 P&CR 625; [1983] EGD 273; 268 EG 1035, CA
Lee-Verhulst (Investments) Ltd v Harwood Trust [1973] QB 204; [1972] 3 WLR 772; [1972] 3 All ER 619; (1972) 24 P&CR 346; [1973] EGD 467; 225 EG 793, CA
Morrisons Holdings Ltd v Manders Property (Wolverhampton) Ltd [1976] 1 WLR 533; [1976] 2 All ER 205; (1975) 32 P&CR 218; 238 EG 715, CA

This was an application by the defendants, Royal Insurance plc, in proceedings which had commenced by an originating summons by the plaintiffs, the Department of the Environment, seeking an order for the grant of a new tenancy of premises consisting of the first floor and part of the ground floor of 5-7 Chancery Lane, London WC2. There had also been a summons by the defendant landlords for the determination of an interim rent. The main proceedings had, however, ended in a consent order whereby the applications for a new tenancy and for determination of an interim rent were withdrawn. Under this order the parties were at liberty, failing agreement, to apply for directions as to the amount of statutory compensation under section 37 of the Landlord and Tenant Act 1954; hence the present application by the defendant landlords.

J F Mummery (instructed by the Treasury Solicitor) appeared on behalf of the plaintiffs; David Neuberger (instructed by Linklaters & Paines) represented the defendants.

Giving judgment, FALCONER J said: In this matter I am asked to determine on an application of the defendants, Royal Insurance plc, who were the landlords of the premises in question, the amount of compensation payable to the former tenants, the Department of the Environment, under section 37 of the Landlord and Tenant Act 1954 in respect of the premises known as the first floor and part of the ground floor of 5-7 Chancery Lane, London WC2. Those premises were at one time occupied by the Industrial Relations Court and subsequently, I think, by the Lands Tribunal.

It is not in dispute that compensation is payable under the provisions of section 37, but the question is whether it is payable at the higher rate under the provisions of section 37(2)(a) or at the lower rate under section 37(2)(b). The defendants, Royal Insurance plc, the landlords, of course contend that it is to be payable at the lower rate and that would be a sum of £161,665. It is common ground that that has been paid and accepted. That is a sum of three times the rateable value, three being the multiplier to which I will refer in a moment; that is the sum which is payable if compensation is payable under para (b) of section 37(2). I will come to the provisions in a moment. If the Department of the Environment are right, as they contend, then it would be payable at the higher rate under para (a), which will be three times two, that is six times the rateable value, which would be a sum of £333,330. There is no dispute, as I understand it, about the arithmetic at all. The matter really turns on the construction of section 37 and in particular section 37(3)(a) as applicable to the facts of this case.

As to the facts, there is really no dispute as I understand it. The matter arises in this way: there were negotiations between the department and Royal Insurance in 1971, which resulted in an offer being made by letter dated August 16 1971 by the Royal Insurance to let to the department the premises in question. As I have indicated, they comprised part of the ground floor and the first floor of 5-7 Chancery Lane. That offer is evidenced by a letter to be found in exhibit WJP1 to the affidavit of Mr Purdie on behalf of the department. In that letter the last paragraph is of some relevance. The agents acting for the insurance company said that their clients, that is the Royal Insurance:

. . . are prepared to allow your building works to commence immediately, upon receipt of a satisfactory undertaking from the Department, stating that the Department will be prepared to complete the lease substantially in the form made available to you on August 5, and subject to the Royal's approval of the works which you wish to undertake.

The offer, of course, was subject to contract. There was to be a lease and the lease was to commence on August 23 1971 according to the letter, and it was to be for a 14-year period. There was a reply on August 19 1971 in which, on behalf of the Secretary of State, the senior estate surveyor dealing with the matter said he accepted the offer to let to the Secretary of State the whole of the first floor and part of the ground floor of the premises on the terms and conditions specified in the offer. He went on to state that two sets of drawings had been handed over as to some of the alterations and adaptations the department wished to make. The final paragraph of the letter stated:

As you know, we wish to start our adaption (*sic*) work on August 25 next and I shall be obliged if you will kindly let me have by return your approval of the works and consent to their being put in hand at that date.

In fact, on August 25 the works of adaptation started on the first floor, in the sense that the contractors went in to start work, or to survey and see what was to be done. There is no dispute that they went in, so far as the first floor was concerned, on August 25.

It is convenient to say at this point that Mr Neuberger for the defendants, the Royal Insurance, takes no point that at that stage, so far as occupancy is concerned, the part of the ground floor which was to be taken over was not immediately available. No point is taken about that. On the evidence, Mr Mummery frankly pointed out that it is not really absolutely clear when the plaintiffs went into the ground floor, but as I say, because of that concession, the point is immaterial.

The actual lease made pursuant to the agreement was executed on June 21 1972 for a term of 14 years and whereas the letter of the 16th offered a lease to commence on August 23 1971 in fact in the lease the expression used is that the term is to start from August 23 1971. It is common ground that so far as the term under the lease is concerned, that commenced as it were immediately after midnight on the night of the 23rd/24th.

On September 11 1984 the defendants, that is to say the Royal Insurance, served the department with a notice under section 25 of the Landlord and Tenant Act 1954. The notice was dated September 11 1984, stating that they proposed to terminate the tenancy on August 23 1985, which it will be remembered is the termination date according to the lease. I need not go to the lease for present purposes. As a result of that notice the short point is whether the period of occupation was such as to entitle the department to compensation at the higher rate in para (a) of section 37(2) or, as the landlords, the Royal Insurance, say, at the lower rate of para (b).

The notice to terminate having been given, the department made an application for a new tenancy and the landlords, Royal Insurance, indicated that they would oppose the grant of a new tenancy and that they would do so under ground (f) in section 30(1)(a), ground (f) being:

That on the termination of the current tenancy the landlord intends to demolish or reconstruct the premises comprised in the holding or a substantial part of those premises or to carry out substantial work of construction on the holding or part thereof and that he could not reasonably do so without obtaining possession of the holding.

I do not think I gave a date for the counternotice by the department that they were unwilling to give up possession. That was actually September 21 1984. The originating summons seeking an order for the grant of a new tenancy was dated December 4 1984.

On December 12 1984 there was a summons by the landlords that there should be determination of interim rent and at some stage thereafter the department realised that they were not going to get a new tenancy, and so a consent order was made on October 1 1985 whereby the department, the plaintiffs, withdrew their application for a new tenancy; the Royal Insurance withdrew its application for the determination of an interim rent; it was agreed that the department, the plaintiffs, would deliver up vacant possession on October 6 1985; it was agreed that following delivery up of vacant possession statutory compensation pursuant to Part II of the 1954 Act should be payable by the defendants to the plaintiffs; there was a provision in that consent order as to a sum by way of interim rent — it is not material for present purposes; finally, under the order, after the provision as to access for purposes of carrying out inspections there was a provision that the parties were to be at liberty to apply for directions to determine the amount of the said compensation in case agreement could not be achieved.

Although there was continuing correspondence about the amount, there was no agreement and so there was an application on the part of the defendants, Royal Insurance, that the court should determine the amount of compensation due to the defendants under section 37 of the Landlord and Tenant Act in respect of the premises which I have mentioned. The master gave directions for evidence and that the application should come in the list, as it has done before me, as a non-witness matter, subject to the right to cross-examination. Neither side has sought to cross-examine — I have read an affidavit on each side.

It is material at this stage just to look at the relevant provisions of the Landlord and Tenant Act 1954. Some point arises on the wording of section 23, the first section in Part II of the Act. Subsection (1):

Subject to the provisions of this Act, this Part of this Act applies to any tenancy where the property comprised in the tenancy is or includes premises which are occupied by the tenant and are so occupied for the purposes of a business carried on by him or for those and other purposes.

The interest is in the words "occupied for the purposes of a business" and so on.

I need not read subsection (2), which deals with the meaning of "business". It is common ground that this is a tenancy under Part II falling within subsection (1) of section 23 by reason of the special provisions dealing with Crown departments to be found in section 56(3).

Section 24 provides for the continuation of existing tenancies to which Part II applies and to the grant of new tenancies. I do not think I need read anything from that for present purposes. Section 25 provides for the way in which the landlord may terminate a tenancy to which this Part of the Act applies by giving an appropriate notice in a prescribed form, and times and so on are provided for. As I have indicated, the Royal Insurance has given appropriate notice to terminate under that provision, the termination to be on August 23 1985.

Section 26 deals with a tenant's request for a new tenancy. I need not read the provisions of that. There was such an application, as I have mentioned, on the part of the department. Subsection (6) of section 26 provides for the landlord, in the event of a tenant's request for a new tenancy, being able to give notice that he would oppose and he has to state on which ground or grounds under section 30 of the Act he will oppose the application.

That takes me to section 30(1), where the grounds are set out, and I have indicated that the ground relied upon was ground (f) and I have already read that.

Coming now to the material section for present purposes, section 37(1), the tenant quitting the premises under the provisions of an appropriate notice by the landlord is entitled to compensation on quitting if he has to quit on certain grounds under section 30(1). I ought perhaps to read subsection (1):

Where on the making of an application under section 24 of this Act the court is precluded (whether by subsection (1) or subsection (2) of section 31 of this Act) from making an order for the grant of a new tenancy by reason of any of the grounds specified in paragraphs (e), (f) and (g) of subsection (1) of section 30 of this Act and not of any grounds specified in any other paragraph of that subsection, or where no other ground is specified in the landlord's notice under section 25 of this Act or, as the case may be, under section 26(6) thereof, than those specified in the said paragraphs (e), (f) and (g) and either no application under the said section 24 is made or such an application is withdrawn, then, subject to the provisions of this Act, the tenant shall be entitled on quitting the holding to recover from the landlord by way of compensation an amount determined in accordance with the following provisions of this section.

Subsection (2) deals with the rate of compensation and provides:

The said amount shall be as follows, that is to say — (a) where the conditions specified in the next following subsection are satisfied it shall be . . .

And then, following the amendment by the Local Government, Planning and Land Act 1980, Schedule 33 para 4, the words are inserted: "the product of the appropriate multiplier and", so the paragraph should now read:

where the conditions specified in the next following subsection are satisfied it shall be the product of the appropriate multiplier and twice the rateable value of the holding,
(b) In any other case it shall be the product of the appropriate multiplier and the rateable value of the holding.

Just pausing there, it is common ground that the appropriate multiplier in this case from the appropriate statutory instrument now presently in force* is three, so that the compensation if it is payable under (a) would be six times the rateable value, if under (b) it would be three times the rateable value. That is how the arithmetic which I dealt with earlier is arrived at. As I say, the department, the plaintiffs, say they come under (a) and the defendants say the plaintiffs come under (b) and they have already paid the appropriate sum under (b) over to the plaintiffs.

That brings me, of course, to subsection (3), where the conditions are set out in order to fall into para (a) of subsection (2). The said conditions are (a) and (b):

(a) that, during the whole of the 14 years immediately preceding the termination of the current tenancy, premises being or comprised in the holding have been occupied for the purposes of a business carried on by the occupier or for those and other purposes.

The similarity of the language used there to that which I quoted from section 23(1) will be observed. I ought to read para (b), I think, for completeness, although it is not really material for present purposes. I merely read it because it was a material provision in one of the cases cited by Mr Mummery.

(b) that, if during those fourteen years there was a change in the occupier of the premises, the person who was the occupier immediately after the change was the successor to the business carried on by the person who was the occupier immediately before the change.

That is taking care of a transfer of the business, the business having been carried on in the same premises; so that the current tenant would have the benefit of his predecessor in business so far as occupancy went.

The only other provision I think I should look at in this section is subsection (7), which provides that:

In this section the reference to the termination of the current tenancy is a reference to the date of termination specified in the landlord's notice under section 25 . . .

So that for the purposes of para (a) of subsection (3) in section 37 the termination is August 23 1985.

It is common ground that the period between then and when they actually quitted, October 6 1985, is not material for present purposes and I need not consider that further.

On the evidence, which I need not read, it is common ground, I think, that the department's tenancy was protected by the Act. It is common ground that the department occupied the premises for the purposes of a government department so as to bring it within Part II of the Act; that the tenancy was terminated on August 23 1985 by the section 30 notice on the ground of para (f). It is common ground on the evidence that the department sent in contractors on August 25 1971 for adaptation work. I think it is common ground that under the lease the department had a tenancy of the premises for a term of 14 years from August 23 1971. As I have already mentioned, but I say it again for completeness, it is common ground that although it is not clear when the department actually went into occupation of the ground-floor part of the holding their going into the first-floor part of the holding on the 25th is regarded as going into the premises. There is no dispute on that aspect of the matter.

Mr Mummery's first submission really is directed to this: that if one goes from August 25 1971 to August 23 1985 inclusive there are not 14 complete years. If one starts from the first instant of August 24 1971 there is one day short; if it is from August 23 1971 there are two days short. The argument in the course of the hearing really went

---

*Editor's note: Landlord and Tenant Act 1954 (Appropriate Multiplier) Order 1984 (SI 1984 No 1932).

upon the assumption, convenient apart from anything else, that it is really one day short of the 14 years that I have to consider.

Mr Mummery's first submission was that, although there are no helpful decisions on this point in issue here, there are cases showing, he submits, that this type of provision, particularly in this Act, is not to be interpreted too literally but is to be a question of fact and degree; a court will look at the substance of the matter and attempt a commonsense conclusion, construing the word "occupy" in its normal and ordinary sense.

I have pointed out, as he did, the similarity in language in subsection (3), condition (a) of section 37, which mirrors the language of subsection (1) of section 23. He pointed out and indeed took me to a case on section 23(1) as to what is meant by the words "occupied by the tenant and are so occupied for the purposes of a business" and so on. He submitted that the approach there should be applied to the construction of subsection (3) of section 37.

I do not think that it is in dispute, I think it was accepted by Mr Neuberger, that the way that the wording in subsection (1) of section 23 is to be construed should be the construction applied to the similar wording to be found in condition (a) of subsection (3) of section 37.

As to the authority to which I have referred, which is *Lee-Verhulst (Investments) Ltd* v *Harwood Trust* [1973] QB 204, that was a case where the tenant carried on in a house the business of letting furnished rooms and services. I need not go into very much of the detail of the case. The tenant at the appropriate time applied under Part II of the Landlord and Tenant Act 1954 for a new tenancy on the expiration of the existing lease. There was a preliminary issue which came before the county court judge, the preliminary issue being whether the house was occupied by the tenant for the purposes of business within section 23 of the Act. The county court judge held that it was and on appeal to the Court of Appeal he was upheld.

In the leading judgment of Sachs LJ the passages relied upon by Mr Mummery are to be found on p 212. After looking at the facts, which were not in dispute, at G on p 212 Sachs LJ said:

It happens that the authorities cited to us on the meaning of the word "occupy" were largely decisions under provisions of the Rent Acts (particularly Part II of the Rent Act 1968) and Rating Acts. It is thus as well at this stage to state plainly that the meaning of this word can vary according to the subject-matter of the statute, as was indeed said by Lord Denning in *Wheat* v *E Lacon & Co Ltd* [1966] AC 577 (a case concerning the Occupiers' Liability Act 1957). In particular, "occupy" and "occupation" in the Rent Act 1968 (eg sections 70, 101 and 102) have not necessarily the same meaning as those words have in the provisions of the Act of 1954. In the latter Act . . .

That is to say the 1954 Act

. . . the word "occupy" has in my judgment a broader and less technical meaning than it has in contexts where it may be necessary to differentiate possession from occupation, to distinguish exclusive from shared occupation, to consider the subtleties of whether a chambermaid enters by right or by courtesy, or to have regard to niceties as between a maid coming in to make and light a fire and a man tending an outside boiler for central heating.

Then, after two paragraphs referring to the particular facts of this case, on p 213 at E, the learned lord justice says:

Is there anything in the Act of 1954 which precludes the court from giving to the word "occupied" in section 23 its natural and ordinary meaning in the context of the subject matter of that Act — a meaning which would in the set of circumstances above described clearly lead to it being held that the tenant did occupy the premises for the purpose of the business? Being unable to find anything in the Act which so precludes the court, I have come to the conclusion that this tenant did so occupy the whole of the premises. For reaching that conclusion it is neither necessary nor desirable to provide a definition of that word which would deal with all the greatly varying sets of circumstances that can exist. As a number of elements have been taken into account, each of a physical nature and each involving a degree of presence on the part of the tenant personally or by goods under his ownership, it is however as well to observe that it could be proper in some other case to reach the same conclusion even if one or more of those elements were subtracted. For instance if the furniture was that of the occupants or if some of the services were not rendered or if the occupancies were not so much controlled, there could still be an occupation by the tenant of the premises as a whole. Much depends on questions of degree. In the end it is necessary to look at the substance of the position as a whole and to seek to apply that common sense to which Roxburgh J in *Narcissi* v *Wolfe* [1960] Ch 10 referred after saying, at p 16: "There is a lot of law about the word 'occupied' but that does not appear to me to be applicable to this Act.''

The only other passage I think which Mr Mummery referred to, and I think rightly, is on p 215, the penultimate paragraph in the learned lord justice's judgment, where he is really stating that he is applying the principles which he has been expounding in the passages which I have read. He said:

It is however on the previously stated basis that the court must look at the substance of the position as a whole, take into account the various elements which have been discussed and then come to a common sense conclusion as to whether the tenant "occupies" the premises for the purpose of his business that this case should in my judgment be determined.

Mr Mummery also referred to the judgment of Stamp LJ at p 218:

It is no doubt correct that if you find the same word appearing in two places in a statute, or in several statutes covering the same subject-matter, it ought in the absence of a controlling context to be given the same meaning.

Mr Mummery relied on that approach which he summarised as follows: he said there are three elements; you have to give the word its ordinary and natural meaning; second, you have to look as a matter of substance, taking all the circumstances into account, whether there was an occupation. I think that is a reference to the actual physical aspect of it. In one part of the passsage I read, Sachs LJ referred to the presence on the part of the tenant personally or by goods under his ownership. That is the physical aspect of it. Third, you have to come to a commonsense conclusion taking all the matters into account.

Taking it a stage further before I come to Mr Mummery's application of that approach, he submitted on a second authority that the intention of the party was also a relevant matter and it was not just a matter of a physical presence either of the tenant or his personnel or his goods, but there was a question of intention too. For that purpose he cited the case of *Morrisons Holdings Ltd* v *Manders Property (Wolverhampton) Ltd* [1976] 1 WLR 533. That was a case, as appears from the headnote, where the tenants held the ground-floor shop and basement of certain premises in Wolverhampton under an underlease, the premises forming part of a larger structure, the Central Arcade. There was a fire, a devastating fire according to the headnote, which occurred in the arcade and burnt it down. It did not, in fact, wholly destroy the premises in question but made them, certainly for the time being, unable to be used. The headnote points out that the day after the fire the tenants wrote to the landlords suggesting that the premises be made weatherproof and suitable for their occupation and generally expressing their desire to resume trading there. They subsequently reasserted their claim to occupation; they never gave up possession of the keys to the premises and they left some fixtures and fittings on the premises after salvaging their stock. Then the landlords purported to determine the tenancy under a clause of the underlease. On a subsequent date the landlords gave notice to terminate under section 25 of the Act, but the tenants applied for a new tenancy under Part II of the Act of 1954. The landlords opposed the application on the ground that they intended to demolish and reconstruct the premises. There was a preliminary issue as to whether the tenants had *locus standi* to apply for a new tenancy, since prior to their application they had ceased to occupy the premises for the purposes of their business in accordance with section 23(1) of the Act. The judge held that they had no *locus standi* to make the application, since the devastating nature of the fire at the premises justified the inference that the tenants' absence from the premises was permanent. On appeal the appeal was allowed, it being held:

. . . the tenants had to show either that they were in physical occupation of the premises for the purposes of a business carried on by them or, if events beyond their control had led to their absence from the premises, that they continued to assert their right to occupancy; . . .

In the leading judgment of Scarman LJ at the bottom of p 539, he quoted from a decision of Cross J, as he then was, in *Caplan (I & H) Ltd* v *Caplan (No 2)* [1963] 1 WLR 1247, where the learned judge said:

I think it is quite clear that a tenant does not lose the protection of this Act simply by ceasing physically to occupy the premises. They may well continue to be occupied for the purposes of the business although they are de facto empty for some period of time. One rather obvious example would be if there was a need for urgent structural repairs and the tenant had to go out of physical occupation in order to enable them to be effected.

Scarman LJ, as he then was, went on to say:

I respectfully agree with the view of the law expressed by Cross J in the two passages to which I have referred. I would put it in my own words as follows: in order to apply for a new tenancy under the Act a tenant must show either that he is continuing in occupation of the premises for the purposes of a business carried on by him, or, if events over which he has no control have led him to absent himself from the premises, if he continues to exert and claim his right to occupancy.

He referred again to *Caplan* and said:

The temporary absence in *Caplan (I & H) Ltd* v *Caplan (No 2)* which did not destroy the continuity of occupation was absence at the volition of the tenant. In the present case the absenting by the tenants of themselves from the premises after the devastating fire was not their choice, but was brought about by the state of the premises created by the fire, which was none of the tenants' making. Nevertheless, they exhibited immediately after the fire, and continued to exhibit, an intention to retain and to claim their right of occupancy, and reminded the landlords from time to time . . .

Then a little lower down on p 540 the learned lord justice said:

It seems to me that, when events such as I have detailed arise and a tenant is faced with the difficulties of occupation that these tenants were, it must be a question of fact to determine whether the tenant intended to cease occupation or whether he was not only, as the judge found these tenants were, cherishing the hope of return, but also making quite clear that he intended to maintain his right of occupancy and to resume physical occupation as soon as the landlords reinstated.

In the third judgment, that of Sir Gordon Willmer, on p 542 he said:

So far as the law is concerned, I think it can be taken as axiomatic that in order to be in occupation one does not have to be physically present every second of every minute of every hour of every day. All of us remain in occupation, for instance, of our houses even while we are away doing our day's work. It follows, therefore, that occupation necessarily must include an element of intention as well as a physical element.

He then went on to give some examples and a little lower down on the same page, between E and F, he said:

It seems to me that the tenants, who had been in continuous occupation up to the fire and immediately after the fire, and who retained the intention to occupy, remained both in fact and in law the occupiers of the premises at the relevant time.

From that authority, and rightly so, Mr Mummery submitted that the intention of the parties, particularly the tenants I suppose but certainly the parties, was a real matter to consider when one is considering whether or not a certain state amounts to occupation for the purposes of either section 23(1) or, as in this particular case, section 37(3) para (a).

Looking first of all to the first aspect, what I might call the physical aspect, in the individual circumstances of this case: it is accepted and common ground that there was physical presence on the premises from August 25 1971 until the termination date August 23 1985. What happened thereafter is common ground and is not material. What is said by the defendants, Royal Insurance, is that that did not amount to a period of 14 years.

I do not think there can be any question that as a fact the department were not in occupation and physical occupation for the complete period of 14 years; assuming that the 14 years in question starts at the beginning of the first instant of the 24th, there was a whole day on the 24th when there was no physical occupation. Mr Mummery submitted on the authorities, and I do not think that this was disputed, that occupying premises does not require there to be physical presence every day of the period. He referred to some of the examples which have been referred to in the two authorities which I have looked at and which he cited, such as when a tenant moves out for structural work to be carried out or where a tenant is prevented from continuing immediate and physical occupation because the place has been burned down, as in the *Morrisons* case, or where, as suggested by Sir Gordon Willmer, a tenant goes away on holiday or where the tenancy may be occupied for seasonal periods because there is a seasonal business.

It seems to me that all those sorts of cases are different from the present case in that they were all examples of cases where there had been physical occupation prior to the gap or break which occurred and the real question to be determined every time by the courts was: had the absence for that period, for whatever reason, affected a cesser of the occupation which had already been in existence? In the present case, as I say, it is common ground that as a physical fact the initiation of the occupation by the contractors going in did not commence until August 25.

It is at this stage, I think, that Mr Mummery, as I understood his argument, prays in aid the intention of the parties as a relevant matter. Of course, from the last authority to which I referred one sees how the courts have regarded intention as important and a relevant consideration where the question is whether, after a period of occupation, there is a period of absence, say on holiday or because of structural change or, as in the *Morrisons* case, of a fire destroying the premises. Whether there is an intention to retain the occupancy or return to the occupancy would be a factor to be taken into account. But what Mr Mummery submits here, as I understand him, is this: he submitted that the intention of the parties, as to be gathered from the two letters initiating the relationship of August 16 and August 19 1971, was such as to be totally consistent with occupation by the department for the whole of the 14 years; the 14 years meaning thereby 14 years preceding the determination, that is to say from immediately after midnight on the 23rd, starting at the beginning of August 24 1971. He argued that the landlords were prepared to allow the building works to commence immediately and therefore there was no intention on the part of the landlords to retain occupation themselves after August 23. He pointed, in support of that view, of course, to the subsequent execution of the lease when the term under the lease provided for the tenancy to begin from August 23 1971 and that on that date, according to the terms of the lease, the department were entitled to exclusive possession. He submitted that the landlord might be estopped from denying that. That seems to me to be a different point, as to whether there was in fact occupancy from August 23. He went on to submit that it was a common intention of both the Royal Insurance and the department, from that correspondence, that there should be occupation on the 23rd, but I do not accept that on my view of the correspondence. I cannot spell out of that correspondence, and I have looked at the relevant parts of the letters, or really on the terms of the lease which was ultimately executed which provides for a term on August 23 1971, any intention that the department should enter into occupation on August 24 1971. With no actual physical occupation, as I have said, until the 25th and, as I see it, nothing to show there was an intention of either party that the occupation should start on the 24th as distinct from the 25th, it seems to me that Mr Mummery does not get over the hurdle that at best the period of occupation was 13 years 364 days; that is to say August 24 1971 was a date when the department were not in occupation.

That view was reinforced by a submission of Mr Neuberger as follows. He pointed out that the effect of the correspondence, the two letters in question, was that there was "an arrangement" between the parties that there would be an agreement subject to contract for a lease and an understanding that subject to the contract the department would come into the premises — the workmen would anyhow — on August 25, but the work that was undertaken at that time was still subject to the approval of the landlord. He submitted on the proper construction of that position that at that time, on the exchange of those letters, there was no binding agreement between the parties for a lease and that there was no binding agreement under which the workmen went in. He said the situation did not really change until August 25 when the workmen went in. He conceded that after the workmen went in, by consent on the 25th, either side might be estopped from saying that there was not an agreement to grant a lease on the terms in the letter, but the powerful part of the submission was that at least until that happened the landlords were in the position to withdraw if they received a better offer and let the premises to another tenant. Indeed Mr Neuberger pointed out that that position was true both ways. Either side could have withdrawn at that stage.

That submission seems to me to have considerable force and I regard it as supporting the view I have already indicated, that on his first argument Mr Mummery does not get over the hurdle that at best the period of occupation was one day short of the 14 years required by condition (a) of subsection (3).

To get over the difficulty of the shortness of the period of occupation because there was no actual physical occupation on August 24, and, as I hold, no intention of the parties that the defendants should enter into occupation before the 25th, Mr Mummery put forward a second argument. This was a *de minimis* argument. He pointed out that 14 years (and I accept his arithmetic without having checked the figure) is 5,106 days, and if one regards one day or two days as the time when there was no occupation, that is to say the 23rd and 24th, or just August 24, that is only one or two in five thousand odd when in occupation. He submits that the court should apply a *de minimis* rule to say that in substance that period of time, that is to say 13 years 364 days, was 14 years.

On this aspect of the case, which was really his second main submission, Mr Mummery submitted that there must be a presumption that Parliament intended to produce a reasonable solution to the question of compensation for disturbance of a business tenancy. He submitted that it would be unreasonable to deprive the department of 14 years' compensation on the grounds that they were not in occupation for one or two days as the case may be. The intention of Parliament, he submitted, was that the longer period qualified for, if it were qualified for, should get the larger compensation.

He referred me to two cases in support of his view that the courts dealing with this question of compensation for a termination of a business tenancy under the 1954 Act have adopted, and should adopt, a favourable approach to the tenant in dealing with cases that arise under the provisions of section 37. The first one was that of *Cardshops Ltd* v *John Lewis Properties Ltd* [1983] QB 161. That was a case, however, in which the construction to be given to the concluding words of subsection (1) of section 37 was in issue. Put shortly what had happened was that there had been a termination pursuant to the Act of a business tenancy. Between the date of termination and the date when the tenant actually quitted the premises there had been a change in the extent of the value of the multiplier so there was a difference in value between what the tenant would be entitled to as compensation if it was to be determined as at the date of the termination of the tenancy, or at the time at the date of actually quitting. The majority view decision of the Court of Appeal was that the relevant date was the date when the tenant actually quitted, so the tenant had the benefit, as it turned out, of a higher rate of compensation, the relevant multiplier being a larger one at that date than at the date of the termination.

This was simply a case, in my judgment, of a question of the construction to be given to the concluding words of subsection (1) of section 37:

. . . shall be entitled on quitting the holding to recover from the landlord by way of compensation an amount determined in accordance with the following provisions of this section.

Waller LJ, who gave the leading judgment, said, having found in favour of the tenant, allowing the appeal:

In my judgment the law to be applied in cases such as the present is the law at the date at which the tenant is obliged to quit, in this case June 29. I might add that I do not see any injustice to the landlord in such a conclusion. The policy of the Landlord and Tenant Acts is to hold the balance between landlord and tenant with an obligation on the landlord to pay compensation (and I assume fair compensation) when the tenant is dispossessed. If the unamended law was not achieving fairness I do not see that the landlord suffers injustice by having to pay what Parliament views as proper compensation.

I was also referred by Mr Mummery to the observations of Ackner LJ in the concluding part of his judgment on p 179:

Moreover, if the date specified in the landlord's notice for the termination of the tenancy is to be the appropriate date for the assessment of compensation, or the earlier date of the notice itself, then it could properly be argued that cases will occur where the apparent intention of the legislation would be frustrated and hardship would be suffered by the tenant. Parliament intends that the tenant should be properly compensated for the disturbance in having to vacate the premises, and the clear inference is that Parliament considered that by the beginning of 1981, if not earlier, the compensaton which was then available was far too low. Accordingly, if for example the tenant succeeded at first instance in obtaining a new tenancy, but in a Court of Appeal it was established that the court was precluded from granting that new tenancy on ground (e), (f) or (g) of section 30(1), such a decision might well have only been made a year or more after a new multiplier had come into force. When the tenant is then obliged to quit, is his compensation to be on the out-of-date scale?

A choice had to be made by Parliament as to whether the tenant was to be entitled to the tariff operating on the date when he lawfully quits (or should have quitted, if he wrongfully stayed on), or whether the new tariff should not apply in cases where the landlord's section 25 notice had been previously served, or where the date for the termination of the tenancy specified in that notice had passed. Since Parliament could have, but did not make any special provision for suspending the operation of the new tariff once it came into force I agree with Slade J and Walton J . . .

that was a reference to quotations from two judgments

. . . that the tenant is entitled to the prevailing rate of compensation upon his quitting the premises.

Mr Mummery, after reading those passages to me, said that was a signpost as to how section 37(2) should be interpreted; that is to say, not absolutely literally. I do not accept that submission on the basis of that case. In section 37(3) (a), as I think I have already indicated, it seems to me that Parliament has made its intention perfectly clear. It provides for a period of 14 years and not only does it provide for a period of 14 years immediately preceding the termination to be the qualifying period for the higher rate under para (a) of subsection (2), it says: "during the whole of the fourteen years immediately preceding", emphasising in my mind that there must be a complete 14 years. Cases have arisen, of course, where the occupancy had been broken in the ways I have indicated; they give rise to the question of whether the break that occurs causes a cesser of the occupation. But that question does not arise when the occupation has not yet commenced.

I was also referred, again in support of the submission that the courts adopt a favourable approach to the tenants in dealing with these provisions of this Act, to the case of *Edicron Ltd* v *William Whiteley Ltd* [1984] 1 WLR 59. I do not propose to go over that case for present purposes. That was a case in which again a question of construction arose, the question whether or not the word "premises" in condition (b) of subsection (3) has the same meaning or was to be read in the same way as "premises" in condition (a). I did not find that I got any assistance in the present case from that case.

In my judgment, Parliament having made very clear, as I see it, in its language that its intention was that to qualify for the higher rate of compensation under para (a) of subsection (2) of section 37 a tenant must establish occupancy for the purposes of a business for the whole period of the 14 years immediately preceding the termination, it seems to me that it would not be right for a court in those circumstances to apply, as Mr Mummery submitted that the court should, if the court came to the conclusion that there was no occupancy on the 24th, the *de minimis* rule in a case such as this where there has, in fact, not been physical occupation for the whole of the 14 years, and a case incidentally where, as I have already held, there was nothing to indicate that it was the intention of either party that there should be an occupation before August 24.

In the result, in my judgment the department, the plaintiffs, do not qualify for compensation at the higher rate provided for in para (a) of section 37(2) because their occupancy of the premises in question failed to meet condition (a) of subsection (3) of section 37. Accordingly the compensation payable is £161,665, a sum which has in fact already been paid, and is not £323,330 contended for by the defendants and I so hold.

Judgment for defendants with costs.

Court of Appeal

March 10 1987

(Before Lord Justice FOX and Mr Justice SHELDON)

ORIANI AND OTHERS v DORITA PROPERTIES LTD

Estates Gazette May 23 1987

282 EG 1001-1005

Landlord and Tenant Act 1954, Part II — Application for new tenancy of café-restaurant — Application not opposed but issue as to assessment of rent — Appeal by landlords from decision of county court judge — Premises in East Precinct of St George's Walk, Croydon — Agreed valuation approach was by means of zone A equivalent method — East Precinct a less favoured area than West Precinct — Landlords gave evidence of rents of two open market lettings in the East Precinct and rents expected from two prospective lettings there — Tenants relied on evidence from nine renewals of leases in the West Precinct and two rent reviews — The county court judge added together the zone A equivalent figures of the nine renewal transactions relied on by the tenants (but with a 20% reduction on account of the difference between the East and West Precincts) and of the two open market lettings put forward by the landlords, and divided the sum by eleven — This produced a figure of £12.59 per sq ft, resulting in an annual rent of £6,345 — On appeal the landlords made various criticisms of the judge's assessment, contended that she had misdirected herself and applied the wrong test — A new trial was sought — It was suggested that by describing the figure of £6,345 as "the fair and reasonable rent" the judge had not

been determining an open market rent as required by the 1954 Act — The judge was also criticised for giving too much weight to the nine renewal rents in the West Precinct; for not evaluating the worth of each of these nine cases; for distorting the figures by her method of dealing with the East Precinct comparables; and for ignoring altogether the evidence of the two prospective lettings — These criticisms were rejected by the Court of Appeal, although the judge's words about "the fair and reasonable rent" were "unhappily chosen"; she meant that her figure was a fair and reasonable assessment of the open market rent — The judge had, however, erred in one respect — The figures relating to the nine comparables in the West Precinct had not been adjusted for increases in market rents down to April 1986, when the application was heard — To that extent the appeal must be allowed

No cases are referred to in this report.

This was an appeal by the landlords, Dorita Properties Ltd, from a decision of Her Honour Judge Graham Hall, at Croydon County Court, on an application by the tenants, the present respondents, Stefano Oriani, Rina Oriani and Nevisia Oriani, for a new tenancy of premises used as a café-restaurant at 47 St George's Walk, Croydon.

Nicholas Dowding (instructed by Grangewoods) appeared on behalf of the appellants; J R T Rylance (instructed by Streeter Marshall & Wilberforce Jackson, of Croydon) represented the respondents. Mr Dowding did not appear in the county court proceedings.

Giving judgment, FOX LJ said: This is an appeal from the decision of Her Honour Judge Graham Hall given at the Croydon County Court upon an application by tenants for a new lease of business premises under Part II of the Landlord and Tenant Act 1954. The premises are used as a café-restaurant and are situate at 47 St George's Walk, Croydon. Part of St George's Walk is called the West Precinct and the other part is the East Precinct. The premises with which we are concerned in this case are in the East Precinct.

The premises were demised for a term of 21 years from June 24 1964. At all material times the reversion to the lease has been vested in the respondents and the residue of the term in the applicants.

On October 8 1984 the applicants applied for a new lease under the 1954 Act. That application was not opposed. The application was heard on April 29 1986, when the only issues were (a) the rent and (b) the interim rent.

The surveyors for both parties were agreed that the correct approach in the matter of the rent was what was called the zone A equivalent method of evaluation and that the relevant area of the premises in issue was 504 sq ft.

The landlords' surveyor's evidence given at the trial as to open market lettings of premises concerned premises which were in the East Precinct of St George's Walk, being nos 31 and 45A — 45A in fact adjoins the premises with which we are concerned. Both premises were let by the landlords in 1985 on the open market at rents which equated to zone A equivalent of £18.99 per sq ft and £22.92 per sq ft respectively. The judge accepted that evidence.

The landlords' surveyor also gave evidence on a prospective letting of nos 28 and 30 St George's Walk, which premises are also in the East Precinct. That was a prospective letting on the open market by the landlords at a rent equated to the zone A equivalent of £18.49 per sq ft. The judge stated that that transaction was likely to become what she called "a completed negotiation".

The tenants relied upon 11 alleged comparable transactions, two of which were rent reviews and the remainder were renewals of leases, at premises in the West Precinct, agreed in each case between the Legal & General Assurance Society Ltd and the sitting tenant of the respective premises.

The rent for the subject premises was fixed by the judge and equated to a zone A equivalent of £12.59 per sq ft. The judge arrived at that figure by adding together the zone A equivalent rates of (a) nine of the transactions relied upon by the tenants but making a 20% reduction to reflect her finding that there was a difference between the premises in the West Precinct and those in the East Precinct and (b) the two open market lettings relied upon by the landlords, and then dividing the resultant figure by 11. The figure of £12.59 was accordingly an average rent, which the judge then applied to the subject premises. The result was that the annual rent should be fixed at £6,345.

The landlords contend that the judge misdirected herself and that there should therefore be a new trial. It is said, first, that she applied a test which was wrong in principle.

Section 34 (1) of the Landlord and Tenant Act 1954 (so far as material) is in these terms:

The rent payable under a tenancy granted by order of the court . . . shall be such as may be agreed between the landlord and the tenant or as, in default of such agreement, may be determined by the court to be that at which, having regard to the terms of the tenancy (other than those relating to rent), the holding might reasonably be expected to be let in the open market by a willing lessor . . .

In the course of her judgment the judge put the matter thus:

I find that all 11 comparables shall be taken into my calculation of a fair and reasonable rent with 20 per cent deduction of any that have been in the West Precinct by first adding the two which the respondents have been able to freely negotiate. This brings the overall figure to £12.59 per sq ft authorising: — (1) £504 = £6,345 per annum and I find that this sum is the fair and reasonable rent.

It is said that "fair and reasonable" is not a test imposed by the statutory provisions of section 34 of the 1954 Act. The test under the statute is the open market rent, and "fair and reasonable" may involve considerations which play no part in the determination of an open market rent.

The language used by the judge is not, I think, happily chosen, but I find it impossible to believe that the judge, who had had a full day's evidence and argument as to the rental this property would fetch, can have had in her mind anything other than the open market rental. Indeed, previously in her judgment she refers to "the open market values agreed by them" — that is to say the landlords — "and two willing tenants". Then previously to that, she referred to the updating of comparables by the use of the Hillier Parker guide "to property values in this area".

I do not think by the use of the words "fair and reasonable" that the judge was introducing extraneous factors outside market considerations. She was merely saying, I think, that on the evidence before her as to values, the figure upon which she decided was a fair and reasonable assessment of the open market rental of the subject premises.

The next point is this: It is said that the judge gave too much weight to the nine rentals in the West Precinct. Those rentals were tendered as comparables. The precise circumstances in which they were agreed (and the terms of the leases in question) were not investigated before the judge. But that was the choice of the parties — in particular the choice of the landlords — and if they had wanted further details, they could have asked for them from the other side's surveyor, who was co-operating in providing information relevant to the comparables tendered.

In my view the judge was never asked to examine the details of the West Precinct comparables. Therefore she was entitled, in my judgment, to assess the evidence as it had been tendered to her. She said that she found it helpful, and one cannot quarrel with that — it obviously was helpful. They were all premises in St George's Walk and in so far as the West Precinct was a more favoured area than the East Precinct (in which these premises were situated) the judge made an allowance of 20% (as opposed to the 30% which had been suggested in the expert evidence), which took that fact into account.

It is true that the nine West Precinct comparables were all renewals of leases; but businessmen who are taking renewals of leases of business premises protected by the Landlord and Tenant Act 1954 are normally going to pay the market rental of the premises — or something near to it. They will be professionally advised and will, in the nature of things, pay more or less what the market requires them to pay. There might very well be scope for cross-examination to show, in individual cases, that some special factors were operating, but the landlords do not appear to have run their case in that way. They may have thought that there was not much to be derived from such an approach.

It is then said that the judge's method of calculation distorted the figures by giving too much weight to the West Precinct figures. But, if the West Precinct figures were comparables (and, on the evidence, I see no reason why the judge was not entitled to treat them as comparables), then it was reasonable and proper to take them into the calculation.

It is said that the judge failed to evaluate the worth of each of the nine properties in the West Precinct. No doubt that would have been an excellent course, but the landlords' case did not involve such an individual evaluation; and, so far as one can judge, there was no material before the judge to enable that to be done upon an individual basis. The tenants' witness does not appear to have been cross-examined in that way, and, as I have already mentioned, that may well have been a perfectly understandable decision upon their part.

A further complaint is made that the judge was wrong in ignoring altogether (as she did) the case of nos 28 and 30 St George's Walk. The prospective letting was, however, subject to planning permission, and there were objectors to the grant of planning permission. If planning permission were to be refused, the transaction itself might very well go off and, if it did, one did not know what might happen or what rent would be obtained in the open market for premises in that condition.

Further, the letting was for a high-class restaurant for a 20-year term and was for a double unit which had two frontages. It was not, in the circumstances, a similar letting to that which would be involved in the subject premises.

There were in fact some differences between the subject premises and the two comparables in the East Precinct which were relied upon by the landlords. Both had initial rent-free periods and both had stood empty before being let. Doubt was therefore cast by the tenants' counsel upon the value of regarding them as being comparable at all. On the facts as we have them I do not feel able to make much of those considerations, and although the position of those comparables is not free from doubt, they do stand.

However, there is, as it seems to me, one matter upon which the judge plainly erred. She took the figures of the nine comparables in the West Precinct from column (6) of the schedule at p 74 of the appeal bundle. However, those figures had not been adjusted for increases in market rents down to April 1986. The latest figures were the index of the Hillier Parker figures to November 1985 — but some figures, being a little before then, were not adjusted to the November 1985 figures.

In my judgment, the figures have to be brought down to April 1986; the precise form of order whereby that is achieved is something which can be considered with counsel.

There is one other matter: It was said on behalf of the landlords that any rents decided upon in this case would be of consequence in determining the rentals for letting of other premises in the precincts. I do not think that is so. Other cases will have to be decided upon their own facts and upon their own evidence — and upon market conditions prevailing at the time when the rents have to be determined.

In the circumstances, I can see no basis for interfering with the decision of the judge, save to the extent which I have mentioned, regarding the adjustment of the figures down to April 1986. To that extent, therefore, I would allow the appeal.

There was a further question as to interim rent. When Mr Dowding opened the case, he indicated that while he was asking for a new trial on the main issue, if that were not ordered he would not be seeking a reconsideration of the interim rent — as he would do so if a retrial were ordered on the main issue. Having regard to the limited extent to which the court is interfering with the judge's decision, it does not seem to me that it would be proper to alter the figure for interim rent.

Accordingly, in my view, the appeal will be allowed to the limited extent which I have mentioned.

SHELDON J agreed and did not add anything.

The appeal was allowed in part as indicated in the judgment of Fox LJ. No order was made for costs in the Court of Appeal; costs below to stand.

Court of Appeal
February 27 1987
(Before Lord Justice DILLON and Lord Justice GLIDEWELL)

BAR v PATHWOOD INVESTMENTS LTD

Estates Gazette June 20 1987

282 EG 1538-1542

Landlord and Tenant Act 1954, Part II — Time-limit under section 29 (3) for application for new tenancy — Application within time but covering part only of holding — Effect of amendment after expiry of time-limit — Tenancy comprised a ground-floor shop with basement storage area, a ground-floor studio and three flats on first, second and third floors — Tenant occupied shop and basement storage for a ladies' boutique and the ground-floor studio and first-floor flat as her home — The second- and third-floor flats were sublet to persons who had no connection with her business — Tenant's originating application for a new tenancy, following the usual preliminary notices, was made within time but referred only to "the ground-floor shop premises at 32 Rosslyn Hill, Hampstead" — Subsequently, but outside the section 29 (3) time-limit, the description was amended to read "Shop and premises situate at 32 Rosslyn Hill, Hampstead" — This was an appropriate description for an application, although any order made for a new tenancy would have to exclude the second- and third-floor flats — Landlords opposed the amendment and the county court judge ruled that it must be disallowed as amounting to a new originating application made out of time — The application must therefore be treated as limited to the ground-floor shop premises — Held on appeal that there was power under Ord 15 of the County Court Rules 1981 to amend an originating application and that the court would itself exercise the discretion which the judge had failed to exercise — G Orlik (Meat Products) Ltd v Hastings & Thanet Building Society followed — A submission by the landlords that to allow the amendment would be to deprive them of a vested right acquired as a result of the expiry of the time-limit was rejected — Appeal allowed

The following cases are referred to in this report.

Beardmore Motors Ltd v *Birch Bros (Properties) Ltd* [1959] Ch 298; [1958] 2 WLR 975; [1958] 2 All ER 311
Davies v *Elsby Bros Ltd* [1961] 1 WLR 170; [1961] 1 All ER 672, CA
Fernandez v *Walding* [1968] 2 QB 606; [1968] 2 WLR 583; [1968] 1 All ER 994; (1967) 19 P&CR 314; [1968] EGD 9; 205 EG 103, CA
Olley v *Hemsby Estates* (1965) CLYB 2205
Orlik (G) (Meat Products) Ltd v *Hastings & Thanet Building Society* (1974) 29 P&CR 126; [1975] EGD 104; 234 EG 281, CA
Williams v *Hillcroft Garage Ltd* (1971) 22 P&CR 402; [1971] EGD 430; 218 EG 1163, CA

This was an appeal by the tenant of a shop and premises at 32 Rosslyn Hill, Hampstead, London NW3, Mrs Hancock, who carried on business there under her maiden name of Nurit Bar, from the decision of Judge Brooks, at Bloomsbury County Court, in regard to her application for a new tenancy under Part II of the Landlord and Tenant Act 1954. The respondents were the landlords, Pathwood Investments Ltd.

Paul Morgan (instructed by Herbert Oppenheimer Nathan & Vandyk) appeared on behalf of the appellant; Norman Primost (instructed by Jacobson Ridley) represented the respondents.

Giving the first judgment at the invitation of Dillon LJ, GLIDEWELL LJ said: The first and major question which arises on this appeal is: can an originating application for a new tenancy under section 24 of the Landlord and Tenant Act 1954 be amended after the time-limit provided by section 29(3) of the Act for making an application has expired?

The facts which give rise to this application are these: by a lease dated May 28 1964 the respondent landlords demised a shop and premises at 32 Rosslyn Hill, Hampstead, for a term of 21 years from May 28 1964.

The applicant, Mrs Hancock, who carries on business in her maiden name of Nurit Bar, and to whom I shall refer as the tenant, took an assignment from the original lessee on November 4 1977.

The premises, the subject of the lease, comprise a ground-floor shop with a basement storage area; a ground-floor studio, and three flats on the floors above. The tenant occupies the shop and the storage for her business of a ladies' boutique, and the ground-floor studio and the first floor as her home. The second- and third-floor flats are sublet by her to persons unconnected with her business.

On November 19 1984 the landlords gave notice to terminate the tenancy of what is described as "the shop and premises" — that termination taking effect on the expiry of the lease on May 28 1985. The notice indicated that the landlords would not oppose an application for a new tenancy. That notice was given in the time provided by section 25 (2) of the Act.

On January 10 1985, again in time, the tenant gave a counternotice and indicated that she was not willing to give up possession of the premises.

Section 32 (1) of the Landlord and Tenant Act 1954 provides:

Subject to the following provisions of this section, an order under section 29 of this Act for the grant of a new tenancy shall be an order for the grant of a new tenancy of the holding; and in the absence of agreement between the landlord and the tenant as to the property which constitutes the holding the court shall in the order designate that property by reference to the circumstances existing at the date of the order.

The phrase "the holding" is defined in section 23 (3) as:

... the property comprised in the tenancy, there being excluded any part thereof which is occupied neither by the tenant nor by a person employed by the tenant and so employed for the purposes of a business by reason of which the tenancy is one to which this Part of this Act applies.

In this case the exclusion in that definition had the effect of excluding the flats on the second and third floors from "the holding" — since they were occupied by persons totally unconnected with the business. It follows, and it is agreed by the parties, that "the holding" within section 23 (3) comprised the ground-floor shop with basement storage, and the ground-floor studio and first-floor flat occupied by the tenant.

On March 8 1985 an originating application to the county court for the grant of a new tenancy was made by solicitors acting on behalf of the tenant. This begins:

In the matter of the Landlord and Tenant Act 1954 and In the matter of ground floor shop premises at 32 Rosslyn Hill Hampstead London NW3.

It then names the tenant and the landlords and indicates that the tenant is applying for the grant of a new tenancy pursuant to Part II of the Landlord and Tenant Act 1954, and continues:

The premises to which this application relates are the ground floor shop premises at 32 Rosslyn Hill Hampstead London NW3. The rateable value of the premises is £597

— which the parties are agreed is the rateable value attributable to the shop alone, and does not include the rateable value of the studio behind or any of the flats.

Under the particulars of the current tenancy of the premises, the last subpara reads: "Whether any, and if so what, part of the property comprised in the tenancy is occupied neither by the tenant, nor by a person employed by the tenant for the purposes of the business carried on by the tenant in the premises" and the answer is "None", which in the circumstances is not merely wrong but astonishingly so.

Why the application in terms was expressed to apply only to the ground-floor shop premises when the holding, as a moment's thought makes clear, applies to the shop premises and to the flat occupied by the tenant is not clear and to my mind does not matter.

That application was made within the time limited by section 29 (3) and that time expired on March 19 1985.

There was then correspondence between solicitors with regard to rent which was inconclusive, and it was not until July 1985, as a result of some indication from the tenant, that her solicitors woke up to the fact that the application did not cover the whole of the holding.

On July 29 1985, without leave, they filed what purported to be an amended application, and a yet further amended application on August 19 1985. It is the first amendments which are material in the present case, because, by those amendments, the description of the premises is altered so that instead of reading "ground-floor shop" it reads (after amendment) "shop and premises situate at 32 Rosslyn Hill Hampstead", which is of course an accurate description of the premises contained in the lease and is appropriate to an application for a new tenancy of the holding.

After the second set of amendments, it became apparent that what the solicitors were then doing was seeking to apply for a new tenancy of the whole premises; but it is now conceded that the order for a new tenancy cannot include the second- and third-floor flats — and indeed that was made clear by a telex sent by the tenant's then solicitors on December 5 1985.

Since that date, the nature of the respective contentions of the parties has been clear. The tenant has been seeking to amend her original application (or her claims to have amendments made to the original application) so that the court will be dealing with an application for a new tenancy of the holding, including her own flat. The landlords have been taking the attitude that she is not entitled so to amend.

In pursuance of that attitude, on November 25 1985 the landlords applied to the court for an order that the amendments to the originating application should be disallowed. Mr Registrar Wakefield made an order to that effect; the tenant appealed, and on May 21 1986 His Honour Judge Brooks dismissed the appeal. The tenant now appeals to this court against the judge's decision, with his leave.

The critical part of the judge's judgment is contained in para 8 of the approved note of his judgment which we have before us, where he said this — after setting out the respective contentions of counsel for the parties:

In my judgment an originating application under the Act cannot be treated as a pleading. Suffice it to say that pleadings have their own set of rules. An originating application under the Landlord and Tenant Act 1954 is governed by the strict provisions of the sections I have already referred to.

The amendments sought by the appellant amount to what can only be described as a "new originating application" in which the appellant seeks to include four separate flats, in addition to the shop premises. It is crucial in my judgment that an application for a new tenancy under the Act should contain the *precise* nature and extent of the premises the tenant wishes to hold as a new tenancy and the application must be lodged within the time-limit prescribed. (I perhaps ought to make it clear that provided the time-limit has not expired an applicant would be entitled to amend his application.)

In this case the appellant chose to limit her application to the ground floor of the premises and in my judgment it is now not open to her to amend her application to include other premises albeit comprised in the same tenancy. To do so would be contrary to the provisions of section 29 of the Act.

It is quite plain, therefore, that the basis of the judgment was that the judge had no power to allow the amendments to stand. The amendments have been made — or are purported to have been made — under Ord 15 r 2 of the County Court Rules 1981, which provide in subrule (1):

Subject to Order 9, rule 2 (3), and the following provisions of this rule, a party to an action or matter may, without an order, amend any pleading of his at any time before the return day by filing the amended pleading and serving a copy on the opposite party.

Then by subrule (3) it is provided:

The court may, of its own motion or on the application of the opposite party, disallow an amendment made under paragraph (1) and shall do so where it is satisfied that, if an application for leave to make the amendment had been made under rule 1, leave would have been refused.

Ord 15, r1, applies to amendments made with leave.

Mr Morgan, who appears for the appellant today — not having previously represented her, just as Mr Primost, who appears for the landlords, had not previously represented them — submits that the judge was wrong. There is power to amend an originating application, under the rules to which I have just referred, and indeed under r 1 of Ord 15.

In that submission he relies particularly on a decision of this court in *G Orlik (Meat Products) Ltd* v *Hastings & Thanet Building Society* (1974) 29 P & CR 126. That was a case in which the tenant had made an application for a new tenancy under the 1954 Act within the appropriate time-limit, but which did not include in the proposed terms of the lease the right to use land adjoining the building, the subject of the tenancy but owned by the landlord, for the purpose of parking motor cars owned by the tenant.

After the expiry of the four-month period within which the application could be made, the tenant sought leave to amend the application by including a claim that the new lease should permit such parking. That was refused on the ground that there was no power to amend, but an appeal against that decision was allowed.

Giving the judgment of the court, Stamp LJ said, towards the bottom of p 128:

When the application came on for hearing before the judge, the tenants sought to claim rectification of the pre-existing lease so that it should be altered in such a way as to provide for the parking rights. It was pointed out that the court did not have jurisdiction to entertain such a claim. The tenants then, it would seem, abandoned any suggestion of rectification of the lease, but sought instead leave to amend the proposals contained in their application so as to include amongst them a term relating to the parking of their vehicles on, in part, the landlords' adjacent land. The application to amend was opposed by the landlords, one ground being the ground urged in this Court, namely, that because of the time-limits in section 29 (3) of the Act, which had by then expired, there was no jurisdiction to give leave to amend. It does not appear to have been suggested that the landlords were taken by surprise, but, of course, if there were lack of jurisdiction the absence of any prejudice created by the granting of leave would not be relevant.

In our judgment no ground has been shown for holding that there is no jurisdiction to allow an applicant to amend, after the four months' period, the proposals contained in an application to the court made within that period. Counsel for the landlords is right in saying that the Act is strict, and uses strict language, in its time provisions. We are, however, unable to see anything in the Act which deprives the court of jurisdiction to grant leave to the applicant to amend the detail of his proposals after the expiration of the four months' limit. The decision of this Court in *Williams* v *Hillcroft Garage Ltd*, if not a direct authority for that proposition, provides, at the least, most persuasive indication that it is correct. Accordingly, we reject this second ground put forward in support of the appeal.

Apart from *Williams* v *Hillcroft Garage Ltd* (1971) 22 P & CR 402 there referred to, Mr Morgan also referred us to another decision of this court, *Olley* v *Hemsby Estates* only reported (as far as he is aware) in the *Current Law Year Book* for 1965, para 2205, where landlords, on serving a notice under section 25 of the 1954 Act on a tenant, stated that they would oppose the grant of a new tenancy. However, when the tenants applied to the county court for a new tenancy, the landlords filed an answer saying that they would not oppose the grant — apparently they had changed their minds in the interim as the result of some redevelopment proposals which they had in mind. They then changed their minds again and sought leave to amend their answer so as to indicate that they would oppose the grant of a new tenancy. The county court judge, in effect, allowed the amendment — he did so in a slightly different way — which led to an appeal against his decision, in which it was held that the amendments were properly allowed under Ord 15 r 1 of the County Court Rules.

In my judgment, this submission of Mr Morgan is correct. I take the view that even if strictly we are not bound to conclude that, in this case, the judge was wrong to say that he had no jurisdiction, nevertheless the authorities to which I have referred are the clearest indication that there is power to amend and that he should have considered whether or not to exercise it. It is, however, only fair to the judge to say that, so far as one can ascertain, he was not referred to any of the authorities on this point to which reference has been made before us.

One then turns to consider, since the judge did not exercise any discretion himself, how we should exercise the discretion which he had. Mr Primost urges us to follow the general principle that the courts will not deprive a party of a vested right acquired as the result of the expiry of a time-limit. For that proposition, which is undoubted, he referred us to *Beardmore Motors Ltd* v *Birch Bros (Properties) Ltd* [1959] Ch 298, a decision of the late Harman J, which was a decision under the 1954 Act and was specifically approved by this court in *Davies* v *Elsby Brothers Ltd* [1961] 1 WLR 170 — a case not concerned with the Landlord and Tenant Act.

I need not make any further reference to the authorities, because the principle is undoubted. The question is whether it applies here.

Mr Morgan's answer is that it does not. He reminds us that under section 32 (1) the court's power is only to grant a new tenancy of the holding. What constitutes the holding is for the court, at the hearing, to decide — unless there is agreement between the parties. Once a valid application for a new tenancy had been made — and in this case it is conceded it was made — then, Mr Morgan submits, the landlords had no right to resist a new tenancy of the whole of the holding. Thus, they were not deprived of any vested right, since they had no vested right to resist an application for a new tenancy, including the flat occupied by the tenant. All the amendment does is to make clear what had hitherto not been clear and which had been wrongly stated — that is to say, what constituted the holding which was misdescribed in the original application.

Mr Primost seeks to answer that by saying that there was indeed an agreement between the landlords and the tenant as to what did constitute the holding. His argument is that the tenant's originating application which described the premises of which the new tenancy was sought as "the shop" was a description of the holding and was an offer to the landlords to agree that that — that is the shop only — constituted the holding for the purposes of the Act. That offer, he submits, was accepted when, as they did, the landlords made an application for an interim rent which they, in turn, limited to the interim rent of the shop premises.

Quite apart from the fact that I am satisfied that the application for the interim rent should have related to the premises comprised in the original tenancy, and not a part, I cannot construe those two documents as constituting an agreement between the landlords and the tenant as to what constituted the holding for the purposes of section 32 (1) of the Act. Certainly, I cannot find, and do not find, that the tenant was bound thereafter in a way which prevented the court from exercising what was its undoubted duty to grant a new tenancy of the premises constituting the holding.

That brings one back to the question whether this court should exercise its discretion in favour of the tenant. Once it is clear that the landlords were not being deprived of any vested right, but merely of what might be described as a possible windfall, then it is equally clear that while there must be some detriment to the landlords, it is limited in extent.

On the other hand, if this appeal fails, there will be substantial detriment to the tenant, because she will not have the protection of the Act in relation to her occupation of her home; and even if she managed to negotiate a new tenancy of that flat — and one knows not if she will be able to do so — it may well be at a rent which will not be that for which the Act itself would provide.

Accordingly, in my judgment, it is quite plain that the detriment to the tenant substantially outweighs that to the landlords — that is to say detriment to the tenant if the appeal fails.

I would therefore allow the appeal. I should say that we did allow Mr Morgan to put in an affidavit from his client by way of further evidence in relation to one point in the judgment, but although we did consider it, I for my part do not find it necessary to refer to it because it did not seem to me to add anything of value to the matters to which I have referred.

For those reasons, I would allow the appeal.

Agreeing, DILLON LJ said: Judge Brooks in the court below did not have the benefit of the very clear and helpful argument and skeleton which we have had from Mr Morgan — who, as my lord has said, did not appear for the tenant in the court below.

It is not in dispute that the original form of application made by the tenant in March 1985 was a valid application to the court for a new tenancy under Part II of the 1954 Act. It is said that it was only a valid application in respect of the ground-floor shop premises and basement storage which are part only of the premises at 32 Rosslyn Hill comprised in the tenant's lease.

It is said — and found favour with the judge in the court below — that any amendment to include a further part of the property comprised in the lease in the application for a new tenancy is tantamount to a fresh application under the Act, which is made out of time. However, as Mr Morgan has pointed out to us, under section 32 of the Act an order under section 29 of the Act for the grant of a new tenancy is to be an order for the grant of a new tenancy for the holding, and in the absence of agreement between the landlord and the tenant as to what constitutes the holding, the court must designate that property in the order.

In this case there is no possibility of arriving at the view that there has been any agreement or contract between the landlords and the tenant that the holding is only constituted by the ground-floor shop and basement storage. On the facts it is common ground that the holding as defined in section 23 (3) of the Act includes the ground-floor shop and basement storage, the studio on the ground floor at the rear of the shop and the first-floor flat which is occupied by the tenant.

As was pointed out by Winn LJ in *Fernandez* v *Walding* [1968] 2 QB 606 at p 616B, it is only the holding in respect of which the court has jurisdiction to grant a new tenancy. There is no jurisdiction to grant a new tenancy of part of the holding, except in the circumstances not relevant to the present case, covered by section 31 A (1) (a) and (b) of the Act. Therefore, when the application which was launched in March 1985 comes on for trial, the court's duty and power will only be to grant a new tenancy for the holding as truly defined.

The amendment which the tenant desires and which the registrar and the judge have refused is thus one which correctly describes the holding to which the application must necessarily relate, and to allow such an amendment is, in my judgment, well within the reasoning of Stamp LJ in *G Orlik (Meat Products) Ltd* v *Hastings & Thanet Building Society*, to which Glidewell J has referred.

I agree entirely with the judgment of Glidewell J and I would allow the appeal accordingly.

The appeal was allowed and the order below set aside; the respondents' application to disallow the amendments was dismissed. The appellant was awarded costs in the Court of Appeal (other than costs of application to adduce fresh evidence) and the costs below.

LANDLORD AND TENANT
LEASEHOLD REFORM

Court of Appeal
November 5 1986
(Before Lord Justice SLADE and Mr Justice WAITE)

DIXON v ALLGOOD

Estates Gazette January 17 1987

281 EG 183-190

Leasehold Reform Act 1967 — Right to enfranchisement Appeal by landlord from decision of county court judge in favour of tenant — Tenant had a term of 51 years from May 13 1964 at £52 per annum of 6.11 acres of land with, originally, two derelict cottages thereon — Over a period of years the cottages were rebuilt, in accordance with a covenant in the lease, and a number of garages built — A small part of the land was acquired by compulsory acquisition and the rent marginally reduced — In 1981 the tenant served notice on the landlord of his desire to acquire the freehold under the 1967 Act — The claim was resisted by the landlord on various grounds — It was submitted that the two cottages did not constitute a "house", that the tenancy was not a long tenancy at a low rent, and that the tenant was not entitled to claim the freehold of the whole acreage leased to him — The county court judge reduced the area to which the tenant was entitled but otherwise rejected the landlord's submissions — On appeal to the Court of Appeal the issue resolved itself into the question of whether the tenancy was a long tenancy at a low rent; if it was not, the other points raised became academic — The county court judge had assumed that the garages built on the land should be included in determining the rateable value on the appropriate day, thus arriving at a rateable value of £100, of which the reserved rent of £52 was less than two-thirds — In this the judge erred, failing to distinguish between the "house and premises" for the purpose of section 2(3) and the "dwelling-house consisting of the house in question" for the purpose of section 4(1)(a) — For the latter purpose the garages should not be included — As a result, the rateable value was either £42 or £76, according to what was considered to be "the appropriate day", but in either case the reserved rent of £52 was more than two-thirds — Held accordingly that the tenancy was not a long tenancy at a low rent and the tenant did not qualify for enfranchisement — Landlord's appeal allowed

No cases are referred to in this report.

This was an appeal by the landlord, Lancelot Guy Allgood, from the decision of Judge Hall, at Hexham County Court, in favour of the claim by the tenant, James Dixon, that he was entitled under the Leasehold Reform Act 1967 to enfranchise the house and premises known as Riverside Cottages, Acomb, Hexham, Northumberland, subject to a restriction as to the area of the property. There was a cross-appeal by the tenant, the present respondent, contending that he was entitled to acquire the whole of the premises comprised in the lease.

J H Fryer Spedding (instructed by Wilkinson Marshall Clayton & Gibson, of Newcastle upon Tyne) appeared on behalf of the appellant; the respondent appeared in person.

Giving judgment, SLADE LJ said: This is an appeal by Mr Lancelot Guy Allgood from a judgment of His Honour Judge H G Hall given in the Hexham County Court on October 24 1985. By his order it was in effect decided that Mr James Dixon was entitled to enfranchise a house and premises known as Riverside Cottages, Acomb, Hexham, Northumberland, subject to a limitation that the house and premises should be restricted to the house and gardens "as defined in the decision of the Lands Tribunal dated January 28 1981".

There is also a cross-appeal by the respondent to the appeal, Mr Dixon. Mr Allgood (whom I will henceforth call "the landlord") has appeared before us by counsel, Mr Fryer Spedding. Mr Dixon has appeared in person and has, if I may say so, assisted us ably by clear oral and written submissions.

The history of the matter begins with the execution of a lease dated December 31 1964 by which there was granted to Mr Dixon a term of 51 years at a yearly rent of £52 from May 13 1964 of property defined in the lease as

ALL THAT land situate in the Parish of St John Lee and forming part of the Lessors Hermitage Estate coloured blue on the plan annexed hereto and containing in area Six acres and eleven one-hundredth parts of an acre or thereabouts TOGETHER with two derelict cottages situate thereon.

Clause 4(c) contained a tenant's covenant in the following terms:

within five years from the [13th] day of May [1964] in accordance with plans previously submitted to and approved of by the Lessors or their Agents to rebuild the said two derelict cottages to the reasonable satisfaction of the Lessors or their Agents. . . "

By clause 4(d) the tenant covenanted

from the [13th] day of May [1969] to maintain the said two rebuilt cottages in a good and tenantable condition and properly painted and decorated inside and outside and in such condition to hand over the demised premises to the Lessor at the end or sooner determination of the said term.

By clause 4(i) the tenant covenanted

not to build on the said land any building other than the said two new cottages and domestic office used therewith and not to use the demised premises for the purposes of any trade profession or business.

As to the subsequent events in this matter, I turn to the learned judge's judgment, of which I would like to express my appreciation, because it is a most careful one and it has set out the facts very clearly for our assistance. I quote from para 7:

(a) At the time of the grant [of the lease] the two cottages were mostly unroofed and without doors or windows. The walls remaining were as shown on the "plan as existing" produced and marked "A1".

(b) From this plan it appears that the two cottages were separated by an unpierced party-wall albeit in "dog-leg" form.

(c) A plan was drawn up showing the proposed reconstruction which is stamped Dec 19 1963 which he [Mr Dixon] produced and is marked "A2".

(d) At the time he already planned to occupy the whole structure for himself and his family but was advised by the landlord's agent to put in plans for separate cottages as these would attract two improvement grants. As can be seen from these plans the party wall was to be pierced in two places on the ground floor and parts of the first floor occupied with one cottage were situated over parts of the other cottage on the ground floor.

(e) The work of reconstruction was carried out by the Applicant assisted by local labour over an extended period. The first part to be completed was that now identified in the valuation as Riverside Cottage with an entry dated May 9 1966. He took up occupation of this part but retained his principal home in Newcastle.

(f) When the second cottage, identified as Riverside Cottage East with the entry February 6 1967, was completed he let it, furnished, to a subtenant.

(g) In 1970 he relinquished his home in Newcastle and Riverside Cottage became his home.

(h) The range of garages was completed, as appears from the valuation list by December 22 1971. It is a single open building with 5 doors and in addition to being used for garaging cars is used as a domestic storeplace.

(i) In the course of construction he departed from the original plans without informing the lessors. He was unable to provide plans showing these departures. No plans have been produced showing the present layout and I have therefore carried out an inspection of the property. From this it is apparent to me that there have been piecemeal alterations and additions as new ideas have occurred to the Applicant during the prolonged course of construction.

(j) In 1977 the subtenancy was terminated and the Applicant's daughter and her husband moved in to the premises. The doorway at the end of the ground floor passage which had been temporarily sealed was unsealed and they lived together as a family unit from then on, eating at the same table and each family having free access over the whole premises notwithstanding that the daughter, her husband and now their adopted child have their bedrooms in the part identified as Riverside Cottage East.

(k) The Applicant's reason for this arrangement is to make provision for the declining years of himself and his wife. The property is in a very isolated situation. Living alone would become a heavy burden on them as they advanced in years and the comfort of having younger members of the family living with them was much to be desired. This is a solution to the problems of old age which is becoming well recognised and is often pursued to the mutual advantage of both generations.

(l) After completion of the cottages and garages the Applicant built an extensive walled garden with greenhouses from which he provides the household needs for fruit and vegetables.

In or about 1974 a Government department acquired a small part of the 6.11 acres demised by the lease by means of a compulsory acquisition. Mr Dixon made an application to the Lands Tribunal to apportion the rent between the land acquired by the department and the residue of the leased land. A chartered surveyor, Mr J M Clark, gave evidence to the Lands Tribunal on behalf of the landlord, in the course of which he produced a plan and schedule purporting to give particulars of the property let to Mr Dixon. It divided the property into four parts, consisting in the whole of 6.11 acres. One of these four parts was referred to by Mr Clark as "cottages and gardens"; it was said to comprise 0.761 acre and was shown hatched brown on the attached plan.

The order of the Lands Tribunal dated January 28 1981 was that the amount of rent to be apportioned to the land acquired was the small sum of 56p. The result was to reduce the annual rent payable under the lease to £51.44.

By a notice of February 18 1981, later amended on June 22 1981, Mr Dixon gave the landlord notice of his desire to acquire the freehold of the land demised in exercise of his rights under the Leasehold Reform Act 1967. The notice, as amended, asserted *inter alia* that the rateable value of the cottages and premises at the "appropriate day" was £100. The significance of this assertion will appear later in this judgment.

At this point I think it will be convenient to refer to a few of the relevant provisions of the Leasehold Reform Act 1967 (to which, as subsequently amended, I will refer as "the 1967 Act").

Section 1(1) provides:

This Part of this Act shall have effect to confer on a tenant of a leasehold house, occupying the house as his residence, a right to acquire on fair terms the freehold or an extended lease of the house and premises where —
(a) his tenancy is a long tenancy at a low rent and subject to subsections (5) and (6) below the rateable value of the house and premises on the appropriate day is not (or was not) more than £200. . . ; and
(b) at the relevant time (that is to say, at the time when he gives notice in accordance with this Act of his desire to have the freehold or to have an extended lease, as the case may be) he has been tenant of the house under a long tenancy at a low rent, and occupying it as his residence, for the last three years or for periods amounting to three years in the last ten years; and to confer the like right in the other cases for which provision is made in this Part of this Act.

It will thus be seen that, in order to qualify for a right of acquisition under this subsection, a tenant has to satisfy both the conditions set out in subparas (a) and (b); and one of those conditions is that his tenancy must be "a long tenancy at a low rent" within the meaning of the Act.

Section 1(2), so far as material, provides:

In this Part of this Act references, in relation to any tenancy, to the tenant occupying a house as his residence shall be construed as applying where, but only where, the tenant is, in right of the tenancy, occupying it as his only or main residence (whether or not he uses it also for other purposes); but
(a) references to a person occupying a house shall apply where he occupies it in part only;. . .

Part occupation therefore may suffice to provide a qualification.

Section 2(1), so far as material, provides:

For purposes of this Part of this Act, "house" includes any building designed or adapted for living in and reasonably so called, notwithstanding that the building is not structurally detached. . . ; and —
(a) where a building is divided horizontally, the flats or other units into which it is so divided are not separate "houses", though the building as a whole may be; and
(b) where a building is divided vertically the building as a whole is not a "house" though any of the units into which it is divided may be.

Section 2(3) provides:

Subject to the following provisions of this section, where in relation to a house let to and occupied by a tenant reference is made in this Part of this Act to the house and premises, the reference to premises is to be taken as referring to any garage, outhouse, garden, yard and appurtenances which at the relevant time are let to him with the house and are occupied with and used for the purposes of the house or any part of it by him or by another occupant.

The provisions of section 2(1)(b) which I have quoted show that the division of a building vertically will prevent its being treated as a "house" for the purposes of the 1967 Act.

By a notice of April 8 1981 the landlord refused to admit Mr Dixon's right to enfranchise the property on three grounds, which were substantially those eventually argued before the learned judge. Accordingly, on July 1 1981 Mr Dixon issued an originating application in which he asked for an order

declaring that the applicant, the said James Dixon, is the tenant of a house and premises known as Riverside Cottages, Acomb, Hexham, in the county of Northumberland, within the meaning of section 1 of [the 1967 Act] and as such is entitled to acquire the freehold of the said house and premises in accordance with the provisions of the said Act.

This was followed by a defence in which all the allegations of fact in the originating application were admitted, save for, first, the description of the Riverside Cottages as "house and premises", secondly, the allegation contained in the originating application that the rent of £52 per annum included £25 per annum to fish at least two rods, and, thirdly, the allegation that the net annual value for rating purposes of "the said house and premises" was £100. I should say at this point that Mr Dixon's assertion that the rent of £52 per annum includes £25 per annum to fish at least two rods is no longer pursued.

The learned judge heard evidence from Mr Dixon and from Mr Clark on behalf of the landlord. Mr Clark, as is recorded in the judge's notes of evidence, gave evidence to the effect that the curtilage of the property did not include the garages but was where the cottages were situated together with an "obvious back yard". The judge reacted adversely to this suggestion, as will appear.

The first ground of opposition to Mr Dixon's application argued on behalf of the landlord was in effect that the two cottages did not constitute together a "house" for the purpose of the 1967 Act. This argument essentially depended on the construction and application of section 2(1)(b) of the 1967 Act, which I have already quoted. As to this, the learned judge said this:

He [the landlord] relies upon the evidence of Mr Clark, a chartered surveyor and the respondent's land agent since 1971 who said here were two cottages with two rateable values, two telephones, two staircases, two kitchens and "they appear to be vertically divided". The question of vertical division appears to me to be the crucial one having regard to the combined effects of section 1(2)(a) and section 2(1)(b) of the Act of 1967. Mr Clark did not go on to describe where he considered that vertical division to be. He produced no drawing to illustrate it.

Then the learned judge referred to various authorities which had been cited to him and distinguished them thus:

(iii) The case before me seems quite different. The lease of 1964 envisaged not just the repair of the two derelict cottages but their modernisation including the provision of "internal sanitation" which would be difficult to provide within the confines of the old layout. The solution to this problem was the destruction of the old vertical division between the cottages and the fitting in at different levels of the accommodation of each to accommodate the new facilities. For instance the bedroom of the westerly cottage was to extend over the bathroom of the easterly cottage and the bedroom of the easterly cottage was wholly over the kitchen of the westerly cottage.

(iv) although, as I have said, these plans had been changed and extended in the course of the building, I was unable, during my inspection, to find any evidence of a vertical division within section 2(1)(b) of the Act and was not assisted by Mr Clark's generalisation.

The learned judge on these grounds rejected the submission that the premises comprised two houses vertically divided, so that this ground of opposition failed.

The second ground of opposition was in effect that Mr Dixon's tenancy is not a "tenancy at a low rent". I will revert to what the learned judge said in this context in due course.

The third ground was in effect that, in any event, Mr Dixon was not entitled to claim the freehold of *the whole* 6.11 acres leased to him. The learned judge accepted this submission. On the other hand, he did not think that Mr Dixon was to be restricted to the area described by Mr Clark in his evidence as "where the cottages are situated and the obvious back yard". He accordingly decided that Mr Dixon was entitled to the declaration which he sought, subject to a limitation of the term "house and premises" so that it was to be co-extensive with the cottages and gardens which had been defined by Mr Clark in his evidence to the Lands Tribunal and had been hatched brown on the plan he produced, comprising, I understand, 0.761 acre.

In an amendment notice of appeal now before us, two main grounds of appeal are relied upon by the landlord in support of the submission that the judge reached the wrong decision:

(1) The learned judge erred in law or alternatively drew an incorrect conclusion of fact or of mixed fact and law from the primary facts found by him in holding that the above mentioned property constituted "a house" reasonably so called for the purpose of section 2 of the above Act.
(2)(i) That there was no evidence to support the learned judge's finding that the rateable value of the said property was £100 on the appropriate day which day was May 9 1966 or alternatively February 6 1967.
(ii) That the learned judge erred in law in including the rateable value of the five garages now situate upon the said property in the rateable value of the property for the purposes of section 4 of the Act.

In a respondent's notice Mr Dixon, for his part, submits in effect that the learned judge erred in holding that he was entitled to acquire only the 0.761 acre, but should have held that he was entitled to acquire the whole of the premises still comprised in his lease. However, this submission only becomes relevant if it is established that he possesses all the necessary qualifications to entitle him to acquire the freehold under section 1(1) of the 1967 Act. For present purposes, as I have already indicated, the most important of the necessary qualifications is that his tenancy must be a "long tenancy at a low rent" (see subsection 1(a)). If his tenancy is not a "long tenancy at a low rent" within the relevant definition, his claim must inevitably fail at the outset.

With the sensible agreement of Mr Dixon and counsel for the landlord, we have, in all the circumstances, begun by hearing argument from both sides on the second of the two main grounds raised by the notice of appeal, which relates to the existence or not of a "long tenancy at a low rent". It has also been agreed that, if we were to decide the point in favour of the landlord, we should give judgment on the appeal to this effect and not proceed to hear argument on the other points raised in the notice of appeal or the respondent's notice, which on this footing would have become academic.

In dealing with the point raised by the second main ground in the notice of appeal, I will assume in favour of Mr Dixon, though without deciding the point, that the learned judge was right in deciding that the building comprising the two cottages is one house, notwithstanding section 2(1)(b) of the 1967 Act and notwithstanding the submissions which the landlord has made, and would make, to the contrary.

Section 4 of the 1967 Act, to which I have not yet referred, contains a definition of a tenancy at a "low rent" for the purposes of the 1967 Act. Section 3 defines a "long tenancy". There is no doubt that Mr Dixon's tenancy is a long tenancy within this definition. It is section 4 which presents the difficulties so far as he is concerned.

Section 4(1) begins as follows:

For the purposes of this Part of this Act a tenancy of any property is a tenancy at a low rent at any time when rent is not payable under the tenancy in respect of the property at a yearly rate equal to or more than two-thirds of the rateable value of the property on the appropriate day . . .

The subsection therefore obliges one to see what was the rateable value of the property "*on the appropriate day*". The ascertainment of "the appropriate day" is therefore of crucial importance.

I interpose a reference to section 4(2), which, so far as material, provides:

Where on a claim by the tenant of a house to exercise any right conferred by this Part of this Act a question arises under section 1(1) above whether his tenancy of the house is or was at any time a tenancy at a low rent, the question shall be determined by reference to the rent and rateable value of the house and premises as a whole, and in relation to a time before the relevant time shall be so determined whether or not the property then occupied with the house or any part of it was the same in all respects as that comprised in the house and premises for purposes of the claim . . .

I now revert to section 4(1)(a), which provides that, for the purpose of this subsection:

"appropriate day" means the 23rd March 1965 or such later day as by virtue of section 25(3) of the Rent Act 1977 would be the appropriate day for purposes of that Act in relation to a dwelling-house consisting of the house in question.

I stress that there can be only one "appropriate day" in relation to any one claim under section 1 of the Act, such as that now made by Mr Dixon. We have, therefore, to ascertain what was the one "appropriate day" for the purpose of his claim.

There was no entry in the rating valuation list on March 23 1965 in respect of any of the buildings on the land in question. There is a letter addressed to Mr Dixon from the local council dated September 26 1985, which confirms that the relevant buildings were assessed for rating purposes and were first included in the valuation list on directions issued by the valuation officer on the following dates: "Riverside Cottage, Howford, Acomb" on May 9 1966 with a rateable value of £42; "Riverside Cottage East, Howford, Acomb" on February 6 1967 with a rateable value of £34; "5 Garages adj Riverside Cottage, Howford, Acomb" on December 22 1971 with a rateable value of £24. The total of those three rateable values represents the figure of £100 referred to by Mr Dixon in his pleadings.

Since none of the buildings appeared in the valuation list on March 23 1965, section 4(1)(a) of the 1967 Act makes it necessary to ascertain what would, by virtue of section 25(3) of the Rent Act 1977, be the appropriate day for the purposes of that Act in relation to "a dwelling-house consisting of the house in question". These references in subpara (a) to "a dwelling-house" and to "the house in question" are in striking contrast with the earlier references in the earlier sections to "the house and premises". Though Mr Dixon forcefully submitted to the contrary, they seem to me to oblige us, in applying section 4(1)(a) to the facts of the present case, to confine our attention to the building comprising the two cottages and not to take into account the separate building comprising the five garages, for the purpose of ascertaining "the appropriate day".

I now turn to section 25(3) of the Rent Act 1977, which provides:

In this Act "the appropriate day" —
(a) in relation to any dwelling-house which, on 23rd March 1965, was or formed part of a hereditament for which a rateable value was shown in the valuation list then in force, or consisted or formed part of more than one such hereditament, means that date, and
(b) in relation to any other dwelling-house, means the date on which such a value is or was first shown in the valuation list.

Subpara (a) can have no application on the facts of the present case. It inevitably follows from subpara (b) that, in relation to the dwelling-house said to comprise Riverside Cottage and Riverside Cottage East, "the appropriate day" must be either May 9 1966 or February 6 1967, being the dates on which the two cottages first appeared, respectively, in the valuation list.

Mr Fryer Spedding submits as his first argument that the 1966 date rather than the 1967 date is "the appropriate day". I do not find it necessary to decide this point, because it seems to me as clear as crystal that "the appropriate day" in respect of the house (assuming for this purpose that the two cottages are to be regarded as one house) must be one or the other of the two dates which I have last mentioned. The rateable value of the house on "the appropriate day" was either £42, if "the appropriate day" is taken as being May 9 1966, or £76, if "the appropriate day" is taken as being February 6 1967. £52 is more than two-thirds of the figure of £42 or £76, whichever figure is taken. In these circumstances I see no answer to Mr Fryer Spedding's submission that Mr Dixon's tenancy is not a tenancy "at a low rent" within the relevant definition contained in section 4(1) of the 1967 Act.

The learned judge dealt with the point now under discussion in his judgment as follows:

Although the rent reserved of £52 is less than ⅔rds of the rateable value of £100 on the appropriate day it is submitted that the rateable value of the garages should be excluded thus reducing the total rateable value to £76 of which the rent of £52 is marginally more than ⅔rds. I reject this submission. It is made on the basis of Mr Clark's evidence that the garages are not within the curtilage of the property. It is distressing to find that this evidence is in direct contradiction of his statement of evidence before the Lands Tribunal. In this he clearly shows an area of 0.761 acres described as cottages and gardens, hatched in brown on his plan, which covers the garages.

Section 2(3) of the Act provides that "house and premises" is to be taken as referring to any garage, outhouse, garden, yard and appurtenances and section 4(2) states that the question of (a low rent) shall be determined by reference to the rent and rateable value of the house and premises as a whole. Section 4(6) also provides for a just apportionment of an entire rent should the need arise.

I pause to say that section 4(6) of the 1967 Act appears to me to have no relevance to the facts of the present case, because there has been no severance of the property in so far as it consists of the two cottages and the five garages.

As Mr Dixon pointed out to us, the learned judge appears, from what he said, to have actually decided that the rateable value of the property on "the appropriate day" was £100, though he gave no reasons for that decision. If that decision had been correct, he would of course have been correct in deciding that this was a tenancy at a low rent. With all respect to him, however, it seems to me that he must have misunderstood the argument that was being put to him on this point, the argument being (a correct one in my view) that the appropriate day was either May 9 1966 or February 6 1967. Mr Fryer Spedding has explained to us that this particular submission was not based on Mr Clark's evidence at all; it was based on the figures shown in the letter from the rating authority. Mr Dixon submitted to us that the learned judge was justified in regarding December 22 1971 as being "the appropriate day", since it was the first date upon which both the two cottages *and* the five garages appeared in the rating list. I would have had some sympathy with this argument before making detailed reference to the statutory provisions, but, when such detailed reference is made, in my opinion they do not permit it. In particular, section 4(1)(a) of the 1967 Act appears to me to preclude it.

It may well be that, as Mr Dixon has submitted to us, the five garages fall to be included in the definition of the "house and premises" contained in section 2(3) of the 1967 Act, but, for reasons which I hope I have sufficiently explained, this definition does not in my opinion assist in ascertaining "the appropriate day". Its relevance lies in ascertaining the extent of the land and buildings which a tenant, who has all the qualifications required by section 1(1), is entitled to acquire by virtue of that subsection. Unfortunately so far as he is concerned, Mr Dixon does not in my judgment possess all those qualifications.

I conclude that his tenancy is not a tenancy "at a low rent" and that accordingly he does not possess the necessary qualifications to entitle him to invoke section 1(1) of the 1967 Act. It follows that I would allow this appeal. I would set aside the learned judge's order and I would make an order refusing the order for enfranchisement sought by the originating application.

Agreeing, WAITE J said: Although we are differing from the views of the learned judge, I do not think it necessary to add any detailed grounds of my own for taking the same view as Slade LJ, because it seems clear to me that, as my lord has said, the learned judge had not fully appreciated the significance of the differences in wording between "a dwelling-house consisting of the house in question", as that expression is used in section 4(1)(a) of the 1967 Act, and the phrase "the house and premises" referred to in section 2, and elsewhere in section 4, of the same Act. I would therefore agree in allowing the appeal and would concur in the order proposed.

The appeal was allowed with costs in the Court of Appeal and below. The order below was set aside and the application for an order for enfranchisement refused. Leave to appeal to the House of Lords was refused.

Court of Appeal

December 4 1986
(Before Lord Justice LAWTON, Lord Justice DILLON and Lord Justice RALPH GIBSON)

RENDALL v DUKE OF WESTMINSTER AND OTHERS

Estates Gazette March 14 1987

281 EG 1197-1198

Leasehold enfranchisement — Leasehold Reform Act 1967 — Appeal by leaseholder from decision of county court judge that leaseholder was not entitled to the freehold under Part I of the Act — Whether rateable value on the appropriate day was not more than £1,500, so as to qualify — Rateable value of appellant's house had been £1,597, but as a result of a claim under Schedule 8 to the Housing Act 1974 a reduction (attributable to improvements) of £88 had been certified — This still left the rateable value above the limit, but, pursuant to a proposal on June 8 1984, the valuation list was further amended on January 14 1985 by the alteration of the description of "house" to "house and garage" and by a reduction of rateable value from £1,597 to £1,547 — If this reduction could be dated back to April 1 1973, then, having regard also to the certified reduction of £88, the value would be below £1,500 — Held, however, that, as the county court judge decided, there was no provision which would enable the reduction to £1,547 to be back-dated to April 1 1973 — Whether under section 79 of the General Rate Act 1967 or under section 37(6) as amended, the alteration made in January 1985 to £1,547 could be related back only to April 1 1984 — This result was in line with the decisions in *MacFarquhar* v *Phillimore* **and** *Rodwell* v *Gwynne Trust Ltd* **— An alternative argument by the appellant, based on taking the appropriate day as January 14 1985, when the rateable value was reduced to £1,547, was also rejected — Appeal dismissed**

The following cases are referred to in this report.

MacFarquhar v *Phillimore* [1986] 2 EGLR 89; (1986) 279 EG 584, CA
Rodwell v *Gwynne Trusts Ltd* [1970] 1 WLR 327; [1970] 1 All ER 314, HL

This was an appeal by John Lewis Rendall from a decision of Judge Oddie, at West London County Court, declaring that he, as leaseholder of the house known as 74 Eaton Terrace, London SW1, was not entitled to the freehold under the provisions of the Leasehold Reform Act 1967. The respondents were the Duke of Westminster, John Nigel Courtenay James and Patrick Geoffrey Corbett.

The appellant appeared in person; Nigel Hague QC (instructed by Boodle Hatfield & Co) represented the respondents.

Giving the first judgment at the invitation of Lawton LJ, DILLON LJ said: Mr Rendall appeals against the decision of His Honour Judge Oddie in the West London County Court whereby it was ordered and declared that he, as leaseholder of the house and premises known as 74 Eaton Terrace, London SW1, was not entitled to the freehold pursuant to Part I of the Leasehold Reform Act 1967.

The Leasehold Reform Act 1967 provided in its original form by section 1(1) that that Part of the Act should:

have effect to confer on a tenant of a leasehold house, occupying the house as his residence, a right to acquire on fair terms the freehold or an extended lease of the house and premises where —
(a) his tenancy is a long tenancy at a low rent and the rateable value of the house and premises on the appropriate day is not (or was not) more than £200 or, if it is in Greater London, than £400; . . .

As the Act originally stood, it could not have applied to 74 Eaton Terrace because the rateable value was too high. However, the section was amended by the Housing Act of 1974. That provided:

In subsection (1)(a) above, "the appropriate day", in relation to any house and premises, means the 23rd March 1965 or such later day as by virtue of section 25(3) of the Rent Act 1977

— I interject that that replaced an earlier Act referred to in the amendment of 1974 —

would be the appropriate day for purposes of that Act in relation to a dwelling-house consisting of that house.

Then by subsection (6) it was provided:

If, in relation to any house and premises, —
(a) the appropriate day for the purposes of subsection (1)(a) above falls before 1st April 1973, and
(b) the rateable value of the house and premises on the appropriate day was more than £200 or, if it was then in Greater London, £400, and
(c) the tenancy was created on or before 18th February 1966,
subsection (1)(a) above shall have effect in relation to the house and premises as if for the reference to the appropriate day there were substituted a reference to 1st April 1973 and as if for the sums of £200 and £400 specified in that subsection there were substituted respectively the sums of £750 and £1,500.

There is no doubt that the tenancy under which Mr Rendall holds 74 Eaton Terrace was a tenancy granted before February 18 1966 and was a long tenancy at a low rent within the meaning of the Leasehold Reform Act. The crucial question is whether, to satisfy section 1(6), Mr Rendall can say that at April 1 1973, or at the appropriate day if that is later, the rateable value was not more than £1,500.

In point of fact, if the new valuation list which came into force on April 1 1973 had been looked at at that date, it would have shown 74 Eaton Terrace under the description of "house" with a rateable value of £1,597. Subsequently, on a claim made by Mr Rendall under Schedule 8 to the Housing Act 1974, the valuation officer in May 1981 certified a notional reduction in rateable value for the purposes of the 1967 Act of £88.* That means that it was established that, of the rateable value of £1,597, £88 is to be treated as attributable to tenant's improvements previously made and is, therefore, for the purposes of the 1967 Act, to be disregarded in considering whether the rateable value exceeded £1,500. However, the £88 reduction from a rateable value of £1,597 is not enough to bring Mr Rendall below £1,500.

Then, pursuant to a proposal made by Mr Rendall on June 8 1984, the valuation list was further amended on January 14 1985: first by altering the description "house" to "house and garage" and, second, by reducing the rateable value to £1,547. If Mr Rendall can pray that reduction in aid as dating back to April 1 1973, he can, in view of the £88 which I have already mentioned, say that the rateable value at April 1 1973 was below £1,500 and that he qualifies. His alternative argument is to say that, because of the alteration of description in 1985 from "house" to "house and garage", the appropriate day for the purposes of his case is January 14 1985 when the rateable value was reduced to £1,547 from which the £88 has to be taken away and not April 1 1973.

I will deal with the second point first because it can be dealt with very shortly. The property demised to Mr Rendall by the lease was described as a piece of land delineated and coloured in the plan together with the messuage garage and buildings erected thereon and known as 74 Eaton Terrace. The plan shows that the property was at the junction of Eaton Terrace and Graham Terrace. The house was the last house in a terrace of houses. The garage was at the rear with its entry from Graham Terrace; it did not have a second entrance from the house and garden of 74 Eaton Terrace, though it was contiguous with the house and garden. The house was in fact built, Mr Rendall has told us, as long ago as 1820, and the garage was originally described as a "cart shed".

To see what is meant by "house" in the valuation list as it stood at April 1 1973 it is necessary to have recourse to extrinsic evidence to explain the entry and, even though evidence of the intention of the valuation officer is not admissible, it is palpable, from looking at what was occupied by the then lessee, because Mr Rendall was not at that time on the scene, and from looking at the site, that the description "house" included the garage. In fact, when the alteration was made to refer to "house and garage" in 1985, that was referred to as a mere alteration in the description of the same hereditament. Mr Rendall seeks to say that the appropriate day was the day when the entry was made and he refers to section 25(3) of the Rent Act 1977 because that is what is mentioned in section 1(4) of the 1967 Act as amended. Section 25(3) provides:

In this Act "the appropriate day" —

a) in relation to any dwelling-house which, on 23rd March 1965, was or formed part of a hereditament for which a rateable value was shown in the valuation list then in force, or consisted or formed part of more than one such hereditament, means that date, and

b) in relation to any other dwelling-house, means the date on which such a value is or was first shown in the valuation list.

I have no doubt at all that, in relation to this dwelling-house consisting of the hereditament being the house and garage, the rateable value was shown in the valuation list in force on March 23 1965. Therefore, that is the appropriate day for the purposes of section 1(1) and, whether or not the rateable value exceeds £1,500 has to be tested as at April 1 1973 under section 1(6). I would, therefore, reject Mr Rendall's second point.

As to his first point, he says that the reduction made pursuant to his proposal of June 8 1984 was made because he had submitted, and it was accepted, that the original valuation in the valuation list which took effect on April 1 1973 was excessive when his house and garage was compared with various other comparable properties in the neighbourhood. Therefore, he says that the alteration should be retrospective to when the valuation list came into force.

His difficulty is that the statutory provisions which have been made to deal with this situation do not so provide.

Section 79 of the General Rate Act 1967 provides that, subject to various provisions which are not material to the present case, where an alteration is made in a valuation list, then, in relation to any rate current at the date when the proposal in pursuance of which the amendment so made was served on the valuation officer, or where the proposal was made by the valuation officer current at the date when notice of the proposal was served on the occupier of the hereditament in question, that alteration shall be deemed to have had effect as from the commencement of the period in respect of which the rate was made and shall, subject to the provisions of this section, have effect for the purposes of any subsequent rate. That means that, for rating purposes at any rate, the alteration made in January 1985 reducing the rateable value to £1,547 only related back to April 1 1984.

The position does not, however, rest there because section 37(6) of the 1967 Act as amended provides that section 25(1), (2) and (4) of the Rent Act 1977 shall apply to the ascertainment, for purposes of this Part of the 1967 Act, of the rateable value of a house and premises or any other property as they apply to the ascertainment of that of a dwelling-house for purposes of that Act. Section 25(4) of the 1977 Act provides that where, after the date which is the appropriate day in relation to any dwelling-house, the valuation list is altered so as to vary the rateable value of the hereditament of which the dwelling-house consists, or forms part, and the alteration has effect from a date not later than the appropriate day, the rateable value of the dwelling-house on the appropriate day shall be ascertained as if the value shown in the valuation list on the appropriate day had been the value shown in the list as altered. Subsection (1) provides that, in effect, the rateable value of a dwelling-house is to be ascertained for the purposes of the Rent Act 1977 on the basis of what is shown as the rateable value in the valuation list.

It follows — and this is in line with the decision of this court in the case of *MacFarquhar* v *Phillimore** as decided on May 19 1986, and the decision of the House of Lords in *Rodwell* v *Gwynne Trusts Ltd* [1970] 1 WLR 327 — in view of the interrelation of the Rent Act 1977 and the provisions of section 79 of the General Rate Act 1967, that the reduction in the rateable value by the amendment of January 1985 can, for purposes of the Leasehold Reform Act as amended and the Rent Act of 1977, relate back only to April 1 1984, and that is not good enough for Mr Rendall's purposes.

Mr Rendall seeks to pray in aid the provisions of earlier legislation, and particularly of section 38 of the Local Government Act 1948, subsection (6) of which says that the list for any rating area, settled, signed and sent to the rating authority, shall, as from the date when it comes into force and subject to any alterations made in accordance with the relevant Part of that Act, be the valuation list for the rating area. Any failure on the part of a valuation officer to complete any proceedings with respect to the preparation, revision or settling and signing of the list within the time required by the Part of the Act, or any omission from the list of any matters required by law to be included therein, shall not of itself render the list invalid and, until the contrary is proved, the list shall be deemed to have been duly made in accordance with the provisions of the Part of the Act.

Mr Rendall points to a note in *Halsbury's Statutes* to the effect that the reference to alterations made would seem to refer to all alterations made during the life of the list rather than to alterations made between the date of settling the list and the date of its coming into force. But subsection (6) is not concerned at all to deal with when an alteration made subsequently to the settlement of the list is to be treated as coming into force. That was dealt with by section 42 of the Local Government Act 1948. Nor is it concerned with the position under the Rent Act 1977 or the Leasehold Reform Act 1967 as amended. It cannot help Mr Rendall.

I agree entirely with the conclusions of Judge Oddie and the substance of the reasons which he gave in his very careful judgment and, accordingly, I would dismiss this appeal.

I should mention that there was a further reduction of the rateable value from £1,547 to £1,538 later than the amendment of January 14 1985, but that is not material to this appeal. It is not of itself enough to bring the rateable value below £1,500 at April 1 1973, and, in any event, for the same reasons that the reduction to £1,547 cannot relate

---

*Editor's note: This matter was the subject of a decision of the Court of Appeal reported t [1986] 1 EGLR 163; (1986) 278 EG 1090.

*Editor's note: Reported at [1986] 2 EGLR 89; (1986) 279 EG 584.

back to April 1 1973, so that to £1,538 cannot relate back.

LAWTON and RALPH GIBSON LJJ agreed that the appeal should be dismissed for the reasons given by Dillon LJ, and did not add anything.

The appeal was dismissed with costs. Leave to appeal to the House of Lords was refused

LANDLORD AND TENANT
RENT ACTS

Court of Appeal
November 27 1986
(Before Lord Justice KERR and Mr Justice SWINTON THOMAS)

SWANBRAE LTD v ELLIOTT

Estates Gazette February 28 1987

281 EG 916-920

Rent Act 1977, Schedule 1, Part I, para 3 — Claim to statutory tenancy by succession — Claimant was daughter of deceased statutory tenant and only question in dispute was whether she was residing with her mother at the time of and for the period of 6 months before her mother's death — County court judge decided on the evidence that, although the claimant visited and cared for her mother during her last illness, she was not "residing with" her mother during the 6 months preceding the latter's death — Claimant appealed to the Court of Appeal — A number of authorities considered — The facts were that the claimant had a home of her own about two miles distant from her mother's house where she lived with her son, aged 21, her husband having left in 1978 — She visited her mother regularly when the mother became seriously ill and for more than 6 months before her mother's death moved into a spare room in her mother's house in order to look after her, sleeping there, according to the judge's finding, three or four nights a week — The claimant throughout retained the tenancy of her own house, paying the rent and other outgoings and the son continued to live there — Held by the Court of Appeal that the claimant did not satisfy the requirement of "residing with" her mother within the meaning of the Act — The claimant had retained her settled home in her own dwelling-house and had not "made her home" with her mother, whom she had visited on a regular basis to give care in her illness — The fact that the claimant had a permanent home in which she intended to live in the foreseeable future was not *ipso facto* fatal to her claim, but it was a matter to be taken into account in a case which was essentially one of fact and one of degree — The claimant here moved in for a limited time and for a limited purpose and the judge, who had heard all the evidence, was entitled to conclude that the claim had failed — Appeal dismissed

The following cases are referred to in this report.

Collier v Stoneman [1957] 1 WLR 1108; [1957] 3 All ER 20, CA
Foreman v Beagley [1969] 1 WLR 1387; [1969] 3 All ER 838, CA
Hampstead Way Investments Ltd v Lewis-Weare [1985] 1 WLR 164; [1985] 1 All ER 564; [1985] 1 EGLR 120; (1985) 274 EG 281, HL
Morgan v Murch [1970] 1 WLR 778; [1970] 2 All ER 100, CA
Neale v Del Soto [1945] KB 144; [1945] 1 All ER 191
Peabody Donation Fund Governors v Grant (1982) 264 EG 925, CA

This was an appeal by Mrs Sheila Elliott, defendant in an action for possession of a dwelling-house at 49 Wellington Road, East Ham, London E6, from a decision of Judge Dobry at Bow County Court in favour of the landlords, Swanbrae Ltd, present respondents.

T Gallivan (instructed by Wiseman Greenman & Lee) appeared on behalf of the appellant; Roger McCarthy (instructed by Wallace Bogan & Co) represented the respondents.

Giving the first judgment at the invitation of Kerr LJ, SWINTON THOMAS J said: This is an appeal against a judgment of His Honour Judge Dobry given at the Bow County Court on February 18 1986. On that day Judge Dobry ordered that the plaintiff, Swanbrae Ltd, do recover against the defendant, Mrs Sheila Elliott, possession of a dwelling-house, 49 Wellington Road, East Ham, London E6. The plaintiffs are the owners of the freehold of the premises and in October 1948 their predecessors in title granted a tenancy to a Mr Binns, the defendant's father, and Mrs Elizabeth Mary Wood, the defendant's mother. Mr Binns died on September 20 1981 and Mrs Wood died on April 20 1985. The plaintiffs issued their particulars of claim in the Bow County Court claiming possession on August 12 1985. The defence was filed on September 3 1985. In the defence and the further and better particulars thereto dated December 9 1985 the defendant claimed that, by reason of the provisions of Part I of Schedule 1 to the Rent Act 1977, she was the successor to her mother, Mrs Wood, and was accordingly entitled to a statutory tenancy of the premises. It was and is common ground that she is entitled to such a tenancy if, pursuant to para 3 of Part I of Schedule 1 to the Rent Act 1977, she was a member of the original tenant's family who "was residing with him at the time of and for the period of six months immediately before his death". There was no dispute that the defendant was a member of Mrs Wood's family or that, to use a neutral term, she had been staying at the premises for a period in excess of six months prior to her mother's death on April 20 1985. There was no dispute that Mrs Wood was the statutory tenant of the premises. The sole dispute was whether Mrs Elliott "was residing with" her mother for the requisite period.

Mrs Elliott lived at 49 Wellington Road as her home with her parents until she married at the age of 21. In about 1971 or 1972 she and her husband went to live at 4 Gainsborough Avenue, Manor Park, London E12. Mr Elliott was the tenant of those premises. He left in about 1978, leaving Mrs Elliott and their son, who is now aged 21, living at the premises. In 1983 Mrs Wood became ill with cancer. 4 Gainsborough Avenue and 49 Wellington Road are about two miles apart from one another and after her mother became seriously ill Mrs Elliott visited her regularly. In September 1984 she moved into 49 Wellington Road, at least on a part-time basis, in order to look after her mother. She retained the tenancy of 4 Gainsborough Avenue and her son continued to live there.

At this point the judge's findings of fact are, in my view, important. In relation to 4 Gainsborough Avenue the judge said: "No doubt she has a secure home there but I am not dealing with Gainsborough Avenue except indirectly." He found Mrs Elliott to be a truthful witness and said that she looked at 49 Wellington Road as her home in the sense that her mother and her parents had lived there and she had lived there since the age of 3. I think that the judge clearly meant that Mrs Elliott regarded 49 Wellington Road as her home prior to her marriage. In relation to the crucial period between September 1984 and Mrs Wood's death in April 1985, the judge said:

Mrs Elliott tells me and I entirely accept it that in September 1984, she "moved" to stay and sleep upstairs in the spare bedroom at 49 Wellington Road. I find as a fact that she slept there regularly from September until her mother's death at least three to four nights a week but her evidence puts it a little higher, five or six nights a week over the material period.

We have notes of the evidence given, but it has to be borne in mind that they are notes only. Mrs Elliott is recorded as having said in evidence that from September 1984 onwards she spent three or four nights per month only at Gainsborough Avenue. The amount of time which Mrs Elliott spent at 49 Wellington Road was very much in issue and a Mrs Savage, who is a warden employed by the local authority, said that on a visit which she made to Mrs Wood Mrs Elliott had told her that she lived at Manor Park and did not live at Wellington Road. That evidence is recited by the learned judge and was clearly accepted by him. Although the learned judge accepted Mrs Elliott's evidence as a matter of generality, he was, in my judgment, clearly entitled on the evidence to come to the conclusion, having heard all the evidence, that Mrs Elliott slept at 49 Wellington Road from September 1984 onwards three or four nights a week as opposed to the five or six nights a week stated by her. Mrs Elliott continued to pay the rent and the outgoings on the home at Gainsborough Avenue. Her post continued to be sent to 4 Gainsborough Avenue. When giving the relevant information to the Registrar of Births and Deaths she gave her "usual address" as 4 Gainsborough Avenue, Manor Park, E12. It was in these circumstances, put shortly, that the learned judge had to resolve whether or not for the material period Mrs Elliott was "residing with" her mother at 49 Wellington Road.

I now turn to consider the relevant law. Section 2(1)(a) of the Rent Act 1977 provides:

After the termination of a protected tenancy of a dwelling-house the person who, immediately before that termination, was the protected tenant of the dwelling-house shall, if and so long as he occupies the dwelling-house as his residence, be the statutory tenant of it.

I have already recited the relevant provisions of para 3 of Part I of the First Schedule.

In *Collier* v *Stoneman* [1957] 1 WLR 1108, the headnote reads:

L, who was the grandmother of the plaintiff, was the statutory tenant of a flat consisting of two rooms and a kitchen on the first floor of a dwelling-house. In 1950 L allowed the plaintiff and her husband, who had no matrimonial home, to live in the back room of the flat, in which they had been allowed to store their furniture since 1947. The arrangement was that the plaintiff and her husband had the back room to themselves. L had the front room to herself and the kitchen was shared between them. L, who was 94 at the time of her death in October 1956, kept very much to her own room and did most of her own shopping and cooking and only occasionally shared a meal in the kitchen with the plaintiff and her husband.

In an action by the plaintiff against the defendant landlord claiming a declaration of her title to a statutory tenancy of the flat on the same terms as that held by L at the time of her death, the county court judge held that, on the facts as found, the plaintiff was not "residing with" L at the time of her death within the meaning of section 12(1)(g) of the Increase of Rent and Mortgage Interest (Restrictions) Act, 1920, and dismissed the action.

On appeal:—

Held, that inasmuch as L had not granted a sub-tenancy of the back room to the plaintiff, the plaintiff was at all material times "residing with" her grandmother within the ordinary meaning of those words, which was the meaning to be placed upon them for the purposes of section 12(1)(g), and that, accordingly, she was entitled to the relief claimed.

Jenkins LJ cited a passage in a judgment of Lord Evershed MR in *Neale* v *Del Soto* [1945] KB 144 in which the Master of the Rolls said that the words "residing with" must be given their ordinary popular significance. They do not involve any technical import or have some meaning only to be defined by lawyers. The fact that the words must be given their ordinary natural meaning does not necessarily make the task in a case such as the present case any easier. At p 1113 Jenkins LJ said:

The arrangement being of this nature, it seems to me that the Colliers were plainly residing with Mrs Langshaw according to the ordinary meaning of that expression.

Sellers LJ at p 1118 said:

The grandmother, as tenant, had control of the premises, and I find it difficult to see how, without a tenancy of their own, the plaintiff and her husband, making their home there, could be said not to be residing with grandmother up to the date of her death.

There is, in my view, an important factual distinction between that case and the present case in that, unlike the Colliers, Mrs Elliott does have a tenancy of her own. Further, in my judgment, the words used by Sellers LJ "making their home there" are important words.

In *Foreman* v *Beagley* [1969] 1 WLR 1387, the headnote is as follows:

A widow, the statutory tenant of a dwelling-house, was in hospital for the last three years of her life. Her son first came to her flat to air the premises, but eventually appeared to take up residence there for the last year of her life. After she died he claimed to be entitled to remain in possession of the flat as statutory tenant and second successor under Schedule 1 paragraph 7 to the Rent Act 1968. The judge made an order for possession against him, on the ground that he was not "residing with" his mother at the time of and for the period of six months immediately before her death.

On the defendant son's appeal:—

Held, dismissing the appeal, that the defendant had not been "residing with" his mother within the meaning of para 7 of Schedule 1 to the Rent Act 1968, and, therefore, he could not be regarded as the second successor and statutory tenant of the dwelling-house.

At p 1391 Russell LJ (as he then was) said:

It is never very wise in these cases to generalise; but at the least it seems to me that in the phrase in this context the alleged second successor must be able to point to his situation as being a member of the tenant's household.

Then, a little later:

. . . there is insufficient evidence to show that the defendant moved in and was permitted by his mother to move in with a view to establishing a household.

Sachs LJ at p 1392 said:

To my mind the words "residing with" import some measure of factual community of family living and companionship.

Then later:

"residing with" is something more than "living at" even when the premises become a person's normal postal address.

In *Morgon* v *Murch* [1970] 2 All ER 100, the headnote is:

The defendant's father was the statutory tenant of a dwelling-house owned by the plaintiff. When the father died, his widow, the defendant's mother, became first successor to the tenancy. The defendant had originally lived in the house with his parents but in 1955 after his marriage he had moved to a council house. On 4th November 1967, the defendant left his wife and children and went to live with his mother. In April 1968, the wife obtained an order from the justices who found that the defendant had been guilty of cruelty to the wife and had deserted her on 4th November 1967. Despite requests by his wife to do so, the defendant took no steps to have the tenancy of the council house transferred to her name. On 27th June 1968, the defendant's mother died and the defendant continued to live in the house. The plaintiff brought proceedings in the county court and obtained an order for possession of the house. The defendant appealed. On the question whether the defendant had been residing with his mother for the period of six months immediately preceding her death so as to enable him to succeed to the statutory tenancy by virtue of para 7 of Sch I to the Rent Act 1968,

Held — The word "reside" in the 1968 Act must be given its ordinary, natural, common language meaning (see p 103 d, post); the defendant was living with his mother in the sense that he was making his home with her, during the material period, during that time he had made no move to become reconciled with his wife and there had been no immediate prospect of his return to the matrimonial home; accordingly he was entitled to succeed to the statutory tenancy.

At p 103 Winn LJ said that in his view the word "reside" was synonymous with the words "live at". That is, of course, contrary to the view expressed by Sachs LJ in *Foreman* v *Beagley* and, if it is necessary for me to do so, and with the greatest respect, I agree with the view of Sachs LJ that "residing with" is something more than "living at". Winn LJ then continues:

In my opinion there is no doubt at all that the learned judge here made a finding, or drew an inference, which he was only qualified to draw, or purported to be qualified to draw, as a lawyer, since he posed to himself a test which he was regarding as a test having validity in law. He was not merely finding an inference of fact from primary facts established by his own findings. In my personal opinion he erred in that he took, as his crucial test, and the dominant criterion, the question whether or not the defendant had abandoned, and intended no longer to return to, his own former living place or — I have to use the word — residence.

The judge posed to himself correctly, and then later, in my respectful opinion, incorrectly, the problem which he had to answer. He said:

". . . the problem now is whether the defendant is entitled to claim that he was residing with her at the time and had been for six months before her death . . ."

That is a correct statement of the problem. Later he said ". . . the problem is whether for six months he has been occupying the dwelling as his residence" He said again, still later in the judgment:

"The important word is residing which requires reasonably permanent residence — seven months' residence was not enough unless he gave up his own home. Had the defendant given up his home?"

In my respectful opinion, with all respect to the learned county court judge that is a misdirection since a man may, in law, have more than one place in which he resides as well as more than one place in which he lives. He may have a house in London, he may have a house in the country, he may have a house in New York, he may have a house also at Florida, he may have a house in New England, and it may well be that it would be right to say of him in fact and in law that he resides in each of those places, if he spends time in them of which it can be postulated that the time which he spends there, and the intention which he has when he spends it, is more than is comprised in and directed to the paying of a temporary visit.

Peabody Donation Fund Governors v *Grant* (1982) 264 EG 925 is case decided under the provisions of section 30 of the Housing Act 1980, which provides that a person is qualified to succeed the tenant under a secure tenancy if he occupied the dwelling-house as his only or principal home at the time of the tenant's death and either (a) he is the tenant's spouse or (b) he is another member of the tenant's family and has resided with the tenant throughout the period of 12 months ending with the tenant's death. The defendant lived with her mother and stepfather in Salisbury. Her father became ill and the defendant went up to live with him for a large proportion of the week. She said in evidence, which was accepted, that she came to regard the flat in London as being "her home". It should be noted that she did not have a home of her own. The judge found that the defendant occupied the flat as her home at the time of her father's death. He was satisfied that there was a sufficient measure of, as he put it, factual community of family living and companionship to constitute residence with the father. The Court of Appeal held that the judge had not erred in any way in the conclusions which he reached on the facts of that case.

In *Hampstead Way Investments Ltd* v *Lewis-Weare* [1985] 1 WLR 164, the headnote reads:

In 1970 the statutory tenant of a flat married, and his wife and stepchildren came to live with him there. In 1978 they purchased and moved into a house nearby. The tenant, however, retained one room in the flat for the sole purpose of sleeping there five times a week on his return from work at a night-club in the early hours of the morning, so as not to disturb his family. He paid the rent and all the outgoings apart from the gas bill which was paid by his adult step-son who occupied the remainder of the flat. The tenant kept his clothes in his room and had his mail addressed to the flat but never had any meals there. The landlord claimed possession of the flat on the ground that the tenant no longer occupied the flat as his residence within the meaning of section 2(1)(a) of the Rent Act 1977. In the county court, the judge dismissed the application but the Court of Appeal allowed the landlord's appeal and made an order for possession.
On appeal by the tenant and stepson it was:
Held, dismissing the appeal, that while a person could occupy two dwelling-houses "as his residence" within the meaning of section 2(1)(a) of the Act of 1977, where he occupied one of them only occasionally or for limited purposes, it was a question of fact and degree whether he occupied it "as his residence"; that the tenant's limited use of the flat was insufficient to amount to occupation of it as a residence and accordingly his tenancy was not protected.

When considering an appeal from a judge of first instance who has heard the evidence it is very important, in my view, for this court to bear in mind that questions of "residence" and "residing at" are very much ones of fact and degree. A judge must view the quality of the residence alleged and come to a conclusion upon the totality of it as to whether in truth it falls within the proper usage of the term "residing with".

In his notice of appeal on behalf of the defendant/appellant Mr Gallivan makes a number of criticisms of the learned judge's judgment. On p 3 of his judgment the judge quotes a passage from *Morgon* v *Murch (supra)* and poses what is, in my view, the correct test, namely whether the defendant is entitled to claim that she was residing with her (the mother) at the time and had been for six months before her death. He then goes on in the next passage to say:

I should add that it was submitted by counsel for the plaintiff in my view correctly that if a person intends to return to her abode or does not make a decision as to her future while living at her mother's house she cannot be classified as a person "residing with" her mother.

Mr Gallivan criticises that passage and says that it is wrong and that therefore the judge posed the wrong test. Taking the passage alone as it stands, it is open to some criticism. If a person intends to return to her abode, in this case 4 Gainsborough Avenue, that may well be a relevant factor in deciding whether she is "residing with" her mother at 49 Wellington Road. It may also be a factor that she had not "made a decision as to her future while living at her mother's house". However, as was made clear in *Morgon* v *Murch,* the defendant does not have to establish that she has been residing with her mother with an intention to reside there indefinitely in the sense that she would not intend ever to move away. She has to establish merely that for the relevant period she was residing with her mother. I shall say a word or two more about intention below. As I have said, the learned judge posed the correct test in the immediately preceding paragraph and I am satisfied that what the judge meant in the passage which is criticised is that Mrs Elliott did indeed have a settled abode and residence at 4 Gainsborough Avenue, and that if in those particular circumstances she intended to return to her abode and had not made a decision as to her future then she could not be classified as a person residing with her mother.

On the next page of his judgment there is a further passage which is criticised by Mr Gallivan. The learned judge said: "I cannot disregard the fact that she had another home which she maintained as her sole residence." I think, with respect, that some criticism might be made of the word "sole". However, as was pointed out by my lord Kerr LJ in the course of submissions, it is quite correct that there was one house only which Mrs Elliott maintained in the sense of paying the rent, rates, and outgoings, namely 4 Gainsborough Avenue.

Mr Gallivan further criticises the learned judge's finding that "residing with" connotes an element of intention. Winn LJ in *Morgon* v *Murch* specifically refers to "the time which he spends" there and the intention which he has when he spends it". The words of Sellers LJ, "making their home there", the words of Russell LJ, "being a member of the tenant's household" and the words of Sachs LJ, "some measure of factual community of family living and companionship" all suggest an element of intent. Although intent is, in my judgment, a relevant factor, none the less the question of whether or not it has been established that the defendant was or was not residing with her mother over the relevant period must be judged objectively on all the facts of the case.

The learned judge said on p 4 of his judgment:

I have to decide on the evidence of Mrs Elliott whether the test of intention and residence with her mother is satisfied. I regret that I have to find Mrs Elliott was not a resident within the meaning of the relevant paragraph. I find that she came to live there with intention of helping her mother as a dutiful daughter would. At that stage and at no time until after her mother died did she form an intention to reside in those premises.

Then a little later:

The question whether for the purpose of the relevant paragraph a person "residing with" can have another residence does not arise as I have found that Mrs Elliott did not reside with her mother within the meaning of the Act because she did not have the necessary intention.

Was the learned judge right? I have not found this case an easy one and I confess that my mind has wavered from time to time in the course of argument. I have no doubt that this was in part caused by the clarity and, if I may say so, ability with which the submissions on both sides were made.

The *Oxford English Dictionary* defines "reside" as "having one's home, dwell permanently". Clearly, for reasons already canvassed, the words "reside with" in the context in which they are used in the Rent Act 1977 do not mean dwell permanently in the sense of dwell indefinitely. They certainly mean something more than dwell transiently and to my mind they have the connotation of having a settled home. A person may reside with a relevant relative for the requisite period but none the less have an intention to move away at some later stage. However, I do think, with Sellers LJ in *Collier* v *Stoneman,* that the words "have one's home" are very helpful. A person may well, of course, have more than one home, although he does not usually do so. As Winn LJ said in *Morgon* v *Murch,* a person may have more than one residence. In my view the person claiming the statutory tenancy must show that he or she has made a home at the premises which they are claiming and has become in the true sense a part of the household. In this case Mrs Elliott had lived at 49 Wellington Road for a limited period. She did not spend all her time, by any means, at that address. She had a settled home at 4 Gainsborough Avenue. She went to 49 Wellington Road for the purpose of caring for her mother who was ill. Her son remained at Gainsborough Avenue. Having considered this case with great care during the submissions and for some period of time since, I have come to the conclusion that it was not established by Mrs Elliott that she had made a home at 49 Wellington Road or that in any true sense she had become part of her mother's household there. Accordingly, in my judgment the judge was right to conclude that she was not residing with her mother within the meaning of the relevant paragraph. Further, it always has to be borne clearly in mind in a case of this nature that a judge sitting in the county court has heard all the evidence and neither the notes which he takes nor his judgment can encompass the totality of the evidence. In a case such as this, which is essentially one of fact and one of degree, the judge of first instance has an opportunity of making a judgment on the evidence which is denied to this court. Despite the formidable arguments put forward by Mr Gallivan, I do not think it has been shown that the learned judge was wrong and I would dismiss this appeal.

Agreeing, KERR LJ said: The feature which distinguishes this case from all the authorities to which Swinton Thomas J has referred is that Mrs Elliott undoubtedly had her own home at 4 Gainsborough Avenue, Manor Park, throughout the period that she claims that she was residing with her mother so as to satisfy the statutory requirement on which she relies. Her mother's house, 49 Wellington Road, East Ham, was not her home in any relevant sense of the word, but only in the sense that it had been her childhood home, that some of her belongings were still there, and that she was a constant visitor, since the two addresses were only about two miles apart.

The fact that Mrs Elliott had a permanent home of her own, in the sense that she evidently could, and intended to, live there for the foreseeable future, that her son and furniture were there, and that she had no plans at any relevant time for giving it up, is not *necessarily* fatal to her claim that she was nevertheless "residing with" her mother during the six months before her mother's death. As pointed out by Winn LJ in *Morgon* v *Murch,* a person can have more than one home and more than one residence. But the existence and continuing availability of Mrs Elliott's "permanent" home, simultaneously with the claimed residence elsewhere, distinguishes the present case drastically on its facts from *Morgon* v *Murch*. Such a state of affairs is bound to render it far more difficult for a defendant to satisfy the test of having "resided with" a member of his or her family for the necessary period at the same time. The reason is that "residence" must connote more than physical presence during the required period, albeit as a member of the household. This is consistent not only with the dictionary definition of "reside" which Swinton Thomas J has cited but also with the social purpose of the legislation. I think that the *Peabody* case is much closer to the present case on its facts than *Morgon* v *Murch,* since the defendant was there to some extent based in two premises at the same time, though neither was owned or rented by her, and she was effectively in the process of transferring herself from one to the other. That was a borderline case (see *per* Sir David Cairns), but the facts of the present case are much less favourable to Mrs Elliott. While I certainly do not exclude the possibility that a defendant who has his or her own home — in the fullest sense of the word — elsewhere may nevertheless satisfy the test of this legislation, such cases will inevitably be rare.

The judge regarded Mrs Elliott as a "visitor", for want of a better word. One of the dictionary meanings of "visit" is "temporary residence with person or at place". For want of a better word, I think that Mrs Elliott was a visitor, a temporary resident, but without having made her home with her mother, within the ordinary and dictionary meaning of "residing with" her. Her position can hardly be put better than she did herself, entirely frankly, in her evidence, when she said: "I moved in with my mother for so long as was necessary." Having regard to the existence and availability of 4 Gainsborough Avenue, to which she returned for odd days and nights throughout the six-month period, that sentence describes no more than what one would usually refer to as "staying with her mother" in order to look after her. She moved in for a limited time and for a limited purpose. In my view, that is clearly not sufficient and I would equally dismiss this appeal.

The appeal was dismissed with costs, not to be enforced without further order; legal aid taxation of defendant's costs.

Queen's Bench Division

January 28 1987

(Before Mr Justice MCCULLOUGH)

R v BRISTOL RENT ASSESSMENT COMMITTEE, EX PARTE DUNWORTH

Estates Gazette April 11 1987

282 EG 214-218

Rent Act 1977 — Fair rent — Decision of rent assessment committee challenged by tenant — Application for judicial review — Rent officer had determined a rent of £1,068 per annum plus £168 for services — Committee subsequently fixed a rent of £2,000 per annum, with the same amount of £168 for services — It was the intervening events and procedure which gave rise to the issue in this case — Communications from the tenant to the committee were ambiguous as to whether she wished to withdraw her objection or was merely stating that she did not wish to attend a hearing — Unknown to committee the tenant had written a letter which did indicate an intention to withdraw, but this letter never reached the committee, having been given to a friend who failed to post it — Committee made efforts to clarify the tenant's intentions but eventually decided that she did not intend to withdraw and notified her of the date and arrangements for the hearing — Even at this stage there was some ambiguity — Tenant said on a reply card that she did not intend to be present but was willing to make arrangements for an inspection — She also wrote a letter which repeated that she would not attend, made comments on increases of rent, but was otherwise ambiguous — Committee proceeded with the hearing in the absence of the tenant — Despite the fact that the landlord supported the rent officer's determination, the committee increased the fair rent to the figure mentioned above, stating in their reasons that the flat had been substantially undervalued — Committee had been unable to inspect the flat as the tenant was not there to admit them — McCullough J, after referring to the principles governing a withdrawal, as established in *Hanson* **v** *Church Commissioners for England***, said that the essential question to be decided in the light of** *Associated Provincial Picture Houses Ltd* **v** *Wednesbury Corporation* **was whether the committee's decision to proceed to a determination in the absence of the tenant and without an inspection was one which no reasonable rent assessment committee in the circumstances could reasonably have arrived at — This question had to be answered by considering whether the only conclusion which the committee could reasonably have reached was that the tenant had indicated an intention to withdraw her objection, or that what she had written was so ambiguous that, without further inquiry, she could no reasonably be taken not to have indicated such an intention — Held by McCullough J, having reviewed the correspondence, that these questions could not be answered in the tenant's favour so as to enable the committee's decision to be quashed — It was hoped that the tenant would appreciate that the court had no power to consider the fairness of the rent — Nor could it be swayed by its considerable sympathy over the mishap of the letter which was never posted showing the tenant's intention to withdraw — Application dismissed**

The following cases are referred to in this report.

Associated Provincial Picture Houses Ltd v *Wednesbury Corporation* [1948] 1 KB 223; [1947] 2 All ER 680, CA
Attorney-General of Hong Kong v *Ng Yuen Shiu* [1983] 2 AC 629; [1983] WLR 735; [1983] 2 All ER 346, PC
Hanson v *Church Commissioners for England; R* v *London Rent Assessment Committee ex parte Hanson* [1978] QB 823; [1977] 2 WLR 848; [1977] 3 All ER 404; (1976) 34 P&CR 158; 241 EG 683, CA

This was an application by Mrs Coral Marjory Dunworth, tenant of Flat 13a at 13 Lansdown Crescent, Cheltenham, for judicial review of a decision of a Bristol rent assessment committee determining the fair rent of the flat under the Rent Act 1977. The applicant sought to have the decision quashed or, alternatively, declared unlawful.

R Gordon (instructed by Willans & Gyles, of Cheltenham) appeared on behalf of the applicant; N P Pleming (instructed by the Treasury Solicitor) represented the respondent committee.

Giving judgment, MCCULLOUGH, J said: This is an application by Mrs C M Dunworth for judicial review of the decision of a Bristol rent assessment committee made on August 9 1985, and communicated to her by a letter of August 12 1985, determining a fair rent in respect of premises occupied by her at Flat 13a, 13 Lansdown Crescent, Cheltenham.

On February 11 1985 the rent officer, upon application by her landlord, decided that a fair rent would be £1,068 per annum plus £168 per annum for services, to be effective from April 10 1985. The applicant objected to those figures and said so.

The issue fell to be determined by the respondent committee. The original hearing was to have been on June 5 1985 in the Montpellier Room, Municipal Offices, Cheltenham. On May 22 1985 Mrs Dunworth was notified of the date and of the fact that the committee would visit her premises on the same day.

She wrote to the clerk to the committee on May 27 saying *inter alia*:

I am unable to attend the hearing on Wednesday June 5 1985 as I will be away, and I wish not to leave the key of the premises to any other person, whilst being on vacation.
Also I do not have anyone to represent me, as I cannot afford it; knowing full well I don't stand a chance of winning this case, so there's no point wasting your's and my time.

It not being apparent to the committee's officers whether or not the applicant wished to withdraw her objection, she was written to on June 20 1985 in the following terms:

I refer to your letter of May 27 1985.
I should be grateful if you could advise me whether or not you want the rent assessment committee to proceed with this case? If you want to withdraw your objection to the rent officer's registration, the committee will not consider the matter, and the rent will remain as registered by the rent officer.
If you do want the rent assessment committee to consider this case, then a date will be arranged for the inspection and hearing in July.
Please let me know as soon as possible whether or not you wish to withdraw your objection.

Mrs Dunworth then wrote a letter dated July 25 1985 to the committee which said:

Received your letter June 21 1985 asking me if I want to withdraw my objection to the rent officer's registration.
I do not wish to pursue this case, I didn't expect it to go this far, all I wanted was an explanation.

Her reference to the letter of June 21 1985 must be to the one written to her by the committee dated June 20 1985. Mrs Dunworth then gave the letter which she had written to a friend and asked her to post it. Unfortunately her friend lost it in her handbag and did not post it; she forgot all about it and so it never reached the committee.

Not having received any reply to their letter of June 20, the committee wrote again on July 10 1985 in the following terms:

In the absence of a reply to my letter to you of June 20 1985, I assume that you still require the rent assessment committee to consider this matter.
I will advise you in due course of the arrangements for inspection of the property, which will now take place in August.

The next thing that happened was that on July 26 1985 the clerk to the rent assessment committee wrote to Mrs Dunworth as follows:

The Rent Assessment Committee considering the fair rent for the premises named above will hold a hearing. It will be held at No 2 Magistrates' Court, St George's Road, Cheltenham, on Friday, August 9 1985 at 2.45 pm. You have a right to speak at the hearing or to be represented by someone else. If you intend to be represented please let me know as soon as possible.
The committee will visit the premises on the same day at 11.15 am. If you are the tenant would you please make sure that someone is there to let them in. You are welcome to attend but they cannot hear any new arguments or evidence during the visit. Its purpose is simply to let them see the premises for themselves.

At the foot of the page was written:

Please complete and return the enclosed card, in the reply envelope provided, as soon as possible.

Mrs Dunworth did complete the card, which reads:

acknowledge receipt of the notice of inspection/hearing and will arrange for the members to be shown over the premises when they arrive. I intend/I do not intend to attend the hearing. Signed C Dunworth. Date 31-7-85.

The whole of the card, apart from the applicant's signature and the date, is printed. There are asterisks against the words "I intend" and "I do not intend" and a note below indicating "delete as appropriate".

The card was received by the committee on August 2 and on the same day the clerk to the committee received a further, undated, letter from Mrs Dunworth which said:

Dear Mrs Collins, Reference to your Rent Assessment Form, saying you wish me to be at the court hearing Friday, August 9.
I am unable to attend, as this wasn't my intention for this case to go as far as the Magistrates' Court, my nerves couldn't take it.
I was just trying to state that I couldn't understand why the rent go up so much every year, when I was told when I was visited before moving from my old flat to a new flat, that the rents would go up for aleast 5 years.
This is what was lead to understand.

I understand that rents do go up from time to time, but surely not every year. I trust you understand what I am trying to say.
Yours sincerely, C Dunworth (Mrs).

That was the material before the committee on August 9 1985 when they decided to go ahead and make a determination of the fair rent. The figure which resulted from that determination was £2,000 per annum with £168 per annum for services. The landlord was in fact content with the figure which had been determined by the rent officer but the committee decided of its own volition, for reasons set out in its determination, that this significantly higher figure should be substituted. The committee wrote and told Mrs Dunworth so and a statement of reasons was sent in due course.

In fairness to the committee I will read those reasons:

Rent assessment committee was unable to inspect the flat which was the subject of the objection by the tenant due to the tenant's failure to be present to admit the committee at the time arranged for inspection. As this was the second occasion that this has happened and furthermore in spite of intervening correspondence sent by the BRAP Office to the tenant the committee decided to proceed with the application without an inspection and in the absence at the hearing of the tenant.
The committee was, however, able to inspect a similar flat on the ground floor of 13 Lansdown Crescent in the presence of a representative of the landlord who indicated that that flat was similar in all respects to the subject flat save that as it was on the top floor the ceilings were at a reduced level.
As a result of the said inspection it was concluded that the subject flat was a self-contained flat of well above average proportions providing above standard accommodation comprising a living room, three bedrooms, a kitchen, bathroom, separate WC, two stores, hall and linen cupboard. The standard of fittings throughout the flat and, in particular, in the kitchen were of high quality.
The committee considered the representations made on behalf of the landlord who supported the rent officer's assessment. However, the committee considered that his figure of £1,068 per annum was a very substantial undervaluation of this flat particularly having regard to comparable flats where committee decisions had been registered and especially having regard to rents registered the same day by the committee in Lansdown Crescent. This was a high quality flat in a top residential area of Cheltenham but convenient to the centre of Cheltenham and its shopping facilities.
By applying the requirements of section 70 of the Rent Act 1977 and using its own knowledge and experience the committee determined a fair rent of £2,000 per annum.
The committee included in that sum an amount for services of £168 per annum being fully satisfied that such a sum represented quite a modest charge for the services provided by the landlord which were listed in the schedule supplied and which were described in some detail by the landlord's representative.

That is the decision which this court is invited to quash or, alternatively, declare to have been unlawful.

The first point made by Mr Gordon, who appeared on behalf of Mrs Dunworth, is that the letter from the committee of June 20 1985 created a legitimate expectation that if she were to tell the committee that she wanted to withdraw her objection the registered rent would stand. Although he did not go on to spell it out, the effect of his submission was that if, with this expectation, she thereafter told the committee that she did want to withdraw her objection and the committee none the less went ahead with its determination that decision could be successfully challenged in this court.

Mr Pleming for the committee was inclined to accept that this letter did create such an expectation. However, I, for my own part, am hesitant to accept that a determination following an indication of withdrawal would inevitably be open to a successful challenge in this court. I have in mind the words in form RR 102 (Rev 1982), which was sent to Mrs Dunworth by the committee on March 5 1985. This is a standard form running to four pages. One section, headed F, reads as follows:

Withdrawing an Objection. Once you have made an objection to the rent fixed by the rent officer you do not have an automatic right to withdraw it if you change your mind. You will normally be allowed to withdraw if the other party agrees, unless the committee think that there are good reasons why they should continue with the case. Even when you have told the committee that you wish to withdraw your objection, you should not assume that this has been agreed to until they tell you.
If you try to withdraw your objection a short time before a hearing is due to take place, there may not be enough time to consult the other party. The committee will continue with the hearing but they may allow you to withdraw your objection at the beginning of the hearing if the other party agrees.

These words reflect the decision of the Court of Appeal in *Hanson v Church Commissioners for England* and *R v London Rent Assessment Committee, ex parte Hanson* [1978] QB 823, and in particular the words of Lord Denning MR at pp 832 G to 833 E, about

A withdrawal by one party from a dispute in which there is a public interest in addition to the interest of the parties. (See also Roskill LJ at p 836 B to E and Lawton LJ at p 839 C to F.)

In this connection it is material also to note a passage in the judgment of the Privy Council in *Attorney-General of Hong Kong* v *Ng Yuen Shiu* [1983] 2 AC 629 delivered by Lord Fraser of Tullybelton at p 638 E to F, namely:

When a public authority has promised to follow a certain procedure, it is in the interest of good administration that it should act fairly and should implement its promise, so long as implementation does not interfere with its statutory duty.

B My doubt stems from those last eleven words, for it may be that what was said in the letter of June 20 1985 would have interfered with the statutory duty cast upon the committee. However, it is not necessary to say more than I have about that submission by Mr Gordon.

The essential question which I have to decide is one to be determined according to the well-known *Wednesbury** principles: was the decision of the committee to proceed to a determination on August 9 1985 in the absence of Mrs Dunworth and without an inspection one which no reasonable rent assessment committee could in the circumstances reasonably have reached?

Since the point at issue concerns whether or not the tenant had indicated or had tried to indicate that she wished to withdraw her objection to the registered rent, the question is to be answered by
C considering whether the only conclusion which could reasonably have been drawn by the committee was that this tenant had indicated an intention to withdraw her objection by August 9 1985, or that what she had written was ambiguous in the sense that, without further inquiry of her, she could not reasonably be regarded as someone who had not so indicated. I phrase the matter in that way because if a tenant who has set in train the process of determination by a rent assessment committee wishes that process to be halted short of such a determination it is for the tenant to indicate that such is his or her wish.

No separate argument was addressed to this court by Mr Gordon on the basis that, although it might have been proper for the
D committee to make a determination, they could not properly have made one without an inspection of the flat.

Looking at the correspondence, the position seems to me to be as follows. Mrs Dunworth's letter of May 27 1985 gave rise to the question of whether or not she was indicating her wish to withdraw her objection; hence the letter from the committee of June 20 1985. No reply having been received by the committee, they wrote on July 10 1985 telling her what they assumed to be the position, in other words, that her objection had not been withdrawn. It is to be noted that she did not write in reply and say, "Contrary to your assumption, I do want to withdraw it". Nor did she write and say, "But did you not get my letter of June 25 1985 telling you that I did want to withdraw?" What followed from her was silence.
E Then on July 26 she was told that the hearing and the visit were both to be on August 9 1985. Her subsequent response was in the form of two documents. The first was the card which indicated two things: her preparedness to have her flat inspected and the fact that she would not attend the hearing. The second document was her undated letter which was received on August 2 1985. That letter also says, effectively, "I am not going to attend". It then goes on to set out her point of view about an increase in her rent.

I do not regard it as unreasonable for the committee not to have taken her statement that she had not intended the case to go as far as the Magistrates' Court as an indication that she did not want it to be determined by the rent assessment committee. Nor do I regard it as unreasonable for the committee not to have read the word "was"
F (the second word in the third paragraph of her letter) as indicative of a consideration which she previously had desired to have taken into account but no longer did.

In all the circumstances I am not able to answer the relevant question in the way which would enable this determination to be quashed.

I do hope that the applicant realises that this court has no power to consider the fairness of the rent which was determined. Nor could this court be swayed by its considerable feeling of sympathy for her because, unknown to the committee, she did in fact want to withdraw

*Editor's note: *Associated Provincial Picture Houses Ltd* v *Wednesbury Corporation* [1948] 1 KB 223.

her objection and wrote a letter to that effect intending that it should be posted. It was no fault of hers that it was not. Having written it, she believed that her objection would be withdrawn and that the registered rent of £1,068 per annum would stand. She then found that, despite the fact that her landlord had asked for no further increase, the committee determined that her registered rent was to rise by a further £932 from £1,068 per annum to £2,000 per annum. If that is an unfair result, I can see no assistance for her other than that she might try to prevail upon her landlord's kindness and consideration in not exacting the full measure of what the law entitles them to demand of her.

The application must be dismissed.

No order for costs except legal aid taxation.

Court of Appeal
February 23 1987
(Before Lord Justice DILLON and Lord Justice GLIDEWELL)
BOSTOCK v TACHER DE LA PAGERIE

Estates Gazette May 23 1987
282 EG 999-1001

Rent Act 1977 — Case 9 in Schedule 15 — Appeal by tenant against county court judge's order for possession and rent arrears — Tenant also complained of the refusal of an adjournment and of a refusal to stay execution in regard to the rent arrears pending the separate trial of a counterclaim for illness alleged to be due to a defectively installed boiler — The main ground of appeal was a point concerning the landlord's title and the application of *McIntyre* v *Hardcastle* — The landlord was the owner of the flat in question but he held it under a declaration of trust as trustee for himself and a daughter as joint tenants in equity in equal shares, with an undertaking to transfer the property into their joint names when the daughter attained the age of 18 — His claim for possession was based on Case 9 on the ground that the flat was reasonably required by him as a residence for his daughter, who had attained the age of 18 before the commencement of the proceedings for possession — It was submitted by the appellant tenant that if, before the claim for possession came on, the property had been transferred into the joint names of the respondent and his daughter, then, as a result of *McIntyre* v *Hardcastle*, they could not have obtained an order for possession for occupation of the flat for the daughter alone — It was suggested that Case 9 should be read subject to some such qualification as "required for his daughter in some capacity other than as equitable owner or for his daughter, his daughter not being a beneficial joint tenant" — Held rejecting this submission, that there was no justification for reading the wording with any such gloss — The respondent was the landlord; there was evidence that the flat was reasonably required for occupation by the daughter, who had been medically advised to have a flat of her own, away from the paternal or maternal roof; and it had been amply shown that the greater hardship was on the side of the landlord's daughter — In all the circumstances it had been reasonable to make the order — The court also rejected the appellant's complaints about the judge's refusal of adjournment and refusal to stay execution of judgment for arrears of rent pending the trial of the counterclaim — Appeal dismissed

The following case is referred to in this report.

McIntyre v *Hardcastle* [1948] 2 KB 82; [1948] 1 All ER 696, CA

This was an appeal by Monique Tacher de la Pagerie, the tenant, from a decision of Judge Vick, at Westminster County Court, awarding the landlord, the present respondent, David Ashton Bostock, possession of Flat 5, 91 St George's Drive, London SW and also a sum of £3,800 in respect of arrears of rent. Judge Vick had upheld a ruling given by Judge McDonnell that a counterclaim for damages by the present appellant should be tried separately from the proceedings in regard to possession and rent arrears.

Miss D I Romney (instructed by Simmonds Church Smiles & Co) appeared on behalf of the appellant; Kim Lewison (instructed by Piper Smith & Basham) represented the respondent.

Giving judgment, DILLON LJ said: This is an appeal by the defendant in the original action against a judgment of His Honour Judge Vick, given in the Westminster County Court on September 26 1986, whereby he awarded the plaintiff in the action (the respondent to this appeal) possession of a flat called Flat 5, 91 St George's Drive, London SW1; and also judgment for a sum of £3,800 arrears of rent. The history of the matter is somewhat complicated and it is not helped by the way the papers are included in different bundles. It appears, however, that the respondent bought a leasehold interest in the flat on October 30 1983. In February 1984 he let the flat for two months to the appellant. That was renewed in April 1984 for six months from April 27 1984 and renewed again in October 1984 for a further six months from October 1984 to April 1985.

The respondent, Mr David Bostock, has a daughter, Sophia Ashton Bostock, who attained the age of 18 on September 2 1985. The respondent brought proceedings for possession of the flat under Case 9 of Schedule 15 to the Rent Act 1977 on the ground that the flat was reasonably required by him as landlord for occupation as a residence for his daughter, Sophia.

The court on such an application has to be satisfied that the circumstances are made out and has to consider matters of greater hardship and whether it is reasonable to make such an order. Proceedings to that effect were begun in 1985. The defendant ceased paying any rent for the flat in October 1985. The basis for this was that, it is said on her behalf in a counterclaim subsequently launched, she suffered illness because a boiler in the flat had not been properly installed by the respondent's contractors, and the boiler had further been boxed in, in a way that concentrated escaping fumes, by either those contractors or servants or agents of the respondent. It was said that the result of that was that carbon monoxide gas escaped into the flat and, as a result, she suffered very unpleasant symptoms, migraine and nausea, and intolerance to fumes and smoke, so that she was unable to take up any employment. She therefore ceased to pay rent, claiming by way of counterclaim unquantified damages for alleged negligence and breach of duty on the part of the respondent and also on the part of the contractors who had installed the boiler.

At a fairly advanced stage of the pleadings in the county court it was claimed that there was a right to set off the sums due under the counterclaim against the accruing rent. She had paid no rent since October 27 1985. The pleadings were amended to include a claim for possession under ground 1 on the basis that rent had not been paid. However, in July 1986 there was an order by the registrar of the county court that the claim for possession, that is to say the action, be tried separately from the counterclaim. There was an appeal by the present appellant against that order, which came before His Honour Judge McDonnell on September 15 1986. He dismissed the appeal and upheld the ruling that the counterclaim be tried separately. He was also asked to direct an adjournment of the claim on the ground that witnesses for the present appellant were not available, but he refused to allow any adjournment. The claim had by then been fixed for hearing on September 25 1986. When it came on for hearing on that date, the solicitors who up to then had been acting for the appellant applied to be removed from the record, and that application was granted. The appellant then renewed her application for an adjournment of the hearing, but that was refused and at the end of the trial, on September 26, the judge made the order which I have mentioned.

Since then, there was an application to this court for a stay of the possession order, which I granted in December, over the first week of this term, on the footing that the case would come on for hearing then. It did not. The stay was then discharged and we are told that the possession order has since been executed.

The first ground of appeal which has been put before us today by Miss Romney is that the judge exercised his discretion wrongly when he refused to grant the appellant an adjournment on September 25, after her solicitors had been discharged from the record. It appears that the reasons why she sought a stay then were, first, that the solicitors had ceased to act for her and she was in person; second, that she then intended to apply for legal aid; and, third, that she had not got the witnesses available.

So far as legal aid is concerned, that is a matter which could perfectly well have been considered long before. As to the witnesses' availability, Judge Vick pointed out that that seemed to be the same point that Judge McDonnell had dealt with on September 15. As for the fact that she was acting in person and no longer had solicitors, that was obviously appreciated by the judge. He said:

> The judge is limited in what assistance he can give to a litigant in person, but I will assist if I can.

Whether or not there should have been an adjournment was a matter entirely for the discretion of the judge. For my part I do not take the view that this court should interfere.

The main ground of this appeal arises from the wording in the provisions of the transfer to the respondent of the property. That document transferred a registered title of the flat to the respondent. There was then a declaration that the respondent held the property as trustee for himself and Sophia Ashton Bostock as joint tenants in equity in equal shares; and he undertook to transfer the property into joint names at the date on which Sophia attained the age of 18 years.

It is said that if, before the claim for possession came on, the property had been transferred into the joint names of the respondent and Sophia, then, under the decision of this court in *McIntyre v Hardcastle* [1948] 2 KB 82, the joint landlords could not, under Case 9, have obtained an order for possession for the property to be occupied by Sophia alone. Whether or not that would have been so if the facts had been that way round, however, it is conceded that as no transfer had been executed the respondent is the sole landlord. The appellant as lessee is not concerned with the trust relationship between the respondent and Sophia. She is concerned only in dealing with her own landlord.

Case 9 provides for possession, subject to hardship and reasonableness, where the dwelling-house is reasonably required by the landlord for occupation as a residence for himself or for any son or daughter of his over 18 years of age. That wording is clear in its application to the facts as they are. The respondent does require possession for occupation by Sophia. Indeed, there was ample evidence to support the judge's findings on Sophia's need for this flat. She has suffered from anorexia, there is a history of a broken marriage of her parents, and strong medical recommendations that she needs a flat of her own away from the paternal or, for that matter, the maternal roof. The case of hardship, therefore, was amply made out; and so was the case of reasonableness, subject to the point which is argued on what was called the *McIntyre v Hardcastle* point. It was indeed urged that the judge's finding of fact failed to take proper account of certain documents produced and certain offers of a flat made to the appellant herself. But, as to that, there was evidence to support the findings of the judge: he accepted the evidence of the respondent and this court could not interfere with his findings of fact.

It is urged that the wording of Case 9 should be read subject to a qualification "required for his daughter in some capacity other than as equitable owner or for his daughter, his daughter not being a beneficial joint tenant". I see no justification for reading in any gloss on the wording of the provisions in the Schedule to the Act. She is not the landlord. She is the daughter of the landlord. Therefore, the requirements are satisfied. Equally, I do not for my part see that the fact that she would have been the landlord if a transfer had been executed can be prayed in aid as a ground for saying that it is not reasonable to grant the order for possession under Case 9 in the circumstances of this case.

The case has to be dealt with on the facts as they are, not on the facts as they might have been. There is no evidence, so far as I am aware, to establish that there was any deliberate taking advantage of a loophole in the Act, even if it had been relevant if that had been the case. Therefore, I would reject that main ground of appeal. It is not fully dealt with in the judgment of the learned judge: he does not refer to *McIntyre v Hardcastle,* but he does refer to authorities establishing that, despite any question of a trust element, the respondent is the landlord. He then deals with the provisions of the deeds and says that they do not raise any reason why he should not make the possession order.

It is said that the judge erred on reasonableness because he mentioned that there were the arrears of rent, and he did not take into account the counterclaim which was ultimately pleaded as a set-off; and it is said also that though he may have entered judgment for the arrears of rent, he should have stayed execution of the judgment pending the trial of the counterclaim. Those points, in a way separate, I would take together. One thing to which the judge drew attention is the appalling obscurity of the pleading of the counterclaim. It is very difficult to discern what it is saying, and the damages claimed are wholly unparticularised. It is said there was evidence from the appellant at the trial that she had been unable, because of her illness, to take jobs which would have paid her £10,000 per year. But the pleading does not give any particulars at all, even of special damages, and this is a case where special damage would be the nub of the claim in the matter of damages.

Furthermore, exactly how the claim is put on the matter of liability is obscure, because it is accepted in this court, at any rate by Miss Romney for the appellant, that the respondent could not have been liable for the defective installation of the boiler if he did not know of the defect and left it to independent contractors to install the boiler. There is the further factor that the case appears to be by no means ready for trial, in that even now, we were told, legal aid had not been obtained to prosecute that counterclaim. In all the circumstances, while it may very well be the normal rule and practice that any judgment on a claim which was given before the hearing of the counterclaim should be stayed pending the hearing of the counterclaim, I take the view that in the particular circumstances of this case the judge was entitled to grant judgment on the claim for the arrears of rent, without staying that pending the hearing of this highly nebulous and ill-pleaded counterclaim. It must equally follow that he was entitled to have in mind, though he does not seem to have attached great importance to it in considering the reasonableness of making the possession order, that there were the admitted arrears of rent to the liability for which there was no defence apart from the counterclaim: there were those on the one hand, while the counterclaim was so nebulous and unparticularised that he could not take it into the balance at all.

I would therefore dismiss this appeal.

GLIDEWELL LJ agreed and did not add anything.

The appeal was dismissed, with costs not to be enforced without leave; legal aid taxation and further directions as to appellant's costs.

Queen's Bench Division

November 19 1986
(Before Mr Justice HODGSON)

R v AGRICULTURAL DWELLING-HOUSE ADVISORY COMMITTEE FOR BEDFORDSHIRE, CAMBRIDGESHIRE AND NORTHAMPTONSHIRE, EX PARTE BROUGH

Estates Gazette June 20 1987
282 EG 1542-1546

Rent (Agriculture) Act 1976 — Application for judicial review, challenging proceedings of agricultural dwelling-house advisory committee established under section 29 of the Act — Committee appointed to advise local housing authority on applications for possession — Cottage claimed by applicant to be required to house a person to be employed by him in forestry — Questions as to regularity of proceedings before advisory committee — Question of administrative law as to whether certiorari could go to quash an advisory report when the power to make a binding determination resided in another body — Applicant for judicial review was the owner of some forestry land — Respondent had been in applicant's employment as a forester and occupied a cottage in that capacity — He had been (unfairly) dismissed and given notice to quit, becoming a statutory tenant under the 1976 Act — Applicant sought possession of the cottage for his replacement — There were irregularities during the proceedings of the advisory committee — Each party was heard in the absence of the other and the respondent had made allegations of bad faith against the applicant which the latter had no opportunity to rebut — In addition, the committee's report failed to state adequately why they rejected the application, although required to give reasons for their decision by section 28(6) of the 1976 Act — After considering various authorities, the judge decided that certiorari could go to quash the committee's report — Although the report was advisory and not determinative, there was sufficient proximity between it and the decision of the local authority, which would be likely to be strongly influenced by it, to justify the court in granting relief at this stage — It would be wrong to allow the proceedings to go further and require the applicant to wait until the decision of the local authority was made against him — Committee's report quashed

The following cases are referred to in this report.

R v Boycott [1939] 2 KB 651
R v Electricity Commissioners, ex parte London Electricity Joint Committee [1924] 1 KB 171
R v Metropolitan Police Commissioner, ex parte Parker [1953] 1 WLR 1150; [1953] 2 All ER 717, DC
R v St Lawrence's Hospital Statutory Visitors, Caterham, ex parte Pritchard [1953] 1 WLR 1158; [1953] 2 All ER 766, DC
R v Winchester Area Assessment Committee, ex parte Wright [1948] 2 KB 455; [1948] 2 All ER 552, CA

This was an application for judicial review by Reginald Arthur Brough, with a view to challenging the report of the Agricultural Dwelling-House Advisory Committee for Bedfordshire, Cambridgeshire and Northamptonshire, advising the South Bedfordshire District Council against his application for vacant possession of a cottage occupied by the respondent, Mr Atkinson.

I A B McLaren (instructed by Turner Kenneth Brown, agents for Brown Jacobson & Roose, of Nottingham) appeared on behalf of the applicant; I Karsten (instructed by Geoffrey Leaver & Co, of Milton Keynes) represented the respondent; the advisory committee were not represented and took no part in the proceedings.

Giving judgment, HODGSON J said: It may be that at the heart of this case there is a point of administrative law, to do justice to which would require elaborate and extended investigation of the authorities. However, I feel reasonably confident that without the citation of authorities with which the point may be concerned I have come to the just decision.

In this case the applicant seeks judicial review of a decision of the Agricultural Dwelling-House Advisory Committee for Bedfordshire, Cambridgeshire and Northamptonshire, whereby on February 13 1985 it resolved to advise the South Bedfordshire District Council that:

The interests of agricultural efficiency do not call for provision of suitable alternative accommodation in this case, principally on the grounds that the forestry operations being carried out on the estate under the supervision of the applicant's forestry adviser do not warrant the presence of a forester living on the estate.

The decision under attack by the committee — I shall speak of the committee and the local authority from now on — is pursuant to and under the provisions of the Rent (Agriculture) Act 1976, as amended by the Rent (Agriculture) Amendment Act 1977.

The background facts can be very briefly stated. The applicant is the owner of forestry land. He does not live on the estate and nor does his son. He employs a forestry adviser, a Mr Acott, who lives some 2 miles away.

Between January 1973 and September 28 1984 the forestry land was cared for by a forester, Mr Atkinson, who is the respondent in these proceedings. He occupied a small house situated near the main entrance to the estate as, I suppose, the licensee of the applicant to begin with.

Under the provisions of the 1976 legislation Mr Atkinson became a protected tenant. The applicant and Mr Atkinson fell out, the applicant dismissed Mr Atkinson and his employment terminated on September 28 1984. He brought proceedings for unfair dismissal which were successful. The applicant then served notice to quit and when that expired on January 2 1985 Mr Atkinson became a statutory tenant under the legislation. The applicant wished to use the cottage occupied by Mr Atkinson for his replacement as forester. Mr Atkinson being, not unnaturally, unwilling to depart, the applicant attempted to avail himself of the provisions of the 1976 Act.

At this stage of my judgment it will be convenient to examine the statutory background against which this application for judicial review is brought.

The 1976 legislation made a very substantial alteration in the security of tenure of agricultural workers occupying premises and it was, no doubt, appreciated that it might not be in the interests of agriculture generally if houses were tied up with statutory tenancies where the tenant was no longer working on the land to which the property belonged. Accordingly, provision was made for the estate owner who was himself unable to provide alternative accommodation to make application to the housing authority concerned: in this case the South Bedfordshire District Council, to whom I shall refer as "the authority".

Section 27 of the Act provides that an application may be made by the occupier of land used for agriculture to the housing authority concerned on the ground that:

(a) vacant possession is or will be needed of a dwelling-house which is subject to a protected occupancy or statutory tenancy . . . in order to house a person who is or is to be employed in agriculture by the applicant, and that person's family, (b) the applicant is unable to provide, by any reasonable means, suitable alternative accommodation for the occupier of the dwelling-house, and (c) the authority ought, in the interests of efficient agriculture, to provide the suitable alternative accommodation.

Accordingly, the applicant in this case makes such an application.

Section 28 of the Act is concerned with the duty of the housing authority to which such an application is directed. Section 28(1) provides that the application shall be in writing. Section 28(2) provides for notification to the occupier of the dwelling-house. Section 28(3) provides:

The authority, or the applicant, or the occupier of the dwelling-house, may obtain advice on the case made by the applicant concerning the interests of efficient agriculture, and regarding the urgency of the application, by applying for the services of a committee under section 29 of this Act.

The applicant, it appears, made such an application. Section 28(4) provides:

The committee shall tender its advice in writing to the authority, and make copies of it available for the applicant and the occupier of the dwelling-house.

Section 28(5) provides:

In assessing the case made by the applicant and in particular the importance and degree of urgency of the applicant's need, the authority shall take full account of any advice tendered to them by the committee in accordance with section 29 of this Act . . .

Section 28(6) provides for notification of the authority's decision, with a generous timetable: three months after receiving the application or, if an application is made for the services of a committee, within two months of their receiving the committee's advice.

Section 28(6A) provides:

The notification shall state — (a) if the authority are satisfied that the applicant's case is substantiated in accordance with section 27 above, what action they propose to take on the application; (b) if they are not so satisfied, the reasons for their decision.

Therefore, the housing authority have a duty to take full account of any advice tendered to them and a statutory duty, if they find against the applicant, to give reasons for their decision.

Section 28(7) places upon the authority a duty to use their best endeavours to provide suitable alternative accommodation. Section 28(8) provides, rather unusually, for an action by the applicant against the local authority for breach of statutory duty, if they are in breach of that duty. For the purposes of this case I think I need not refer to any of the other subsections of that section.

In response to the request of the applicant to have a committee to assess the agricultural need and urgency of this application one was appointed, and it is against the decision of that committee that this application for judicial review is brought.

Section 29 of the Act makes provision for advisory committees and for the composition of the committee. Section 29(8) provides for the committee to act in accordance with any directions given by the Minister; I understand none have been given. Section 29(9) empowers the minister to make regulations; I understand that none have been made. Section 29(10) provides:

Subject to regulations, or any direction, under subsection (9) above the procedure of any committee shall be such as the chairman of that committee may direct.

Section 29(12) provides:

The Minister may with the consent of the Minister for the Civil Service make payments to persons other than members of a committee by way of fees or compensation for expenses incurred and time lost by them in or in connection with their giving, at the request of the committee, any advice or information.

The committee having been appointed, they, in their turn, as they were entitled to do, sought the advice of the Forestry Commission. That advice was contained in a three-page document. The advice concluded:

The estate's policy at present is for the forester to carry out all the work himself. Considering the annual forestry work programme, the Christmas tree production and sales, the work being requested by the Nature Conservancy Council and the work involved in the production of semi-mature trees it is my opinion that the employment of a skilled forester is justified. On an estate where public footpaths are in the vicinity of young plantations, and with an intensive Christmas tree operation, and especially considering the distance at which the owner, his son and their forestry adviser live, it is necessary for the forester to live in a strategic place in the vicinity for security reasons.

Therefore, the opinion of the Forestry Commission, through its official, was that the forestry land justified the employment of a skilled forester, and that because of the distance from the estate at which the applicant, his son and his forestry adviser lived the forester should live in a strategic place in the vicinity for security reasons.

The committee met on February 13 1985 at an hotel. They had before them the advice of the Forestry Commission. What happened on that day seems to me quite plainly to have amounted to procedural impropriety, allowing for the fact that the chairman was entitled to be the master of his own procedure. In my judgment, having devised an unsatisfactory procedure, the committee then proceeded to put it into operation in an unfair way. The committee, the applicant and Mr Atkinson, the tenant, all had a copy of the Forestry Commission report. The committee decided to see the parties separately. First they saw the applicant, his solicitor and his expert Mr Acott, whose views agreed with those of the Forestry Commission.

At that first meeting the committee were told of something which did not appear in the Forestry Commission report: that the applicant was unable to obtain the services of a skilled forester unless he could offer him housing. The committee were told that advertisements had been put in the trade papers unsuccessfully. It is true that later on the applicant was asked whether he had copies of the advertisements actually with him and said that he did not. I do not see that very much turns on that.

Having seen and heard the applicant's side the committee then saw Mr Atkinson and his solicitor. As all the expert evidence was one way, and as Mr Atkinson had seen the independent report of the Forestry Commission, he clearly knew the case he had to meet. The applicant could be forgiven for thinking that there was little or nothing which Mr Atkinson could say about the technical matters on which the committee had to be satisfied in order to advise the local authority. What Mr Atkinson did do, and what now appears from his affidavit, was to throw doubt upon the *bona fides* of the application.

In his affidavit Mr Atkinson says:

There were a number of points which needed to be made at the hearing, and which I arranged for my solicitor to make on my behalf:
(1) Mr David Brough, the applicant's son, had indicated to me in early 1984 that he would like my cottage for renting out, which suggested that perhaps they were not being straightforward about applying on the basis of agricultural need.

That is a plain allegation of *mala fides* against the applicant.

Mr Atkinson then went on to deal with two other properties on the estate. He then instructed his solicitor to allege inaccuracies of fact in the expert report from the Forestry Commission, set out in para 4(3) of his affidavit, which I do not cite in full. Then at para 4(4) he said:

Since my dismissal in September 1984, the estate had been looked after by a forestry worker who had worked previously on the Woburn estate and who lived in Woburn, some four miles distant from Rushmere, and this had apparently worked out perfectly satisfactorily for several months following my dismissal. Mr Marshall, the Woburn forestry worker, had his own home in Woburn, and he would have been very surprised indeed to hear that my cottage was required for his occupation, indeed he informed me himself that he did not wish to move.

Those are all relevant averments made in the proceedings, and made at a time when, on the basis of the procedure adopted by the committee, the applicant and his solicitor were given no proper opportunity to rebut them. True it is that, after the proceedings — if you can call them that — at the hotel were completed, there was a visit to the site, during which certain matters were discussed as the inspection took place. It is suggested, forcefully and well, by counsel for Mr Atkinson that any defect in procedure at the hearing — or should I say hearings — in the hotel were put right by what happened during the inspection. I do not accept that for one moment.

It is true that there is uncontradicted evidence that there were relevant conversations as the party moved from place to place around the estate. But one does not need to have had the experience of such inspections oneself to realise that the sort of thing which goes on during such an inspection cannot take the place of the opportunity to give proper replies to allegations made. Nor were the allegations of *mala fides* against the applicant and his son and as to the other property on the site ever put to the applicant so that he could, if he wished, refute them.

Whether or not any of those things had any effect on the deliberations and determinations of the committee I do not and cannot know, because they have not filed evidence in this case and we are therefore left in ignorance as to what did and did not affect their judgment. All I can really be concerned about is to look at what happened and then, allowing as I said for the fact that the chairman was entitled to devise his own procedure, decide whether it was fair within that procedure. I unhesitatingly conclude that it was not.

After the inspection and the conversations which took place there the committee provided their advice to the local authority. That advice is contained in an advice sheet of which there is a copy. It is perhaps of some interest to know that section 1 of that form provides for the committee to make observations on factual data as follows: "With particular reference to any instances where, in the committee's view, the data furnished by the applicant need to be amplified or qualified so as to give what they would regard as a full and fair picture of the relevant facts." Nothing was written there in the report.

Then, in their assessment, after the printed words "in the committee's view the interests of agricultural efficiency do not call for provision of suitable alternative accommodation in this case, principally on the grounds that," the committee have inserted the words:

The forestry operations being carried out on the estate under the supervision of the applicant's forestry adviser do not warrant the presence of a forester living on the estate.

On behalf of the applicant it is submitted that those reasons are wholly inadequate for the purpose for which they are required, namely, the advice to the local authority so that it can decide whether it has any duty under the Act.

It must be remembered that the local authority is required by statute to take full account of the advice that it receives. There is some slight evidence that when advice is received it is, if not conclusive, then very nearly so; as one would expect with an authority faced with the possibility of having to fulfil a statutory obligation to find accommodation.

In the context of the need of the local authority for the advice, I have come to the conclusion that, particularly in view of the overwhelming expert evidence with which the committee were faced and no hint of which appears in their advice, the reasons given were, in the context of this case and this legislation, wholly inadequate reasons.

That being so, as long as the decision of the committee to give the advice they did is subject to judicial review and quashing by way of certiorari, and if I am satisfied in my discretion that certiorari should go, then that, on the view I have formed of the case, is the end of the matter.

The question of law which may be at the heart of this decision, and to which I referred at the beginning of my judgment, is whether certiorari will go to quash an advisory decision of this nature when the determination is itself to be that of another body.

I have looked with counsel at a number of passages from the fifth edition of *Wade on Administrative Law* and the fourth edition of *de Smith*.* I dare say that we could have gone on for hours and hours finding passages both in the textbooks and in the authorities which would support either view. I should particularly refer to *R v Boycott* [1939] 2 KB 651 and *R v Winchester Area Assessment Committee, ex p Wright* [1948] 2 All ER 552, together with various passages from *Wade* and *de Smith*; *R v St Lawrence's Hospital, Caterham,*

*Editor's note: *Judicial Review of Administrative Action*, by S A de Smith.

Statutory Visitors, ex p Pritchard [1953] 1 WLR 1158, the contiguity of which with *R v Metropolitan Police Commissioner, ex p Parker* [1953] 1 WLR 1150 is perhaps worth noticing; and also the case of *R v Electricity Commissioners* [1924] 1 KB 171.

In my judgment, particularly when one is considering the procedural impropriety or otherwise by which a decision of this nature — that is, one which is not finally determined — can be subject to judicial review, one has to pay great regard to a consideration which appears in a sentence of *de Smith* at p 234:

The degree of proximity between the investigation in question and an act or decision directly adverse to the interests of the person claiming entitlement to be heard may be important.

I think that is right. Merely because a decision to give advice, or the advice itself, is not finally determinative of a question is not in my view the determining factor. I think it is important to look at all the facts and see in general terms what part that subdecision, if I can coin a phrase, plays in the making of the decision as a whole.

If it is only a decision to give evidence one way or the other, then plainly it would not be subject to judicial review. But where that advice is sought by the determining authority from a committee of whose decision the authority is required by statute to take full account, and where there is some evidence that in practice the advice is — to put it no higher — highly likely to be followed, then I think it would be wrong to allow the proceedings to go further and require the applicant to wait until the decision of the local authority is made against him, if it is, before attacking that decision on the basis that the material upon which it was based was flawed.

That would seem to me to be a wholly unnecessary requirement, and I have no doubt on the facts of this case and within the context of this legislation that the court has power to interfere at this stage and that it is a power which it ought to exercise if it is satisfied that there has been a procedural impropriety. I am satisfied that there has been that procedural impropriety. I think that in my discretion I ought not to refuse the relief sought at this stage and the consequence of that is that this decision of the committee must be brought up to this court and quashed.

Certiorari ordered to go to quash committee's decision. Respondent to pay applicant's costs.

Court of Appeal

February 2 1987

(Before Lord Justice STEPHEN BROWN and Lord Justice RUSSELL)

FOWLER v MINCHIN

Estates Gazette June 20 1987

282 EG 1534-1538

Rent Act 1977, Case 16 in Schedule 15 — Landlord sought possession of a cottage occupied by the tenant which had formerly been occupied by the landlord's cowman and which the landlord now required for a person to be employed by him in agriculture — The conditions required by paras (a) and (c) of Case 16 were satisfied, but there was an issue as to para (b) under which it was necessary for the tenant to be given "notice in writing that possession might be recovered under this Case" — The questions were whether a document which the landlord claimed to be such a notice had been given to the tenant and, if so, whether it complied with para (b) — The course of events was "remarkable" — Until the actual hearing in the county court the case had proceeded on the basis that no notice complying with para (b) had been given — This had appeared from the pleadings and in a letter from the landlord's solicitor — Unexpectedly, during the hearing evidence was given by the landlord that a written agreement, unfortunately since destroyed, had been entered into under which the tenant undertook to vacate the cottage on 28 days' notice if the landlord required it for a farm worker — The county court judge accepted this as evidence that a notice complying with para (b) had been given and he held that the landlord was

entitled to possession — The Court of Appeal expressed disquiet about the proceedings in the county court so far as this last-minute disclosure was concerned — However, they held that, even if a written agreement in the terms mentioned had come into existence, it did not satisfy para (b), because it was not a notice "that possession might be recovered under this Case" — Such a notice must state quite specifically that possession might be recovered under the provisions of the Act, ie must make it clear that a situation would obtain which would be a compulsory situation so far as the tenant was concerned — The alleged term of the so-called agreement that the tenant would vacate on 28 days' notice, if the landlord required it for a farm worker, was no more than a voluntary undertaking that he would do so — Tenant's appeal allowed

No cases are referred to in this report.

This was an appeal by the tenant, David Richard Minchin, from a decision of Judge Hutton, at Gloucester County Court, in favour of the landlord, John Phillips Fowler, whereby the landlord was granted possession of a dwelling-house, 2 Riddlers End Cottages, Tirley, Gloucestershire. This was formerly a tied cottage occupied by a cowman employed by the landlord. The cowman left and the cottage was subsequently let to the tenant, Mr Minchin, who had never at any time been employed in agriculture by the landlord.

D R Lewis (instructed by Langley-Smith & Sons, of Gloucester) appeared on behalf of the appellant; Simon Buckhaven (instructed by Leslie J Slade & Co, of Newent) represented the respondent.

Giving judgment, STEPHEN BROWN LJ said: This is an appeal by the defendant, David Richard Minchin, from the judgment of His Honour Judge Hutton given at Gloucester County Court on June 27 1986. The learned judge gave judgment in favour of the plaintiff and ordered that the plaintiff should be entitled to recover possession of a dwelling-house, 2 Riddlers End Cottages, Tirley, Gloucestershire, within 28 days.

The facts giving rise to the claim for possession can be shortly stated. The plaintiff, Mr Fowler, is a farmer who owns cottages in connection with his farm. No 2 Riddlers End, Tirley, was formerly a tied cottage occupied by a cowman employed by Mr Fowler. The cowman left in 1978 and the cottage became vacant. The plaintiff then let the cottage to the defendant, Mr Minchin. The circumstances in which he let it have been the subject of dispute. The claim for possession was brought under Case 16 of Schedule 15 to the Rent Act 1977. Case 16 provides:

Where the dwelling-house was at any time occupied by a person under the terms of his employment as a person employed in agriculture, and
a) the tenant neither is nor at any time was so employed by the landlord, and is not the widow of a person who was so employed, and
b) not later than the relevant date, the tenant was given notice in writing that possession might be recovered under this Case, and
c) the court is satisfied that the dwelling-house is required for occupation by a person employed, or to be employed, by the landlord in agriculture.

When the case was pleaded the particulars of claim made no reference to any notice in writing having been given to the tenant not later than the relevant date in accordance with subpara (b) of Case 16. The particulars of claim were then amended to allege that prior to the defendant's taking possession the premises had always been used for the purpose of housing agricultural workers, but again there was no reference in the amended particulars of claim to a notice in writing in accordance with the provisions of subpara (b) of Case 16. Para 3 of the amended particulars of claim merely alleged:

At the commencement of the said tenancy it was expressly agreed between the parties that the Defendant would vacate the said premises if it should at any time become necessary to provide accommodation for farm labourers required for the proper management and control of the Plaintiff's adjoining farm.

The particulars of claim did, however, specify that possession was claimed under Case 16 of the 15th Schedule in the prayer for relief.

The defence denied that any relevant notice had been given in writing and the defendant sought further and better particulars of any relevant document, if the agreement was alleged to be a written agreement. At p 28 of the bundle of documents before this court the request and the answer to the relevant request for further and better particulars is set out at para 5. The request was:

5) If the agreement was written, the date and the identity of the document.

Answer — Insofar as the agreement was evidenced in writing the document was a written undertaking that the rent should be paid weekly and would be equal to that of council houses in Tirley and would also be increased yearly on a par with inflation. The Defendant will have a copy of this undertaking in his possession. The written agreement was made on or about August 26 1978.

There is no reference, it will be noted, to any notice in writing indicating that possession might be recovered under this Case. That was the answer in the further and better particulars.

On the day before the hearing a reply was delivered. It is dated June 26 1986. Para 2 is in these terms:

The Defendant is estopped from alleging that no notice in writing was given to him that possession might be recovered under Case 16 of Schedule 15 to the Rent Act 1977 by reason of the following matters:
(1) At the date of the letting agreement on or about August 26 1978 the Defendant represented to the Plaintiff that if the Plaintiff should require the property for a farm worker he would surrender up vacant possession on the giving of one month's notice.
(2) In reliance upon the Defendant's said representation, the Plaintiff let the premises to the Defendant without giving him notice in writing pursuant to Case 16 aforesaid.
(3) In the premises it is inequitable that the Defendant should now seek to rely on the absence of written notice.

That was the reply, and it will be seen that it quite specifically states that there was no notice in writing pursuant to Case 16 and seeks to raise what appears to be an issue of estoppel.

The case came on for hearing on June 27 1986. The ensuing course of events was, in the view of this court, remarkable. The case was opened by counsel on behalf of the plaintiff. The same counsel does not appear in this court today for reasons which will no doubt become apparent. The issue was raised, quite plainly at the very beginning, by the defendant that he was contending that there was no notice in writing sufficient to satisfy subpara (b) of Case 16 in this case.

The plaintiff gave evidence — we have the judge's note of the evidence — that he decided that he had to give up the farm work side because of ill health and would therefore need another person to work the farm. He gave evidence about his ownership of the cottage and its previous occupation by an agricultural worker employed by him, and he gave evidence about the letting. The note at the top of p 11 about seven lines down reads:

We verbally agreed to

the defendant's initials

putting interior right and keeping in good repair and if I gave a month's notice he would vacate if I needed it for farm labour.
It was a weekly term and rent in line with council houses in Tirley.
He was to start paying rent one month after he had the key.
Rent book first date August 26 1978.
The meeting was about five weeks previously.
He was happy to agree.
At the meeting I agreed to get the term written out in duplicate so that when

the defendant's initials

collected the key we would sign it.
I wrote them out myself in duplicate.
It said those three items.
He signed the agreement when he had the key at Tirley Hill.
It was two or three days after we had the verbal discussion.
We both signed both copies and I handed one copy to him and retained one.
It went with other documents to our solicitors when we had difficulty in collecting rent.
It got lost then.

The defendant had produced a document, D1, which purported to set out the terms of the agreement. It is at p 17 of the bundle and reads:

An agreement between J P Fowler, the owner of Riddlers End Cottage (No 2) and Mr P Minchin the tenant of the above cottage.
Riddlers End cottage will be let to Mr Minchin from August 26 1978 at a rent of £14.00 per week exclusive of rates.
The rent to be reviewed annually in relation to the percentage rate of inflation.

The plaintiff said during his evidence in chief that that note, and another note produced as D2 headed: "Riddlers End Cottage (No 2). August 26 1978 — Amanda Fowler. One week's rent received until September 2", was written by his daughter when she got the first rent, but he said that that was not the document he had originally written out. He then went on to deal with his reason for requiring the cottage, to enable him to obtain good labour. He said, according to the judge's note:

My wife and I knew about it when the original got lost. The solicitors were and he names them and names a particular partner.

Documents sent

it is not clear just what the note should read at that point, but it may be "at times when we were trying to get arrears paid".

I think 1982 or 3. Discovered immediately I got documents back.
I brought some back but not this.

In the course of cross-examination he said:

She said

that is a reference to the partner in the firm of solicitors

it was worthless and she threw it in the waste paper basket. I saw her do it.

The position at the trial was this. Until the actual hearing commenced or was about to commence the defendant's counsel had had no notice that reliance was to be placed upon a specific written document which it would be alleged complied with the requirement of subpara (b) of Case 16. During the course of the case, therefore, additional voluntary further and better particulars of the amended particulars of claim were given by the plaintiff. They are handwritten and we have a photostat copy of them:

1. Date of written notice
During the week prior to July 29 1978 ie in the week before four weeks prior to the date shown in the Rent Book.
2. When and where it was given to the Defendant
At Tirley Hill Farm at the time the Defendant was given the key during the week prior to July 29 1978, a day or two after the oral agreement.
3. Exactly what it said
That the Defendant was to keep the premises in a good state of repair. That the Defendant would vacate on 28 days' notice if the Plaintiff required it for a farm worker

and lastly

That the rent was to be paid weekly and was to be kept on a level with Tirley council house rents.

In the course of the hearing the plaintiff's counsel found himself in an embarrassing position. Until the plaintiff had answered in cross-examination that his solicitor had thrown the alleged written document into the waste paper basket saying it was worthless, he had no knowledge of any such allegation. All he had, apparently, was an indication on the same morning shortly before the hearing commenced from his lay client that a document had been lost by his solicitors.

Clearly the pleadings to which I have referred, including the reply, which is dated June 26, implied that there was no written notice at all, so that counsel was in a difficult position. We are told by the defendant's counsel (the appellant's counsel today) who was present at the court of trial that the plaintiff's counsel asked the judge for an adjournment and asked to be relieved of his position as the plaintiff's counsel because of an apparent conflict between his solicitor on an important factual matter and his lay client. We are told by Mr Lewis, who appears for the appellant, that he did not oppose that application, but the judge did not grant an adjournment and the case proceeded. The defendant gave evidence that there never was a written notice in accordance with subpara (b) of Case 16. In due course the learned judge gave judgment.

The judgment is a short judgment. We have a note of it approved by the judge. He set out the fact that it was not disputed that this was formerly a tied cottage and that the requirement (a) in Case 16 was not disputed. The issues are as to requirements (b) and (c).

Having heard the evidence of Mr Fowler I have no difficulty about requirement (c). I find as a fact that he is genuine about his need for labour and that he is more likely to obtain a man of suitable quality if he can offer him a cottage. As a matter of law it is only necessary to satisfy me that the plaintiff wishes to have possession for this purpose — not that it is necessary.

That leaves requirement (b). This is hotly in dispute. It is a pure question of fact. On the face of it the defendant has a strong case. Up until today the case was conducted on the basis that no notice was given — see the further and better particulars of the amended particulars of claim and exhibit D3. When counsel for the plaintiff saw his client it became apparent that his case was not as pleaded and that the notice had been lost. He therefore sought leave to file further and better particulars of his claim. In cross-examination of Mr Fowler the position became even clearer, ie, the solicitors had not just lost the notice but had deliberately thrown it away (this was Miss Bennett). Mr Fowler said that he accepted that he had been mistaken in not raising the matter before but did not wish to upset his solicitors.

I then heard evidence from Mrs Fowler and Amanda Fowler. Both gave evidence in support that there was a written agreement. In the course of the hearing documents — P2 and P3 — were produced to show that in November 1983 there was an issue between the Fowlers and the solicitors as to the document.

Having heard the evidence of all three Fowlers I have decided to accept their evidence despite the obvious difficulties in their way. I accept that Mr Fowler is telling me the truth about the notice and the signing of it.

I heard Mr Minchin's evidence which was largely credible, although I am surprised to hear him say that he did not raise with Mr Fowler the question of the cottage being required in the future for an agricultural labourer. Reluctantly I conclude that the defendant is not telling the truth and I find that due notice was given. All the elements of Case 16 are made out and the plaintiff is entitled to possession.

In this court today Mr Lewis submits first of all that the learned judge's finding of fact that a written document purporting to be or answering the requirement of subpara (b) of Case 16 was given was not a finding which any reasonable tribunal could have reached on the evidence adduced at the hearing taken in conjunction with the pleaded case and all the circumstances which I have indicated. Second, he submits that in any event if there was a written document, there was not a notice which satisfied the requirements of subpara (b) of Case 16. The plaintiff relies on a document alleged to be in the terms of the additional further and better particulars of the amended particulars of claim, but that would not satisfy the requirements of subpara (b). At best it is merely an undertaking that the defendant would vacate on 28 days' notice if the plaintiff required it for a farm worker. That would be unenforceable.

Quite plainly this case took a very unusual course. Here was a plaintiff with a pleaded case which specifically proceeded upon the basis that no written notice had been given. It does not only rest with the pleadings, for at p 18 of the bundle there is a letter from the plaintiff's solicitors dated March 26 1986 addressed to the defendant's solicitors. Para 2 is in these terms:

In respect of supplying you with a copy of the original agreement between Mr and Mrs Fowler and Mr Minchin in 1978 we again must say that we are unable to supply you with a copy as the agreement was handed over to Mr Minchin immediately after it was signed and no copy was retained by Mr and Mrs Fowler. Having said that the agreement in writing only relates to such matters as rent and repairs and we accept that it makes no reference to the grounds on which possession may be recovered, as to be capable of comprising written notice as required by the relevant section of the Rent Act 1977.

Accordingly it is quite appropriate for Mr Lewis to categorise the course of proceedings as being "remarkable".

I must say, speaking for myself, that I am troubled by the way this case proceeded because clearly there was a dramatic and fundamental last-minute change in the plaintiff's entire case alleging that there was a written notice which would comply with subpara (b) of Case 16. Moreover it involved a clear conflict between the plaintiff himself and his solicitor, who was a partner in a very responsible firm of solicitors. The solicitors had also assured the defendant on the plaintiff's behalf that no notice in writing had existed.

These facts alone, it seems to me, would provide a reason for this court to be so disquieted by the course of events as to consider very seriously ordering a rehearing. It is quite plain that counsel for the plaintiff was placed in a personal position of great embarrassment and that the defendant was taken by surprise although his counsel did not himself seek an adjournment. There are in my judgment real grounds for questioning the finding of fact of the learned judge.

Of course this court does not see and hear the witnesses: the learned judge is an experienced judge, and this court does not lightly interfere with findings of fact. However, there is in this case a ground of appeal which in my judgment is well founded, and it is the ground which forms the second matter which Mr Lewis argued in the course of his address. That is that, accepting for the purposes of this argument that a written agreement in the terms of the additional further and better particulars was brought into existence even though no copy was produced to the court, it would still not comply with the requirements of subpara (b) of Case 16, which I will read again:

(b) not later than the relevant date, the tenant was given notice in writing that possession might be recovered under this Case.

The argument is very simple. It is that even if there was a term or undertaking in the tenancy agreement in the terms set out at p 37 of the bundle, it could not satisfy subpara (b) because there it is not a notice that possession might be recovered under this particular provision.

It is quite true that the notice does not have to follow any particular form, but in my judgment it must state quite specifically that possession might be recovered under the provisions of the Act: that is to say, to make it clear that a situation would obtain which would be a compulsory situation so far as the tenant was concerned. In m judgment the alleged term of the so-called agreement that the defendant would vacate on 28 days' notice if the plaintiff required it for a farm worker is no more than a voluntary undertaking that he would do so. It does not indicate or suggest that the plaintiff would be entitled to go to court to require him to give up possession and indeed would do so if he required the dwelling for a farm worker. However informal the phraseology might be, it would have to satisfy that requirement. Accordingly, in my judgment, even if a written agreement was brought into existence in the terms of the additional further and better particulars, it did not comply with subpara (b).

Mr Lewis argued in addition that subpara (c) was not satisfied because the plaintiff had in effect merely said that it would be more attractive to a prospective employee to be able to make him the offer of a cottage. However, it seems to me that so far as that ground of appeal is concerned it fails: the learned judge dealt with that satisfactorily and indeed the note of the evidence shows that the plaintiff was saying in terms that he now required to have a labourer because of his own ill health. He said in terms that he would have to offer the cottage in order to be able to get the type of labour that he needed. I do not think there is anything in that ground of appeal.

However, for the reasons that I have given in relation to the requirement for a notice in writing under subpara (b) this appeal should be allowed.

Agreeing, RUSSELL LJ said: For the reasons rehearsed by my lord I, too, am very unhappy about the course which these proceedings took prior to and during the course of the hearing. As to subpara (b) of Case 16, I agree that, even on the basis that a notice in the terms set out in the further and better particulars of the amended particulars of claim was served, about which I have grave reservations, it would not comply with subpara (b). I agree with my lord, therefore, that this appeal must be allowed.

The appeal was allowed with costs in the Court of Appeal and below; legal aid taxation of appellant's costs.

For a further case on this subject see p 224

LANDLORD AND TENANT
RENT REVIEW

Chancery Division
June 13 1986
(Before Mr Justice HARMAN)

GENERAL ACCIDENT FIRE & LIFE ASSURANCE CORPORATION PLC v ELECTRONIC DATA PROCESSING CO PLC

Estates Gazette January 10 1987
281 EG 65-68

Landlord and tenant — Rent review clause in lease — Construction — Rejection by Harman J of reliance on the "Commercial commonsense" approach — Nourse LJ's comments in *Philpots (Woking) Ltd* v *Surrey Conveyancers Ltd* preferred as "the classic and correct method of approach" — Lease granted in 1980 for a term of 25 years at a rent of £42,025 per annum subject to upward review at 5-year intervals — Review clause contained no provision that the hypothetical lease was to be on the terms of the actual lease and there was no express reference, among eight matters to be assumed, to any assumption as to rent review in the hypothetical lease — The arbitrator had assumed a hypothetical lease without rent reviews — On that basis he determined a rent at the first review of £55,500, with an alternative finding of £48,500 if rent reviews were to be assumed — On an appeal against the arbitrator's decision, the tenants' representative, citing *British Gas Corporation* v *Universities Superannuation Scheme Ltd*, submitted that the review clause should be construed in the light of an underlying commercial purpose and that one of the necessary provisions to be implied was a provision for rent review — Landlord's representative argued that the lease directed that eight specific assumptions should be made in respect of the hypothetical lease and that there was no justification for any implications, whether as to commercial purpose, true market value, or otherwise — The review clause was workable as it stood without any such implications, although it resulted in a rent 15% higher than a letting in the real market would command — There was no basis for overriding the express terms even if the result seemed a little surprising — Judge's comments on dates when gales of rent fall due — Appeal against arbitrator's award dismissed

The following cases are referred to in this report.

British Gas Corporation v *Universities Superannuation Scheme Ltd* [1986] 1 WLR 398; [1986] 1 All ER 978; [1986] 1 EGLR 120; (1986) 277 EG 980
Philpots (Woking) Ltd v *Surrey Conveyancers Ltd* [1986] 1 EGLR 97; (1985) 277 EG 61
Prenn v *Simmonds* [1971] 1 WLR 1381; [1971] 3 All ER 237, HL
Sterling Land Office Developments Ltd v *Lloyds Bank plc* (1984) 271 EG 894

This was an appeal by originating notice of motion against the decision of an arbitrator on a rent review clause in a lease of premises at 1-3 Tapton Park Road, Sheffield. The parties were the landlords, General Accident Fire & Life Assurance Corporation plc, and the tenants, Electronic Date Processing Co plc.

Kim Lewison (instructed by Benson Burdekin & Co) appeared on behalf of the tenants; David Neuberger (instructed by Edwin Coe & Calder Woods) represented the landlords.

Giving judgment, HARMAN J said: I have before me an appeal by originating notice of motion against the decision of an arbitrator on a rent review clause. Such matters are part of the daily bread of the Chancery Division these days.

The parties to the motion are the present landlord and the present tenant of premises at 1-3 Tapton Park Road, Sheffield. The lease was granted on November 12 1980, so that it is a modern lease. It was granted of those premises to hold for a term of 25 years at a yearly rental of £42,025 subject to upward revision as provided in clause 3(2), and upon covenants by the lessee set out in clause 2(1) to (20). Clause 3 provides in subclause (1) an irrelevant set of obligations, and in subclause (2) the rent review provisions to which I will return later. It then sets out a provision in subclause 3(2)(v) providing for a review of the review dates, which is an unusual provision, such as I have not met before.

The lease goes on to set out in clause 4 the landlord's covenant for quiet enjoyment and a covenant for insurance by the landlord, there being an insurance rent reserved as part of the reddendum. By clause 5 there is granted an option to the tenant to take a new lease for a further term of 25 years at the expiration of the term hereby granted. The rent review dates are specified as within the six months immediately preceding the expiry of the 5th, 10th and 15th years of the term granted and thus the first rent review became due in the six months immediately prior to August 1985.

The rent review was before an arbitrator, the parties being unable to agree, and Mr Watson, a chartered surveyor, acted as arbitrator and made an award on February 25 1986. The award is in a classic form and is very clear and well laid out. It sets out in clause 5 the rent review provision in clause 3(2)(i) of the lease. It refers to valuation issues which do not arise before me, and in para 7 of the award it refers to the legal issue which was discussed in documentation before the arbitrator, he having only written submissions since no hearing was thought necessary.

The contentions set out on the landlord's surveyor's side are that:

Because there is no mention in the rent review clause of any review in the 20-year term that is to be assumed, the rent is to be assessed as if there were no such reviews.

He therefore added 15% to his valuation. The tenant's surveyor did not accept this view and made no such addition; the point was not closely argued by that surveyor.

The arbitrator goes on to set out what is in fact contained in the rent review clause and to define the matters in it. At the foot of p 6 he says:

There are two conspicuous omissions from this list of provisions of the hypothetical lease. The first omission is the direction almost universally found in modern rent review clauses to assume a lease: "On the terms of this lease." The second omission is any reference to rent reviews. The inclusion of either would have had the effect that the hypothetical lease should contain rent reviews as they are in fact contained in the existing lease.

He goes on to consider the two contentions and he holds at p 8: "I have reluctantly come to the conclusion that I must assume a lease without rent reviews," and he values accordingly, fixes the figure at £55,500 as his award, and then makes an alternative award on the basis that there should be assumed to be rent reviews in the hypothetical lease which he determines at £48,500.

Before me counsel have analysed the rent review provisions in very much closer terms. Clause 3(2)(i) requires the lessor and the lessee within six months before the end of the 5th, 10th and 15th years of the term to: "endeavour to agree the annual rack rental market value then prevailing." I break off to observe that the draftsman is not the most frugal of users of words, the phrases "annual rack rental" and "market value" are tautologous. The draftsman then defines that as: "Hereinafter called the current rental value of the whole of the demised premises assuming," and he then sets out a series of assumptions.

The assumptions are, as I construe the lease, and as Mr Neuberger submitted to me for the landlord, eight in number. The hypothetical lease is to be assumed, first, to be a lease of the whole; second, it is between a willing landlord and a willing tenant; third, it is without payment of any fine or premium; fourth, it is for the balance of the term of years then remaining unexpired under this lease; fifth, it is with vacant possession; sixth, it is for use for the purpose hereby authorised; seventh, it is on a tenant's full repairing and insuring basis as hereinbefore provided; eighth, one is directed to disregard the factors in section 34(a), (b) and (c) of the Landlord and Tenant Act 1954. As the arbitrator observed, there is no stated hypothesis that the hypothetical lease is upon the terms of this lease, or upon the terms contained herein or any such phrase, and, as he also observed, there is no express reference in the assumptions to any rent review.

I think it right to go on to notice, as I have recited above, that the lease contains in clause 5 an option to the tenant. That grants an option in these terms:

If the lessee shall be desirous of taking a lease for a further term of 25 years from the expiration of the term hereby granted at the rent and on the terms and conditions hereinafter mentioned

I now paraphrase, "and gives proper notice of its desire and has performed its covenants, then"; I resume the lease

The lessors will at the cost of the lessee grant a lease of the demised premises to the lessee for the further term of 25 years from the 1st August 2005 at a rent to be determined in the manner provided by the schedule hereto commencing and payable as therein provided and subject in all other respects to the same stipulations as are herein contained except this clause for renewal and subject to the amendments of subclause 3(2) of this lease which are specified in the said schedule.

Thus the draftsman of this lease was familiar with the concept of importing the same stipulations as are herein contained and chose to so do in the option clause; he did not do so in the rent review clause, whether as a deliberate choice or by some inadvertence I cannot at present determine.

Mr Lewison, for the tenant, submitted that in accordance with a recent decision of the present Vice-Chancellor in *British Gas Corporation* v *Universities Superannuation Scheme Ltd* [1986] 1 WLR 398,* the rent review clause should be construed in the light of an underlying commercial purpose. The Vice-Chancellor expressed himself as laying down a correct approach to rent review clauses at p 403 just above the letter (B), he having said that he felt himself free to adopt his own approach, as follows:

In my judgment the correct approach is as follows: (a) words in a rent exclusion provision

a term he had defined earlier

which require *all* provisions as to rent to be disregarded produce a result so manifestly contrary to commercial commonsense that they cannot be given literal effect; (b) other clear words which require the rent review provision (as opposed to all provisions as to rent) to be disregarded (such as those in the *Pugh* case†, 264 EG 823) must be given effect to however wayward the result; (c) subject to (b), in the absence of special circumstances it is proper to give effect to the underlying commercial purpose of a rent review clause and to construe the words so as to give effect to that purpose by requiring future rent reviews to be taken into account in fixing the open market rental under the hypothetical letting.

He then beseeches the help of higher courts and endeavours to restrict the practice of the Bar in these cases by deploring the number of cases, a view which I confess I do not share.

The Vice-Chancellor's decision is one which I, with respect to him, would have come to myself upon the terms in that lease. They were, as he plainly sets out, words involving a rent exclusion provision. That is to say that the rent review clause itself contains some references which are apparently directed to excluding some parts of the rental provisions of the lease. The question then is the meaning of those words in that particular context. With all respect to the Vice-Chancellor, I do not think he could, and in my judgment he did not in fact attempt to, lay down that every rent review provision should be approached with some prior conception that its only purpose is to establish the true market rent of the premises from time to time upon the terms of the actual lease then in issue before the court. To so say would be to assert a power to shape all the clauses to a particular purpose whatever they may say. I do not believe the Vice-Chancellor said that, intended to say that, or that, if he had said that, it would have been correct.

In my judgment the correct approach is exemplified by Nourse LJ, giving the decision of the Court of Appeal in *Philpots (Woking) Ltd* v *Surrey Conveyancers Ltd,* reported in (1985) 277 EG 61‡ again a decision on a rent review clause. It was cited by Mr Neuberger, not I think for the reason that he succeeded in that case but because the reasoning does seem to give some assistance in approaching these matters.

The learned lord justice said, in the middle of p 62 in the left-hand column, the fourth complete paragraph:

A court of construction can only hold that they intended it

"it" obviously means the rent review clause

to have that effect if the intention appears from a fair interpretation of the words which they have used against the factual background known to them at or before the date of the lease, including its genesis and objective aim. It is not in this case suggested that there is any material background beyond the fact that the lease is a commercial lease between commercial parties.

In my view that is the classic and correct method of approach. The court is to take the words the parties have used, to consider those words in the context of the whole of the document before the court and to consider those words in the context of such facts as are proved to be what Lord Wilberforce in *Prenn* v *Simmonds* [1971] 1 WLR 1381 called the matrix of fact surrounding the genesis of the transaction. I am sure that Nourse LJ intends a reference to Lord Wilberforce's concept when the learned lord justice uses the word "genesis" in *Philpots*' case.

In this present case, as in *Philpots*' case, I have no evidence of background material, no matrix of fact beyond the fact that the lease is a commercial lease between commercial parties. As Nourse LJ says in *Philpots*' case, one should construe the words as they stand and it is only if they produce a result which can be characterised as absurd that one should hold that the words cannot mean what they apparently say. Such a holding must be based on a conclusion that, to quote Knight-Bruce LJ in a rather different context, the draftsman was one of those who considered that the office of language was to conceal thought. Unless one finds absurdity one must give effect to the words.

The result in *Philpots*' case was that Nourse LJ rejected the proposition, at the foot of the left-hand column on p 62, in these words:

Inherent in Mr Smith's principal argument is the proposition that commercial parties to a commercial lease *invariably*

his emphasis

intend that the rent should not in real terms fall below the market rent initially agreed upon. I cannot accept that proposition. It is perfectly possible for them to assume that the fair rack rental value will always increase, but none the less to intend that the rent payable from the second review onwards shall always be less.

That is less than the fair rack rental value.

There is nothing

the lord justice goes on:

in the words of the lease to exclude that possibility, unless it be that the tenant's construction of them leads to results so absurd that the parties cannot be credited with an intention to achieve them. I do not think that it does.

That set of observations was made in the context of an illustration given by Nourse LJ earlier in his judgment on p 61, where he demonstrates that upon the tenant's construction, which he held to be correct, at the end of the 10th year you might find the following facts. The aggregate of £8,000, which was the initial rent, plus the amount by which the fair rack rental value, assumed to be £17,000 at that date, exceeds the yearly rent of £11,000 fixed at the previous date of review — that produces a figure of course, by deducting £11,000 from £17,000, of £6,000 — creates a calculated amount of £14,000. Thus on the tenant's construction, the fair rack rental value was £17,000 but the rent payable under the rent review was only £14,000. Nourse LJ held that that was the result of the words; it was not an absurd result, it was not one which should compel a court to force

---

* Editor's note: Also reported at [1986] 1 EGLR 120; (1986) 277 EG 980.
† Editor's note: *Pugh* v *Smiths Industries Ltd.*

‡ Editor's note: See also [1986] 1 EGLR 97.

language into a form other than that which it expressed and the language was effective to produce that result. The motives for negotiating a term in those words were not known to the court or material to the decision. The private intentions of the party are not only unknown but inadmissible on construction, and the court cannot and should not speculate about matters which would be inadmissible to be proved by evidence before it. The court must just take the document and try to understand it.

Reverting to this present case, Mr Neuberger's contention is that the rent review clause contains within itself the whole of the terms which are to be assumed by the valuer. It directs him to assume a hypothetical lease and to make eight specific assumptions as to what is contained therein. Nothing else, says Mr Neuberger, is to be contained in it, the words are entirely adequate to create a perfectly simple lease which a valuer can value. It excludes some matters, maybe, although one cannot know and perhaps should not try to speculate, in order to avoid such difficulties in fixing a rent, as attributing a value to an option to take a further 25-year term in the year 2005, which is still fairly remote.

Mr Neuberger submits that the eight assumptions are effective to determine the length of the term; they establish the fact that there is no premium; they fix the user; they specify the tenant's covenants, which are expressed to be a tenant's full repairing and insuring basis as hereinbefore provided, that is the actual covenants in this lease for the purposes of repair and insurance; and those provisions are all that there is. There is no *need*, necessity being the mother of, or the governor of, implication, to imply anything further by way of covenant, says Mr Neuberger.

Mr Neuberger went through the tenant's covenants in this case for me. He pointed to those which he said were included by reason of the express terms of the rent review clause. He agreed that a considerable number of covenants actually included in the existing lease would not, on his contention, be in the hypothetical lease which the valuer has to value. The most significant missing covenant was, I think, any covenant against assignment, there being no reference whatever to such a covenant in the assumptions expressly set out in the hypothetical lease indicated by the rent review clause. That might well have an effect upon value, but, he said, there we were, that was what was provided and that was what resulted.

Mr Neuberger further submitted that the absence of many of the other covenants in the existing lease — for example, the covenant against committing nuisance; the covenant against acting in breach of planning law; the covenant not to make repairs — may not really matter in terms of valuation. As he observed, most of them may in fact be governed by different obligations. Not obligations in covenant, and not obligations necessarily owed to the landlord, but obligations which in fact govern the premises because the general law applies them, so that the valuer will be able to say: "Quite true I am not to assume a covenant against doing something without planning permission, but that will not make any difference to the value and I need not pause to think in any detail about it." That would be a proper valuation exercise because, in fact, if the tenant makes an alteration, for example builds an extra storey, without planning permission, he will almost certainly be subject to enforcement proceedings by the local authority and nobody would add to or detract from the rental value of a lease because of the absence of that covenant from it.

Thus, says Mr Neuberger, although many covenants are not included in the covenants in the hypothetical lease laid down by clause 3(2), yet in fact the hypothetical lease is on perfectly workable terms because the really important covenants, the repairing and insurance covenants, are incorporated. He submits that contrary to Mr Lewison, his, Mr Neuberger's, contention results in only those terms being included which are specified to be included, whereas Mr Lewison's contention results in there being an imputed intent to have the repairing and insurance covenants plus an actually expressed intent to have the repairing and insurance covenants. Such a construction, he says, is a forced, unnatural, and pointless way of construing the document.

Lastly, Mr Neuberger prays in aid the terms of the option in clause 5. Why, he asks, should this document contain within its own four corners a provision referring to the grant upon the same stipulations as are herein contained in the option clause but mentions a grant in quite different terms in the rent review clause, unless there is to be a difference derived from the contrasting terms in these two provisions.

Mr Lewison urged me that the law has always taken a different view of terms which are to be supplied by the court, or can be supplied by the court, in aid of an existing relationship between two parties, and terms which will not be supplied by the court for the purpose of creating a relationship between two parties. He submitted that the court in construing the rent review terms was considering supplying or implying terms into an existing and continuing relationship of landlord and tenant. By way of contrast, the question of implying or supplying any terms in the option would be implying terms into a relationship which would be created in the future. Thus, he said, it was natural for the draftsman to be more careful in drawing the option clause because the court cannot supply terms for creating relationships and the draftsman, knowing that, will be more careful in drawing the provisions relating to future relationships to be created. On the other hand, the draftsman could be more slapdash in dealing with the terms in the rent review clause, since these apply to an existing relationship, and the draftsman can rely, in a sort of a sense, on the court making good his deficiencies.

I cannot accept that the court should construe a document on an assumption that there has been slapdash drafting. Although I entirely agree with Mr Lewison and accept his analysis of the difference in the court's approach in supplying terms to an existing relationship and refusing to supply terms in creating a relationship, I do not accept that the conclusion which Mr Lewison postulates follows from the analysis. In my judgment the analysis is a valuable way of recognising the difference between a true option which creates future relationships and a rent review clause which deals with an existing relationship and rights, yet I cannot accept that the contrast between the option clause and the rent review clause is not of any significance and can be ignored in regarding the terms of clause 3(2).

Mr Lewison has urged me to hold that for the court to imply *no* terms produces a foolish result. It produces a rental 15% higher than a letting in the real market would command in the present case according to the arbitrator's decision. He submits that I should proceed on some basis which he derives from the decision in *British Gas*. He submits that the Vice-Chancellor has there laid down a guideline that one should always try to construe a rent review provision as producing the current true market value of the letting which is in fact being made. Mr Lewison in his argument accepts that the result of saying that one should imply (as he says I implied in a rather different case, *Sterling Land Office Developments Ltd* v *Lloyds Bank plc**) that the terms not written out are to be the terms of the present lease is here to write out in the hypothetical lease all the covenants of the present lease including the provision for rent review.

I cannot accept that submission. The result postulated produces a duplication of clauses expressly provided for in the rent review clause and impliedly provided for by implying all the clauses of the present lease into the hypothetical lease. The submission must result in doing violence to language in order to produce what is assumed to be an intended result, ie that the revised rent must always be the open market rent. In my judgment no such assumption is justified. The parties have expressed themselves in certain terms; those terms are workable; there is no basis in law for overriding the terms actually expressed even if the result seems a little surprising. People are entitled to make their own bargains, even if the results surprise some observers.

For all these various reasons I am persuaded that the award was rightly made and that Mr Neuberger's argument is correct. There is only one oddity which I will mention so that it may be considered in future; I do not believe it arises directly today. That is the question of when the gales of rent shall fall due. In this lease the rent is payable under the reddendum quarterly in advance as in many commercial leases. The rent review clause makes no reference to the reddendum, nor to gale dates at all. It merely says that there is an annual rack rental value to be fixed. That seems to provide for an annual rent and, absent any provisions at all about gale dates, an annual rent is payable, as I understand the law, yearly in arrear.

It might be that there would be a difference in market valuation between a lease with rent on gale dates payable quarterly in advance and a lease with a rent payable yearly in arrear. No such suggestion seems to have been made in this case, but, says Mr Neuberger, it would be of assistance if I indicated whether in my view the hypothetical lease created by the rent review clause is to be a lease

---

*Editor's note: See (1984) 271 EG 894.

with a reddendum as in this lease or is to be on the basic common law term now thoroughly out of date.

In my view it is right to construe this rent review provision between a willing landlord and a willing tenant as upon the terms which are likely to be agreed between such people. In my judgment it is most improbable that anyone today would agree a rent of a substantial commercial building as a rent payable annually in arrear and the landlord could not properly be described as a willing landlord to grant such a term. In my view a reddendum, which is not mentioned at all by express term in the rent review clause, must have been intended to be included in the hypothetical lease and must have been intended to be included in the terms of the present lease. That, as it seems to me, is a matter of construction of the terms of this document, not a matter of implying any term and not a matter of contradicting any provision of it. Thus, I come to the conclusion that the arbitrator was right, although my reasons for it, I fear, are rather more lengthy than his. I dismiss this appeal.

Chancery Division

October 2 1986

(Before Mr Justice VINELOTT)

BISSETT v MARWIN SECURITIES LTD

Estates Gazette January 10 1987

281 EG 75-76

Landlord and tenant — Rent review clause in lease — Construction point — Lease for 30 years of premises used for the operation of amusement machines at rent of £12,500 per annum, with upward-only rent review provisions at mainly four-year intervals — Rental value for purposes of review was defined as being the highest of four figures — These were full market value of the premises let with vacant possession, or the previous rent plus an addition related to either the retail price index or the cost of living index, or the previous rent plus an addition "calculated at 10 % per annum on a compound basis from the date of the immediately preceding rent review" — The machinery for determination was a trigger notice by the lessors specifying the rental value, with provision for the lessee's counternotice and a reference to a valuer — The lessors at the initial review at first specified a rent of £18,300, based on the 10 % compound addition, but this was resisted by the lessee on the ground that the 10% compound formula operated only if there had been a "preceding rent review", whereas there had been no preceding review — Lessors replied that if the 10% formula was inappropriate they relied on the price or cost index basis, which produced a figure of £15,635 — On the plaintiff lessee's summons for construction, the lessors argued that the 10% formula should be read as if, after the words "the immediately preceding rent review", there were added "or the commencement of the term (as the case may be)" — Held that, construing the lease as a commercial document, it would be an odd result if the parties intended that the 10% escalation should start from the commencement of the term — Declaration that the rent payable from the first review date was £15,635 per annum (based on the indices formula)

The following cases are referred to in this report.

Bailey (CH) Ltd v Memorial Enterprises Ltd [1974] 1 WLR 728; [1974] 1 All ER 1003; (1973) 229 EG 613, CA
Wight v Dicksons (1813) 1 Dow 141

This was an originating summons taken out by the plaintiff, Paul Anthony Bissett, seeking the construction of the rent review provisions in a lease of premises in Peckham, for a term of 30 years from September 29 1981, of which the lessors were the defendants, Marwin Securities Ltd.

W Blackburn (instructed by Nicholson, Graham & Jones) appeared on behalf of the plaintiff; Norman Primost (instructed by Lithgow Pepper & Eldridge) represented the defendants.

Giving judgment, VINELOTT J said: This originating summons raises a short but by no means easy question as to the construction and effect of a rent review clause in a lease dated January 21 1982 and made between the defendant Marwin Securities Ltd, the lessor of the first part, the plaintiff Paul Anthony Bissett, the lessee of the second part, his wife Mrs Dorothea Elizabeth Bissett of the third part, and Alfred Marks Bureau Ltd, the former lessee who entered into a limited guarantee, of the fourth part.

By the lease, the lessor demised premises in Peckham to the lessee for a term of 30 years, from and including September 29 1981, at a rent of £12,500 per annum "subject to upward review as hereinafter provided" and payable by quarterly instalments in advance.

The lease contains in clause 2(14) a covenant by the lessee not to use the premises otherwise than for amusement machines with prizes. Then by clause 4(6) the lessor was given the right on the 15th, 20th or 25th anniversary of the term to determine the lease without compensation if it wishes to demolish or reconstruct the building or any adjoining, adjacent or neighbouring property.

The rent review provisions are contained in clause 5. Subclause (1) defines the expression "the review date" as meaning: "the 29th September 1985, 29th September 1989, 29th September 1992, 29th September 1997, 29th September 2001, 29th September 2005, 29th September 2009". Subclause (2) defines the expression "rental value" as meaning: "whichever is the highest of the following figures". There follow three paragraphs. Under para (a) the figure is the full market value of the premises let with vacant possession on the same terms "except as regards rent" and disregarding any effect on rent of the lessee being in occupation, the break provision and the restriction on user.

Para (b) starts with the words "the rent payable for the period immediately preceding the review date plus by way of additional rent". There follow two subparas joined by the disjunctive "or". The first defines "a sum bearing the same relation to the rent payable for the period immediately preceding the review date as shall be borne by any increase in the index of retail prices to the figures shown therein for the month of September 1981 or as at the review date for the preceding three-year period"; the second is in the same terms except for the substitution for the reference to the index of retail prices of a reference to the cost of living index. Para (c) I must read in full: "a rent equivalent to the rent payable for the period immediately preceding the review date plus an additional rent calculated at 10% per annum on a compound basis from the date of the immediately preceding rent review".

The next provision is numbered (d) and starts with the words "the rental value shall be determined as follows". The machinery provided is shortly the service of a notice by the lessor "at any time", specifying the rental value, the service at the option of the lessee of a counternotice within 28 days from receipt of the lessor's notice and, if the parties are unable to agree, a reference to a valuer. If no counternotice is served, the sum specified in the lessor's notice is to be the new rent. Para (d) should, I think, have been numbered subclause (3) (it is not part of the definition of the rental value but relates to the machinery for its ascertainment) and indeed the next provision is numbered subclause (4). That provides for the signature and annexation to the lease of a memorandum of any agreement or determination of the rental value.

Subclause (5) provides that:

pending such agreement or determination of the rent as aforesaid, the Lessee shall pay rent at the rate of the rent payable in respect of the period immediately preceding the review date on every date due for payment until the amount of the said rent shall be agreed or determined (hereinafter called "rent at the revised rate"). And forthwith upon such agreement or determination the Lessee shall pay in addition to the Quarter's rent at the revised rate such sum as with the rent already paid in respect of the period to which such agreement or determination relates down to such Quarter day shall be equal to rent at the revised rate for such a period down to such a Quarter day.

What happened was this. On April 25 1985 (the first review date being then six months ahead) the lessor's surveyors served a notice specifying the new rent as £18,300. That figure was arrived at by adding to the rent payable immediately before the forthcoming review date (the initial rent of £12,500) an additional rent calculated at 10% per annum compound on that rent. On May 9 the lessee's surveyors wrote objecting to the revised rental. The lessor's surveyors responded on May 22 1985 to say that the figure of £18,300 had been calculated in accordance with subclause (2) (c) and that there was accordingly no ground for disputing it.

On June 14 1985, the lessee's surveyors wrote to say that they had inspected the premises and that they were satisfied that the premises were let at full market rent and that no increase was called for. That of course did not answer the point made in the lessor's surveyors' letter of May 22. That point was answered by the lessee's solicitors on November 14, when they wrote to say that they had been advised by leading counsel that the lessor could not rely on para (c) "since it refers to a preceding rent review and there has been no such review to date".

The lessor's surveyors riposted with a further notice claiming, in effect, that if para (c) was inappropriate, the rental value ascertained in accordance with both subparas (i) and (ii) of para (b) was £15,635. There is no dispute about the accuracy of that figure. There was initially a dispute whether the lessor was entitled to have a second bite. The lessee now accepts that it was.

The question therefore is whether the rent during the first review period should be the £18,300 arrived at in accordance with para (c) or the £15,635 arrived at in accordance with para (b). That of course turns on whether para (c) applies to the determination of the rent at the first review date, there being at that date no "immediately preceding review date".

As I have said, I have not found this question an easy one. There can be no doubt that literally construed the formula in para (c) produces a rent of £12,500, that being the rent payable for the period immediately before September 29 1985. No addition falls to be made because the additional rent is to be calculated at 10% per annum compound (*sub silentio* on that initial rent) "from the date of the immediately preceding rent review", and on September 25 1985 there had been no preceding rent review.

The case for the lessor is that the parties must have intended that para (c) (like paras (a) and both subparas of para (b)) would produce a figure larger than £12,500 (or, I think, more accurately in the case of paras (a) and (b) would be capable of producing a figure larger than £12,500) so that on each review date the lessor would be able to choose from four different ways of recalculating the rent that one which produced the largest rent. To give effect to that intention requires only the addition after the words "the immediately preceding rent review" of the words "or the commencement of the term (as the case may be)". The question is whether, on construction, these words can be supplied.

It is common ground between the parties (as it was in *C H Bailey Ltd* v *Memorial Enterprises Ltd* [1974] 1 All ER 1003) "that this is a commercial document between commercial parties, which ought to be construed so far as possible to give effect to commercial good sense". (See *per* Megaw LJ at p 1007.) The clear commercial purpose of para (a) is to give the lessor on each review date the right to require the rent to be increased to the figure that the premises would have commanded if offered on the market for the residue of the term on the same terms as those in the lease, save in so far as expressly varied.

Para (a) thus ensures that the lessor is not prejudiced by the grant of a 30-year term if rental values of similar premises in the same area escalate. The clear commercial purpose of para (b) is to give the lessor the right to require the rent to be increased in line with changes in either of the two indices mentioned. There is in fact a slip in drafting in both limbs of para (b), which refers to "the review date for the preceding three-year period"; the review dates are for the most part separated by four, though in one instance five, years. However, the commercial purpose being once discerned, this error (and also the inelegant definition of "*the* review date" — my emphasis) can readily be corrected on construction.

Together paras (a) and (b) give the lessor both belt and elastic braces. If rental values increase ahead of inflation, he can call for rent to be assessed by reference to para (a); if rental values lag behind inflation (measured by either of the two indices) he can rely on para (b), notwithstanding that in the real world he has retained the premises and could not have let them at a rent which would reflect depreciation in the real value of the rent.

It is less easy to discover the commercial purpose of para (c). Under that para, the lessor is entitled to demand an arbitrary increase, which may bear no relationship either to the rent which the lessor might have been able to obtain, if he had let the premises for a shorter period and relet them on each review date on similar terms, or to the income he might have received if he had sold the premises and invested the proceeds in a way which would yield an income which would increase in step with increases in either of the two indices (assuming that that magical result was capable of achievement) and which, even on the lessee's construction, would yield a massive rent by the year 2009.

It is odd to find that (on a literal construction) this automatic escalation will not start until the first review date, but it is not, it seems to me, so odd as to compel the conclusion that the parties must have intended that the escalation would start at the commencement of the lease. There is a striking contrast between the language of subparas (i) and (ii) of para (b) (which refer in terms to the relevant index figure in September 1981 or, as I read para (b), the last preceding review date) and para (c), which omits any reference to September 1981 or to the commencement of the term. It may be that the lessee was willing to accept the unusual provision in para (c) only on the footing that it would not operate until September 29 1989 (the effect of compounding is that a reduction of four years in the period over which the rent is automatically increased makes a substantial difference towards the end of the term). This is not a case where (as in *Wight* v *Dicksons* (1813) 1 Dow 141) words have to be supplied in order to make sense of other provisions of the lease. It is true that, a point stressed by Mr Primost, subclause (5) is drafted on the assumption that at each review date *an* additional rent will be payable and that in theory there might be no increase at the first review date if para (c) does not then operate. But I think that the answer to this point is that the parties assumed that rental values generally would continue to increase and that the value of the pound sterling, measured by the indices, would continue to decline.

In sum, in my judgment, it cannot be said that it is clear from the terms of the lease, construed as a commercial document and with a view to giving effect to the commercial purpose to be discerned from the terms of the lease as a whole, that the parties must have intended that the automatic escalation in para (c) would start from the commencement of the term. If the existence of such a common intention can be inferred from other material, the lessor's remedy is an action for rectification.

On this summons, therefore, I propose only to declare that the rent payable from September 29 1985 is £15,635 per annum.

Court of Appeal

November 17 1986

(Before Lord Justice FOX, Lord Justice PARKER and Lord Justice GLIDEWELL)

YOUNG AND OTHERS v DALGETY PLC

Estates Gazette January 31 1987

281 EG 427-430

**Landlord and tenant — Rent review clause in lease — Whether certain items were landlord's fixtures which should be taken into account in the determination of the open market rent under the review clause or whether they were tenant's fixtures which should not be taken into account — Appeal from decision of Mervyn Davies J, who held that the items in question were tenant's fixtures — The items were light fittings, in the form of fluorescent tubes contained in glass boxes fixed securely to the plaster of the ceiling, and floor coverings, in the form of carpeting fixed to the floor with gripper rods, the rods being fixed to the floor with pins which were themselves attached to the carpet: the rods were laid on a screeded floor — Both the floor coverings and the light fittings had been installed in pursuance of obligations binding the tenants to do so in an agreement entered into prior to the execution of the lease — The effect of this was a main issue in the case — Decision in *Mowats Ltd* v *Hudson Bros Ltd* cited in support of the tenants' contention that the fact that the items had been installed in pursuance of the tenants' contractual obligations did not prevent them from being removable tenant's fixtures — Landlords argued that this was in effect allowing the tenants to take advantage of a breach of their obligations and that the items should be treated as landlord's fixtures — Held by the Court of Appeal that the judge had reached the correct conclusions — The decision in *Mowats'* case was indistinguishable in principle — The landlords obtained no title to the items in question — On the assumption that they were fixtures (there might be a question as to the carpeting, but it was a matter for the judge), they were tenant's fixtures and not to be taken into account in the determination of rent under the review clause — Appeal dismissed

The following cases are referred to in this report.

Mowats Ltd v *Hudson Bros Ltd* (1911) 105 LT 400
Spyer v *Phillipson* [1931] 2 Ch 183
Webb v *Frank Bevis Ltd* [1940] 1 All ER 247, CA

This was an appeal by the lessors, Stuart Young, Alasdair David, Gordon Milne and Harold Paul Hughes, acting as trustees of the BBC New Pension Scheme, from a decision of Mervyn Davies J holding, for the purpose of a rent review clause in a lease of premises at 19 Hanover Square, London W1, that certain items of floor covering and light fittings were to be treated as tenants' fixtures. The respondent lessees were Dalgety plc. Stuart Young died prior to the hearing and the action was continued by the remaining appellants.

John Colyer QC, Paul de la Piquerie and Wayne Clark (instructed by Bennetts & Partners) appeared on behalf of the appellants; Miss Hazel Williamson (instructed by Speechly Bircham) represented the respondents.

Giving judgment, FOX LJ said: This is an appeal from a judgment of Mervyn Davies J, in which he dealt with various points arising upon the construction of a rent review clause in a lease. The lease in question is dated July 9 1973 and concerns business premises at 19 Hanover Square, made between Land Securities Investment Trust as lessors of the first part, Schlesinger Insurance and Institutional Holdings Ltd as lessees of the second part, and Eagle Star Insurance Co Ltd as surety of the third part.

The lease was for a term of 30 years at a rent of £240,000 per annum, subject to a review.

Prior to the lease, there was an agreement entered into in April 1973 between the same parties and it was agreed that as soon as the premises (which were then in course of construction) had been completed in a manner therein specified the lessors would grant the lease. It was also agreed that the lessors would complete the premises by June 15 1973 in accordance with certain plans and a schedule of finishes prepared by the lessors' architects.

Clause 4 of the agreement provides as follows:

(a) THE Lessees when so requested by the Lessors' Architects shall forthwith at their own cost prepare and submit to the Lessors' Architects all suitable drawings and specifications (in triplicate) of the works mentioned in sub-clause (b) of this Clause and any other works which the Lessees desire to carry out and within seven days of the receipt by them of such Architects' approval thereof on the Lessors' behalf the Lessees shall thereupon with all due diligence apply for all requisite permissions consents licences and approvals of the Town Planning Local or other interested authorities as may be necessary for such works

(b) So soon as the Lessors' Architects shall certify in writing (such certificate being hereinafter referred to as "the Architects Certificate") that the said premises have been practically completed in accordance with the said plans and schedule of finishes (except that the said Certificate shall not be withheld or delayed on account of the Lessors having been unable to complete the Fire Escape arrangements for the said premises as hereinafter mentioned) or such earlier date as shall be agreed between the parties hereto the Lessees in a good and workmanlike manner shall proceed forthwith to carry out and complete without delay the following works, namely:

(i) lay adequate and suitable floor finishes to the offices on the first second third fourth and fifth floors of the said premises and the showroom premises on the ground and part of the basement floors of the said premises

(ii) decorate the interior of the said premises and install light fittings therein and

(iii) fit up and complete or cause to be fitted up and completed the said premises with all necessary services and installations including (where necessary) suitable shop fronts and relative shop front fittings in keeping with a first class shopping centre

PROVIDED (i) that all such Lessees works shall be carried out in accordance with the said specifications and drawings previously approved or amended by the Lessors and with the best materials available of their kind and in all respects to the reasonable satisfaction of the Lessors.

The rent review provisions, which gave rise to these proceedings, initially hinge upon the determination of the open market rent, which is defined in para 8 (c) of the Third Schedule to the lease in the following terms:

"open market rent" means the yearly rent for which the demised premises might reasonably be expected to be let on the open market with vacant possession by a willing lessor to a willing lessee for a term equal to the unexpired residue on the relevant rent review date of the term hereby granted and otherwise upon the terms and conditions (save as to the amount of rent payable but including this Schedule) contained in this Deed there being excluded:

(i) Any effect on rent of the fact that the Lessees have been in occupation of the demised premises

(ii) Any goodwill attached to the premises by reason of the carrying on thereat of the business of the Lessees (whether by him or by a predecessor of his in that business)

(iii) Any effect of an improvement carried out by the Lessees otherwise in pursuance of an obligation to the Lessors.

The lessees complied with the agreement to carry out the obligation contained in clause 4 (b) of the agreement, but doubts arose as to precisely what those obligations were since the relevant documents have been lost.

The identification of the works was necessary for the operation of the rent review clause — and in particular to subparagraph (b) to which I have referred. That gave rise to various questions before Mervyn Davies J which he determined at the specific request of the parties, that request being as to what can and must be assumed to have been done by the lessees in pursuance of the provisions of Clause 4 (b) of the agreement. The judge in fact had doubts as to whether he should embark upon that course, but as he was requested by the parties to do so, understandably he did so.

The matter was dealt with by reference to a number of items contained in a Scott Schedule, and the issue related to two things: floor coverings and light fittings. The learned judge found that floor coverings and light fittings were provided by the lessees, pursuant to the provisions of clause 4 (b), and by the originating summons (as amended) he was requested to determine whether the various items of floor coverings and light fittings were (a) tenant's fittings or (b) landlord's fixtures or (c) tenant's fixtures.

It was common ground that tenant's fittings or tenant's fixtures would not fall to be taken into account in determination of the open market rent; in my view that was a correct basis on which to proceed.

With regard to the light fittings, which the learned judge found to have been assumed to be provided, these consisted of fluorescent tubes contained in glass boxes fixed securely to the plaster of the ceiling. However, during the course of time, the lessees had in part substituted for those light fittings standard lamps which provided diffused lighting in an upwards direction.

However, what originally ought to have been was the fluorescent tubes contained in glass boxes which were securely affixed to the plaster of the ceiling.

The other item concerned the floor covering. That was carpeting fixed to the floor by gripper rods, such rods being fixed to the floor with pins which were themselves attached to the carpet. The rods were laid on a screeded floor.

The learned judge held that the carpeting was a fixture as were the light fittings. He held that both the carpets and the light fittings were tenant's fixtures and were therefore, under the ordinary principles of the law of landlord and tenant, removable by the tenants during the term or within a reasonable time thereafter. Accordingly (and this was not in dispute), since they were tenant's fixtures, they did not fall to be taken into account for the purposes of the rent review.

I now refer to various findings of the judge concerning the state of the premises. He said:

Question (1) is whether the 4 (b) obligations are to be related to a fitting out of the ground floor and basement as showrooms or as a banking hall. It is common ground that the ground floor and basement have never been used for showroom purposes. From the outset they were used as a banking hall by Western Bank Limited. There is an application by the bank for planning permission to change the "use of part of ground floor and basement to bank". The application is dated April 11 1973, that is two days before the date of the Agreement. On August 16 1973 planning permission was given for "use of the basement and ground floor (shown outlined in red on the submitted plan) of 19 Hanover Square, W1 as a bank". In the meantime on July 20 1973 there is a letter wherein the landlords state:

"The alterations to the ground and basement floors are the subject of our consent in principle dated the 21st June and we appreciate that you will be in touch with us as soon as you can with the further information so that these also can be formally approved".

One then looks at one of the plans produced. I refer to plan 271 dated June 8 1973 which refers to "The Western Bank Limited project". It shows a ground floor with a "banking hall" and a basement plan with "bank vaults" and "safe deposit". All these considerations suggest that the parties had in mind from April onwards the possibility that the ground floor and basement would be fitted out for banking rather than as a showroom. One then turns to the Agreement and the Lease to see what was agreed as to the ground floor and basement. The Agreement does not refer to banking; but in Clause 4 (b) (i) it refers to "the showroom premises on the ground and part of the basement floors of the said premises"; as well Clause 2 refers to the "Schedule of Finishes" and when one inspects that schedule one sees that it is headed "Schedule of Finishes relating to the new showroom and office block". The Lease has the User Clause 2 (18) that I have already quoted. Clause 2 (18) confines the user of the ground floor to showroom use with the proviso that "if the Lessee shall obtain the necessary permission it may use that part of the premises hereby authorised to be used as a showroom or parts thereof as a banking hall with ancillary offices".

In their argument before this court, the landlords place great weight on the fact that the tenants installed the carpets and the lights in pursuance of the contractual obligations contained in clause 4 of the agreement. It is submitted that to construe the tenant's covenant in clause 4 as requiring the tenants to install lights and carpets which could (if the tenants' case is correct) constitute tenant's fixtures which could be removed immediately by the tenants as soon as the lease had been entered into would (in commercial terms) make nonsense of the bargain made between the landlords and the tenants.

It is further submitted that it would, in effect, permit the tenants to benefit from a breach of their own covenant undertaken in clause 4 of the agreement and deprive the landlords of the benefit of the tenants' covenant which the landlords bought as part of the bargain for the lease.

It is said that the tenants' works consisted of applying "finishes", including carpets, and adding "fittings" without either of which the building would be incomplete and which were necessary to the convenient occupation of any tenant.

In terms the covenant merely was to provide and install the lights and the carpets. The argument that the conclusion of the learned judge would, in effect, allow the tenants to take advantage of a breach of the covenant seems to me to be begging the question and does not advance this case.

I turn then to consider further what the effect of the covenant was. The judge referred to the decision of the Court of Appeal in *Mowats Ltd v Hudson Bros Ltd* (1911) 105 LT 400, which the judge said showed that any item regarded as having been installed by the tenant pursuant to his obligation in clause 4 (b) is nevertheless removable if held to be tenants' fixtures.

The headnote to *Mowats* reads as follows:

By the lease of an unfinished shop the lessees covenanted at their own expense to "complete and finish . . . all necessary fittings for the carrying on of the trade of a provision merchant" and also to deliver up the demised premises in good repair at the end of the term. In pursuance of their covenant the lessees affixed certain fittings to the premises which became "trade fixtures" and they removed them shortly before the end of the term.

Held (allowing the appeal, Vaughan Williams LJ dissenting), that the covenants in the lease did not take away the right of the lessees during the term to remove the fittings as trade fixtures.

The covenant in question is set out in the judgment of Vaughan Williams LJ at p 402, with the preliminary observation:

The building thus described as a messuage or shop was, before the shop front and shop fittings were added, a mere shell and not a shop at all. Then follow the covenants . . .

which were to pay the rent and

(3) And also will within three calendar months from the date of these presents at their own expense erect and complete and finish in good, substantial, and workmanlike manner under the inspection and to the satisfaction of the lessors or their surveyor a suitable shop front to the said premises and all necessary fittings for the carrying on of the trade or business of a provision merchant. And also will at all times during the said term keep the said premises in good and sufficient repair and the same in good and sufficient repair deliver up to the lessors at the expiration or sooner determination of the said term of the tenancy.

The work which the tenants performed was to put in a new shop front. They erected a cashiers' office inside the building and installed other fixtures which were necessary for the covenient carrying on of the business. What they did in fact complied with the covenant.

At the end of the lease the tenants left the shop front as it was, but they removed other fixtures. The issue before the court was whether they were entitled to do so — there was disagreement as to that.

Fletcher Moulton LJ, delivering one of the majority judgments, said:

It would seem that at the time that the lease was granted the premises were in a somewhat unfinished state. For example, they possessed no shop front, although they were so constructed as to be suitable for and evidently intended to be used as a shop. This accounts for the presence of a covenant in the lease, which is substantially the only portion of the lease which we are asked to construe. In fact, its presence constitutes the only peculiarity in the document. In all other respects the lease is an ordinary lease for twenty-one years, terminable upon notice at the end of seven or fourteen years at the option of the lessees, containing the usual provisions and no others. The covenant reads as follows (speaking of the lessees)

and the learned lord justice read the covenant and continued:

I can see no reason for ascribing to the words of this covenant any other or further meaning than that which they naturally bear. The covenant provides that the lessees will at their own expense put in a shop front and fittings to the satisfaction of the lessors, and it says no more. It makes no reference to their being or becoming the property of any person, and, in the absence of any such provisions, the only legitimate conclusion is, in my opinion, that it leaves the question of property to be settled by the ordinary rules of law. There can be no doubt as to the result of so doing.

Then Buckley LJ said:

The learned judge has found as a fact that all the articles in question are things which the defendants would have a right to take away if it were not for the wording of the lease. They were all trade fixtures; they were articles, such as counters and shelves and a pay desk or office placed in the shop, which were brought upon the demised premises and fixed (so far as they were fixed) under such circumstances as that, but for anything to the contrary contained in the lease, the lessees would before the expiration of the term have been entitled to detach them and take them away.

Two propositions were urged before us. First, that by operation of law these chattels having been affixed to the freehold became part of the land, not *sub modo* and subject to the tenant's right of removal during the term, but absolutely because it was part of the consideration for the lease that they should be supplied and fixed; and, secondly, that upon the construction of the lease they had, as between lessor and lessee, been made the property of the lessor. If chattels are brought upon demised premises, whether in pursuance of a covenant to bring them there or not, the mere fact that they are brought upon the demised premises confers no title to them upon the owner of the freehold.

In the result the tenant was held entitled to remove them. The judge's view of the *Mowats* decision is challenged by the landlords, who submit that in *Mowats* it was conceded that the items were tenant's fixtures. It is correct that that concession was made. But in my view the concession cannot have been regarded by anybody as determining the matter which was then before the Court of Appeal. In effect, the case turned upon the meaning to be given to the covenant and, in my view, the concession which was made was merely that under the general law the items would, other things being equal, be regarded as tenant's fixtures.

The question under consideration was: What was the effect of the covenant on what would otherwise have been the position of the general law of landlord and tenant?

For myself, I cannot regard the concession which was made in relation to the classification of the items as trade fixtures or tenant's fixtures as being of materiality of the decision itself — in relation to the fundamental point concerning the effect of the covenant.

The decision of the majority in *Mowats* was that the covenant was merely a covenant to install items and was nothing more. In particular it said nothing about ownership or permanent attachment to the premises of the items in question. The matter was left — as appears from the judgments of Vaughan Williams and Buckley LJJ — to the operation of the ordinary rules of law. It was in effect a question of the construction of the covenant.

I can see nothing in terms of the commercial content of the agreement between the parties to compel me to the view that the landlords' contention that these were landlord's fixtures or fittings was correct. One must have regard to the covenant and its commercial implications, but the commercial reality of the matter is simply this: the tenants having taken a lease of this property at a high rental and having installed the carpets and the light fittings at their own cost for their own business purposes are not going to remove them the next day. They were chattels which, subject to the agreement of the landlords, were left to their choice and they were obviously chosen as being suitable for the tenants' purposes.

In those circumstances, the tenants having installed them in compliance with the covenant are going to leave them and obtain the maximum use from them. The landlords, for their part, have the benefit of the installation of carpets and light fittings of a quality which satisfy them as being suitable to the maintenance of the repute of the building. The landlords know that, in business terms, the tenants are likely to leave them there in order to obtain the fullest value from them.

In the same way there will be other chattels which will be brought into the premises by the tenants, though without obligation, being kept there by the tenants for use to maximum capacity. Of course with the passage of time, if the tenants feel that what has been installed is no longer satisfactory to them, then as a matter of convenience for the running of their organisation they will be replaced with something as good if not better.

In the present case the tenants substituted the standard up lighters for the lights which were required to be installed under the provisions of the agreement. That is the kind of situation which is satisfactory both to landlords and tenants.

Landlords require the building to be properly fitted out and so do tenants.

In my judgment, the commercial reality does not require one to go beyond that. It does not require the assumption that the tenants are bound to replace items under the repairing covenants or upon yielding up the premises at the end of their term.

The landlords stipulated for the benefit of a covenant to install the items; they did not stipulate for a covenant to give them title to the items themselves. In particular, this covenant (as the covenant in *Mowats*) said nothing about ownership and nothing about permanent attachment to the hereditament. The matter was left to the ordinary rules of law.

In my view this case is no different from *Mowats*. However, it is submitted that it does differ (1) because the *Mowats* agreement contained a break clause, but the existence of that break clause is irrelevant, having no practical impact on the question which this court has to decide. (2) It is said that *Mowats* lease envisaged fittings would be provided for the purposes of a trade to be carried on in the premises, the fittings being brought into the premises for that purpose. But in my view so does clause 4 (b) (3) in the present case.

In my judgment, therefore, the position is this: as with *Mowats*, here is a simple agreement to install certain items of equipment, without an indication as to their eventual ownership. The items were paid for by the tenants — and there is nothing in the documents which deals with their eventual ownership, so that *prima facie* it could be said that they belonged to the tenants.

What then is the impact of the general provisions of the law pertaining to landlord and tenant upon that situation? First of all, were they fixtures? The judge decided that both the carpets and the light fittings were fixtures. The light fittings were securely fitted to the plaster of the ceiling. On the other hand, the fitting of the carpets could be said to be more tenuous because such fitting depended upon a gripper rod and pin arrangement. However, in my view it is not necessary for the court to decide whether the learned judge was right or wrong in that connection.

The lessees have contended that these items were not fixtures, but for the purposes of this case I am prepared to assume that the learned judge was correct in his conclusion that they were fixtures. Therefore, on the assumption they were fixtures, the next question is: Were these items tenant's fixtures or landlord's fixtures? If they were landlord's fixtures, then they must be left for the landlords and it would therefore be proper to bring them into account in relation to a rent review. However, the position is otherwise and it is not in dispute if they were tenant's fixtures. In general, fixtures attached by tenants for the purposes of their trade or business will be tenant's fixtures. In that connection we were referred to *Megarry and Wade's Law of Real Property* 5th ed, p 735, where the general rule is stated.

It is said on behalf of the landlords that the fixtures were installed for the completion and fitting out of the building itself and that they were not there for the convenience of the tenants. Fundamentally, they were there for the benefit of *any* tenant who came into the building.

It is said, further, that the carpets were "floor finishes" within the terms of clause 4 (b), and that the term "floor finishes" indicates a greater degree of absorption in the hereditament than would an ordinary carpet.

The learned judge found that the items were attached merely to render the premises convenient for the *tenants*' occupation of the building as business premises. In my view that conclusion was perfectly open to the learned judge on the facts before him. I see no reason to depart from it. The fact that installation was in pursuance of a contractual obligation by the tenants to the landlords under clause 4 (b) does not prevent such a conclusion any more than did the covenant in *Mowats'* case.

I accept that the attachment must be such that, on removal, it would not lose its essential character. In that connection the learned judge referred to *Webb v Frank Bevis Ltd* [1940] 1 All ER 247. In my view neither the light fittings nor the carpets would, upon removal, lose their essential character, or such value as they then had. The carpets in fact would only need to be freed from the rods; and as to the light fittings, there is no reason to suppose that by freeing them from the plaster on the ceiling, that would destroy either their character or any value placed upon them. The judge took that view, and on the evidence before him he was entitled so to do.

It is accepted of course that an article must be capable of being removed without irreparable damage to the demised premises. In that connection the judge referred to *Spyer v Phillipson* [1931] 2 Ch 183 and to a passage from the judgment of Romer LJ which he set out.

In my judgment the removal of these items could not in any sense of the word cause irreparable damage to the demised premises — as I have said, the carpets would simply have to be removed from the grip rods. The removal of the light fittings from the plaster of the ceiling would no doubt cause some damage to the plaster itself, but that is a minor matter, being easily remedied.

The position is, therefore, that in my view the judge was right in the conclusion to which he came; these were matters of fact for him; I can see no reason to depart from the conclusion which he reached, and I would therefore dismiss this appeal.

PARKER and GLIDEWELL LJJ agreed and did not add anything.

The appeal was dismissed with costs.

Court of Appeal

January 22 1987

(Before Lord Justice FOX, Lord Justice DILLON and Lord Justice RUSSELL)

WOLFF AND ANOTHER v ENFIELD LONDON BOROUGH

Estates Gazette March 21 1987

281 EG 1320-1321

Landlord and tenant — Rent review clause in lease — Construction — Effect of use clause in lease and planning permission — Lease provided for use of the demised premises for any purpose within Class III in the Town and Country Planning (Use Classes) Order 1972 or any other class or classes within which fell the use or uses permitted by the planning authority from time to time — Last use before present lease was granted was use for light industrial purposes, but shortly before the grant of the lease planning permission was given for change of use to that of a non-teaching service unit for the Middlesex Polytechnic — The question arose whether the permitted purpose of the non-teaching service unit for the polytechnic was a composite use which did not fall into any of the use classes, or whether the wording was shorthand for listing all the various uses which the polytechnic might seek to make of the premises, such as office use, light industrial use for printing, storage use for warehousing, or anything else which might be considered relevant — The latter construction would enable any subsequent user (not the polytechnic) to take advantage of a whole range of uses — Held, affirming the judgment of Whitford J, from which the landlords had appealed, that the use authorised by the planning permission was a composite use which did not fall within any of the prescribed use classes — The consequence was that there was nothing in the rent review clause to permit the rent to be assessed by reference to any user authorised by a use class except for light industrial purposes within Class III of the Use Classes Order — Appeal dismissed

No cases are referred to in this report.

This was an appeal by the plaintiff landlords, Werner Wolff and Esther Wolff, as trustees of the Wolff Charity Trust, from the decision of Whitford J [1985] 1 EGLR 75; (1984) 273 EG 1121 in proceedings to determine the true construction of a rent review clause in a lease of premises known as the Chase Side Works, in Chase Side, Southgate. The defendants (the present respondents) were the tenants, the London Borough of Enfield.

Michael Barnes QC and J C Harper (instructed by Rabin Leacock & Partners) appeared on behalf of the appellants; Anthony May QC and S A Furst (instructed by W D Day, London Borough of Enfield) represented the respondents.

Giving the first judgment at the invitation of Fox LJ, DILLON LJ said: This is an appeal by the plaintiffs in proceedings against a decision of Whitford J given on December 13 1984. The appeal raises a question of the true construction and effect of a rent review clause in a lease. The point raised lies in a very narrow compass, and the clause is an unusual one to be applied in unusual circumstances.

The appellants, the trustees of the Wolff Charity Trust, are the landlords by assignment of the reversion; the respondents, the London Borough of Enfield, are the tenants.

There are in fact two leases, one a lease in possession and the other a reversionary lease, both dated August 22 1977. Nothing turns on the fact that there are two leases.

The lease in possession demised the premises (which are premises known as the Chase Side Works, in Chase Side, Southgate) for a term from August 22 1977 to March 23 1983. The reversionary lease demised the same premises from March 25 1983 to August 21 2002.

The rent under the lease in possession was at a fixed yearly rate of £39,000 for the first five years of the term, but for the remainder of the term (which is called "the last period") it was to be either a yearly rent of £39,000 or the fair market rent of the demised premises at the commencement of the last period, whichever was the higher.

Under the reversionary lease, the rent for the period commencing on March 25 1983 and continuing to August 21 1987 was to be a yearly rent equal to the yearly rent payable immediately prior to the commencement of the term of the reversionary lease under the previous lease, the lease in possession. There were then provisions in the reversionary lease for later rent reviews.

The court has to consider the rent review at the commencement of the last period of the lease in possession. The wording which prescribes that is to be found in clause 1(2) of that lease, which states:

> The said fair market rent shall be the amount which shall be agreed between the Landlord and the Tenant to be the best annual rent for the time being obtainable as between a willing landlord and a willing tenant in respect of the demised premises on a letting thereof as a whole with vacant possession for use for any purpose within Class III of the Town and Country Planning (Use Classes) Order 1972 or any other class or classes of the said Order within which falls the use or uses of the demised premises permitted by the planning authority from time to time . . .

The Use Classes Order prescribed a number of use classes. Under the Town and Country Planning Act it was provided that in a case of buildings or other land which are used for a purpose of any class specified in an order made by the Secretary of State for the use thereof for any other purpose of the same class would not involve development of the land.

Among the use classes prescribed by the order, those said to be material to the present appeal are Class II (use as an office for any purpose); Class III (use as a light industrial building for any purpose); and Class X (use as a wholesale warehouse or repository for any purpose).

When the premises were last occupied before this lease was granted, they had been used for light industrial purposes within Class III. Shortly before the lease was granted, however, a planning permission was granted for a change of use. That was granted by the planning authority of the London Borough of Enfield (who are the prospective tenants) on July 29 1977 on an application made by their chief education officer.

The permission reads as follows:

> Whereas in accordance with the provisions of the Town and Country Planning Act 1971 and the Orders made thereunder you have made application dated 28 6 77 and illustrated by plans for the permission of the Local Planning Authority to develop land situated at Chase Side Works, Chase Side, Southgate N14 by change of use of existing premises from light industry to non-teaching service unit for the Middlesex Polytechnic Now therefore THE COUNCIL OF THE LONDON BOROUGH OF ENFIELD, the Local Planning Authority, HEREBY GIVE YOU NOTICE pursuant to the said Act and the Orders made thereunder that permission to develop the said land in accordance with the said application is HEREBY GRANTED.

There was a condition as to the beginning of the development, but nothing turns on that.

In view of the reference to the application, and to elucidate the words "non-teaching service unit for the Middlesex Polytechnic" it is convenient — and in my judgment permissible — to look at the planning application referred to. That gives as brief particulars of the proposed development "Non-teaching Service Unit for the Middlesex Polytechnic, primarily office and light industrial work such as printing" and then, a little further down, in indicating the areas of each aspect of that use, it sets out "Administrative offices, 600 square metres, Special Research/Workshop storage, 2,180 square metres".

Therefore one can obtain an idea of what was meant by a "Non-teaching Service Unit for the Middlesex Polytechnic", but it is clear that the planning permission does not tie the Middlesex Polytechnic to the use of precise areas for offices, or precise areas for workshop or storage, or other purposes within the general description.

We have in evidence plans supplied by both sides indicating roughly what the premises, which consist of a building having a ground floor and a lower-ground floor, are used as. It appears that a substantial amount is used as offices, including a filing area; there is also a warehouse or storage area, a boiler house and a printing room. There would also appear to be another form of storage area or stationery store — although the precise nature of that use is not clear. What is clear, however, is that any printing activities are not ancillary to the office activities. Likewise, the office activities are not ancillary to the storage or printing activities. Each activity is ancillary to the general purposes of the Middlesex Polytechnic, and that is why it is referred to as a "Non-teaching Service Unit" for that institution.

The question then is: how is the rent review clause to be applied to that grant of planning permission? I have already referred to the clause. One has to see what use is permitted by the planning authority from time to time. It is a question of looking not at how the buildings are actually used on the site but what is authorised by the planning permission. One then has to consider whether what is authorised falls into any one or more of the use classes within the Use Classes Order which I have mentioned.

A search of the Use Classes Order would not disclose any use class which applies specifically and in terms to non-teaching service units for the Middlesex Polytechnic, or any other polytechnic or educational establishment.

The question therefore is: whether the permitted purpose of the non-teaching service unit for the Middlesex Polytechnic is a composite use which does not fall into any of the use classes; or whether that wording is a shorthand phrase for listing all the various uses which the Middlesex Polytechnic might seek to make of the premises — such as office use, light industrial use for printing, storage use for warehousing, or anything else which might be considered relevant.

It is established that there are permitted planning uses which do not fall within any of the classes in the Use Classes Order. One instance established of such a use by a decision binding on this court is use as a builders' yard. Another instance, treated as such a use by a decision of the Divisional Court, is use as a sculptors' studio. I am not concerned to consider whether that decision was right or wrong.

The difficulty I feel about treating the description "non-teaching service unit for the Middlesex Polytechnic" as a form of shorthand describing uses within the classes is that, as I read this permission, it was leaving it to the Middlesex Polytechnic to decide which of its ancillary or non-teaching service activities were to be carried on in the Chase Side Works. If, for instance, they had decided initially to use them merely for the printing works and offices, then they could do so; if they decided subsequently to close the printing works and use the space they had occupied for storage of books and equipment for use in the lecture rooms of the polytechnic elsewhere, they could do so under the legend "non-teaching service unit for the Middlesex Polytechnic" without going back for any further planning permission.

On the other hand, I am not attracted by the notion that this is to be regarded as, in some sense, a list of uses, authorising offices, light industrial use, storage use, or anything else which might be regarded as being relevant to a service unit for the Middlesex Polytechnic in a general way, so that any subsequent user (not the Middlesex

Polytechnic) would be able to develop its own activities within the use classes referred to which comprehend those particular activities of the Middlesex Polytechnic which have supposedly been listed as separate permitted activities.

I prefer, therefore, the view that what is authorised by this planning application is a composite use which does not fall into any of the prescribed use classes. If that be right, there is nothing in the rent review clause to permit the rent to be assessed by reference to any user authorised by a use class except use for light industrial purposes within Class III of the Use Classes Order. This is the conclusion to which Whitford J came. I agree with him and would dismiss this appeal.

RUSSELL LJ said: I agree with the judgment delivered by Dillon LJ and have nothing to add. For the reasons he has given, I, too, would dismiss this appeal.

FOX LJ also agreed and did not add anything.

The appeal was dismissed with costs. Leave to appeal to the House of Lords was refused.

Chancery Division

November 20 1986
(Before Mr Justice MILLETT)

POWER SECURITIES (MANCHESTER) LTD v PRUDENTIAL ASSURANCE CO LTD

Estates Gazette March 21 1987

281 EG 1327-1331

Landlord and tenant — Construction of rent review clause — Unusual provisions raising an old problem in a new form — Machinery of review formula was complicated, but the issue was whether time was of the essence in so far as it concerned a requirement that a component part of the review formula should be agreed by a named date — The component part was a figure, essential to the calculation of the reviewed rent, of the total income receivable from the demised premises — It was in the interests of the landlords to contend that time was of the essence for this purpose and that the time-limit had passed without the required agreement having been reached — Judge's statement of four principles derived from the case law following the decision in United Scientific Holdings Ltd v Burnley Borough Council **— Various arguments by landlords in present case in favour of displacing the presumption that time was not of the essence rejected by judge — Main argument was based on the express provision as to what was to happen in case of default, namely, that the figure to be taken for total income was £225,000 — Held that the strength of such a provision as a contra-indication depended on its nature and purpose, there being considerable differences between default clauses in this respect — In the present case it was understandable and necessary that some time-limit should be fixed by which the tenants were to satisfy the landlords as to the true amount of income from the demised premises and that some provision on default should be included — But in all the circumstances this did not imply that time was of the essence — The default provision was entirely consistent with the assumption that time was not of the essence — It was still open to the parties in the present case to agree as to the total income receivable for the premises — Declaration to be framed accordingly —** *Dictum* **as to possibility of making time of the essence by serving a notice**

The following case is referred to in this report.

United Scientific Holdings Ltd v Burnley Borough Council [1978] AC 904; [1977] 2 WLR 806; [1977] 2 All ER 62; (1977) 33 P&CR 220; [1977] EGD 195; (1977) 243 EG 43 & 127, HL

This was an originating summons by which the plaintiff tenants, Power Securities (Manchester) Ltd, sought a declaration in regard to provisions in a rent review clause of the lease of a shopping centre at the Royal Exchange, Manchester, of which the defendants, Prudential Assurance Co Ltd, were the landlords.

Michael Barnes QC and Jonathan Brock (instructed by Ellison & Co) appeared on behalf of the plaintiffs; Derek Wood QC and Edward Cole (instructed by J T Flanagan, of the legal department of the Prudential Assurance Co) represented the defendants.

Giving judgment, MILLETT J said: This case raises a familiar question on the construction of an unusual form of rent review clause in a lease. The lease in question is dated August 3 1981 and demises a high-class shopping centre at The Royal Exchange, Manchester, for a term of 99 years from December 25 1979.

Until December 25 1986 the rent payable under the lease is fixed at £150,000 per annum. Thereafter the term is divided into 18 successive rental periods of five years each. During each rental period the rent is to be the greater of the highest rent previously payable under the lease and such rent as shall be equal to an amount obtained by multiplying 95% of the fair yearly rent for the demised premises at the beginning of the relevant rental period by a fraction of which the numerator is £150,000 and the denominator is the amount of the total income receivable for the demised premises. The total income receivable for the demised premises is defined (omitting detailed qualifications not material for present purposes) as the aggregate of (i) the yearly rents which the lessee was entitled to receive as at December 25 1981 under the provisions of all underlettings and agreements for underletting made with the consent of the landlord, and (ii) the fair yearly rent as at that date for all such parts of the demised premises as were not then the subject of such underlettings or agreements for underletting. Forsaking strict accuracy in the interests of brevity, I shall refer to these as "voids".

There are thus two components in the formula which require to be ascertained: the fair yearly rent (both of the demised premises as a whole and of the voids) and the total income receivable. The lease contains elaborate provisions for the ascertainment of both. Before I turn to these, however, three features call for comment.

First, of the two components in the formula, one (the fair yearly rent for the demised premises as a whole) is a variable which requires to be ascertained afresh for each successive rental period of the term, whereas the other (the total income receivable for the demised premises) is a constant which is to be ascertained once and for all as at December 25 1981 and is applicable for the determination of the rent payable under the lease throughout the whole of the residue of the term after December 25 1986.

Second, the constant component, the total income receivable for the demised premises as at December 25 1981, itself has two components: one (the rents receivable from underlettings) being an actual figure and the other (the fair yearly rent for the voids) being a notional one.

Third, and unusually, since the total income receivable constitutes the denominator of the fraction in the formula, the higher the amount the less is the rent payable under the lease. The lease provides for a maximum and a minimum figure of £275,000 and £225,000 respectively for the denominator, so that the proportion represented by the fraction can vary within a limited range from 54% at £275,000 to 66⅔% at £225,000. £225,000 was evidently the figure which the parties contemplated could be achieved on the first underletting; if the lessee succeeded in obtaining more, it was evidently to have the added incentive of a reduction throughout the residue of the term in the proportion of the revised fair yearly rent which was payable to the landlord.

The machinery for the ascertainment of the components in the formula may be summarised as follows: *First, the fair yearly rent of the demised premises as a whole.* This may be agreed between the parties at any time. If not agreed, it may be determined, at the option of the landlord, either by an arbitrator or by an expert valuer to be appointed by the president for the time being of the Royal Institution of Chartered Surveyors on the application of the landlord made not more than two quarters before, or at any time after, the commencement of the relevant rental period. If by the first day appointed for the payment of rent for any rental period the landlord has not made an application for the appointment of an arbitrator or expert, then the lessee may at any time thereafter serve on the landlord a notice in writing containing a proposal as to the amount of the fair yearly rent for that rental period. Then, unless the landlord duly makes an application for the appointment of an expert or arbitrator within three months after service of such notice, the amount so proposed is to constitute the fair yearly rent.

Second, the fair yearly rent for the voids as at December 25 1981 (a component of the total income receivable). This, too, may be agreed at any time between the parties. In the absence of agreement, it is expressly provided that it is to be determined by the same means and upon the same terms (*mutatis mutandis*) as are provided for the determination of the fair yearly rent for the whole of the demised premises. It is common ground that that incorporates the provision by which the lessee may serve a notice upon the landlord setting out the lessee's proposals and the landlord, if it has not already applied for the amount to be determined, may thereafter within three months make the application for determination, failing which the amount in the lessee's notice is to be taken as binding.

Third, the actual rents receivable from the underlettings (the other component of the total income receivable). This is not, in fact, provided for separately, or at all. Instead, provision is made for the ascertainment of the total income receivable: that is to say, the figure of which one component has already been dealt with. Since it is in relation to this that the question of construction arises, I shall read it in full:

Provided that the total income receivable for the demised premises shall be agreed between the landlord and the lessee within six months after the expiration of the second year of the said term, and if not so agreed shall be the sum of £225,000. And when the amount of the total income receivable for the demised premises has been ascertained, memoranda thereof shall thereupon be signed upon behalf of the landlord and the lessee and annexed to this lease and counterpart thereof and shall bind the landlord and the lessee including their respective successors in title and also the guarantors. The lessee shall provide the landlord with such information and assistance, including an accountant's certificate, as the landlord shall reasonably require to enable it to verify the total income receivable for the demised premises and without prejudice to the generality of the foregoing permit any person authorised on their behalf by the landlord to inspect and take copies of the counterparts of any under-leases agreements for under-leases and duplicate licences affecting the whole or any part of the demised premises and of the relevant books and accounts of the lessee and the lessee will afford such explanations thereof as that person or the landlord may require.

No attempt to agree the total income receivable was made before June 25 1982 and no agreement has even now been concluded. In fact, of course, if the strict requirements of the lease are for the moment laid aside, there is no need for the total income receivable for the demised premises to be ascertained before December 25 1986 at the earliest, this being the first review date. Until then, the rent is fixed at £150,000 per annum. The landlords, however, contend that, no agreement as to the amount of the total income receivable having been concluded by June 25 1982, it must now be taken to be £225,000.

The lessee accepts that this is indeed the inevitable result of a literal construction of the lease. It is, however, submitted that this is not the correct result in law, since time is not of the essence of the requirement that the total income receivable should be agreed by June 25 1982, nor has time since been made of the essence of that date by notice served by either party. Accordingly, it is submitted, it is still open to the parties to agree, if they can, the true figure and displace the figure of £225,000 which is to apply in default of agreement.

This alone, however, is not enough for the lessee. On its behalf it was contended that, there being no obligation on the part of the landlord to agree any figure, either within the period stipulated or later, and the default figure being the most favourable result which the landlord could possibly achieve, some term must be implied in order to give business efficacy to the contract. Otherwise, whatever the lessee does, and whatever the true position, the landlord, simply by refusing to agree a higher figure, could always procure a figure of £225,000 for the total income receivable. This would make the elaborate provisions for the ascertainment of the fair yearly rent of any voids (a component in the total income receivable) and for the provision by the lessee of information and assistance to enable the landlord to verify the total income receivable mere surplusage.

The lessee originally submitted (though in his reply Mr Barnes, for the lessee, resiled from this submission) that the words "or otherwise determined" should be implied after the words "and if not so agreed" in the proviso, so that it read:

Provided that the total income receivable for the demised premises shall be agreed between the landlord and the lessee within six months after the expiration of the second year of the said term and if not so agreed or otherwise determined shall be the sum of £225,000 . . .

This, it was hoped, would avoid altogether the question whether time was of the essence of the requirement that agreement be reached within the six months' period, since that would not apply to the alternative.

I do not agree that the question can be avoided by this means. The addition, expressly or by implication, of a second negative contingency in the concluding part of the sentence requires the implication of a corresponding requirement in the earlier, so that in full it would read:

Provided that the total income receivable for the demised premises shall be agreed between the landlord and the lessee, or be otherwise determined, within six months after the expiration of the second year of the said term and if not so agreed or otherwise determined . . .

However, I would in any case reject the implication of the proposed words. In the context of this lease, in which elaborate machinery is provided for the resolution of potential disputes in the ascertainment of each component in the calculation, I would hesitate long before implying a provision requiring a dispute to be resolved otherwise than by agreement and yet omitting any machinery for its determination. Moreover, the suggested term is, in my judgment, repugnant to the concluding paragraph of the Schedule, which shows that the process by which the total income receivable is to be determined, and the only process which is envisaged, is that the lessee is to put forward a figure and support it by such information and assistance as the landlord may reasonably require in order to enable it to verify the figure. This is the process which is described as "agreement" and it is this process which must be completed within the six months' period.

But there is a more fundamental reason — which in the end I think Mr Barnes in his reply accepted — for rejecting the words proposed to be implied: they are not necessary to give business efficacy to the contract. In my judgment, on the true construction of this contract, the landlord cannot arbitrarily or unreasonably withhold its agreement to the proper figure if it is established to the landlord's reasonable satisfaction. The process described as "agreement" is not one of negotiation or bargaining but simply of the acceptance by both parties of facts capable of being objectively established. In so far as the total income receivable depends on the fair yearly rent for the voids, it can be established by obtaining a determination by an arbitrator or expert in accordance with the procedure prescribed. In so far as it depends on the actual rent received from underlettings, it can be established by documentary and accountancy evidence. It is for the lessee to establish that the total exceeds £225,000 and to provide information and assistance to enable the landlord to verify the figure. But, in my judgment, the landlord is bound to agree the figure if established to its reasonable satisfaction within the stipulated period of six months, and the landlord, to be fair, never contended to the contrary. I should add that, since one of the elements in the ascertainment of the total income receivable is the fair yearly rent for any voids, and express provision is made for this to be determined in case of dispute by an arbitrator or expert, it necessarily follows, in my judgment, (i) that the landlord cannot be said to be acting unreasonably if it insists on invoking the procedure prescribed for the ascertainment of this element of the total income receivable by arbitration or by expert report, and (ii) that the landlord is bound by any such determination.

This brings me to the real question in the case: whether time is of the essence of the requirement that the total income receivable be agreed, within the meaning of the clause, within the period of six months. The cases on this topic are not easy to reconcile, but, in my judgment, the following principles emerge:

(1) The correct approach to a rent review clause is to begin with a presumption that time is not of the essence of the time-limits laid down for the various steps to be taken for the determination of a revised rent or, it may be added, of any component element in its calculation.

(2) This presumption will be displaced if, on a consideration of the lease as a whole, and in particular of the provisions of the rent review clause as a whole, it appears that the parties have evinced a contrary intention.

(3) Where the parties have not only required a step to be taken within a specified time but have expressly provided for the consequences in case of default, this provides an indication, of greater or less strength, that time is to be of the essence, but it is not necessarily decisive. Whether it is so or not must depend on all the circumstances of the case, including the context and wording of the provision, the

degree of emphasis, the purpose and effect of the default clause and any other relevant consideration.

(4) In the end, the matter is one of impression to be derived from a consideration of the rent review clause as a whole, together with any other relevant considerations, avoiding fine distinctions, but giving effect to every provision in the lease.

The landlord's first submission was that this was not a case in which the presumption against time being of the essence applied at all, since the step in question was not a step in a process towards the determination of a figure (in the present case not the rent but the total income receivable) but the determination itself. The date prescribed is the date by which the relevant amount is to be determined in whatever manner. In my judgment, this cannot exclude the presumption. In *United Scientific Holdings Ltd* v *Burnley Borough Council* [1978] AC 904 the presumption was applied to just such a case, where the only time-limit was that the rent for each successive rental period was to be determined by agreement or, failing agreement, by arbitration during the year immediately preceding the period to which the rent would relate.

The landlord next submitted that the present case is simply one of contingency. If the total income receivable is agreed at a higher figure than £225,000 within six months, then that higher figure is to be taken; if not, the figure to be taken is £225,000. There is, it was said, no room for the court to rewrite the contract which the parties have made. It is, it was argued, as if they had expressly agreed that the figure should be £X if all the shops were to be let by the end of the period and £Y if they were not. I do not accept the submission or the appropriateness of the analogy. The distinction, in my judgment, is not between the final step in the determination of the figure and earlier steps in the process leading to its determination but between the completion of the process or of a step in that process, on the one hand, and the happening of an external event on the other, which, though determinative of the figure, is not itself a step in its determination.

On the footing, therefore, that the normal presumption applies, is it displaced in the present case? To displace it the landlord relied on the following features:

(i) The rent review clause in the present case is not of the normal kind found in any of the precedent books or in any of the cases but clearly represents a special bargain negotiated at length between the parties.
(ii) The relevant time is not for the completion of a stage in the determination of the figure but for its final determination.
(iii) The time provision is in stark contrast to other steps in the timetable which may by express provision be taken at any time, and
(iv) There is an express provision for what is to happen in case of default, ie that the figure to be taken for total income receivable is £225,000.

In my judgment, there is nothing at all in the first or second of these features. If anything, the first might be said to tend in the opposite direction. Since 1978 the parties or their legal advisers must be taken to know that, unless they evince an intention to the contrary, the presumption is that time is not of the essence of any limit specified in the rent review clause. Failure to negate that presumption by express words in a specially negotiated clause might well be thought by some to indicate an intention to accept the presumption; but whether this be so or not, it certainly cannot be taken to evince an intention to displace it.

The second, as I have already pointed out, was a feature of *United Scientific Holdings Ltd* v *Burnley Borough Council* and was not considered by any of their lordships in that case to constitute a contrary intention which might be held to displace the presumption. I can see no possible reason why it should.

The third, in my judgment, is explained by considering the purpose of the time provision in question. It would be ridiculous to have no time-limit within which the lessee must satisfy the landlord of the true amount of the total income receivable. It is essential to have some date after which the landlord can say to the lessee "You have had enough time to satisfy me that the figure is in excess of £225,000 and you have failed to do so — the figure must now be taken to be £225,000". But it does not follow that time is necessarily of the essence of the date specified. If the period is exceeded and time is not of the essence, the landlord can at any time make time of the essence by the simple expedient of serving a notice giving the lessee some further but limited time to provide the landlord with the evidence required.

Mr Wood, on behalf of the landlord, submitted — in relation, it is true, to the three months' period in the machinery for the determination of the fair yearly rent, but the submission is equally applicable to the point that I am now considering — that where a party has an express right to do something at any time, the other can always make time of the essence by serving a notice stipulating a final date beyond which there should be no further opportunity given and making time of the essence of that date. I disagree. Neither party is entitled to abridge the time given by the contract to the other. The most that he can do is to make time of the essence for the taking of a step for which a time-limit has been prescribed in the contract and has been exceeded.

It was the fourth feature which gave rise to most of the argument. In my view, the strength of the contra-indication which is provided by an express provision for what is to happen in case of default depends upon a number of considerations, but primarily upon the nature and purpose of the default provision itself. There is, in my judgment, a considerable difference between a default clause which merely expresses what would be implied anyway and a default provision which provides for an entirely new figure for the rent, or for some component element in the rent, which would not otherwise be implied. The former can have little, if any, weight. The latter must have considerably greater force and may in some circumstances be decisive. Finally, there may be a default clause which provides that if a notice is not served within the time stipulated it shall be of no effect or shall be void. It is difficult to see how more clearly parties could express an intention to make time of the essence of the serving of a notice in such a case.

The present clause does not come within any of these categories, although, perhaps, the nearest that it comes to is the second. But I have already explained its purpose. It is simply not a case in which it would be sensible or appropriate to give the lessee an indeterminate length of time in which to satisfy the landlord of the true figure. It was essential to specify some time-limit within which that should be done, and it was also essential to provide for what was to happen in default. It does not follow in the slightest that time was intended to be of the essence of the date specified.

In my judgment, neither singly nor cumulatively do any of the considerations to which I have referred constitute enough to displace the presumption; but in any event there are three considerations which point in the opposite direction. First, there was no great urgency about the determination of the total income receivable within the six months stipulated. There was no need for the total income receivable to be determined before the first rent review date, and if the landlord was prepared to give the lessee further time in which to satisfy it of the amount of the true income receivable right up until the first review date then that was a matter for it. If for any reason commercial considerations rendered it convenient or desirable that the landlord should know the value of its investment and, for that purpose, should know the proportion of the fair yearly rent of the premises to which it would be entitled throughout the rest of the term, then there was nothing simpler than to call upon the lessee to satisfy it as to the true income receivable within some stated time. The true effect, as it seems to me, of the six months' provision was to prevent the landlord from abridging that time, but I see no reason why it should be thought that the landlord was entitled to require strict compliance without serving a notice to make time of the essence.

Second, it must be remembered that the total income receivable consists of two components, of which one is, or may be, the fair yearly rent for the voids, and that a means for determining this element in the total has been prescribed. As I have already pointed out, the landlord could not be said to be acting unreasonably if it insisted upon that element in the total income receivable being determined by the prescribed means. Since the existence of any void could not be known for certain before December 25 1981 the application for the appointment of an expert or arbitrator could not reasonably be made before that date. If the landlord should fail to make that application the lessee's remedy was at any time thereafter to serve a notice stating what in its view was the true figure, and the landlord then had a further opportunity to apply for the appointment of the expert or arbitrator. Only if it failed to do so within three months would the tenant's figure become binding. It is clear that, first, there could be no application for the appointment of the arbitrator or expert before December 25 1981 and, second, that even if the lessee served a notice on the next day there could be no certainty that there would be any application for the appointment of the expert or arbitrator for a further three months. The parties must be taken to have contemplated at least the possibility — and I would have thought the strong probability — that if the procedure were invoked, then there would be no determination of the fair yearly rental for the voids within the six months' period stipulated for the ascertainment of the total income receivable. This is a strong indication that time was not intended to be made of the essence of that six months' period. Moreover, however swiftly or efficiently the lessee might act, and however quickly the expert or arbitrator might be appointed, his determination would be outside the control of the parties. That is a further indication supporting the presumption that time was not intended to be of the essence of the time-limit laid down.

Finally, I take the view that the very nature and purpose of the time provision is in the end the strongest reason for the view that time was not intended to be of the essence. I have already explained what I take to be its nature and purpose, that is to say to provide a time-limit which the lessee was to have to satisfy the landlord of the true figure; and that it was essential for some time-limit to be specified but that it did not involve any necessity for making time of the essence. In my judgment, when it can be seen that some time-limit must be specified, and that some default provision must also be included, and that this is entirely consistent with time not being of the essence, it is impossible to rely on the default clause to exclude the presumption.

I have reached the conclusion that the six months' period is not one which has to be strictly complied with, and that it is still open to the parties to agree, if they can, by the processes I have described, what is the total income receivable in the present case.

Declaration to be drawn up accordingly. The plaintiffs were awarded costs.

Court of Appeal
March 19 1987
(Before Lord Justice FOX, Lord Justice DILLON and Lord Justice RUSSELL)

EQUITY & LAW LIFE ASSURANCE SOCIETY PLC v BODFIELD LTD

Estates Gazette March 28 1987
281 EG 1448-1452

Landlord and tenant — Rent review clause in lease of industrial estate — Construction — Appeal from decision of Gibson J — Confirmation of strict approach to construction of individual lease without any presumption in favour of commercial realism or otherwise — Guidance of Vice-Chancellor in *British Gas Corporation* v *Universities Superannuation Scheme Ltd* not to be understood as entitling the court to apply, not the actual clause which the parties have entered into, but the different clause which they probably would have entered into if fully advised as to the practical consequences of their scheme — In the present case there was a 70-year term with rent reviews at the end of the 14th, 28th, 42nd and 56th years — The rent review provisions included a somewhat complicated formula with the unusual feature that a discount on the "net rental value" was allowed to the tenants — The "net rental value" was defined as the best open market rent on the assumption (*inter alia*) that the hypothetical lease should be on the terms of the actual lease "other than as to duration and rent" — Landlords had sought a declaration that the true construction was that the hypothetical lease should contain no rent review provisions and Gibson J had so decided — Tenants appealed, but Court of Appeal have affirmed Gibson J's decision — It was common ground that the words "other than as to . . . rent" did not exclude the covenant to pay rent or the power of re-entry for non-payment — It was also clear that they did exclude the yearly rents of fixed amounts payable before the first review — The difficulties arose as to other provisions — The somewhat complicated provisions as to a discount from the net rent could not, however, be imported into the hypothetical letting; that became clear on construction — It followed, in the court's opinion, that the current rent review provision must be covered by the words of exclusion, not as a matter of implication, but of the meaning of the express words — It was a necessary consequence that future reviews should also be excluded from the hypothetical letting — Comments by court on the unhelpfulness of decisions on the construction of other leases as an aid to the construction of the particular lease which has to be considered — Appeal dismissed

The following cases are referred to in this report.

Aspden v *Seddon* (1875) 10 Ch App 394
British Gas Corporation v *Universities Superannuation Scheme Ltd* [1986] 1 WLR 398; [1986] 1 All ER 978; [1986] 1 EGLR 120; (1986) 277 EG 980
Guys 'n' Dolls Ltd v *Sade Brothers Catering Ltd* (1983) 269 EG 129, CA

This was an appeal by the tenants, Bodfield Ltd, from the decision of Gibson J (reported at [1985] 2 EGLR 144, (1985) 276 EG 1157) in favour of the submission of the landlords, Equity & Law Life Assurance Society plc, on the construction of the rent review provisions of the lease of premises on an industrial estate known as Commerce Estate, Raven Road, South Woodford, Essex.

Michael Barnes QC and Christopher Priday QC (instructed by Baker & McKenzie) appeared on behalf of the appellants; Kim Lewison (instructed by A C McIntosh, Legal Department, Equity & Law Life Assurance Society plc) represented the respondents.

Giving the first judgment at the invitation of Fox LJ, DILLON LJ said: This is an appeal from a decision of Peter Gibson J regarding the construction of a rent review clause in a lease. Briefly, the problem is this. The lease provides that each rent review shall be on the basis of a hypothetical lease on the same terms as those of the lease other than as to duration and rent. What is meant by rent? Does the reference to rent require that the terms as to future rent reviews contained in the existing lease should be included in the hypothetical lease or excluded?

The lease is dated April 28 1968 and was made between the plaintiff, Equity & Law Life Assurance Society (as lessors), of the one part and Town & Commercial Properties Ltd (as lessees) of the other part. The term has since become vested in the defendants. The demised property was an industrial estate in Essex. The term granted was 70 years from the date of the lease at rents of:

> £22,500 for the first 3 years
> £27,000 for the next 7 years
> £28,500 thereafter.

The lease further provides for payment of "such increased rent as may be agreed or determined in manner hereinafter provided".

Clause 4 (1) of the lease conferred on the lessors the right to have the rent reviewed as from the end of all or any of the 14th, 28th, 42nd and 56th years of the term. The present question has arisen on the first rent review.

How the reviews are to work is set out at considerable length in clause 4 (2) of the lease. That in turn depends on the definition of "Net Rental Value" in clause 4 (4). Rather than attempting a paraphrase, I set out the relevant provisions of clause 4 (2) as follows:

If a sum equal to eighty-five per cent of the net rental value so agreed or determined on a review under sub-clause (1) hereof as from the end of the Fourteenth year of the term hereby granted exceeds the sum of Twenty-eight thousand five hundred pounds there shall become payable by way of additional rent hereunder an annual sum equal to the amount of such excess. If at the time of a review as from the end of any one or all of the Twenty-eight Forty-second or Fifty-sixth years of the term hereby granted no additional rent is payable under the provisions of this Clause then if on any such review a sum equal to eighty-five per cent of the net rental value is found to exceed Twenty-eight thousand five hundred pounds there shall become payable by way of additional rent hereunder an annual sum equal to the amount of such excess. If at the time of a review as from the end of any one or all of the Twenty-eight Forty-second or Fifty-sixth years of the term hereby granted an additional rent is payable under the provisions of this Clause then if on any such review a sum equal to eighty-five per cent of the net rental value is found to exceed the sum of Twenty-eight thousand five hundred pounds plus the additional rent already payable the additional rent then payable shall be increased by the amount of such excess.

There are then provisions as to how the additional rent or increased additional rent is to be paid. Clause 4 (4) of the lease is, so far as material, as follows:

The "net rental value" means the best rent which the premises hereby demised might reasonably be expected to fetch on the open market upon the following assumptions that is to say
(i) that they are vacant and to let as a whole without a premium or other capital payment for the residue unexpired of the term hereby granted upon the terms of this lease other than as to duration and rent
(ii) that the premises have been kept in good repair and condition in all respects in accordance with the Lessees' obligations hereunder.

The lessors on March 14 1983 took out an originating summons asking for a declaration that:

upon the true construction of the above-mentioned Lease and in the events which have happened, any valuer appointed to determine the net rental value of the . . . property (under Clause 4 (4) of the said Lease) ought to determine the same upon the assumption that the said property is let upon a hypothetical lease which contains no provision for revising the rent payable thereunder.

The judge accepted the plaintiffs' case and made an order as asked in the originating summons. From that decision the lessees appeal.

The dispute turns on the meaning of the words in clause 4 (4) ". . . upon the terms of this Lease other than as to . . . rent".

There is no doubt that the general object of a rent review clause, which provides that the rent cannot be reduced on a review, is to provide the landlord with some measure of relief where, by increases in property values or falls in the real value of money in an inflationary period, a fixed rent has become out of date and unduly favourable to the tenant. The exact measure of relief depends on the true construction of the particular rent review clause. It is, however, in my view, a very unsatisfactory procedure for arriving at a decision on the true construction of the particular clause that counsel should, as counsel for the appellants did in the present case, feel constrained to read aloud in open court the judgments from start to finish in 10 or more other cases decided on differently worded clauses. These other cases serve in truth to illustrate the great variety of different forms of words which draftsmen may use in preparing rent review clauses. For my part I would accept the point made by Mr Lewison (by reference to the judgment of Sir George Jessel MR in *Aspden* v *Seddon* (1875) 10 Ch App 394 at p 397) that to refer to authorities on other documents merely for the purpose of ascertaining the construction of a particular document is to be deplored as a wrong approach and likely to lead to confusion and error.

However, I welcome and approve the rough guidelines stated by the Vice-Chancellor in *British Gas Corporation* v *Universities Superannuation Scheme Ltd* [1986] 1 WLR 398 at p 403* where he said:

In these circumstances, there are in my judgment conflicting decisions as to the correct approach to the construction of these clauses. I am accordingly free to adopt the approach I prefer. In my judgment the correct approach is as follows: (a) words in a rent exclusion provision which require all provisions as to rent to be disregarded produce a result so manifestly contrary to commercial common sense that they cannot be given literal effect; (b) other clear words which require the rent review provisions (as opposed to all provisions as to rent) to be disregarded (such as those in the *Pugh* case, (1982) 264 EG 823) must be given effect to however wayward the result; (c) subject to (b), in the absence of special circumstances it is proper to give effect to the underlying commercial purpose of a rent review clause and to construe the words so as to give effect to that purpose by requiring future rent reviews to be taken into account in fixing the open market rental under the hypothetical letting.

I am conscious that such an approach is perilously close to seeking to lay down mechanistic rules of construction as opposed to principles of construction. But there is an urgent need to produce certainty in this field. Every year thousands of rents are coming up for review on the basis of clauses such as the one before me: witness the growing tide of litigation raising the point. Landlords, tenants and their valuers need to know what is the right basis of valuation without recourse to lawyers, let alone the courts. The question cannot be left to turn on the terms of each lease without the basic approach being certain. It is in my judgment most desirable that this, or some other case, should at an early stage be taken to the Court of Appeal so as to resolve the conflicting judicial approaches that have emerged.

These were of course no mechanistic rules of construction, to be applied rigidly in every case. They were only guidelines, and, however valuable guidelines are, the function of the court in each particular case is to construe the particular rent review clause which is in issue in that case.

Where lawyers have prepared a particularly complicated rent review it is likely to be difficult if not impossible for the parties to discover the right basis of valuation without recourse to lawyers. Indeed, one of the most fruitful sources of difficulty in the construction of such clauses is that parties leave it to their lawyers to devise appropriate wording to, in the Vice-Chancellor's words, "give effect to the underlying commercial purpose of a rent review clause" and the lawyers, in devising complicated review provisions, fail to appreciate the less immediately apparent commercial implications of what they have drafted. Guidelines such as the Vice-Chancellor's cannot entitle the court to construe and apply not the clause which the parties have entered into but the different clause which they might have, or probably would have, entered into if their lawyers had thought rather more deeply about how the intricate scheme they were setting up would work in practice.

In the present case, the rent review clause in its definition of "net rental value" requires the assumption of a hypothetical letting of the premises as a whole for the residue unexpired of the term granted by the actual lease "upon the terms of this lease other than as to duration and rent". It is common ground that the words "other than as to . . . rent" do not exclude the covenant in the actual lease to pay rent and the power of re-entry in the actual lease for non-payment of rent; any other conclusion would, in the Vice-Chancellor's words, be manifestly contrary to commercial common sense. The words "other than as to rent" do, however, exclude the yearly rents of fixed amount reserved by clause 1 of the lease so as to be payable at various times before the first rent review. So much is easy. The difficulty arises over clause 4 (2) of the lease set out above.

Clause 4 (2) is a complicated clause, with its provisions for additional rents and increased additional rents to become payable on successive reviews. What is clear to me, however, is that the provisions of clause 4 (2) for calculating 85% of the net rental value on a review, and calculating the additional rent, if any, payable from that review date on the basis of that 85%, cannot be imported into the hypothetical letting which is postulated for that particular review date by clause 4 (4), because what has to be determined for that hypothetical letting, as the "net rental value" as defined in clause 4 (4), is the best rent which the premises might fetch on the open market on the prescribed assumptions, and not 85% of the best rent.

It follows, in my judgment, that the provisions of clause 4 (2) as to the current review which cannot be included in the hypothetical letting must be covered by the words of exclusion in the phrase "upon the terms of this lease other than as to . . . rent" in the formula for determining the "net rental value" — ie these provisions of clause 4 (2) are excluded because they are terms of the actual lease as to rent. This is emphasised by the fact that the whole cumbrous edifice of clause 4 (2) is based on the specified rent of £28,500, which is itself excluded from the hypothetical letting by the words of exclusion just quoted.

I do not for my part regard the exclusion, from the hypothetical letting on a review, of the provisions of clause 4 (2) in relation to that current review as being merely a matter of necessary implication, because to include them would be absurd. Exclusion by necessary implication does not arise where there is an express exclusion provision, as there is in the words "upon the terms of this Lease other than as to rent". Those words exclude the yearly rents of fixed amount reserved by clause 1 of the lease, payable before the first rent review, which it would be equally absurd or impracticable to include in the hypothetical letting. In my judgment (as indicated), those words also exclude the parts of clause 4 (2) applicable to the current review. But clause 4 (2) is a long and detailed clause providing for what is to happen in relation to each successive rent review. If any of its terms are "terms of this lease as to rent", then all of them are.

Peter Gibson J dealt with this in a passage which is fundamental to his judgment. Referring to Mr Barnes, counsel for the present appellants, he said:

He submitted on the authority of the decision of the Court of Appeal in *Guys 'n' Dolls Ltd* v *Sade Brothers Catering Ltd* (1983) 269 EG 129, that where a lease provides for a rent review on the basis of a hypothetical lease on the terms of an actual lease and the actual terms provide for the payment of rent at a discount to or a premium over the market value ascertained on the rent review, the court will as a matter of necessary implication exclude that term from the terms of the hypothetical lease. I of course accept that is so where the terms of the lease are explicit in requiring the hypothetical letting to be on the terms of the actual lease. But in the case of a lease such as this which does not contain such explicit terms to my mind it is a very odd process of construction to construe the words "other than as to rent" as in the first place limited only to the rents initally reserved, but then to exclude clause 4 (2) by a process of necessary implication. The more natural way to approach the matter is, in my view, to seek a meaning for "other than as to rent" which comprehends all the

---

* Editor's note: Also reported at [1986] 1 EGLR 120 at p 121.

terms to be taken as excluded from the hypothetical letting.*
I agree and I would therefore dismiss this appeal.

In addition, if, by whatever process of construction, only the provisions of clause 4 (2) as to the current review are excluded from the terms of the hypothetical letting, but the provisions of clause 4 (2) as to future reviews are not excluded, those provisions as to future reviews would have to be redrafted — probably drastically, since clause 4 (2) is such an involved clause — to base them on a datum rent other than the rent of £28,500. I do not for my part believe that such redrafting is permissible, given the terms of the definition of "net rental value" in clause 4 (4).

Fox and Russell LJJ agreed and did not add anything.

The appeal was dismissed with costs; leave to appeal to the House of Lords was refused.

---

* Editor's note: See [1985] 2 EGLR 144 at p 145.

Chancery Division

December 8 1986
(Before Mr Justice HOFFMANN)

UNITED CO-OPERATIVES LTD v SUN ALLIANCE & LONDON ASSURANCE CO LTD AND ANOTHER

Estates Gazette April 4 1987

282 EG 91-92

Appointment of arbitrator or expert by president of the RICS — Procedure questioned but held to be "sensible and right" — The case arose out of a rent review clause in a lease which, after giving the landlord the right to serve a six months' notice requiring a review, went on to provide that, in default of agreement between the landlord and tenant within three months of the expiration of such notice, the question of the fair rack rental value of the demised premises should be determined by an independent surveyor to be appointed, in the absence of an agreed nomination, by the president — The appointee was to act as an expert, not as an arbitrator — Tenants, on hearing of landlords' intention to apply to the president for an appointment, objected on the grounds (1) that they were contemplating court proceedings for the construction or rectification of the lease, and (2) that the appointment would be premature — The second ground was based on a submission that, as a matter of construction, it was correct for a party to apply to the president for an appointment only if the parties had failed to reach agreement on the rent before the expiry of the three months following the expiry of the landlords' six months' notice — The president replied that it was not his practice to enter into disputes between the parties over the construction of a lease and that he would proceed to make the appointment unless restrained by an order of the court — The tenants then issued a writ, joining the landlords and the president as parties, seeking an injunction restraining the president from making the appointment — During the trial of this action various different constructions of the review clause were canvassed, including suggestions that there was no reason why negotiations between the parties should not run concurrently with the investigation by the expert and that "within three months of the expiration of such notice" meant three months *before* the expiration — Hoffmann J, however, pointed out that it was not for him in these proceedings to decide between competing constructions, but to determine whether an injunction should be granted against the president — Held, dismissing the motion, that there was no cause of action against the president — He owed no duty to the parties to a rent review clause in these circumstances to refrain from making an appointment — A person who considered an appointment void could apply to the court for a declaration in proceedings to which the president would not have to be a party — If a party, fearing prejudice, wished to prevent a valuer from proceeding immediately with a valuation he could request him to wait and, on a refusal, take proceedings in which the valuer, but not the president, would be a party — The suggestion that the president should, in case of objection, await the outcome of court proceedings before making an appointment would make him a pawn in the tactical moves between the parties — The present procedure was correct — Motion dismissed

No cases are referred to in this report.

This was a motion by United Co-operatives Ltd, tenants of a department store in Lancaster, as plaintiffs, seeking an injunction against the president of the RICS restraining him from appointing a surveyor as expert to determine the fair rack rental value under the rent review clause of a lease of which the Sun Alliance & London Assurance Co Ltd were the landlords. The Sun Alliance were named as the first defendants and the president, Donald Troup, as the second defendant to the proceedings.

Edward Cole (instructed by Addleshaw Sons & Latham, of Manchester) appeared on behalf of the plaintiffs; Kim Lewison (instructed by Maples Teesdale) represented the first defendants; R J Furber (instructed by Linklaters & Paines) represented the president.

Giving judgment, HOFFMANN J said: This motion raises an important practical point concerning the functions of the president of the Royal Institution of Chartered Surveyors in appointing an arbitrator or an expert pursuant to a rent review clause.

The plaintiffs are the tenants of a department store in Lancaster. They hold under a lease from the first defendant — the Sun Alliance & London Assurance Co Ltd — which was granted for a term of 150 years less 10 days from June 24 1980. It contained provision for five-yearly rent reviews. The original rent was £201,875 pa, but the rent review clause gave the landlords the right to serve a notice requiring a review in the first instance from January 5 1987. That notice had to be given not less than six months earlier and the landlords duly gave notice on June 17 1986.

The review schedule provided that the new rent was to be the fair rack rental value of the premises at the review date. This was defined as the greater of two figures arrived at in alternative ways. The first was the application of a formula to the letting values of the zone A parts of shop properties in the immediate neighbourhood of the premises. The second was the fair rack rental value of a department store of comparable size and location in a comparable town or city in England.

It appears from the correspondence between the surveyors that those two methods of calculating the fair rack rental value are likely to produce very different results. The rental values of shops in the neighbourhood in Lancaster seem to have gone up a great deal more than the rental values of department stores in the country generally. According to the tenants' surveyors, a valuation according to the shop formula will produce a rent in the neighbourhood of £1m whereas a valuation according to the department store formula will result in a rent of only £200,000 or £300,000.

The tenants consider that there must be something wrong with these provisions of the lease and are said to be contemplating proceedings in the High Court for either construction or rectification of the schedule.

On October 21 the landlords made an application to the president of the Royal Institution of Chartered Surveyors for the appointment of an independent expert to determine the fair rack rental value in accordance with the schedule. The provision for the appointment of an independent expert is contained in the schedule in the following words. After providing for the six months' notice requiring a review, the schedule goes on to say:

And thereupon, in default of agreement in writing between the landlord and the tenant within three calendar months of the expiration of such notice, the question of what shall be such fair rack rental value of the demised premises aforesaid for the time being shall be determined by an independent surveyor to be nominated by the landlord and the tenant jointly, or if they shall fail to agree upon a nominee, then at the request of either of them, by the president for the time being of the Royal Institution of Chartered Surveyors, and such independent surveyor shall act as an expert and not as an arbitrator and his decision shall be final and binding on both parties.

When the tenants were notified of the landlords' intention to apply for the appointment of an independent surveyor, they wrote to the president of the Royal Institution of Chartered Surveyors and asked him not to proceed. The tenants said that the appointment would, for two reasons, be premature: first, because they were contemplating proceedings for construction or rectification, and it would not be possible for an expert to determine the rent until those proceedings had been concluded. Second, they said that the appointment was premature in accordance with the timetable of the schedule which I have read. The tenants submit that upon its true construction the schedule allows the parties three months *after* the expiration of the six months' notice to arrive at an agreement as to the rent. It is only after that period has expired without agreement that it is proper for one or other of the parties to apply for the appointment of an expert. In this case, the application was made on October 21 1986, well before the notice had expired.

The president replied that it was not his practice to enter into disputes between the parties over the construction of the lease. If, according to the terms of the lease, he was the party who had the power to appoint a surveyor, he would, at the request of either party, proceed to do so. The president therefore said that unless the parties otherwise agreed, or unless he was restrained by an order of this court, he would proceed to appoint a surveyor.

The tenants responded by issuing a writ joining both the landlord and the president as parties, and by this motion seek an injunction against the president restraining him from proceeding to appoint a surveyor.

The question of whether, on the true construction of the schedule, a valid application for the appointment of a surveyor can be made before three months after the expiration of the notice is not an easy one. Mr Cole, for the tenants, submits that although the language of the schedule does no more than postpone the determination by the surveyor until after the three-month period, it must also be implied that the surveyor shall not be appointed or act within that period. Otherwise, he says, the parties may be put to the expense of making submissions to the surveyor which turn out to have been unnecessary because they are able within the three-month period to agree upon a rent.

Mr Lewison, for the landlords, on the other hand, says that there is no reason why negotiations for the agreement of the rent should not run concurrently with the investigation of the matter by the expert. The knowledge that the expert is ready to make his determination at the expiry of the three-month period may concentrate the minds of the parties and achieve an earlier agreement. Furthermore, Mr Lewison says that the language of the schedule, which postpones the determination by the expert until a period "within three calendar months" of the expiration of such notice, does not mean three months *after* such expiration. In his submission, "within three calendar months of the expiration of such notice" means three months *before* it expires. That, he says, makes commercial sense because it leaves three months for the expert to do his work before the rent review date. On the tenants' submission, a disagreement between the parties will inevitably produce a period after the rent review date during which it is not known what the review rent is going to be. It is true that the landlords will subsequently be able to recover the difference in rent retrospectively as from the review date, but they will not be entitled to interest on the money.

I do not, as I have said, find these questions of construction easy, and therefore do not propose to determine them today. The important question is whether in circumstances such as this, which must frequently occur, a party to a rent review clause should be entitled to injunctive relief against the president of the Royal Institution of Chartered Surveyors.

The president has filed evidence as to the practice which he follows in making such appointments. He says that a great number of applications are made each year. In 1985 there were 7,664. He has prepared standard documentation for dealing with the applications. There are standard forms of application and standard forms of reply. When an application is received, his staff check the lease to see whether it gives the president a power of appointment. If it does, his practice is to look no further and make an appointment if either party asks him to do so. He does not consider it his function to determine any legal questions which may arise in the course of a rent review. Indeed, one can see that if he were to attempt to do so, it would add a great burden to the administration of his office.

Mr Furber, who appeared for the president, submitted that there is no legal relationship between the tenant and the president which would entitle the tenant to an injunction. There is certainly no contract between them. It may be — though I express no opinion on this point — that the president would owe some duty of care in selecting the person appointed to be valuer. But, Mr Furber says, he owes no duty as to whether to make an appointment or not. If he makes an appointment it will be either valid or void. If it is valid, the tenant can have no complaint. Nor will the tenant be prejudiced if the appointment is void. The standard form of application for appointment requires the applicant to pay a non-returnable fee of £70 plus VAT and to undertake responsibility for payment of the professional fees and costs of the surveyor appointed. Appointment does not, therefore, in itself involve the non-applicant party in any expense.

If the non-applicant party considers that the appointment is void, he can apply to the court for a declaration to this effect in proceedings to which the president would not have to be a party. If he considers that it would involve him in unnecessary expense or cause him prejudice if the valuer were to proceed immediately with the valuation, it would be open to him to ask him to wait and if he should refuse, to apply to the court in proceedings to which the valuer may be a party, but not the president. In practice, that would seem to me a much more desirable state of affairs. The president, considering the number of cases passing through his office, is not in a position to enter into the merits of whether the determination should or should not be delayed. The valuer, on the other hand, will make himself acquainted with the facts of the individual case in which he has been appointed and can form a view as to whether the request for delay is bona fide or merely frivolous and vexatious. The president says in his evidence:

In my experience, the appointed surveyor will usually await the determination by the court of any genuine and important legal disputes before proceeding himself, if any party wishes him to do so.

Mr Cole submitted that the president should change his procedure. He acknowledged that it would be difficult for the president to enter into the merits of each case, but argued that he should reverse the policy which he presently adopts. Instead of making an appointment and leaving it to the parties to deal with the valuer, he should in case of objection refrain from making an appointment until the matter has been determined by the court.

This practice would, in my judgment, make the president simply a pawn in tactical moves between the parties. It would mean that the valuation procedure would inevitably be brought to a halt by any objection, whether frivolous and vexatious or not. The other party would be left to take the initiative by going to court to get it started again. If, on the other hand, it is left to the valuer, subject to any legal proceedings, to decide whether to proceed or not, he will be in a position to make an informed judgment as to whether there are genuine grounds for holding up the valuation or not.

In my judgment, the president's procedure is sensible and right, and he owes no duty to the parties to a rent review clause not to make an appointment. The plaintiff therefore has no cause of action against him and the motion must be dismissed.

May I conclude by saying that I thought that this motion was extremely well argued by all three counsel and I am very grateful to them.

Chancery Division

October 24 1986
(Before Mr Justice VINELOTT)

BRITISH RAIL PENSION TRUSTEE CO LTD v CARDSHOPS LTD

Estates Gazette April 18 1987
282 EG 331-332

Landlord and tenant — Rent review clause — Whether letter from tenants headed "subject to contract" constituted a valid counternotice in accordance with provisions which, in the judge's view, were unusual in the sense of not having been considered hitherto in any reported case — The machinery

was for the landlord to serve a notice stating his opinion as to the market rent and for the tenant within eight weeks of receipt (time to be of the essence) to serve a counternotice stating his opinion as to the market rent — If the tenant did not serve such a counternotice within the time-limit the amount stated in the landlord's notice was to be the market rent — If the tenant did serve such a counternotice within the time-limit the two parties were then to seek to agree the market rent within seven weeks of the landlord's receipt of the counternotice — If they failed to agree within that period the landlord would apply to the president of the RICS to appoint a surveyor as expert — Landlords served a notice proposing a rent of £36,000 per annum — After some correspondence about comparables, tenants sent a letter headed "subject to contract" stating that they were prepared to agree a rent of £24,000 per annum, this letter being sent a short time before expiry of the time-limit — The issue was whether this letter was a valid counternotice bringing the subsequent machinery into operation, as the tenants contended, or whether it was a step in negotiation intended to avoid serving a counternotice, as the landlords contended — Landlords relied on *Shirlcar Properties Ltd* v *Heinitz* and *Sheridan* v *Blaircourt Investments Ltd* in support of their argument — Held, distinguishing these cases and rejecting landlords' submission, that the letter constituted a valid counternotice — In the present case the letter was not part of machinery to "trigger off" the review, as in the *Shirlcar* case or to exercise an election to have a rent determined by arbitration, as in the *Sheridan* case — It merely started a period during which the parties were required to negotiate in good faith and which resulted in a reference to an expert only if they failed to agree — It would be unreasonable to read the letter as an offer to which the tenant expected a reply before deciding whether to serve a counternotice in the few days remaining before the expiry of the time-limit — The heading "subject to contract" could be explained either as a mistake (such mistakes are not uncommon) or as intended to ensure that the letter was not taken as an offer capable of acceptance — Declaration in favour of tenants

The following cases are referred to in this report.

Sheridan v *Blaircourt Investments Ltd* [1984] EGD 176; (1984) 270 EG 1290
Shirlcar Properties Ltd v *Heinitz* (1983) 268 EG 362, CA

These were proceedings between the plaintiff landlords, British Rail Pension Trustee Co Ltd, and the defendant tenants, Cardshops Ltd, to determine whether a letter written by the tenants was a counternotice served in accordance with the rent review provisions in an underlease of shop premises at 101 Queen Street, Cardiff.

Terence Cullen QC and Miss Hazel Williamson (instructed by the Solicitor, British Railways Board) appeared on behalf of the plaintiffs; N J Patten (instructed by Stuart Hunt & Co, of Croydon) represented the defendants.

Giving judgment, VINELOTT J said: The short but by no means easy question in this case is whether a letter written by a tenant constitutes a valid counternotice given under the terms of a rent review clause.

The facts can be shortly stated. On October 29 1982 the plaintiff, British Rail Pension Trustee Co Ltd, the landlord, granted an underlease to the defendant, Cardshops Ltd, of shop premises in Cardiff for a term of 20 years from March 25 1981, at a rent of £15,000 per annum (subject to review). The provisions for review are contained in a schedule. The schedule is not, in fact, referred to in the body of the underlease, but nothing turns on that; it is common ground that the relevant schedule, the third schedule, must be read into the lease, the words "subject to review" being read as "subject to review as provided by Schedule 3".

The rent review provisions are unusual, at least to the extent that they are unlike any form of rent review that has so far been considered in any reported case. They may, for all I know, be in very common use. They are certainly well and clearly drawn. Schedule 3 starts in para 1 with definitions of a number of phrases, including "Review Notice", a notice served by the landlord pursuant to para 2; "Review Notice Period", each of the periods of five years immediately following the commencement of each fifth year of the term; "Market Rent", the rent at which the demised premises might reasonably be expected to be let on certain assumptions, and the "Review Date", the commencement of the second year of any Review Notice Period, or the date of service of a Review Notice within that Review Notice Period, whichever is the later date.

Para 2 provides that the landlord may once only at any time during a Review Notice Period serve on the tenant a notice in writing stating the amount which, in the opinion of the landlord, is the market rent. Para 3 provides that within the period of eight weeks from and exclusive of the date of service of the Review Notice "(time to be of the essence hereof)":

The tenant may serve on the landlord a Counter Notice in writing stating the amount which in the opinion of the tenant is the market rent, but if on the expiration of eight weeks from and exclusive of the date of service of the relevant Review Notice (time to be of the essence hereof), the tenant shall not have served on the landlord such a Counter Notice in writing, the amount stated in the relevant Review Notice shall be deemed to be the market rent.

Para 4 I also read in full:

If the tenant does serve such a Counter Notice, then the landlord and the tenant shall seek to agree the market rent and record that agreement in writing within seven weeks of the date that the tenant's Counter Notice was received by the landlord.

Para 5 provides that if the landlord and the tenant fail to reach agreement within that seven-week period, the landlord will apply to the president of the RICS for the appointment of an independent surveyor to determine the market rent, the surveyor acting as an expert not as an arbitrator.

I can pass over provisions governing the basis on which the market rent is to be ascertained, the endorsement of a memorandum on the underlease, and the date on which the new rent is to be paid.

Para 9 provides that the fees of the surveyor are to be borne in the following way: if the market rent is equal to or greater than that stated in the landlord's review notice by the tenant, if it is equal to or less than that stated in the tenant's counternotice by the landlord, and in any other case equally.

What happened was shortly this. On February 18 the landlord served a formal notice of his intention to review the rent with effect from March 25 1986, and of the landlord's opinion that the market rent of the premises was £36,000 a year. On the following day the tenant wrote to the landlord to say that they were not in agreement with the rent proposed. The tenant commented: "Before commencing further on this matter, I would be pleased if you would let me have details of comparables which you are using to justify the rent we are being asked." That letter was answered on March 3, when the landlord's estate manager put forward a rent of £70 per sq ft as that at which shops in the neighbourhood were currently let. He added: "I look forward to hearing from you if you have any counter proposals as to the market rent". There was a formal reply on March 6 to the effect that the director of the tenant company concerned was on holiday until the end of March. Then on April 9 the director in question on behalf of the tenant wrote the following letter:

Your letter of March 3 has been put before me on my return and I am surprised at the contents in view of the fact that you recently agreed rent reviews on numbers 99 and 97, which reflected rentals considerably below the £70 you appear to be looking for. I can only assume from this figure of £70 was referring to premises in the peak section of the street. In view of the agreement reached and the level of rentals in the area, we are prepared to agree a rental of £24,000 per annum, which would appear to be in line with agreements on premises in the block in better positions than our own.

That letter was headed in large type "SUBJECT TO CONTRACT". The question is whether that letter was a counternotice within para 3 of Schedule 3.

It is trite law that a lease must be construed in the same way as any other commercial document; that is, it must be given the meaning in which ordinary businessmen would understand it. The rent review provisions in turn are intended to be operated by the parties and not always by their lawyers. There is no specific requirement as to the form a counternotice should take. There is no magic formula. The question whether a letter constitutes a counternotice is like any other question of construction and must be answered in the light of the particular provisions of the lease and of the surrounding circumstances, including any other correspondence of which the letter forms part. The test is: would a reasonably sensible businessman in the light of all the surrounding circumstances have been left in no real doubt that the tenant wished to bring paras 4 and

5 of the third schedule into operation; or could a sensible businessman in the position of the landlord in this case have been reasonably mistaken as to what was meant? Mr Cullen submitted that even if the letter of April 9 had not been headed "subject to contract", it would not have been free from ambiguity and could sensibly have been read as a last offer by the tenant, made in an endeavour to agree the rent and to avoid having to specify what in his opinion was the market rent. He pointed out that it was important to the landlord to know not merely the figure which the tenant would agree to accept by way of negotiation but also the figure which, in the opinion of the tenant, was the market rent. The latter figure would govern the liability of the parties to the costs of the ascertainment of the market rent if, during the period specified in para 4, the parties failed to agree the new rent.

I have no hesitation in rejecting this submission. In my judgment, looking at the sequence of letters and the date when the letter of April 9 was written (it was written six or seven days before the expiry of the period of eight weeks specified in para 3), it would be only a most unreasonable landlord who would read it as an offer to which the tenant required an answer before deciding whether to specify what in his opinion was a market rent, thereby exposing himself to a possible risk as to the costs if the market rent had to be determined by an expert.

The only question to my mind is whether the words "subject to contract" in effect converted what any sensible businessman would otherwise read as a counternotice into something else — a step in a negotiation and intended to avoid the necessity of serving a counternotice.

Mr Cullen relied on the decision of the Court of Appeal in *Shirlcar Properties Ltd* v *Heinitz* (1983) 268 EG 362 where a landlord's notice triggering a rent review clause was similarly headed "subject to contract". In that case the Court of Appeal took the view that, in the words of Dillon LJ:

these time-hallowed words "subject to contract" would leave the tenant in real doubt as to whether the figure of £6,000 a year was being put forward as a firm figure specified by the landlords under the rent review clause, or was merely being put forward as a provisional figure, which, if not agreed by a binding contract (that is, as envisaged by the words "subject to contract"), the landlord might reserve the right to revise.

In *Sheridan* v *Blaircourt Investments Ltd* (1984) 270 EG 1290, a letter was relied on as a counternotice which, under the rent revision clause there under consideration, was, in effect, an election by the tenant to have the rent determined by an expert. It was headed "Without Prejudice and Subject to Contract". Nichols J (as he then was) held that that letter was wholly insufficient to alert the landlord to the fact that the tenant wished to exercise his right of election. He held that, apart from the heading, it would have been quite inadequate for that purpose; that if it had been otherwise adequate, it would have been made inadequate by the heading. He said:

I would, in the context of the correspondence in this particular case, and having regard to the heading of the letter, have come to the conclusion that for this reason alone it would not be possible to construe the wording in this letter as an unequivocal intimation to the landlord's advisers that the tenant was requiring the ascertainment of the substituted rent be referred to a referee.

However, there is nothing in these cases which binds me to hold that the letter of April 9 cannot be a valid counternotice. I have to consider the effect of that letter in the context of a rent review clause in which the document so headed does not trigger a rent review clause, as it did in the *Shirlcar* case, and which does not exercise a right to elect to have a rent determined by arbitration or by an expert, as it did in the *Sheridan* case, but which starts a period during which the parties are required to enter into negotiations in good faith and which results in a reference to an expert if, and only if, they fail so to agree. I think it is unreal in this context to regard the words "subject to contract" as stamping this letter as unequivocally an offer in a round of negotiation antecedent to the commencement of, as it were, mandatory negotiations, and necessarily antecedent to a step which might, in the event of there being a reference to an expert, expose the tenant to a risk as to costs. The words would be taken, I think, by a sensible businessman, as either a mistake (by no means an unfamiliar mistake — the words "subject to contract" are frequently used in quite inappropriate circumstances) or, a suggestion put forward by Mr Patten, as designed to ensure that the letter was not taken by the landlord as an offer capable of acceptance — a construction which is wholly consistent with the operation of the letter as a counternotice; that is, as a statement of the opinion of the tenant as to the market given on the footing that the expression of that opinion is not to be taken as an offer. Under the rent review clause it would be open to the expert to determine a rent at a figure below that given in the counternotice, and the tenant might have had this possibility in mind.

It would, in my judgment, be unreasonable in all the circumstances for the landlord to have read it as an offer to which the tenant expected a reply before determining whether or not to serve a counternotice in the day or so that would remain after receipt of the tenant's letter and before the expiry of the eight-week period. In these circumstances I shall make the declaration sought by the tenant and, unless I hear argument to the contrary, I will say that the normal consequences as to costs will follow.

Costs were awarded to the defendants.

Chancery Division

February 3 1987
(Before Mr Justice WARNER)

PHIPPS-FAIRE LTD v MALBERN CONSTRUCTION LTD

Estates Gazette April 25 1987

282 EG 460-464

Landlord and tenant — Rent review provisions in lease — Construction — Whether time was of the essence for the purpose of a particular paragraph — This paragraph provided that, subject to certain conditions (which were satisfied), the lessee might after a particular date serve on the lessor a notice proposing the amount of the revised rent and this amount "shall be the revised rent" for the relevant period unless the lessor within three months after service of such a notice by the lessee applied to the president of the RICS for determination of the rent by a valuer — The lessees served a notice under this paragraph proposing that the rent should be the same figure as during the previous five years, namely £11,000 per annum — Lessors did not apply to the president within the three months and the lessees sought a declaration that the rent for the new period should remain at £11,000 — Lessors counterclaimed for a declaration that their application to the president of the RICS after the expiry of the three months was valid — The issue before the court was whether time was of the essence for the purpose of the three months' time-limit — It was submitted on behalf of the lessees that there were "contra-indications" sufficient to displace the presumption established by *United Scientific Holdings Ltd* v *Burnley Borough Council* that time is not of the essence for the purpose of steps in a rent review clause — Reliance was placed on a dictum by Griffiths LJ in *Trustees of Henry Smith's Charity* v *AWADA Trading & Promotion Services Ltd* and on the presence of a "default provision" in the relevant paragraph — Held that the contra-indication required to rebut the presumption must be "a compelling one" and that those put forward were not compelling — Declaration in favour of lessors

The following cases are referred to in this report.

Cheapside Land Development Co Ltd v *Messels Service Co* [1978] AC 904; [1977] 2 WLR 806; [1977] 2 All ER 62; (1977) 33 P&CR 220; [1977] EGD 195; 243 EG 43 & 127, HL
Greenhaven Securities Ltd v *Compton* [1985] 2 EGLR 117; (1985) 275 EG 628
Lewis v *Barnett* (1981) 264 EG 1079
Mecca Leisure Ltd v *Renown Investments (Holdings) Ltd* (1984) 49 P&CR 12; [1984] EGD 200; 271 EG 989, CA
Smith's (Henry) Charity Trustees v *AWADA Trading & Promotion Services Ltd* (1983) 47 P&CR 607; [1984] EGD 103; 269 EG 729, CA
Taylor Woodrow Property Co Ltd v *Lonrho Textiles Ltd* [1985] 2 EGLR 120; (1985) 275 EG 632
United Scientific Holdings Ltd v *Burnley Borough Council* [1978] AC 904; [1977] 2 WLR 806; [1977] 2 All ER 62; (1977) 33 P&CR 220; [1977] EGD 195; (1977) 243 EG 43 & 127, HL

The plaintiffs in these proceedings, lessees of two units on the Huncoat Industrial Estate at Accrington in Lancashire, sought declarations that the rent payable under the lease for the period April 14 1984 to April 13 1990 was £11,000 per annum and that any application made by the lessors after August 31 1985 to the president of the RICS for the nomination of a valuer under the rent review clause was invalid. The lessors counterclaimed for a declaration to the opposite effect.

Joanne Moss (instructed by Bower Cotton & Bower, agents for Becke Phipps, of Northampton) appeared on behalf of the plaintiff lessees; P W Smith (instructed by O'Collier, Littler & Kilbeg, of Stockport) represented the defendant lessors.

Giving judgment, WARNER J said: This is yet another case raising the question whether time is of the essence of a particular paragraph in the rent review provisions of a lease. The lease is dated June 18 1980. The property thereby demised consists of two units on an industrial estate at Accrington in Lancashire known as the Huncoat Industrial Estate. The demise was for a term of 25 years from April 14 1980. The rents reserved by the lease were, apart from an "insurance rent", a peppercorn for the first month of the term (from April 14 1980 to May 13 1980), a rent at the yearly rate of £11,000 for the remainder of the first five years of the term (from May 14 1980 to April 13 1985) and thereafter — I quote from clause 2.1 of the lease:

during each of the successive periods of five years beginning on 14th April 1985 14th April 1990 14th April 1995 and 14th April 2000 respectively a yearly rent equal to the rent payable under this Lease immediately prior to the commencement of the relevant period or such revised yearly rent as may be agreed or determined under the provisions of Schedule 2 below (whichever shall be the greater).

Clause 2.1 goes on to provide, so far as material, that the rent shall be payable in advance by equal quarterly payments on the usual quarter days in each year and that:

until any revised rent is agreed or determined under the provisions of Schedule 2 below the rent payable for the relevant period shall (subject to paragraph 10 of Schedule 2 below) be the rent payable immediately prior to the commencement of such period.

Schedule 2 is entitled "Rent Review Provisions". It has eleven paragraphs, each with its own heading. Paras 1, 2 and 3 are as follows:

DETERMINATION BY AGREEMENT
1 The revised rent referred to in Clause 2.1 above may be agreed at any time between the Lessor and the Lessee.
DETERMINATION BY A VALUER: LESSOR'S OPTION
2 In the absence of agreement under paragraph 1 above the revised rent may at the option of the Lessor be determined not earlier than two months before the commencement of the period to which it relates (the "Relevant Period") by a Valuer (the "Valuer") to be nominated in the absence of agreement by the President of the Royal Institution of Chartered Surveyors ("the President") on the application of the Lessor made not more than six months before or at any time after the commencement of the Relevant Period.
DETERMINATION BY A VALUER: LESSEE'S OPTION
3 In the absence of agreement under paragraph 1 above and provided that the Lessor has not applied to the President in accordance with paragraph 2 above the Lessee may at any time after the commencement of the Relevant Period serve on the Lessor a notice containing a proposal as to the amount of such revised rent (not being less than the rent payable immediately before the commencement of the Relevant Period) and the amount so proposed shall be the revised rent for the Relevant Period unless the Lessor shall apply to the President for determination by a Valuer within three months after service of such a notice by the Lessee.

Para 4 provides that the valuer shall act as an expert and not as an arbitrator and that his determination of the revised rent shall be binding on the lessor and the lessee unless the revised rent has been agreed prior to such determination in accordance with para 1.

Para 5 lays down the principles on which the revised rent is to be determined by the valuer. It is to be such rent as in his opinion

shall be the yearly rent of the demised premises having regard to market rental values current at the commencement of the Relevant Period

on a number of assumptions. I need not go into the details of those assumptions. They are of a familiar kind.

Para 6 deals with the procedure to be adopted by the valuer.
Para 7 provides for the appointment of a substitute valuer in certain events.
Para 8 provides that the cost of the reference to the valuer

shall be in the award of the Valuer whose decision shall be final and binding on the Lessor and Lessee.

Paras 9 and 10 are important and I must read them:

CONTINUATION OF EXISTING RENT UNTIL DETERMINATION OF REVISED RENT
9 If and so often as a revised rent in respect of any Relevant Period has not been agreed or ascertained pursuant to the above provisions before the first day appointed for the payment of rent for such Relevant Period rent shall continue to be payable during that Relevant Period at the rate equal to the rent payable immediately before the commencement of that Relevant Period (the "Old Rent") until fourteen days after that revised rent has been agreed or ascertained or until the expiration of that Relevant Period (whichever happens first)
PAYMENT OF ARREARS
10 Within fourteen days after any revised rent has been determined by a Valuer in accordance with the provision of this Schedule or agreed under paragraph 1 above or ascertained under paragraph 3 above the Lessee shall (provided the revised rent is greater than the Old Rent) pay to the Lessor an amount equal to the difference between the Old Rent and the revised rent together with Interest upon such difference calculated from the commencement of the Relevant Period to the actual date of such payment

By virtue of clause 1 of the lease, which is a definition clause, such interest is calculable at the rate of 2 per cent per annum above Barclay's Bank base rate from time to time.

Finally, para 11 of Schedule 2, which is headed "Rent Review Memoranda", provides:

When the amount of any rent to be ascertained as provided above has been so ascertained memoranda of it shall thereupon be signed by or on behalf of the Lessor and the Lessee and annexed to this Lease and the counterpart of this Lease each party bearing its own costs.

On January 22 1985, a firm of chartered surveyors acting for the landlord (Malbern Property Holdings Ltd) wrote to the tenant (Phipps-Faire Ltd) saying that they had been instructed by the landlord to negotiate with the tenant the new rent to take effect from April 14 1985 and proposing that that rent should be £19,920 per annum. They received a reply dated January 26 1985, from a chartered surveyor instructed by the tenant, questioning whether there should be any increase in the rent at all. After a further exchange of letters between the surveyors in March 1985, the tenant's surveyor wrote to the landlord's surveyors on May 30 1985 maintaining that there was no case for an increase in the rent. He added:

If your client does not agree with this I would ask that you accept this letter as formal notice under paragraph 3 of Schedule 2 of the lease dated 18th June 1980 that the lessee proposes that the review rent for the period 14th April 1985 to 13th April 1990 be £11,000 per annum.

Though there is no evidence as to when that letter was received by the landlord's surveyors, counsel have treated it as common ground that the period of three months prescribed by para 3 of Schedule 2 ended on or about August 31 1985. It was not until October 7 1985 that the landlord's surveyors wrote to the president of the Royal Institution of Chartered Surveyors applying for a valuer to be nominated. There had been some correspondence between the parties' surveyors in the meantime, but I do not think that anything in that correspondence is relevant to the question that I have to decide.

This action is brought by the tenant against the landlord for declarations that the rent payable under the lease for the period April 14 1984 to April 13 1990 is £11,000 per annum and that any application to the president of the Royal Institution of Chartered Surveyors made after August 31 1985 for the purposes of Schedule 2 to the lease is out of time and invalid.

There is a counterclaim by the landlord for, *inter alia*, a declaration that the application to the president of the Royal Institution of Chartered Surveyors made on October 7 1985 was valid.

The whole question, of course, is whether time was of the essence of the provision in para 3 of Schedule 2 that the lessor should apply to the president within three months after service of the lessee's notice under that paragraph. If the answer to that question is "Yes", the landlord's application to the president of October 7 1985 was out of time and the rent payable under the lease for the period April 14 1985 to April 13 1990 is now fixed at £11,000 per annum.

On that question I was referred to a number of authorities including the *United Scientific* and *Cheapside* cases [1978] AC 904, *Lewis* v *Barnett* (1981) 264 EG 1079, *Trustees of Henry Smith Charity* v *AWADA Trading and Promotion Services Ltd* (1983) 4 P&CR 607*, *Mecca Leisure Ltd* v *Renown Investments (Holdings) Ltd* (1984) 49 P&CR 12†, *Greenhaven Securities Ltd* v *Compton*

---

*Editor's Note: Also reported at (1983) 269 EG 729.
†Also reported at (1984) 271 EG 989.

[1985] 2 EGLR 117 and *Taylor Woodrow Property Co Ltd* v *Lonrho Textiles Ltd ibid* p 120.

Miss Moss, on behalf of the tenant, takes as the starting point of her argument the statement of principle by Lord Diplock in the *United Scientific* and *Cheapside* cases ([1978] AC 904 at p 930):

So upon the question of principle which these two appeals were brought to settle, I would hold that in the absence of any contra-indications in the express words of the lease or in the interrelation of the rent review clause itself and other clauses or in the surrounding circumstances the presumption is that the time-table specified in a rent review clause for completion of the various steps for determining the rent payable in respect of the period following the review date is not of the essence of the contract.

Miss Moss submits that in the present case there are three "contra-indications" in the express terms of the lease.

First, she points to the contrast between the phrase "at any time", which is used in each of paras 1, 2 and 3 of Schedule 2, and the phrase "within three months" used in para 3. I do not regard that contrast as a material contra-indication. The phrase "at any time" appears to me to be used in each of the places where it is used because it is the appropriate phrase to use there to convey the intended meaning clearly. Miss Moss suggests that the phrase could have been left out in each of those places but, in my opinion, to have left it out would have amounted to slipshod drafting. Of course the phrase "within three months", if read literally, imports a strict time-limit. But the very essence of the principle laid down in the *United Scientific* and *Cheapside* cases is that there is a presumption that such a phrase is not to be taken literally. The use of the phrase cannot of itself serve to rebut the presumption.

Second, Miss Moss points to the purpose of para 3, which is to enable the lessee, after the commencement of the "Relevant Period", to shorten to three months the period within which the lessor must apply to the president of the Royal Institution of Chartered Surveyors for the nomination of a valuer, a period which, by virtue of para 2, would otherwise be indefinite. Such a provision is necessary for the protection of the lessee, not only because, as was recognised in the *United Scientific* and *Cheapside* cases, there are circumstances in which a tenant needs to know without undue delay what his liability for rent is going to be, but also because of the provisions of para 10 of Schedule 2 for the payment of the so-called "arrears" of rent with interest. Thus, says Miss Moss, para 3 is concerned with the imposition of a time-limit which was not there before, with the object of achieving finality. By the very nature of such a provision, time must be of the essence of it. In support of that submission, Miss Moss relies on a dictum of Griffiths LJ (as he then was) in the *A WADA* case. Discussing the clause there in question, he said (at 47 P & CR 607 at p 617):

By clause 1(f) market rent is to be assessed as at the date of the landlords' rent review notice. Clause 2 allows the landlords to serve a rent review notice at any time during a five-year review period; so if rents are rising fast the landlords may think it to their advantage to delay serving their notice. However, the tenants are given the chance of protecting themselves against this tactic by serving a notice on the landlords requiring them to serve a notice within six months, and if the landlords do not do so, clause 2 expressly prohibits the landlords from serving another notice during that review period. This clause must be read with time being of the essence otherwise the protection of the tenant is either destroyed or greatly reduced.

As reinforcing that contra-indication, Miss Moss points, thirdly, to the fact that para 3 contains a "default provision", that is to say the provision that, if the lessor does not apply to the president within three months after service of the lessee's notice, the amount proposed in that notice "shall be the revised rent". Miss Moss recognises, in view of the decision of the Court of Appeal in the *Mecca Leisure* case, that the presence of such a default provision does not necessarily indicate that time is of the essence, but she rightly submits that it may well do so, particularly in conjunction with other factors pointing to the same conclusion.

Miss Moss' arguments based on her second and third "contra-indications" are attractive. I have, however, come to the conclusion that I ought to reject them. The authorities seem to me to show that the presumption that time is not of the essence of a provision in a rent review clause is strong and that it will not be rebutted by any contra-indication in the express terms of the lease unless it is a compelling one. I do not find the contra-indications relied upon by Miss Moss compelling. The effect of holding that time is of the essence of the provision here in question is to entitle the lessee, once the three months' period has elapsed, to bring matters to a head by serving on the lessor a notice making time of the essence, that is a notice specifying a reasonable period within which the lessor is to exercise its right to apply to the president of the Royal Institution of Chartered Surveyors for the nomination of a valuer or lose that right. Since the act of writing to the president is a fairly simple one, the period need not be long. Mr Smith, who appeared for the landlord, suggested 28 days at the most. I should have thought that 14 days would be ample. It does not seem to me that such an extension of the prescribed period of three months, which may be necessary to prevent para 3 from being what Eveleigh LJ in the *Mecca Leisure* case called "a trap", would substantially impair the protection intended to be afforded to the lessee by that paragraph.

I do not overlook the dictum of Griffiths LJ on which Miss Moss relies but, as she fairly recognises, the view taken by a judge of the effect of a particular document in one case does not always afford safe guidance as to the effect of a different document in a subsequent case. I am not sure that I understand why Griffiths LJ thought that unless time was held to be of the essence of the clause with which he was concerned the protection of the tenants would be "either destroyed or greatly reduced". Probably it was because there the market rent was to be assessed as at the date of the landlord's notice. There is no equivalent of that here. Also the time within which in that case the landlords were to act in response to the tenant's notice was six months, not three.

Nor do I overlook that, by holding time not to have been of the essence here, I do violence to the language of para 3 to the extent that the amount proposed in the tenant's notice of May 30 1985 will not necessarily be "the revised rent" for the period April 14 1985 to April 13 1990. But, in every case where time is held not to be of the essence of a provision in a rent review clause, it follows that the new rent may not be what it would have been if the words of the clause had been given literal effect. So that is not an objection.

In the result I will declare that the application made by the landlord to the president of the Royal Institution of Chartered Surveyors on October 7 1985 was valid. It does not appear to me that the other declaration sought by the counterclaim is appropriate, but I will hear counsel on that.

Chancery Division

January 21 1987
(Before Mr John MOWBRAY QC, sitting as a deputy judge of the division)

PANTHER SHOP INVESTMENTS LTD v KEITH POPLE LTD

Estates Gazette May 2 1987
282 EG 594-597

Landlord and tenant — Rent review clause — Construction and effect — Lease for 20 years with five-yearly rent reviews — Material part of review clause provided for the disregard of any effect on rent of any improvement carried out by the tenants or any person deriving title under them otherwise than in pursuance of an obligation to the landlords — Two improvements, a back extension and a storage building, had in fact been carried out, not under the present lease, but under a previous lease between the same parties — By the time the present lease was executed these structures, which had not been erected under any obligation to the landlords or their predecessors in title, had become landlords' fixtures — The question at issue was whether the effect on rent of these structures, erected as improvements before the present lease began, was to be disregarded — Held, following the decision of the Court of Appeal in *Brett* **v** *Brett Essex Golf Club Ltd,* **that "improvements" for the purpose of the rent review clause meant improvements to the demised premises — The structures here were not improvements to the demised premises; they were part of them — The** *Brett* **case was indistinguishable from the present case —** *Hambros Bank Executor & Trustee Co Ltd* **v** *Superdrug Stores Ltd,* **on which**

the tenants relied, was distinguishable on the facts — Declaration accordingly that the effect of the improvements in question did not fall to be disregarded for the purpose of the rent review clause

The following cases are referred to in this report.

Brett v Brett Essex Golf Club Ltd [1986] 1 EGLR 154; (1986) 278 EG 1476, CA
Hambros Bank Executor & Trustee Co Ltd v Superdrug Stores Ltd [1985] 1 EGLR 99; [1985] 274 EG 590
"Wonderland", Cleethorpes, Re [1965] AC 58; [1963] 2 WLR 1426; [1963] 2 All ER 775, HL

This was an originating summons by which the plaintiff landlords, Panther Shop Investments Ltd, sought the court's determination of the true construction of the part of a rent review clause in a lease of premises at 7 Symes Avenue, Hartcliffe, Bristol, which concerned the disregard of the effect on rent of improvements. The defendants were the tenants of the premises, Keith Pople Ltd.

Mrs P M Lucas (instructed by Lehrer Segal) appeared on behalf of the plaintiffs; Gordon Bennett (instructed by Burges Salmon, of Bristol) represented the defendants.

Giving judgment, MR JOHN MOWBRAY QC said: This originating summons raises questions about the interpretation and effect of a rent review clause in a lease of 7 Symes Avenue, Hartcliffe, Bristol. The lease is dated October 5 1979. It is a 20-year lease with five-yearly rent reviews. The review clause is numbered 6. It provides that if the rent for the next five years is not agreed it is to be determined by an arbitrator, and I quote from clause 6:

disregarding:
(i) any effect on rent of the fact that the Lessees or any person deriving title under them have been in occupation of the demised premises;
(ii) any goodwill attached to the demised premises since the commencement of the term hereby granted by reason of the carrying on thereat of the business of the Lessees or of any person deriving title under them; and
(iii) any effect on rent of any improvement carried out by the Lessees or any person deriving title under them otherwise than in pursuance of an obligation to the Lessors.

That is all I need to read.

The parties to the lease are the predecessor of the plaintiffs as landlord, and the defendant company — I now read from the beginning of the lease:

hereinafter called "the Lessees" which expression shall, where the context so admits, include their permitted assigns.

There had been a previous lease between the same parties as the 1979 lease before me. During the previous term the defendant company erected in 1960 a back extension, and in 1966 a separate storage building beyond. This is referred to in the user covenant, clause 2(9)(b), and the covenant against assignment, clause 2(18) of the 1979 lease, as an existing part of the demised premises. It is common ground that both erections were in existence when the 1979 lease was granted and were by then landlord's fixtures; also that they were erected by the defendant company and not pursuant to any obligation to the then lessor, let alone the present landlord.

In these circumstances the question arises whether what I may perhaps call "disregard no (iii)" in the rent review clause requires the arbitrator to disregard buildings erected before the lease began as an improvement within that disregard. Mrs Lucas, for the present landlord plaintiffs, argued that the disregard does not operate on those buildings; that would increase the rent on the review now pending. Mr Bennett, for the tenant, argued the contrary.

I say at once that in my view the buildings are not excluded from the valuation by disregard (iii) because they are not improvements as the word is used in the clause. In my view, this means improvements to the demised premises. These buildings were not improvements to the demised premises; they were part of them.

That is the ground on which the Court of Appeal recently decided in Brett v Brett Essex Golf Club Ltd (1986) 278 EG 1476.* I refer especially to the middle of the left-hand column on p 1483, where Slade LJ, with whose judgment Croom-Johnson LJ and Sir John Megaw agreed, said that in the clause before him "improvements" must mean improvements to the demised premises.

Mr Bennett for the tenant argued that I should distinguish the Brett case because the clause there was drafted by incorporating the three similar disregards in section 34 of the Landlord and Tenant Act 1954, and those in turn had to be interpreted with regard to the whole of that Act, including the definition of "the holding" in section 23. In the course of his able address, which lost nothing by its concision but rather gained from it, he pointed out quite rightly that the Court of Appeal in the Brett case relied on the speech of Lord Morris of Borth-y-Gest in the House of Lords in the "Wonderland", Cleethorpes case [1965] AC 58 at p 74. That was a case on section 34 of the Act as it then stood, and Lord Morris referred to "the holding". But as I understand Slade LJ, he was not bringing into the interpretation of the lease any influence from provisions of the Act which were not incorporated into it. At the top of p 1481 he said that the incorporated words should be construed in the context of the lease, not the Act; and at p 1483 he said that the reference in the Act to "the holding" must be read in the lease as if it were a reference to the demised premises.

In my view, then, I can take the conclusion of Slade LJ about the meaning of "improvements" in the lease before the Court of Appeal as directly helpful in interpreting the same word in the lease before me. If I can add this without disrespect, I would think without any such guidance that inherent in the meaning of the word "improvements" in a lease is the sense of an alteration to the premises demised, rather than an existing part of them. And putting the word "any" in front of it does not extend its meaning. Things that are not improvements are not any kind of improvements.

Mr Bennett relied strongly on Hambros Bank Executor & Trustee Co Ltd v Superdrug Stores Ltd [1985] 1 EGLR 99, where Scott J held that shopfitting works carried out by the tenant just before the lease was granted were an improvement carried out by the tenant, required to be disregarded on any rent review. About this case Slade LJ in Brett simply said that it was plainly distinguishable on its facts. That is at the bottom of the left-hand column of (1986) 278 EG 1483†. It is equally distinguishable here.

Scott J found the approach of the landlord, in his case, to verge on the unconscionable. I will not read the passages from his judgment, but they are both on p 101 of the report, one in the lower part of the left-hand column and the other towards the bottom of the right. There is nothing like that here, so Scott J's decision would fall to be distinguished on the facts in the present case. I notice, too, that the argument about the meaning of "improvements" does not seem to have been addressed to Scott J. I think there may have been reasons of discretion and tact for that.

In the present case there is no such strong factual matrix as there was in Scott J's. The parties here may have agreed the starting rent on the footing that the buildings were to be disregarded. I make no finding about that. But assuming they did, it was only what the 1954 Act, as by then amended, required, and an originating summons had been issued under the Act. The buildings were erected, one 18 and the other 24 years, before the first review. The period under the Act as now amended after which an improvement falls out of disregard is 21 years. For these reasons I am not applying the decision of Scott J.

I had better end by entering my own disregard. I have read and considered the correspondence of 1977 and 1978 which is exhibited to one of the affidavits before me, but I say nothing more about the circumstances surrounding the grant of the 1979 lease. One reason is that I do not need to. The other is that I am anxious not to say anything which might prejudge any proceedings to rectify the lease which might be brought by the tenant, as mentioned in evidence. Nothing I have said should be taken as reflecting one way or the other on the questions of intention and so forth which might arise in such proceedings.

For the reasons I have given I will make the two declarations sought in the originating summons, subject only to any discussion there may be about the precise terms.

The judge made a declaration that upon the true construction of the lease the improvements in question did not fall to be disregarded for the purpose of the rent review clause. The summons was transferred to Bristol District Registry for the purpose of a counterclaim for rectification of the lease. Plaintiffs were awarded costs, but the order was stayed upon the defendants' undertaking to pursue the counterclaim with all due diligence.

---

*Editor's note: See also [1986] 1 EGLR 154 at p 158 (H).

†Editor's note: See also [1986] 1 EGLR 154 at p 158 (L).

Court of Appeal

February 25 1987

(Before Lord Justice FOX, Lord Justice DILLON and Lord Justice RUSSELL)

DENNIS & ROBINSON LTD v KIOSSOS ESTABLISHMENT

Estates Gazette May 16 1987

282 EG 857-862

Landlord and tenant — Construction of rent review clause — Appeal from decision of Mr Michael Wheeler QC, sitting as a deputy judge of the Chancery Division — Underlease for 25 years with four five-yearly reviews — A "full yearly market rent" to be determined, meaning the yearly rent at which the property might reasonably be expected to be let in the open market — There was no express reference to "a willing lessor" or to "a willing lessee" — Tenants contended that there should be no assumption that there would in fact be a willing lessee to whom the property might reasonably be expected to be let, it being a matter for consideration whether, if the property were offered in the open market, anyone would wish to take a lease on the terms offered — Landlords submitted that the independent valuer was required to determine the rent which would be expected to be agreed between a willing lessee and a willing lessor — The deputy judge came to the conclusion that to say that it must be assumed that there will be a letting (whatever may be the reality of testing the appropriate open market) is to incorporate into this hypothetical exercise a situation which the language of the rent review clause does not warrant and which is completely unreal — Held by the Court of Appeal, allowing the landlords' appeal, that the language of the review clause required the following assumptions: (1) there will be a letting of the property, (2) there is a market in which that letting is agreed, (3) the landlord is willing to let the premises, (4) equally, the supposed tenant is willing to take the premises — Although there was no express reference here to a willing lessor and a willing lessee, such an implication was necessary to achieve an open market letting — The fact that the above assumptions are artificial is irrelevant; that is the bargain which the parties have made — It is, however, a matter for the valuer, using his experience and judgment, to determine the strength of the market — Although it is assumed that there is a market there is no assumption as to how lively that market is — The court could give only limited guidance to the valuer — It was not appropriate to make any formal declaration — The judge's order would be set aside but no order would be made on the originating summons — A subsidiary point, as to the date of commencement of the 25-year term, had been decided by the judge, but he had not been asked to decide it and it was not in issue; the order in so far as it purported to determine this point would have to be set aside for that reason also

The following cases are referred to in this report.

Evans (FR) (Leeds) Ltd v English Electric Co Ltd (1977) 36 P&CR 185; [1978] EGD 67; 245 EG 657
Law Land Co Ltd v Consumers' Association (1980) 255 EG 617, CA
Wallersteiner v Moir [1974] 1 WLR 991; [1974] 3 All ER 217, CA

This was an appeal by the landlords, Kiossos Establishment, from an order of Mr Michael Wheeler QC, sitting as a deputy judge, on an originating summons by the tenants, Dennis & Robinson Ltd, underlessees of premises on an industrial estate at Lancing, Sussex, seeking a declaration as to the construction of the review clause in the underlease. The judgment of Mr Michael Wheeler is reported at [1986] 2 EGLR 120.

David Neuberger (instructed by Teacher Stern & Selby) appeared on behalf of the appellants; Michael Barnes QC and J M Male (instructed by Speechley Bircham) represented the respondents.

Giving judgment, FOX LJ said: This is an appeal by the landlord from an order of Mr Michael Wheeler QC, sitting as a deputy judge of the Chancery Division. The case is concerned with the construction of a rent review clause in a lease.

The lease is an underlease dated October 20 1982 and was made between Smiths Industries plc, as landlords of the one part, and the plaintiffs, Dennis & Robinson Ltd, as tenants of the other part. The reversion has since been assigned to the defendants, Kiossos Establishment, who are now the landlords.

The demised property consists of premises on an industrial estate at Lancing in Sussex. It comprises a large warehouse built in about 1906.

The lease was for a term of 25 years from December 31 1981. The rent payable was £30,000 per year but was subject to increase as provided by clause 5.

Clause 5 (1), (2) and (3) provide as follows:

IT IS HEREBY AGREED BY THE LANDLORD AND TENANT that the rent hereby reserved (or such increased sum as results from the application of this clause) shall be subject to review in the manner following namely:
(1) The Landlord may give notice in writing to the Tenant at any time not more than six months before nor more than twelve months after all or any of the following dates namely the thirty-first day of December in the years one thousand nine hundred and eight-five one thousand nine hundred and ninety one thousand nine hundred and ninety-five and two thousand (each of such dates being called a "Material Date") calling upon the Tenant to negotiate with the Landlord as to what represents the full yearly market rent (as hereinafter defined) of the property at the date three months before the Material Date in question and after the giving of such notice the following provisions of this clause shall take effect for the purposes of reviewing the rent reserved by this Lease.
(2) For the purposes of this clause the expression "full yearly market rent" shall mean the yearly rent (exclusive of rates and other payments (if any) to be made by the Tenant by virtue of this Lease) at which the property might reasonably be expected to be let in the open market three months before the Material Date for a term of twenty-five years with vacant possession and otherwise on the same terms and conditions as this Lease (including the provisions for rent review at the intervals herein specified) there being disregarded those matters referred to in paragraphs (a) (b) and (c) of subsection (1) of Section 34 of the Landlord and Tenant Act 1954 PROVIDED ALWAYS and It is hereby agreed and declared that Firstly the expression "Improvement" in the said paragraph (b) of the said subsection 34 (1) shall only be deemed to cover any improvement carried out to the property on or after the thirty-first day of July one thousand nine hundred and eighty-two and Secondly that if the full yearly market rent as aforesaid shall be less than the Existing Rent then and in such case the Existing rent shall continue to be payable as from the relevant Material Date.
(3) If the Landlord and the Tenant have not reached an agreement as to the full yearly market rent by a date three months after the service of a Notice pursuant to sub-clause (1) of this Clause 5 the review of rent may at any time thereafter be referred to an independent valuer to be appointed in default of agreement between the parties by the President for the time being of the Royal Institution of Chartered Surveyors upon the application of either party such valuer to determine (as an expert and not as an arbitrator) what sum in his opinion represents the full yearly market rent of the property at the date three months before the Material Date in question.

Under these provisions there is protection for the landlords in that no rent review is to result in the tenants' paying a rent of less than the existing rent (ie at present £30,000 per year).

On August 14 1985 Smiths Industries plc, the then landlords, gave notice under clause 5 (1) of the lease calling upon the tenants to negotiate with the landlords as to what represents the full yearly market rent. The parties were unable to reach agreement by the expiration of the period of three months referred to in clause 5 (3). On May 9 1986 the president of the Royal Institution of Chartered Surveyors appointed an independent valuer under the provisions of clause 5 (3). A question has arisen as to the basis on which he should make his valuation.

I need not refer further to the terms of the lease, except to mention the tenants' covenant not to assign, underlet or part with possession of the demised premises without first making an offer to the landlords to surrender the lease for a premium: ". . . representing . . . its fair market value between willing buyer and willing seller at the date of the offer".

The originating summons was taken out by the tenants and asked for a declaration:

. . . that on the true construction of the Underlease . . . the independent valuer in determining the full yearly market rent thereunder in accordance with Clause 5 (2) is not required or entitled to make an assumption that there would be a willing lessee to whom the property might reasonably be expected to be let but must consider and determine whether if the property were offered to be let in the open market anyone would wish to take a lease of the property on the terms offered and must determine the rent, if any, which would be agreed in the open market in respect of such a lease.

The judge, by the order appealed from, declared that upon the true construction of the lease ". . . the independent valuer therein referred to shall determine the reviewed rent thereunder in accordance with the attached judgment of Mr Michael Wheeler QC . . ."

The judgment appealed from is in two parts. The first part dealt with the meaning, in clause 5 (2), of the words ". . . for a term of 25 years". The judge stated that he had "no hesitation in concluding that the reference to 25 years is a reference to a term of a lease which commenced (as did the existing underlease) on December 31 1981". The tenants cross-appeal against that decision.

Mr Barnes for the tenants submitted that the construction adopted is wrong and that the clause requires the assumption of a lease for 25 years from three months before the relevant material date. He also contended that the judge was not entitled to decide the point, as it was not in issue before him.

Mr Neuberger for the landlords offered no argument in opposition to the tenants' contention.

In my opinion the judge's order, so far as it purports to determine this point, must be set aside. The judge was not asked to decide the point, which was never in issue in the proceedings, and he heard no argument upon it. I do not, however, think it would be appropriate for this court to make a declaration as to the interpretation of the material words. Since we heard no argument from the landlords we would, in effect, be making a declaration by consent. The court does not normally make declarations by consent or upon admissions (see the *Annual Practice 1986*, at p 218, and *Wallersteiner* v *Moir* [1974] 1 WLR 991 at pp 1029 and 1030).

Nor is there any reason to do so in the present case. Either the parties are agreed upon the meaning of the lease or they are not. If they are agreed, they do not need any declaration. They can give effect to their own agreement. If they are not agreed, the matter must be raised in proceedings and fully argued.

The second part of the judgment dealt with the issue raised in the originating summons. The declaration thereby sought by the tenants I have already set out. That sought by the landlords was as follows:

. . . is required to determine the rent which would be expected to be agreed between willing lessee and willing lessor three months before the material date and otherwise on the assumptions set out in the said sub-clause.

For present purposes, the material parts of the judgment appear to be the following: on p 6, after referring to the decision of Donaldson J in *F R Evans (Leeds) Ltd* v *English Electric Co Ltd* (1977) 36 P & CR 185, the deputy judge said:

. . . it appears to have been argued that the arbitrator's concern was with the attitude of the hypothetical willing lessee and that the rent review clause therefore assumed that there was such a person so that it was *nihil ad rem* to prove that there was no such willing lessee. I am not sure how far Donaldson J (as he then was) accepted the last part of the statement. The learned judge accepted that it was to be assumed that there was at least one tenant bidding for the lease. So be it. But he also accepted: "That the rent must be determined at an amount that the hypothetical tenant — the willing lessee — would agree" and that: "The willing lessee has to agree this rent with a hypothetical landlord — the willing lessor."

On p 192, Donaldson J said: "The parties will reach agreement." With much of this I respectfully agree. But if examination of the appropriate open market were to reveal no willing lessee at any price, I do not see for myself why the formula should be construed as requiring one to be invented. It may well be, of course, that Donaldson J did not have in mind the admittedly remote possibility of the open market turning up no "willing lessee" at all. In any event, there is no express reference in the present case to "a willing lessor" or to "a willing lessee"; and for my part I would if necessary find no difficulty in implying after "willing lessee" the words "if any".

At p 8 of his judgment, the deputy judge said:

I conclude, therefore, that the fact that even if some such expression "as between a willing lessor and a willing lessee" were found in clause 5 (2) in this present case it would add nothing to the requirement that the rent was to be that at which the property might reasonably be expected to be let "in the open market". An open market transaction necessarily proceeds on the hypothesis of a willing seller/lessor and a willing buyer/lessee. But if examination of the open market were to reveal a total or almost total absence of hypothetical willing lessees, the result would, as it seems to me, necessarily be that the best rent that the hypothetical lessor could get in the open market would either be a very low rent or (in an extreme case) a peppercorn rent. Of course, in the present case, the landlord is protected against that sort of hardship by the £30,000 "safety net" provided at the end of clause 5 (2). But to say that the formula must be applied on the hypothetical assumption that there will be a letting (whatever may be the reality as a result of testing the appropriate open market) is in my judgment to incorporate into this hypothetical exercise a situation which the language of the rent review clause does not warrant and which is completely unreal.

What the lease requires the valuer to determine is "the full yearly market rent". That is defined as:

the yearly rent . . . at which the property might reasonably be expected to be let in the open market three months before the Material Date for a term of 25 years with vacant possession and otherwise on the same terms and conditions as the Lease.

The following assumptions in relation to that provision appear to me to be correct:

(1) There will be a letting of the property. The judge, as I read his judgment (p 8 of the transcript), was not prepared to accept that in general terms. But in my opinion it must be so. The language of clause 5 (2) expressly contemplates a letting on the open market.

(2) There is a market in which that letting is agreed.

(3) The landlord is willing to let the premises. Equally, the supposed tenant is willing to take the premises. The notion of a letting in the open market between an unwilling lessor and an unwilling lessee (or between a willing lessor and an unwilling lessee) for the purpose of determining a reasonable rent makes no sense.

These assumptions seem to me to follow from the language which the parties chose to use.

I refer to the judgment of Templeman LJ in *Law Land Co Ltd* v *Consumers' Association Ltd* (1980) 255 EG 617 at p 617 (right-hand column) where he said:

Clause 3 (5) (a) (ii) defines the market rent. It means the yearly rent at which the demised premises, fully repaired in accordance with the provisions of the lease, might reasonably be expected to be let in the open market with vacant possession by a willing lessor . . . That clause envisages the existence of an open market, an offer with vacant possession, a willing lessor and, by implication, a willing lessee.

In the present case, although there is no express reference to a willing lessor, that has to be implied if an open market letting is to be achieved. A willing lessee must be implied in consequence.

Mr Barnes, for the tenants, says that the purpose of the rent review clause is to determine the amount that would actually be paid in the open market if the property were actually offered to be let on the relevant terms at the relevant time. The valuer, it is said, must therefore investigate whether there would in reality be any willing lessee or lessees in the market if the property were so offered.

The important fact is that clause 5 (2) requires assumptions to be made. The fact that those assumptions are artificial is irrelevant. That is the bargain which the parties have made. For the reasons which I have given, the clause requires the assumption of a willing lessor and a willing lessee. I do not think that the assumption of a willing lessee, any more than the assumption of a willing lessor, can be qualified by the addition of the words "(if any)".

But the dispute which has arisen has, I think, some unreality about it. The assumptions are only a part of the process of computation of the full yearly market rent. It is assumed that there is a willing lessee. But the willing lessee is not going to pay more than the market requires him to pay. It is essentially a matter for the valuer to inquire into and determine the strength of the market. He is, for example, entitled, if such is his expert opinion on the facts, to say that, having regard to the state of the market and the condition of the property, a tenant, though a willing tenant, could not be expected to take the stipulated lease save at a low or nominal rent and that the full yearly market rent must be determined accordingly. Further, any determination below £30,000 will leave the existing rent unaltered anyway.

The judge obviously found difficulty in formulating any specific declarations to deal with the matter. So do I. I do not think that the tenants are entitled to a declaration that clause 5 does not contemplate a willing lessee. On the other hand, I think that the declaration sought by the landlords is unsatisfactory in that it might divert attention from the central fact that it is for the valuer to determine, in the light of his own expertise and judgment of the market, at what rent the property might reasonably be expected to be let at the time and in the circumstances specified in clause 5.

I would discharge the order made by the judge and make no further order.

Agreeing, DILLON LJ said: In these proceedings, the parties ask the court to express a view on the finer nuances of the legal interpretation of the rent review clause in the lease for the guidance of the independent valuer appointed by the president of the Royal Institution of Chartered Surveyors, who is, by the lease, charged with determining, as an expert, the "full yearly market rent" of the premises. I do not find this a very satisfactory or desirable procedure: first, because the determination of the full yearly market rent of the premises depends primarily on the practical experience and common sense of the independent valuer; second, because the nuances of legal interpretation which are thought by the parties to be in issue are very difficult to express in language which will be helpful and not confusing to the independent valuer when the court does not know how this particular problem of valuation strikes the valuer or what particular difficulties he perceives in it.

The learned deputy judge commented that when he was first presented with the rival contentions of the parties he was hard put to it to see what differences of substance there were between them. As I see it, the differences are essentially differences of emphasis. The learned deputy judge decided that the best he could do to help the parties was, hopefully — as he put it — to give sufficient indications in his judgment of the factors which he considered to come within the true construction of the rent review formula so as to enable them to prepare a revised form of declaration which they could agree and which the expert could then proceed to implement. He therefore made an order declaring that the independent valuer should determine the rent in accordance with his judgment. It is, however, somewhat difficult to discern precisely what he has meant in some parts of his judgment.

The most that I feel I could say to the independent valuer by way of guidance is as follows:

The lease uses the phrases "full yearly market rent" and "rent . . . at which the property might reasonably be expected to be let in the open market". These phrases assume that there is a market in which agreement will be reached for a hypothetical letting of the premises to a hypothetical tenant. That necessarily imports a hypothetical landlord who is willing to let the premises and a hypothetical tenant who is willing to take the premises on the terms prescribed by the rent review clause, ie a willing lessor and a willing lessee. But though it is assumed that there is a market, there is no assumption required as to how lively that market is. The strength of the market and the rental value of the premises in the market are matters for the valuer's discretion based on his own knowledge and experience of the letting value of such premises.

The learned deputy judge dealt in his judgment with a further question on which there was no dispute between the parties and on which he had not heard argument. This was a question as to the duration of the hypothetical lease which the valuer had to consider. The judge considered this as part of the background to his consideration of the question he was asked to rule on; his conclusion was that the hypothetical lease, which was said by clause 5 (2) of the actual lease to be for a term of 25 years, was to be a lease for 25 years from the date of commencement in 1981 of the actual term of 25 years of the actual lease. By the form of the deputy judge's order as drawn up — declaring that the valuer should determine the rent in accordance with the deputy judge's judgment — that conclusion of the deputy judge as to the term has become binding on the parties.

The respondent tenants have given a notice of cross-appeal against the deputy judge's conclusion as to the term of the hypothetical lease. The appellant landlords have declined to argue the point. Since there is no *lis* on it between the parties, this court should, in my judgment, decline to decide the point. It should simply set aside the deputy judge's order, so that each party will have complete freedom to argue the point in other proceedings, if the point ever does become a live one.

For the rest, I have indicated the limited extent of the guidance which I feel able to offer to the independent valuer. I do not think, however, that it is appropriate for this court to make any formal declaration. It should set aside the deputy judge's order and simply make no order on the originating summons.

RUSSELL LJ agreed with both judgments and did not add anything.

The appeal was allowed and the order of the judge below discharged. No order was made as to costs in Court of Appeal or below.

Chancery Division

December 16 1986

(Before Mr Justice WARNER)

FACTORY HOLDINGS GROUP LTD v LEBOFF INTERNATIONAL LTD

Estates Gazette May 23 1987

282 EG 1005-1010

Landlord and tenant — Rent review clause — Construction and application — Effect of notice "making time of the essence" — Warehouse in trading estate — In this case a rent review clause provided for a "trigger" notice by the landlords followed by machinery for reaching agreement on a fair market rack rental and, in default of agreement, determination of the rent by a surveyor appointed, failing agreement, by the president of the RICS — Surveyor to act as arbitrator — Apart from the disputed effect of the tenants' letter mentioned below, time was not of the essence for the purposes of the review clause — After some correspondence between the parties the tenants wrote to say that they could not agree with a revised rent proposed by the landlords and gave formal notice to the latter that they were required within 28 days of the date of the letter to refer the calculation of a fair market rent to arbitration — Landlords did not comply with this time-limit and tenants claimed that, as their letter had made time of the essence, it was now too late to proceed with the rent review and that the appointment of an arbitrator by the president was invalid — Landlords sought declarations that the appointment was valid and that the tenants would be liable to pay the rent determined by the arbitrator — Apart from a preliminary point decided in favour of the tenants, the issue before the court was whether the tenants were entitled to give an effective notice making time of the essence — Observations in the speeches in *United Scientific Holdings Ltd* **v** *Burnley Borough Council* **considered — It was not correct to say that in every case where equity treated time as not being of the essence of a stipulation there was a countervailing right for the party affected to serve a notice making time of the essence — In the context of rent reviews a notice making time of the essence can be used in appropriate cases, but not all time-limits can be the subject of such a notice — It was submitted on behalf of the landlords that a tenant was not entitled to serve a notice making time of the essence of a step in the procedure that was open to the tenant himself to take — In the present case, as in the** *United Scientific* **case, it was equally open to the landlord and the tenant to apply to the president of the RICS for the appointment of an arbitrator — Held, accepting this submission, that in the circumstances of the present case the tenants were not entitled to serve a notice making time of the essence — Declarations claimed by landlords granted**

The following cases are referred to in this report.

Amherst v *James Walker (Goldsmith & Silversmith) Ltd* [1983] Ch 305; [1983] 3 WLR 334; [1983] 2 All ER 1067; (1983) 47 P&CR 85; [1985] EGD 157; 267 EG 163, CA
Cheapside Land Development Co Ltd v *Messels Service Co* [1978] AC 904; [1977] 2 WLR 806; [1977] 2 All ER 62; (1977) 33 P&CR 220; [1977] EGD 195; 243 EG 43 & 127, HL
London & Manchester Assurance Co Ltd v *G A Dunn & Co* [1983] EGD 86; (1982) 265 EG 39 & 131, CA
Mecca Leisure Ltd v *Renown Investments (Holdings) Ltd* (1984) 49 P&CR 12; [1984] EGD 200; 271 EG 989, CA
Smith's (Henry) Charity Trustees v *AWADA Trading & Promotion Services Ltd* (1983) 47 P&CR 607; [1984] EGD 103; 269 EG 729, CA
Thorn EMI Pension Trust Ltd v *Quinton Hazell plc* (1983) 269 EG 414
Touche Ross & Co v *Secretary of State for the Environment* (1982) 46 P&CR 187; 265 EG 982, CA
United Scientific Holdings Ltd v *Burnley Borough Council* [1978] AC 904; [1977] 2 WLR 806; [1977] 2 All ER 62; (1977) 33 P&CR 220; [1977] EGD 195; (1977) 243 EG 43 & 127, HL

In this case the plaintiff landlords, Factory Holdings Group Ltd, sought by originating summons declarations (a) that the appointment of Mr G J Calver FRICS as arbitrator was valid in relation to the dispute with the defendant tenants, Leboff International Ltd, as to the terms of the rent review clause in the lease of a warehouse on a trading estate at Yate, Gloucestershire, and (b) that the defendants were liable to pay to the plaintiffs with effect from June 12 1985 such rent as might be awarded by Mr Calver.

Kim Lewison (instructed by Helder Roberts & Co, of Epsom) appeared on behalf of the plaintiffs; David Neuberger (instructed by Linklaters & Paines) represented the defendants.

Giving judgment, WARNER J said: By a lease dated July 25 1969, the plaintiff, Factory Holdings Group Ltd (which I will call "the landlord"), demised to the defendant, whose name was then S Leboff (Fobel) Ltd and is now Leboff International Ltd (I will call it "the tenant"), what I understand to be a warehouse on a trading estate at Yate in Gloucestershire which is owned by the landlord. The lease is for a term of 25 years from June 12 1969. The rents initially reserved were a yearly rent of £8,930 and an insurance rent. However, the lease contains, in clause 5 (2), provisions for rent reviews, the relevant parts of which are as follows:

(a) If the Landlord shall desire to review the rent hereinbefore reserved at or after the expiration of the eighth and sixteenth years of the term hereby granted (or either of them) and of such desire shall give to the Tenant not less than six calendar months previous notice in writing then as from the date of the expiration of the said notice or of the expiration of the relevant year of the term hereby granted in respect of which such notice shall have been given (whichever shall be the later) for the residue of the term hereby granted or until any subsequent review as herein provided the rent first hereby reserved shall be revised and shall be such an annual sum as may be agreed between the Landlord and Tenant or as may be determined as provided by the following sub-clauses of this present clause to be the fair market rack rental of the demised premises PROVIDED that in no circumstances shall the rent payable hereunder following any such review be less than the yearly rent payable by the Tenant at the date of the Landlord's notice calling for the review in addition to the insurance rent.
(b) If within two calendar months after the service by the Landlord of the notice referred to in sub-clause (a) of this present clause the Landlord and the Tenant have been unable to agree upon a fair market rack rental (as defined by sub-clause (c) of this present clause) then the question of what is a fair market rack rental of the demised premises shall as soon as practicable and in any event not later than four months before the expiration of the said notice or of (sic) the expiration of the relevant year of the term hereby created in respect of which such notice shall have been given (whichever is the later) be referred for decision to a Surveyor to be mutually agreed between the Landlord and the Tenant or in default of agreement to be nominated by the President for the time being of the Royal Institution of Chartered Surveyors and such Surveyor whether agreed or nominated as aforesaid shall act as an Arbitrator in accordance with the Arbitration Act 1950 or any statutory enactment in that behalf for the time being in force and the decision of such Surveyor shall be binding on both the Landlord and the Tenant but subject to the proviso contained in sub-clause (a) of this clause.
(c) The expression "fair market rack rental" shall for the purposes of this clause mean the amount which would in addition to the insurance rent be the annual amount obtainable at the date of agreement or determination as aforesaid as between a willing Landlord and a willing Tenant in respect of the demised premises on a letting thereof as a whole with vacant possession for a term of eight years and subject to similar covenants and conditions as those contained in this Lease but ignoring any goodwill value attaching to the Tenant's business or any improvements carried out by the Tenant solely at its own expense and not pursuant to any obligation to the Landlord and ignoring the provisions of this present clause for revision of the rent but subject to the proviso contained in sub-clause (a) of this clause.

There was, I understand, a rent review under those provisions at the eighth year of the term as a result of which the yearly rent of £8,930 was increased to £26,000.

On November 26 1984, the landlord duly served notice under clause 5(2)(a) of its desire to review that rent at June 12 1985, the expiration of the 16th year of the term. I will for convenience, as is customary, refer to that notice as a "trigger notice".

Nothing then seems to have happened until April 16 1985, when the landlord's group estates manager wrote to the tenant in these terms:

Further to my letter of November 26 1984, I have now considered the matter of rental values and I am of the opinion that the exclusive annual rental value of the above premises with effect from June 12 1985 should be not less than £52,000 per annum. I look forward to hearing from you that this is acceptable.

To that, the tenant's secretary replied on May 23 1985 as follows:

Thank you for your letter of April 16. Having consulted with my colleagues I regret to say the revised rental proposed in your letter is not acceptable to the company. Please treat this letter as formal notice by the company that they require you within 28 days of the date of this letter to refer the calculation of a fair market rental to arbitration.

That letter was received by the landlord on May 24 1985. On May 28 1985 the landlord's group estates manager wrote to the tenant saying:

Thank you for your letter of May 23 1985, the contents of which I note. I am in the course of preparing an application to the President of the RICS for the appointment of an independent expert, but perhaps in the meantime you would place me in touch with your surveyors so that we may establish the basic facts before the appointment is known. I look forward to hearing from you.

That evoked no response from the tenant. On June 26 1985, that is to say after the expiration of the period of 28 days mentioned in the tenant's letter of May 23 1985, the landlord's group estates manager wrote to the president of the RICS asking him to appoint a member of that institution to determine the rent. On June 26 1985, he wrote to the tenant, enclosing a copy of his letter to the president and asking again to be put in touch with the tenant's surveyors so that they could agree basic facts.

The tenant's response to that took the form of a letter from its solicitors dated July 25 1985 contending that, whether or not the timetable contained in the lease made time of the essence, the tenant's letter of May 23 1985 had certainly done so, with the result that, the landlord having failed to apply to the president of the RICS for the appointment of an arbitrator within the period of 28 days stipulated in that letter, it was now too late for the landlord to seek to proceed with the proposed rent review.

In October 1985 the parties were informed by Mr G J Calver, a Fellow of the RICS, that he had accepted an appointment by the president of that institution to act as arbitrator in their case. The tenant's solicitors at once took the point that, the landlord having (as they contended) lost its right to review the rent, the appointment of Mr Calver was invalid.

On April 14 1986, the landlord issued the present originating summons, seeking declarations that the appointment of Mr Calver was valid and that the tenant is liable to pay to the landlord with effect from June 12 1985 such rent as may be awarded by Mr Calver.

I can dispose shortly of the question whether, apart from the tenant's letter of May 23 1985, time was of the essence of the provision in clause 5(2)(b) of the lease for referring the matter of the fair market rack rental of the demised premises to an arbitrator. Mr Neuberger, who appears for the tenant, while not formally conceding the point, very properly accepts that, in the light of the decisions in *Touche Ross & Co* v *Secretary of State for the Environment* (1982) 46 P & C R 187* and *Thorn EMI Pension Trust Ltd* v *Quinton Hazell plc* (1983) 269 EG 414, it would be a waste of time for him to argue that the answer to that question was in the affirmative.

The next question is one of construction of clause 5(2)(b) and it is whether that time-limit expired on June 12 1985, the date of "the expiration of the relevant year of the term", or on February 12 1985, four months earlier. Mr Neuberger concedes that, if the time-limit did not expire until June 12 1985, the tenant's letter of May 23 1985 seeking to make time of the essence was premature. It is common ground that the phrase "and in any event not later than four months before the expiration of the said notice or of the expiration of the relevant year of the term . . . (whichever is the later)", which is the crucial phrase in clause 5(2)(b), is ungrammatical. The word "of" after the word "or" should not be there. Mr Neuberger suggests that it probably crept in as a result of blind copying of the corresponding phrase in clause 5(2)(a), where its presence is appropriate. Be that as it may, Mr Lewison, on behalf of the landlord, submits that the time-limit did not expire until June 12 1985, because, he says, the words "four months before" relate only to the first limb of the alternative, "the expiration of the said notice". He submits that if it had been intended to make them relate also to the second limb, "the expiration of the relevant year of the term", the natural thing to have done would have been not to repeat the words "the expiration of" so that the phrase would have read "and in any event not later than four months before the expiration of the said notice or of the relevant year of the term . . . (whichever is the later)".

On this point, I prefer Mr Neuberger's argument. He points out that under clause 5(2)(a) the landlord was free either to serve its trigger notice not less than six months before the end of the relevant year of the term or to serve it later. In the former case, the new rent would become payable as from the end of that year. In the latter case, it would become payable as from the end of six months from the date

---

*Editor's note: Also reported at (1982) 265 EG 982.

of the notice. In clause 5(2)(b), as in clause 5(2)(a), the first limb of the alternative relates to a late notice while the second limb relates to a timeous notice. In either case, by virtue of clause 5(2)(c), the "fair market rack rental" would fall to be ascertained as at the date of agreement between the parties on its amount or of the determination of that amount by the arbitrator. There is no sensible reason why the reference to the arbitrator should be required to be made four months before the date when the new rent becomes payable in one case but not in the other. In either case there is sense in having the new rent fixed before it becomes payable or as soon thereafter as possible. As I have indicated, I accept that argument.

That brings me to the difficult question in this case which is: was the tenant, in the circumstances, entitled to give notice making time of the essence?

On that question, Mr Lewison submitted *in limine* that there was no authority binding me to hold that the concept of a notice making time of the essence was applicable in the sphere of rent review clauses; that the only authorities to that effect were *obiter dicta*; and that, on principle, I should hold that that concept was not applicable in that sphere. Mr Lewison pointed out that the concept was in origin an equitable one applicable in the sphere of contracts for the sale of land. (As to that there is no doubt: see *per* Lord Diplock in *United Scientific Holdings Ltd* v *Burnley Borough Council* and *Cheapside Land Development Co Ltd* v *Messels Service Co* [1978] AC 904 at p 928.) In the sphere of contracts for the sale of land, Mr Lewison's argument continued, its effect was to entitle a party to the contract, once the contractual date for completion had passed or, if the contract fixed no date for completion, once a reasonable period had elapsed, to serve on the other party a notice requiring him to complete within a reasonable further period specified in the notice, failing which the party serving the notice, if not himself in default, would have the option of either suing for specific performance of the contract or treating it as wholly at an end. In no case could a party, by serving such a notice, entitle himself both to affirm the contract and relieve himself of some of his obligations under it. It would be inequitable, said Mr Lewison, to extend the scope of the application of the concept in such a way as to enable a party to a lease to achieve such a result. Alternatively, if the concept were to be applied in the sphere of rent review clauses, it must be applied to the full, that is to say on the footing that a tenant availing himself of it would, if the landlord failed to comply with the tenant's notice, have the option of either proceeding with the rent review none the less or treating the lease as at an end and accordingly giving up possession of the demised premises.

I cannot accept those submissions. In the first place, I do not think that what Lord Diplock and Lord Fraser said on the point in their speeches in the *United Scientific* and *Cheapside* cases (as I will call them for short) was *obiter*. I think that it formed part of the reasoning that led them to their decision, at least in the *Cheapside* case, and I observe that Lord Simon of Glaisdale expressed, at the beginning of his speech, his agreement with Lord Diplock's arguments and conclusions.

Second, I have been referred to numerous *dicta* in the Court of Appeal, in cases subsequent to the *United Scientific* and *Cheapside* decisions, supporting the view that a notice making time of the essence may be used in the course of a rent review. There are *dicta* to that effect by Slade LJ in *London & Manchester Assurance Co Ltd* v *G A Dunn & Co* (1982) 265 EG 39 at pp 134-135; by Oliver LJ (as he then was), by Ackner LJ (as he then was) and by Lawton LJ in *Amherst* v *James Walker Goldsmiths & Silversmiths Ltd* [1983] Ch 305 at pp 315, 318 and 319 respectively*; by the Master of the Rolls and by Slade LJ again in *Trustees of Henry Smith's Charity* v *AWADA Trading & Promotion Services Ltd* (1983) 47 P & C R 607 at p 609-610 and p 619 respectively†; and by Eveleigh LJ and May LJ in *Mecca Leisure Ltd* v *Renown Investments (Holdings) Ltd* (1984) 271 EG 989 at pp 990 and 992 respectively. Cumulatively, those *dicta* represent a formidable body of authority.

Third, even if there were no authority to that effect, I would have no doubt that, in principle, once it is accepted, as it must be accepted in view of the decisions of the House of Lords in the *United Scientific* and *Cheapside* cases, that there is a presumption against time being of the essence of provisions in rent review clauses, it must follow that the party who is thereby adversely affected should be entitled, in appropriate circumstances, to invoke the remedy of serving a notice making time of the essence. The problem is to determine what those circumstances are. As Mr Lewison pointed out, if the concept of a notice making time of the essence is to be imported into the sphere of rent reviews, it must be adapted so as to operate in that sphere with due fairness to both parties. I cannot think that it would be an appropriate adaptation to hold that where such a notice was given by a tenant and the landlord did not comply with it, the tenant was put to his election either to proceed none the less with the rent review or to give up the lease altogether.

I turn to what was Mr Lewison's main argument on this part of the case. On the basis of an analysis of the speeches in the House of Lords in the *United Scientific* and *Cheapside* cases, Mr Lewison submitted that there were three reasons why it should be held that the tenant in the present case was not entitled to give the notice on which it relies.

The first reason was that, in the case of a rent review initiated by a landlord's trigger notice, the tenant was not entitled to make time of the essence of any subsequent step in the procedure.

The second reason was that a tenant was not entitled to make time of the essence of a step in the procedure the carrying out of which was not within the exclusive control of the landlord. The only relevant time-limit here was the time-limit in clause 5(2)(b) for taking the step of referring the matter of the fair market rack rental value to an arbitrator. Adherence to that time-limit depended in part on the speed with which the president of the RICS appointed an arbitrator and on the speed with which the appointed arbitrator accepted the appointment.

The third reason was that a tenant was not entitled to serve on a landlord a notice making time of the essence of a step in the procedure which it was open to the tenant himself to take. If the relevant time-limit in clause 5(2)(b) was to be interpreted as one within which an application should at least be made to the president of the RICS for the appointment of an arbitrator, the tenant did not need the remedy of being able to serve a notice on the landlord making time of the essence of it because the tenant was, under the clause, entitled equally with the landlord to make the application. Mr Lewison went so far as to say that the tenant was not only entitled but under an obligation to make the application. I was not, however, persuaded by his argument to that effect.

With those submissions in mind, I turn to the speeches in the House of Lords in the *United Scientific* and *Cheapside* cases. Lord Diplock, in the first part of his speech, deals generally with the question of principle, whether there is a presumption that time is not of the essence of a rent review clause and comes to the now well-known conclusion that there is. In so doing, he mentions the concept of a notice making time of the essence only in the passage that I referred to earlier, in which he states its historical origin.

In the second part of his speech, Lord Diplock turns to the actual facts of the *United Scientific* case. The lease there was for 99 years and the rent reviews were to take place every 10 years during the term. The only stipulation as to time in the rent review clause was that the rent for each successive period of 10 years or, in the case of the last period, nine years, was to be determined in the year immediately preceding that period. The case had two features in common with the present case. The first was that at any rent review, failing agreement between the landlord and the tenant as to what the new rent should be, that matter was to be referred to arbitration at the initiative either of the landlord or of the tenant. The second was that the arbitrator was to be nominated by the president for the time being of the RICS. In reaching the conclusion that the presumption that time was not of the essence applied in that case, Lord Diplock said this at pp 931-932:

> The Court of Appeal took the view that it was a detriment to the tenant not to know what his new rent was going to be in advance of the date when it started to accrue, as he might not be able to afford the additional rent and might feel compelled to assign the residue of the term to someone else. For my part, I find this unrealistic, if only because under this particular clause the tenant can initiate the review procedure himself and unless there is some unforeseen delay on the part of the arbitrator, has it in his power to ensure that the new rent is determined before the stipulated date. Apart from this, delay in the determination of the new rent until after the first rent day following the stipulated date works to the economic benefit of the tenant since until the higher rent has been determined he has the use of the money representing the difference between the former rent and the new rent which he would otherwise have been compelled to pay.

Lord Diplock made no mention in that part of his speech of any possibility that the tenant might serve a notice making time of the essence.

---

*Editor's note: Also reported at (1983) 267 EG 163: see pp 166-167.
†Editor's note: Also reported at (1983) 269 EG 729: see pp 729, 732 and 735.

In the third part of his speech, Lord Diplock turned to the facts of the *Cheapside* case. The relevant provisions of the rent review clause there were as follows:

3 The market rent may be determined and notified to the lessees in the manner following:
(a) the proposed rent shall be specified in a notice in writing ("the lessors' notice") served by the lessors or their surveyor on the lessees not more than twelve months nor less than six months prior to the review date. (b) the lessees may within one month after service of the lessors' notice of the proposed rent serve on the lessors a counter-notice ("the lessees' notice") either agreeing the proposed rent or specifying the amount of rent which the lessees consider to be the market rent for the period in question. (c) in default of service of the lessees' notice or in default of agreement as to the market rent to be payable for the period in question the rent shall be valued by a Fellow of the Royal Institution of Chartered Surveyors agreed between the lessors and the lessees or in default of agreement to be appointed not earlier than two months after service of the lessors' notice on the application of the lessors by the President for the time being of the said Institution whose valuation shall be made as an expert and not as an arbitrator and shall be final and binding upon the lessors and the lessees and shall be given in writing to the lessors and the lessees not less than fourteen days before the review date.

In fact, the lessors had given a trigger notice under clause 3 (a) in due time. Negotiations between the parties followed, but no agreement was reached either as to the new rent or as to a valuer to determine it. The delay that occurred was in the lessors' application to the president of the RICS to appoint a valuer. They did not make that application until after the review date. In commenting on the clause, Lord Diplock said this at pp 933-934:

These provisions contain an elaborate timetable as to what is to be done in various eventualities, not only by the landlord and tenant but also by persons over whom neither has any control — the President of the Royal Institution of Chartered Surveyors and whatever Fellow of the Institution may be appointed as valuer.

Leaving out a paragraph, he goes on:

In two respects under the terms of the review clause the progress of the procedure for determining the new rent is, or may become, within the exclusive control of the landlord. He alone can initiate the procedure; and he alone can apply to the President of the RICS if negotiations with the tenant do not result in an agreement as to the rent or upon the person who is to value it. The tenant's position under this clause thus differs from that of the tenant under the rent review clause that is the subject of the first appeal inasmuch as he has no right under his contract to initiate the procedure or to apply for the appointment of a valuer if the landlord himself fails to do so within the stipulated times. But this difference has not in my view any significant practical consequences so far as concerns any detriment to the tenant from the landlord's failure to do either of these things within the stipulated times. If the tenant reckons that the advantage of knowing before the review date exactly how much higher his new rent will be outweighs the economic benefit of having the use of the money representing the difference until the new rent has been determined, he has the remedy in his own hands. Quite apart from the fact that he can get a pretty good idea of what the market rent is from his own surveyor or can himself offer to enter into negotiations with the landlord before the stipulated time for serving a lessor's notice has expired, so soon as that time has elapsed he can give to the landlord notice specifying a period within which he requires the landlord to serve a lessor's notice if he intends the market rent to be determined and payable instead of the former rent for the ensuing seven years. The period so specified, provided that it is reasonable, will become of the essence of the contract.

Lord Diplock did not go on to say that the tenant could do the same thing if the landlord failed to apply in good time to the president of the RICS for the appointment of a valuer.

I need not trouble with the rest of Lord Diplock's speech, which contains nothing directly material for the present purposes. The same is true of the speeches of Viscount Dilhorne, Lord Simon of Glaisdale and Lord Salmon. It is true that there is a passage in Lord Simon's speech at p 946 where he discusses making time of the essence by notice, but, as is common ground between counsel, he does so only by way of comment on the historical survey in the first part of Lord Diplock's speech and, apart from confirming that the concept is a creature of equity, what he says has no direct bearing on the questions arising in the present case.

Lord Fraser of Tullybelton drew the same distinction between the *United Scientific* case and the *Cheapside* case as Lord Diplock. In commenting on the facts of the *United Scientific* case he said this at p 960:

There is provision for the arbitrator to be nominated, failing agreement between the parties, by the President of the Royal Institution of Chartered Surveyors and it is left open to either party to request the President to make a nomination. If the stipulation in the schedule requiring the rack rent to be ascertained "during the year" is to be strictly enforced the result would be that if, owing to some accident for which the landlord was not responsible or to the illness or dilatoriness of the arbitrator, the rack rent had not been ascertained until a month or even a day after the end of the year, the review would be abortive and the former rent would continue in force for another ten years. That result would seem to be inequitable . . .

Nowhere did Lord Fraser suggest that in the *United Scientific* case, or a case like it, the tenant might in any circumstances resort to a notice making time of the essence. Lord Fraser began his discussion of the *Cheapside* case in this way:

A more difficult question is raised in cases where the clause is in a form giving the landlord the sole right to initiate a review provided he does so by a certain time. Provisions of this sort are conveniently described as "triggering" provisions. A typical triggering provision is found in the *Cheapside* case.

He then summarised the facts of that case. In the course of doing so, he commented as follows:

No time limit for the application to the President or for the appointment of the valuer was stated, but, as the valuation had to be made not less than fourteen days before the review date, it was implied that the application and the appointment must be made in reasonable time to enable that to be done. In fact, the lessors did not apply to the President until more than two months after the review date and the President declined to make the appointment until its validity had been decided by the court. Hence these proceedings.

The landlord's right to operate the trigger and to apply to the President are both unilateral rights. The former might be described as an option. The latter would not I think normally be so described but, in my opinion, it is for the present purpose indistinguishable from the former in that both are unilateral rights which the landlord is under no obligation to exercise.

Lord Fraser went on to discuss an argument advanced on behalf of the tenants which was based on the view that the landlord's rights were options. He then, at p 962, said this:

It was also argued on behalf of the tenants that the lessors in a case such as *Cheapside* are not under any obligation to initiate a review and that there is therefore no room for applying the equitable rule so as to release them from the consequences of failure to perform an obligation. But the equitable rule originated in relieving a mortgagor from the consequence of failure to redeem his property by the stipulated date although he had no obligation to do so. The mortgagor, like the landlord here, had a unilateral right which might be described as an option, yet he was able to rely on the equitable rule to relieve him from the consequences of failure to exercise his right in time. There seems no reason in principle why the landlord should not be able to do the same and in my opinion he can. If a tenant felt himself prejudiced by the landlord's delay in serving a triggering notice, it would be open to him after the time for serving it had expired to give notice prescribing a further time within which the triggering notice must be served. Provided that the further time was reasonable, he could thus make time of the essence.

Like Lord Diplock, Lord Fraser did not say that the tenant could do the same if the landlord delayed his application to the president of the RICS.

From the way in which Lord Diplock and Lord Fraser expressed themselves in the *United Scientific* and *Cheapside* cases, I deduce that not every time-limit in every rent review clause of which time is not of the essence can be the subject of a notice making it so.

Mr Neuberger contended that so to hold would be contrary to basic principles. He submitted that it was never a rule of equity that a time limit could be ignored. Equity never went further than to say that a time-limit was not of the essence of the contract and, when equity said that, it gave by way of *quid pro quo* to a party who was thereby prejudiced the right to make time of the essence by serving an appropriate notice. To deny him that right would amount to treating the time-limit as not being in the contract at all.

In my judgment, that argument is ill founded. As was observed by Lord Diplock in his historical survey ([1978] AC 904 at p 927), by Lord Simon in his comments thereon (*ibidem* at pp 941-942) and by Lord Fraser in the last passage that I have read, the equitable rule about not treating time as being of the essence of a contract had their origin in the law of mortgages. It has never been part of those rules that a mortgagee can defeat the mortgagor's equity of redemption by serving a notice making time of the essence. It is therefore not correct to say that, whenever equity treats time as not being of the essence of a stipulation in a contract, it confers on a party thereby affected the countervailing right to serve such a notice. There is nothing contrary to principle in saying that the right to serve such a notice is non-existent in the sphere of mortgages, unlimited (unless by the contract itself) in the sphere of contracts for the sale of land and limited in the sphere of rent review clauses. I do not overlook that some of the *dicta* in the Court of Appeal to which I referred earlier are wide enough, read literally, to suggest that in the field of rent review clauses also the right is unlimited but, as Mr Neuberger very properly conceded, in none of the cases in which those *dicta* were uttered was the present question even remotely before the court and I cannot attribute to their authors an intention to express a view on it.

I turn to Mr Lewison's three reasons for saying that the tenant was not entitled to serve a notice making time of the essence in the present case.

The first (that, in the case of a rent review initiated by a trigger notice given by the landlord, the tenant is not entitled to make time of the essence of any subsequent step in the procedure) was, of course, based on the omission by both Lord Diplock and Lord Fraser, when dealing with the *Cheapside* case, to say that the tenant could serve a notice making time of the essence of the exercise by the landlord of its right to apply to the president of the RICS for the appointment of a valuer. At first sight that omission is puzzling, but I think that the reason for it must be that, as was pointed out by Lord Fraser, there was there no express time-limit for the application to the president. It would therefore have been necessary for Lord Diplock and Lord Fraser, if they were going to deal with the point, to express views *obiter* on questions that had probably not been argued as to the effect of the absence of an express time-limit. Lord Diplock, it seems, preferred to say nothing about it, while Lord Fraser went no further than to say that a reasonable time was implied. As is made clear by both Lord Diplock and Lord Fraser, the purpose of conferring on the tenant a right to serve on his landlord a notice making time of the essence is to afford him a remedy against inaction by the landlord which may delay the ascertainment of the new rent. I see no reason why the tenant should be denied that right at any stage of the rent review procedure at which inaction by the landlord may delay progress, if that right can be conferred on the tenant without unfairness to the landlord. I derive some assistance in that regard from the *Trustees of Henry Smith's Charity's* case, where the Court of Appeal saw nothing unfair in holding that time was, on the construction of the rent review clause there in question, of the essence of a provision in it imposing on the landlord a time-limit for applying to the president of the RICS for the appointment of a surveyor. I therefore reject Mr Lewison's first reason.

On Mr Lewison's second reason (that a tenant is not entitled to serve a notice making time of the essence of a provision concerning a step in the procedure the carrying out of which is not within the exclusive control of the landlord) I need not express a view and I refrain from doing so, because Mr Neuberger in effect conceded that the period of 28 days mentioned in the tenant's letter of May 29 1985 was unreasonably short if the letter meant that within that period not only must the landlord have applied to the president of the RICS for the appointment of an arbitrator but the president must have appointed one and the person appointed must have accepted the appointment. Mr Neuberger's contention was that the letter meant no more than that, within the 28 days, the landlord must have written to the president of the RICS applying for the appointment of an arbitrator.

So I turn to Mr Lewison's third reason (that a tenant is not entitled to serve on the landlord a notice making time of the essence of a step in the procedure that it is open to the tenant himself to take). This case shares, of course, with the *United Scientific* case, the characteristic that it was equally open to the landlord and the tenant to apply to the president of the RICS for the appointment of an arbitrator. It seems to me to be implicit in the speeches of both Lord Diplock and Lord Fraser, that in those circumstances, the tenant is not to have the alternative remedy of serving on the landlord a notice making time of the essence, because he does not need it.

Mr Neuberger urged upon me that the tenant ought to have it because he ought to be entitled in effect to say to the landlord: "I do not want to go to arbitration, because I think it will be an expensive waste of time, but I do not want the threat of arbitration hanging over my head. If you want to go to arbitration, you must do so within a reasonable time."

It seems to me, however, that, in a case where it was true that the relevant facts and figures were such that an arbitration would be an expensive waste of time, the landlord and the tenant ought to be able to settle the matter by agreement. It is failing agreement between them that the rent review clause provides for arbitration. The purpose of conferring on a tenant the right to make time of the essence of a provision in a rent review clause is not to give him an opportunity of avoiding fortuitously his obligations under the lease.

In saying that, I have in mind the point made by Lord Diplock at [1978] AC 904 at p 930 where he said:

> The determination of the new rent under the procedure stipulated in the rent review clause neither brings into existence a fresh contract between the landlord and the tenant nor does it put an end to one that had existed previously. It is an event upon the occurrence of which the tenant has in his existing contract already accepted an obligation to pay to the landlord the rent so determined for the period to which the rent review relates. The tenant's acceptance of that obligation was an inseverable part of the whole consideration of the landlord's grant of a term of years of the length agreed. Without it, in a period during which inflation was anticipated, the landlord would either have been unwilling to grant a lease for a longer period than up to the first review date or would have demanded a higher rent to be paid throughout the term than that payable before the first review date. By the time of each review of rent the tenant will have already received the substantial part of the whole benefit which it was intended that he should obtain in return for his acceptance of the obligation to pay the higher rent for the succeeding period.

That point was echoed by Viscount Dilhorne at p 938, by Lord Simon at p 946, by Lord Salmon at p 948 and by Lord Fraser at p 958.

For that reason, I do not think that the tenant in the present case was entitled to serve a notice making time of the essence in the circumstances in which it claims to have done so.

That is enough to dispose of the case and I need not express any view on Mr Lewison's further submission that, in any event, the tenant's letter of May 23 1985 was inadequately worded because it did not clearly state what the landlord must do in order to comply with it or what the consequences would be if the landlord did not comply with it.

I will grant the declarations claimed by the originating summons.

Court of Appeal

March 16 1987

(Before Lord Justice MAY, Lord Justice BALCOMBE and Lord Justice BINGHAM)

JAMES AND ANOTHER v BRITISH CRAFTS CENTRE

Estates Gazette June 6 1987

282 EG 1251-1255

Landlord and tenant — Rent review clause in lease — Construction — Appeal from part of decision of Scott J — The difficulties in this case had arisen in determining what exactly were the terms of the hypothetical lease which had to be assumed in arriving at the "commercial yearly rent" for the purposes of the rent review clause — After providing that such rent should be the open market rent of a lease for a term equal to the residue of the actual term of the current lease, the review clause stated that the hypothetical lease was to be "in the same terms in all other respects as these presents" — The construction problems were due to the fact that both the user covenant and the restriction of assignment covenant contained a specific reference to the lessee by name, "the British Crafts Centre" — The user covenant permitted the use of part of the premises for storage, sale and display of craftsmen's work while the lessee was the British Crafts Centre — The covenant restrictive of assignment permitted the sharing of occupation with a holding or subsidiary company, "but only whilst the lessee is the British Crafts Centre" — The relevance was that the broader the class of potential lessees the higher the rent that might be expected to be obtained — The tenants had accordingly contended that the provisions should be construed narrowly, directed to the particular personal position of the British Crafts Centre as such, whereas the landlords had argued that they should be taken to refer to the lessee who took the hypothetical lease, whoever that lessee might turn out to be — Scott J decided that the two covenants should be construed as applying differently in this respect — He held that in the case of the user covenant it was intended to grant a personal privilege to the Centre and

that the hypothetical lease should be assumed to contain a similar restriction — He held, however, that in the case of the assignment covenant the hypothetical lease should be treated as leaving blank the name of the lessee, the intention being to prevent the privilege of sharing occupation to be enjoyed by anyone other than the original lessee, whoever the original lessee might be — Before the Court of Appeal the landlords challenged the narrow construction placed by Scott J on the user covenant — The tenants did not, however, serve a respondent's notice with a view to appealing against the judge's wider construction of the assignment covenant — The Court of Appeal agreed with the judge's construction of the user covenant — That construction involved less departure from the actual language of the covenant and gave effect to what appeared to be the intention of the parties — As there was no cross-appeal by the respondent tenants in regard to the assignment clause, the correctness of the judge's construction of that clause did not arise — However, if the matter had come before the court May LJ would have reached a different conclusion from that of Scott J — Balcombe and Bingham LJJ also expressed doubts about Scott J's conclusion — Appeal dismissed

The following case is referred to in this report.

Law Land Co Ltd v *Consumers' Association* (1980) 255 EG 617, CA

This was an appeal by the plaintiff landlords, Robert Hedley James and Christopher Brian Carr, from part of the decision of Scott J (reported at [1986] 1 EGLR 117; (1986) 277 EG 976) in regard to the construction of the rent review clause in a lease to the respondent tenants, British Crafts Centre, of premises at 43 Earlham Street, Covent Garden, London WC2.

J P Whittaker (instructed by Hempsons) appeared on behalf of the appellants; Paul de la Piquerie (instructed by Sacker & Partners) represented the respondents.

Giving judgment, MAY LJ said: This is an appeal from a decision of Scott J on the hearing of an originating summons between the present appellants as plaintiffs and the respondent as defendant on October 30 1985. That originating summons sought declarations as to the proper construction of a rent review clause in a lease dated August 12 1977 entered into between the appellants' predecessors in title as lessors and the present respondent as lessee.

The demised premises were 43 Earlham Street, Covent Garden, London WC2. The term granted by the lease was one of 14 years from September 29 1976. The rent originally reserved by the lease was £10,500 per year.

The premises comprised four floors; basement, ground floor and two floors above. It seems that at the date of the lease the defendant respondent was already in occupation of the basement and ground floor. The first floor was let to a company known as High Vision Ltd.

The learned judge in his judgment referred to two planning permissions that were in existence in relation to the premises, but for my part I do not think they take the matter of proper construction of the lease any further and I do not propose to refer to them in detail.

I turn to the lease itself in which in particular the respondent, the British Crafts Centre, is referred to in this way:

hereinafter called "the Lessee" which expression where the context so admits includes the persons or corporate body in whom the term hereby granted may from time to time be vested.

For the purposes of the present appeal I turn to clause 2 (16) of the lessee's covenants which, in so far as material, reads as follows:

not to use or permit or suffer the demised premises or any part thereof to be used for any purpose other than
(i) for high class business commercial or professional offices . . . or
(ii) in respect of such part of the demised premises as shall for the time being be occupied and used by the Lessee (here meaning The British Crafts Centre party hereto) for storage sale and display of craftsmens work and ancillary offices and in respect of the first floor of the demised premises (whilst not occupied and used by British Crafts Centre) as an office and studio for the trade or business of designers advertising and press agents.

I then move to the next of the lessee's covenants, that is the one against alienation of the term, and, again omitting immaterial parts, it reads in this way:

Not to assign . . . demise underlet or otherwise part with possession of any part of the demised premises (here meaning a portion only and not the whole thereof) or (subject to the provisions of paragraph (ii) of this sub-clause) . . . to share occupation of the whole or any part thereof for all or any part of the said term . . .
(ii) Notwithstanding the provisions of paragraph (i) of this sub-clause . . . but only whilst the Lessee is The British Crafts Centre the Lessee may share occupation of the demised premises or any part thereof with a holding or subsidiary company of it or a subsidiary of such holding company . . .

That covenant went on to contain a number of provisos. In my opinion, however, they do not take the matter any further.

I then turn to the rent review clause, which was clause 5 in the lease. In general terms that provided for a rent review upon notice in writing by the lessors at five-year intervals and for the assessment of what was described in the lease as "the commercial yearly rent" at the quinquennial review date, and for the payment of that assessed rent from that review date, if when assessed it was greater than the rent then actually being paid.

Finally, in so far as the lease is concerned, I turn to the definition in clause 5 of the commercial yearly rent. That reads as follows:

"The commercial yearly rent" means the clear yearly rent at which the demised premises, assuming the due performance and observance of the covenants on the part of the lessee and conditions contained in these presents, might reasonably be expected to be let at the review date by a willing landlord in the open market with vacant possession and without premium or any other consideration than that evidenced by execution of a lease thereof to a willing tenant for a term equal to the residue then unexpired of the term hereby granted by a lease in the same terms in all other respects as these presents (including this sub-clause).

There then followed various sub-clauses, but again these are immaterial for the purposes of the present appeal.

The basic issue in the latter is whether in the hypothetical new lease on offer as contemplated in the definition of the commercial yearly rent which I have just quoted, the user for storage, sale and display of craftsmen's work and the ancillary offices referred to in the "user clause" is limited to the respondent, the British Crafts Centre, and no other body or person, as the respondent contends, or is to be permitted to the hypothetical new lessee, whoever he or it may be, which is the contention put forward by the appellant plaintiffs. The learned judge held that on that particular issue the correct view to take was the former. It is against that decision the lessors now appeal.

A similar question also arose below in respect of the proper construction of the covenant against alienation. On this point the learned judge held differently from his view on the proper construction of the user covenant. He accepted the lessors' contention that the reference to the British Crafts Centre in that part of the existing lease was not intended to be personal to the respondent but that in the hypothetical new lease, as it were, a blank would appear in the relevant clause to be filled in with the name of the new and also hypothetical lessee. The learned judge's decision on this second point has not been the subject of a cross-notice by the respondent.

The appellants' argument before us was four-fold. First, that in any event on the lease as it presently stands, one cannot give a wholly literal meaning to the definition of the "commercial yearly rent". On the judge's finding and the respondent's contention, there is to be substituted in the hypothetical new lease for the phrase "the lessee (here meaning the British Crafts Centre party hereto)" the phrase "The British Crafts Centre". This is inconsistent, it was submitted, with the concept of a hypothetical new lease in which the British Crafts Centre may be neither lessee nor a party to the conveyance. Second, Mr Whittaker argued that when one looks at the actual drafting of the relevant part of the user covenant in clause 2(16) of the lease it is not easy to follow, but if the respondent and the judge are correct it involves giving a new meaning to the word "lessee" in that clause different from that in the parties' clause of the lease, with which I started this judgment, and in which sense the word "lessee" is used in numerous other places throughout the lease. The appellants' suggestion that there should merely be a blank in which the name of the hypothetical new tenant could be substituted in this clause, as in the covenant against alienation, involves substantially less alteration to the wording actually used and for that reason also is to be preferred. Third, counsel submitted that the learned judge was inconsistent in reaching different conclusions on effectively the same point in the two covenants, the one in respect of permitted user and the one against alienation. It was submitted that, of the two, the latter was correct and should have been the conclusion reached on the user covenant also.

Finally, Mr Whittaker referred the court to *The Law Land Co Ltd v Consumers' Association Ltd* (1980) 255 EG 617. In that case there was a lease which contained a similar review clause to that for consideration in the instant appeal. There was, however, a more limited user clause by which

the tenants covenanted not, without the prior written consent of the landlord, to use or permit the demised premises or any part thereof to be used, other than as offices of the Consumers' Association and its associated organisations.

Counsel relied in particular on this passage from the judgment of Buckley LJ at p 623:

When one considers that the hypothesis upon which the clause is to operate is that the premises are vacant and that they are being offered on the market to a lessee who is prepared to accept them upon the terms of a lease tendered by the lessor, it is reasonable to suppose that the lease so hypothetically tendered will be a lease in which the name of the lessee will not be stated, because the assumption is that the lessee has not yet been identified. Also if the lease is to be a lease in the form of that with which we are concerned [the user clause] will necessarily be a clause in which the user covenant does not yet specify the name of the tenant which is to be inserted in that clause, although it will be drawn in such a way as to suggest that, when the identity of the hypothetical tenant has been identified, the name of that tenant will be inserted in the clause as the name of the Consumers' Association is inserted in the clause in the actual lease.

Mr Whittaker submitted that Buckley LJ was there suggesting an approach to that particular case which involved leaving blanks in the appropriate places in the respective covenants in which the name of the new hypothetical lessee would be inserted, as he contended should also be the approach in the present case.

Against those contentions and on behalf of the respondent, Mr de la Piquerie argued, first, that the learned judge was correct in the conclusion to which he came about the proper construction of the user covenant (clause 2 (16) of the lease). It was clear, he suggested, that the benefit of that clause and of the clause permitting sharing in certain circumstances of the demised premises was, and was intended to be, personal to the British Crafts Centre only. It was not intended to inure to the benefit of some wholly different hypothetical tenant. That solution required, he contended, substantially less rewriting of the actual wording of the clauses in the lease than would the contention of the appellants.

As to the alleged inconsistency between the learned judge's decision on the two clauses, the one as to user and the one against alienation, I think counsel in the end accepted that it was not easy to reconcile the learned judge's decision on the two points and that perhaps it would have been better had there been a respondent's notice in this appeal, seeking to have the learned judge's decision on the non-alienation clause varied so as to conform with his decision on the user clause.

In so far as *The Law Land Co Ltd* case is concerned, Mr de la Piquerie submitted that when one looks at the facts of that particular case, without a radical redrafting of the relevant parts of the lease, and in particular of the rent review and user clauses, there could not in the event have been the hypothetical open market which those clauses postulated. Therefore, in that case one has to do a certain amount of injury to the literal wording of the lease in order to enable just that open market to be looked at as and when the rent review dates came up. That difficulty will not arise in the instant case. The hypothetical open market can well exist and can easily be considered in the terms of the instant lease as they stand.

So much for the arguments on each side. For myself I respectfully agree with the respondent's submission and the learned judge's conclusion as to the proper construction of the user covenant, clause 2 (16). Although I think that the drafting of that covenant could have been clearer, in my opinion its intention is clear. It was to provide that the lessee should, first, be entitled to use the demised premises for high-class business, commercial or professional offices; and second, but only in respect of such part of the demised premises as are for the time being occupied and used by the "lessee", which in this instance and for this purpose is to be restricted to the British Crafts Centre, the party to the lease itself, for the further purposes mentioned. I prefer the learned judge's view about the proper construction of this clause because I think that it involves the least alteration to the actual language of the lease and covenants. It gives effect to what I think, on careful consideration of the covenant, was its clear intention, particularly because of the use of the word "here" immediately after the first bracket in the sub-clause. Finally, I can see in the circumstances a good policy reason for drafting the covenant in that way. In so far as *The Law Land Co Ltd* decision is concerned, I agree that that can and should be distinguished on the basis contended for by Mr de la Piquerie on behalf of the respondent.

In so far as the suggested inconsistency in the learned judge's decision on the two sub-clauses is concerned, we need not consider his decision on the second because there is no respondent's notice. If one is satisfied that his view on the first is correct, then the fact that his view on the second was different is not sufficient to make me change my mind on the first. I would add, however, that if the second had indeed been for our consideration in this appeal, then for my part I would with respect have reached a different conclusion upon it from that of the learned judge below. Nevertheless it is sufficient for present purposes to say that, for the reasons which I have sought briefly to indicate, I think this appeal should be dismissed.

Agreeing, BALCOMBE LJ said: The party of the second part to the lease of August 12 1977 is the British Crafts Centre. After the description of its registered office come these words: "hereinafter called the Lessee' . . ." If this definition clause had stopped there, it would have been no more than the parties making their own dictionary and substituting two words for four, so that wherever in the lease you find the words "the Lessee" you simply read "The British Crafts Centre". In fact the definition clause goes on: ". . . which expression where the context so admits includes the persons or corporate body in whom the term hereby granted may from time to time be vested."

When one comes to consider clause 2 (16) (ii) one finds these words:

. . . in respect of such part of the demised premises as shall for the time being be occupied and used by the Lessee (here meaning the British Crafts Centre party hereto) . . .

As my lord has said, with the benefit of the close attention this clause has received in two courts, the draftsman might have found a happier form of wording, but it seems to me clear that, in the light of the definition clause, what that subparagraph means is "in respect of such part of the demised premises as shall for the time being be occupied and used by the British Crafts Centre", ie the context does not admit a reference to anyone else, and indeed that meaning is strengthened by the later part of the same subparagraph where there is a reference in parenthesis to "(whilst not occupied and used by British Crafts Centre)". On that construction (which seems to me unanswerable), the way in which the learned judge approached the terms of the hypothetical new lease imposes no strain on the language used. It is merely saying what the lease already says, although using more words.

I agree that the case of *The Law Land Co Ltd v Consumers' Association Ltd* is distinguishable. In that case, unless some change was made to the wording of the new hypothetical lease, there was no potential open market. That problem does not arise here, because there is here under clause 2 (16) (ii) a user "for high-class business commercial or professional offices" which is already available to the world at large.

Finally, I would add that I, too, am uncertain about the construction which the learned judge gave to clause 2 (17A) (ii) of the lease (the proviso about sharing occupation), but since there has been no respondent's notice I need say no more about that.

I agree that this appeal should be dismissed.

Also agreeing, BINGHAM LJ said: This rent review clause has the same purpose as every rent clause, to provide for the rent reserved under the lease to be increased at the stated intervals in line with rises in rental levels in the open market. Thus the operation of any rent review clause involves a fusion of the actual and the hypothetical. The rent to be determined is that actually to be paid by the actual lessee under the lease in question or his successor in title, but the measure of that rent is determined by reference to what would be paid by a hypothetical willing lessee to a hypothetical willing lessor if the premises were available for letting on the open market, which of course they are not. Depending on the wording of the clauses in question, difficulties may arise (as they do here) in determining where the actual ends and the hypothetical begins.

The lessee's covenant in clause 2 (16) of this lease prohibits a long string of different users. By way of exception from the prohibition, three users are, as I construe the clause, permissible: (a) use for high-class business, commercial or professional offices; (b) use for storage, sale and display of craftsmen's work and ancillary offices; (c) use as an office and studio for the trade or business of designers, advertising and press agents. But each of these permitted users is subject to an express restriction.

User (a) is permitted subject to the obtaining of appropriate planning permission. This restriction is expressly stated in the language of the sub-clause, is quite clear, and causes no problem of construction.

User (c) is permitted in respect of the first floor of the demised premises "whilst not occupied and used by British Crafts Centre". The lease was expressly granted subject to but with the benefit of an existing tenancy of the first floor and this provision reflects that fact. This permitted user is subject to two restrictions. First, it is permitted on the first floor only. There can be no doubt about that. Second, in my view, it is permitted so long as the first floor is not occupied and used by the British Crafts Centre. In this part of the sub-clause there is no reference to the lessee. The reference is to the British Crafts Centre alone. In my judgment, therefore, the hypothetical open market rent is to be determined on the assumption that design, advertising and press agency user is permitted on the first floor in all circumstances, save when the floor is occupied and used by the British Crafts Centre.

User (b) is permitted "in respect of such part of the demised premises as shall for the time being be occupied and used by the lessee (here meaning The British Crafts Centre party hereto) . . ." Plainly there is a restriction of the part of the premises in which this user is permitted. This is defined by reference to actual events. There is also a restriction of the party by whom such user is permitted. Under the actual lease, as is agreed on both sides, this user is not permitted by a successor or assign of the British Crafts Centre. If the sub-clause had simply referred to "the Lessee" without amplification or qualification, there would be no problem. This user would be permitted, subject to actual occupation by the British Crafts Centre and any persons or corporate body in whom the term granted might from time to time be vested. That follows from the definition of "the Lessee" at the outset of the lease. The open market rent payable by the hypothetical lessee would be determined on the assumption that such user would, subject to occupation, be permitted. But the parties have stipulated that "the Lessee" shall here mean the British Crafts Centre. There is, in my view, no escaping from that, either in the actual lease or, because the hypothetical lease is to be in the same terms, in the hypothetical lease. Effect must be given as closely as possible to what the parties have agreed; otherwise there would be a disparity between the effect of the actual lease under which the rent is after all to be paid and the effect of the hypothetical lease, which is to provide a measure of that rent. It would seem to be anomalous if an assignee of this term were, following a rent review, obliged to pay rent based on an assumption that a user was permitted because permitted to a hypothetical lessee although not actually permitted to him. Such an anomaly should, I think, be avoided unless one is driven to it.

I share my lords' doubts on the judge's construction of subclause (17A)(ii), but this does not affect the outcome of this appeal, which I also would dismiss.

The appeal was dismissed with costs. An application for leave to appeal to the House of Lords was refused.

Chancery Division

November 21 1986

(Before Mr Justice SCOTT)

NORTH EASTERN CO-OPERATIVE SOCIETY LTD v NEWCASTLE UPON TYNE CITY COUNCIL AND ANOTHER

Estates Gazette June 13 1987

282 EG 1409-1414

Landlord and tenant — Rent review clause in lease — Construction — Ambiguous provisions as to capacity of independent surveyor appointed by agreement of parties to determine rack-rental value — Lease provided that such value at review date should be as agreed by the parties or, in default of such agreement, as determined by an independent surveyor agreed by the parties or, in default of such agreement, by an arbitrator nominated by the president of the RICS "and this lease shall be deemed for this purpose to be a submission to arbitration within the Arbitration Act 1950" — Whether the independent surveyor, who was in fact appointed by agreement of the parties, and not by the president, was to be regarded as acting as an arbitrator under the 1950 Act or simply as an expert valuer — The matter came before Scott J both by way of appeal by originating motion, on the assumption that the surveyor was an arbitrator, and by way of originating summons asking the court to decide whether or not he was an arbitrator — The plaintiff lessees wished to challenge the surveyor's assessment of the rack-rental value at £17,865 per annum, which appeared to them excessive as they had failed for some years before the review to find anyone willing to purchase the lease at the existing rent of £5,725 — The term was 42 years from December 1 1969 and the first rent review was on December 1 1983 — It was argued for the plaintiffs that it would be strange if the person appointed by agreement between the parties was not to act in the same capacity as the person to be appointed by the president of the RICS — On the other side it was pointed out that the lease drew a clear distinction between an independent surveyor and an arbitrator and that parties might be content to have their own appointee act as a valuer, but might wish to have the president's nominee subject to the arbitral machinery with its judicial features — The main authorities considered were *Sutcliffe* v *Thackrah*, *Arenson* v *Casson Beckman Rutley & Co* and *Palacath* v *Flanagan* — Held, on balance, although the indicia were not very strong, that it was intended that the independent surveyor should act as an expert if appointed (as he was) by agreement of the parties, but that he should act as an arbitrator if appointed, in default of such agreement, by the president of the RICS — This decision necessarily disposed of the motion by way of appeal under the 1950 Act — The judge refrained from making any declaration as to whether the surveyor, having been found not to be an arbitrator, was amenable to a suit for negligence — That issue had not been raised on the summons and it involved serious questions which required considered submissions

The following cases are referred to in this report.

Arenson v *Casson Beckman Rutley & Co* (on appeal from *Arenson* v *Arenson*) [1977] AC 405; [1975] 3 WLR 815; [1975] 3 All ER 901; [1976] Lloyd's Rep 179, HL
Langham House Developments Ltd v *Brompton Securities Ltd* (1980) 256 EG 719
Palacath Ltd v *Flanagan* [1985] 2 All ER 161; [1985] 1 EGLR 86; (1985) 274 EG 143
Safeway Food Stores Ltd v *Banderway Ltd* [1983] EGD 213; (1983) 267 EG 850
Schuler (L) AG v *Wickman Machine Tool Sales Ltd* [1974] AC 235; [1973] WLR 683; [1973] 2 All ER 39; [1973] 2 Lloyd's Rep 53, HL
Sutcliffe v *Thackrah* [1974] AC 727; [1974] 2 WLR 295; [1974] 1 All ER 859 [1974] 1 Lloyd's Rep 319, HL

The plaintiffs both in respect of the originating motion by way of appeal under the Arbitration Act 1950 and the originating summons were North Eastern Co-operative Society Ltd, lessees of supermarket premises in Newcastle upon Tyne. The respondent to the originating motion was Newcastle upon Tyne City Council and the council was also the first defendant to the originating summons. The second defendant to the summons was David Kendrick FRICS, the independent surveyor.

Grant Crawford (instructed by Punch Robson, of Middlesbrough) appeared on behalf of the plaintiffs; K Hornby (instructed by R A Brockington, director of administration, Newcastle upon Tyne) represented the city council; Miss Erica Foggin (instructed by Hadaway & Hadaway, of Newcastle upon Tyne) represented Mr Kendrick.

Giving judgment, SCOTT J said: I have before me two matters which are connected. Both arise as a result of a rent review carried out in connection with the rent payable under a lease whereunder the plaintiff, North Eastern Co-operative Society Ltd, is the lessee, and the first defendant, Newcastle upon Tyne City Council, is the lessor. The lease is a lease of supermarket premises in Newcastle upon Tyne. The supermarket is located in a shopping precinct in a housing estate which, I understand, has suffered urban decay. The lease granted a term of 42 years from December 1 1969. The lease was dated January 12 1971. It reserved the rent of £5,725 per annum for the first 14 years of the term and provided for rent reviews thereafter. The first rent review date, therefore, was December 1 1983.

I must refer in detail to the contents of the lease so far as concerns the rent review, but, put shortly, it provided a scheme under which the lessor and lessee were first to try to agree the new rent and, in default of agreement, the rent was to be fixed either by an independent surveyor agreed upon by the two parties or by an arbitrator to be appointed by the president of the Royal Institution of Chartered Surveyors.

The new rent was to correspond with what was called in the lease the "rack-rental value" of the demised premises. The rent was not, however, to fall below the initial £5,725 per annum.

Some three years or so before the rent review date arrived, the lessee ceased using the premises as a supermarket and offered its lease for acquisition. It was offered, of course, at the rent reserved of £5,725 per annum.

For three years up to the time of the rent review there were no takers; virtually no interest seems to have been shown by would-be purchasers in acquiring the plaintiff's lease. The plaintiff regards that as an indication that the rack-rental value of the premises could not have exceeded £5,725 per annum.

The lessor, the city council, did not agree that that was so. It contended that the rack-rental value of the premises for the seven-year period following December 1 1983 ought to be a sum in the region of £25,000.

As the parties could not agree, the machinery prescribed by the lease came into effect. The parties agreed on an individual to be appointed as the independent surveyor who was to determine the new rent. The person concerned was Mr David Kendrick, who is the second defendant before me. He was appointed and, after considering representations in writing from both sides, he gave what he called an "award" in which he assessed the rent to be paid for the seven years from December 1 1983 at the sum of £17,865.

The plaintiff regarded that decision as quite unwarranted. In argument the plaintiff contends that a conclusion to that effect could not for a moment be supported in view of the fact that nobody was willing to take the property at a rent of £5,725 per annum. It seems, from the material before me, that Mr Kendrick was particularly influenced by the rent that had been agreed on a rent review for another supermarket in broadly the same area as the supermarket with which I am concerned. The rent review in that case had proceeded, apparently, on the basis of the rent that might be expected to be commanded for the property if offered with vacant possession. Influenced by that, Mr Kendrick, with various adjustments, arrived at his figure of £17,865 per annum. The plaintiff contends that Mr Kendrick has applied quite the wrong tests. He has taken his eye off the required criterion of rack-rental value and allowed himself to be diverted, first, by the rent agreed, or fixed, on the rent review of the other supermarket and, second, by what he regarded as a reasonable rent for the tenant to be required to pay.

The plaintiff, having formed the view, rightly or wrongly, to which I have referred, naturally wished to challenge Mr Kendrick's decision. The form to be taken by the challenge became, however, complicated by doubt as to whether Mr Kendrick was acting as an arbitrator under the Arbitration Act 1950 in making his decision, in which case the appropriate remedy for the plaintiff would be to obtain leave to appeal against his decision and then to prosecute such appeal by application to the High Court, or whether Mr Kendrick was, in truth, not an arbitrator but simply acting as an expert in making his valuation decision. In the latter case an appeal under the provisions of the Arbitration Act 1950 would not be available and the plaintiff's only remedy would be by way of a negligence action against Mr Kendrick.

Mr Kendrick's award was given on December 5 1985. On February 5 1986 the plaintiff issued a summons for leave to appeal out of time against the award. That was, of course, on the footing that Mr Kendrick had been acting as an arbitrator and that the relevant provisions of the Arbitration Act 1950 applied. The complication caused by the doubt as to whether Mr Kendrick was or was not an arbitrator had led to the expiry of the time within which leave to appeal ought normally to have been sought and so leave to apply out of time was necessary.

On March 26 1986 Hoffmann J gave leave to the plaintiff to appeal out of time, but in his order he expressed that leave to be without prejudice to the contention of the city council that Mr Kendrick was not an arbitrator and that, for that reason, an appeal did not lie.

On February 7 1986, in optimistic anticipation of the order of Hoffmann J that was eventually made, notice of originating motion was issued by the plaintiff by way of appeal. That notice would be an effective notice if, but not unless, Mr Kendrick was, in truth, an arbitrator. That notice of originating motion is before me.

Then, on April 18 1986, the plaintiff issued an originating summons asking the court to decide, in effect, whether or not Mr Kendrick was an arbitrator. That originating summons is before me. The city council is first defendant and Mr Kendrick is second defendant. The city council is, of course, respondent to the originating motion whereby the appeal is brought. Mr Kendrick is not a party to the appeal.

The issues arising out of the originating summons have been argued before me. That was right and logical, because they are by way of being preliminary issues to the appeal itself. If Mr Kendrick was not an arbitrator the appeal falls away.

Para 1 of the originating summons seeks a declaration that Mr Kendrick was acting as arbitrator. Para 2 asks, in the alternative, for a declaration as to the capacity in which Mr Kendrick was acting in determining the new rent to be paid.

It was, I think, in the mind of Mr Crawford, who has appeared before me for the plaintiff, that, if not satisfied that Mr Kendrick was acting as arbitrator, I might make a declaration that he was acting as quasi-arbitrator or, perhaps, as an expert. At an early stage in the arguments I indicated that "quasi-arbitrator" was not, to my mind, a term of art. A declaration that somebody was acting as a "quasi-arbitrator" would be likely itself to be the subject of a future application to the court as to what the declaration meant. A declaration that a person was acting as an expert would be merely a preliminary to some further proceedings. It is quite clear to me what the plaintiff has in mind. If the plaintiff cannot prosecute the appeal because Mr Kendrick was not an arbitrator, it would then wish to sue him in negligence. The plaintiff wants to establish that he is amenable to being sued in negligence. The plaintiff has in mind that Mr Kendrick might contend, as a defence to a negligence action, that, in determining the new rent, he was acting, if not as arbitrator, then at least in such a quasi-judicial capacity as entitled him to immunity from suit. The plaintiff wants that possible defence to be dealt with under the originating summons. Hence para 2 of the prayer. I will return to this aspect of the summons later.

The main point is whether or not in determining the rent Mr Kendrick was acting as arbitrator.

I must now turn to the lease and refer in more detail to its relevant provisions.

Para 3 of the lease reserves the rent. The rent is expressed to be £5,725 for the first 14 years of the term. The clause then proceeds:

and thereafter such other annual rent as shall be agreed between the parties hereto in accordance with the provisions contained in the third part of the schedule hereto.

I turn to part III of the schedule. Para (1) of part III entitles either party to give a notice in writing requiring a rent review.

Para (2) I should read in full.

From the relevant date of review the yearly rent shall (in default of agreement) be the said yearly rent of £5,725 or the yearly rent of an amount equal to the rack rental value of the premises as at that date as agreed by the parties hereto or (in default of such agreement) as is determined by an independent surveyor agreed between the lessors and the lessee or (in default of agreement) by an arbitrator to be nominated by the President of the Royal Institution of Chartered Surveyors on the application of either party and this lease shall be deemed for this purpose to be a submission to arbitration within the Arbitration Act 1950 or any statutory modification or re-enactment thereof for the time being in force and the assessment fixed by the independent surveyor or arbitrator as the case may be shall be communicated to the parties hereto in writing and immediately upon such communication the rent so assessed as the reasonable rent for the ensuing period of the term granted by the lease immediately following the date of review until the next date of review shall be the rent payable for the said period under the terms of the lease and an endorsement to that effect shall be made on the lease and counterpart thereof and executed by the lessors and lessee respectively.

Para (3) more accurately defines what is meant by "rack-rental value". I think I need not read that, although it would be highly relevant to the question of an appeal, if appeal there is to be.

Para (4), too, I need not read; but para (5) is relevant. It is in these terms:

The fees payable to the independent surveyor or arbitrator hereinbefore mentioned for such assessment as aforesaid shall be borne by the parties hereto equally.

There is one other provision of the lease to which I ought to refer. It is to be found in clause 6(f) thereof. Clause 6(f) is dealing with the event of the premises becoming damaged or destroyed by fire so as to be unfit for occupation or use. It provides for a suspension of the rent or some part thereof for an appropriate period until the premises shall have been reinstated and once more be rendered fit for occupation and use. At the end of the clause there is this:

any dispute concerning this clause shall be determined by a single arbitrator in accordance with the Arbitration Act 1950 or any statutory enactment in that behalf for the time being in force.

Those are the relevant provisions of the lease.

The question for me is whether Mr Kendrick, having been appointed by the parties as the independent surveyor and having then proceeded to determine the rent in the sum I have mentioned, was acting as an arbitrator under the Arbitration Act 1950 or was simply an independent surveyor acting as a valuer or expert.

The arguments put forward by Mr Crawford for the plaintiff and Mr Hornby for the first defendant were on both sides, I thought, very cogent and very simple.

Mr Crawford referred to the provisions for an arbitrator to be nominated by the president of the Royal Institution and to the provision that the lease "shall be deemed for this purpose to be a submission to arbitration within the Arbitration Act 1950". It would be strange, he pointed out, if the parties had not contemplated that the same function would be discharged by the independent surveyor, if they could agree on one, as by the person to be appointed by the president of the Royal Institution of Chartered Surveyors. It is clear from the clause that the person to be appointed by the president is to be an arbitrator and so it should follow, said Mr Crawford, that the person appointed by the parties should be an arbitrator: both would be discharging the same function.

As against that, Mr Hornby drew attention to the contrast in part of the schedule between the independent surveyor, on the one hand, and the arbitrator, on the other hand. The contrast is not simply to be found in the first mention of "independent surveyor" and "arbitrator" respectively, namely, in the part of the clause in which it is said that the rent shall be determined by "an independent surveyor agreed between the parties or, in default of agreement, by an arbitrator". The distinction is drawn again in the passage which refers to "the assessment fixed by the independent surveyor or arbitrator", and also in clause 5, which refers to "The fees payable to the independent surveyor or arbitrator . . .". Mr Hornby relied on those references as indicating that the parties were not regarding the function of the independent surveyor appointed by the parties and of the arbitrator appointed by the president as the same.

I was referred to the correspondence which led to the appointment of Mr Kendrick. The correspondence commenced with a letter of September 4 1985 from the city council to Mr Kendrick. He was by this letter asked to:

act as independent surveyor for the purpose of assessing the annual rental in accordance with the terms of the lease.

That is language consistent with Mr Kendrick's simply being a valuer. However, in the last line of the letter Mr Kendrick was asked for "your confirmation that a reasoned award will be given". That is language more consistent with arbitration than simply with valuation. Mr Kendrick wrote on September 10 1985 to chartered surveyors acting for the plaintiffs. He said, among other things: "I would be pleased to act as an independent surveyor in respect of this matter. I do, however, feel that it is prudent to invite written submissions from the parties", and he then set out a suggested timetable for submissions.

Mr Hornby pointed out — I think rightly pointed out — that that language was more consistent with Mr Kendrick's not being an arbitrator than with his being one. He was not regarding it as inevitable that written submissions would be invited from both parties: he simply put the suggestion forward as something he thought a good idea as a matter of prudence.

On September 10 1985 Mr Kendrick wrote also to the city council in answer to the city council's letter of September 4. The letter is in broadly the same terms as his letter to the plaintiff's surveyors; but he rather underlined the point that he was acting as an independent surveyor in saying this: "Although acting as an independent surveyor I do . . . feel that it is prudent to invite written submissions from the parties."

The point made by Mr Hornby arising out of the previous letter is, therefore, accentuated in the letter to the city council.

That is consistent with an affidavit sworn by Mr Kendrick on May 28 of this year in which he sets out his own understanding of the position. In para 4 he says: "It was clear to me that my appointment was as an independent surveyor and not as an arbitrator."

However, it does not seem that Mr Kendrick's understanding of the position, nor the understanding of the position by any other of the parties, has been entirely consistent. In a letter of September 26 1985, Mr Kendrick wrote again to the plaintiff's surveyor and said: "I would also confirm that I would be willing to grant a reasoned award in this arbitration."

The correspondence leaves it, in my view, at large whether the parties thought that they were setting in train an arbitration as opposed simply to a valuation procedure.

I have had submissions from both counsel as to the criteria that ought to be applied in order to identify an arbitration, strictly so-called, from a mere valuation. I have been referred in this connection to two cases in the House of Lords, *Sutcliffe* v *Thackrah* [1974] AC 727 and *Arenson* v *Casson Beckman Rutley & Co* [1977] AC 405. Neither case directly raised the question whether an individual was acting as valuer or as arbitrator. *Sutcliffe* v *Thackrah* concerned the position of an architect employed under an RIBA contract who had issued certificates from time to time. The question for decision was whether the architect was amenable to being sued for negligence as a consequence of the certificates issued. It was suggested, on the strength of some earlier authority, that an architect, acting under an RIBA contract, was in a judicial or quasi-judicial position — in a position, if not that of an arbitrator, then that of a quasi-arbitrator — and that there was a public interest immunity from suit enjoyed by persons acting in a judicial or quasi-judicial capacity. There had apparently been Court of Appeal authority in the early years of this century to that effect. That Court of Appeal authority was overruled by the House of Lords in *Sutcliffe* v *Thackrah*. In *Arenson* v *Casson Beckman Rutley & Co* the defendants were chartered accountants who had valued shares in a private company for the purpose of fixing the "fair value" at which the shares were to be sold. *Sutcliffe* v *Thackrah* was followed. In both cases there are dicta bearing upon the indicia of arbitral judicial proceedings as opposed to proceedings not of that character. In *Sutcliffe* v *Thackrah* Lord Salmon at p 763 said this:

In *In re Hopper* Cockburn CJ, with whom Blackburn and Lush JJ agreed, was in effect saying that the question as to whether anyone was to be treated as an arbitrator depended upon whether the role which he performed was invested with the characteristic attributes of the judicial role. If an expert were employed to certify, making a valuation or appraisal or settle compensation as between opposing interests, this did not, of itself, put him in the position of an arbitrator. He might, eg, do no more than examine goods or work or accounts and make a decision accordingly. On the other hand, he might, as in *In re Hopper*, hear the evidence and submissions of the parties, in which case he would clearly be regarded as an arbitrator. Everything would depend upon the facts of the particular case. I entirely agree with this view of the law.

In the *Arenson* case Lord Simon at p 423 expressed the issue in this way:

The main issue in this part of the case was whether it was of the essence of a judicial decision that it answers a question (the respondents' contention) or decides a dispute (the appellant's contention). The latter seems to me to be the right view both in principle and on authority

Then, a little further down the same paragraph, he says:

The general judicial role in society is to resolve disputes which the parties themselves cannot resolve by conciliation, compromise or surrender

At p 424 Lord Simon said this:

There may well be other indicia that a valuer is acting in a judicial role, such as the reception of rival contentions or of evidence, or the giving of a reasoned judgment. But in my view the essential prerequisite for him to claim immunity as an arbitrator is that, by the time the matter is submitted to him for decision, there should be a formulated dispute between at least two parties which his decision is required to resolve. It is not enough that parties who may be affected by the decision have opposed interests — still less that the decision is on a matter which is not agreed between them.

Lord Wheatley at p 428 set out what, in his view, were the indicia of the judicial proceedings as opposed to that of mere valuation. He said:

> The indicia are as follows: (a) there is a dispute or a difference between the parties which has been formulated in some way or another; (b) the dispute or difference has been remitted by the parties to the person to resolve in such a manner that he is called upon to exercise a judicial function; (c) where appropriate, the parties must have been provided with an opportunity to present evidence and/or submissions in support of their respective claims in the dispute; and (d) the parties have agreed to accept his decision.

Mr Crawford, in his submissions, asserted that each one of these indicia was to be found in the present case. Mr Hornby, on the other hand, submitted that the indicia were inconclusive. He submitted that a valuation exercise would, in rent review cases at least, be likely to involve a dispute or difference between the parties; it would certainly involve the parties having agreed to accept the decision; it might well involve the parties being entitled to place evidence before the tribunal; it might very well involve the tribunal exercising some element of judgment upon the material placed before it.

Before returning to the particular submissions and the issue which I must decide, I think I should refer to the last of the authorities placed before me. This was a decision of Mars-Jones J in *Palacath Ltd* v *Flanagan*, reported in [1985] 2 All ER 161.* The question in this case, like that which arose for decision in *Sutcliffe* v *Thackrah* and *Arenson* v *Casson Beckman Rutley & Co*, was whether a particular individual was liable to be sued for negligence arising out of his function in acting to resolve a question or dispute between the parties. In *Palacath* v *Flanagan* the person concerned was a surveyor who had determined the rent under a rent review clause in a lease. The relevant provision of the lease whereunder he was appointed provided as follows:

> The surveyor: (1) will act as an expert and not as an arbitrator; (2) will consider any statement of reasons or valuation or report submitted to him as aforesaid but will not be in any way limited or fettered thereby; (3) will be entitled to rely on his own judgment and opinion; (4) will within two months after his appointment or within such extended period as the landlord and the tenant may agree give to the landlord and to the tenant written notice of the amount of the rent as determined by him and his determination will be final and binding on the landlord and on the tenant.

The point before Mars-Jones J for decision was not whether the surveyor had acted as an arbitrator. It would, I think, have been impossible to have contended that he did, in the face of the express statement in the lease that the surveyor "will act as an expert and not as an arbitrator". The question before the learned judge was whether the surveyor was amenable to suit for negligence. He held that he was. At p 166, however, Mars-Jones J analysed what, in his view, were the particular distinctions of importance between the role of a person acting as arbitrator — acting judicially — and the role of a person acting simply as valuer. He said:

> the ultimate test is: how was he to arrive at his decision? Was he obliged to act wholly or in part on the evidence and submissions made by the parties? Or was he entitled to act solely on his own expert opinion? If the answer to the question is the latter, then the defendant could not be exercising a judicial function or a quasi-judicial function, if there is any such distinction. In the instant case, the defendant was specifically enjoined in clause 8 of the second schedule to act as an expert, and was not to be limited or fettered in any way by the statement of reasons or valuations submitted by the parties, but was entitled to rely on his own judgment and opinion. In the light of those express provisions it is impossible for me to hold that the parties intended that the defendant should act as an arbitrator or quasi-arbitrator in determining the revised rent. I am satisfied that the provisions of clause 8 were not intended to set up a judicial or quasi-judicial machinery for the resolution of this dispute or difference about the amount of the revised rent. Its object was to enable the defendant to inform himself of the matters which the parties considered were relevant to the issue. He was not obliged to make any finding or findings, accepting or rejecting the opposing contentions.

In that case, as it seems to me, the argument against the surveyor having been an arbitrator was a good deal stronger than the corresponding argument in the case before me.

I return to the language of the lease in order to try to discern from it what the parties must have contemplated would be the role of the independent surveyor. It was commented by Mr Hornby that the reference in clause 6(f) to arbitration was relevant in that it showed that the draftsman of the lease, and, accordingly, the parties, knew well how to make clear that an arbitration procedure was intended. I find myself unimpressed by that as a guide. In para 2 of part III of the schedule there is a clear arbitration provision as well as a reference to the appointment of an independent surveyor. If the contrasting references to "arbitrator" and to "independent surveyor" in that paragraph do not suffice to justify the distinction between the intended function of the independent surveyor and the intended function of the arbitrator, the contents of clause 6(f) cannot, in my view, do so.

The relevant provisions in part III of the schedule follow upon the failure of the parties to agree on the new rent. The first provision is that "in default of such agreement the rent shall be determined by an independent surveyor agreed between the lessors and the lessee". If the clause had stopped there, there would, I think, have been little difficulty in concluding that the independent surveyor was intended to act simply as a valuer and was not acting as an arbitrator. That was the decision come to by Sir Robert Megarry V-C in *Langham House Developments Ltd* v *Brompton Securities Ltd* (1980) 256 EG 719. It was the decision come to also by Goulding J in *Safeway Food Stores Ltd* v *Banderway Ltd* (1983) 267 EG 850. If the provision in the lease with which I am concerned had stopped at the place I indicated, there would have been no valid grounds of distinction, in my view, between this lease and those leases. The provision, however, goes on: "In default of agreement," that is to say, in default of agreement as to the identity of the independent surveyor, the rent is to be determined "by an arbitrator . . . ".

The question then is whether the parties, having clearly intended an arbitral function, to that extent a judicial function for the arbitrator, must be taken to have intended the same function for the independent surveyor.

Mr Crawford's argument rested very heavily on the improbability of the parties having intended a different machinery for the independent surveyor than that which they must clearly be taken to have contemplated for the arbitrator. He referred in this connection to the dictum of Lord Reid in *L Schuler AG* v *Wickman Machine Tool Sales Ltd* [1974] AC 235 to the effect that the court should lean against a construction which would attribute to the parties an unreasonable or perverse intention.

Mr Hornby, however, had what to my mind was a fair answer to that submission. He pointed out that the independent surveyor would have to be a person in whom both parties had confidence and who was known to both parties. He would almost certainly be a person practising in the area in which the demised premises are to be found. It is understandable, said Mr Hornby, that the parties might be content that their own approved appointee should act as a valuer and as an expert, while wishing, none the less, to have arbitral machinery, with its judicial characteristics, for the nominee of the president of the RICS. The president's nominee would not necessarily be known to them or necessarily practise in the same area as that in which the demised premises are to be found. Whether thinking of that character was in fact part of the reasons why the parties accepted para 2 in the form in which it stands I know not; but the reasonableness of the possibility, in my view, deprives Mr Crawford's argument of much of its force.

I turn again to the language of the paragraph. After the reference to the appointment of the arbitrator by the president the paragraph continues: "and this lease shall be deemed for this purpose to be a submission to arbitration within the Arbitration Act 1950". I take this language to be a slight indication that the "purpose" is limited to that of the arbitrator. The phrase could have been "this lease shall be deemed for these purposes to be a submission to arbitration". That might have been more apt if it had been intended to be a submission to arbitration not simply for the purpose of the arbitrator and his function, but also for the purposes of the independent surveyor and his function.

Then there is the reference to the "assessment fixed by the independent surveyor or arbitrator as the case may be". That language is not wholly inconsistent with the view that the independent surveyor was to have an arbitral function, but it points a contrast between "independent surveyor", on the one hand, and "arbitrator", on the other. That point is underlined and given a little more substance than it would otherwise have by the wording of para 5 relating to fees — "the fees payable to the independent surveyor or arbitrator hereinbefore mentioned". Again there is the contrast.

I have to say that none of these points seems to me to be very strong. The parties have left it unclear, to my mind, as to whether the independent surveyor was to be acting simply as valuer or was to be, like his colleague appointed by the president, an arbitrator. But, on

---
*Editor's note: Also reported at [1985] 1 EGLR 86 and (1985) 274 EG 143.

balance, I have concluded that the correct construction is that the independent surveyor was not intended to be an arbitrator and was intended to act as an expert.

Returning to the indicia, there was, of course, a dispute between the parties at the time in question in the sense that they could not agree on the amount of the yearly rent; but it was not a dispute in which each had formulated a view which was then placed for decision before the independent surveyor. The independent surveyor asked for their submissions in order to assist him in his task. He did not proceed on the footing that he was obliged to have their submissions. He was not appointed in order to arbitrate between £5,725 per annum on the one hand and £24,000-odd on the other hand. I am not clear, therefore, that there was a formulated dispute in quite the sense that, for instance, Lord Wheatley had in mind in the *Arenson* case. Second, to ask whether the dispute or difference was to be resolved by the independent surveyor in a judicial manner is to beg the question. If he was an arbitrator, he would have to resolve the dispute in a judicial manner: if he was simply a valuer, he would be able to call for such assistance as he might think desirable but would deal with that assistance and with the submissions and evidence the parties might think fit to place before him as a valuer. He would not be bound to confine himself to that evidence. He could go beyond it and, of course, could reject it. I am not clear that that indicium is satisfied in the present case.

The third indicium was that the parties must have been provided with an opportunity to present evidence and/or submissions in support of their respective claims. Although, in fact, they were presented with that opportunity, that machinery was not written in as an essential procedure to be followed by the independent surveyor. It was suggested by him in his letters of acceptance of office.

Finally, the parties are bound to accept the independent surveyor's decision. This is the only one of Lord Wheatley's four indicia that is clearly present.

On balance, therefore, I conclude that Mr Kendrick was not an arbitrator and I do not propose to make the declaration sought by para 1 of the originating summons. That conclusion means that the originating motion by way of appeal pursuant to the provisions of the Arbitration Act 1950 cannot proceed.

I now return to para 2 of the originating summons. It raises, not in its terms but in its intended effect, the question whether Mr Kendrick, not having been an arbitrator, is amenable to being sued for negligence. Whether, of course, there was any negligence is a matter on which I say nothing at all.

The only question for consideration, if it is a question I can decide on this application, is whether he is amenable to being sued or whether there may not be some public interest amenity behind which he is entitled to shelter.

In finding that he was not an arbitrator I am finding, in a negative sense, something about his capacity. I am not prepared to make a declaration in a positive sense as to his capacity because I do not understand that any accurate meaning can be ascribed to such an expression as "quasi-arbitrator". Nor am I prepared to make a declaration that he was acting in the capacity of expert because I am not clear what that would encompass either. Mr Kendrick was employed, and accepted office, to apply his expertise to the question before him for determination. That must be common ground between the parties.

If the question had been clearly raised on the originating summons whether Mr Kendrick was entitled to public interest immunity from suit, I would have dealt with it. It is a proper matter, as it seems to me, to be raised by way of preliminary point and it might as well have been raised by the originating summons as in a summons taken out in an action actually commenced against Mr Kendrick. Mr Kendrick is a party to this summons. But that issue has not been clearly raised on the originating summons. It arises only incidentally, it seems to me, in trying to think through what lies behind para 2 of the summons.

Neither counsel is able to refer me to any authority to the effect that a person in Mr Kendrick's position is entitled to public interest immunity. Mars-Jones J's decision in the *Palacath Ltd* case was that the surveyor with whom he was concerned was not entitled to that immunity. That decision, of course, is not binding upon me and I do not know, because it has not been explored, whether there are any distinguishing features between that case and this. There is a sense, of course, in which, though not an arbitrator, Mr Kendrick was exercising judgment on rival contentions which he had invited and which, no doubt, he had considered. Whether that would be sufficient to allow him on public policy grounds to immunity from suit raises, in my view, very serious questions indeed. I am loath to decide that question without counsel having had a full opportunity of researching any relevant case law that there may be and making considered submissions. I think it is too important a topic to be satisfactorily dealt with as it were by a side-wind arising out of para 2 of the originating summons in its present form. I therefore propose to say nothing more about it.

The plaintiffs were ordered to pay the costs of both defendants.

Chancery Division

February 4 1987
(Before Mr Justice SCOTT)

CORNWALL COAST COUNTRY CLUB v CARDGRANGE LTD

Estates Gazette June 27 1987

282 EG 1664-1675

Landlord and tenant — Questions arising in rent review arbitration — Arbitration Act 1979, section 2 — Determination of matters of law by the court — Premises in Curzon Street, Mayfair, used as a gaming club or casino — Rent review clause in sublease — Gaming activities conducted by Crockford's, a licensee of the sublessee — Questions submitted to court with a view to clarifying the situation after sublessor's expert put forward an open market rental value of £3,000,000 per annum and sublessee's expert a figure of £180,000 — There were eight questions designed mainly to elucidate the hypotheses and assumptions on which the rental value was to be assessed — Many of the complications arose from the fact that Crockford's held a gaming licence in respect of the premises and the effect of a requirement in the review clause that there should be disregarded any addition to the value attributable to the gaming or justices' licence which might be so held if it appeared, having regard to the terms of the tenancy and any other relevant circumstances, that the benefit of the licence belonged to the subtenant or any associate — The judge was asked to consider such questions as whether it should be assumed that Crockford's had never been in occupation or that someone other than Crockford's held the gaming licence or that Crockford's did not in fact hold the gaming licence, or whether it should be assumed that the number of casino licences in the London casino circuit was one less than the number actually in existence — In considering in detail the answers to the questions posed the judge emphasised the importance of adhering strictly to the limits of the hypotheses required by the review clause and of refraining from deducing elaborate consequences from them as if the original hypotheses had an independent existence as actual features of the real world — This would result in the creation of fictitious situations or circumstances not within the contemplation of the review clause — After detailed consideration the judge summarised the assumptions on which the arbitrator should regard the parties to the sublease as "higgling" — A separate question put to the judge was whether the arbitrator was entitled to value the demised premises by reference to, or in reliance on, profits made by the sublessee or an associate carrying on a business therein — This question was associated with an appeal by the sublessee against a refusal by the arbitrator to order specific discovery of the full management accounts and supporting basic records of the casino undertaking — After considering authorities, the judge held that the admissible evidence as to profit-earning capacity would be that which is available to prospective tenants in the hypothetical open market — Whether that would extend to Crockford's private trading records would be

a matter for the arbitrator — Unless it did so extend in the arbitrator's opinion the documents would not be discoverable — As regards the discovery appeal itself, the arbitrator had refused discovery for the wrong reasons (no evidence of absence of comparables, expense and unlikelihood of saving costs) and the appeal would have to be allowed on that ground — The judge did not, however, propose to remit the discovery application for rehearing, as it was extremely unlikely that an arguable case could be made that these trading records would have been available to prospective lessees in the hypothetical open market

The following cases are referred to in this report.

Barton (WJ) Ltd v *Long Acre Securities Ltd* [1982] 1 WLR 398; [1982] 1 All ER 465; [1982] EGD 265; (1981) 262 EG 877, CA
Evans (FR) (Leeds) Ltd v *English Electric Co Ltd* (1977) 36 P&CR 185; [1978] EGD 67; 245 EG 657
Harewood Hotels Ltd v *Harris* [1958] 1 WLR 108; [1958] 1 All ER 104, CA
Lynall v *Inland Revenue Commissioners* [1972] AC 680; [1971] 3 WLR 759; [1971] 3 All ER 914, HL
Norwich Union Life Insurance Society v *Trustee Savings Banks Central Board* [1986] 1 EGLR 136; (1986) 278 EG 162
Scottish & Newcastle Breweries plc v *Sir Richard Sutton's Settled Estates* [1985] 2 EGLR 130; (1985) 276 EG 77

In these proceedings, in which the plaintiffs were Cornwall Coast Country Club, sublessors of 30 Curzon Street, Mayfair, London W1, and the defendants were Cardgrange Ltd, the sublessees, the court was asked to determine a number of questions of law pursuant to section 2 of the Arbitration Act 1979. There was also an appeal from the arbitrator's refusal to order discovery of certain documents.

Michael Barnes QC and Ian Glick (instructed by Stilgoes) appeared on behalf of the plaintiffs; M A F Lyndon-Stanford QC and Kim Lewison (instructed by Cameron Markby) represented the defendants.

Giving judgment, SCOTT J said: I have before me for decision a number of questions which have arisen in the course of a rent review arbitration. The plaintiff, Cornwall Coast Country Club Ltd, is the sublessor of 30 Curzon Street, Mayfair. The defendant, Cardgrange Ltd, is the sublessee.

The headlease of 30 Curzon Street, dated January 27 1978, was granted by Daejan Investments Ltd to Ladbroke Rentals Ltd. The term granted was 26 years from December 8 1977. The rent reserved was £130,000 pa. The headlease provided for rent reviews to take place on December 8 1983 and thereafter at five-yearly intervals.

The sublease, dated October 9 1980, was granted by Ladbroke Rentals Ltd to Ladup Ltd, also a member of the Ladbroke Group of companies. The sublease granted a term of 26 years from December 8 1977 less three days. The rent reserved was £130,000 with rent reviews on the same dates as were provided for under the headlease.

The rent review provisions in the headlease and in the sublease were identical. They required the "Rack Rental Value" of 30 Curzon Street to be ascertained by a chartered surveyor acting as an arbitrator.

"Rack Rental Value" was defined in subclause (5) of clause 6 as follows:

The highest rent at which at the rent review date the premises might reasonably be expected to be let in the open market by a willing lessor, with vacant possession, for the residue of the term hereby granted, and upon the terms and conditions including provision for rent review of this lease, it being assumed that the demised premises have been put into a state of repair and condition consistent with full performance of the obligations of the tenant under this lease, but there being disregarded

(a) any effect on rent of the fact that the tenant or his predecessor in title or some associate of the tenant has been in occupation of the holding;

(b) any goodwill attached to the holding by reason of the carrying on thereat of the business of the tenant or some associate of the tenant, whether by him or by a predecessor of his in that business;

(c) any effect on rent of any improvement carried out by the tenant;

(d) any addition to the value of the demised premises attributable to the gaming or justices licence which may be held in respect of such premises if it appears that having regard to the terms of the current tenancy and any other relevant circumstances the benefit of the licence belongs to the tenant or some associate of the tenant.

I need not read the rest of the rent review provisions. They provided broadly for upwards adjustments of rent, with the rack rental value at the relevant review date being substituted for the rent previously payable. The new rent to be payable as from the first review date, December 8 1983, was to be at least £170,000.

The user covenants in the headlease and in the sublease are to the same effect. They authorise user of the property or any part thereof as a gaming club or casino, as a non-residential club of some other description, as offices, as a restaurant, as licensed bars, as staff rooms, as showrooms and as residential premises.

By assignment dated August 26 1983 the headlease was assigned by Ladbroke Rentals Ltd to the plaintiff, also a member of the Ladbroke Group.

By assignment dated November 3 1982, the sublease was assigned by Ladup Ltd to the defendant. The defendant is a member of the Lonrho Group of companies.

No 30 Curzon Street comprises a basement, a ground floor and four upper floors. The top floor consists of a residential flat. The rest of the property consists of a gaming club or casino.

Conversion of the premises to enable its use as a gaming club or casino took place over the period 1978 to 1980. The gaming use did not, however, commence until after the assignment of the sublease to the defendant and the defendant had carried out additional alterations and refurbishments. A gaming licence for 30 Curzon Street was obtained by Crockford's Club Ltd (also a member of the Lonrho Group) on June 29 1982. A liquor licence was obtained in February 1983. Use of 30 Curzon Street as a gaming club and casino began on March 1 1983 and has continued ever since.

The gaming activities at 30 Curzon Street have been and are being conducted by Crockford's, the holder of the gaming licence, rather than by the defendant, the sublessee. Nothing turns on this. Crockford's, as a gaming club, is a household name. Before moving to 30 Curzon Street, Crockford's had conducted a gaming club from premises in Carlton House Terrace.

The first rent review date both under the headlease and under the sublease was December 8 1983. The headlease rent review has been completed. The arbitrator was Mr J G Powell, chartered surveyor. He was appointed on January 18 1984. A number of questions of law arose in the course of the arbitration. Those questions were, with the consent of the parties, Daejan Investments Ltd and the plaintiff, submitted to the High Court for determination. The case came before Peter Gibson J, who gave judgment on December 14 1984. I have been supplied with a transcript of his judgment. One of the questions raised the point whether in disregard (d) of subclause (5) the reference to "the tenant or some associate of the tenant" should be read as including a sublessee or a licensee of a sublessee. Peter Gibson J held that "tenant" in the headlease meant the tenant under the headlease and did not include a sublessee.

This point is significant because it has the consequence that the respective rent review clauses in the headlease and the sublease, although in identical terms, will produce different, perhaps very different, results. In the sublease rent review, the effect on rent of the occupation of 30 Curzon Street by Crockford's is required to be disregarded. Not so in the headlease rent review. Crockford's is not, for the purposes of the headlease, "the tenant or his predecessor in title or some associate of the tenant". Nor is the defendant. In the sublease rent review, any addition to the value of the demised premises attributable to Crockford's gaming licence is to be disregarded. Not so in the headlease rent review.

The arbitrator in the headlease rent review gave his award dated May 9 1985. He found the rack rental value, for the purposes of the headlease, to be £900,000 pa.

The arbitration procedure, for the purposes of the sublease rent review, has not yet been completed. The arbitrator, Mr V D Revell, chartered surveyor, was appointed on December 19 1985. On January 7 1986 a preliminary hearing took place at which pleadings were directed. These have been served. It was directed that proofs of evidence of the expert witnesses be exchanged. This has been done. On the plaintiff's side Mr R John Stephenson, a chartered surveyor and partner in the firm Grant & Partners, has submitted a proof. It ends with Mr Stephenson's conclusion that on December 8 1983 the rental value of 30 Curzon Street for the purposes of the sublease rent review was the sum of £3,000,000 pa exclusive.

On the defendant's side there are two expert witnesses. Mr J R Trustram Eve, senior partner of J R Eve, chartered surveyors, has submitted a proof setting out the manner in which, in his view, a valuer should proceed to calculate the rental value of 30 Curzon Street in accordance with the terms of the sublease rent review clause. Mr Leslie Aarons, a chartered surveyor and partner in Baker Lorenz, has done the actual valuation in accordance with Mr Trustram Eve's view as to the correct approach. Mr Aarons has arrived at the conclusion that the rental value as at December 8 1983 would be represented by the sum of £180,000 pa.

The valuers on either side may be assumed to be experienced and competent professional valuers. The startling discrepancy in their conclusions is attributable to the different assumptions and hypotheses used by the respective sides. Very sensibly the parties have endeavoured to resolve these differences by submitting a number of questions of law to the court for determination pursuant to section 2 of the Arbitration Act 1979. These are quite different questions from those submitted to Peter Gibson J in connection with the headlease rent review arbitration.

There are eight questions in all. Counsel before me, Mr Barnes for the plaintiff and Mr Lyndon-Stanford for the defendant, have rightly treated questions 1 to 6 as associated. This group of questions endeavours to clarify the hypotheses and assumptions on which the estimate of the rental value ought to be based.

Questions 1 to 6 are in these terms:

1 Does the stipulation in clause 6(v) of the underlease, that the premises are to be assumed to be let with vacant possession, require the arbitrator to assume that the respondents as Crockford's Club Ltd have never been in occupation of the premises;

2 Does the stipulation in clause 6(v)(a) of the underlease, that there is to be disregarded any effect on the rent of the fact that the tenant or his predecessor in title or some associate of the tenant has been in occupation of the holding, require the arbitrator to assume that the respondents, as Crockford's Club Ltd, have never been in occupation of the premises;

3 Is the arbitrator entitled to assume that some person other than the respondents and/or Crockford's Club Ltd held the gaming and justices licence in respect of the premises and carried on gaming there up to the review date;

4 Is the arbitrator required to assume that the respondents and Crockford's Club Ltd did not hold a gaming or justices licence in respect of the premises;

5 Is the arbitrator required to assume or entitled to find that in considering for the purposes of the rent review the London casino market as a whole, the number of gaming licences in existence on the rent review date was one less than the number of such licences actually in existence on that date;

6 Is the arbitrator entitled to assume and/or find that the respondent was a possible hypothetical tenant in the market for the property on the rent review date.

Counsel are agreed that I should not regard myself as bound strictly by the wording of these questions. In the course of the hearing before me, areas of common ground have emerged and the real issues between the two sides have seemed in many cases to be more narrowly defined than in the questions. Counsel have invited me not only to answer the actual questions but also to try to resolve the other issues of principle which have emerged from the arguments before me. It is recognised that the valuation process is a matter for the arbitrator, but it is hoped that the result of this reference to the court will be to settle the outstanding points of principle as to the basis on which the valuation should be made.

For reasons that I need not take time to explain, question 7 cannot usefully be proceeded with at this stage.

Question 8 is in these terms:

Is the arbitrator entitled to value the demised premises by reference to or in reliance upon profits made by the tenant or an associate of the tenant in carrying on a business therein.

This question is associated with an appeal by the plaintiff against a refusal by the arbitrator to order certain specific discovery. General discovery took place between the parties following the close of pleadings. The defendant's list of documents did not include any documents relating to the profits earned by the gaming business carried on at 30 Curzon Street. The plaintiff applied to the arbitrator for specific discovery of the "full management accounts and supporting prime or other basic records of the casino undertaking". The application was not limited to documents in respect of the period March 1 1983 to December 8 1983. As I understand it, all documents and records from March 1 1983 to date were sought. The plaintiff's contention was that an estimate of rental value could be based on or supported by evidence of the profitability of 30 Curzon Street as a casino. The documents of which discovery was sought were, it was said, relevant to that issue. The overlap with question 8 is patent.

By an interim award given on June 18 1986 the arbitrator (*inter alia*) refused the application for this specific discovery.

The plaintiff, leave having been given by Millett J on November 18 1986, has appealed against the arbitrator's refusal of discovery. That appeal, too, is now before me. It has been argued by counsel in conjunction with their arguments on question 8. Before me the plaintiff has limited its discovery claim to documents in respect of the period March 1 1983 to December 8 1983.

Many of the complications in the rent review arise, directly or indirectly, from the fact of Crockford's gaming licence in respect of 30 Curzon Street. The licence was first granted on June 29 1982. Gaming licences are annual licences. They must be renewed each year. On the review date, December 8 1983, Crockford's held a current licence due to expire in May or June 1984. The exact date does not for present purposes matter.

Gaming licences are governed by the Gaming Act 1968, as amended by the Gaming (Amendment) Act 1982. I have been supplied by counsel with an agreed statement of facts which sets out, under para 5, the relevant practice regarding the application for and grant of gaming licences. I think I ought to read the whole of that paragraph:

5.1 – The Gaming Act 1968 Schedule 2 lays down a two-stage procedure for the obtaining of a new gaming licence;

5.2 – The first stage is for application to be made to the Gaming Board for Great Britain for a certificate of consent, consenting to the applicant applying for a gaming licence in respect of the particular premises. If and when this is obtained the applicant is then enabled to make application to the Licensing Justices for the appropriate area for a gaming licence to be granted in respect of the particular premises;

5.3 – Since August 28 1982 an application could be made at any time to the Gaming Board for a certificate of consent in respect of particular premises;

5.4 – Subject to the above, in determining whether to issue a certificate the Gaming Board shall "have regard only to the question whether in their opinion the applicant is likely to be capable of and diligent in securing that the provisions of the Gaming Act and of any regulations made under it will be complied with, that gaming on those premises will be fairly and properly conducted and that the premises will be conducted without disorder or disturbance";

5.5 – For this purpose the Gaming Board in particular take into consideration "the character reputation and financial standing" of (*inter alia*) the applicant, but may also take into consideration any other circumstances appearing to them to be relevant in determining whether the applicant "is likely to be capable of and diligent in securing the matters" mentioned in the above paragraph;

5.6 – A certificate of consent was granted to Crockford's Club Ltd as dated February 15 1982;

5.7 – It will be an issue in the arbitration whether and if so to what extent the number of gaming licences granted by the relevant justices is limited. Pursuant to para 19 of Schedule 2 to the 1968 Act, the Gaming Board give advice to the justices as to the extent of demand at least once a year. The justices are required to take account of such advice, but are not bound by it.

There were, I have been told, in the south Westminster area, on December 8 1983, 19 gaming establishments including 30 Curzon Street. There were 21 current gaming licences held in respect of premises in this area, including Crockford's 30 Curzon Street licence. The holders of two of these licences were, I have been told, seeking transfer of their licences from the premises in respect of which the licences had been granted to new premises. This explains the discrepancy between 19 gaming establishments and 21 licences. I have been told, also, that the policy of the Gaming Board and of the justices with jurisdiction in south Westminster was that the number of gaming establishments in this prime area should be strictly limited. I have been told that these authorities were not prepared to allow any increase in the number of licensed gaming establishments. These details may or may not be established by evidence before the arbitrator. They are by no means all common ground. But if the situation has been correctly described, the position of Crockford's and of Crockford's current gaming licence becomes highly significant.

It is common ground that use of 30 Curzon Street as a casino would justify a higher rental value than would be justified by any other use permitted under the sublease. It is common ground that a casino use would not be possible without a consent from the Gaming Board and a licence from the justices. So, on what basis ought the valuer to proceed so far as use of 30 Curzon Street as a casino is concerned? The rack rental value for the purposes of the rent review is required by the terms of the rent review clause to be ascertained on a hypothetical basis. The hypothesis is that 30 Curzon Street is available on the rent review date for letting in the open market by a willing lessor to a willing lessee for a term of years equal to the residue of the term of the sublease and with vacant possession given on the rent review date to the new lessee. The terms of the hypothetical letting are all, bar one, supplied by the relevant provisions of the rent review clause. The missing term is that of the rent. The rent is to be the highest rent that the hypothetical lessee would be willing to pay. But here the four "disregards" come into play.

The rent that the hypothetical lessee would be prepared to pay would depend upon the use to which the property could lawfully and profitably be put. A casino use is authorised by the sublease. So should it be assumed that the hypothetical lessee would succeed in obtaining the necessary certificate of consent from the Gaming Board and the necessary gaming licence from the justices? This was a question put to Peter Gibson J for the purposes of the headlease rent review. He answered "No" and it is common ground before me that the same answer must be given to that question for the purposes of the sublease rent review. Peter Gibson J expressed his conclusion thus, and I read from p 15 of the transcript:

I have no doubt that clause 6(5) does not require or permit the arbitrator to assume that the hypothetical lessee will take possession with a gaming licence already in his possession. That does not however mean that the arbitrator cannot take into account evidence tending to show the existence in the market of persons who might stand a good chance of obtaining a certificate of consent and a gaming licence, if they obtained an interest in the premises. The arbitrator's valuation will no doubt reflect that fact if established. It will no doubt take account of the uncertainty that attaches to any applications for a certificate of consent and a gaming licence.

Counsel before me have not directed any criticism to the correctness of my learned brother's analysis. They have, I think, accepted it as correct. So, it follows that for the purposes of the sublease rent review, the hypothetical lessee will be a person anxious to use 30 Curzon Street as a casino and anxious to obtain as soon as possible the necessary consent and gaming licence that, on the rent review date, December 8 1983, he does not yet hold. The uncertainty, of which the hypothetical lessee will be well aware, as to whether he will be able, lawfully, to use 30 Curzon Street as a casino is critical. The existence of this uncertainty is common ground. The arbitrator will have the task of assessing its extent and its effect on the rent that the hypothetical lessee would be prepared to offer for the hypothetical lease. But the parties are miles apart as to the basis on which the arbitrator should approach this critical question.

On the defendant's side it is said that the arbitrator should consider the hypothetical lessee's chances of obtaining a gaming licence on the footing that there were 21 current licences and 19 current gaming establishments in south Westminster. Since 30 Curzon Street is to be available to the hypothetical lessee with vacant possession, the hypothesis requires that Crockford's right to be in occupation has come to an end. So, says the defendant, Crockford's must be assumed to have obtained the necessary consents to the transfer of its gaming licence to premises elsewhere in south Westminster. Thus the number of current licences and current gaming establishments is unaffected by the vacant possession hypothesis. The defendant, relying on the very restrictive policy of the Gaming Board and of the justices to which I have referred, will seek to satisfy the arbitrator that the hypothetical lessee would have, and would know he had, little chance of obtaining a gaming licence for 30 Curzon Street. That being so, the hypothetical lessee would not be willing to pay a rent appropriate to casino premises but only a rent applicable to premises to be used for, say, offices or, perhaps, a restaurant.

This is the basis, broadly, on which Mr Aarons' assessment of the rental value was made.

The plaintiff's approach is quite different. The fact of Crockford's occupation of 30 Curzon Street and of Crockford's gaming licence must, it is said, be ignored — disregards (a) and (d) are relied on. So, contrary to the fact, it must be assumed that there were only 18 gaming establishments in south Westminster and only 20 gaming licences. Consistently with the alleged policy of the Gaming Board and of the justices, there was clearly room for one more gaming establishment and one more gaming licence. Moreover, the suitability of 30 Curzon Street for casino use is known and established. So the hypothetical lessee would have a very strong expectation of succeeding in obtaining the necessary gaming licence. There would, it is accepted, remain some unavoidable uncertainty. But it would be very small and would not justify much, if any, reduction in the rental value of 30 Curzon Street from that which would reflect an established lawful casino use. This seems to be the basis on which, broadly, Mr Stephenson proceeded.

Mr Barnes and Mr Lyndon-Stanford helpfully prefaced their respective arguments with a general review of authority. I will do the same.

The rent provisions in the sublease are based upon the terms of section 34 (1) of the Landlord and Tenant Act 1954 (as amended). Slight alterations have been made to the wording of the four "disregards" but thereapart there is no material difference between the relevant language of the section and that of the sublease.

Guidance as to the right approach to the operation of rent review provisions in this form may be found in the judgment of Donaldson J in *F R Evans (Leeds) Ltd* v *English Electric Co Ltd* (1977) 36 P & CR 185*. On appeal Donaldson J's judgment was affirmed by the Court of Appeal without significant comment.

A particular feature of the case was that, having regard to the peculiar nature of the demised premises, a willing lessee would, in reality, have been very difficult to find. The actual lessee would have been delighted to be rid of the premises and, in the assumed market, would not have made any offer at all. Against this factual background Donaldson J emphasised that the particular circumstances of the actual lessee were irrelevant. He said at pp 190-191:

The arbitrator's concern is with the attitude of the hypothetical *willing* lessee, who is not in occupation of the premises. The clause assumes that there is such a person, and it is nothing to the point to prove that there was not.

At p 192 Donaldson J said this:

I accept that the rent has to be agreed in the light of all the circumstances which in fact affect the property and, in theory, affect the hypothetical lessor and lessee. Any circumstance which affects the actual landlord and the actual tenant, but which would not affect the hypothetical lessor and lessee, is irrelevant. I agree that these circumstances include, but stress that they are not limited to, the fact that it is unlikely that there will be more than one willing lessee, and that in October 1976, which was the relevant date, there was no other property on the market which provided the accommodation and facilities provided by the Walton Works. I also agree that the possibility of the parties failing to reach agreement is to be disregarded. To borrow and adapt an immortal phrase, "We are not interested in the possibilities of a failure to reach agreement. They do not exist." As the negotiations proceed, however, each will be considering whether it would not be better at a given level of rent to break off the negotiations. True it is that they will resist these temptations, but the extent to which they will operate on their respective minds will be reflected in the rent which will notionally be agreed in the end.

In *Norwich Union Life Insurance Society* v *Trustee Savings Banks Central Board* [1986] 1 EGLR 136, Hoffmann J said this at p 137:

There is, I think, a presumption that the hypothesis upon which the rent should be fixed upon a review should bear as close a resemblance to reality as possible.

In my view, these passages justify, and require, an approach in which, first, the hypotheses required by the rent review provisions are strictly adhered to but, subject to that, the real circumstances of the case are taken into account. This approach requires, I think, a clear distinction to be drawn between, on the one hand, hypothetical assumptions directed by the language of the rent review provisions and, on the other hand, allegedly consequential assumptions which, it is argued, must follow the former assumptions. May I try to explain the distinction I have in mind?

The hypothetical lease will carry with it the right to vacant possession. It must, therefore, be assumed that on December 8 1983 30 Curzon Street will be available for occupation and use by the hypothetical lessee. It is a necessary assumption that Crockford's right of occupation, as licensee of the defendant, came to an end on that date. It must also be assumed that Crockford's actual occupation came to an end on that date.

Mr Lyndon-Stanford argued that if the hypothetical assumption were true, Crockford's would, in the real world, have established themselves elsewhere and have obtained a transfer of the 30 Curzon Street gaming licence to other premises. This, in my opinion, is to confuse reality and hypothesis. Crockford's departure from 30 Curzon Street has not happened. It is a hypothetical assumption, demanded by the rent review provisions. The rent review provisions do not demand any assumptions at all about Crockford's situation on or after December 8 1983 other than its departure from 30 Curzon Street. Crockford's establishment elsewhere is nothing to do with the real world. It is simply another hypothesis. And it is not a hypothesis required by the rent review provisions.

A similar approach should, in my view, be applied to the

---

* Editor's note: Also reported at (1977) 245 EG 657.

"disregards". The "disregards" are required to be left out of account. They are to be assumed to be disregarded by the hypothetical lessee in deciding what rent to offer and by the hypothetical lessor in deciding what rent to accept.

Disregard (a) requires any effect on rent of the fact that the defendant or Crockford's have been in occupation of 30 Curzon Street to be disregarded. Disregard (b) requires any goodwill attached to 30 Curzon Street by reason of the gaming club user since March 1 1983 to be disregarded. Disregard (d) requires any addition to the rental value of 30 Curzon Street attributable to Crockford's gaming licence to be disregarded. None of these disregards requires, in terms at least, any positive hypothetical assumption to be made about the defendant or about Crockford's.

But, it is said, hypothetical assumptions have to be made because Crockford's is a real company, an important operator in the London gaming club world, and, in argument, part of the open market in which the hypothetical lessee is competing for the lease of 30 Curzon Street. The hypothetical lessee must outbid Crockford's as well as the other companies in the market.

So some assumptions about Crockford's must, it is said, be made. Crockford's cannot, for the purposes of this hypothetical competition for the lease, be invested with its actual characteristics. To do so would be to ignore disregards (a), (b) and (d). So hypothetical circumstances have to be attributed to Crockford's, eg the transfer to and conduct of a gaming business from other premises.

I reject this approach. It was made clear by Donaldson J in the *English Electric* case that the hypothetical tenant must not be invested with the qualities of the actual tenant. The qualities of the actual tenant cannot be attributed to the hypothetical lessee either to suppress the rent (as the lessee sought to do in the *English Electric* case) or to inflate the rent. The hypothetical lessee, and the hypothetical lessor for that matter, is an abstraction. One of the questions I am asked is whether the arbitrator should assume that Crockford's is a possible hypothetical tenant in the market. Mr Barnes has asked me to answer that question "Yes". But if the answer is "Yes" then Crockford's must be invested with fictitious qualities and fictitious circumstances. Crockford's becomes, I suppose, the hypothetical underbidder. Mr Lyndon-Stanford invited me to answer the question "No". But although he did not want Crockford's in the market as a bidder, he did want Crockford's in the market as a disincentive to other bidders. For that purpose, he invested Crockford's with a number of fictitious qualities and circumstances, ie the transfer to other gaming premises.

I reject both approaches. Reality must be adhered to so far as possible. It would, in my view, be wrong to regard the market as containing a hypothetical Crockford's, whether as a bidder or as a gaming presence elsewhere.

With those preliminary observations I turn to the six questions:

Question 1

Both counsel are agreed that I should answer this question in the negative. I agree. The "vacant possession" hypothesis requires that Crockford's right to occupy and its actual occupation have come to an end on December 8 1983, but requires no other assumption.

Question 2

Mr Barnes invites me to answer this question "Yes". He relies on the literal meaning of disregard (a). Crockford's occupation of 30 Curzon Street from March 1 1983 to December 8 1983 is a fact. It is also a fact, but one on which nothing turns, that either the defendant as sublessee or Crockford's as its licensee was in occupation from the date of the sublease to the commencement of the gaming user on March 1 1983. If Crockford's occupation from March 1 1983 to December 8 1983 is relevant at all on this rent review, its relevance must be attributable to some argument relating to the rental value of 30 Curzon Street on December 8 1983. But disregard (a) requires "any effect on rent" of that occupation to be disregarded. So, it is argued, the arbitrator must assume that Crockford's has never been in occupation.

Mr Lyndon-Stanford, on the other hand, submits that the answer should be in the negative. He relies on authority which, he submits, establishes that the broad literal meaning should not be given to disregard (a). First, there is *Harewood Hotels Ltd* v *Harris* [1958] 1 WLR 108, a decision of the Court of Appeal. The question was whether, for the purpose of fixing the rent of a new tenancy under section 34 of the Landlord and Tenant Act 1954, evidence of the tenant's business trading accounts during the previous five years had been rightly admitted. It was held by the Court of Appeal that the trading accounts were probative of the earning capacity of the premises and had been rightly admitted. The argument that to admit the accounts would permit the fact of the tenant's previous occupation of the premises to have an effect on the new rent was dealt with by Lord Evershed MR in this way:

It has been Mr King-Hamilton's main argument that the terms of paragraph (a) upon their ordinary sense must exclude any evidence at all about the occupation of the tenant, because its only possible admission could be to affect the rent, and it is said, according to the argument, that one cannot allow the quantum of rent to be affected in any way by considerations derived from the tenant's occupation. If that argument is right then I think it would follow that the judge here must be treated as having in that regard misdirected himself and even though, in the end of all, the right sum should turn out not to be very different from that awarded, still I think that Mr King-Hamilton would be entitled to have the matter remitted for reconsideration. But I am not, for my part, able to go as far with Mr King-Hamilton as to produce a total exclusion of any evidence based upon or derived from the tenant's previous occupation.

After referring to a passage from *Woodfall's Landlord and Tenant*, Lord Evershed said:

If the evidence was led for the purpose of showing that these tenants ought to be granted some concession because of some particular hardship that they had suffered, that might be another matter. And I agree also that the terms of the paragraph serve to exclude the consideration that a tenant might be expected to be willing to pay rather more than an outsider, because he would not wish to be disturbed in his occupation. But, in my judgment, it is plainly legitimate for a judge to hear evidence which bears upon the question which he has to decide, namely, what would the particular holding reasonably be expected to be let at in the open market? Plainly I should have thought in arriving at a conclusion upon that question it is legitimate to hear evidence of what similar premises which are being let for a particular purpose, as the one in suit is, can be expected to earn for a potential lessee in the market in the place where the premises are. And if so, then similar evidence is in my judgment admissible for proving the same point about the premises in suit. In other words, if the purpose of the evidence of the figures was for that limited objective, then I think for my part that they were perfectly admissible and that no objection can be made to them.

Romer LJ at p 114 said:

It seems to me that paragraphs (a) and (b) are really directed to saying that, for example, the fact that the sitting tenant has been in occupation for some time past and has built up a goodwill is to be disregarded in assessing the rent which he is to pay under the new lease. Normally, of course, a man who is in the position of sitting tenant and has built up a business and has been there for some years and established himself, would be prepared to pay a higher rent than anybody else then coming in for the first time. It is that kind of thing, in my view, to which paragraph (a) and (b) are directed.

Harewood Hotels Ltd v *Harris* was given careful attention in *W J Barton Ltd* v *Long Acre Securities Ltd* [1982] 1 WLR 398, also a decision of the Court of Appeal. This case, too, involved a section 34 new tenancy. The question was whether discovery should be ordered of the tenants' business accounts and records for the previous three years. The Court of Appeal distinguished *Harewood Hotels Ltd* v *Harris* on grounds which I will examine when dealing with question 8 and the discovery appeal, and refused to order the discovery. But the Court of Appeal accepted in principle that there might be cases where trading documents were relevant and ought to be produced on discovery. Oliver LJ, who gave the judgment of the court, described the problem at p 401. He said:

So what the court has to look for is the open market rent of the premises, simply as premises at which a business of the type carried on by the tenants can be carried on, but that rent is not to be enhanced, reduced, or otherwise affected by the tenant's own actual occupation of the holding, or by any goodwill created as a result of the business which the tenant has carried on. To put it broadly, the rent is to be arrived at on the hypothesis that the premises are empty and without regard to the tenant's previous trading.

This immediately raises the question in one's mind of what relevance to such an inquiry are the tenant's trading results. The court is not concerned with the tenant's ability to pay rent but the rent which a willing lessor could command for these premises in the hypothetical open market and there is a perfectly well recognised way of arriving at that, by reference to the rents payable for similar premises in the vicinity.

Indeed if one is to take into account the results of the tenant's trading as a relevant factor in arriving at the open market rent, the elimination from that consideration of any effect on rent from the tenant's occupation and from the goodwill involves an extraordinarily difficult practical exercise.

In his judgment the judge observed:

It is highly material to know in order to ascertain the open market rent what the trading position is.

A little later he said:

In a case of this sort where the open market rental value is in dispute evidence of trading is relevant and admissible to consider and show what the open market is.

We confess that, for our part, we are entirely unable to follow this in the case of a property such as this where there are, as it is conceded that there are, plenty of comparable premises in the vicinity from which the open market value of premises of this type can be deduced. No doubt evidence of the tenant's trading would indicate whether his business had been successful or unsuccessful, and so might be a pointer to the rent which this particular individual tenant might be prepared to pay in order to spare himself the disruption of moving to other similar premises in the area, but that has nothing to do with the open market rent which the court is directed by the Act to ascertain.

Of the *Harris* decision Oliver LJ said at p 403:

The landlord appealed on the grounds that the evidence was irrelevant and that, in any event, the judge was precluded from considering it by the provisions of section 34 (a) and (b) to which reference has been made above. The court dismissed the appeal. We cannot, however, read the decision as supporting any general proposition that evidence of this type is relevant and admissible in every application under the Act. It was relevant in that case because of the absence of any comparable premises and of the nature of the business under consideration; and the effect of the decision appears to us to be only this, that where such evidence is required in order to establish the open market rent, there is nothing in section 34 which prohibits its reception for this limited purpose.

At p 404 Oliver LJ goes on, having cited from the judgment of Romer LJ in the *Harris* case:

Certainly the decision shows that there may be cases where the production of trading figures may be both relevant and admissible and that where that is so section 34 does not inhibit the consideration of such evidence for the narrow limited purpose described above, even though the exclusion of any consideration of the effect of the tenant's occupation and of goodwill may present the judge with a very difficult task.

In *Scottish & Newcastle Breweries plc* v *Sir Richard Sutton's Settled Estates* [1985] 2 EGLR 130, His Honour Judge Paul Baker, sitting as a judge of the High Court, said this about disregard (a) at p 136 at E:

Then it was suggested to me that with regard to the first disregard, he has got to disregard any effect on rent of the fact that the tenant or any person deriving title under it has been in occupation of the demised premises, and if that requires him to assume that it is with vacant possession it ignores any sort of occupation that has been or is going on. That, in my judgment, is not a legitimate use of that disregard, indeed not the purpose of it, which is limited to negating the special effect of the tenant's own occupation, because that might either enhance the value in that he is likely to make a special bid and thereby increase it, or his occupation might diminish the value of the premises in that he had been in any way unsatisfactory in his occupation and thereby the premises had deteriorated. It is really directed at those sorts of considerations and not to conclude the question as to whether it is with or without vacant possession that the arbitrator is to review the rent.

Disregard (a) requires that in considering the rent the hypothetical lessee would offer, in considering what rent any other bidders on the open market might offer, the arbitrator must ignore any effect of the fact of Crockford's or the defendant's past occupation. But the authorities, to which I have referred, establish, in my judgment, that disregard (a) does not necessarily require the exclusion of all evidence "based upon or derived from the tenant's previous occupation" (*per* Lord Evershed at p 111).* The assumption expressed in question 2 is in wider terms than is justified by the language of disregard (a) or is consistent with these authorities. I would therefore answer the question "No".

If, however, the question had asked whether the arbitrator should assume that "the defendants and Crockford's have never been in occupation of the premises if and to the extent that the occupation had an effect on rent", I would have answered the question in the affirmative.

Question 3

I answer this question "No". No one, other than Crockford's, has ever held a gaming licence in respect of 30 Curzon Street. So why should the arbitrator proceed on the basis of a false assumption?

Mr Barnes asked me to answer this question in the affirmative. The reason was that he wanted the arbitrator to take account of the fact that from March 1 1983 to December 8 1983 30 Curzon Street had been used as a casino. But if it had been used as a casino, the operator/occupant must have had a gaming licence. Disregards (a),

---

* Editor's note: This refers to p 111 of *Harewood Hotels Ltd* v *Harris* [1958] 1 WLR 108.

(b) and (d) together require that any enhancement of rental value attributable to Crockford's gaming activities or gaming licence should be disregarded. So, it was argued, the arbitrator should proceed on the assumption that some other innominate person had held the necessary gaming licence. In that way, it was submitted, the arbitrator would be able to take account of the fact that 30 Curzon Street had been used as a casino while none the less honouring the disregards.

The real question, in my view, is whether the arbitrator ought to assess the rental value on the basis of a gaming user of the premises from March 1 1983 to December 8 1983 or whether the disregards require that fact to be ignored.

On what basis are the hypothetical lessee and the hypothetical lessor to be taken to be assessing the rent they are prepared to pay and to receive respectively? They must be taken to leave out of account the fact of Crockford's occupation (disregard (a)). They must be taken to leave out of account any goodwill attached to 30 Curzon Street by reason of its use as a casino (disregard (b)). And they must be taken to leave out of account any enhancement of rental value attributable to the fact that Crockford's held a gaming licence in respect of 30 Curzon Street (disregard (d)). If all this is to be left out of account, what is left of the fact that the premises were used as a casino from March 1 to December 8 1983? In my view, nothing. If and to the extent that the hypothetical lessee would be prepared to pay more rent, or the hypothetical lessor would expect to receive more rent, on account of the fact that for the nine months or so before the review date 30 Curzon Street had been used as a casino, they would, I think, thereby be taking into account the combined effect of Crockford's occupation, gaming activities and gaming licence. It follows, in my judgment, that the arbitrator is not entitled to assess the rent on the basis that the hypothetical parties are taking into account that 30 Curzon Street was used as a casino for the period March 1 to December 8 1983.

So far as this point is concerned, the hypothetical parties must, in my view, be taken to be contemplating a lease of premises suitable for use as a casino but which have not yet been used as a casino.

Question 4

Mr Barnes invited the answer "Yes". An affirmative answer is required, he submitted, by disregards (a) and (d). Mr Lyndon-Stanford, on the other hand, said that the answer should be "No". Disregard (a) requires only that the effect on rent of Crockford's gaming licence be disregarded. As to disregard (d), he pointed out that it was only an *enhancement* of rental value attributable to a gaming licence that must be disregarded. If Crockford's possession of a gaming licence would tend to depress the rental value, disregard (d) does not apply.

Both of Mr Lyndon-Stanford's points are, in my view, good ones. But the disregards must be applied consistently. The hypothetical lessee is deciding what rent to offer. He is "higgling" with the hypothetical lessor about rent. What facts will influence his thinking? Will he take into account that 30 Curzon Street is the subject of a current gaming licence held by Crockford's? If he does take that into account, the fact that the premises have been regarded as suitable for gaming by the Gaming Board and the justices may tend to reduce the uncertainty as to whether he, the hypothetical lessee, will succeed in obtaining a gaming licence and, accordingly, to enhance the rental value. This would be contrary to disregard (d). On the other hand, the fact that Crockford's current licence in respect of 30 Curzon Street will not expire until May or June 1984 may tend to increase the relevant uncertainty and to depress the rental value. This would not be contrary to disregard (d). It is, in my view, a question for the arbitrator whether the fact of Crockford's current gaming licence in respect of 30 Curzon Street would, on balance, enhance or depress the rental value. I would, however, emphasise that it is the gaming licence alone which is here in point. The use, pursuant to that licence, of 30 Curzon Street as a casino must, as I have held, be disregarded. There is, I think, no inconsistency in so holding. The proprietor of a gaming licence may or may not exercise the right granted by the licence. From June 1982 to March 1983, for example, Crockford's held a gaming licence, but 30 Curzon Street was not being used as a casino.

In my view, therefore, Question 4 cannot be answered with a definite "Yes" or "No". The arbitrator must make up his mind whether the fact of Crockford's current gaming licence in respect of 30 Curzon Street will, on balance, increase or reduce the rent which the hypothetical lessee would be prepared to pay. If it would increase the rent then the arbitrator must assess the rent the hypothetical lessee would offer without regard to it. But if not, then, since the licence is a fact and part of the real world, and since none of the disregards would require it to be disregarded, the licence's effect on the rent the hypothetical lessee would be prepared to pay should be taken into account. If, however, the licence is taken into account it must be taken into account for its rent-enhancing as well as its rent-depressing implications, although it must not have, on balance, a rent-enhancing effect.

Question 5

The answer to this question depends, in my opinion, on the answer to Question 4. The fact is that on December 8 1983 there were 21 current gaming licences in respect of premises in the south Westminster area. The "higgling" between hypothetical lessee and hypothetical lessor is assumed to be taking place in the real world, except to the extent that the rent review provisions require otherwise. Mr Barnes relied on disregards (a) and (d) as requiring the hypothesis of only 20 gaming licences. His reliance on disregard (a) is, in my judgment, misplaced. It is the effect on rent of occupation, not of gaming licences, that must be disregarded. As to disregard (d), it requires any addition in rental value attributable to the licence to be disregarded. But Mr Barnes wants the licence disregarded in order to increase the rental value. This is topsy-turvy.

It may be, however, that, pursuant to the approach I have outlined under Question 4, the arbitrator will conclude that Crockford's gaming licence in respect of 30 Curzon Street would tend to increase the rental value of the premises. In that event, disregard (d) requires that increase to be disregarded. The rent the hypothetical lessee would be willing to pay for the lease of 30 Curzon Street would have to be assessed on the footing that 30 Curzon Street was not on December 8 1983, and never had been, the subject of a gaming licence.

As I have already remarked, the hypotheses required by the rent review provisions must, in my view, be applied consistently. If the hypothetical lessee, in considering what rent he will pay, is to be taken to be contemplating 30 Curzon Street without any gaming licence, he cannot, in my view, be taken to be contemplating a south Westminster area with 21 gaming licences. The 20 other gaming licences and the premises to which they respectively relate are facts of the real world. Nothing in the rent review provisions requires them to be left out of account. Crockford's gaming licence in respect of 30 Curzon Street is a fact of the real world. If disregard (d) requires that fact to be left out of account, it must, in my judgment, be left out of account for all purposes of the assessment of rental value. If that were not so, to what premises would the 21st licence in the south Westminster area be taken to relate? There is no such thing as a gaming licence at large. If the gaming licence in respect of 30 Curzon Street is to be left out of account, then it must follow, in my judgment, that the hypothetical circumstances of the "higgling" involve 20 and not 21 gaming licences in the south Westminster area.

There are, therefore, two possible hypotheses, so far as Crockford's gaming licence is concerned. One is that the "higgling" is taking place on the footing that Crockford's held a gaming licence in respect of 30 Curzon Street and that there were 21 current gaming licences in the south Westminster area. The other is that Crockford's did not hold a gaming licence in respect of 30 Curzon Street and that there were 20 gaming licences in the south Westminster area. The first hypothesis accords with reality and, *prima facie*, should be adopted. But if it would result in a higher rental value being attributed to 30 Curzon Street than the second hypothesis, the arbitrator must, in my judgment, pursuant to disregard (d), adopt the latter.

There is another point thrown up by question 5. The number of gaming establishments in the south Westminster area on December 8 1983 was, in fact, 19. One was 30 Curzon Street. 30 Curzon Street is, however, assumed to be available on December 8 1983 for letting with vacant possession. In my view this "vacant possession" hypothesis reduces, for the purposes of the hypothetical higgling, the number of gaming establishments on December 8 1983 by one.

The 18 other gaming establishments are part of the real world that the hypothetical lessee must take into account. The 19th, 30 Curzon Street, is assumed to be vacant on December 8 1983. So, in my view, the hypothetical lessee must be taken to be assessing the rent he is prepared to pay on the basis that on December 8 1983 the south Westminster area includes the 18 gaming establishments and no more. It will be for the arbitrator to assess the effect of this on the view the hypothetical lessee would take of his chances of obtaining a gaming licence for 30 Curzon Street.

Question 6

Crockford's is a real company. Its stature in the gaming world is well known. The "vacant possession" hypothesis requires it to be assumed that Crockford's right of occupation and actual occupation of 30 Curzon Street have ended on December 8 1983. So is Crockford's to be assumed to be a potential bidder in the open market for the hypothetical lease of 30 Curzon Street. If it is, then the hypothetical lessee must outbid Crockford's. The presence of Crockford's in the market without any right of occupation of 30 Curzon Street and without any other gaming premises would tend to push up the rent the hypothetical lessee would have to pay. No surprisingly, Mr Barnes submitted that I should answer the question in the affirmative.

In my judgment, however, the right answer to this question is "No". The open market must be, so far as is possible, a market which corresponds with reality. The vacant possession hypothesis necessarily introduces an element of unreality. But as I have already tried to emphasise, the various hypotheses must, in my view, be taken no further than their terms make strictly necessary. It is not necessary for the purpose of giving effect to the "vacant possession" hypothesis that Crockford's should be treated as a possible hypothetical tenant. On the other hand, so to treat Crockford's introduces hypothesis upon hypothesis. It requires Crockford's to be invested with hypothetical qualities and surrounded by hypothetical circumstances that do not correspond with reality. Thus the proposition that Crockford's is a possible hypothetical tenant requires the hypothesis that it would be in the market for 30 Curzon Street if 30 Curzon Street were vacant. That latter hypothesis is not a necessary consequence of the vacant possession hypothesis. It is no more than an arguable one. The proposition that, if it had to leave 30 Curzon Street, Crockford's would try to negotiate a transfer of its gaming licence to the new lessee of 30 Curzon Street is another arguable hypothesis. So is the hypothesis that Crockford's would have established itself in other premises and obtained a transfer of its licence to those premises. None of these hypotheses is the fact. None is demanded by the terms of the rent review provisions. Each is no more than an arguable consequence of the vacant possession hypothesis. I reject them all. It is impossible to treat Crockford's as a possible hypothetical tenant without inventing for it hypothetical circumstances. So, in my view, Crockford's cannot be assumed to be a possible hypothetical tenant.

Summary

In my view, therefore, the arbitrator, in considering what rent the hypothetical lessee would be prepared to pay and the hypothetical lessor would be prepared to accept, should regard them as higgling on this footing:

(i) Vacant possession of 30 Curzon Street can be given on December 8 1983;
(ii) 30 Curzon Street is suitable for casino use but has not previously been used for that purpose;
(iii) There are, on December 8 1983, 18 licensed gaming establishments operating in the south Westminster area;
(iv) The competitors for the new lease do not include Crockford's
(v) Crockford's have not established themselves elsewhere;
either (vi) Crockford's hold a gaming licence in respect of 30 Curzon Street in which case 21 licences are, on December 8 1983, held in respect of premises in the south Westminster area;
Or (vii) Crockford's do not hold a gaming licence in which case 20 licences are held in respect of premises in the south Westminster area. The correct alternative as between (vi) and (vii) will be that which would, on balance, result in the lower rental value.
(viii) In any event, no one else has held a gaming licence in respect of 30 Curzon Street.

Question 8 and the discovery appeal

Basic to this question is the proposition that the earning capacity of commercial property is relevant to the level of rent that the property will command. This proposition is, in my view, self-evident. But although the earning capacity may be relevant to the rent level, it is not easy to translate that relevance into a sum of money, nor is it easy to be certain what would be the perception of a hypothetical tenant as to the earning capacity of a particular commercial property.

A common, perhaps the most common, tool used by valuers in order to form an opinion, or to justify an opinion, on the value of a property takes the form of comparables, that is to say, evidence of the rent or the price obtained on the letting or selling of comparable property. But the weight of comparables is obviously variable and will depend upon the extent to which the circumstances of the comparable transaction match the circumstances of the hypothetical transaction under review.

A valuer may or may not be satisfied with the evidence of the available comparables. He may look for some other tool by which to assess the rental value. He may look for evidence of the profit-earning capacity, either in an income or in a capital sense, of the property with which he is concerned. He may look at the development potential of the property. He may try to assess the trade turnover of which the property is capable. Evidence on these lines may tend to confirm or to invalidate the evidence of the available comparables.

There is not, in my view, any doubt but that the arbitrator in valuing 30 Curzon Street is entitled to take into account the income-earning capacity of the premises. The question is whether the arbitrator is entitled, in forming a view of the premises' income-earning capacity, to take into account the actual profits made by Crockford's. Mr Barnes submitted that he is. A possible objection is that that would be to allow Crockford's occupation of 30 Curzon Street to have an effect on rent, contrary to disregard (a). In answer to that objection Mr Barnes relied on *Harewood Hotels Ltd v Harris* and *W J Barton Ltd v Long Acre Securities Ltd*. In the former case the Court of Appeal held that the tenant's trading accounts for the preceding five years had been properly admitted in evidence. The objection based on disregard (a) was not accepted. I have already cited the relevant passages from the judgments of Lord Evershed MR and Romer LJ. In the *Barton* case the Court of Appeal refused to uphold an order for discovery of the tenant's accounts for the three previous years. But it did so not on the ground that the trading accounts were in principle inadmissible but on the narrow ground that they were not necessary because adequate comparables were available. As Oliver LJ said at p 402:

It is conceded that there are plenty of comparable premises in the vicinity from which the open market value of premises of this type can be deduced.

These two Court of Appeal authorities establish the proposition, binding on me, that disregard (a) is no reason for excluding from evidence the trading accounts of Crockford's at 30 Curzon Street.

Mr Lyndon-Stanford submitted that trading accounts should be admitted as evidence of value only where adequate comparables were not available. He relied on the *Barton* case as authority for this submission. But in *Barton* the adequacy of available comparables seems to have been conceded. In the present case, it is certainly not conceded by Mr Barnes that adequate comparables are available. The probative weight of the available comparables will, in due course, have to be decided by the arbitrator. He may or may not regard them as constituting compelling evidence of the rental value of 30 Curzon Street. His decision cannot be prejudged and the plaintiff is, in my view, entitled to place before him evidence of the income-earning capacity of 30 Curzon Street in order to supplement, discredit or qualify the evidence of the available comparables. The *Barton* case is not authority to the contrary.

Mr Barnes naturally relied on *Harewood Hotels Ltd v Harris* as authority for the admissibility into evidence of Crockford's trading accounts. He pointed out that the Court of Appeal had upheld the reliance by the trial judge on the tenant's trading accounts.

Mr Lyndon-Stanford, however, mounted a full-scale attack on *Harewood Hotels Ltd v Harris*. The case was, he submitted, wrong in principle and was decided *per incuriam*. His attack was based on the nature of the "open market" in which the "higgling" between hypothetical lessee and lessor is to be assumed to be taking place. The hypothetical lessee must be taken to be in possession of all information about the premises that either would be available to the public at large or would be supplied to prospective lessees by the hypothetical lessor. The previous tenant's statutory accounts would fall into the first category. Prospective lessees could obtain these from Companies House. But the previous tenant's trading accounts and records underlying the statutory accounts or additional to the statutory accounts would not be available to the public. They would be private documents. Nor would they be available to the lessor. The proposition that an outgoing tenant would make his trading accounts and records available to the lessor so as to enable the lessor to supply them to prospective new tenants has only to be stated to be seen to be devoid of any reality. These trading documents would be neither public nor available to the lessor for supply to prospective lessees. In the real world they would play no part in the "higgling" between lessor and prospective lessees.

These arguments seem to me, I am bound to say, almost self-evidently right. The "higgling" is to be assumed to be taking place in the real world. A previous tenant's private trading records would not, in the real world, be available either to lessor or to prospective lessee. So how can it be right that they should be admitted into evidence for the purpose of fixing rental value whether for the purposes of a section 34 new tenancy or for the purposes of rent review provisions such as those in the present case?

Mr Lyndon-Stanford's point does not seem to have been argued either in *Harewood Hotels Ltd v Harris* or in the *Barton* case. It does, however, have support from *Lynall v IRC* [1972] AC 680, a decision which post-dated *Harewood Hotels Ltd v Harris* and was not cited to the court in *Barton*.

In *Lynall v IRC* the House of Lords was dealing with a question which had arisen on the valuation for estate duty purposes of shares in a private company. Section 7 (5) of the Finance Act 1894 required the shares to be brought into account at the value "such property would fetch if sold in the open market at the time of the death of the deceased". The shares were a minority holding and their value on an open market sale would have been increased if a public issue were likely. Private documents of the directors in existence at the date of death indicated that a public flotation was under active consideration. The question was whether in assessing "the open market" value of the shares regard should be had to these private documents. Plowman J held that regard should only be had to published information and to information which the directors would in fact have given in answer to reasonable questions. He held that the directors would not have disclosed the private documents and that the documents were not, therefore, admissible in evidence. The Court of Appeal reversed his decision, but the House of Lords restored it. All five members of the House of Lords agreed that, for the purposes of the hypothetical sale in the open market, the hypothetical purchaser could not be treated as having access to confidential information.

Lord Reid at p 695 said this:

If the hypothetical sale on the open market requires us to suppose that complete competition has been invited, then we have to suppose that steps have been taken before the sale to enable a variety of persons, institutions or financial groups to consider what offers they would be prepared to make. It would not be a true sale in the open market if the seller were to discriminate between genuine potential buyers and give to some of them information which he withheld from others, because one from whom he withheld information might be the one who, if he had the information would have made the highest offer.

The respondent's figure of £4 10s per share can only be justified if it must be supposed that these reports would have been made known to all genuine potential buyers, or at least to accountants nominated by them. That would only have been done with the consent of Linread's board of directors. They were under no legal obligation to make any confidential information available. Circumstances vary so much that I have some difficulty in seeing how we could lay down any general rule that directors must be supposed to have done something which they were not obliged to do. The farthest we could possibly go would be to hold that directors must be deemed to have done what all reasonable directors would do. Then it might be reasonable to say that they would disclose information provided that its disclosure could not possibly prejudice the interests of the company. But that would not be sufficient to enable the respondents to succeed.

Not all financiers who might wish to bid in such a sale, and not even all the accountants whom they might nominate, are equally trustworthy. A premature leakage of such information as these reports disclose might be very damaging to the interests of the company, and the evidence in this case shows that in practice great care is taken to see that disclosure is only made to those of the highest repute. I could not hold it right to suppose that all reasonable directors would agree to disclose information such as these reports so widely as would be necessary if it had to be made available to all who must be regarded as genuine potential bidders or to their nominees. So in my opinion the respondents fail to justify their valuation of £4 10s.

Lord Morris at p 699 said:

The somewhat limited issue as between the two figures of £3 10s or £4 10s mainly depends upon the question whether knowledge of the category B documents and of the information which they contained would be "open market" knowledge. The conclusion of the learned judge was that as such knowledge was not published information and as (on Mr Alan Lynall's evidence, which the learned judge accepted) it would not in fact have been elicited on inquiry, it ought not to enter into the calculation of price and value.

A The differing view of the Court of Appeal was based on the evidence, above referred to, of the practice of boards of directors to answer reasonable questions in confidence to the advisers of an interested potential purchaser. If this is the practice, and even if the sought for information may be given in confidence to an interested potential purchaser himself, I cannot think that this equates with open market conditions. It was said that it should be assumed that a purchaser would make reasonable inquiries from all available sources and that it must further be assumed that he would receive true and factual answers. If, however, the category B documents and the information contained in them were confidential to the board, as they were, the information could not be made generally available so that it became open market knowledge.

Lord Dilhorne at p 701 said:

B On a sale in the open market is it to be assumed that possible purchasers would have information as to the contents of the reports of McLintocks and Cazenoves? They were confidential to the directors. All the shareholders in Linread were directors, but it is not to be assumed that they would disclose confidential information they possessed to the public without the consent of the board; nor is it to be supposed that the board would have given its consent to the disclosure of the contents of those reports. In the light of the evidence given by Mr Alan Lynall, whose evidence was tendered and accepted as being the evidence of the board, and accepted by Plowman J . . . it is clear that that would not have been given.

It was agreed that if it were held that it is to be assumed that purchasers would have knowledge of those reports in a sale on the open market, the shares were to be valued at £4 10s a share, but that if no such assumption was to be made, their value was £3 10s a share.

C Lord Donovan agreed. He said at p 702:

I concur in the view that confidential information ought not to be regarded as available to a hypothetical purchaser under section 7 (5) of the Finance Act 1894; though I would think it right not to treat as confidential information for this purpose accounts of the company already prepared and awaiting presentation to the shareholders. I have in mind the accounts of the present company for the year to July 31 1961.

Lord Pearson at p 705 said:

The crucial question, therefore, is whether this information should be deemed to be available to participants in the hypothetical market.

He answered the question at p 706. He said:

D In the present case, however, the company's board of directors had received reports and advice which were obviously of a confidential character and the board had come to no decision as to whether they would act on the advice or not but were maintaining their cautious and uncommitted attitude. It is reasonable to imagine that in that situation the board would have kept these matters confidential and would have been unwilling to disclose the reports and advice which they had received, and in particular unwilling to make them available to participants in the open market. *Prima facie* the information would not have been available.

It is, however, suggested that it would have been available in two ways: first it is said that the likely purchasers might have included a director of the company and he would have had the information *ex officio*. But unless others also knew it, his possession of the information would not materially affect the market price, which he or any other purchaser would have to pay.

E Then a little later on he said:

In my opinion the reasonable supposition is that the information would not be available in the hypothetical open market, and so the assessment should be £3 10s and not £4 10s.

The ratio of each of the judgments delivered in *Lynall* v *IRC* was that confidential information was not to be admitted into evidence unless it represented information which would be available in the hypothetical market. The principle applies, in my judgment, in the present case. If and to the extent that the decisions in *Harewood Hotels Ltd* v *Harris* and *W J Barton Ltd* v *Long Acre Securities Ltd* are to the contrary effect, I agree with Mr Lyndon-Stanford that they should be regarded as decided *per incuriam*. I would observe, however, that the trading accounts admitted into evidence in *Harewood Hotels Ltd* v *Harris* may have been the statutory accounts or directors' accounts which subsequently became the statutory accounts. It is not in dispute that accounts of this character are admissible (see Lord Donovan in *Lynall* v *IRC* at p 702C). And the decision of the Court of Appeal in *W J Barton Ltd* v *Long Acre Securities Ltd* was to exclude the trading records in question. The *Lynall* v *IRC* point represents an additional ground upon which that decision could have been based.

I must now return to question 8. It is not, in my view, a question to which at this stage a firm answer can be given.

First, there is nothing necessarily contrary to principle in an assessment of rental value being based on the profit-earning capacity of 30 Curzon Street. Evidence of its profit-earning capacity may, therefore, be admissible in evidence.

Second, the only admissible evidence as to the profit-earning capacity of 30 Curzon Street will be evidence available to prospective lessees in the hypothetical open market. This will include evidence available to the lessor and which the lessor would be likely to make available to prospective lessees.

Whether Crockford's trading records would, in the real world, be made available by Crockford's to all prospective lessees is, strictly, a matter to be decided by the arbitrator. I cannot at the moment think of any reason at all why Crockford's should be willing to release its records to all or any prospective lessees and Mr Barnes, not surprisingly, could not suggest a reason. But unless the arbitrator can come to the conclusion that Crockford's private trading records would be available to prospective lessees in the hypothetical open market, the documents would not, in my judgment, be admissible in evidence and would not be discoverable.

Third, it will be for the arbitrator, having considered all the admissible evidence, both of the profit-earning capacity of 30 Curzon Street as well as of the available comparables, to decide what, if any, weight ought to be attributed to the evidence of the property's profit-earning capacity.

I must now deal with the discovery appeal. The arbitrator refused discovery of Crockford's trading records on the ground, first, that no evidence of the absence of available comparables had been put forward and, second, that the discovery sought would be time-consuming, expensive and unlikely to save costs. The arbitrator did not have the advantage of the arguments on law that have been addressed to me. Counsel for the defendant submitted to the arbitrator that trading records would be admissible only in the absence of adequate evidence of comparables. The arbitrator accepted this submission. In doing so he was, I think, in error in that the adequacy of the available comparables could not be gauged at that stage. But the point that could have been, but was not, made was that the accounts would not have been available in the hypothetical open market. If that point had been made and if the arbitrator had accepted it the accounts would have been inadmissible whether or not adequate comparables were available.

Since, in my judgment, the arbitrator misdirected himself I must, I think, allow the appeal. I do not, however, propose to remit the discovery application to the arbitrator for rehearing, since, as at present advised, I think it unlikely in the extreme that any arguable case could be put forward that these trading records would have been available to prospective lessees in the hypothetical open market. If the plaintiff thinks otherwise, it can renew its discovery application to the arbitrator.

NEGLIGENCE

Court of Appeal
February 3 1987
(Before Lord Justice PURCHAS and Mrs Justice BUTLER-SLOSS)

SUTCLIFFE AND ANOTHER v SAYER

Negligence — Valuation or survey — Action by purchasers of house in Southport against estate agent — Appeal by purchasers from decision of county court judge rejecting their claim — House was purchased in reliance on a report by the estate agent, described as a "valuation", which drew attention to a settlement on the gable, dampness entailing a new damp-proof course, electrical wiring needing overhaul at a future date and plumbing old but seeming to be in reasonable condition — A purchase at £8,650 was recommended and purchasers in fact paid asking price of £8,745 for the house, intended to be their retirement home — Subsequently, owing to husband's state of health, they wished to sell the house and move to the south of England, but found difficulty in reselling when they put the house on the market (about $3\frac{1}{2}$ years after the purchase) at prices varying in the region of £19,000 reducing to £17,000 — Owing to the differential between house prices between the north and south of England they found themselves unable to move, and they sued the estate agent in contract and tort for breach of duty and negligence — The statement of claim was based on the premise that the estate agent would provide a full survey and it was contended that the report should have contained a warning as to probable difficulty in reselling — The county court judge found that the instructions were to carry out a valuation, not a full survey, the question being "was the price right?" He also found that the estate agent's report was not misleading; it drew attention to the settlement, which was in fact obvious — As a valuation, the judge found that the report was accurate — Held by the Court of Appeal that the judge was fully entitled on the evidence to find that there had been no breach of contract or of the general duty of care by the estate agent, who had been instructed as a valuer only — He was not instructed as a surveyor to carry out a structural survey, which he was not qualified to do — The court did not consider that there was a duty on a valuer to warn a purchaser as to the difficulties of resale — Appeal by owners dismissed

No cases are referred to in this report.

This was an appeal by Harry Sutcliffe and his wife, Margaret Ethel Sutcliffe, from a decision of Judge Edward Jones at Southport County Court, rejecting their claims for alleged breach of duty and negligence against J Leslie Sayer, an estate agent, in respect of a report on a house at 17 Southport Road, Southport.

Miss Jane Shipley (instructed by Park Nelson & Doyle Devonshire, agents for Gordons, of Bradford) appeared on behalf of the appellants; S W Baker (instructed by Mawdsley Hadfield & Lloyd, of Southport) represented the respondent estate agent.

Giving the first judgment at the invitation of Purchas LJ, BUTLER-SLOSS J said: This is an appeal from an order of His Honour Judge Edward Jones made on July 2 1986 when he found judgment for the defendant together with costs upon scale 3. It is an appeal by the plaintiffs, who are the appellants in this matter, in respect of the purchase of a house in July 1979. The short facts are that the plaintiff appellants had lived in Bradford for many years and had come to the conclusion that they would like to retire and settle in Southport. They had limited means and had little knowledge of the area, except that it appeared attractive to them. They visited in May 1979 a house at 17 Southport Road and decided to buy it. The price was within their price range. The appellants appreciated that the house was somewhat lower in price than some others in the area and decided to buy it subject to its being, according to them, surveyed. They walked along the road in Southport which is the main shopping street and walked into the premises of the defendant, who was an estate agent, having seen the notice and decided that he was a suitable person to advise them on the purchase price of this house. They did not see the defendant himself; they saw somebody who turns out to have been his wife. At no time did they see the defendant. However, they left a message, and in due course there was some telephone call, according to the appellants, which was not remembered by the defendant. After that a report was sent to them which was headed "Valuation". The defendant respondent to this appeal is not qualified, and he put on the outside of his valuation, as no doubt was on the board outside his premises "J Leslie Sayer, MNAEA., Surveyors, Valuers & Estate Agents", and it would appear that he had been in the business of estate agents for many years, and indeed retired, according to his evidence, rather later that year.

The valuation set out short particulars of the property and its rateable value, the rates payable, and the final paragraph on the first page said:

The property, about 45 years old has settlement on the gable and there is dampness, which entails a new damp proof course. The plumbing, although old seems to be in reasonable condition, but we can give no warranty whatsoever and though the electrical wiring has been improved in part it will need an overhaul at some future date. *Taking into account the shortcomings we can recommend a purchase.*

Then on the second page of this short report it says:

WE HEREBY CERTIFY that we have this 30th day of May 1979 inspected and appraised the value of the aforementioned property to be in our opinion the sum of £8,650 . . . with vacant possession.

He also separately sent his fee in the sum of £50. After receipt of that valuation the appellants went ahead and purchased the house some time in July 1979. It is clear (and the learned judge so found) that it was bought as a house for their retirement, with no immediate view in their minds of the prospects of resale. However, beyond their immediate contemplation they had the misfortune that the husband suffered an illness, it appears largely from overstraining himself in the do-it-yourself repairs that he did to the house, and within a comparatively short time of their settling there he found himself in hospital and thereafter advised to move to a better climate, and wishing, apart from anything else, having a heart condition, to live on one floor instead of on two. Again it is clear from the evidence and the finding of the learned judge that, had he not had this misfortune, it is unlikely that they would have contemplated leaving for no doubt most of the period of their retirement life.

When they came to sell they discovered that the house was extremely difficult to sell. It became clear that it was situated on substrata of peat and there was settlement which had not moved, it appears, since 1979. But this sort of house with these defects has proved in the ensuing years to create very considerable difficulties of resale. There was evidence before the learned judge as to the difficulties of resale and the views of three experts based upon the building society's view of mortgages on this type of properties and the limited appreciation in value of this property compared with other properties without the defects of settlement.

The appellants, having the desire to move from Southport to the south of England and having placed the house upon the market at a price which was considerably above that which would have been advised by any of those who gave evidence as experts in the case, found themselves unable to purchase the property that they would like to purchase in the south of England in the climate in which they have now been hoping to live. In these circumstances they bring an action against the defendant respondent based upon breach of the contract between them and him and in tort for the giving of negligent advice. I should have said that the price which was recommended by the respondent to this appeal, of £8,650, was in fact exceeded in that the appellants paid the price which was asked by the vendor, that is to say, £8,745.

The points of complaint in the statement of claim are based upon the premise that the respondent would provide a full survey of the property to be bought, that it was an implied term of that survey that it would be carried out with skill and care, as was reasonable to expect from a surveyor, valuer and estate agent, and that he was in breach of such duty of care for a number of reasons.

The respondent in his defence set out that he was instructed to provide a valuation and not a survey. I am very much indebted to the helpful and comprehensive submissions of counsel. The notice of appeal can conveniently be summarised in that the first two grounds relate to the implied terms as to breach of contract, and the third ground relates to the giving of negligent advice. But all of those points really can be summarised into three matters which arise in both claims, that is to say: (1) what was the defendant instructed to do; (2) were the plaintiffs misinformed; and (3) was the valuation accurate?

The learned judge made a number of findings of fact, and, in relation to the first one — what was the defendant instructed to do — he found that he was instructed to prepare a valuation rather than to prepare a full survey, and that the questions that were uppermost in the minds of these appellants were: "Was the price right?" and "What defects, if any, would affect the price that was being asked for this property?". He says at p 53 of his judgment:

The plaintiffs claim they asked for, and indeed contracted for, a full survey. Plaintiffs' counsel accepts [that] this would not have included a structural survey. His real contention is — and this is in fact the point of the case — that however limited the survey is to be construed, it would include warning a prospective purchaser that the house would be difficult, indeed for people with the financial resources of the plaintiffs, impossible, to sell.

He said at p 56:

The plaintiffs wanted at the time to know if the price was right for the property. They wished to have their attention drawn to any defects which affected that valuation. They got just that.

At p 57 he said:

I find the defendant was no more than an unqualified estate agent, but with a knowledge of property values at the material time in this particular area. He provided a report about which no criticism can be levelled as to the qualities of the house.

The judge, upon facts that were available to him, found as a fact that the instructions of the plaintiff appellants were as to a valuation and not as to a full survey, and he was entitled to come to that view.

The second point was: Were these appellants misinformed? As I have already read out, there were certain defects which were referred to on the valuation: "*The property*, about 45 years old, has settlement on the gable", and then various other matters. There was therefore an indication to the purchasers of settlement. Much criticism is made by counsel for the appellants that the report provided a partial, misleading and inaccurate view of the very real defects that there were of a house which was standing upon peat and where there was settlement already obvious, as indeed it would be obvious to anybody, as I understand it, walking along the road.

Counsel on behalf of the respondent points out that the settlement was obvious and that, if there was settlement to the gable (and there was only one gable in this house), it was clear it would be to the wall, and, if there were settlement to that, there would bound to be settlement and movement of other parts of the house.

The learned judge had the evidence before him of a number of experts and he took the view (at p 55):

As I mentioned earlier, the plaintiffs believed they were going to have a full survey. As far as the physical condition of the property is concerned, they have not been misinformed. All the defects, actual or possible, are mentioned in the report. The point at issue here is referred to in the words "about 45 years old, has settlement on the gable".

That again is a finding upon which the learned judge was entitled to come.

Was the valuation accurate? At p 56 he said:

I do not accept that the property, at the time of the sale, was indeed overpriced. There seems no reason to believe that the defendant, despite his lack of professional qualification, was out of touch with the level of house prices at the time he wrote his report. His valuation was £8,650. The plaintiffs, indeed, paid £8,745 for it.

Then at p 57 he said:

His valuation was, within limits, completely accurate.

The respondent to this appeal said at p 30 in his valuation: "Taking into account the shortcoming we can recommend a purchase", and it is said by counsel for the appellants that that was recommending a purchase and providing advice and that should have taken into account advice as to the prospects of resale. I for my part take the view that this was a recommendation for the purchase of this property at the price which on the following page is put forward, that is to say, £8,650; and that there was no greater duty of care upon this respondent than to provide a valuation, which was a valuation within acceptable limits, taking into account various matters which would have an effect upon the valuation, that is to say, that there were defects and shortcomings. There was no means for this respondent to know how long these appellants were proposing to live at the house. He may or may not have known that they were buying it for retirement purposes; he never met them. He was not asked to advise on it as an investment; nor indeed was that the purpose of the purchase of the house. On the evidence before the learned judge and before this court, there was nothing to suggest that in 1979 the property that was on the market and sold to these appellants could not have been resold.

Undoubtedly the resale factor is a consideration, and it is a consideration which has an effect upon the price of the property to be bought. There was before the learned judge some conflicting evidence as to whether or not a building society would have given a mortgage in 1979 and whether or not this was a property that was peculiarly difficult to sell rather than rather more than usually difficult to sell.

The learned judge put his mind to that at p 57 and said:

I am not satisfied the house had at that time, or in the foreseeable future, from that time, a blight upon it. To find the last matter would entail the whole area suspect,

and I for my part cannot shut my eyes to the fact that this is a very experienced judge who has sat in this area for many years and was listening to experts in this area and would have considerable ability to sift the evidence that was provided on houses in an area with which he was particularly well acquainted. As it happens, this house, when put up for purchase, has been put on the market, as I have already said, at a price considerably in excess of that which any of the experts would have recommended.

I do not consider that there is any duty upon a valuer to warn a purchaser as to the difficulties of resale. The plaintiffs appellants in this case cannot have it both ways. The complaint is that the defendant respondent should have given to the appellants many more details as to the defects and the consequences of such defects, in particular the consequences as to the difficulties of resale. But it is worthy of comment that the details which the appellants were given were at no time investigated by the appellants; they neither requested the respondent to explain what he meant by "settlement of the gable" nor asked what were the consequences of that settlement; nor indeed did they ask whether they should get a structural survey or consider further what other information it might be necessary to have upon which they might judge whether they should in fact buy this property.

The learned judge rightly found, in my view, that the purpose of the valuation to these appellants was to know if the price was right. The price appears to have been reasonably accurate. The appellants were not misled. There was no breach, in my judgment, of any implied condition in the contract. I can find no breach of the general duty of care to take proper steps to do the best that he can as a valuer and not as a qualified surveyor, and I for my part would dismiss this appeal.

Agreeing, PURCHAS LJ said: I of course have sympathy for the plaintiffs, who have had their plans for retirement dashed by an unfortunate heart attack sustained by Mr Sutcliffe in April 1982, with the result that instead of remaining at 17 Southport Road in their retirement, which was their plan, they now seek to move elsewhere as a result of Mr Sutcliffe's change in health. They put 17 Southport Road on the market on December 15 1982 at prices varying in the region of £19,000 reducing to £17,000, but they had been unable to find a buyer at this price. They cannot move without selling the house first because of the differential in house prices between Southport and the area to which they hope to move.

The learned judge, as has been said in the judgment just delivered, applied his mind to the material issue in the case. Understandably looking back with hindsight now, Mr and Mrs Sutcliffe contend that they expected much more from the defendant than they received, but the learned judge, having considered their evidence, the evidence of a number of experts as to value and property, and, as he was entitled to do, taking into account his own experience in the area in which he sits, came to conclusions which were adverse to the appellants in this case. I read only one passage to which my Lady has already referred: "As I mentioned earlier, the plaintiffs believed they were going to have a full survey." The learned judge's finding as to their attitude at the time was this:

The plaintiffs wanted at the time to know if the price was right for the property. They wished to have their attention drawn to any defects which affected that valuation. They got just that. I am not satisfied that [at] any time a mortgage would not have been obtainable.

On the question of evidence as to the value, the learned judge found in these terms:

I do not accept that the property, at the time of the sale, was indeed overpriced. There seems no reason to believe that the defendant, despite his lack of professional qualification, was out of touch with the level of house prices at the time he wrote his report. His valuation was £8,650. The plaintiffs, indeed, paid £8,745 for it.

The case made for the appellants was that, if they had been properly warned about settlement and the difficulties that might attend a resale, they then might have taken other steps or made other decisions in relation to the purchase of the property. I find no evidence or finding by the learned judge to support such a contention. In my judgment the learned judge properly applied his attention to the issue in this case and reached conclusions which he was fully entitled to reach on the evidence before him.

For those reasons and for the reasons that have already fallen from my Lady, I agree that this appeal must fail.

The appeal was dismissed with costs, not to be enforced against appellants without leave of the court. Legal aid taxation of appellants' costs ordered.

Court of Appeal

March 13 1987

(Before Lord Justice DILLON, Lord Justice GLIDEWELL and Sir Edward EVELEIGH)

SMITH v ERIC S BUSH (a firm)

Estates Gazette April 18 1987

282 EG 326-331

Negligence — Action against surveyors in respect of mortgage valuation — Appeal by defendant surveyors from assistant recorder's decision awarding damages to house buyer — Application of Unfair Contract Terms Act 1977 — Plaintiff, respondent to present appeal, purchased a terraced house, at the lower end of the market, in Norwich — The asking price was £17,250, but respondent agreed, subject to contract, to pay £17,500 and in fact subsequently paid £18,000 — She required a mortgage, but only to the extent of £3,500 — The Abbey National Building Society instructed the appellant firm of surveyors to carry out the mortgage valuation — It was common ground that what was required was not a structural survey but a reasonably careful visual inspection — The report was reassuring, valuing the house for mortgage purposes at £16,500 and stating that it was readily saleable, had been modernised and that no essential repairs were needed — Respondent, who had paid a fee of £36.89, was given a copy of the report, expressed satisfaction with it and proceeded with the purchase — Both the mortgage application form and the report itself contained clear and comprehensive disclaimers of liability — Unfortunately, the inspecting surveyor overlooked a serious defect — Although aware that chimney breasts in two first-floor rooms had been cut away, he did not check whether the chimneys above had been left with adequate support — He could easily have done so by "putting his head and shoulders through the trap door into the roof space" — Eighteen months later one of the flues collapsed through ceilings, bringing part of another with it, causing extensive damage but luckily no personal injuries — In the respondent's action against the appellants the assistant recorder found the appellants liable for negligence and awarded the respondent £4,379.97 damages — On appeal the appellants did not dispute the existence of a duty of care or their failure to carry out a reasonably careful visual inspection or the amount of damages if liability was established — They contended, however, that the disclaimers absolved them from liability for negligence and that they satisfied the requirement in section 11(3) of the Unfair Contract Terms Act 1977 that it was fair and reasonable to rely on the disclaimers — Held by the Court of Appeal that, apart from the 1977 Act, the disclaimers would have provided the appellants with an effective defence, but that, having regard to all the circumstances, it would not be fair and reasonable to allow the appellants to rely on them and thus avoid liability — The onus was on the appellants, under section 11(5) of the 1977 Act to establish reasonableness, and they had not done so — It was emphasised by the court that the appellants were dealing with a property at the lower end of the market, that they knew that the respondent would rely on their report and that to their knowledge she was unlikely to instruct another surveyor to inspect the property — It might be different in cases where the applicant was "a surveyor or a lawyer who understands these matters" — In the circumstances of the present case the reasonableness requirement was not satisfied — Appeal dismissed

The following cases are referred to in this report.

Alderslade v *Hendon Laundry* [1945] KB 189
Hedley Byrne & Co Ltd v *Heller & Partners Ltd* [1964] AC 465; [1963] 3 WLR 101; [1963] 2 All ER 575; [1963] 1 Lloyd's Rep 485, HL
White (Arthur) (Contractors) Ltd v *Tarmac Civil Engineering Ltd* [1967] 1 WLR 1508; [1967] 3 All ER 586, HL

This was an appeal by surveyors, Eric S Bush, a firm carrying on business as surveyors and valuers in Norwich, against a decision of Assistant Recorder Gerald Draycott QC, at Norwich County Court, awarding the plaintiff, Mrs Jean Patricia Smith, the present respondent, damages for professional negligence. The matter concerned a valuation for mortgage of a house at 242 Silver Road, Norwich.

Nigel Hague QC and Miss Jane Davies (instructed by Barlow Lyde & Gilbert) appeared on behalf of the appellants; Robert Seabrook QC and Philip Havers (instructed by Hood Vores & Allwood, of Dereham, Norfolk) represented the respondent.

Giving judgment, DILLON LJ said: The defendants in this action, a firm of surveyors, appeal against a judgment against them for damages for professional negligence awarded to the plaintiff by Mr Gerald Draycott, sitting as an assistant recorder, at the trial of this action in the Norwich County Court on April 17 1986. The appeal raises a question of some general importance as to the application of the Unfair Contract Terms Act 1977 (the "Act").

The story begins in 1980. At that time the plaintiff, Mrs Smith, lived at Coulsdon in Surrey in a house which she owned and which was subject to a mortgage in favour of the Abbey National Building Society. She decided, however, to move to Norwich. Her personal circumstances were that she was a state enrolled nurse, but had been working for some months as a clerk for a well-known firm; she had divorced or was separated from her husband and she had two children, both living with her, a girl of 15 and a boy of 12.

She found a purchaser for her house in Surrey, and she also offered for a house at Horsham St Faiths, near Norwich. But that purchase fell through. By the beginning of October 1980 she was, as the judge's note of her evidence records, desperate to get another house in Norwich, because she had been offered a good price for her house in Surrey and did not want to lose her purchaser. She then found the house with which this action was concerned, 242 Silver Road, Norwich. It was on offer at an asking price of £17,250, but the vendor said that he already had someone else interested at that price, and so she agreed to buy the house, subject to contract, for £17,500. It was a terraced house, some 70 years old, at the lower end of the housing market, but the plaintiff regarded it as suitable for herself and her children as it was well decorated, had central heating and she could move in without doing any work to it.

To raise the price she needed a mortgage, and so on October 6 1980 she applied to the Abbey National for a mortgage and completed their standard mortgage application form. In fact she only required a mortgage for £3,500, which was well below the price of the house, but nothing turns on that.

Under section 25 of the Building Societies Act 1962 (now section 13 of the Building Societies Act 1986) a building society which makes an advance on the security of a freehold or leasehold property has to get a report from an appropriately experienced person as to the value of that property and as to any matter likely to affect the value thereof. Building societies satisfy this requirement either by instructing firms of surveyors in general practice or by using their own in-house employed surveyors. In the present case, the Abbey National instructed the appellants, a firm of surveyors and valuers carrying on practice in Norwich; they acted by their senior partner, Mr Cannell*, a chartered surveyor who was amply qualified for the job.

It has always been the case that applicants for mortgages are required by the building society to pay the fees for the valuation reports which the societies are by the statute required to get. It used for many years to be the practice that building societies refused to disclose the reports to the applicants, on the footing that they were only obtained for the building societies' own purposes. But this practice gave rise to widespread discontent in that applicants could not see why they should not have copies of reports for which they had paid. Before October 1980, therefore, the building societies had adopted instead a new practice of making copies of the reports available to the applicants for mortgages, but with extensive disclaimers of liability.

In accordance with this new practice, there was included in a somewhat lengthy set of declarations, just above the place where the plaintiff signed the mortgage application form, the following:

I/We have read the Notes for the guidance of Mortgage Applicants and accept that if the loan is approved the Mortage Guarantee Policy Premium will be added to the Loan.
I/We accept that the Society will provide me/us with a copy of the report and mortgage valuation which the Society will obtain in relation to this application. I/We understand that the Society is not the agent of the Surveyor or firm of Surveyors and that I am making no agreement with the Surveyor or firm of Surveyors. I/We understand that neither the Society nor the Surveyor or the firm of Surveyors will warrant, represent or give any assurance to me/us that the statements, conclusions and opinions expressed or implied in the report and mortgage valuation will be accurate or valid and the Surveyor(s) report will be supplied without any acceptance of responsibility on their part to me/us.

Just below her signature it is recorded that she had paid the inspection fee, which was £36.89. There is a somewhat similarly worded disclaimer of liability in printed notes for the guidance of mortgage applicants which it was the practice of the Abbey National to supply to all mortgage applicants, but I need not refer to those notes because the disclaimer there adds nothing to what appears in the mortgage application form, as above set out, and in the copy of the appellants' report which was supplied to the plaintiff as mentioned below.

Mr Cannell inspected the property on October 8 1980. It is common ground that he was not required to carry out a structural survey. What he was required to do was to make a reasonably careful visual inspection of the property and to fill in the Abbey National's standard printed form headed "Report and Mortgage Valuation". Mr Cannell said that he would usually spend about half an hour on an inspection for a building society; that was generally in line with the evidence, as to their own practice, given at the trial by two other experienced surveyors, Mr Manwaring and Mr Wreford. It seems that thousands of such inspections and reports are made by surveyors for the purposes of building society mortgage applications every year over England and Wales as a whole.

A copy of Mr Cannell's report was sent by the Abbey National to the plaintiff under cover of a letter from the Abbey National dated October 9 1980. The report gives the value of the property for mortgage purposes, with vacant possession, as £16,500. It answers affirmatively the printed question whether the property was readily saleable for the purposes of owner-occupation and was likely to remain so at or about the mortgage valuation. The report states that the property had been modernised in recent years to a fair standard and that no essential repairs were required. The report amply warrants the plaintiff's reaction to it, as expressed in her evidence at the trial:

The impression I got from the report was that I was very pleased with it — it did not show anything seriously wrong with the property.

On the third page of the copy of the report sent to the plaintiff there was the disclaimer on which the appellants particularly rely, distinctively printed in red, whereas the rest of the form is printed in black. It is as follows:

TO THE MORTGAGE APPLICANT(S):
IMPORTANT
1 THIS DOCUMENT IS NOT A MARKET VALUATION. IT IS NOT, AND SHOULD NOT BE TAKEN AS, STRUCTURAL SURVEY. It has been obtained by the Society from a qualified surveyor or firm of surveyors to comply with Section 25 of the Building Societies Act 1962.
2 If you are purchasing the property, you will receive a notice that the Society does not warrant that the purchase price is reasonable.
3 This is a report to the Society by its surveyor(s) and neither the Society nor the surveyor(s) give any warranty representation or assurance to you that the statements, conclusions and opinions expressed or implied in the document are accurate or valid.
4 The surveyor(s) has/have made this report without any acceptance of responsibility on his/their part to you.

The plaintiff admits that she read this.

The plaintiff then proceeded to buy the property. She was forced by her vendor to increase her offer to £18,000, but nothing turns on that. She did not have any other survey of the property done, because, as she said in evidence and the judge accepted, she relied on the building society survey.

Unfortunately, in his inspection of the property Mr Cannell overlooked a serious defect. He noticed that the chimney breasts in two of the first-floor rooms, including the main bedroom, had been cut away, but it did not occur to him to check whether that had been done in a way which left the chimneys above adequately supported. Had it occurred to him to check, it would have been easy to have done so (since there was a trapdoor through which he could have looked into the roof space with the aid of the ladder he had with him) and it would have taken him no more than 10 minutes. Had he checked he would have seen at once that the brickwork of the chimneys had been left unsupported by the removal of the chimney breasts. The absence of adequate supports meant inevitably, as explained by Mr Manwaring at the trial, that the flues would collapse at some time. The house was thus in truth a dangerous structure, and unfit for habitation until adequate support for the flues had been provided.

Some 18 months later, at about 10.30 pm on June 1 1982, the inevitable happened. One of the flues collapsed through the main bedroom and landing ceilings, bringing down part of another flue with it. Rubble came down the stairs into the lounge. Fortunately the plaintiff was downstairs at the time; had she been in her bed she might have been killed or seriously injured. The building work to get the place right was put in hand at once by the plaintiff, but it was five weeks before she could move back into her bedroom.

By this action the plaintiff claimed damages from the appellants for the negligence of Mr Cannell in failing to check whether the removal of the chimney breasts had left the chimneys without adequate support and consequently failing to give in his report correct advice as to the condition of the property. The assistant recorder upheld the plaintiff's claim, and awarded her £4,379.97 damages, including interest.

---
*Editor's note: The surveyor, Richard S Cannell FRICS, was partner in charge of the surveys and professional services department of the firm of Eric S Bush and not the senior partner.

On this appeal two points only are raised, to which I shall come. The appellants do not dispute the assistant recorder's findings: (i) that in preparing his report Mr Cannell owed a duty of care to the plaintiff, apart from the disclaimers which I have mentioned, since he knew that she was likely to rely on his report; (ii) that she did indeed rely on his report; (iii) that he was negligent in the sense that in carrying out a reasonably careful visual inspection of the property he ought to have looked to see whether the chimneys had been left with adequate support after the removal of the chimney breasts; and (iv) that there was no contributory negligence on the plaintiff's part. The appellants do not in this court dispute the amount of damages awarded in the court below, if liability is established.

The appellants submit, however:
1 that the disclaimers, ie the provisions set out above in the form of mortgage application signed by the plaintiff and in the notice in red on the copy of Mr Cannell's report which was supplied to the plaintiff, have the effect of absolving the appellants from all liability for negligence which would otherwise have fallen on them; and
2 that the Act does not preclude the appellants from relying on the disclaimers because, they submit, the requirement of reasonableness under section 11(3) of the Act is satisfied in relation to the disclaimers.

The assistant recorder held against the appellants on issue (1) because he found that it was an implied condition of the disclaimers that if the disclaimers were to apply, there should have been a complete visual inspection of all parts of the fabric of the house. In other words, the disclaimers would have applied if Mr Cannell had looked into the roof space but had then negligently failed to see what was plain before his eyes there, but they did not apply, as he negligently failed to look into the roof space at all. Counsel for the plaintiff in this court have — rightly — felt unable to support the reasoning of the assistant recorder on this issue, but they have submitted that the wording of the disclaimers is not sufficiently explicit to exempt the appellants from liability for negligence. They say that so far as the appellants' position is concerned (as opposed to the Abbey National's position) the disclaimers merely told the plaintiff that the appellants had not carried out a full structural survey of the property and that she had no contract with the appellants.

Reference was made in a general way in the course of argument to the well-known line of authorities, stemming from *Alderslade* v *Hendon Laundry Ltd* [1945] KB 189 in which the courts have considered the principles which are applicable to clauses which purport to exempt one party to a contract from liability to the other party for the consequences of the first party's own negligence. Assuming, however, that those are the principles which ought to be applied to the construction of these disclaimers, I find it impossible to conclude that the disclaimers do not cover tortious liability of the appellants for negligence. The final words in the notice on the copy of the report supplied to the plaintiff

The surveyor(s) has/have made this report without any acceptance of responsibility on his/their part to you

and the corresponding words at the end of the declaration in the form of mortgage application are really meaningless if they do not extend to responsibility or liability in negligence. If authority is required, it is to be found in the speeches of Lord Upjohn and Lord Pearson in *Arthur White (Contractors) Ltd* v *Tarmac Civil Engineering Ltd* [1967] 1 WLR 1508 at 1526B-F and 1529 C-G.

At common law, as is recognised in *Hedley Byrne & Co Ltd* v *Heller & Partners Ltd* [1964] AC 465, where there is no contract the assumption of a duty of care can by appropriate words be avoided — eg where a reference is given "without responsibility" — and where a report is supplied subject to stipulations disclaiming responsibility the recipient cannot accept the report and ignore the stipulations: see *per* Lord Reid at p 492 and *per* Lord Morris at p 504. Apart, therefore, from the Act, the disclaimers relied on would, in my judgment, have provided an effective defence for the appellants against the plaintiff's claim.

I turn, therefore, to consider issue (2), the effect of the Act. The relevant sections of the Act are as follows:

2(1) A person cannot by reference to any contract term or to a notice given to persons generally or to particular persons exclude or restrict his liability for death or personal injury resulting from negligence.
   (2) In the case of other loss or damage, a person cannot so exclude or restrict his liability for negligence except in so far as the term or notice satisfies the requirement of reasonableness.
11(3) In relation to a notice (not being a notice having contractual effect), the requirement of reasonableness under this Act is that it should be fair and reasonable to allow reliance on it, having regard to all the circumstances obtaining when the liability arose or (but for the notice) would have arisen.

Under section 11(5) the onus is on those claiming that a contract term or notice satisfies the requirement of reasonableness to show that it does.

The assistant recorder includes in his judgment a finding that he had not been satisfied by the appellants that in all the circumstances of this case it would be fair and reasonable to allow reliance on the disclaimers. But this is tacked on to the section of his judgment which deals with the quantum of damages and it may relate back to the assistant recorder's view that the disclaimers were inadequate because Mr Cannell had never looked into the roof space at all.

The question is whether the requirement of reasonableness, referred to in section 2(2) of the Act is satisfied, ie whether, on the words of section 11(3), it is fair and reasonable to allow reliance on the disclaimers, having regard to all the circumstances obtaining at the relevant time. In view of the generality of the words "all the circumstances", Mr Hague for the appellants concedes that the court is not limited to considering merely the questions whether the notice referred to in the section was sufficiently clearly brought to the attention of the person concerned or was sufficiently clearly expressed for that person to have been able to understand it.

Whether in any particular case it is fair and reasonable in all the circumstances to allow reliance on a notice excluding liability for negligence must, of course, depend on the circumstances of the particular case. In that sense no case is a precedent for any other case. But the circumstances of the present case are very ordinary, and we were told that there were many other cases awaiting the outcome of this appeal. That is not surprising considering the number of building society surveys that are carried out. It appears that it is the practice of very many if not all building societies to include as a matter of course in their printed forms disclaimers to protect the society itself and its surveyors from all liability, in substantially the same terms as the disclaimers in the present case.

It is common ground that it was no part of Mr Cannell's duty to carry out a full structural survey of the house, and this was understood by the plaintiff. She knew that she was taking a chance on there being no hidden defects in the house which would not have been apparent on a reasonably careful visual inspection of the house. To hold that the surveyor in such a case as the present cannot rely on the disclaimers does not involve extending the surveyor's obligations so as to make it incumbent on him to detect such hidden defects. His obligation is still merely to carry out a reasonably careful visual inspection.

But where he is dealing with a property at the lower end of the market and he knows that the purchaser is likely to rely on his report, and not instruct his or her own surveyor, I find it very difficult to see why it should be fair and reasonable to allow him to rely on an automatic blanket exclusion of all liability for negligence if his visual inspection of a property turns out not to have been reasonably careful.

The appellants have urged various considerations as showing that it is fair and reasonable that they should be allowed to rely on the disclaimers. They are all general considerations, unrelated to the particular facets of this case, and they are said to warrant an automatic blanket exclusion of all liability to mortgage applicants for negligence in respect of building society surveys. They say, for instance, that the surveyors never meet the mortgage applicants and so have no opportunity of clarifying or explaining anything they may have put in a report for a building society. They say also that it is of great benefit to the mortgage applicants that the report required by a building society under the statute, before it can agree to make a mortgage advance, should be available quickly and cheaply, and they suggest that neither of these ends will be attained if the surveyor has always in his mind the fear of litigation and the need to satisfy his insurers. They also naturally point to the fact that the interests of the building society, merely as mortgagee, and the interests of the purchaser, as owner of the equity of redemption, are not the same; but this point is greatly weakened when it is seen that the report covers such questions, in which the purchaser is vitally interested, as the marketability of the property and whether urgent repairs to the property are required.

Giving full consideration to all that has been urged by Mr Hague, I cannot see that it is fair or reasonable that a professional surveyor making a mortgage report at the lower end of the property market, when he knows that the would-be purchaser who is applying for a mortgage on the property has paid the fee for the report, will be supplied with a copy of the report and is likely to rely on the report and so is not likely to instruct any other surveyor, should be able to rely on any general disclaimers, such as those in the present case, unrelated to any special factors affecting the particular property, to exempt him from liability to the purchaser for negligence if it should happen that he, the surveyor, carries out his visual inspection of the property without due care.

It may be different — I express no opinion — where the mortgage applicant who chooses to rely on a building society's surveyor's report despite having read such disclaimers is himself a surveyor or a lawyer who understands these matters. So far as the plaintiff was concerned, — and I fancy this would go for most purchasers — I do not think she would have seen any reason to incur the additional cost of instructing a second surveyor after she had read the copy of Mr Cannell's report.

I do not doubt that Mr Cannell, when he makes a visual inspection of a property with a view to making a mortgage report to a building society, intends to act with reasonable care. I do not see why he should not be liable in damages like any other professional man who fails to show reasonable care if on a particular occasion he has failed to take reasonable care in his inspection.

I would accordingly dismiss this appeal.

Agreeing that the appeal should be dismissed, GLIDEWELL LJ said: I have had the advantage of reading in draft the judgment of my Lord Dillon LJ. I gratefully adopt his recital of the facts, including the relevant documents.

There are two issues in this appeal:
(1) On their face, did the words of the disclaimer suffice to prevent any duty of care being owed by the defendants to Mrs Smith?
(2) If so, did the disclaimer satisfy the requirement of reasonableness — Unfair Contract Terms Act 1977, sections 2(2) and 11(3).

(1) *The effect of the disclaimer*

The documents which are particularly important are the declaration on the mortgage application form signed by Mrs Smith and the notice at the end of the defendants' report of October 8 1980, below their valuation. Dillon LJ has set this out in his judgment. Mrs Smith said in evidence that she read this notice. We are thus not concerned with the question whether the disclaimer was brought to her attention. Clearly it was.

The reason which the assistant recorder gave for holding that the disclaimer was not sufficient to avoid responsibility, ie that there was an implied condition that, if the disclaimer is to apply, there should have been a complete visual inspection, is unsound. Counsel for Mrs Smith did not attempt to rely upon it.

Mr Seabrook's argument is that the disclaimer did not cover liability for negligence. If that is correct, the disclaimer did not cover anything. Mrs Smith was never in a contractual relationship with the defendants. If they were or could be liable to her, that liability could only be in the tort of negligence — which is how the plaintiff's claim is framed

The payment by Mrs Smith of £36.89 was not a payment of a fee for the provision of a report to her. The building society required the inspection, report and valuation by a surveyor both for their own purposes and to comply with section 25 of the Building Societies Act 1962. At October 6 1980, the only contract was a promise by the building society to obtain a report and valuation and then consider whether to make to Mrs Smith the mortgage loan she required in consideration of the payment of £36.89 to them.

In my view, the wording of the disclaimer set out in the notice on the defendant's report and valuation did suffice to prevent the defendants owing any duty of care to Mrs Smith. In other words, it was so worded as to relieve them of all liability in negligence.

(2) *The requirement of reasonableness*

Did that notice satisfy the test of reasonableness in the Act of 1977? In other words, is it "fair and reasonable to allow reliance on it, having regard to all the circumstances obtaining at" October 8 1980, ie the date of the report? This is a question of policy, which could sensibly be answered either way.

The argument for the plaintiff, as it eventually emerged, is that it is necessary to distinguish between a disclaimer of liability for not doing something which the defendant was not purporting to do, ie carrying out a full structural survey, and doing incompetently that which Mr Cannell was purporting to do, ie making a visual inspection sufficient to reveal any state of disrepair which would affect the value of the property, to ascertain the general condition of the property and to place a value upon it. It is submitted that it is unfair to exclude liability for negligence in the latter category.

Mr Hague for the defendants has two main arguments on this:
(a) The lack of support for the chimney flues in the roof space and the chimney stack above would only have been revealed by a full structural survey, not by a visual inspection of the sort Mr Cannell was carrying out. In my view this is not correct. The judge accepted the evidence of Mr Manwaring, a chartered building surveyor called as an expert witness for the plaintiff, that the fact that the chimney breasts had been removed (which Mr Cannell observed) should have put him on inquiry as to whether the flues and stack above had been left unsupported. That inquiry would have involved no more than putting his head and shoulders through the trapdoor into the roof space, when the lack of support would at once have been apparent. The judge said that it would have taken no more than 10 minutes to check, and that Mr Cannell should have looked through the trapdoor but did not. I agree with him. In my view Mr Cannell was guilty of lack of proper care in doing the sort of inspection he was purporting to do.

(b) That Mr Cannell was carrying out the inspection for, and reporting to, the building society and nobody else. Mr Hague argues that it is not fair or reasonable that the defendants should be under any duty of care to a potential purchaser, with whom they had no contractual relationship, simply because the building society chose to supply a copy of their report to Mrs Smith. This is the nub of the case. If a surveyor's inspection is competently carried out, a potential purchaser who chooses not to commission his own inspection, despite the warning notice, will obtain the benefit of the report. But Mr Hague argues that he cannot justifiably complain if the inspection failed to reveal some defect which a reasonably competent inspection should have revealed.

It is my opinion that this argument fails because its basic premise is incorrect. While it is true that the surveyor carried out his inspection and wrote his report and valuation because he was commissioned to do so by the building society, he knew that a copy of his report would be sent to Mrs Smith and that most probably she would rely upon his report and valuation without obtaining any other advice. In my judgment, a professional man who knows that a potential purchaser of a house (herself possessed of no special skill) will most probably rely upon his skill and competence should not be allowed to relieve himself of liability if he fails to exercise a reasonable degree of skill and competence in the task on which he is engaged.

In agreement with Dillon LJ, I would thus hold that, in so far as the disclaimer excluded liability for negligence in the carrying out of the limited sort of inspection Mr Cannell was engaged to do and was purporting to do, the notice of disclaimer did not satisfy the requirement of reasonableness within sections 2(2) and 11(3) of the Act of 1977. For this reason I, too, would dismiss the appeal.

Also agreeing, SIR EDWARD EVELEIGH said: Section 2(1) of the Unfair Contract Terms Act 1977 prevents the defendant firm from excluding "liability for negligence except in so far as the term of notice satisfies the requirement of reasonableness".

In the present case it is first necessary to decide whether or not there would be liability for negligence apart from the notice which seeks to exclude that liability.

The words in red displayed prominently in the report are not simply a disclaimer of liability. They also constitute a warning that the report is of limited value as a survey report. If the plaintiff had relied upon the report to an extent which would be unreasonable, bearing in mind the warning, there would be no liability on the defendants and no need to consider the efficacy of a notice excluding liability.

However, in so far as the words constitute a warning, they do not, in my opinion, warn a person to place no reliance at all upon the report. They do not say that it is dangerous to rely upon the report for what it purports to be or for what it says.

The report did not reveal what a valuation report would have revealed if properly prepared, albeit for its limited purpose. On the contrary it asserted that the property was readily saleable for the purpose of owner-occupation. The plaintiff relied upon it in this respect, but the report did not merit even that degree of confidence. The warning, in my opinion, was not such as to make it unreasonable for the plaintiff to rely upon the report as she did. Therefore the defendant will be liable to the plaintiff unless the notice is effective to exclude liability.

In this court, counsel has not sought to argue that the defendant would not have been liable in the absence of a notice of disclaimer. None the less, I have analysed the position because I think it is important to bear in mind the basis of the defendant's liability when considering whether or not it would be fair and reasonable to allow reliance upon the notice having regard to all the circumstances obtaining when liability arose.

A relevant part of the circumstances obtaining when liability arose was the fact that the report stated that the house was readily saleable for the purpose of owner-occupation and was likely to remain so. There was no warning that this assertion was unreliable. It would have been perfectly simple to discover the dangerous state of the premises. No structural survey was needed for that. The least attention to the task that the plaintiff was entitled to expect from the defendant would have revealed the fault. The defendant's fault can hardly be regarded as a mere accidental error or omission.

I, therefore, am of the opinion that it would not be fair and reasonable to allow reliance upon a disclaimer which might, I know not, be effective in other circumstances.

The appeal was dismissed with costs. Leave to appeal to The House of Lords was refused.

Queen's Bench Division

January 23 1987

(Before Mr P J COX QC, sitting as a deputy judge of the division)

WESTLAKE AND ANOTHER v BRACKNELL DISTRICT COUNCIL

Estates Gazette May 16 1987

282 EG 868-872

Negligence — Action against local authority on the ground of negligence by an employee in authority's surveyor's department in reporting on a house with a view to a loan by way of mortgage to a young couple about to be married — Report stated that structural and decorative condition of house was good and valued it at £11,750 — On the basis of this report the authority offered a loan of £10,900 provided that a few minor defects were remedied — At time of inspection house was occupied by vendor and was furnished, with furniture and fitted carpets — Plaintiffs did not have an independent survey and did not see the report, but relied upon a letter from the authority representing that the property afforded security for a loan of £10,900, and proceeded to purchase the house — Shortly after moving in, plaintiffs became aware of a gap between the skirting board and floor of a front room downstairs; subsequent evidence indicated that the gap was about 20mm wide at that time — At the request of the plaintiffs, who had become worried by contemplation of the gap, a further inspection was made on behalf of the authority, probably by the person who made the original inspection and report — He reassured the plaintiffs that it was purely a matter of settlement and that there was nothing to worry about — After about five years or so the husband's employers moved their premises and the plaintiffs, with the aid of a bridging loan, purchased another house, putting the subject property on the market at a price of about £31,000 — Unfortunately, it proved to be unsaleable and the plaintiffs had to move back into it, selling their new house — A thorough examination now showed progressive settlement of the original house — The concrete floor slab had been laid on a bed of rejects, uncompacted and of excessive depth — Plaintiffs sued the local authority, their writ being issued in June 1983, nearly eight years after the completion of the purchase — Held (1) that the local authority owed a duty of care to the plaintiffs, knowing that they would rely on the valuation made by the authority; (2) that the authority's surveyor had been negligent in failing to examine properly the state of the ground floor, which would have shown that the house was unsuitable for mortgage; that the plaintiffs' cause of action accrued at the completion of the purchase in July 1975; but that their cause of action was not statute-barred, as the words of reassurance by the surveyor on his second visit amounted to deliberate concealment of the facts within the meaning of section 32(1)(b) of the Limitation Act 1980; alternatively these words raised an estoppel against the defendants, preventing them from averring that the plaintiffs' cause of action was out of time — In the event the plaintiffs were awarded damages of £5,000 (together with interest thereon from date of purchase) on the basis of the difference in price between what the plaintiffs paid in 1975 and the true market value at that time — They were also awarded general damages of £1,500 for distress and discomfort suffered

The following cases are referred to in this report.

Kaliszewska v *Clague (J) & Partners* [1984] CILL 131; (1984) Const LJ 137
Perry v *Sidney Phillips & Son* [1982] 1 WLR 1297; [1982] 3 All ER 705; [1982] EGD 412; (1982) 263 EG 888, CA
Yianni v *Edwin Evans & Sons* [1982] QB 438; [1981] 3 WLR 843; [1981] 3 All ER 592; [1981] EGD 803; (1981) 259 EG 969

This was an action by Mr and Mrs Westlake alleging negligence against the Bracknell District Council in relation to the valuation for mortgage purposes of the house, 11 Yorktown Road, Sandhurst, Berkshire, purchased by the plaintiffs.

Roy Lemon (instructed by W Bradly Trimmer & Son, of Alton, Hampshire) appeared on behalf of the plaintiffs; Robert Gaitskell (instructed by Barlow Lyde & Gilbert) represented the defendant local authority.

Giving judgment, MR P J COX QC said: This is a claim by Mr and Mrs Westlake for damages against the Bracknell District Council by reason of the negligence of one of its employees, a surveyor, described as a technical services officer. There is virtually no dispute between the parties as to the history of the case, which I find as follows.

The plaintiffs married on August 11 1975. Mr Westlake was then aged 22. He was employed as an electronics repair technician and, as first-time buyers, Mr and Mrs Westlake had decided to buy a house at 11 Yorktown Road, Sandhurst in Berkshire. This was a semi-detached house described as being at the lower end of the market and priced at £11,500. In common with many young couples, they needed to borrow some 95% of the purchase price on mortgage. They were inexperienced in house purchase and were unaware of the problems of subsidence and the like. They had found the house by May 1975 and they applied for a loan by way of mortgage to the local authority, Bracknell District Council, some three months before their wedding.

Before granting the loan, in accordance with the usual practice the defendants instructed their surveyor to inspect the house for the purpose of satisfying themselves that it represented adequate security for the loan. As a result of this inspection, a report dated May 13 1975 was prepared. It was signed by a Mr Robert King who, at the material time, was employed by the council as a section head in the surveyor's department of the defendants. The actual inspection of the property was carried out by a member of Mr King's staff whose identity has not been established in this case. The report stated that the structural and decorative condition of the house was good and, with vacant possession, was valued on a freehold basis at £11,750.

At the time of this inspection the house was occupied by the plaintiffs' vendor and was furnished with furniture and fitted carpets. On the open market, in reasonable structural and decorative condition, the house was worth £11,500. The plaintiffs did not see a copy of the defendants' surveyor's report but, by a letter dated June 6 1975 from the defendants to the plaintiffs, an offer of a loan of £10,900 by way of mortgage was made to the plaintiffs. This letter stated that the advance would be subject to the terms set out in six numbered paragraphs contained in the letter, and para 6 required the plaintiffs to remedy certain minor defects within one year of becoming the owners of the property. In making this offer the defendants clearly relied upon the report dated May 13.

The plaintiffs, in common with the great majority of first-time buyers, did not have survey of their own. They accepted the offer, relying upon the defendants' representations contained in the letter that the property afforded security for a loan of £10,900. They completed the purchase of the house some time, I think, in July 1975 and they moved into occupation immediately after their marriage.

Within a very short time they had become aware of the existence of a gap between the skirting board and the floor of the front downstairs room, particularly below the window. There is a plan, which we have called "P2" (which is marked "RT1"), exhibited to the affidavit of Mr R T Treadwell, which has been put in evidence in this case. The plan gives various measurements, showing the gap between the skirting board and the concrete floor at various points on the ground floor of the house at a time when Mr Treadwell carried out the survey. This gap had also been noted by Mrs Westlake's father, Mr Pritchard, when he visited the property before the plaintiffs moved in. Mr Pritchard told me in evidence that when he saw the gap he commented upon it and he thought it appeared to be, at that time, between three-quarters and one inch wide. I am satisfied upon the evidence in this case that this gap was 20mm wide below the window in May and in August 1975. The plaintiffs had not noticed the gap when they viewed the house to begin with, prior to purchase, probably because of the fact that a settee was concealing the skirting board beneath the front window; but it is quite clear that three or four weeks after the wedding they had become somewhat worried because of the presence of this gap. It was not only under the window but it was visible in the hall and also down the side walls of the front room, narrowing towards the back of the house.

Being worried in this way, they telephoned the defendants' surveying office and asked for a surveyor to visit the house for the purpose of inspecting this problem. In response to this, a man, described by the plaintiffs as being fairly short with grey hair and in his 50s, visited the house and looked at the gap, told the plaintiffs that it was purely settlement and assured them that there was nothing to worry about. He was there for about five minutes only and then he left. This reassured the plaintiffs. Some time before Christmas, Mr Westlake filled the gaps with some type of plaster or cement filler, primarily to stop the draught and, I suppose, also to improve the appearance of the decoration. He told me that he had also had to construct some sort of small ramp from the front door into the hall because of the dropping of the hall floor.

Mr King, the surveyor who was called by the defendants and who had been a section head in the defendants' surveying department in 1975 (the material time), told me that he considered it likely that the man described by the plaintiffs was the same person who had done the original survey for the defendants. In the absence of any evidence to the contrary I am satisfied on balance of probability that this was so.

The plaintiffs lived at 11 Yorktown Road until late 1980 or early 1981, when Mr Westlake's employers moved their premises to Basingstoke. The plaintiff was encouraged to move house by the offer of a free bridging loan from his employers. The house was then put in the hands of agents for sale at a price of about £31,000, which reflects the rise in house prices between 1975 and 1981, and the plaintiffs purchased a house at Four Marks in Hampshire. They were happy at this house as it was better than their first, being described by Mrs Westlake as a country place with a huge garden and where she could enjoy her hobby, I think, of keeping and showing dogs.

Despite the efforts of the agent, 11 Yorktown Road remained unsold. After about a year, Mr Westlake's employers, not unnaturally, became restive about the continuing burden of the bridging loan and they commissioned a firm of estate agents, surveyors and valuers, Messrs Poulters (the firm in which Mr Treadwell, the surveyor witness for the plaintiffs, is a partner), to report upon the property. The report is dated December 7 1981. I quote from that report under the paragraph "General Condition". The inspection revealed:

Evidence of significant downward deflection within the ground floor concrete slab which may be attributed to either settlement of the hard core base or, alternatively, subsidence due to adverse ground conditions. In our opinion this movement has been progressive over a number of years as gaps between the floor surfaces and fitted skirtings have been filled with plaster or similar material. We very much doubt whether any building society would be prepared to grant a mortgage on the property in its existing state and we therefore strongly advise that your company obtains professional advice as regards the structural condition of the floor in order that any necessary remedial works may be undertaken prior to sale.

This report was shown to Mr and Mrs Westlake in early 1982 and it came as a considerable shock to them: this was the first time they had become aware of the structural problem. Despite some further effort to persuade local builders to buy — and this would have to be, of course, at a reduced price — there was no success as the property proved unmortgageable. As a consequence of this the plaintiffs were obliged to sell their house at Four Marks and to move back into 11 Yorktown Road. They consulted their present solicitors in February 1982 and the writ in this action was issued on June 1 1983.

In March 1982 Mr Treadwell, who is a chartered surveyor and, as I have said, a partner in the firm of Poulters, carried out a careful examination of the floor slab. In the course of this he made trial holes through the floor in the hall and also in the front garden to determine the underlying strata. His findings are set out in an affidavit (exhibit P4) to which were exhibited two drawings, numbered RT1 (that is exhibit P2 to which I have already referred) and RT2, now exhibited as P7, respectively. Mr Treadwell impressed me as a careful and knowledgeable witness and I have no hesitation in accepting his evidence both as to his findings and his opinions.

The trial hole through the hall floor revealed that the concrete floor slab had been laid upon a bed of rejects — this being the description of the layer of large stones 1½in sieve size found beneath the concrete. Exhibit P7 (which is a drawing of the trial holes) shows more precisely what Mr Treadwell found. He said that these rejects should have been firmly compacted at the time of construction and their depth should not have exceeded 600mm, being the maximum recommended by the National Housebuilding Council's guidelines. The depth found by Mr Treadwell of these rejects was 760mm and the stones were found to be loose under the concrete. He explained, in the course of his evidence, the difficulty of assessing the state of the rejects throughout the subfloor — because, of course, the very act of using an auger to bore into the rejects would of necessity disturb them and loosen them.

At the time of this examination in March 1982 the maximum downward movement of the floor was 28mm and plan exhibit P2 shows the depth of this movement at various points on the plan. The plaster or cement filling done by Mr Westlake in 1975 was also clearly seen by Mr Treadwell and measured a maximum of 20mm, thus confirming Mr Westlake's evidence and also the evidence of Mr Pritchard as to the approximate size of the gap in 1975. It is clear, therefore, that the floor slab had moved down 20mm by 1975 when the Westlakes purchased. I am satisfied that a movement of this magnitude should have been noticed by a reasonably competent surveyor and it was an indication of faulty construction and certainly required further investigation.

This was not only Mr Treadwell's view but was confirmed by Mr King who, in his evidence, said that he would have been concerned about a gap of 20mm and would have expected it to have been found by a surveyor despite the presence of fitted carpets. This was also the view of Mr Elliott, the independent chartered surveyor called by the defendants. I am also satisfied, in the light of the evidence I have heard in this case, that, having found evidence of downward movement of this magnitude in 1975, a reasonably competent surveyor should have advised that the property was unsuitable for mortgage purposes by reason of the substantial cost involved in putting the property into a reasonably saleable condition.

Perhaps I should pause here to state that, on the evidence, the increase in subsidence from 20mm in 1975 to 28mm in 1982 when Mr Treadwell made his inspection was probably due, certainly in part, to the fact that in 1975 and 1976 there had been two very dry summers. This had caused a drying out of the subsoil which, in this particular part of the country, caused contraction of a special kind; it was a contraction which was irreversible. In the light of my findings in this case, the fact that this additional cause of subsidence existed in 1975 and 1976 does not materially affect the judgment I have reached.

There was some difference of opinion between Mr Treadwell and Mr R F Elliott (the independent surveyor called on behalf of the defendants) as to the proper method of remedying this defective floor in 1975. Mr Treadwell said that the only satisfactory solution at that time would have been to grout the rejects and that would have involved pumping a cement slurry into the reject layer, which would then solidify and prevent further movement. This work would mean that the house would remain uninhabitable for some three months while the work was being done. Such work would only probably be done by a contractor who would execute the repairs and then put the house back on to the market. This would involve, of course, expenses of solicitors, estate agents and the expense of interest on capital and would also, perhaps, include a measure of profit. Mr Treadwell's estimate was that the cost of all this in 1975 would have been £5,000. Mr Elliott agreed that this figure was about right for the work that Mr Treadwell contemplated. However, Mr Elliott said that he did not think that such work was necessary in this type of house being, as it was, at the bottom end of the market. Mr Elliott thought that the problem could be adequately overcome simply by rescreeding the surface of the concrete slab at a cost of about £1,250.

I am satisfied upon the evidence that this would not have been an adequate way of correcting the defect. Mr Treadwell accepted that such a solution would have been adequate and permissible had the movement simply been up to about 10mm. With a 20mm settlement, he said that he would expect that there had been cracking of the underlying concrete slab and to rescreed in such circumstances would simply have been covering up the fault and inviting later problems. I am satisfied, therefore, that in 1975 the total expense of reparation of the floor would have been £5,000 including the incidental expenses — fees, interest and the like.

I now turn to the question as to what duty was owed by the defendants to the plaintiffs in this case. As to this the evidence seems to me to be quite clear. These plaintiffs were inexperienced first-time buyers of a property pretty well at the bottom of the market. They were in need of a 95% mortgage as their means were very limited. In 1975 such purchasers rarely, if ever, instructed their own surveyors and instead relied upon the valuation made by those who were advancing the money. In this case the defendants made a charge against the plaintiffs for the cost of the valuation and I would refer to condition 3 of the mortgage offer letter. Furthermore, the defendants did not seek to exclude their responsibility in this regard and did not invite the plaintiffs to have their own survey. The offer letter, by requiring certain minor matters to be rectified, carried the implication that the house was structurally sound. Mr King, in his evidence, very frankly accepted this position. He said:

In 1975 it was the case that purchasers did not usually have their own surveyors. It was common that purchasers relied upon the mortgage offer as indicating no major structural problems.

Thus, this case bears many similarities to the case of *Yianni* v *Edwin Evans & Sons* [1981] 3 All ER 592, to which I have been referred. I am satisfied that the defendants owed a duty of care to the plaintiffs knowing that the latter would rely upon the accuracy of the valuation for mortgage purposes. I am further satisfied that the defendants' surveyor who examined and reported upon the house in the terms of the report signed by Mr King was negligent in that he failed properly to examine the state of the ground floor of this house. Had he done so he would, as a reasonably competent surveyor, have noticed the 20mm gap and the other lesser gaps and ought to have been put on further inquiry as to the structural stability of this floor. It was negligent of him to report that the house was structurally sound and to value it at £11,750. No mention is made in the report of the fact that the ground floor was of concrete and this, in itself, points to a very superficial examination of this property.

It is clear that the plaintiffs have suffered damage by reason of this negligence because they have bought a house which has proved to be unsaleable by reason of the defective floor. This state of affairs came into existence as soon as they completed their purchase in July 1975, which is the date at which I find that the cause of action accrued.

The defendants say that the plaintiffs' claim for damages is barred by statute. The writ was not issued until June 1983 — some eight years later — and thus by reason of the provisions of section 2 of the Limitation Act 1980 it is said that this action, being founded on tort, was started too late.

The plaintiffs seek to answer this plea by praying in aid the provisions of section 32(1)(b) of the Act. This reads as follows:

Subject to subsection (3) below [which is not relevant to this case], where in the case of any action for which a period of limitation is prescribed by this Act . . . (b) any fact relevant to the plaintiff's right of action has been deliberately concealed from him by the defendant . . . the period of limitation shall not begin to run until the plaintiff has discovered the concealment . . . or could with reasonable diligence have discovered it. References in this subsection to the defendant include references to the defendant's agent . . .

I also read from subsection (2) of section 32:

For the purposes of subsection (1) above, deliberate commission of a breach of duty in circumstances in which it is unlikely to be discovered for some time amounts to deliberate concealment of the facts involved in that breach of duty.

It is the plaintiffs' case that there was a deliberate concealment of facts, within the meaning of that expression, by the defendants when their surveyor (as I find) revisited the house at the plaintiffs' request some four weeks after they moved into occupation and who, on being shown the 20mm gap, simply said: "It is purely settlement, nothing to worry about." Coming as this did from a surveyor who ought to have known better, I can only construe it as reckless in the extreme unless it was done deliberately. I have to bear in mind that (as I find) this surveyor had already negligently reported to his employers that the house was sound when it was not, as a consequence of which a loan of £10,500 had been advanced. Thus, when he saw the 20mm gap, probably for the first time, he hoped that there would be no adverse consequences.

My attention has been drawn to the commentary in relation to section 32 in the 1980 volume of *Current Law Statutes*. The words "deliberate concealment" in the 1980 Act replaced the words "fraudulent concealment" in section 26 of the Limitation Act 1939 which it superseded. It is not necessary that I should enter into an analysis of the meaning of this expression in the light of the decided cases, but it is reasonably clear that the words "deliberate concealment" involved something more than mere neglect of duty. Conduct involving recklessness or turning a blind eye or unconscionable conduct would all seem to fall within the ambit of the words "deliberate concealment".

Whatever phrase is used, I have come to the conclusion that the conduct of the defendants' surveyor on this short five-minute visit amounted to a deliberate concealment of facts relevant to the plaintiffs' right of action. Thus, the period of limitation did not begin to run until early 1982 when the plaintiffs first learned of the reason why the house would not sell.

If I am wrong about this, then I consider that the words and conduct of the defendants' surveyor on the short visit clearly led the plaintiffs into thinking that the gap was nothing to worry about. This raises an estoppel against the defendants from averring that the plaintiffs' cause of action accrued more than six years before the commencement of these proceedings. In this connection, I have found the reasoning of His Honour Judge White in the case of *Kaliszewska* v *J Clague & Partners* (1984) (vol 5 of *Construction Law Reports* at p 62) helpful. I am very much indebted to Mr Gaitskell for searching out that case in the course of his researches. It was not particularly helpful to his case, but it was plainly his duty to bring it to the attention of this court.

The plaintiffs are, therefore, entitled to damages. Upon the authority of the Court of Appeal decision in *Perry* v *Sidney Phillips & Son* [1982] 1 WLR 1297 I conclude that the proper measure of damage in this case is the difference in price between what the plaintiffs paid for the property in 1975 and its true market value at that time. Accepting, as I do, Mr Treadwell's evidence as to this, I assess damages under this head at £5,000, together with interest thereon from the date of purchase.

There is also a claim for damages for the distress and discomfort suffered by these plaintiffs as a consequence of this negligence. The authorities suggest that sums awarded under this head should be modest. One has only to consider the history of this matter to realise that the plaintiffs, once they became aware of the problem in early 1982, have suffered much inconvenience and heartache by having to move back into a poorer house in a less desirable locality together with the attendant upheaval of an unwanted move. Doing the best I can, I award general damages in respect of that part of the claim at £1,500.

I note, on rereading the amended defence in this case, that there was, in fact, an allegation of contributory negligence on the part of the plaintiffs, but no argument has been addressed to me under this head and I do not think that calls for any further comment.

For further cases on this subject see p 231

RATING

Court of Appeal
November 13 1986
(Before Lord Justice WATKINS, Lord Justice PURCHAS and Lord Justice GLIDEWELL)

IMPERIAL COLLEGE OF SCIENCE AND TECHNOLOGY v EBDON (VALUATION OFFICER) AND ANOTHER

Estates Gazette January 31 1987

281 EG 419-427

Rating — "Contractor's basis" — Appeal by rating authority from decision of Lands Tribunal — Hereditament occupied by Imperial College of Science and Technology — Complaint by appellants that, although the contractor's basis was the correct method of valuation to use in the case of such a hereditament, it had been incorrectly applied — The values determined by the Lands Tribunal, including the Huxley Building, were GV £846,000, RV £704,972 — Of the five steps in the approach to the contractor's basis valuation described in *Gilmore (VO) v Baker-Carr*, no issue arose before the Court of Appeal as to the first three, estimated replacement cost, adjustment to arrive at effective capital value, and capital value of the site — The main issue was as to the fourth step, the decapitalisation rate to arrive at annual rental and there was a secondary issue as to the final adjustment at stage five — The Lands Tribunal discounted inflation entirely in fixing a rent for a year certain and arrived at a real interest rate of between 1.4% and 2.4% — To this was added a borrower's premium of 1% and a figure of 1% for depreciation and repairs, resulting in a total of between 3.4% and 4.4% — In the light of agreed settlements of assessments of other universities throughout the country at a decapitalisation rate of $3\frac{1}{2}$%, the Lands Tribunal adopted the figure of $3\frac{1}{2}$% — As a final step the tribunal made a deduction of $7\frac{1}{2}$% as an adjustment to reflect the actual characteristics of the hereditament not already taken into account — The Court of Appeal rejected most of the criticisms of the tribunal's decision on the ground that the court's jurisdiction was confined to dealing with errors of law, while the criticisms had been directed against matters of fact and valuation which were reserved for technical decision by the tribunal — The tribunal had not been shown to have erred in law — Appeal dismissed

The following cases are referred to in this report.

Baker Britt & Co Ltd v *Hampsher (VO)* [1976] RA 69; (1976) 19 RRC 62; [1976] EGD 566; 239 EG 971
Cardiff City Council v *Williams (VO)* [1973] RA 46; (1973) LGR 221; [1973] EGD 780; 226 EG 613, CA
Dawkins v *Ash Brothers & Heaton Ltd* [1969] 2 AC 366; [1969] 2 WLR 1024; [1969] 2 All ER 246; (1969) 67 LGR 499, HL
Dawkins (VO) v *Royal Leamington Spa Corporation* (1961) 8 RRC 241; [1961] RVR 291 178 EG 293, 365 & 461 LT
Edwards v *Bairstow* [1956] AC 14; [1955] 3 WLR 410; [1955] 3 All ER 48; (1955) 48 R&IT, HL
Gilmore (VO) v *Baker-Carr (No 2)* (1963) 10 RRC 205; [1963] RA 458; [1964] RVR 7; 188 EG 977, LT

Humber Ltd v *Jones (VO)* (1960) 6 RRC 161; 53 R&IT 293, 175 EG 1195, CA
Metropolitan Water Board v *Chertsey Assessment Committee* [1916] AC 337
Westminster City Council v *American School in London and Goodwin (VO)* [1980] RA 275; [1980] EGD 684; (1980) 255 EG 999 & 1107, LT

This was an appeal by case stated by Westminster City Council, the rating authority, against the decision of the Lands Tribunal (Mr C R Mallett FRICS) on appeals from the Greater London (Central) Valuation Court in respect of the gross and rateable values of the Imperial College of Science and Technology in South Kensington. The Lands Tribunal's decision is reported at [1985] 1 EGLR 209; (1984) 273 EG 81 & 203.

Graham Eyre QC and Richard Hone (instructed by the City Solicitor, Westminster City Council) appeared on behalf of the appellants; Alan Fletcher QC and G N Huskinson (instructed by the Solicitor of Inland Revenue) represented the valuation officer, Mr Ebdon; Guy Roots and N King (instructed by Lovell White & King) represented the Imperial College.

Giving the first judgment at the invitation of Watkins LJ, GLIDEWELL LJ said: This is an appeal by case stated against the decision of the Lands Tribunal (C R Mallett FRICS) given on October 24 1984, on eight related appeals from the Greater London (Central) Valuation Court. By its decision the Lands Tribunal determined the gross value and rateable value of the hereditament occupied by the second respondents, the Imperial College of Science and Technology, in the valuation list for rating which came into force on April 1 1973. The appellants in this court, Westminster City Council, are the rating authority. They contend that in reaching his decision the learned member of the Lands Tribunal has made fundamental errors in his application of a method of valuation known as the contractor's basis. Mr Ebdon, the valuation officer, supports the decision of the Lands Tribunal.

The facts
The history of the matter and the relevant facts are set out in an agreed statement of facts which was placed before the Lands Tribunal, and in the tribunal's decision. It is therefore not necessary for me to refer to more than a brief outline. Imperial College occupies a site of some $15\frac{3}{4}$ acres, on which stand substantial buildings, in South Kensington. The college is a school of the University of London. One of the buildings now on the site, the Huxley Building, was not completed or occupied when the valuation list came into force on April 1 1973. The list was therefore amended to include the Huxley Building as a result of a proposal made by the valuation officer on March 29 1977.

Lands Tribunal's decision
The values determined by the Lands Tribunal are as follows:

	Gross value	Rateable value
At April 1 1973	£767,000	£639,138
With addition of Huxley Building	£846,000	£704,972

The points raised in this appeal affect both sets of values, but in the same manner. The matter was therefore argued before us, and can be treated, as a single appeal.

Contractor's basis of valuation
By section 19(6) of the General Rate Act 1967 the gross value of a hereditament is defined as:

the rent at which the hereditament might reasonably be expected to let from year to year if the tenant undertook to pay all usual tenant's rates and taxes and the landlord undertook to bear the cost of the repairs and insurance and the other expenses, if any, necessary to maintain the hereditament in a state to command that rent.

It is the function of the rating surveyor to ascertain that rent.

Normally the principal weapon in his armoury is evidence of recent lettings, or failing that, sales of similar or comparable properties. For some categories of property, however, such evidence is rarely if ever to be found. University colleges and their constituent buildings fall into such a category. The valuer called upon to assess the gross value of such a property has no evidence of comparable transactions to guide him. If the hereditament itself is let at a rent, or if it is occupied by a commercial enterprise which trades at a profit, evidence of the actual rent or profit may assist towards the necessary valuation. But Imperial College is not let at a rent and is not a trading concern. The valuer must therefore seek some other method.

The "contractor's basis" is a method of calculation which has been devised by valuers, with the approval of the Lands Tribunal and the courts, to meet this kind of situation. In the present case all parties are agreed that it is the correct method of valuation to be applied. In *Cardiff City Council* v *Williams (VO)* [1973] RA 46, Lord Denning MR quoted as the "classic explanation" of this method a passage from the address of Sir Jocelyn Simon QC, Solicitor-General, in *Dawkins (VO)* v *Royal Leamington Spa Corporation* (1961) 8 RRC 241, which the members of the Lands Tribunal had adopted in that case:

As I understand it, the argument is that the hypothetical tenant has an alternative to leasing the hereditament and paying rent for it; he can build a precisely similar building himself. He could borrow the money, on which he would have to pay interest; or use his own capital on which he would have to forgo interest to put up a similar building for his owner-occupation rather than rent it, and he will do that rather than pay what he would regard as an excessive rent — that is, a rent which is greater than the interest he forgoes by using his own capital to build the building himself. The argument is that he will therefore be unwilling to pay more as an annual rent for a hereditament than it would cost him in the way of annual interest on the capital sum necessary to build a similar hereditament. On the other hand, if the annual rent demanded is fixed marginally below what it would cost him in the way of annual interest on the capital sum necessary to build a similar hereditament, it will be in his interest to rent the hereditament rather than build it.

In his judgment in the *Cardiff* case, Lord Denning MR added a qualification to this explanation. He said at pp 50-51:

To that statement, however, I would make this qualification. The annual rent must not be fixed so as to be only "marginally below" the interest charged. It must be fixed much below it, and for this reason: By paying the interest charged on capital cost, he gets not only the use of the building for its life, but he gets the title to it, together with any appreciation in value due to inflation: whereas, by paying the annual rent, he only gets the use of the building from year to year — without any title to it whatsoever — and without any benefit from inflation.

Mr Eyre, for the city council, is critical of this qualification, describing it as a matter for valuation evidence rather than a statement of principle. However, he accepts that a deduction should properly be made from the rate of interest at which the hypothetical building owner could borrow to finance his building, to take account of the two factors to which Lord Denning referred. Indeed in his evidence to the Lands Tribunal on behalf of the city council Mr W A Hampsher ARICS made such a deduction. As I shall explain later, Mr Eyre's major criticism of the Lands Tribunal's decision is that too great a deduction has been made in respect of the element of inflation.

Lands Tribunal's approach
Mr Mallett divided the valuation into five stages, thus following the approach described in the decision of the Lands Tribunal in *Gilmore (VO)* v *Baker-Carr (No 2)* (1963) 10 RRC 205, which is now generally adopted. These stages are:
1 To estimate the replacement costs of substituted buildings.
2 To adjust this cost to take account of the actual state of the buildings comprising the hereditament. The resulting figure is often known as "effective capital value", but Mr Mallett in his decision prefers to call it "adjusted replacement cost".
3 To add the capital value of the site comprising the hereditament.
4 To adopt and apply a rate at which to decapitalise the total capital value, so as to achieve an annual rental.
5 To make any adjustment which is necessary in order to reflect the actual characteristics of the hereditament, but which has not already been taken into account at an earlier stage.

The issues
Before the Lands Tribunal the parties agreed the estimated replacement cost of substituted buildings at April 1 1973 at a total of £20,340,438. There was disagreement between the valuers as to the appropriate adjustments at stage 2, as to the value of the site (stage 3) and as to the cost of the Huxley Building, but the Lands Tribunal's findings on these matters are not challenged in this court. The total adjusted capital figure for land and buildings at April 1 1973 is £23,125,971. Thus before us there is no issue as to stages 1, 2 or 3.

The principal issue in this appeal relates to the decapitalisation rate adopted by the Lands Tribunal at stage 4. There is also a secondary issue as to the adjustment at stage 5.

An appeal to this court from a decision of the Lands Tribunal lies only if the decision "is erroneous in point of law": Lands Tribunal Act 1949, section 3(4). Mr Eyre submits that the member of the Lands Tribunal departed in several respects from the accepted method for arriving at a decapitalisation rate and as a result made errors which resulted in his adopting too low a rate. In addition Mr Eyre points to one alleged error at stage 5. We have to decide whether and to what extent Mr Eyre's criticisms are justified and, if so, whether the Lands Tribunal's error is on a point of law.

Stage 4 — the decapitalisation rate
The tribunal's decision on this issue begins by setting out four propositions which are said to be agreed, namely:
(i) that the object is to arrive at the annual equivalent of the adjusted cost in terms of the hypothetical tenancy;
(ii) that the decapitalisation rate is the rate at which the hypothetical tenant could expect to borrow the capital required over the term of the hypothetical tenancy at a rate of interest fixed for one year certain;
(iii) that inflation must be taken into account in converting a fixed-interest rate on long-term borrowing to an interest fixed from year to year; and
(iv) that the decapitalisation rate must reflect the ability of the particular hypothetical tenant . . . to borrow money at a preferential rate.

Mr Eyre asserts that none of these propositions was agreed. Moreover, he submits they are wrong, and so fundamentally wrong as to vitiate the decision. Mr Roots for Imperial College and Mr Fletcher for the valuation officer accept that the propositions were not agreed but do not accept that they are wrong, though Mr Fletcher suggests that they might have been differently expressed.

Mr Roots describes the first proposition as "unimpeachable", and I agree that it is a brief but accurate summary of the valuer's objective. However, it is an introduction only.

The first major issue between the parties relates to the phrase "a rate of interest fixed for one year certain" in the second proposition. Mr Eyre submits that, if the hypothetical landlord and tenant are negotiating a rent for a tenancy from year to year, they will in practice reach agreement on the basis that, though the landlord will not allow the rent to run unaltered indefinitely, he will allow a reasonable (though undefined) time to elapse before giving notice to increase the rent, which is in effect notice to terminate the existing tenancy. Although in theory the landlord could give such a notice to increase the rent at the end of each year, in practice he would not do so. Thus it is an error to regard the rent, or the rate of interest from which it is to be derived, as "fixed for one year certain".

Mr Roots and Mr Fletcher accept that a tenancy from year to year is a tenancy for an indefinite time, though terminable by notice, and that the hypothetical tenancy must be so regarded — see *Dawkins* v *Ash Brothers & Heaton Ltd* [1969] 2 AC 366, *per* Lord Pearce at p 383 F-G, Lord Wilberforce at p 387 and Lord Pearson at p 392 G-F. Nevertheless, they submit, the effect of continuing inflation on the negotiations between hypothetical landlord and tenant cannot be disregarded. The hypothetical landlord is bound to enter into a tenancy from year to year; he cannot protect himself against inflation by negotiating a rent for a fixed term longer than a year or by inserting a rent review clause in the lease. Thus it must, or can properly, be assumed that he would give notice to terminate at the end of the first year of the tenancy and repeat the process yearly thereafter.

In my judgment, the learned member's decision to assess the rent, and the rate of interest on which it was based, as if they were to be fixed for a year certain was a finding of fact and valuation opinion to which he was entitled to come on the evidence before him. In *Baker Britt & Co Ltd* v *Hampsher (VO)* (1976) 19 RRC 62, the House of Lords held that an appeal by case stated against a decision of the Lands Tribunal, being on a point of law, can only succeed within the principle laid down in *Edwards* v *Bairstow* [1956] AC 14, *viz* if the decision contains material which is on its face wrong in law or if on the evidence no tribunal acting properly could have reached the decision arrived at. Since there was material upon which the learned member could base his second proposition, no error of law in it is disclosed. Mr Eyre's challenge on this point therefore fails.

His major criticism of the third proposition is also based on the reference there to "an interest rate fixed from year to year". I would therefore also reject this argument. However, Mr Eyre argues the point relating to inflation in a somewhat different way.

A rate of interest to be paid on a long-term loan is, in theory, made up from several elements. The first is the lender's profit, his desired return on his money. Secondly, the lender will add an element to take account of the risk that his loan may not be repaid or will be repaid late. This is relevant to the fourth proposition, and I will return to it later. Thirdly, in times of continuing inflation the lender expects that his capital will have depreciated in value when it is returned to him, and thus he will add to the rate of interest he requires in order to protect himself against future inflation.

However, if the loan is not for a long term of years, but for a period equivalent to a tenancy from year to year, the effect of inflation over the period will be much less. Moreover, as Lord Denning said, in the passage I have quoted from his judgment in the *Cardiff* case, the tenant from year to year is not protected against inflation. Thus a rate of interest appropriate for a long-term loan must be reduced in order to relate it to a yearly term.

In the Lands Tribunal Mr Jackman, an economist giving evidence on behalf of Imperial College, put in evidence a table showing the annual rate of inflation and the minimum lending rate over the 20 years from 1952 to 1972. This showed that, on average, minimum lending rate over the period was only 1.42% higher than the rate of inflation. Mr Jackman called this the "real rate of interest". Over the same period inflation averaged just over 4%. Mr J R Trustram Eve MSc FRICS adopted in his valuation for Imperial College this "real rate of interest", rounded up to 1.5%. Professor Ilersic, an economist who gave evidence for the city council, did not disagree with this approach but preferred to take the average of a shorter period, which gave a "real rate" of 2.4% and a rate of inflation of 3.3%.

In his decision the member of the Lands Tribunal followed Mr Eve's approach by starting with a minimum lending rate of 5.5% and deducting the whole average figure for past inflation. This gave him a real interest rate of between 1.4% and 2.4%. To this he added the "borrower's premium" and an addition for depreciation and repairs, giving a total of 3.4% to 4.4%. He then referred to evidence given on behalf of Imperial College that assessments in relation to other universities and university colleges outside London had been agreed at figures based on a decapitalisation rate of 3.5% and adopted this figure to apply at stage 4.

Mr Eyre submits that this approach was wrong in three respects. His main objection is that, though it was correct to make some discount to reflect inflation, it was wrong, and not in accordance with past practice, to seek to eliminate the effect of inflation entirely from the rate of interest to be applied. He refers to the most recent previous decision of the Lands Tribunal in relation to the contractor's basis, namely, *Westminster City Council* v *American School in London and Goodwin (VO)* [1980] RA 275. In that case the Lands Tribunal (J H Emlyn Jones FRICS) adopted 5% as the decapitalisation rate.

Mr Roots points out that a major difference between the present case and the *American School* case is that, in the latter, the Lands Tribunal had no clear evidence as to the effect of inflation, whereas in the present case there was the evidence of the economist to which I have already referred. This, in my view, justified the tribunal in making a greater discount for the effect of inflation than had been applied in the *American School* decision.

I have already said that the member of the Lands Tribunal was entitled to discount inflation to a figure appropriate to fixing a rent for a year certain. Should such a figure nevertheless include some element for inflation? If so, was it an error of law to discount for the whole effect of inflation?

If I had been deciding this matter as a question of fact, I might well have concluded that in the inflationary climate of April 1973 a rent negotiated for a year certain would probably have included an element to take account of inflation. But Mr Mallett considered this matter expressly; he recorded Mr Eve's opinion that "where the rent was subject to annual review, the effect of inflation on the rent would be imperceptible"; having no contrary evidence, he adopted this opinion and thus discounted for the whole effect of inflation. I cannot say that, in taking this course, he made any error of law.

Mr Eyre's second criticism is that the figure adopted for the "buyer's premium", ie the 1% added to minimum lending rate to compensate for the lender's risk, was too low. This point relates to the last of the four propositions. Mr Ebdon had suggested 2%. It was agreed by all parties that the hypothetical tenant would be, or to be more precise would have the characteristics of, Imperial College. The tenant would thus inevitably be an institution with government backing, providing the hypothetical landlord with as sound a covenant as he could find. In my judgment, the Lands Tribunal was entitled to adopt a 1% "borrower's premium" as a matter of opinion based on evidence. Since it was agreed that the hypothetical tenant, and thus the hypothetical borrower, would have the characteristics of Imperial College, he was not debarred from this finding by the observations of Earl Loreburn in *Metropolitan Water Board* v *Chertsey Assessment Committee* [1916] AC 337, where at pp 347-8 he made it clear that a rate of interest used to ascertain a hypothetical rent may not be fixed by taking into account the financial position of the particular occupier.

Lastly, Mr Eyre submits that the Lands Tribunal should not have taken into account the evidence of the decapitalisation rate adopted in settlements of assessments at other universities throughout the country without detailed evidence to show how far those hereditaments were truly comparable to Imperial College. The answer to this, in my view, is that the member used this evidence as his final stage in choosing a rate in the range he had already established between 3.4% and 4.4%, not as primary evidence of allegedly comparable transactions.

In summary, were I deciding this matter at first instance I might have come to a different conclusion on one aspect of the discount for inflation, but I cannot say that in any part of his decision to adopt a decapitalisation rate of $3\frac{1}{2}$% the member of the Lands Tribunal made any error of law. I emphasise that this is not a decision that $3\frac{1}{2}$% is necessarily the correct rate to be applied for decapitalising a capital value in the application of the contractor's basis to the rating assessments of all universities or colleges. Each case depends upon its particular facts and the evidence called. My judgment is that the facts established and the opinion evidence called justified the Lands Tribunal's decision in this case.

Stage 5 — the final adjustment

At this stage the Lands Tribunal made a deduction of $7\frac{1}{2}$%. The submission is that the evidence only justifies 5%. Mr Eve in his valuation deducted $7\frac{1}{2}$%, but of this $2\frac{1}{2}$% related to the alleged effects of a district heating scheme for which the Lands Tribunal made no deduction. On the other hand, Mr Ebdon accepted that there were disadvantages for which he allowed 5%, expressing them more generally than did Mr Eve. The member said in his decision, "whether or not these items have any effect and to what extent is a subjective opinion based upon the information available." This is clearly correct. The information available to him included that gained from a view of the hereditament. His decision to deduct $7\frac{1}{2}$% is a matter of opinion which cannot be challenged as being wrong in law.

For these reasons I would dismiss the appeal.

Agreeing, PURCHAS LJ said: Out of acknowledgement for the able and interesting submissions made by Mr Eyre, I propose to deliver a short judgment of my own. The facts and background to the appeal have already been fully set out in the judgment of Glidewell LJ, a draft of which I have had the privilege of reading, and need not be repeated here.

The appeal is brought under the provisions of the Lands Tribunal Act 1949, section 3, the relevant part of which provides:

(4) A decision of the Lands Tribunal shall be final: Provided that any person aggrieved by the decision as being erroneous in point of law may . . . require the Tribunal to state and sign a case for the decision of the court . . .

Earlier tiers of review and appeal are provided by section 76 of the General Rate Act 1967 (appeals to local valuation courts against objections to proposals) and section 77 of that Act (appeals from decisions of local valuation courts to the Lands Tribunal). There are, therefore, two levels of appeal at which technical questions of valuation and other matters of expertise may be fully rehearsed before review bodies specially qualified for this purpose. Although in his submissions Mr Eyre commented that a review of the gravity of that involved in the present appeal would in former years almost certainly have been considered by a panel consisting of two members of the tribunal each possessing differing qualifications and expertise rather than a tribunal of one member, I do not find any force in the implied criticism, if that was the purpose of the submission in the present case. On this appeal we are concerned solely with a consideration of the case stated under the Act of 1949 to see whether the appellant establishes that the decision was erroneous in a relevant point of law.

By section 19(6) of the General Rate Act 1967 the gross value of a hereditament is defined as:

the rent at which the hereditament might reasonably be expected to let from year to year if the tenant undertook to pay all usual tenant's rates and taxes and the landlord undertook to bear the cost of the repairs and insurance and the other expenses, if any, necessary to maintain the hereditament in a state to command that rent.

The valuation of this "rent" is the statutory duty imposed upon the valuation officer against whose valuation appeal lies on questions of technical expertise and approach to the valuation court and to the Lands Tribunal. Beyond that the appeal lies, as I have already stated, on a point of law only. The exercise, therefore, is to determine whether the question of complaint raised in the case stated relates to matters of technical method and expertise or whether it can be shown that the Lands Tribunal acted, as a matter of law, so that its approach lay outside the statutory duty imposed by the section.

Subject to the qualification in the oft-cited passage from the speech of Viscount Radcliffe in *Edwards* v *Bairstow* [1956] AC 14 at p 36, namely, that where the facts found are such that no person acting judicially and properly instructed as to the relevant law could have come to the determination under appeal, provided the method and approach of the valuation officer is one acceptable to the highest tier of technical appeal, namely the Lands Tribunal, then criticisms either of the method of approach or the application of expertise in the valuation field cannot, in my judgment, found a successful appeal from the Lands Tribunal on a case stated. Lord Diplock in his short speech in *Baker, Britt & Co Ltd* v *Hampsher (VO)* (1976) 19 RRC 62 criticised the framing of the case stated which used the terms "whether the tribunal misdirected itself in holding that greater weight ought to be given to the evidence of the rents of the comparables" on the basis that this invited the Court of Appeal to express its view as to the comparative weight to be attached to two kinds of evidence adduced at the hearing, *viz* (1) the terms of the actual letting to ratepayers of the premises to be valued, and (2) the rents at which comparable premises were let, both of which approaches were accepted as relevant in the ascertainment of the rateable value of the premises.

In this case it was common ground that the proper approach in the case of the hereditament in question was that known as "the contractor's basis of valuation". This has received the approval of the courts (see *Cardiff City Council* v *Williams (VO)* [1973] RA 46). The method is based upon the "classic explanation" of the Solicitor-General in *Dawkins (VO)* v *Royal Leamington Spa Corporation* (1961) 8 RRC 241, which has been set out in the judgment of Glidewell LJ. I venture only the comment that the qualification of that method made by Lord Denning MR, also already cited by Glidewell LJ, purported to do no more than indicate that the valuer should as a matter of general principle make an adjustment to the interest charged in order to arrive at the annual rent within the meaning of the section. Otherwise, the precise degree to which a reduction should be made is more appropriate for the techniques and expertise of the valuer than for the judicial decision of the judge.

In the judgment of Glidewell LJ the submissions made by Mr Eyre attacking in detail the technical method of application of the valuation on the contractor's basis are considered at length. In particular, Mr Eyre's attack was that the decapitalisation rate adopted by the Lands Tribunal at stage 4 was too low. I agree with the conclusion of Glidewell LJ that these are matters appropriate to the expertise of the economist or valuer and not of the lawyer. The same criticism applies to the figure adopted in the evidence accepted by the Lands Tribunal for what is known as "buyer's premium". And, thirdly, a reflection of the criticism of Lord Diplock in *Baker, Britt's* case, the use made of and the approach to the decapitalisation rate by the Lands Tribunal in considering other comparable assessments which had been accepted by way of settlement could not be challenged on appeal to this court.

Finally, so far as stage 5, the final adjustment, is concerned, again the application of this factor is one essentially in the province of the valuer and therefore is not susceptible to challenge on this appeal.

In summary, therefore, notwithstanding the able and attractive arguments of Mr Eyre, and without, I hope, doing a disservice to the detail of his argument and the careful analysis of it by Glidewell LJ, Mr Eyre's submissions really amounted to an invitation to the court to leave the territory of judicial decision on points of law and to move into the arena reserved for technical decision by the Lands Tribunal, the valuation court, and the experts in valuation, be they economists, land agents, surveyors, or other specialists in this field. Inviting as the invitation was made to appear by Mr Eyre, I agree with my lord that it is an invitation which this court would be wrong to accept, quite apart from the opinions expressed by Glidewell LJ on the detailed merits of the valuation or decision of the Lands Tribunal, to which I do not find myself able to make any useful contribution.

Accordingly, for the reasons given by Glidewell LJ and the reasons given in this judgment, I would dismiss the appeal.

Also agreeing, WATKINS LJ said: The first and main question arising out of the stated case for our decision is expressed by Mr Mallett, the member of the Lands Tribunal, in these terms:

Whether the approach of the tribunal was correct in law in failing to find the appropriate market rate and make adjustments reflecting the nature, extent and terms of the hypothetical tenancy, but instead quantified the element of inflation and adopted a rate of interest (representing some form of borrowing rate free of all inflation) when determining a decapitalisation rate for the purpose of Stage IV of a contractors' basis valuation.

We were informed by the appellant that this is the first case in respect of a university hereditament to reach the Court of Appeal under the current 1973 valuation list, and has the widest possible implications in respect of the rates which university hereditaments throughout England and Wales will be liable to pay. If the decision of the member can be said to be founded on an error of law, and this matter has to be remitted for a hearing based upon what is contended is the proper law, that may very well be so, seeing that in agreed arrangements for other universities throughout the country $3\frac{1}{2}$% has been regarded as the appropriate decapitalisation figure. If that is wrong then obviously the implications are liable seriously to affect rating of universities generally.

It is that figure which is under attack, that is to say, the decapitalisation rate, or in other words the application of the appropriate rate of interest to the effective capital value.

Agreed assessments for other universities undoubtedly influenced the member in coming to his final conclusions, which were:

In summary it seems to me that the real interest rate is not less than 1.4% and not more than 2.4%, that the borrower's premium should be not more than 1%, and that the addition for depreciation and repair should be not more than 1%. This would result in a lowest possible figure of 3.4% and a highest of 4.4%. In the absence of any evidence to the contrary, the evidence of agreed assessments on other universities throughout the country is overwhelmingly in favour of $3\frac{1}{2}$% as the appropriate decapitalisation figure and this is the percentage I adopt.

In achieving the bracket of 3.4% and 4.4% within which to settle upon the rate of $3\frac{1}{2}$% the member clearly relied upon the evidence preferred by him to all other of experts called on behalf of Imperial College. That evidence, which he set out in his illuminating reasoned decision, cannot in my judgment be said to contain an approach erroneous in point of law to what to the layman is a perplexing problem and to the expert one which admits of varying opinion some of which may be of equal validity. The choice of which expert opinion is to be preferred seems to me unquestionably to be for the tribunal of fact, in this case the member.

The member is himself a very experienced surveyor, who manifestly comprehended the law he had to adopt as expounded in a number of leading authorities referred to him. Moreover, I detect no arbitrary approach to the conflicting expert evidence he heard such as would enable this court to say that his factual findings were against the weight of the evidence. In my judgment, they clearly were not.

Where, then, was the error of law complained of? In common with Glidewell LJ, with whose judgment I agree, I see none. The member, in my view, was in a like position to that explained by Willmer LJ, as to another member of the Lands Tribunal, in *Humber Ltd* v *Jones (VO)* (1960) 6 RRC 161 at p 171:

The fact is that it is impossible to get away from the situation that the statute postulates not only a hypothetical tenant but also a hypothetical landlord, and, as the Lands Tribunal said in the passage read, in the context of a hypothetical world in which the hypothetical tenant cannot become the owner of the premises and cannot get a lease for a term of years. Moreover, one has to postulate a world in which not only this hypothetical tenant is in that position, but everybody else is in the same position. In the end, therefore, we are in a world of make-believe. What the value of premises in such imaginary circumstances would be seems to me to be very much a question of fact. It is a question of fact for the tribunal, guided by the expert evidence of valuers and such skilled persons, to say how much help can be derived from the terms of actual tenancies negotiated in the real world for a term of years. The learned member of the Lands Tribunal treated the question as one of fact; he had the evidence of valuers before him, and he came to his conclusion treating the matter as one of fact. I am quite unable to see that he was wrong and, for the reasons which I have already stated, I cannot see that he has misdirected himself with regard to the test to be applied.

The factual issues involved in that and the present case differ, but the approaches to settling upon the value of premises and so forth do not. To that I would add that, once the member here had found the bracket as a matter of fact, his decision to settle upon a $3\frac{1}{2}\%$ capitalisation rate, which coincidentally fell neatly within the bracket, cannot be faulted. By agreement it has been of general application in similar institutions, which doubtless have benefited from expert advice.

For these and the reasons provided by Glidewell LJ as to the first and second questions, I would dismiss this appeal.

The appeal was dismissed with costs of both respondents; an application for leave to appeal to the House of Lords was refused.

Court of Appeal

December 18 1986
(Before Lord Justice O'CONNOR, Lord Justice WOOLF and Sir George WALLER)

ADDIS LTD AND OTHERS v CLEMENT (VALUATION OFFICER)

Estates Gazette February 14 1987
281 EG 683-688

Rating — General Rate Act 1967, section 20(1)(b) — "Tone of the list" — Proposals for reductions in rates of industrial or commercial hereditaments situated just outside enterprise zone — Adverse effects of zones on values of premises situated outside — Whether it was correct to take into account, under section 20(1)(b), the designation of an enterprise zone in arriving at the ceiling provided for by the section — Appeal by valuation officer from decision of Lands Tribunal dismissing his appeal and holding that the adverse effects of the existence of the zone should properly be taken into account in accordance with section 20(1)(b) — Consideration of the statutory provisions and of the judgment of Sir Patrick Browne in *K Shoe Shops Ltd* v *Hardy (Valuation Officer) and Westminster City Council* — It was submitted on behalf of the appellant valuation officer that section 20(1)(b) was concerned with physical factors or at least with factors which affect the physical use and enjoyment of the hereditament, an interpretation disputed by the respondent ratepayers — The Court of Appeal accepted in general the valuation officer's approach, but with important guidance as to the implications of their decision — At the material time there were no changes shown with a physical connotation consequential on the designation of the zone — The Lands Tribunal had been in error in taking account of the mere designation or existence of the zone — However, the zone might in due course create physical changes which could be capable of being taken into account for the purpose of section 20(1)(b) — If the existence of a zone affects the prosperity of an area in a manner which is manifest and can be observed, this should be taken into account — Appeal by valuation officer allowed

The following cases are referred to in this report.

Baker Britt & Co Ltd v *Hampsher (VO)* [1976] RA 69; (1976) 19 RRC 62; [1976] EGD 566; 239 EG 971
Dawkins v *Ash Brothers & Heaton Ltd* [1969] 2 AC 366; [1969] 2 WLR 1024; [1969] 2 All ER 246; (1969) 67 LGR 499, HL
K Shoe Shops Ltd v *Hardy (VO) and Westminster City Council* [1983] RA 26; [1983] EGD 749; (1982) 266 EG 119, CA

Ladies Hosiery & Underwear Ltd v *West Middlesex Assessment Committee* [1932] 2 KB 679, CA
Sheerness Steel Co plc v *Maudling (VO)* [1986] RA 45, LT

This was an appeal by the valuation officer (Mr P J Clement BSc ARICS) from a decision of the Lands Tribunal (Mr J H Emlyn Jones CBE FRICS) holding that the ratepayers' hereditaments situated outside the Lower Swansea Valley Enterprise Zone were adversely affected by the existence of the zone and that this was a matter which should be taken into account in their valuations pursuant to the provisions of section 20(1)(b) of the General Rate Act 1967. The ratepayers were Addis Ltd and the four others named in the table of gross and rateable values on the two bases set out in Woolf LJ's judgment. The Lands Tribunal's decision is reported at (1984) 271 EG 291.

Alan Fletcher QC and David Mole (instructed by the Solicitor of Inland Revenue) appeared on behalf of the appellant; Guy Roots (instructed by McKenna & Co) represented the respondents, Addis Ltd; Matthew Horton (instructed by Roy Thomas Begley & Co, of Swansea) represented the remaining respondents.

Giving the first judgment at the invitation of O'Connor LJ, WOOLF LJ said: This is an appeal by case stated from a decision of the Lands Tribunal given on June 21 1984. It relates to proposals for the reduction in rates of five hereditaments which are industrial or commercial premises situated just outside the Lower Swansea Valley Enterprise Zone ("the zone").

The appeal, which is by the valuation officer against the dismissal of his appeal to the Lands Tribunal, raises an issue of general importance as to the method of determining valuations for the purpose of inserting or altering an entry in the current valuation list. The proposals have been made on the grounds that the establishment of the zone has adversely affected the value of the hereditaments. The issue is whether, when making valuations for this purpose, it is right to take into account changes in rental value caused by events of this character.

Section 179 of and Schedule 32 to the Local Government, Planning and Land Act 1980 authorised the setting up of enterprise zones and the zone was established by the Lower Swansea Valley Enterprise Zone Designation Order 1981 with effect from June 11 1981. With the exception of one proposal made by Addis Ltd on March 18 1981 all the proposals were made in 1981 after the zone had been designated.

The object of establishing enterprise zones is to encourage industrial and commercial activities in run-down areas. This is achieved by providing fiscal and administrative incentives within an enterprise zone for a period of 10 years from the date when the zone comes into effect. The incentives apply both to new and existing industrial enterprises within the zone. The incentives which are available are described in the decision as follows: (1) Exemption from development land tax. (2) Exemption from rates on industrial and commercial property. (3) 100% allowances for corporation and income tax purposes for capital expenditure on industrial and commercial buildings. (4) Applications for certain customs facilities for firms within the zones are processed as a matter of priority and certain criteria are relaxed. (5) Industrial development certificates are not needed. (6) Employers are exempt from industrial training levies and from the requirement to supply information to industrial training boards. (7) A greatly simplified planning regime; developments that can conform with a published scheme in each zone will not require individual planning permission. (8) Those controls remaining in force are to be administered more speedily. (9) The Government's requests for statistical information are reduced.

It is not surprising that it is accepted that these substantial attractions of operating a business in an enterprise zone have had an adverse effect on values of premises situated outside, but in the same locality as, the zone. This adverse effect will commence when it is known that an area is going to be designated even though the designation has not yet come into effect, and so it is accepted that the issue raised on this appeal applies equally to the proposal which was made by Addis Ltd on March 18 1981 before the zone came into effect.

The current valuation list came into force on April 1 1973 and it is not in dispute that irrespective of the outcome of this appeal the values of the hereditaments at the date of the proposals would be higher than their values on April 1 1973 before there was any question of there being an enterprise zone in the locality. Having regard to the scale of inflation between 1973 and 1981 this is also not surprising. It is, however, important because if this were not the position the valuation would be carried out solely under section 19 of the General Rate Act 1967 and it would not be necessary to consider section 20 of the Act, which is the section which has created the difficulties of interpretation on which the outcome of this appeal depends.

Section 19 provides, so far as relevant:

(1) Subject to the provisions of this part of this Act . . . the rateable value of a hereditament shall be taken to be the net annual value of that hereditament ascertained in accordance with subsections (2) to (4) of this section. (2) In the case of a hereditament consisting of one or more houses or other non-industrial buildings . . . the net annual value of the hereditament shall be ascertained by deducting from its gross value such amount, or an amount calculated in such manner, as may for the time being be specified by the Minister by order in relation to the class of such hereditaments to which the hereditament in question belongs. (3) The net annual value of any other hereditament shall be an amount equal to the rent at which it is estimated the hereditament might reasonably be expected to let from year to year if the tenant undertook to pay all the usual tenant's rates and taxes and to bear the cost of the repairs and insurance and the other expenses, if any, necessary to maintain the hereditament in a state to command that rent . . . (6) In this section, the following expressions have the following meanings respectively, that is to say — . . . "gross value", in relation to a hereditament, means the rent at which the hereditament might reasonably be expected to let from year to year if the tenant undertook to pay all usual tenant's rates and taxes and the landlord undertook to bear the cost of the repairs and insurance and the other expenses, if any, necessary to maintain the hereditament in a state to command that rent;

The general approach to valuations under section 19 is now well established. In *Dawkins* v *Ash Brothers & Heaton Ltd* [1969] 2 AC 366, at p 381, Lord Pearce said:

Rating seeks a standard by which every hereditament in this country can be measured in relation to every other hereditament.

Later Lord Pearson said (p 388) that the aim is

to produce a just and true result, attributing to the hereditament its actual . . . value — the real value of the beneficial occupation to the occupier.

Having regard to this general approach, it is accepted by Mr Fletcher, on behalf of the valuation officer, that on a valuation under section 19 it would be necessary to take into account the effect of the zone being designated and the individual benefits which I have set out which follow from the designation. It is also accepted that under section 19 it is the valuer's duty to take into account the present effect of an anticipated event so that under section 19 the proposed designation could be taken into account on the valuation following the proposal made by Addis Ltd.

However, under section 19 it is also necessary to carry out the valuation as at the date of the proposal, and at times when rents are increasing rapidly this can create substantial differences between the values originally inserted in the valuation lists and subsequent valuations resulting from subsequent proposals. This resulting lack of uniformity in valuations would be aggravated because it has been established for many years that the fact that other hereditaments have been assessed at a lower figure cannot be used to justify a reduction of an assessment which would otherwise be appropriate. See *Ladies Hosiery & Underwear Ltd* v *West Middlesex Assessment Committee* [1932] 2 KB 679.)

It will be readily appreciated that at times of rising rents this inability to correct later assessments by comparing them with assessments of similar hereditaments at lower values at an earlier date can cause unfairness to those assessed at a later date, since their burden in relation to rates would be higher than those assessed at an earlier date. If section 19 stood alone, this unfairness could be corrected only by frequent revaluations of all hereditaments, which in itself would be administratively undesirable. Similar problems could arise at times of falling values. Section 20 of the 1967 Act is designed to deal with this mischief. Its provisions are as follows:

(1) For the purposes of any alteration of a valuation list to be made under Part V of this Act in respect of a hereditament in pursuance of a proposal, the value or altered value to be ascribed to the hereditament under section 19 of this Act shall not exceed the value which would have been ascribed thereto in that list if the hereditament had been subsisting throughout the year before that in which the valuation list came into force, on the assumptions that at the time by reference to which that value would have been ascertained — (a) the hereditament was in the same state as at the time of valuation and any relevant factors (as defined by subsection (2) of this section) were those subsisting at the last-mentioned time; and (b) the locality in which the hereditament is situated was in the same state, so far as concerns the other premises situated in that locality and the occupation and use of those premises, the transport services and other facilities available in the locality, and other matters affecting the amenities of the locality, as at the time of valuation.
(2) In this section, the expression "relevant factors" means any of the following, so far as material to the valuation of a hereditament, namely — (a) the mode or category of occupation of the hereditament; (b) the quantity of minerals or other substances in or extracted from the hereditament; (c) in the case of a public house, the volume of trade or business carried on at the hereditament; and in paragraph (c) of this subsection the expression "public house" means a hereditament which consists of or comprises premises licensed for the sale of intoxicating liquor for consumption on the premises where the sale of such liquor is, or is apart from any other trade or business ancillary or incidental to it, the only trade or business carried on at the hereditament.
(3) References in this section to the time of valuation are references to the time by reference to which the valuation of a hereditament would have fallen to be ascertained if this section had not been enacted.
(4) This section does not apply to a hereditament which is occupied by a public utility undertaking and of which the value falls to be ascertained on the profits basis.

Section 20 has since been amended, but it is not necessary for the purposes of this judgment to refer to those amendments.

In a judgment which was subsequently described by Lord Templeman as "impeccable" when the case went to the House of Lords, Sir Patrick Browne in *K Shoe Shops Ltd* v *Hardy (Valuation Officer) and Westminster City Council* [1983] RA 26 at p 36* said:

The object of section 20 and its predecessors was clearly to remedy this unfairness by providing a ceiling which valuations on proposals made during the currency of the list were not to exceed.
Parliament dealt with this unfairness in stages. The Valuation for Rating Act 1953 applied only to dwelling-houses and private garage or private storage premises and applied only to "the making or altering of the first valuation list made after the passing of this Act", which turned out to be the 1956 revaluation. The effect of section 2 was to impose as a ceiling of value the rent at which the hereditament might reasonably have been expected to let on or about June 30 1939. The assumptions required by section 2(3) to be made correspond closely with those required by section 20(1) and (2) of the 1967 Act, but the 1953 Act, unlike the 1967 Act, specified a date by reference to which the valuation was to be made. Then came section 17 of the Local Government Act 1966, re-enacted by section 20 of the 1967 Act. As the Lands Tribunal said [1980] RA 333, 344: "Section 20 of the General Rate Act 1967 is not one of those statutory provisions that yield up their meaning at a glance". But we have no doubt that its general intention and effect were and are to protect ratepayers against the effect of inflation since the coming into force of the list current at the date of the proposals; at the date of the proposals in this case the relevant list was the 1973 list. That figure is a ceiling — the value could be reduced below it. Section 20 has no application in the making of a new valuation list. In *Ryde on Rating*, 13th ed, p 479, it is said: "It cannot be doubted that the intention of the legislature was to put an end to the rule in the *Ladies Hosiery* case."
We do not agree. The effect of the section is not that correctness must now be sacrificed to uniformity but that if the correct value of the hereditament as at the date of a proposal is higher than the value which would have been ascribed to it in the 1973 list, the extent to which effect can be given to the correct value is limited; the date of the proposal is no longer to be the governing factor. If the correct value is lower than the value ascribed in the list, effect can be given to it. In *Baker Britt & Co Ltd* v *Hampsher (VO)* [1976] RA 69, 85, Viscount Dilhorne at p 93 cited the *Ladies Hosiery* case with approval (see also Lord Morris of Borth-y-Gest). In our view section 20 has no application (except as to the date of valuation) to the ascertainment of the value which would have been ascribed to a hereditament in the 1973 list; for example, *Ladies Hosiery* still establishes that it would be no reason for reducing the assessment of shops in Regent Street to show that shops in Oxford Street or Bond Street were under-valued. The section deals merely with proposals made for the valuation of new or altered hereditaments in the context of the current valuation list, or the revaluation of hereditaments included in the current list. As mentioned below, it seems that it was primarily the former category which Parliament had in mind.
It is fair to say that increases in value are now limited by the "tone" of the 1973 list, in the sense of the level of values appearing in that list, and the side note is therefore a convenient label to the section.
Our impression from the wording of section 20 is that Parliament had primarily in mind hereditaments newly created after the passing of the Act, or so altered after that date as to become new hereditaments, but having regard to the words "*any* alteration" it must in our view apply to all hereditaments in respect of which a proposal is made after that time, as counsel for the valuation officer does not dispute.

Before this court this extract from Sir Patrick Browne's judgment has been accepted as correct. However, before the tribunal the valuation officer submitted that the agreed objective of section 20, having regard to the language of the section, had to be achieved by ignoring the existence of the zone and the benefits it offered. The

---

* Editor's note: Also reported at (1983) 266 EG 119 at p 123.

ratepayers argued to the contrary. However, the parties were able to agree what the consequences would be, depending upon which approach was correct, and these are set out in the decision [at (1984) 271 EG 292) as being as follows:

Appeal hereditament	The values properly to be ascribed in the valuation list on the assumption that			
	(a) the benefits available within the zone are not to be taken into account		(b) the benefits available within the zone are to be taken into account	
	Gross value £	Rateable value £	Gross value £	Rateable value £
LVC/122&125/1983 Addis Ltd		45,500		36,500
LVC/222/1982 Coteglade Ltd	790	630	675	534
LVC/223/1982 Coteglade Ltd and Welsh Bakers' Buying Group Ltd	3,325	2,742	2,825	2,326
LVC/224/1982 Ray James Ltd	5,300	4,388	4,500	3,722
LVC/225/1982 Ray James Ltd	8,950	7,430	7,600	6,305

In alternative (b), however, where there are no cross-appeals by the ratepayers, the values which can be determined on these appeals are limited in the last three cases by the determinations of the local valuation court as follows:

	Gross value £	Rateable value £
LVC/223/1982	3,100	2,555
LVC/224/1982	4,950	4,097
LVC/225/1982	8,150	6,763

As it is the effects of the zone upon hereditaments which are agreed to be in the same locality as the zone but outside the zone which it is submitted should be taken into account, it is section 20(1)(b) and not section 20(1)(a) with which this case is directly concerned.

Before the tribunal Mr Fletcher, on behalf of the valuation officer, submitted that the subsection is concerned only with the physical characteristics of the locality and furthermore is not concerned with the prospective changes which could occur in the locality in consequence of the creation of a development zone. The member of the tribunal Mr Emlyn Jones, in a decision of exemplary clarity, rejected this view in accepting the arguments of the ratepayers. He concluded his decision by saying:

... wherever the line is to be drawn between physical factors and other factors and between the *res* and other conditions which have an effect on value, I am satisfied that the enterprise zone benefits are properly to be embraced within the words of section 20(1)(b) when taken as whole, particularly when I have regard to the mischief which the statutory provisions were designed to remedy. In my judgment, the appeal hereditaments were adversely affected at the relevant dates by the designation or proposed designation of the enterprise zone within the locality and that adverse effect is a matter which should properly be taken into account in their valuation in accordance with the provisions of section 20 of the Act.

Before this court Mr Fletcher has modified the submission he made to the tribunal. He submits that section 20(1)(b) is concerned with physical factors, or at least factors which affect the physical use and enjoyment of the hereditament. He gives as examples of such factors new public sewers, the opening of a street market, no waiting restrictions on an adjacent highway or a change in the Heathrow flight path bringing aircraft directly overhead. He further contends that on their natural construction "other facilities" and "other matters affecting the amenities of the locality" do not include benefits or disbenefits that are merely financial or economic.

Mr Roots and Mr Horton, who appeared on behalf of the ratepayers, accept, having regard to the detailed provisions of section 20, that limits have to be placed on the considerations which can be taken into account on a valuation for the purposes of section 20. While seeking to uphold the approach of the tribunal to the designation of the zone and rejecting the distinction which Mr Fletcher still draws between physical and non-physical factors, understandably they prefer not to take on the difficult task of saying where the line is to be drawn between matters which can and cannot be taken into consideration.

I accept that it would be consistent with the language of section 20(1) to adopt either the approach of Mr Fletcher or the approach adopted by the tribunal. Furthermore, notwithstanding Mr Fletcher's arguments to the contrary, I can see nothing inherently difficult in applying section 20 in accordance with the interpretation adopted by the tribunal. However, when the language of section 20 is considered as a whole in the context of the mischief it was designed to cure, I cannot accept the approach of the tribunal. In general, and I emphasise the words "in general", I accept the approach of Mr Fletcher that section 20(1)(b) is limited to physical factors or factors which affect the physical enjoyment of a hereditament. In broad terms the way section 20(1) is intended to operate is that you value the hereditament and any building upon it as it exists at the date of proposal in the setting in which it is situated (with that setting having the actual characteristics of the locality as they would be observed at that date if the locality was to be inspected) on the basis of its 1973 value. For the purposes of carrying out that valuation, it is the economic climate, both local and national, of 1973 which has to be considered and not that at the date of the proposal except to the extent that alterations in the economic conditions result in changes in the locality which are capable of being observed "on the ground" in the locality.

Apart from the fact that the language of section 20(1) gives the strong impression that it is primarily concerned with physical matters, my reasons for adopting this approach to the section are as follows:

(1) As Sir Patrick Browne pointed out, section 20 does not replace section 19 but only provides a ceiling above which a valuation under section 19 cannot go. This provides at least a partial answer to cases such as this where it is argued that a restrictive interpretation results in unfairness. It must be remembered that in the majority of cases the restrictive interpretation will in fact be in the interests of a ratepayer making a proposal.

(2) I consider that there is considerable force in Mr Fletcher's submission that if his limitation is not placed upon the effect of section 20(1) it is difficult to identify any other limitation which can be placed upon the language which would avoid the consequence, which was clearly not intended by Parliament, that on a section 20 valuation all relevant factors have to be taken into account as they exist at the date of the proposal.

(3) While the word "state" can have different meanings in different contexts and can be a word of very wide application, in section 20(1)(a) it applies more naturally to the structural state of the hereditament and in section 20(1)(b) its application is restricted by the words "so far as concerns", with the result that regard can only be had to other premises, the occupation and use of those premises transport services, other facilities and other matters affecting the amenities of the locality.

(4) While the word "amenities" can be of wide ambit and it i capable of applying to the business climate of the locality, which would include its designation as a development zone, I regard "amenities" as being used in a sense where it applies to those aspect of the locality which are capable of affecting all the hereditaments in the locality and not merely a category of hereditaments such a commercial premises. I am cautious about adopting the same approach to the construction of the word "facilities". However, in relation to both amenities and facilities I do recognise that the effect of an area being designated as a development zone, as happens with smokeless zone, can result in changes in the facilities and the amenities of the locality which can be taken into account.

(5) It is interesting to note that section 2(5) of the 1953 Act, to which Sir Patrick Browne referred, limits the effect of the Rent Acts an scarcity of accommodation as factors to be taken into account o carrying out a valuation at the June 30 1939 date. This suggests tha the same language which now appears in section 20(1)(a) and (b) wa then regarded as only having effect at the 1939 date and not the dat of the proposal. This would not be the position if the reasonin adopted by the tribunal were to be applied to the similar provisions o section 2 of the 1953 Act.

On the approach to the interpretation of section 20 which I regar as being correct, the designation of a development zone is not matter which can be taken into account in arriving at the ceilin provided for by section 20. In coming to this conclusion I regard it a proper to look at the effects of the development zone as a whol rather than the individual benefits it confers. However, I would tak the same view if I considered in turn each of the individual benefi which exist in the zone, though I have reservations as to the special customs facilities provided to businesses within the zone. However, by itself I doubt very much whether that benefit could have any material effect upon a valuation.

I should, however, emphasise that I do not accept Mr Fletcher's submission that because a consideration is of a financial nature it cannot be considered as it exists at the date of the proposal because it is incapable of being converted into 1973 values. I would therefore regard it as perfectly appropriate in considering the quality of transport services to take into account the level of fares charged for the services as this could materially affect an assessment of the quality of the service. Likewise, if the existence of a development zone affects the prosperity of an area in a manner which is manifest and can be observed, this should be taken into account. The features which demonstrate a change in prosperity in this way could be properly taken into account as part of the setting in which the valuation at 1973 values is to be made.

There remain two further matters to which I should refer. The first is as to another decision of a tribunal in relation to which it is intended that there should be an appeal which could be affected by the result of this appeal. The second is as to the way the question to be answered has been framed in the case stated for this appeal.

The other decision was that of the tribunal in *Sheerness Steel Co plc v Maudling (Valuation Officer)* [1986] RA 45. That decision was given on February 17 1986 and followed the decision of the tribunal under appeal in this case. The subsection under direct consideration was 20(1)(a) and not 20(1)(b). However, the issue involved is clearly similar to that involved on this appeal. This court therefore considered whether it would be right for our decision to be given before hearing the proposed appeal in relation to the decision in the *Sheerness Steel Co* case. We have decided to do so, in the absence of any application that our decision should be adjourned, because although we felt it would have been desirable to have heard the argument in the other appeal first, it is due to the default of the parties to the later decision that the form of the case has not yet even been settled, so that it is not possible to hear that appeal in the relatively near future.

Mr Mallett, the member who gave the decision in the *Sheerness Steel Co* case, in earlier cases acceded to arguments advanced on behalf of valuation officers in terms similar to the argument which was advanced by Mr Fletcher before the tribunal in this case. However, he rejected those submissions in the *Sheerness Steel Co* case and came to the conclusion that the word "state" in subsection (1)(a) was not limited to the physical state and that consideration of the legal state of the hereditament at the date of the proposal was therefore not excluded. The result was the ratepayer was entitled to have taken into consideration in respect of a steel works a production and delivery quota system which was introduced at the end of 1980 by the European Steel Community which reduced production to a specified percentage of the works' maximum possible output.

This interpretation of subsection 20(1)(a) is in conflict with the views which I have expressed above. However, it is conceivable that the directives which limited the production could be regarded as a "relevant factor" which affected the mode of occupation of the hereditament. As to this we have not heard full argument and I therefore do not express my final view, since I appreciate it could have wide application. For example, it could affect the question of whether a refusal of planning permission which would have extended the lawful use of a hereditament could be taken into account.

The question set out in the case stated on this appeal is in the following terms:

The question upon which the decision of the Honourable Court is desired is whether, in considering what was the "state" of the locality within the meaning of section 20(1)(b) of the General Rate Act 1967 I was correct in law in taking into account the existence, actual or prospective, of premises within the area designated by the Lower Swansea Valley Enterprise Zone Designation Order 1981, and the benefits conferred or expected to be conferred on the owners and occupiers of those premises.

The parties accept that this question is not happily worded and the parties should have sought to have it amended. It was never contended on behalf of the valuation officer that premises already existing within the area designated could not be taken into account. Clearly such a contention would be unsustainable. Furthermore, the question does not, as I feel it should, raise the effect not of the benefits conferred by the zone but of the designation of the zone itself. However, in this judgment I hope I have sufficiently set out my views as to the issue which in fact separates the parties. The position is that the designation itself cannot be taken into account, but the consequences of the designation, when they exist, are capable of being taken into account if they fall within my interpretation as to the effect of section 20 of the Act.

It only remains for me to acknowledge the particularly clear and helpful submissions of counsel on this appeal and to indicate that I would like to hear further argument as to the appropriate order which should be made having regard to the decision to which this court has come.

Agreeing, SIR GEORGE WALLER said: I will briefly state my reasons. The decision depends on the meaning of the word "state". How should the word be construed? Mr Fletcher submitted that it should be limited to physical matters. Mr Roots and Mr Horton supported the decision of the Lands Tribunal, namely that "state" went beyond physical bounds and should include the existence of an enterprise zone with all the tax and other financial advantages for those in business within it. If this were a new application for an assessment of a new hereditament the case would be considered under section 19 of the General Rate Act 1967 and the existence of an enterprise zone would be a matter to be taken into consideration. It would have been a simple matter to apply the same principles to an alteration of a valuation list. But Parliament has not done that but instead has specified the matters which have to be taken into account.

In this case no problem arises on section 20(1)(a); the hereditament had not changed and the mode of occupation was the same as in 1973. It is section 20(1)(b) on which the respondents rely. The member said:

... if the Lower Swansea Valley Enterprise Zone had existed [at the time when the valuation list came into force] or there had been the immediate prospect of its designation, then the effect which such a zone would have on the value of properties immediately outside the zone would have been a factor properly to be taken into account constituting one of the actual conditions affecting the hereditament.

The respondents submit that in considering the state of the locality it is not only the physical state which has to be considered, but also that one of the factors to be considered is the existence of the enterprise zone even though no physical change has taken place.

In *K Shoe Shops Ltd v Hardy* [1983] RA 26 at p 36 Sir Patrick Browne, in a judgment approved by the House of Lords, said, referring to section 20:

But we have no doubt that its general intention and effect were and are to protect ratepayers against the effect of inflation since the coming into force of the list current at the date of the proposals, ...

The section, therefore, is to protect against the effects of inflation, and in this phrase I would include rising rents and not just the retail price index.

In interpreting the section it is necessary to consider the matters which Parliament provided should be taken into consideration. Occupation would be part of the physical state and use would be associated with the physical state. Similarly, transport services and other facilities in the locality, although not a physical part of the locality, would be associated physically with the locality. This would also apply to "other matters affecting the amenities of the locality". The *Oxford English Dictionary* gives a number of meanings of "state", most of which would be difficult to apply to a locality. Those which could apply either specify physical matters or describe matters which are mainly physical. The nearest to a non-physical enterprise zone would be a "combination of circumstances or attributes belonging to a person or thing".

I have come to the conclusion that the Lands Tribunal was in error in taking the mere existence of an enterprise zone into consideration. The enterprise zone in due course may create physical changes which would be relevant to the state of the locality. If so it would be open to the respondents to seek to have their assessments reduced. At the time when this valuation was being made, there were no physical changes whatever, nor were there any changes with a physical connotation. Accordingly I would allow this appeal.

Also agreeing, O'CONNOR LJ said: The appeal should be allowed for the reasons given in the judgments of Woolf LJ and Sir George Waller.

The appeal was allowed, the valuation officer to have the costs of the appeal before the Lands Tribunal and of this appeal. Leave to appeal to the House of Lords was refused.

Court of Appeal
November 5 1986
(Before Lord Justice LAWTON, Lord Justice LLOYD and Lord Justice BALCOMBE)

HEMENS (VALUATION OFFICER) v WHITSBURY FARM & STUD LTD

Estates Gazette March 7 and 14 1987

281 EG 1065-1072 and 1202-1206

Rating — General Rate Act 1967, section 26(3) and (4) — Rating Act 1971, sections 1 and 2 — Exemptions from rating — Premises used as studs for breeding thoroughbred racing stock — Stud premises, including covering yards and paddocks, within or attached to agricultural land — Broad question, complicated by statutory language, was whether for rating purposes the breeding of such thoroughbred horses on premises with agricultural land attached or adjoining should be equated with the breeding of cattle and sheep on agricultural land — Review of English and Scottish authorities — Difference of past treatment in England and Scotland — Decision in *Lord Glanely* v *Wightman* misunderstood, but now distinguished — Issue in the present case, which arose on appeal from the Lands Tribunal, was whether the appellants' stud buildings were agricultural buildings within the definition contained in section 26(4) of the General Rate Act 1967 as extended by section 1(1)(a) of the Rating Act 1971 — This issue turned on two questions of construction: first, whether the buildings, which were admittedly occupied together with agricultural land, were being used "solely in connection with agricultural operations" on that land; and, second, whether, if not, horses bred for riding, hunting or racing were "livestock" so as to bring the case within section 2(1)(a) of the 1971 Act — Held, dismissing the appeal, that the grazing of the thoroughbred stock in question was not an agricultural operation within section 26(4) of the 1967 Act, so that the question whether the buildings were being used, or used solely, in connection with agricultural operations did not strictly arise; and that the thoroughbred stock were not "livestock" within section 2(1)(a) of the 1971 Act — Lloyd LJ, in contrast with Lawton and Balcombe LJJ, would have held that the buildings were used solely in connection with the operations if he had considered the latter agricultural — Appeal dismissed, but leave to appeal to House of Lords granted

The following cases are referred to in this report.

Belmont Farm Ltd v *Minister of Housing and Local Government* (1962) 13 P&CR 417; 60 LGR 319, DC
Cresswell (VO) v *BOC Ltd* [1980] 1 WLR 1556; [1980] 3 All ER 443; [1980] RA 213; [1980] EGD 831; (1980) 255 EG 1101, CA
Crowe (VO) v *Lloyds British Testing Co Ltd* [1960] 1 QB 592; [1960] 2 WLR 227; [1960] 1 All ER 411; (1960) 58 LGR 101; 53 R&IT 54; 5 RRC 371, CA
Derby (Lord) v *Newmarket Assessment Committee* (1930) 13 R&IT 50
Eastwood (W & J B) Ltd v *Herrod* [1971] AC 160; [1970] 2 WLR 775; [1970] 1 All ER 774, HL
Evans v *Bailey (VO)* [1981] EGD 730; (1981) 260 EG 611
Forth Stud Ltd v *East Lothian Assessor* [1969] RA 35
Gilmore v *Baker-Carr* [1962] 1 WLR 1165; [1962] 3 All ER 230; [1962] RVR 486; (1962) 9 RRC 240; [1962] RA 379, CA
Glanely (Lord) v *Wightman* [1933] AC 618
Hardie v *Assessor for West Lothian* 1940 SC 329
Inland Revenue v *Ardross Estate Co* 1930 SC 487
Kidson v *Macdonald* [1974] Ch 339; [1974] 2 WLR 566; [1974] 1 All ER 849; 49 TC 503
McClinton v *McFall* [1974] EGD 16; (1974) 232 EG 707, CA
Malcolm v *Lockhart* [1919] AC 463, HL
Minister of Agriculture, Fisheries and Food v *Appleton* [1970] 1 QB 221; [1969] 3 WLR 755; [1969] 3 All ER 105, DC
Normanton (Earl of) v *Giles* [1980] 1 WLR 28; [1980] 1 All ER 106, HL
Peterborough Royal Foxhound Show Society v *Inland Revenue Commissioners* [1936] 2 KB 497
Sargaison v *Roberts* [1969] 1 WLR 951; [1969] 3 All ER 1072

This was an appeal by Whitsbury Farm & Stud Ltd from a decision of the Lands Tribunal rejecting the appellants' contention that the stud hereditaments occupied by them at Whitsbury, Hampshire, should be exempt from rating. The decision of the Lands Tribunal (V G Wellings QC) is reported at [1985] 1 EGLR 227 and (1984) 274 EG 403.

W J Glover QC and Alun Alesbury (instructed by Ward Bowie, agents for Rustons & Lloyd, of Newmarket) appeared on behalf of the appellants; Alan Fletcher QC and Nicholas Huskinson (instructed by the Solicitor of Inland Revenue) represented the respondent valuation officer.

Giving judgment, LAWTON LJ said: This is an appeal by Whitsbury Farm & Stud Ltd, who are the owners of a stud for the breeding of thoroughbred racing stock, against a refusal by the Lands Tribunal to declare divers buildings on their land exempt from rating. The appeal raises issues which have been discussed and litigated since at least 1930 and which are of interest to all who are engaged in breeding thoroughbred racing stock.

The facts

The facts set out in the case stated by the Lands Tribunal can be stated shortly. The appellants occupy at Whitsbury in Hampshire four separate stud hereditaments. Each stud lies within, or is attached to, land which is agricultural land within the meaning of section 26(3) of the General Rate Act 1967. At all material times the stud buildings were in excellent condition and were essential for accommodating and breeding thoroughbred racing stock. The covering of mares by the stallions was usually accomplished in a covering yard; but sometimes it took place outside in the adjoining paddocks. The stallions were owned by a syndicate of 40 shareholders. The appellants were shareholders in this syndicate. The syndicate made contracts with those owners who brought mares to the stallions to be covered. The income from the service fees went to such of the shareholders as sold the breeding rights. The appellants were paid by the syndicate the cost of keeping the stallions. The mares which were covered were either the appellants' own property or visiting mares. Visiting mares were kept at the stud until such time, usually 60 days, as it took to discover whether they were in foal. Some of the appellants' mares were sent to other studs to be covered. Because of the Jockey Club's rules for deciding the age of racehorses, covering took place between February 15 and July 15 each year. Nearly all the mares produced their foals at night in the foaling boxes, but a few did so in the paddocks in daylight. All the mares and stallions had access to paddocks and, save in frosty weather, the mares spent most of each day in them. Paddocks were essential for the running of the appellants' stud. They afforded space for exercise; and for mares and foals the grass in them provided during the growing season, that is from March until high summer, nourishment of a kind which mares required for providing milk for their foals and bringing them into season and foals for growth. The paddocks were well looked after because producing good-quality grass was important for breeding. From time to time, sheep and cattle were put into the paddocks, the object of doing so being to keep the grass down and to stop seeding. The appellants' stud was run in the way in which studs are normally run. Lord Wright's description in *Lord Glanely* v *Wightman* [1933] AC 618 at pp 634 to 635 of how Lord Glanely's stud was run is much the same as that in the case stated in this case. No more detail is necessary for providing the factual background to this appeal.

The issues

Broadly stated, this court has to ask itself this question: for rating purposes, should breeding thoroughbred racing stock on premises with agricultural land attached or adjoining be equated with the breeding of cattle and sheep on agricultural land? If it should be, those who run studs should have the benefit of the same exemptions from rating as the occupiers of agricultural land enjoy. This broad question, however, has been complicated by the statutory language in which Parliament has given the occupiers of agricultural land their exemptions.

The legislative history

Before 1896 farmers were rated just like other occupiers of land. The Agriculture Rates Act of that year gave for a period of five years a 50% exemption from rates to the occupiers of "agricultural land", which was defined in section 9 as follows:

The expression "agricultural land" means any land used as arable, meadow, or pasture ground only, cottage gardens exceeding one quarter of an acre, market gardens, nursery grounds, orchards or allotments, but does not include land occupied together with a house as a park, gardens, other than as aforesaid, pleasure-grounds, or any land kept or preserved mainly or exclusively for purposes of sport or recreation, or land used as a racecourse.

Thereafter other Acts extended the period and increased the percentage of the exemption. The Local Government Act 1929 derated agricultural land fully. The General Rate Act 1967 consolidated various enactments relating to rating and valuation and by section 26 continued the policy of the earlier statutes of derating "agricultural land" and "agricultural buildings". Agricultural land was defined in the 1967 Act in substantially the same terms as in the 1896 Act: see section 26(3). Such differences as there are have no relevance to this appeal. The relevant part of the definition of "agricultural buildings" in section 26(4) was as follows:

In this section, the expression "agricultural buildings" — (a) means buildings (other than dwellings) occupied together with agricultural land or being or forming part of a market garden, and in either case used solely in connection with agricultural operations thereon; ...

Farmers in Scotland were given the same kind of exemption from rating as those in England and Wales; but, as the rating system there was different, as was the legal terminology for describing interests in land, different wording had to be used. The first Scottish provision granting partial relief from rates was contained in the Agricultural Rates, Congested Districts, and Burgh Land Tax Relief (Scotland) Act 1896. A relevant modern provision was in section 7 of the Valuation and Rating (Scotland) Act 1956. Subsection (2) was as follows:

"agricultural lands and heritages" means any lands and heritages used for agricultural or pastoral purposes only or as woodlands, market gardens, orchards, allotments or allotment gardens and any lands exceeding one quarter of an acre used for the purpose of poultry farming, but does not include any buildings thereon other than agricultural buildings, or any garden, yard, garage, outhouse or pertinent belonging to and occupied along with a dwelling-house, or any land kept or preserved mainly or exclusively for sporting purposes;
"agricultural buildings" means buildings (other than dwelling-houses) occupied together with agricultural lands and heritages, or being or forming part of a market garden, and in either case used solely in connection with agricultural operations thereon;

Since 1945, regulatory statutes have been passed which affect agricultural land, which in the interests of precision and clarity has been defined. One such Act was the Town and Country Planning Act 1947. By section 12(2) it provided as follows:
..
Provided that the following operations or uses of land shall not be deemed ... to involve development of the land ...
(e) the use of any land for the purposes of agriculture ...

Agriculture was defined in section 119(1):

In this Act ... "agriculture" includes ... dairy farming, the breeding and keeping of livestock (including any creature kept for the production of food, wool, skins or fur, or for the purpose of its use in the farming of land) ...

The 1947 Act dealt with a subject-matter different from rating; but it was concerned, just as the General Rate Act 1967 is, with the consequences which follow from carrying on specified activities on land. It follows, in my judgment, that decisions upon the construction of section 7(2) of the 1956 Act and section 119(1) of the 1947 Act provide some, but not conclusive, help in the construction of section 26(3) and (4) of the 1967 Act. I conclude the legislative history relating to the derating of agricultural land and agricultural buildings with the Rating Act 1971, which was, according to its long title:

An Act to extend the provisions relating to the exemption from rating of land and buildings used in connection with agriculture.

This Act was passed following the decision in *W & J B Eastwood Ltd* v *Herrod* [1971] AC 160. The House of Lords had adjudged that broiler houses on a farm were not exempt from rating because they were not "agricultural buildings" as defined by section 2(2) of the Rating and Valuation (Apportionment) Act 1928, which was the statute in force when the dispute arose but which was superseded by section 26(4) of the 1967 Act. The relevant part of the 1971 Act is in section 2 and is as follows:

Subject to subsections (2) to (4) of this section, each of the following is an agricultural building by virtue of this section —
(a) any building used for the keeping or breeding of livestock; ...

(2) A building used as mentioned in subsection (1)(a) of this section is not an agricultural building by virtue of this section unless either
(a) it is solely so used; or
(b) it is occupied together with agricultural land (as defined in the principal section) and used also in connection with agricultural operations on that land, and that other use together with the use mentioned in subsection (1)(a) of this section is its sole use.

The English cases

Shortly after the Local Government Act 1929 derated agricultural land and agricultural buildings, seven occupiers of studs around Newmarket appealed to West Suffolk Quarter Sessions against the refusal of the Newmarket Area Rating Assessment Committee to regard them as exempt from rating on the ground that they were not occupying agricultural buildings within the meaning of section 2(2) of the 1928 Act. The appeals failed, the court finding that "in the circumstances of this case the breeding of livestock is not an agricultural operation": see *Derby (Lord)* v *Newmarket Assessment Committee* (1930) 13 R & IT 50. We were told by counsel that quarter sessions in other parts of England followed this decision.

Then came *Lord Glanely's* case (supra). It was not concerned with rating. The issue was whether the fees which Lord Glanely had derived in respect of a stallion he kept at stud should be assessed to income tax under Schedule D of the Income Tax Acts, notwithstanding the fact that he had already been assessed under Schedule B in respect of the occupation of land as occupied in part for husbandry purposes and in part for stud and racing purposes. The House of Lords adjudged that he should not be further assessed under Schedule D because the fees were profits in respect of the occupation of land and were chargeable to income tax under Schedule B. The decision turned on the construction of the Income Tax Acts and the effect of a number of cases on the issue in dispute (see per Lord Wright at pp 636 to 638). In the course of his speech Lord Wright said at p 640.

On these grounds I think that the service of the stallion is appurtenant to the soil and a profit of the occupation in every case, so that in this regard it is immaterial whether the service is to the appellant's own mares or whether it is sold to strangers; in the latter case the service is sold from the land and as a product of the land, just as much as bullocks, potatoes, fruit or eggs are sold from the land. Without the appellant's stud farm or some other such stud farm the stallions could not live or exercise their generating functions. The value of these functions is inseparably connected with the occupation of land.

Rating authorities in England concluded that *Lord Glanely's* case entitled the occupiers of stud farms of a similar kind to his to claim exemption from rating. The reasoning behind granting exemption must have been that, since the generative powers of stallions were, in Lord Wright's words, "a product of the land", any land on which they grazed was agricultural land and any building in which they were stabled was an agricultural building. This reasoning would have made an enclosure in a zoo which accommodated any grazing animals agricultural land. From about 1933 until 1980 the occupiers of studs in England were not rated. Occupiers in Scotland were not so fortunate.

The Scottish cases

In *Forth Stud Ltd* v *East Lothian Assessor* [1969] RA 35, the Lands Valuation Appeals Court considered a number of cases which had been decided under section 9(11) of the Rating and Valuation (Apportionment) Act 1928, the wording of which had been similar to section 7(2) of the Valuation and Rating (Scotland) Act 1956. The court also considered the relevant English cases, including *Lord Glanely's* case. The Forth Stud was run in much the same way as, but on a smaller scale than, the appellants' stud. The adjoining land was 25 acres in extent and was used for grazing, exercising and taking a crop of hay. In an earlier Scottish case, *Inland Revenue* v *Ardross Estate Co* 1930 SC 487, which was concerned with whether a silver fox farm was exempted from rating as agricultural land, Lord Sands had said at p 490:

The use extends only to the breeding of such animals as are associated with an ordinary farm — horses, cattle, sheep, goats, pigs and poultry.

In yet another Scottish case, *Hardie* v *Assessor for West Lothian* 1940 SC 329, Lord Robertson had said:

As I understand Lord Hunter's opinion, he expressed the view that the breeding of livestock might be an agricultural purpose although there was no pasturage of stock on the land ... In any event I am of opinion that the rearing of stock is, or may be, an agricultural purpose, inferring an agricultural use of the land within the meaning of the Act of 1928, without regard to the question whether the stock is reared to a material extent on the crops raised on the land. But if the rearing of stock is to be regarded in itself as an agricultural purpose the stock reared must be such as produces or directly contributes to produce "the means of human subsistence". I think that this is the ratio of the judgment in *Inland Revenue* v *Ardross Estates Co*.

Lord Fraser, in the *Forth Stud* case, adjudged that racehorses are not animals associated with an ordinary farm and they did not produce or directly contribute to produce the means of human subsistence. Lord Avonside referred to what Lord Robertson had said in the *Hardie* case and went on as follows at p 46:

Further, rearing of stock as an agricultural or pastoral operation must, on any proper view of the words, be rearing of stock which produces or contributes to produce the means of human subsistence. Animals reared for sport or entertainment or for their decorative qualities cannot, in my view, be regarded as being reared in the course of the activity of farming, which in its broad sense covers tillage and pasturing. The whole purpose of farming was and is, from the earliest times until now, to produce the means of human subsistence and there can be no doubt that that is why agricultural land, under legislation, enjoys the substantial financial privilege of being omitted from the valuation roll.

The court adjudged that the Forth Stud was not entitled to exemption from rating as agricultural land and heritages.

I find this case a most persuasive authority as to the overall legislative intention of section 26(3) and (4) of the General Rate Act 1967. The question, however, remains whether Parliament, by the words used in that Act and in the 1971 Act, intended either to qualify or to extend the overall intention. In *Evans* v *Bailey (Valuation Officer)* (1981) 260 EG 611, which was concerned with a stud for breeding hunters, the Lands Tribunal treated the *Forth Stud* case as a persuasive authority. That stud was not given exemption from rating.

Construction of the 1967 and 1971 Acts

Since the intention of the 1971 Act was, *inter alia*, to extend the meaning of the words "agricultural building", it is convenient to consider whether the buildings at the stud come within the extended definition. The appellants, on the facts of this case, had to prove, first, that the buildings were used for the keeping or breeding of livestock, as defined in section 1(3), and, second, that that use came within either para (a) or (b) of subsection (2).

As I have already said, I have no doubt that the overall intention of Parliament in providing for the derating of agricultural land and agricultural buildings was to help farmers. But, since rating is concerned with the occupation of land, Parliament had to define what kind of occupied land or buildings should be exempt from rating. This it did in section 26(3) and (4) of the 1967 Act and sections 1 to 4 of the 1971 Act. The definitions are more extensive than what in the ordinary usage of English would be meant by "farmland" or "farmbuildings"; but, with the exception of the inclusion of use for "a plantation or a wood or for the growth of saleable underwood", all the other uses relate to the production of products directly connected with human subsistence. The words of exclusion at the end of subsection (3) strengthen this construction. Silviculture as a source of fuel and building materials contributes indirectly, if not directly, to human subsistence. In my judgment, the definition of "agricultural land" in section 26(3) was intended to cover land used for purposes contributing to human subsistence. Since in both section 26(4) of the 1967 Act and section 2 of the 1971 Act the adjective "agricultural" qualifies "buildings" as well as land, it follows that a building to be exempt from rating must be one which is used for a purpose contributing to human subsistence and that "livestock" in section 2(1)(a) means mammals or birds which also do so. I cannot accept Mr Glover QC's argument on behalf of the appellants that, because the statutory definition in section 1(3) of the 1971 Act starts with the words "In this part of this Act 'livestock' includes" . . ., mammals such as thoroughbred racing stock which contribute nothing to human subsistence should come within the ambit of an exempting provision which is concerned with land used for purposes which do so contribute. If, as Mr Fletcher QC pointed out, the word "livestock" in the 1971 Act included all mammals, a building in which zebras or giraffes were kept would be an agricultural building. In my judgment, the appellants' buildings at Whitsbury are not used for the keeping or breeding of livestock within the meaning of section 2(1)(a) of the 1971 Act. In *Belmont Farm Ltd* v *Minister of Housing and Local Government* (1962) 13 P&CR 417, which was a planning case, Lord Parker CJ construed a provision of the Town and Country Planning Act 1947, section 119(1), defining "livestock" in an agricultural context, in the same sense as I have done.

The next question is whether the pasturing of racing stock in paddocks is an agricultural operation within section 26(4)(a) of the 1967 Act. As I understood Mr Glover QC's submission, it was as follows. The respondents admitted that the paddocks were agricultural land. This was because horses were grazed in them. Putting horses out to graze so as to increase their breeding potential was an agricultural operation and the buildings were used in connection with it. All this brought the buildings within the ambit of section 26(4) of the 1967 Act. In my judgment, it did not. I accept Mr Glover QC's submission that section 26(3) is concerned with the use of land, not the purpose for which it is used. The fact that the land is used in connection with the breeding of racing stock does not, as Mr Fletcher QC accepted, disqualify it from being "agricultural land". But, in order to qualify for exemption from rating, it is not enough that the buildings should be occupied together with agricultural land. They must be used solely in connection with agricultural operations thereon. The use of the word "operation" is significant. It connotes action for a purpose. In this case, for what purpose? The answer is — the breeding of racing stock. This, in my judgment, in the context of the 1967 Act, is not an agricultural purpose and therefore grazing in connection with breeding is not an agricultural operation. This finding is fatal to the appellants' case.

I should, however, deal shortly with the other two issues in this appeal, namely if the grazing, as part of the breeding activities, was an agricultural operation on the land, were the buildings used in connection therewith? If yes, were they used solely in connection therewith? In what was an otherwise admirably clear and concisely stated case, the learned member did not ask himself the right question when considering whether the buildings were used in connection with the grazing for breeding purposes. He asked himself whether the grazing and the buildings, being, as he found, ancillary to each other, were ancillary to the entire stud operation. I am not satisfied that, in the ordinary use of English, it can be said that the buildings were used in connection with the grazing. The grazing was used in connection with the buildings. Nor am I satisfied that the buildings were used *solely* in connection with the grazing. The buildings were used for a number of purposes connected with breeding. Grazing was but one of them.

I would dismiss the appeal.

Agreeing, BALCOMBE LJ said: It has at all times been conceded by the respondent valuation officer — and rightly conceded — that the paddocks which constitute the land (as opposed to the buildings forming part of the appellants' premises at Whitsbury constitute "agricultural land" within the meaning of section 26(3)(a) of the General Rate Act 1967, as they are used as pasture ground only, for the grazing of thoroughbred horses. The two questions which arise on this appeal are:

(1) Are the stud buildings at Whitsbury used solely in connection with agricultural operations on the paddocks?
(2) Are the buildings used for the keeping or breeding of livestock

If either of these questions is answered in the affirmative, the appellants succeed on this appeal.

I consider these questions in the order in which I have set them out above. In order to answer the first question it is necessary to answer the following subsidiary questions:
(i) What are the operations which are being carried out on the paddocks?
(ii) Are those operations agricultural?
(iii) Are the stud buildings used "in connection with" the operations on the paddocks? If so
(iv) Are the buildings used "solely" in connection with those operations?

Nature of the operations

Mr Glover QC for the appellants submitted that the operations are either: (a) the grazing of thoroughbred horses in the course of breeding and rearing; or (b) the breeding of thoroughbred horses. In my judgment, his first submission is correct. It has been held by the House of Lords that, in considering what are the operations in question, one must look at what is being actually done on the land alone and not at the combined purpose for which the land and buildings are being used. See *W & J B Eastwood Ltd* v *Herrod* [1971] AC 160, 173G-174A; 178F-G; 180H-181B. So I turn to consider the next question on the basis that the operations in question are the grazing of thoroughbreds for the purpose of breeding and rearing. (I prefer to use the word "thoroughbreds" rather than "racehorses", since, although the object of the breeding is to produce horses suitable for racing, that object is not achieved in every case.)

Are the operations agricultural?

Although section 26 of the 1967 Act contains definitions of "agricultural land" and "agricultural buildings", it contains no definition of "agricultural operations". The phrase "agricultural operations" occurs in both subsections (2) and (4) of section 26. Subsection (2) deals with the gross value for rating of a house:

occupied in connection with agricultural land and used as the dwelling of a person who
(a) is primarily engaged in carrying on or directing agricultural operations on that land; or
(b) is employed in agricultural operations on that land in the service of the occupier thereof . . .

Subsection (4) defines "agricultural buildings" as meaning:

buildings . . . occupied together with agricultural land . . . and . . . used solely in connection with agricultural operations thereon.

It is plain that not all operations carried out on agricultural land are themselves agricultural: indeed, if this were not so, it would have been sufficient simply to refer to "operations on agricultural land". Thus, pasture land, which is not preserved mainly or exclusively for purposes of sport, may from time to time be used for shooting. Shooting as such is not an agricultural operation and so a keeper's cottage would not come within subsection (2), nor would any hut used for the purposes of the shooting be within subsection (4). *Cf Earl of Normanton* v *Giles* [1980] 1 WLR 28. Nevertheless, the fact that not all operations carried out on agricultural land are themselves agricultural does not of itself meet Mr Glover's submission that, if the operations in question — in this case the grazing of horses — are such as to make the land on which they are carried on agricultural land (as is here conceded), then they must of necessity be agricultural operations. There is an obvious attraction in this submission; however, I am not convinced that the reasoning behind it is sound.

In the first place, subsections (2) and (4) of section 26 must be read together and in conjunction with the definition of "agricultural land" in subsection (3). When this is done, it is apparent that, in considering whether *land* is agricultural, the court must look only at the actual use of the land and not to the purpose of the occupier. On the other hand, in considering whether the *operations* on that land are agricultural, there is no similar limitation, and the references (in subsection (2)) to a person engaged or employed "in carrying on agricultural operations" and (in subsection (4)) to a building used "solely in connection with agricultural operations" — which, in its turn, contemplates the use of a building *partly* in connection with agricultural operations — suggest to me that in this context the court should look at the facts from the point of view of the occupier of the land and is not limited merely to a consideration of the use to which the land is put.

The next submission of Mr Fletcher QC for the respondent was that the most extensive category of land included in the definition of agricultural land is land "used as arable, meadow or pasture ground *only*". If operations involving the use of land for other purposes are carried out, the land ceases to be agricultural and the question whether these operations are agricultural becomes irrelevant. Accordingly, the draftsman must have envisaged that operations which were not agricultural might be carried out on land used as arable, meadow or pasture ground only. I can see some force in this submission, but its strength is weakened when one considers that other parts of the definition of agricultural land — eg cottage gardens exceeding one quarter of an acre, market gardens, nursery grounds, orchards or allotments — do not refer to the use of the land only for the specific purposes mentioned, and the draftsman was having to provide for operations carried out on all types of agricultural land.

In the end, what persuades me that the meaning of "agricultural" in relation to operations is not confined to the use to which the land is put is a consideration of the consequences of the more limited meaning in the light of the purpose which the section was clearly intended to achieve. The section wholly exempts agricultural land and agricultural buildings from liability to be rated and follows on similar provisions in earlier statutes dating back to 1896. The *Shorter Oxford English Dictionary* definition of agriculture is:

The science and art of cultivating the soil; including the gathering in of the crops and the rearing of live stock; farming (in the widest sense).

Although the definition of agricultural land in the section includes land used for forestry and horticulture, it expressly excludes park land, pleasure grounds, land kept or preserved mainly or exclusively for purposes of sport or recreation, and land used as a racecourse. I am left with the distinct impression that the object of the section is to benefit farming in the widest sense.

If that be right, then in the case of land used as pasture only it is necessary to consider the kind of animals which are pastured upon the land since, if they are not farm animals, their grazing would not be an agricultural operation carried on on the land. In his submissions for the respondent, Mr Fletcher gave as an example the use of land by the occupier of a zoo for grazing zebras or bison: an unlikely, but by no means impossible, use of land. In my judgment, the grazing of such animals could not properly be described as an agricultural operation.

So in the present case the answer to the question turns upon the nature of the animals which graze the paddocks. Of course, some kinds of horse may properly be described as farm animals, in which case their grazing would be an agricultural operation but, in my judgment, a person who carries on the business of breeding thoroughbreds is not to be described as a farmer, and the operation of grazing thoroughbreds for the purpose of breeding and rearing is not properly described as agricultural.

Thus far I have considered the question as if it were free from authority. The majority of the cases support the conclusion at which I have arrived. In *Gilmore* v *Baker-Carr* [1962] 1 WLR 1165, a case which turned on the same definitions of "agricultural land" and "agricultural buildings" in the Rating and Valuation (Apportionment) Act 1928, Lord Denning MR said (at p 1172):

The phrase "agricultural operations" is not defined but I should have thought that it meant operations by way of cultivating the soil or rearing of livestock.

In *Forth Stud Ltd* v *East Lothian Assessor* [1969] RA 35, the Lands Valuation Appeal Court in Scotland was faced with the same question as we are: whether a commercial stud for the breeding of thoroughbred horses was exempt from rating. This turned upon the definition of "agricultural lands and heritages" in section 7(2) of Valuation and Rating (Scotland) Act 1956 as "lands and heritages used for agricultural or pastoral purposes only". The court held that the stud was not exempt from rating, because land "used for agricultural or pastoral purposes only" meant, so far as the breeding of animals was concerned, land used for animals which were associated with an ordinary farm. Although the phrase which the court there had to construe was not the same as that which we have to consider, I cannot identify any material difference. Further, I find myself in complete agreement with the following passage from the judgment of Lord Avonside at p 46:

Further, rearing of stock as an agricultural or pastoral operation must, on any proper view of the words, be rearing of stock which produces or contributes to produce the means of human subsistence. Animals reared for sport or entertainment or for their decorative qualities cannot, in my view, be regarded as being reared in the course of the activity of farming, which in its broad sense covers tillage and pasturing. The whole purpose of farming was and is, from the earliest times until now, to produce human subsistence and there can be no doubt that that is why agricultural land, under legislation, enjoys the substantial financial privilege of being omitted from the valuation roll.

While this case is not strictly binding upon us, it is of considerable persuasive authority and, in considering the effect of statutes in similar terms which apply in different parts of the United Kingdom, I adopt the approach of Foster J in *Kidson* v *Macdonald* [1974] Ch 339 at p 348, following Megarry J in *Sargaison* v *Roberts* [1969] 1 WLR 951 at pp 957-8, viz to give the English statute a construction which looks at the realities of the situation in the two systems of law (English and Scottish) at the expense of the technicalities in any one system.

A similar association of "agricultural" with farming in its ordinary sense is to be found in the speeches in *W & J B Eastwood Ltd* v *Herrod* — see *per* Lord Reid at p 168C-D and *per* Viscount Dilhorne at p 180F-G.

Mr Glover relies on the decision of the House of Lords in *Lord Glanely* v *Wightman* [1933] AC 618, and in particular the speech of Lord Wright at pp 638-9, as leading to a conclusion in the opposite sense. To understand that case, it is necessary to consider carefully what was the issue before the House. Lord Glanely owned and occupied a stud farm near Newmarket for the breeding of thoroughbred racing stock. Not only did he use the stallions he maintained for covering his own mares, but he also charged fees for letting out the services upon his stud farm of those stallions to the thoroughbred mares of other owners. Under Schedule B of the Income Tax Act 1918 tax was chargeable in respect of all lands in the United Kingdom. The rate of tax chargeable under Schedule B depended upon whether or not the land was used only or mainly for the purposes of husbandry. The farm as a whole had been assessed to tax under Schedule B: the agricultural part at one rate, the land used purely for stud and racing purposes at a different rate — see *per* Lord Wright at p 634. So the issue whether that part of the land used purely for stud and racing purposes was used only or mainly for the purposes of husbandry was not before the House: it appears to have been accepted that it was not so used. The Inland Revenue then sought to assess Lord Glanely under Schedule D in respect of the fees he received in respect of the services of his stallions in covering other owners' mares on the stud farm. The question before the House was whether that additional assessment was justified:

that is, whether these fees are or are not covered by the general assessment made on the lands under Schedule B

— *per* Lord Wright at p 634. It was held that the fees were part of the profits in respect of the occupation of the farm and chargeable to income tax under Schedule B, not Schedule D. It was not disputed by the Crown that the occupation of the land for the purposes of the stud farm was an occupation within the meaning of Schedule B — see *per* Viscount Buckmaster at p 629, *per* Lord Tomlin at p 632 — and, as I have already said, the question whether the land was used only or mainly for the purposes of husbandry was not in issue. The reasoning of the majority can be found in the following passages from the speeches:

Now a stud farm is plainly an occupation of the land, and the breeding and sale of foals arises from that occupation, and for that purpose the use of the stallion is as indispensable as the use of the mare. The services, therefore, of the stallion upon the land are as much a breeding operation as the production of the foal by the mare, and I find it difficult to see why, when other people's mares are sent on to the farm, and kept there, the payment for the services of the stallion is not a normal part of the purposes for which the land is occupied and inseparable therefrom

— *per* Viscount Buckmaster at p 629.

Looking at the matter apart from authority, I can see no reason in logic for distinguishing between the profit derived from the reproductive capacity of the female and the profit derived from the reproductive capacity of the male

— *per* Lord Tomlin at p 632.

It was conceded by the Crown, and necessarily conceded, that the normal receipts of a thoroughbred stud farm include stud fees received for the service by the stud farm stallions of mares which belong to other people and which are brought on to the stud farm for that purpose. Those stud fees are therefore in my opinion part of the gains of the appellant in respect of his occupation of this land.

By what right can the Crown then claim to pick out one item from the various gains of the appellant in respect of that occupation, and say that it is not covered with the other gains by the assessment under Schedule B but is available as a separate item for a separate assessment under Schedule D? I can envisage no principle which would justify such a course . . .

— *per* Lord Russell of Killowen at p 633.

Lord Wright delivered a long speech, and Mr Glover relies in particular on the following passages:
At pp 638-9:

If authority were needed, the provisions just quoted do at least show that profits of "occupation" include gains from the animal produce as well as the agricultural, horticultural or arboricultural produce of the soil. And the references to gardens, nurseries and woodlands show a scope of Schedule B beyond the use of the land and its products for the provision of food; equally it is obvious that the rearing of animals, regarded as they must be as products of the soil — since it is from the soil that they draw their sustenance and on the soil that they live — is a source of profit from the occupation of land, whether these animals are for consumption as food (such as bullocks, pigs or chickens), or for the provision of food (such as cows, goats or fowls), or for recreation (such as hunters or racehorses), or for use (such as draught or plough horses). All these animals are appurtenant to the soil, in the relevant sense for this purpose, as much as trees, wheat crops, flowers or roots, though no doubt they differ in obvious respects.

At p 640:

On these grounds I think that the service of the stallion is appurtenant to the soil and a profit of the occupation in every case, so that in this regard it is immaterial whether the service is to the appellant's own mares or whether it is sold to strangers; in the latter case the service is sold from the land and as a product of the land, just as much as bullocks, potatoes, fruit or eggs are sold from the land. Without the appellant's stud farm or some other such stud farm the stallions could not live or exercise their generating functions.

Taken out of context, Lord Wright's remarks do undoubtedly support Mr Glover's submissions but, in my judgment, when properly understood, they are not authority for the proposition that "agricultural operations", in the context of a rating statute, include the rearing of animals for the purposes of recreation. We were told that, as a result of the decision in *Lord Glanely* v *Wightman* (*supra*), stud farms were considered to be exempt from rating for nearly 50 years until the decision of the Lands Tribunal in *Evans* v *Bailey* (LVC/429/1980)*: I can only say that I find it surprising that the misunderstanding of the decision in *Lord Glanely* v *Wightman* (*supra*) lasted so long.

Accordingly, I am of the view that the grazing of thoroughbreds for the purpose of breeding and rearing was not, and is not, an agricultural operation within the meaning of section 26(4) of the General Rate Act 1967.

If I am right in this view, the remaining subsidiary questions under the first question do not arise but, as the matter was fully argued before us, and in case this case goes further I propose to deal with them.

Are the stud buildings used "in connection" with the operations on the paddocks?
The member of the Lands Tribunal found the following facts as to the use of the stud buildings:

I have inspected the buildings and I find as a fact that they are of excellent quality and that they are essential to the respondents' activities. The buildings are particularly important for reasons of security and safety of valuable animals, that is to say, the stallions (in the public stud), the visiting mares and their foals (whether at foot or born at Whitsbury), and for providing warmth in winter and coolness in summer and protection from adverse weather, as premises providing facilities for covering and for hay and concentrates to be dispensed, and for veterinary examinations, weighing and measuring of foals, observation of pregnant mares and grooming of yearlings to take place.

In the light of these findings, it seems to me to be difficult, as a matter of ordinary language, to say that the stud buildings are not partly used "in connection with" the operations on the paddocks, if these operations are, as I have said, the grazing of thoroughbred horses in the course of breeding and rearing. In particular, it seems to me that the use of the stud buildings for providing shelter to the grazing animals must be a use in connection with the operations on the paddocks. However, this phrase has also been the subject of judicial consideration. In *Gilmore* v *Baker-Carr* [1962] 1 WLR 1165 at p 1175 Donovan LJ said:

But the clear impression which I receive from the statutory language is that the buildings exempted were to be ancillary or complementary to the agricultural purpose of the land, and not vice versa.

This definition was approved by Viscount Dilhorne in *W & J B Eastwood Ltd* v *Herrod* [1971] AC 160 at p 181B, who added:

I think that the language of the definition requires that buildings to come within it must be used as adjuncts to the agricultural operations on the land . . .

In the same case Lord Reid said (at p 168G):

I do not foresee serious difficulty if "used in connection with" is held to mean consequential on or ancillary to the agricultural operations on the land which is occupied together with the buildings.

I accept that it would be difficult in the present case to describe the use of the stud buildings as ancillary to the grazing of the thoroughbreds on the paddocks, but I do not see why it cannot properly be described as complementary to those operations. The Lands Tribunal's decision on this matter — that the stud buildings were not used in connection with the operations on the paddocks — is not, as Mr Fletcher submits, a decision of fact which cannot be challenged on appeal. It was an error of law based on a misapplication of the statutory test to the primary facts as found by the tribunal. As such, it is open to review by this court and, in my judgment, the stud buildings are used in connection with the operations on the paddocks.

Are the stud buildings used "solely" in connection with the operations?
Here again, I have the misfortune to differ from the decision of the

---
*Editor's note: Reported at (1981) 260 EG 611.

learned member of the Lands Tribunal. I do not see how the use of the stud buildings as "premises providing facilities for covering" and "for veterinary examinations" can be properly described as use in connection with the operation of grazing the thoroughbreds. Accordingly, I would, if it were material, hold that the stud buildings are not used solely in connection with the operations on the paddocks. I turn now to the second main question:

Are the buildings used for the keeping or breeding of livestock?

The *Shorter Oxford English Dictionary* definition of "livestock" is:

Domestic animals generally; any animals kept or dealt in for use or profit.

If the question fell to be answered without any indication that this dictionary meaning is to be qualified in any way, it is clear that thoroughbred horses fall within the definition of "livestock". However, the extension of the definition of "agricultural buildings" in section 26 of the General Rate Act 1967 to include any building used for the keeping or breeding of livestock is contained in section 2(1)(a) of the Rating Act 1971 and section 1(3) of that Act defines "livestock" for the purposes (*inter alia*) of section 2 as including:

any mammal or bird kept for the production of food or wool or for the purpose of its use in the farming of land.

These words would be unnecessary if it were intended that "livestock" should bear its extensive dictionary meaning, and Mr Glover accepts that in this context "livestock" does not include domestic animals of every description. He submits that it does cover domestic animals found on agricultural premises or domestic animals in an agricultural context. It will be apparent, from what I have already said in relation to "agricultural operations", that, in my judgment, thoroughbred horses would not come within this definition.

Again, the matter is not free from authority. In *Belmont Farm Ltd v Minister of Housing and Local Government* (1962) 13 P&CR 417, a Divisional Court of the Queen's Bench Division held that the breeding and keeping of horses, not intended for use in the farming of land, did not amount to "the breeding and keeping of livestock" and so was not a use of land for the purposes of agriculture within the definition of agriculture contained in section 119(1) of the Town and Country Planning Act 1947. We are not, of course, bound to follow this decision, and I accept that it is a decision on a different definition of "livestock" in a different Act. However, I find the reasoning of Lord Parker CJ in that Act relevant to the question we have to answer. There "livestock" was defined as "including any creature kept for the production of food, wool, skins, or fur, or for the purpose of its use in the farming of land". Apropos of this definition Lord Parker CJ said (at pp 421-2):

Granting that the word "including" has been used in an extensive sense, it seems to me nonsense for the draftsman to use those words "any creature kept for the production of food, wool, skins or fur, or for the purpose of its use in the farming of land", if the word "livestock" was intended to cover the keeping of any creature whether for its use in farming land or not. It seems to me that these words show a clear intention that "livestock", however it is interpreted, does not extend to the breeding and keeping of horses unless it is for the purpose of their use in the farming of land.

In my judgment, the same reasoning applies, *mutatis mutandis,* to the definition of "livestock" in the 1971 Act.

The *Belmont Farm* case was followed by another Divisional Court of the Queen's Bench Division in *Minister of Agriculture, Fisheries and Food v Appleton* [1970] 1 QB 221 — again a case on the meaning of "livestock" within a definition of "agriculture", this time for the purposes of the Selective Employment Payments Tax 1966. The *Belmont Farm* case was also approved by Russell LJ in *McClinton v McFall* (1974) 232 EG 707 at p 709.

Mr Glover seeks to rely on certain passages in the judgments of the Court of Appeal in *Cresswell (VO) v BOC Ltd* [1980] RA 213. In that case the Court of Appeal held that fish were not livestock within the definition contained in the Rating Act 1971. That was the issue before the court and it is in that context that the passages on which Mr Glover relies are to be read. Thus, Eveleigh LJ said (at p 216):

For myself, I would be inclined, at a first look at s2, to say that "livestock" here contemplates domestic animals or birds which are found on agricultural premises and which are supported by the land.

It is clear that no member of the court in that case was considering the question whether thoroughbreds fell within the definition of "livestock" — indeed, Watkins LJ cited the *Belmont Farm* case with apparent approval — and, in my judgment, that case does not support Mr Glover's contention that "livestock" in the 1971 Act includes thoroughbreds.

Accordingly, I answer the second main question also in the negative.

I would dismiss this appeal.

Also agreeing, LLOYD LJ said: In this case we have to determine whether certain buildings are agricultural buildings within the definition contained in section 26(4) of the General Rate Act 1967, as extended by section 1(1)(a) of the Rating Act 1971.

The outcome turns on two short questions of construction: first, whether the buildings, which were admittedly occupied together with agricultural land, were being used "solely in connection with agricultural operations" on that land so as to bring the case within section 26(4) of the former Act; and, second, whether, if not, horses bred for riding, hunting or racing are "livestock" so as to bring the case within section 2(1)(a) of the latter Act.

Before attempting to answer these questions, there are three preliminary observations worth making.

In the first place, I doubt if we get much help in a case such as the present by adopting what is now called a purposive approach. When the legislative purpose is clear, then it is always legitimate, and often very helpful, to have that purpose in mind when approaching questions of construction. But, where the legislative purpose is not so clear, it may be dangerous to speculate. In the present case we are concerned with the derating of agricultural land and buildings. The legislative history goes back nearly 100 years. Partial relief was introduced by the Agricultural Rates Act 1896. Full relief was introduced by the Local Government Act 1929. The same Act also introduced partial relief for industrial hereditaments. It may be that the legislative purpose of Parliament in derating agricultural land and buildings was to encourage the growing of food or the means of human subsistence. But it is not obvious that Parliament may not have had a wider purpose in mind, as is perhaps shown by the inclusion of woodland in the definition of agricultural land.

Second, I doubt if one gets much help by asking what would ordinarily be understood by an agricultural building, still less by asking what sort of building one would expect to find on an ordinary farm. On two occasions in recent years the courts have taken a narrow view of the exempting provisions. On both occasions Parliament has taken prompt action to extend the definition. I have in mind the 1971 Act itself, which was passed in the wake of *W & J B Eastwood Ltd v Herrod* [1971] AC 160. Broilerhouses, which had been held not to fall within section 26 of the 1967 Act on the ground that they could not be said to have been used solely in connection with agricultural operations on agricultural land, are now included in the extended definition. So also are buildings used solely in connection with the keeping of bees. One would not perhaps normally think of such buildings as being agricultural buildings; nor would one expect to find such buildings on an ordinary farm. But they are now specifically included.

The second occasion on which Parliament has extended the definition in recent years followed a decision of this court in *Cresswell (Valuation Officer) v BOC Ltd* [1980] 1 WLR 1556. In that case this court held that a fish farm was not exempt, on the ground that fish were not livestock within the 1971 Act. Within a year Parliament had, by the Local Government, Planning and Land Act 1980, section 31, introduced a new section 26A in the General Rate Act 1967 so as to exempt land and buildings used solely for or in connection with fish farming.

So it would, I think, be dangerous in this case to ask what Parliament must have intended to include or exclude. Our task is to take the words that Parliament has actually used, to note what is in fact included and excluded, and then apply the result to this particular case.

The third general consideration is of a different kind. In 1933 the House of Lords decided *Lord Glanely v Wightman* [1933] AC 618. Although that case was concerned with the liability to income tax, nevertheless the reasoning and, in particular, the speech of Lord Wright, was treated as being applicable in the rating field. Thereafter, for 50 years, it was accepted that stud farms were exempt. Stud buildings were not in fact rated. The courts are always reluctant to disturb such a long-continued practice. Nevertheless, we are bound to do so if we are persuaded that the practice is based on a misunderstanding of *Lord Glanely's* case. It could not be argued —

nor was it argued by Mr Glover — that, by consolidating the law in 1967 without reference to stud farms, and by extending the definition of agricultural land and buildings in 1971, and again in 1980, Parliament has, as it were, given its legislative approval to the practice based, rightly or wrongly, on the decision of the House of Lords in 1933.

With those preliminary observations, I turn to the language of the statute. The first thing to notice is, of course, that "agricultural operations" are not defined. But one thing is beyond dispute. "Agricultural" in the phrase "agricultural operations" must have the same meaning as it has in the phrase "agricultural land". So I would agree with Mr Glover that the approach must be to ask, first, whether the land is agricultural land; second, what are the agricultural operations on that land and, third, whether the buildings which are occupied together with the land are used solely in connection with those agricultural operations. If authority be needed for that approach it is to be found in the speech of Lord Morris in *W & J B Eastwood Ltd* v *Herrod* [1971] AC 160 at p 173. I do not myself regard it as helpful to split up the third question into two sub-questions, ie were the buildings used in connection with agricultural operations on the land? If so, were they used *solely* in connection with such operations? Both sub-questions are best considered together. Failure to do so led the learned member of the Lands Tribunal to include a passage in the reasons for his decision which is not altogether easy to understand.

Is the land agricultural land? Mr Fletcher for the respondents concedes that it is. But it is necessary to examine the reasons for that concession in a little detail. Mr Glover argues that the answer to the first question is crucial in answering the second and the third.

The definition of agricultural land in section 26(3) of the 1967 Act is both positive and negative. In the first half, the subsection tells us what agricultural land is, and in the second half it tells us what it is not. The two halves of the definition are not altogether easy to fit together. There is no difficulty in the case of land used for a plantation or a wood or for the growth of saleable underwood. Such land is agricultural land, unless it is occupied together with a house as a park, or unless it is kept or preserved "mainly or exclusively for the purpose of sport or recreation". So the occasional use of woodland for shooting does not prevent its being agricultural land. But what about arable and pasture? Unlike woodland, arable meadow and pasture are only included in the definition of agricultural land if the land is used as arable meadow or pasture ground *only*. Does that mean that the occasional use of arable for shooting excludes it from the definition? If woodland is included, in spite of its being used as cover for pheasants reared for sport, it would seem artificial to exclude a field of kale on that ground. Occasional use of land as a racecourse prevents its being agricultural land, unless the use is *de minimis*. But the exclusion of land used as a racecourse is not qualified by the words "mainly or exclusively".

So I would think that the occasional use of arable meadow or pasture for purposes of sport or recreation would not prevent its being agricultural land within the definition, despite the word "*only*". But occasional use for other purposes would so prevent it.

I now come to the central issue in the case. Mr Fletcher concedes that the land is agricultural land because it was used by horses for grazing. But he submits that the breeding of horses, whether for use as hunters or in racing, is not an agricultural operation. When you are deciding whether the land is agricultural land, you look only, he submits, at the use made of the land. But, when you are deciding whether the operation is an agricultural operation, you look at the purpose of the occupier.

Mr Glover, on the other hand, submits that there can be no distinction between the use made of land, and the purpose for which land is used, whether as arable or woodland, or for poultry farming or a market garden; and any distinction between the purposes for which land is used and the operations on that land would be highly artificial, at any rate on the facts of this case. Since it is conceded that the use of *this* land for grazing *these* horses made the land agricultural land, then it must follow that that use is an agricultural operation. Of course, there may be operations on land which are not agricultural operations. But here the operation in question, namely the grazing of horses kept for breeding, is the very operation which makes the land agricultural land. It would indeed be strange, he submits, if the operation which makes the land agricultural land is not itself an agricultural operation.

Throughout the hearing before us, I could see no answer to Mr Glover's argument. It seemed to me that Mr Fletcher was in a dilemma. Either the operation carried out on the land which, for brevity, I will call the keeping of brood mares, is an agricultural operation or it is not. If it is, *cadit quaestio*. If it is not, then the land was not being used as arable meadow or pasture ground *only*, in which case the land is not agricultural land within the definition. But it is conceded that it is. So the operation must be an agricultural operation.

In *Peterborough Royal Foxhound Show Society* v *Commissioners of Inland Revenue* [1936] 2 KB 497 Lawrence J (as he then was) seemed to regard it as self evident that the breeding of hunters and racehorses takes place "in the ordinary course of agriculture". That dictum was quoted, without apparent disapproval, by Lord Wilberforce in *Normanton (Earl)* v *Giles* [1980] 1 WLR 28 at p 32. Similarly, the Court of Appeal in *McClinton* v *McFall* (1974) 232 EG 707 held that land let for use as a stud farm was an agricultural holding within the meaning of the Agricultural Holdings Act 1948. Stamp LJ said:

The activities in relation to the stud farm which I have described, so far as they consist of the grazing of horses, pasturing of cattle and making of hay, are clearly agricultural. So far as they consist of the breaking-in of horses for riding, the little schooling that is done, the showing of horses to customers, and the jumps and jumping on the five-acre field I have mentioned, they are not, in my judgment, inconsistent with the agriculture carried on.

Moreover, the distinction between the use made of the land, and the purposes of the occupier — a distinction which lies at the heart of Mr Fletcher's argument — is difficult to reconcile with land which is held for the *purposes* of poultry farming being included within the definition of agricultural land in section 26(3)(a) of the 1967 Act. It is also difficult to reconcile with the language of section 2 of the 1971 Act. Section 2(1)(a) provides that "agricultural building" shall include any building used for breeding livestock. Section 2(1)(b) provides that it shall also include any building used solely in connection with the operations carried on in that building. This suggests that there is no distinction to be drawn between the *use* of the first building and the *operations* carried on in that building. If that is right, why should there be any distinction between the use of the land and operations on the land?

Mr Glover's argument also appeared to be consistent with the approach adopted by the courts in construing other related provisions of the Rating and Valuation (Apportionment) Act 1928, namely those relating to industrial hereditaments. Thus, in *Crowe (Valuation Officer)* v *Lloyds British Testing Company Ltd* [1960] QB 592, a majority of this court held that "industrial purposes" in section 4(2) of the 1928 Act must mean those purposes which make the premises in question a factory within the meaning ascribed by section 3(2) of the Act. It is hard to believe that Parliament intended to draw a distinction between "purposes" in the phrase "industrial purposes" and "operations" in the phrase "agricultural operations".

These are powerful arguments. But, in the end, I have come to the conclusion that they cannot succeed. There is no real dilemma. Parliament must clearly have envisaged that operations which are no agricultural operations might be carried out on land which nevertheless remains agricultural land. Otherwise, there would have been no point in including the reference to "agricultural operations" in section 26(4)(a).

That being so, there are only two possible explanations of the apparent impasse. Either the reference to agricultural operations was intended to exclude, and exclude only, buildings used in connection with sport or recreation for there is no other use of the land which is consistent with the land remaining agricultural land within section 26(3)(a). The alternative explanation is that Parliament did indeed intend to draw a distinction, as Mr Fletcher submits, between the use to which land is put and the operations thereon. I do not find either explanation particularly satisfactory. But of the two I prefer the second. In other words, lands may be *used* as pasture land only and so qualify as agricultural land; but the *operations* carried out on the land, looked at from another point of view, may be non-agricultural operations.

If that be right, then the question whether the operation carried out on this land was an agricultural operation becomes largely, if not entirely, a question of fact. We cannot say, as a matter of law, that the operation was necessarily an agricultural operation by reason of Mr Fletcher's concession that the land is agricultural land. It follows that I would reject, though not without considerable hesitation, M

Glover's central submission.

The learned member has found, after carefully considering all the evidence, that the operation here was not an agricultural operation. I am not persuaded that there is any material on which we could or should disturb that finding.

Before leaving the second question, I should, however, deal briefly with *Lord Glanely* v *Wightman*, the case upon which Mr Glover principally relied. The question in that case was whether the appellant, Lord Glanely, was liable to tax under Schedule D on the stud fees earned by his stallion, Grand Parade. It was held by the House of Lords, reversing a majority of the Court of Appeal, that he was not so liable. There was no distinction to be drawn between the profits earned from the reproductive capacity of the appellant's stallion and the profits earned from the reproductive capacity of the appellant's mares. Since the latter were admittedly profits of the appellant's occupation of the land, and therefore assessable to tax under Schedule B and not under Schedule D, so also were the former. Mr Glover relied in particular on a paragraph of Lord Wright's speech (at p 638):

If authority were needed, the provisions just quoted do at least show that profits of "occupation" include gains from the animal produce as well as the agricultural, horticultural or arboricultural produce of the soil. And the references to gardens, nurseries and woodlands show a scope of Schedule B beyond the use of the land and its products for the provision of food; equally it is obvious that the rearing of animals, regarded as they must be as products of the soil — since it is from the soil that they draw their sustenance and on the soil that they live — is a source of profit from the occupation of land, whether these animals are for consumption as food (such as bullocks, pigs or chickens), or for the provision of food (such as cows, goats or fowls), or for recreation (such as hunters or racehorses), or for use (such as draught or plough horses). All these animals are appurtenant to the soil, in the relevant sense for this purpose, as much as trees, wheat crops, flowers or roots, though no doubt they differ in obvious respects.

But the question in *Lord Glanely's* case, as Mr Fletcher pointed out, was whether the stud fees were severable from the profits of the breeding operation as a whole. There was no issue in the case whether the breeding operation was an agricultural operation. As Lord Wright himself pointed out at p 633, that question did not arise. Mr Glover argued that, though Lord Wright left that question open, the whole tenor of his speech was consistent only with his view being that breeding racehorses is an agricultural operation. I do not agree. Indeed, it is significant that the ground on which Wilfred Greene KC (as he then was) sought to distinguish *Malcolm* v *Lockhart* [1919] AC 463, a previous decision of the House of Lords, was that in *Malcolm* v *Lockhart* the farm was an agricultural farm:

Supplying the services of a stallion is no part of the business of an agricultural farmer, and the fees for those services were easily severable from the profits of the farm. The case of a stud farm is quite different.

That distinction is reflected in Viscount Buckmaster's speech. At p 631 he said:

The whole case

that is to say *Malcolm* v *Lockhart*

was based upon the occupation of the land being for ordinary farm purposes, and there was no reason to displace the finding of the Courts that the sale of the services of the stallion when taken round the countryside formed no part of that business.

So I do not consider that Mr Glover gets any assistance from *Lord Glanely's* case. The case is authority for the proposition (i) that stud fees earned on a stud farm cannot be severed from the other receipts of the stud farm and (ii) that the profits of a stud farm are profits in respect of the occupation of land, and are therefore taxable under Schedule B. It is not authority for the proposition that stud farms are an agricultural operation within the meaning of section 26(4) of the General Rate Act 1967. If, therefore, the practice of exempting the buildings on stud farms was based, as I assume it was, on *Lord Glanely's* case, then I am bound to conclude that it was based on a misunderstanding of that case.

I now turn to the third question. Assuming there were agricultural operations on the land, were these buildings used solely in connection with those operations? If I am right on the second question, then the third question does not arise. But it was fully argued before us. So I mention it briefly.

I have no doubt that, if stud farming as carried on by these appellants was an agricultural operation, then these buildings were used solely in connection with that operation. They serve no other purpose. It is unnecessary to inquire whether the buildings were more important than the land, or vice versa. I agree with Mr Glover's submission that the relative importance of the buildings and the land is irrelevant. There may perhaps be cases where the use of the land is so insignificant compared with the use of the buildings that the proper conclusion on the facts would be that the building was not occupied solely in connection with agricultural operations on the land, even though the buildings have no other use. *W & J B Eastwood Ltd* v *Herrod* (*supra*) was such a case on the facts. But, in general, the question to be asked is not whether the buildings are more or less important than the land, but whether they serve any purpose other than the agricultural operations on the land. If not, then in the great majority of cases they will be used solely in connection with agricultural operations.

Various expressions were suggested in *Eastwood* v *Herrod* as synonyms for the statutory language, including the word "ancillary". It may be that that word has led to the idea that the use of the buildings must be of minor importance only. That would be a misunderstanding. It would be inconsistent with what Lord Morris said in *Eastwood* v *Herrod* at p 174:

The words of the definition of "agricultural buildings" suggest to my mind buildings that are needed as an adjunct or a necessary aid to agricultural operations taking place on agricultural land and used solely in connection with those operations. This does not necessarily involve that the use to which the buildings are put must be of minor or minimal importance but it does involve that no part of the use is unconnected with the agricultural operations on the land.

If a synonym is needed, then I would suggest "complementary" rather than "ancillary". But it is better to apply the statutory language as it stands.

The learned member held that buildings were not occupied in connection with the agricultural operations on the land because, as he puts it, so far from the use of the buildings being ancillary to the use of the land, or the use of the land being ancillary to the use of the buildings, each was ancillary to the enterprise as a whole. I find this hard to understand. If the enterprise as a whole had been an agricultural operation, then the fact that the use of the land was "ancillary" to that operation would not have prevented the use of the buildings being ancillary to the operation on the land. So the learned member must, I think, have misdirected himself as to the correct test. This is confirmed by the learned member's puzzling observation that, if he had found that the buildings were being used in connection with agricultural operations, then he would have found they were being solely so used.

Since the learned member applied the wrong test, I should, had it been relevant, have felt free to reach a different conclusion on the facts. Both counsel agreed that in those circumstances, we should draw our own inference from the evidence. The evidence is very well summarised in the case. It was hardly, if at all, in dispute. Having read the evidence, I am left in no doubt that the use of the buildings was not only essential to the operations on the land, as the learned member himself has found, but was also confined to those operations. In other words, the buildings served no other purpose.

It follows that, if I could have accepted Mr Glover's central submission on the first main question, I should have been in favour of allowing the appeal.

Finally, I turn to the second main question of construction, which I can deal with much more briefly. Sections 1 and 2 of the 1971 Act provide that "agricultural buildings" shall include any building which is used solely for the keeping or breeding of livestock. It was accepted that the buildings were used solely for the keeping and breeding of thoroughbred horses. The question is whether thoroughbreds are "livestock".

The dictionary meaning of "livestock" is "domestic animals generally". This is clearly too wide, since it would presumably include cats and dogs. So the meaning must be confined to domestic animals in an agricultural context. This would cover all domestic animals normally found on a farm, such as cattle, sheep and pigs, and might also include deer other than ornamental deer. But does it include horses?

I have already referred to the dictum of Lawrence J in *Peterborough Royal Foxhound Show Society* v *Commissioners of Inland Revenue*. He clearly regarded horses as livestock. Others might share his view if the words stood alone. But, in the present case, we must have regard to the definition of livestock contained in section 1(3) as including "any mammal or bird kept for the production of food or wool or for the purpose of its use in the farming of land". It is not suggested by Mr Glover that these horses fall within that definition. They are clearly not kept for the production of food or wool, or for use in farming. But Mr Glover points out that the definition is inclusive, not exclusive. It does not prohibit the inclusion of horses if they would otherwise be regarded as livestock.

A similar argument was advanced in *Cresswell* v *BOC,* to which I have already referred. In that case the question was whether fish are livestock. It was argued that, though the fish were not mammals or birds, they were certainly being kept for the production of food. They should therefore be treated as livestock for the purposes of the Act. This court had no hesitation in rejecting that argument. Megaw LJ in particular made it clear that the meaning of "livestock" must take colour from the definition, even if the definition is not exclusive. In *Belmont Farm Ltd* v *Minister of Housing and Local Government* (1962) 13 P&CR 417, Lord Parker CJ, after referring to Lawrence J's dictum, held that thoroughbreds are not livestock. That decision was expressly approved by this court in *McClinton* v *McFall* (*supra*). Though I have doubt on the first main question, for the reasons I have mentioned, I have no doubt on the second. We are obliged to hold by authorities binding on this court that these horses are not livestock within the meaning of the 1971 Act.

For the reasons I have mentioned, I, too, would dismiss the appeal.
The appeal was dismissed with costs. Leave to appeal to the House of Lords was granted.

Court of Appeal
February 13 1987
(Before Lord Justice SLADE, Lord Justice PARKER and Lord Justice MUSTILL)

R v TOWER HAMLETS LONDON BOROUGH COUNCIL, EX PARTE CHETNIK DEVELOPMENTS LTD

Estates Gazette April 25 1987
282 EG 455-460

Rating — General Rate Act 1967, section 9(1) — Claim for refund of overpayments of rates — Appeal from decision of Mann J dismissing application for judicial review seeking relief from refusal of rating authority to repay rates which appellant ratepayers had not been liable to pay — Rates in question had been paid in respect of unoccupied property, but in fact as a result of a condition laid down by the Greater London Council under the London Building Acts 1930 to 1939 occupation of the property during the material period would have been unlawful — The condition required the council's consent to the proposed user and this condition could not be satisfied until a tenant was identified and the user approved — Under para 2(a) of Schedule 1 to the 1967 Act no unoccupied rates were payable during any period when occupation was prohibited by law — Despite section 9(1), the respondent rating authority refused to refund the rates, giving as their main reasons that the payment was made under a mistake of law; that no hardship to the applicants had been alleged; that the scheme and intent of the 1967 Act was to provide for completion notices to prevent owners from avoiding rates by non-completion; and that the applicants could have taken professional advice before making the payments and could have avoided the problem if they had complied with the requirements of the London Building Acts — Mann J, in refusing the application for judicial review, had held that the discretion of the rating authority under section 9 was unfettered, although he accepted that it fell within the *Wednesbury* principles — He considered that the authority's decision did not offend against these principles — Held on appeal, reversing the decision of Mann J, that a rating authority, on receipt of an application under section 9, although not placed under an imperative obligation to refund, was bound to take into account the object which Parliament must have intended to achieve in enacting the section — The object was to enable rating authorities to give redress and to remedy the injustice which would (at least *prima facie*) otherwise arise if they were to retain sums to which they had no right, in cases where persons had paid rates which they were not liable to pay — The court also analysed the four main reasons put forward by the rating authority for their decision and found each one irrelevant — Appeal allowed and orders of certiorari and mandamus to go, quashing rating authority's decision and requiring it to hear and determine the case according to law

The following cases are referred to in this report.

Arsenal Football Club Ltd v *Ende* [1979] AC 1; [1977] 2 WLR 974; [1977] 2 All ER 267; (1977) 75 LGR 483, HL
Associated Provincial Picture Houses Ltd v *Wednesbury Corporation* [1948] 1 KB 223; [1947] 2 All ER 680, CA
Blackpool and Fleetwood Tramroad Co v *Bispham with Norbreck UDC* [1910] 1 KB 592
Council of Civil Service Unions v *Minister for the Civil Service* [1985] AC 374; [1984] 1 WLR 1174; [1984] 3 All ER 935, HL
Meadows v *Grand Junction Waterworks Co* (1905) 69 JP 255
Padfield v *Minister of Agriculture, Fisheries and Food* [1968] AC 997; [1968] 2 WLR 924; [1968] 1 All ER 694, CA and HL
R v *Liverpool City Council, ex parte Windsor Securities Ltd* [1979] RA 159, CA
R v *Rochdale Metropolitan Borough Council, ex parte Cromer Ring Mill Ltd* [1982] 3 All ER 761
Slater v *Burnley Corporation* (1883) 53 JP 70
Slater v *Burnley Corporation (No 2)* (1889) 53 JP 535
Stanley v *Weardale Coal & Coke Co Ltd* (1935) 6 DRA 49

This was an appeal by Chetnik Developments Ltd, a property company, from the dismissal by Mann J of an application for judicial review, seeking relief in the form of orders of certiorari and mandamus against the refusal of the London Borough of Tower Hamlets to refund to the applicants unoccupied property rates which the applicants had paid under a mistake of law. The property in question was a site at 7 Ditchburn Street, London E14, which the applicants were proposing to develop by the provision of warehouse units.

John Taylor QC and John Howell (instructed by Lovell White & King) appeared on behalf of the appellants; Barry Payton and Simon Gault (instructed by H D Cook, solicitors' department, London Borough of Tower Hamlets) represented the respondents.

Giving the judgment of the court, SLADE LJ said: By a judgment delivered on April 3 1985 (reported at (1985) 25 R&VR 87) Mann J dismissed an application by Chetnik Developments Ltd ("the applicants") for judicial review. The respondent to the application was the Council of the London Borough of Tower Hamlets ("the council"). The relief sought was an order of certiorari to quash a decision made by the council on October 26 1983 that rates paid by the applicants in respect of the period from November 16 1976 to March 31 1979 should not be refunded, and an order of mandamus directing the council to hear and determine according to law an application by the applicants for the refund of such rates. The applicants now appeal to this court from Mann J's order.

The application for repayment was dated August 5 1982 and related to an amount of £51,396.92. It was made under section 9(1) of the General Rate Act 1967. That section provides:

9. (1) Without prejudice to sections 7(4)(b) and 18(4) of this Act, but subject to subsection (2) of this section, where it is shown to the satisfaction of a rating authority that any amount paid in respect of rates, and not recoverable apart from this section, could properly be refunded on the ground that —
(a) the amount of any entry in the valuation list was excessive, or
(b) a rate was levied otherwise than in accordance with the valuation list; or
(c) any exemption or relief to which a person was entitled was not allowed; or
(d) the hereditament was unoccupied during any period; or
(e) the person who made a payment in respect of rates was not liable to make that payment,
the rating authority may refund that amount or a part thereof.
(2) No amount shall be refunded under subsection (1) of this section —
(a) unless application therefor was made before the end of the sixth year after that in which the amount was paid; or
(b) if the amount paid was charged on the basis, or in accordance with the practice, generally prevailing at the time when the payment was demanded.

(3) Before determining whether a refund should be made under subsection (1) of this section —
(a) in a case falling within paragraph (a) of that subsection; or
(b) in a case falling within paragraph (c) of that subsection where the exemption or relief was one which ought to have appeared in the valuation list,
the rating authority shall obtain a certificate from the valuation officer as to the manner in which in his opinion the hereditament in question should have been treated for the purposes of the valuation list, and the certificate shall be binding on the authority.

The ground for repayment relied on was ground (e).

The circumstances which gave rise to the application and its refusal are common ground and were summarised clearly and concisely by the learned judge. We gratefully adopt his summary:

The applicant is a property development company. In 1975 and 1976 the applicant developed a site at 7 Ditchburn Street, London E14. The development provided two warehouse units each of which has ancillary accommodation. The applicant secured the requisite consents to the development, including an approval given by the Greater London Council in the exercise of its powers under the London Building Acts 1930 to 1939. That approval was dated the 13th November 1975 and contained 45 conditions, amongst which was one (number 43) which provided that "no part of the building shall be occupied until the consent of the [GLC] had been obtained to the proposed user". A contravention of the condition would have been a criminal offence (see London Building Acts (Amendment) Act 1939, section 148(2)(xiii)). The condition could not be satisfied until a tenant was identified, for until then the usage would be unknown. The applicant tried its best to let the warehouses but it was not until dates in 1978 that a tenant for unit 1 was found, the user was approved and his occupancy commenced. In the case of unit 2 those events did not occur until 1980.

Section 17 of the 1967 Act grants a power to a rating authority (such as is the respondent) to apply the provisions of Schedule 1 with respect to the rating of unoccupied property to their area. The council exercised that power before 1976. The provisions of the Schedule allow the owner of a hereditament to be rated in respect of that hereditament if it is unoccupied for a continuous period exceeding three months (para 1). Where a new building is erected the rating authority may serve on the owner of that building a completion notice stating the date on which the building is to be treated as completed (para 8(1)). There is a procedure whereby the date may be challenged (para 8 (4)), but if there is no successful challenge then if on the stated date the building is unoccupied it is deemed to have become unoccupied on that date (para 7). An owner cannot, however, be rated in respect of an unoccupied hereditament if he is "prohibited by law from occupying the hereditament or allowing it to be occupied" (para 2 (a)).

On July 20 1976 the rating authority served two notices upon the applicant stating that August 16 1976 was to be the date on which units 1 and 2 were to be treated as having been completed. There was no challenge to the date. As from November 16 1976 the applicant was rated as the owner of unoccupied property. It was so rated until the dates on which tenants took occupation of the units. The applicant paid rates in the total sum of £51,396.92 for the period November 16 1976 to March 31 1979. The applicant then declined to pay any further rates in respect of the period in respect of which unit 2 remained without an identified tenant. The rating authority sought the issue of a warrant of distress under section 97 of the Act of 1967, but on April 3 1981 a stipendiary magistrate refused to issue a warrant on the ground that the applicant could not have been rated in respect of unit 2 because the company was prohibited by law from occupying the hereditament or allowing it to be occupied. The prohibition was held to be by reason of condition 42 in the approval of November 13 1975. There was no appeal to the Crown Court against the magistrate's decision. In the light of that decision, the applicant applied for a refund of what it had previously paid whilst trying its best to find tenants for units 1 and 2 and latterly a tenant for unit 2. The application was made on August 5 1982 and was refused by the council's finance committee in the exercise of delegated powers on October 26 1983. It is that decision which is impugned. The application was first before the committee on September 28 1983. The minute for that day is: "Item 4.9 — 7a Ditchburn Street — Application for rates refund under section 9 of the General Rate Act 1967. Details of the decision made by a stipendiary magistrate at Thames Magistrates' Court in refusing the council's application for a distress warrant in respect of this case were tabled. The representatives of the solicitor to the council referred to the counsel's opinion which had been obtained on the application for rates refund made by Chetnik Developments Ltd and indicated that, as this was a discretionary power, the committee would have to make a decision independent of any officer's recommendation. Councillor Charters felt that, in view of the complexity of the application, and the additional tabled information (which had only been received by the solicitor at 4.30 pm on the day of the meeting), the matter be deferred so that members could give adequate consideration to all aspects of the application. Resolved — that the item be deferred until the next meeting."

The minute for October 26 is: "Item 4.6 — 7a Ditchburn Street — Application for refund of rates under section 9, General Rate Act 1967. The representative of the solicitor to the council (Mr A. Tobias, principal legal assistant) reiterated the comments made at the meeting held on September 28 1983, in that it was for the committee to exercise its reasonable discretion on the application, without recommendations from any officer. The members of the committee then entered into a comprehensive discussion on the merits of the application as presented. It was unanimously Resolved — That the application be refused."

On October 28 1983 the applicants' solicitors asked to be given the reasons for the refusal. They were supplied in a letter dated November 8 1983 by the council's solicitor, who said:

When this matter came before the Finance Committee on October 26 last, Members had before them all the representations made by you on behalf of your Client Company, together with the Judgment of the Stipendiary Magistrate and the Advice from Counsel. The Committee were strenuously advised that the decision required pursuant to section 9 of the General Rate Act 1967 was one entirely for their discretion acting fairly in accordance with the principles set out in Counsel's advice. After discussion the Committee decided against a repayment for the following reasons:
(a) the advice that the payment made by the Company was paid under a mistake of law;
(b) on the parallel that an application for relief from empty rates under paragraph 3A of Schedule 1 of the General Rate Act 1967 would require an applicant to demonstrate "hardship" — which was not argued by your Client Company although they were invited to do so;
(c) the scheme and intent of the 1967 Act with regard to completion notices;
(d) that your Clients could have taken professional advice before making such payments and, of course, they could have avoided the problem if they had complied with the requirements of the London Building Acts.

It is thus clear that the committee, in coming to their decision, attached considerable weight to counsel's opinion. A copy of the opinion was also supplied to the applicants. It has been referred to in the course of argument by both sides. The reasons stated in the letter of November 8 1983 to a substantial extent reflect certain features of that opinion, to which it will be necessary to revert.

As to the construction of section 9(1), the learned judge held that the words "may refund the amount or a part thereof" confer on the rating authority "a discretion to repay, rather than impose what is in effect an obligation to repay, subject to its not being equitable to do so". He agreed with a dictum of Forbes J in *R v Rochdale Metropolitan Borough Council, ex parte Cromer Ring Mill Ltd* [1982] 3 All ER 761 who at p 770F had described the discretion as a "true unfettered discretion".

However, in accepting that the discretion of the rating authority under section 9(1) is "unfettered", the learned judge also accepted that it none the less fell within the confines formulated in *Associated Provincial Picture Houses Ltd v Wednesbury Corporation* [1948] 1 KB 223.

Lord Diplock in *Council of Civil Service Unions v Minister for the Civil Service* [1985] AC 374 pointed out at p 410 that *"Wednesbury* unreasonableness" applies to a decision "which is so outrageous in its defiance of logic or of accepted moral standards that no sensible person who had applied his mind to the question to be decided could have arrived at it". He referred to unreasonableness of this nature as "irrationality" and observed (at p 411) that it "by now can stand upon its own feet as an accepted ground on which a decision may be attacked by judicial review".

The learned judge accepted that a rating authority must not make a decision which is irrational and that, in the exercise of its discretion, it must not have regard to a consideration which is irrelevant. However, he held that, in the present case, each of the four factors mentioned in the letter of November 8 1983 manifested a consideration to which a rating authority was entitled to have regard "when considering how to exercise its unfettered discretion". Irrationality was not suggested. Accordingly, he refused the application.

Before turning to the council's decision which is now under attack, it is necessary to consider the construction and effect of section 9(1) itself. According to the wording of the subsection, even in a case where an applicant has clearly established one or other of the grounds (a), (b), (c), (d) and (e), the rating authority, before making repayment, is still obliged to satisfy itself that the relevant sum paid in respect of rates, and not recoverable apart from the section, could, in exercise of the powers conferred by the subsection, *"properly be refunded"* on that ground. There has been some debate in this court as to the force of the word "properly" in this context. Mr John Taylor QC, on behalf of the applicants, at least at one stage in his argument, submitted that the word by itself imports the notion of fairness. This submission we do not accept. In our judgment, the presence of the word "properly" in the subsection is quite adequately explained by the need for the rating authority to satisfy itself that the restrictions contained in section 9(2) are not infringed and that any relevant preconditions set out in section 9(3) are satisfied before determining whether a refund should be made under section 9(1). If the payment sought would involve such an infringement, or if any relevant precondition were not satisfied, they could not properly make the payment. In our judgment, no greater significance than this can be attributed to the word "properly" in the context of subsection (1).

For present purposes, the crucially important word in section 9(1) is the word "may". Though this word, when appearing in a statute, in its ordinary meaning generally falls to be read as conferring an enabling or discretionary power, it has in some contexts been read as bearing an imperative sense; a number of examples are to be found in *Stroud's Judicial Dictionary* (4th ed) vol 3 at pp 1642-1645. However, in the present case, Mr Taylor did not go so far as to submit that the word "may" in section 9(1) should be read as "shall" — and, in our judgment, rightly so. A number of other provisions of the Act of 1967 explicitly impose an imperative obligation on a rating authority to repay rates (for example, sections 6(2)(i), 7(4)(b), 18(4) and 79(3)) in terms which are in contrast with the word "may" in section 9(1). Furthermore (as the opening words of subsection (3) show), section 9 contemplates that the authority will "determine" whether a refund should be made under subsection (1). The whole section, in our judgment, presupposes that, within the limitations imposed by subsections (2) and (3), the authority will have a discretion in determining whether or not to make a refund.

However, though the word "may" suffices to show that the authority has some discretion in determining whether to make repayment, it gives no guide as to the nature or extent of this discretion. "That must be inferred from a construction of the Act read as a whole": see *Padfield* v *Minister of Agriculture, Fisheries & Food* [1968] AC 997 at p 1033G *per* Lord Reid. That was a case concerning section 19 of the Agricultural Marketing Act 1958, which gave the minister power to refer to a committee of investigation complaints of a specified nature. Earlier in his speech (at p 1030B-C) Lord Reid had observed:

Parliament must have conferred the discretion with the intention that it should be used to promote the policy and objects of the Act; the policy and objects of the Act must be determined by construing the Act as a whole and construction is always a matter of law for the court. In a matter of this kind it is not possible to draw a hard and fast line, but if the Minister, by reason of his having misconstrued the Act or for any other reason, so uses his discretion as to thwart or run counter to the policy and objects of the Act, then our law would be very defective if persons aggrieved were not entitled to the protection of the court.

Lord Hodson at p 1045 said that "the discretion must be exercised by the minister in accordance with the intention of the Act" and at p 1046 pointed out that Lord Greene MR in *Wednesbury* itself [1948] 1 KB 223 at p 229 had drawn attention to the necessity to have regard to matters to which the statute conferring the discretion showed that the authority ought to have regard. Lord Upjohn at p 1060 considered that the minister "was entirely misdirecting himself in law based upon a misunderstanding of the basic reasons for the conferment upon him of the powers of section 19".

The principles applied in *Padfield*, in our judgment, reflect merely one facet of the *Wednesbury* principles when applied to an enabling or discretionary power conferred on a minister or public body by statute. By virtue of these principles, it is, in our judgment, clear that a rating authority, on receipt of an application for repayment of rates, is bound to take into consideration the object which Parliament must have intended to achieve in enacting section 9 and not to act in such a way as would frustrate that object.

In *R* v *Liverpool City Council, ex parte Windsor Securities Ltd* [1979] RA 159, Cumming-Bruce LJ, in considering the powers of a rating authority under para 3A of Schedule 1 to the Act of 1967 to reduce or remit rates on the ground of hardship, had this to say at p 179:

Paragraph 3A confers a discretion. In the exercise of that discretion the authority has to behave fairly. It must direct itself properly in law. It must direct its attention to the matters which it is bound to consider. It must exclude from its consideration matters which are irrelevant to what it has to consider. If it does not obey those rules it may be said to be acting unreasonably. See *Associated Provincial Picture Houses Ltd* v *Wednesbury Corporation*. Likewise it must not so act as to frustrate the object and purpose of the statute conferring the power. (*Padfield* v *Minister of Agriculture*.) So a statutory

discretion cannot be accurately described as an unfettered discretion. It is nothing of the sort. . . .

Lord Upjohn in *Padfield* [1968] AC 997 at p 1060 similarly did not approve of the introduction of the adjective "unfettered" to describe the discretion of the minister there in question. As he observed:

. . . the use of that adjective, even in an Act of Parliament, can do nothing to unfetter the control which the judiciary have over the executive, namely that in exercising their powers the latter must act lawfully and that is a matter to be determined by looking at the Act and its scope and object in conferring a discretion upon the Minister rather than by the use of adjectives.

In the present case, as we have said, the learned judge, in describing the discretion of the rating authority under section 9 as "truly unfettered", fully recognised that this was subject to the *Wednesbury* principles. Nevertheless, we think that the introduction of the adjective "unfettered" was potentially misleading. The very existence of the *Wednesbury* principles (as applied in *Padfield*) means that the discretion is by no means an unfettered one. In particular, the learned judge, with great respect to him, seems to us to have overlooked or paid inadequate attention to the fact that it is the duty of the rating authority, in exercising its discretion, to have regard to the purpose for which the relevant power was conferred on it.

The original statutory predecessor of section 9 (1) of the Act of 1967 was section 17 (1) of the Rating and Valuation Act 1961, which was substantially in the same terms. What then was the purpose for which Parliament, by these subsections, conferred on rating authorities the power to refund rates to a person in a case where he has made a payment in respect of rates which he was not liable to make but the amount paid is not recoverable by him apart from the section? The answer to this question is, we think, fairly obvious.

The need for the subsections is illustrated by a brief reference to the position at common law before they were enacted. Though payments of rates made under a mistake of fact involuntarily were recoverable (*Meadows* v *Grand Junction Waterworks Co* (1905) 69 JP 255), rates voluntarily paid under a mistake of law were not recoverable: *Slater* v *Burnley Corporation* (1888) 53 JP 70, *Slater* v *Burnley Corporation (No 2)* (1889) 53 JP 535. This did not mean that at common law a rating authority was or is necessarily entitled to retain the full benefit of rates paid to it which were not in truth owed. Rates paid under a mistake of law may be set off against subsequent demands: see *Blackpool & Fleetwood Tramroad Co* v *Bispham with Norbreck Urban District Council* [1910] 1 KB 592 and *Stanley* v *Weardale Coal & Coke Co Ltd* (1935) 6 DRA 49. As the learned judge rightly observed in the present case at p 88: "The ability so to do is founded on the inequity of allowing the rating authority to retain that to which they have no right without giving credit for it". However, this right of set-off did not and does not avail a ratepayer, such as the applicants in the present case, who has made a payment for which he was not liable, under a mistake of law, but is not in a position to achieve redress by means of a set-off. (Set-off, as the learned judge pointed out, is not available to the applicants because they were rated as an owner and, once each of the two units was occupied, there was no further liability against which credit could be claimed in regard to that hereditament.) Before the Act of 1961 came into force, there were thus a number of situations in which, despite the inequity, a person who had made payments of rates which he was under no liability to make was left entirely without redress.

As Lord Wilberforce observed in *Arsenal Football Club Ltd* v *Ende* [1979] AC 1 at p17:

Uniformity and fairness have always been proclaimed and judicially approved, as standards by which to judge the validity of rates. Indeed I believe that many men feel a more acute sense of grievance if they think they are being treated unfairly in relation to their fellow ratepayers than they do about the actual payments they have to make. To produce a sense of justice is an important objective of taxation policy.

As the law stood before 1961, ratepayers who had made a payment of rates for which they were not liable under a mistake of law (as opposed to fact) but were not in a position to avail themselves of a right of set-off might well have felt a legitimate sense of grievance. We think it clear that, in broad terms, the purpose of section 9 and its predecessor was to enable rating authorities to give redress and to remedy the injustice that would (at least *prima facie*) otherwise ordinarily arise, if they were to retain sums to which they had no right, in cases where persons had paid rates which they were not liable to pay.

If it be asked why the legislature, in enacting section 9 (1) and its statutory predecessor, did not think it appropriate to impose on the rating authority an imperative obligation to make repayment to a ratepayer who brings himself within the wording of the section, we think there is at least one ready answer. Cases could arise in which, on the particular facts, a rating authority could reasonably consider that it was not inequitable to retain the rates paid to it, even though these were moneys to which they had no legal right. Simply, for example (as Mr Taylor suggested in the court below), where the payer had fraudulently represented that he was the occupier when another was in fact the occupier.

In the present case, to quote the wording of the letter of November 8 1983, "the Committee were strenuously advised that the decision required . . . was one entirely for their discretion acting fairly in accordance with the principles set out in counsel's advice". Counsel, in para 8 of his opinion, while saying that the council "should be guided only by consideration of what is reasonable in all the circumstances", went on to refer to four specific factors which he said the council was entitled to take into account in reachng its decision. All these factors could fairly have been read as pointing towards a decision to withhold rather than to make repayment. Nowhere in the course of his opinion did he draw to the council's attention what we conceive to be the statutory purpose of section 9 (1). The reasons given on behalf of the council in the letter of November 8 1983, so far from indicating that they had had regard to this statutory purpose, give a strong indication to the contrary (see in particular reason (a), to which we will revert). We think that this appeal must succeed, if only on the grounds that the council, in reaching its decision, failed to take into account the basic reason why the relevant statutory power had been conferred on it.

However, there are further no less compelling reasons why, in our judgment, this appeal should be allowed. The four factors set out in paras (a), (b), (c) and (d) of the letter of November 8 1983 were regarded and specifically presented on behalf of the council not merely as being background circumstances in which they made their decision but as being the "*reasons*" why "the Committee decided against a repayment". Viewed as reasons for reaching an adverse decision, each one of these factors was, in our judgment, a bad one. We take them seriatim:

(a) "The advice that the payment made by the Company was paid under a mistake of law."

This echoed a factor which counsel advised the council they might take into account, namely "that money paid under a mistake of law is not recoverable as of right". As a statement of the legal position this is true. However, it is an express precondition of making any repayment under section 9 (1) that the money paid should not be recoverable as of right or, in the words of the subsection, "not recoverable apart from this section". The very purpose of the relevant statutory provision, as we have indicated, is to enable inequities to be remedied when rates have been paid under a mistake of law. To regard the existence of a mistake of law as being in itself a good ground for withholding payment would seem to us to frustrate the policy and objects of the section.

(b) "On the parallel that an application for relief from empty rates under para 3A of Schedule 1 [to the Act of 1967] would require an applicant to demonstrate "hardship" — which was not argued by your Client Company although they were invited to do so."

The suggested "parallel" is, in our judgment, a false analogy. Para 3A of Schedule 1, so far as material, provides:

. . . a rating authority shall have power to reduce or remit the payment of any rates payable in respect of a hereditament by virtue of paragraph 1 of this Schedule if they consider that the payment would cause hardship to the person liable for those rates.

The power to remit referred to in para 3A thus relates to rates for which the ratepayer has at all material times been liable in law. We fail to see how the insertion of a statutory requirement of hardship in the context of para 3A can have any bearing at all on the application of a section which gives power to refund rates which were never due to the rating authority — and, we might add, makes no mention of hardship. Quite apart from para 3A, the council would, of course, have been bound to give due weight to the factor of hardship *in the applicants' favour*, if hardship had been shown. A suggested analogy with para 3A, however, did not justify them in regarding the absence of hardship as a positive factor adverse to the applicants. The reference to para 3A was an irrelevant consideration.

(c) "The scheme and intent of the 1967 Act with regard to completion notices."

This somewhat elliptical sentence has to be read and can only be explained in the light of counsel's advice, to which the letter of November 8 1983 referred. Counsel in his opinion (para 1) had stated that the provision in Schedule 1 to the 1967 Act, with regard to completion notices, is designed "for the sole purpose of avoiding the 'mischief' of a developer deliberately keeping a new building unfinished so as to avoid paying rates until he has an acceptable tenant or purchaser". He had continued: "The intent and purpose of the Act is, among other things, to encourage owners to bring new properties into use in the interests of the economy generally and not to hold them unoccupied for ulterior commercial reasons." In para 7 of his opinion he had stated that he inferred that "Chetniks had wished to postpone completion for their own purposes". In para 8 he had listed as the third of the factors which the council were entitled to take into account: "The scheme and intent of the 1967 Act is to provide for completion notices for the purpose of preventing owners from avoiding rates by non-completion."

Whether or not it was so intended, counsel's opinion could reasonably have been read, and, we do not doubt, was read by the council's committee, as suggesting that it would be contrary to the scheme and intent of the provisions of the Act of 1967 relating to completion notices if the council were to accede to the applicant's request for a refund. In our judgment, any such suggestion is misconceived. Para 2(a) of Schedule 1 to the Act of 1967 shows the clear intention of Parliament that no rates shall be payable under para 1 of that Schedule in respect of a hereditament for any period during which "the owner is prohibited by law from occupying the hereditament". At all material times, the applicants *were* prohibited by law from occupying the property. Para 43 of the conditions attached to the approval given by the Greater London Council had specifically prohibited any part of the building from being occupied until the consent of the GLC had been obtained to the proposed user. If the applicants had infringed this condition, they would have acted illegally, by virtue of sections 34(4) and 148(2)(xiii) of the London Building Acts (Amendment) Act 1939 ("the Act of 1939"). It could not fairly be suggested that the applicants had been procrastinating in obtaining the requisite approval of the GLC to a proposed user which would enable the building to be lawfully occupied. As the learned judge commented:

the applicant could have chosen and obtained consent to a particular use in advance of finding a tenant for that user, but such a course would (as counsel for the council accepted) have been a commercially absurd one to adopt in that no sensible person would adapt his building to meet the requirements of the GLC in regard to a particular user until he knew that he could let for that usage.

For practical purposes, therefore, the applicants could neither complete the building until they had found a tenant nor obtain the requisite consent to a proposed user nor permit the building to be occupied. It is now common ground, as the stipendiary magistrate accepted, that the applicants had done their best to let the premises.

In these circumstances, we find it difficult to understand, and impossible to accept, the suggestion that it would be contrary to the scheme and intent of the provisions of the Act of 1967 relating to completion notices if the refund were to be granted. Mr Payton, on behalf of the council, observed that the prohibition appearing in section 34(4) of the Act of 1939 was of a punitive nature, affording a sanction for non-compliance. He pointed out that that sanction did not apply to other areas of the country. He submitted that there is no good reason why "a punitive provision should exempt someone from rates merely because he happens to live in London". The fact remains that, by virtue of the relevant legislation, the applicants had a statutory right to exemption from rates for the material period. The mere fact that they might or would not have been entitled to a similar exemption if the property had been situated outside London, in our judgment, affords no good ground at all for withholding a refund under section 9(1). It is an irrelevant consideration. Reason (c), which influenced the council's decision, was, in our judgment, misconceived.

(d) "That your Clients could have taken professional advice before making such payments and, of course, they could have avoided the problem if they had complied with the requirements of the London Building Acts."

As to the reference to the London Building Acts, we have already said that the council accepts that it would have been commercially absurd to adapt the building to the particular requirements of the GLC in regard to a particular user until it was known that the building could be let for that user. The fact that the applicants could have avoided the problem presented by the absence of consent of the GLC to the proposed user if they had complied with the requirements of the London Building Acts, though no doubt it is a fact, could not, in our judgment, constitute a good reason for the decision to withhold repayment of rates when all the particular facts of this case, and the general purpose of section 9(1), are borne in mind. The fallacy of this proposition is further illustrated by the following consideration. The only way in which the applicants could have avoided any problem created by non-occupation at a time before they had not yet been able to let the building would have been to obtain the necessary consents of the GLC and then to occupy it themselves. Thus, they would have made themselves rateable. In this event, however, they would have had the right to set-off against the rates for which they became liable by virtue of such occupation the rates already paid, for which they were not liable. The fact that the applicants could have "avoided the problem" in the manner suggested was a wholly irrelevant factor, if regarded as a reason for declining to exercise the discretion in their favour.

The same comment applies to the fact that the applicants could have taken professional advice before making the relevant payments. No doubt they could have done so, but we see no good reason why a failure to do so should be taken into the balance against them when the council determines whether or not to make repayment under section 9. The position might be different if, in making its determination, the council could properly regard as an adverse factor the fact that the applicants had made a mistake of law in effecting the payment; but we have already concluded that, contrary to the opinion of the council, it could not. So to hold would be to frustrate the purpose of section 9.

In the result, we conclude that the council, in reaching the determination to withhold repayment of rates for the four reasons which it gave, (1) failed to take into account an important consideration which it was bound to take into account, that is to say, the purpose for which the legislature had conferred on it the relevant power; (2) took into account four factors which could not constitute good grounds for withholding repayment, having regard to the purpose of section 9.

We accordingly allow this appeal. We set aside the learned judge's order and make orders for mandamus and certiorari of the nature mentioned at the beginning of this judgment. As we have already indicated, section 9, in our judgment, gives the council a discretion in determining whether or not to accede to an application for the refund of rates. The discretion is that of the council and we would not think it necessary, appropriate or indeed possible to attempt to list all the factors which they either should or should not take into account on the further determination which they will now have to make. It will suffice to say this. First, in embarking on their new deliberations, they should, in our opinion, take as a starting point the purpose of the relevant statutory provisions — which, in our view, as we have indicated, was to enable a rating authority to give redress and remedy the inequity that would otherwise (at least *prima facie*) ordinarily arise if it were to retain moneys paid in respect of rates which the payer was not liable to pay and to which the authority had no right. Second, they should not regard any of the four matters set out in the letter of November 8 1983, either jointly or severally, as themselves constituting good reasons for withholding repayment of these moneys to which they have no right, though their good faith in initially demanding and accepting them is not in question.

The appeal was allowed with costs and the judgment of Mann J set aside. Orders of certiorari and mandamus were granted quashing council's decision and directing it to hear and determine according to law. Leave to appeal to the House of Lords was refused.

Court of Appeal

March 5 1987

(Before Lord Justice DILLON, Lord Justice GLIDEWELL and Sir David CAIRNS)

TRENDWORTHY TWO LTD v ISLINGTON LONDON BOROUGH

Estates Gazette May 30 1987

282 EG 1125-1131

Rating — General Rate Act 1967, section 17 and Schedule 1 — rating of unoccupied property — Appeal by rating authority from decision of Mervyn Davies J in favour of declaration sought by plaintiff ratepayers that they were not liable to pay unoccupied property rates until the relevant hereditaments, part of the Angel Centre in Pentonville Road and St John Street, Islington, London, and their rateable values had been entered in the valuation list — Cross-appeal as to whether ratepayers were liable to pay unoccupied property rates before their appeal against completion notice had been finally decided — Court of Appeal divided on the appeal (Dillon LJ dissenting from majority decision to dismiss appeal) — It was submitted on behalf of the rating authority that unoccupied property rates could be demanded, as a result of Schedule 1, para 1(1) and section 6(1) and (2), even though the addition of the hereditament to the valuation list had reached only the proposal stage and the hereditament had not yet been entered in the list — Ratepayers argued that, while this might be the position in the case of occupied premises, it was not so in the case of the special code of provisions which governed the liability of owners of unoccupied premises — Held by the majority (Glidewell LJ and Sir David Cairns) that the owners were not liable to pay the unoccupied property rates until the rateable value, and thus the amount to be paid, had been established — Neither of the cases cited by the parties, *Bar Hill Developments Ltd* v *South Cambridgeshire District Council* and *Hastings Borough Council* v *Tarmac Properties Ltd* was an authority binding on the Court of Appeal on the point at issue — Dillon LJ, dissenting, considered that there was no difference in principle discernible in the legislation which would prevent the rule applicable to occupied premises from applying to unoccupied premises in this respect — He agreed with the views expressed by the Divisional Court in the *Bar Hill* case and he would have allowed the present appeal — He agreed, however, with the majority that the cross-appeal by the owners that they were not liable to pay the unoccupied property rates until their appeal against the completion notice had been finally disposed of should be dismissed — In the result both the appeal and the cross-appeal were dismissed, but leave was given to appeal to the House of Lords

The following cases are referred to in this report.

Bar Hill Developments Ltd v *South Cambridgeshire District Council* [1979] RA 379; [1979] EGD 946; (1979) 252 EG 915, DC
Hastings Borough Council v *Tarmac Properties Ltd* [1985] RA 124; (1985) 83 LGR 629; [1985] 1 EGLR 161; (1985) 274 EG 925
Kettle (B) Ltd v *Newcastle under Lyme Borough Council* [1979] RA 223; (1979) 77 LGR 700; [1979] EGD 934; 251 EG 59, CA

This was an appeal by the rating authority, the London Borough of Islington, from the decision of Mervyn Davies J (reported at [1986] 1 EGLR 187; (1986) 277 EG 534) in favour of the plaintiff ratepayers, Trendworthy Two Ltd, in regard to liability to pay unoccupied property rates in respect of the Angel Centre, an office development comprising two buildings at the junction of Pentonville Road and St John Street, Islington, London N1. There was a cross-appeal by the respondent ratepayers in regard to their liability to pay the rates before an appeal by them against the rating authority's completion notice had been decided.

Matthew Horton (instructed by C R Tapp, borough solicitor,

Islington) appeared on behalf of the appellant authority; W J Glover QC and Guy Roots (instructed by Michael Conn & Co) represented the respondents.

Giving the first judgment at the invitation of Dillon LJ, GLIDEWELL LJ said: This appeal concerns two questions relating to the rating of unoccupied property which have not hitherto been decided directly by this court. The appellant council are the rating authority of the London Borough of Islington. They are responsible for making the rate for the borough for each rate year from April 1. The rate is the amount sufficient to provide for such part of the total estimated expenditure to be incurred by the council as will not be met by other means, for example out of grants from central government.

By section 2(4) of the General Rate Act 1967, the general rate:

(a) shall be a rate at a uniform amount per pound on the rateable value of each hereditament in that area

ie in the London borough, and

(b) shall be made and levied in accordance with the valuation list in force for the time being.

The official responsible for the valuation list is the valuation officer. This list contains details of "every hereditament in the rating area and the value thereof" (section 67(2)). The valuation officer may at any time make a proposal for any alteration to a valuation list. He may thus make a proposal to add to the list a newly erected building and to ascribe to it gross and rateable values (section 69(2)).

An owner or occupier of the hereditament so added to the list, or the rating authority, may give written notice of objection to the proposal. Unless the valuation officer, the rating authority, the objector and the occupier then reach agreement, the matter is placed before the local valuation court by way of an appeal against the objection. The local valuation court decides what are the appropriate values to be ascribed to the hereditament and shall give directions for the making of the appropriate entry in the valuation list (sections 70, 72 and 76 of the Act). It is at this stage that a newly erected building first appears in the valuation list. There is a further right of appeal from the local valuation court to the Lands Tribunal (section 77). There may be a further appeal on a point of law by way of case stated to this court.

Normally the person liable to be assessed to rates in respect of any hereditament is the occupier of the property (section 16). The rating authority may resolve, in relation to certain classes of property, to rate owners as opposed to occupiers under section 55, but that is not here relevant. If an existing building became vacant or a newly erected building was left vacant after it was completed, until 1966 no rates were payable in respect of the building, since there was no occupier to be assessed for rates. However, sections 20 and 21 of the Local Government Act 1966 introduced provisions for the rating of unoccupied property which are now contained in section 17 of and Schedule 1 to the 1967 Act.

Section 17 provides:

A rating authority may resolve that the provisions of Schedule 1 to this Act with respect to the rating of unoccupied property —
(a) shall apply, or
(b) if they for the time being apply, shall cease to apply, to their area, and in that case those provisions shall come into operation, or, as the case may be, cease to be in operation, in that area on such day as may be specified in the resolution.

Who, then, is to be liable to the rates? By definition there is no occupier of the property. The answer is to be found when one turns to the First Schedule, in para 1(1), which, so far as material, reads:

Where, in the case of any rating area in which, by virtue of a resolution under section 17 of this Act, this Schedule is in operation, any relevant hereditament in that area is unoccupied for a continuous period exceeding

the standard period

the owner shall, subject to the provisions of this Schedule, be rated in respect of that hereditament for any relevant period of vacancy . . .

Then there is a final clause which is important in this appeal:

. . . and the provisions of this Act shall apply accordingly as if the hereditament were occupied during that relevant period of vacancy by the owner.

In order to decide what value is to be ascribed to the unoccupied hereditament, one turns to para 5 of the Schedule. Under para 5(1) it is provided:

Subject to the provisions of this Schedule, the rateable value of a hereditament for the purposes of paragraph 1 thereof shall be the rateable value ascribed to it in the valuation list in force for the area in which the hereditament is situated or, if the hereditament is not included in that list, the first rateable value subsequently ascribed to the hereditament in a valuation list in force for that area.

With a new building it is obviously necessary to know the date at which it is completed in order to decide whether thereafter it is unoccupied: until it is completed it is not a building much less an unoccupied building. Paras 7 to 10 of the First Schedule provide a procedure by which the decision whether a building has been completed is made. Under para 8(1) it is provided:

Where a rating authority are of opinion—
(a) that the erection of a building within their area has been completed; or
(b) that the work remaining to be done on a building within their area is such that the erection of the building can reasonably be expected to be completed within three months, and that the building is, or when completed will be, comprised in a relevant hereditament, the authority may serve on the owner of the building a notice (hereafter in this paragraph referred to as "a completion notice") stating that the erection of the building is to be treated for the purposes of this Schedule as completed on the date of service of the notice or on such later date as may be specified by the notice.

Under para 8(4) it is provided:

A person on whom a completion notice is served may, during the period of twenty-one days beginning with the date of service of the notice, appeal to the county court against the notice on the ground that the erection of the building . . . has not been or . . . cannot reasonably be expected to be completed by the date specified in the notice.

The last part of para 8(5) provides:

if the notice is not withdrawn and such an appeal is brought and is not abandoned or dismissed, the erection of the building shall be treated for those purposes as completed on such date as the court shall determine.

By para 7 it is provided:

For the purposes of paragraph 1 of this Schedule, a newly erected building which is not occupied on the date determined under the subsequent provisions of this Schedule as the date on which the erection of the building is completed shall be deemed to become unoccupied on that date.

If a completion notice comes into effect or if there is an appeal against it and the county court determines a date which is the appropriate date, then it is deemed to become unoccupied on the date on which the erection of the building is completed.

Finally, para 15, which contains definitions, defines the phrase "relevant period of vacancy" as starting with the unoccupied rating day, which is the next day after the expiry of three months from the date on which, under para 7, the building was deemed to become unoccupied. The relevant period of vacancy ends with the first occupation of the building.

I turn to the facts which have led to this appeal. The respondents are and were the owners of the Angel Centre, which is an office development at the junction of Pentonville Road and St John Street in Islington. It comprises two buildings, one much larger than the other.

On June 1 1983 the borough council served a completion notice under Schedule 1 para 8. This said that the premises could reasonably be expected to be treated as completed by September 1 1983. On June 20 1983 the respondents, Trendworthy, appealed to the county court. On March 24 1984 the valuation officer made proposals with regard to the Angel Centre, attributing to the main building a rateable value of £579,638 and to the smaller building a rateable value of £43,722.

On April 3 1984 Trendworthy objected to both proposals. On July 10 1984 the smaller building was let. On December 21 1984 His Honour Judge Marder QC dismissed the appeal against the completion notice, and on that same day the council issued a demand for rates. In relation to the smaller building that was a demand for the period from December 1 1983 to July 9 1984 (that letter date being, as will be seen, the date before it was let and rateable occupation thus commenced), and in respect of the larger building, which was still unlet, from December 1 1983 to March 31 1985. December 1 1983 is, of course, three months after September 1 1983, that being the date which is the commencement of the relevant period of vacancy based on a completion date of September 1. The total demanded was some £945,000.

On that same day the council amended the rate, that is to say, so far as their own records are concerned, they added these two hereditaments to their rating records and ascribed to them the rateable values contained in the valuation officer's proposals which were still, of course, under objection.

On January 11 1985 Trendworthy appealed against the county

court decision. On January 18 the council started winding-up proceedings against Trendworthy under section 223(a) of the Companies Act 1948 on the basis that Trendworthy had not paid the £945,000 or any part of it. On February 12 1985 Trendworthy obtained an injunction preventing the winding-up on the ground that the debt was bona fide disputed. On February 15 the writ in the present action was issued in the Chancery Division.

This is, at least in my experience, an unusual way of deciding the question which falls to be decided, but it has proved, if I may say so, to be extremely effective. The alternative, I suppose, would have been to wait until the matter came before the magistrates' court. Then either party could have appealed, if they thought right, from the magistrates by way of case stated and so the normal channel of appeal would have been entered.

On April 29 1985 British Telecom took a lease of the main building. On June 11 1985, since the main building was now occupied and the former rate demand had only run until March 31 1985, the council added a demand in respect of the main building for the intervening four weeks between the earlier date and the date of occupation, thus making the total demand now just over £1m, £1,031,000. In June and July 1985 the council, a little optimistically I think, started distress proceedings in the magistrates' court, which were promptly restrained by injunction.

On July 15 there was the first date of hearing before the local valuation court of the appeal arising out of the objections to the valuation officer's proposals. That hearing was adjourned *sine die* pending this court's decision with regard to the appeal against the county court's decision in respect of the completion notice, and those proceedings still stand adjourned. On October 31 1985 Mervyn Davies J gave his judgment in the present proceedings, and on March 24 1986 this court quashed the decision of His Honour Judge Marder in relation to the completion notice and remitted the appeal to that learned judge.

There is one provision of the Act which is relevant to these proceedings to which I have not so far referred, and before I come to submissions I must do so. It is section 6. Reading only the essential parts, section 6(1) provides:

Subject to the provisions of this section, the rating authority may at any time make such amendments in a rate (being either the current or the last preceding rate) as appear to them necessary in order to make the rate conform with the enactments relating thereto, and in particular may —
. . .
(c) make such additions to or corrections in the rate as appear to the authority to be necessary by reason of —
(i) the coming into occupation of any hereditament which has been newly erected or which was unoccupied at the time of the making of the rate.

Subsection (2) provides:

Where the effect of the amendment would be either —
. . .
(b) to charge to the rate a hereditament not shown, or not separately shown, in the valuation list, the rating authority shall not make any amendment of the rate unless either the amendment is necessary to bring the rate into conformity with the valuation list or a proposal for a corresponding alteration to the valuation list has been made by the valuation officer; and if effect, or full effect, is ultimately not given to such a proposal, and the amount of the rate levied in pursuance of the amendment is affected, the difference —
(i) if too much has been paid, shall be repaid or allowed; or
(ii) if too little has been paid, shall be paid and may be recovered as if it were arrears of the rate.

It was under the power contained in subsection (1) and believing that the restriction contained in subsection (2) was satisfied that the council, on December 21 1984, purported to amend the rate.

I turn to the submissions, which raise two main issues. The first and most important is: "Are owners liable to pay unoccupied property rates before the hereditament and the value attributed to it are entered in the valuation list?"

Mr Horton for the borough council submits that under section 6 the rating authority may add a newly erected building to the rates when it comes into occupation (section 6(1)). Rates may then immediately be demanded even though the addition of the hereditament to the valuation list has only reached the proposal stage and the hereditament has not been entered in the valuation list.

I believe that Mr Glover for Trendworthy conceded this in relation to occupied buildings: certainly if he did not concede it he did not oppose the proposition with any great fervour, and in my judgment it is correct.

Mr Horton then comes to the last phrase of para 1(1) of the Schedule. I repeat it: "the provisions of this Act shall apply accordingly as if the hereditament were occupied during that relevant period of vacancy by the owner."

This means, he submits, that, for the application of the principal sections of the Act (the Act preceding the Schedule), the owner is deemed to be the occupier. Thus, for the purposes of section 6(1), the premises are deemed to have come into occupation and under section 6(2) the rating authority can demand payment of the rate once the valuation officer has made a proposal to add the property to the list.

Mr Horton also argues as an alternative that his argument can succeed if the owner is not "deemed" to be the occupier. Mr Glover for Trendworthy submits that Mr Horton, if he is to succeed at all, must establish "deemed" occupation, and with that proposition of Mr Glover I agree.

In reply to Mr Horton's main submission, Mr Glover says that the proper meaning of the last phrase of para 1(1) of Schedule 1 does not make the owner the "deemed" occupier. Other, later, provisions of the Schedule make it clear that the Schedule recognises the actual facts. For instance, para 1(2), to which I have not hitherto referred, provides:

Subject to the provisions of this Schedule, the amount of any rates payable by an owner in respect of a hereditament by virtue of this paragraph shall be one-half of the amount which would be payable if he were in occupation of the hereditament.

Mr Glover says: "When you immediately find a paragraph saying, 'If he were in occupation', you can hardly construe the immediately preceding paragraph as meaning that he is deemed to be in occupation." Moreover, he says, when the statute wants to say "deemed" it says so. For instance, there is the reference which I have already read in para 7 of the Schedule to "deemed" to have become unoccupied.

Mr Glover submits that the effect of the last phrase of para 1(1) is to invest the owner of an unoccupied property with the mantle of an occupier of an occupied property in order to apply those provisions of the Act which are necessary to make the unoccupied property rating effective and which are not covered by, and are not incompatible with, the provisions of Schedule 1. Mr Glover submits that Mr Horton's interpretation is incompatible with para 5(1) of Schedule 1. I remind myself that that provides:

the rateable value of a hereditament for the purposes of paragraph 1 . . . shall be the rateable value ascribed to it in the valuation list in force for the area in which the hereditament is situated or, if the hereditament is not included in that list, the first rateable value subsequently ascribed to the hereditament in a valuation list in force for that area.

This makes it clear, says Mr Glover, that a demand for rates for unoccupied property can be made only when the hereditament first appears in the valuation list. The Schedule does not import, and is incompatible with, a demand for rates based upon the valuation officer's proposal as may be done for occupied property (see section 6(2)).

If there is no need to refer to authority, or assuming that there is no authority binding upon us in respect of this matter, I would agree with and would adopt Mr Glover's submission.

There are, however, two previous relevant decisions. The first is a Divisional Court decision, *Bar Hill Developments Ltd v South Cambridgeshire District Council* [1979] RA 379. There were four separate points argued during the course of that appeal, of which the last was the point which arises here.

Giving a judgment with which the other two members of the court (Lord Widgery CJ and Woolf J (as he then was)) agreed, Eveleigh LJ said at p 389:

It is said that the effect of that paragraph is that if a hereditament is not included in the list there can be no rateable value and there can be no rate demanded in respect of that property.

ie an unoccupied property

. . . the Act itself contains the power to demand rates although the property is not in the list. I see nothing in the Act to indicate that that situation is confined to occupied property as opposed to unoccupied property. After all, section 17 is itself part of the Act and that of course relates to unoccupied property. The Schedule should not be allowed to limit the effect of the sections of the Act itself.

That is accurately summarised in the headnote against subpara (4) at p 382 in the following words:

the rates could be recovered based on the rateable value proposed by the valuation officer notwithstanding the absence of a relevant entry in the valuation list because section 6 of the 1967 Act clearly gave the rating authority the power to demand rates although the property was not in the valuation list and para 5(1) of Schedule 1 should not be allowed to limit the effect of the sections of the Act itself nor did it attempt to do so.

That authority, of course, although not binding on us, is totally contrary to the conclusion to which I would come without authority.

The other decision is a decision of this court in *Hastings Borough Council* v *Tarmac Properties Ltd* [1985] RA 124. In that case a building remained vacant after completion between February 1 1976 and March 1 1979. The hereditament was entered in the valuation list on November 11 1980. A demand for rates had already been made before that, but in that case proceedings were commenced by writ to recover unpaid rates and the writ was issued after the date on which the hereditament was entered in the valuation list.

The issue in the case was: "Could a demand for rates be backdated to the beginning of the relevant period of vacancy?" The present issue did not arise because, as I have said, the writ was not issued until after the hereditament had been entered in the valuation list.

The answer to the question posed was answered by this court in the affirmative, but the relevance of the decision to us is to be found particularly in the judgment of Lawton LJ. At p 127 he said:

Schedule 1 contains detailed provisions for the rating of unoccupied property. In so far as it deals with an owner's liability to be rated, it is a code; but it does not deal with many matters which are incidental to liability without being determinate of liability, such as appeals.

On p 129 in the last para of his judgment he said:

In my judgment the rating authority's construction is correct. The liability of the owners of unoccupied hereditaments to be rated has a statutory origin. It was a new concept in rating law and could not easily be clothed with the existing law based as it was on the concept of occupancy. Situations had to be dealt with which did not arise when there was an occupier of a hereditament. Examples are provided by para 2 of the schedule. Further, there has to be a means for determining the rateable value for the relevant period of vacancy when a newly occupied or altered building has been completed. Schedule 1 deals with all these matters. It is a code enacting when liability to be rated arises and how the quantum of liability is to be determined. On matters within its terms it supersedes any provisions of the Act which determine the liability of occupiers. The concluding words of para 1(1), which were expressed as a subordinate clause, are, in my judgment, intended to make applicable to the owners of unoccupied hereditaments the provisions of the Act which are not concerned with liability. The construction put forward by the defendants would, in my judgment, tend to thwart the policy of the Act.

Mr Glover submits that the statement in that judgment, with which Fox LJ and Kerr LJ agreed, that Schedule 1 provides a code enacting when liabilities to be rated arise and how the quantum of liability is to be determined is binding upon us. I accept in general that that proposition is binding upon us, and indeed if it were not, with great respect, it would be my own opinion. But that does not necessarily mean that section 6 does not apply to the rating of unoccupied property. Whether it does or not depends upon the meaning one ascribes to para 5 of the First Schedule. If para 5 does not have the meaning which I have ascribed to it, the code does not say when a demand for unoccupied property rates may be made, and in that case there is a gap to be filled and the obvious place to look in order to fill it would be section 6. In such a case reference to section 6 would, in my view, be permissible. So, tempting though it is to say that *Hastings* v *Tarmac Properties* already answers the question for us, in my judgment it does not, and we are dealing with this matter without authority binding upon us.

The issue is not whether Trendworthy are liable to pay rates in respect of the Angel Centre for the period before it was occupied. In principle, Trendworthy accept that they are so liable or will be. The issue is when that liability arises. Were the company liable to pay at a time when the rateable value, and thus the amount to be paid, had not been established? In my judgment they were not and are not, and I would therefore dismiss the appeal.

The second issue arises out of the cross-appeal, and I can deal with it more shortly. It is this: "Are Trendworthy liable to pay unoccupied property rates before their appeal against the completion notice has been finally decided?" At the time of the hearing before Mervyn Davies J the decision of Judge Marder had not been quashed. Mervyn Davies J therefore considered it unnecessary to make the second declaration sought, which was a declaration "that the plaintiff is not liable to pay any unoccupied property rates . . . until the plaintiff's appeal against the 'completion notice' . . . has been finally disposed of."

Mervyn Davies J also held that, if it had been necessary to make a decision, he would have declined this declaration. His reason was that on an appeal it was for the county court to determine the date of completion. This the county court had done, and therefore para 8(5) of the Schedule was satisfied. In effect the learned judge must have concluded that an appeal against the decision of the county court to this court was of no effect, at least so far as para 8(5) of the Schedule is concerned.

It is a matter of impression as to whether he was correct in that view or not. My view is, with respect, that on this aspect of the case the learned judge was wrong. However, there is nothing in Schedule 1 to indicate that liability to pay cannot arise until the date of completion of the building has been finally established. If in the end the borough council received an overpayment, they would be under a liability to repay the balance. Although there was argument about this, this in my view is the effect of section 9 of the Act of 1967. There is nothing in the Schedule which in this respect states when liability to unoccupied property rates arises.

I would therefore dismiss the cross-appeal, though for somewhat different reasons from those adopted by the learned judge.

Agreeing with the judgment of Glidewell LJ, both on the appeal and on the matter raised by the respondents' notice, SIR DAVID CAIRNS said: I add a few words on only one point, namely the contention by Mr Horton, on behalf of the appellant, that the effect of the words "the provisions of this Act shall apply accordingly as if the hereditament were occupied during the relevant period of vacancy by the owner" (at the end of para 1(1) of the First Schedule to the General Rate Act 1967) is that, for the purposes of the Schedule, the owner of an unoccupied hereditament is deemed to be the occupier.

It may be that this is not decisive either way of the appeal, but certainly it would be easier to construe section 6 of the Act as applying to a newly erected building if the owner were deemed to be the occupier.

When something is intended to be deemed, certainly so far as this Act is concerned, Parliament says so in terms. An example of that is in para 7 of the Schedule:

For the purposes of . . . this Schedule, a newly erected building which is not occupied on the date determined under the subsequent provisions of this Schedule as the date on which the erection of the building is completed shall be deemed to become unoccupied on that date.

Deeming something to be so means assuming a state of affairs to exist which does not exist. The provision that the Act is to apply as if the owner were the occupier requires no such assumption. I therefore reject that contention on behalf of the appellants, and that is one of the reasons why I, too, would dismiss this appeal.

Dissenting in regard to the appeal, DILLON LJ said: So far as the cross-appeal is concerned, I agree with my lords that it is not appropriate to make the declaration sought that the plaintiff is not liable to pay any unoccupied property rates in respect of the premises until the appeal against the completion notice has been finally disposed of. I find nothing in the Act which precludes the rating of an owner in respect of unoccupied property until after the appeal process in determining the date of completion has been finally disposed of.

The provisions for rating unoccupied property came about because it was found that some properties were, for the commercial reasons of their owners, remaining unoccupied for long periods after completion. But there may be cases where, even though there is a dispute over the completion date, it is clear that on any view the building has been completed before the commencement of a new rating year and has yet remained unoccupied throughout that rating year. I see no reason therefore to include in the Act any mandatory provision that there should be no rating of unoccupied property until the date of occupation has been finally determined, and in the circumstances of the appellate proceedings in this case it is not appropriate to go further into what is likely to be an academic question anyhow.

So far as the appeal is concerned, I regret that I do not find myself in agreement with my lords. I could have wished, therefore, that more time were available to set out a fully reasoned and clearer judgment, but there are circumstances in this case which make it expedient that we should give judgment speedily, and I must express my views as best I can after the short time available.

The question is whether the rating authority can levy the unoccupied rate on the property before the rateable value has been finally established and entered in the valuation list. Can the unoccupied rate be levied on a proposal?

It is clear, at any rate since *B Kettle Ltd v Newcastle-under-Lyme Borough Council* [1979] RA 223, a decision of this court, that the occupied rate can be levied on a proposal. This is as a result of section 6(1) and (2) of the 1967 Act.

In his judgment in *Kettle*'s case Lord Denning MR said at p 226, after citing from section 6(2) of the Act:

Those words came in in 1948 for the first time. It seems to me that the plain intendment of the legislature was to enable amendments to be made when premises came newly into charge for rating and immediately when a proposal was made, before the matter had gone through the various procedures, and before it got into the valuation list.

I find it difficult to discern any reason in principle why that should not also be so with the unoccupied rate, but of course that must depend on the true construction of the Act.

The Divisional Court, consisting of Lord Widgery CJ, Eveleigh LJ and Woolf J (as he then was), in the case of *Bar Hill Developments Ltd v South Cambridgeshire District Council* [1979] RA 379, held that the rating authority could levy an unoccupied rate on a proposal and before the rateable value had been finally established.

Section 17 of the Act and the First Schedule to it are the provisions primarily concerned with the unoccupied rate. Other provisions of the Act come in, if at all, by virtue of the words at the end of para 1(1) of the First Schedule:

the owner shall, subject to the provisions of this Schedule, be rated in respect of that hereditament for any relevant period of vacancy; and the provisions of this Act shall apply accordingly as if the hereditament were occupied during that relevant period of vacancy by the owner.

For myself I would respectfully agree with the interpretation of those words given by Lawton LJ in his judgment in *Hastings Borough Council v Tarmac Properties Ltd* [1985] RA 124 where he said, at p127:

Schedule 1 contains detailed provisions for the rating of unoccupied property. In so far as it deals with an owner's liability to be rated, it is a code; but it does not deal with many matters which are incidental to liability without being determinate of liability, such as appeals.

At p 129, in relation to Schedule 1, he said:

It is a code enacting when liability to be rated arises and how the quantum of liability is to be determined. On matters within its terms it supersedes any provisions of the Act which determine the liability of occupiers.

That tallies, as I understand them, with the references to the Schedule and the Act in the words at the end of para 1(1) of the Schedule. The owner is to be rated subject to the provisions of the Schedule, and the provisions of the Act are to apply accordingly as if mentioned in the Schedule. The Schedule must therefore, so far as it goes, have priority over other provisions of the Act. I do not therefore feel able to support that part of the reasoning of Eveleigh LJ in the *Bar Hill* case where he said, at p 389:

The schedule does not restrict the Act. What it does is to develop the provisions of the Act and to provide ways of determining certain matters which are essential for the implementation of the Act's provisions . . . The schedule should not be allowed to limit the effect of the sections of the Act itself.

It does not, however, follow that the decision in the *Bar Hill* case was wrong. The actual decision in *Tarmac* was on a different point which does not, I think, affect the decision of the present case. The provisions of the Schedule are to prevail if there is conflict with the provisions of the Act, but, save where there is conflict, the provisions of the Act are to apply as if the hereditament were occupied by the owner.

The provisions of the Schedule, though forming a code, form a code which is avowedly incomplete, and there are many provisions of the Act which must apply to the unoccupied property rate even though they are not specifically mentioned by the Schedule. Sometimes this is achieved by some form of general saving in the Schedule of other rights or remedies: sometimes it is not.

Para 13 of the Schedule, for instance, provides:

Any amount due in respect of rates payable by virtue of para 1 of this Schedule shall, without prejudice to the operation of any other enactment under which it is recoverable be recoverable as a simple contract debt in any court of competent jurisdiction.

In my judgment, that gives an alternative remedy, but the provisions of the Act for the recovery of occupied rate through the magistrates' court and by distress must also be applicable to the recovery of unoccupied rate. I did not understand this to be disputed by Mr Glover.

So also the provisions of section 69(2) of the Act as to the making of proposals for the amendment of the valuation list which are a necessary preliminary to any entry being made in the valuation list in respect of a newly erected property or to the alteration of the rate to permit the rating of a newly erected property under section 6 of the Act must be applicable to the unoccupied rate, otherwise no newly erected property would ever achieve a rateable value so long as it was unoccupied.

All this emphasises that the unoccupied property rate is a rate, and a part of the general rate which the rating authority levies under the Act: it is not an entirely separate levy or impost. Many of the provisions of the Schedule or code are enabling provisions giving particular powers to the rating authority which it was thought necessary that they should have in relation to the unoccupied rate, but in my judgment they must also have the powers conferred by the Act in relation to the occupied rate in so far as not excluded by the provisions of the Schedule. *Expressio unius est exclusio alterius* is a limited maxim. This, as I see it, is achieved by the final words of para 1(1) of the Schedule, which I have already quoted. In essence, by those words the owner is deemed to be in occupation of the hereditament during the relevant period of vacancy.

It is objected to this approach that the draftsman of the Schedule has used the word "deemed" where he meant there to be a deeming process. For instance, by para 7 of the Schedule it is provided:

For the purposes of paragraph 1 of this Schedule, a newly erected building which is not occupied on the date determined under the subsequent provisions of this Schedule as the date on which the erection of the building is completed shall be deemed to become unoccupied on that date.

It is said that it is wrong on the scheme of this Act to infer a deeming where the actual word is not used, and it is further said that it would be nonsense and self-contradictory for the one Schedule to provide by para 7 that the hereditament is deemed to be unoccupied and by para 1 that it is deemed to be occupied by the owner.

This, as it seems to me — and with all respect to those who take the different view — is a merely verbal point without substance. The Schedule deems the property to be unoccupied from the completion date (para 7). For the purpose of the Schedule the commencement of the relevant period of vacancy is the unoccupied rating day as defined in the Schedule. Then, for the separate purpose of applying the provisions of the Act to an unoccupied property for the purposes of the Schedule, the provision is made in para 1(1) which I have quoted and which has the effect that the property is, during the relevant period of vacancy, to be treated as, or deemed to be, occupied by the owner. That saves writing out in detail in the Schedule all the provisions of the Act with the amendments *mutatis mutandis* to make them applicable to property which is, in truth and by deeming, unoccupied.

It follows, in my judgment, that, for the purposes of section 6(1) of the Act, a newly erected property has come into occupation. Subject therefore to consideration of para 5 of the Schedule, to which I will come, the provisions of section 6(1) and (2) of the Act are applicable to the unoccupied property rate, ie the rate can be altered subject to a proposal being made, and if a proposal is made under section 69(2) the rate can be collected, as in *Kettle*'s case, on the proposal subject to adjustment when the rateable value is finally established. In this respect I agree with the reasoning of Eveleigh LJ in the *Bar Hill* case.

Does para 5 of the Schedule preclude this? Para 5 provides:

Subject to the provisions of this Schedule, the rateable value of a hereditament for the purposes of paragraph 1 thereof shall be the rateable value ascribed to it in the valuation list in force for the area in which the hereditament is situated or, if the hereditament is not included in that list, the first rateable value subsequently ascribed to the hereditament in a valuation list in force for that area.

In my judgment that does not preclude the collection of the rate on a proposal. It provides that the rateable value shall be the first rateable value ascribed to the hereditament in the valuation list, but that is equally so with the occupied property rate. The rateable value is the value ultimately agreed or established when the proposal has been considered and discussed and, if necessary, the appeal processes have been exhausted. When the rateable value has been so established in accordance with the provisions of the Act, it relates back, as appropriate, either under section 79 of the Act, with occupied property rate, or, with unoccupied property, as explained in *Tarmac*. That does not preclude the interim collection on the proposal as in

Kettle's case subject to adjustment when the rateable value is established and entered in the valuation list. The collection on a proposal, as I see it, is a matter of interim machinery subject to adjustment, and I find nothing in para 5 of the First Schedule which precludes that machinery being applied in relation to unoccupied property.

Accordingly, for my part I would allow this appeal, but as my lords take a different view the appeal will be dismissed.

The appeal and cross-appeal were both dismissed with costs. Leave to appeal to the House of Lords was granted on condition that the appellants' solicitors notify the House of Lords immediately that leave has been granted, so that their Lordships can consider, if they wish, whether they want to hear both appeals together or separately.

For further cases on this subject see pp 201 and 248

REAL PROPERTY AND CONVEYANCING

Court of Appeal
January 19 1987
(Before Lord Justice O'CONNOR, Lord Justice LLOYD and Lord Justice GLIDEWELL)

STROUD v WEIR ASSOCIATES LTD

Estates Gazette March 14 1987
281 EG 1198-1202

Mobile Homes Acts — Dispute between owners of mobile home site and owner of one of the mobile homes as to pitch fee — Agreement provided that in determining amount of pitch fee regard should be had to (1) the Retail Price Index, (2) sums expended by site owner for benefit of occupiers of homes, and (3) "any other relevant factors including the effect of legislation applicable to the operation of the park" — The 1983 Act, unlike the 1975 Act, did not provide for arbitration on disputes but for questions to be determined by the county court — The home-owner, dissatisfied with a proposed increase in the pitch fee from £564 per annum to £705.12, applied to the county court — The judge increased the fee only in accordance with the increase in the Retail Price Index of 3.7 per cent, making the fee £584.88 — Site owners appealed to the Court of Appeal — County court judge had rejected a submission that the court should not determine actual figures but consider only whether the appropriate factors had been taken into account — He had construed the words "applicable to the operation of the park" as referring to the whole of the phrase quoted above — He had ruled that evidence as to pitch fees on other sites was not relevant and that he was not required to determine a "fair market rent" — Held by the Court of Appeal that the judge had approached the matter correctly and that his decision should be upheld — He had dealt with all the factors and the only one which provided scope for an increase was the Retail Price Index, which he applied — The case of *Beer* v *Bowden*, where the term "fair" was implied, was distinguishable — Appeal dismissed

The following case is referred to in this report.

Beer v *Bowden* [1981] 1 WLR 522; [1981] 1 All ER 1071; (1976) 41 P&CR 317, CA

This was an appeal by Weir Associates Ltd, leasehold owners of a mobile home site at Havenwood Park, near Arundel in Sussex, from a decision of Judge Slot at Chichester County Court in favour of George Albert Stroud, the present respondent, in a dispute mainly concerning the pitch fee in respect of the pitch occupied by the respondent.

R J F Gordon (instructed by Davies, Thomas & Cheale, of Littlehampton) appeared on behalf of the appellants; Charles S Taylor (instructed by Hubbard & Co, of Chichester) represented the respondent.

Giving the first judgment at the invitation of O'Connor LJ, GLIDEWELL LJ said: The present appellants, Weir Associates Ltd, are the lessees of a mobile home site called Havenwood Park, which is near Arundel in Sussex. The freehold is owned by Mr and Mrs Weir. Mr Weir is a director of and shareholder in the company, which holds under a lease, originally held by him and then assigned to the company in 1982, which was a 25-year lease and thus runs until the year 2007.

Mr Stroud is the owner of one of the mobile homes on that site. He has lived in that home on a pitch at Havenwood Park since August 1978. At that time the company had not been formed and Mr and Mrs Weir themselves were the operators of the site. Mr Stroud at that time occupied the pitch under a written agreement between him and Mr and Mrs Weir.

At that time the relevant legislation governing residential mobile home sites was the first legislation of its kind, the Mobile Homes Act 1975. Among other matters, that agreement (ie the 1978 agreement) provided for arbitration as to disputes, and clause 8(b), the arbitration clause, contained the following provisions:

If an arbitrator is called upon to determine the amount of the annual pitch fee he shall take into account in determining the same:-
(a) the Index of Retail Prices
(b) sums expended by the Owner for the benefit of the occupiers of Mobile Homes on the Park
(c) any other factors which he shall consider relevant.

The 1978 agreement provided for annual review of the pitch fee, which I shall from now on call the rent. Generally the evidence given and accepted was that until 1983 the rent was increased by a percentage equal to the increase in the Retail Price Index in the preceding year. In May 1983 the 1975 Act, which had always been intended by Parliament as an interim measure, was repealed and replaced, save as to certain ancillary provisions, by the Mobile Homes Act 1983. The 1975 Act had for the first time introduced some security of tenure for the occupiers of pitches on mobile homes sites. Generally speaking, that Act provided for security of tenure for five years with the opportunity to extend for a further three.

The 1983 Act gives greater security. Part I of Schedule 1 to the Act contains a number of implied terms which the Act incorporates into all agreements from the time when the Act comes into force. In so far as those implied terms are in conflict with express terms they override those express terms. The effect of those implied terms so far as security of tenure for the pitch occupier is concerned is that, provided he continues to pay his rent and complies with the other terms of his agreement and provided that he is occupying the mobile home as his main or principal residence, and finally that the condition of the home is not such as would be detrimental to the amenities of the site, he has security which is either indefinite or lasts until the end of the owner's interest, which in this case, as I have already said, means until the year 2007.

Under the 1975 Act the home-owner could sell his mobile home to another potential occupier with the right to assign the right to occupy the pitch. That required the consent of the site owner, but the 1975 Act provided that such consent should not be unreasonably withheld. Moreover, under the 1975 Act, if a home-owner wished to sell his home he had first to offer it to the site owner, who was entitled to purchase it at a discounted price of 85 per cent off the market price. Alternatively, if the site owner did not buy it and it was resold to another person, the site owner was entitled to a commission on the sale which, under the agreement in force in this case, was 15 per cent.

Almost identical provisions apply under the 1983 Act, but with this important practical alteration. The commission which the site owner may take is at a rate which is to be fixed by a statutory instrument made by the Secretary of State. It has been fixed at 10 per cent, so the percentage which the site owner was allowed to take was reduced as a result of the coming into force of the 1983 Act: a benefit, of course, to the pitch occupier. The pitch occupier thus gained two major benefits from the 1983 Act: one was the increased security of tenure, which not merely gave him an intangible but important benefit but also increased the value of his home itself; and second, as I have said, his right to assign, if he wished to do so, was subject to a reduced maximum commission payable to the site owner.

The 1983 Act also requires that the site owner shall, within six months of the Act coming into force in relation to existing occupiers or three months in relation to new occupiers of pitches, serve upon the pitch occupier a written statement which shall specify a number of matters. It has to set out the terms implied under Schedule 1, Part I, to the Act, and then it has to set out the express terms of the agreement. From that recital of the legislation and its effect I go back to the facts.

The lease under which Weir Associates Ltd hold this site commenced on November 30 1982, ie before the 1983 Act came into force. At some date after May 1983 (the precise date is not clear), Weir Associates served on Mr Stroud a written statement in accordance with the 1983 Act in a form provided by the National Federation of Site Operators and the National Caravan Council, neither of whom, we are told, though it may not be strictly relevant, is a body representing pitch occupiers. The Federation of Site Operators represents site owners — as the name makes clear.

The written statement, as it was required to do, contained, among other things, the express terms of the agreement between the parties. The statement said at its commencement that the agreement commenced on August 7 1978, so in terms it was referring back to the original agreement between Mr Stroud and Mr Weir. In my view, the express terms set out should then have been the terms of the 1978 agreement. That is what the statute provides, subject only to the alterations made by those terms which were required to be implied under the 1983 Act.

Nevertheless, Mr Stroud accepted the new agreement in so far as it differed from the old one (I will come in a moment to one way in which it differed) and did not, as he might have done, object, by going to the court, about the alteration of the terms of the new agreement, so nothing turns on that so far as these proceedings are concerned.

The differences that are relevant to these proceedings between the 1978 and the 1983 agreements are these: first, in the 1983 agreement there is no arbitration clause. The effect of that is that by section 4 and section 5(1) of the 1983 Act questions arising under the agreement are to be determined by the county court. Second, there is an alteration in the matters to which regard is to be had when the pitch fee is reviewed. The relevant clause in the 1983 agreement is clause 7(a). It does not talk about what the arbitrator is to have regard to because, of course, there is no provision for arbitration. It reads:

On the review date namely the first day of January in each year the amount of the annual pitch fee shall be reviewed and in determining the amount of the reviewed pitch fee regard shall be had to:
(i) the Index of Retail Prices
(ii) sums expended by the owner for the benefit of the occupiers of mobile homes on the park
(iii) any other relevant factors including the effect of legislation applicable to the operation of the park.

It will at once be seen that the first and second of those factors are exactly the same as the factors to be taken into account by the arbitrator under the 1978 agreement. The third factor has been altered in two ways which may be significant: first, by adding the words "including the effect of legislation" and, second, by adding the words "applicable to the operation of the park".

In 1983 the rent for Mr Stroud's pitch was £564 per annum or, to be very precise, £47.01 per month, plus £66 in round figures for water and sewerage. On August 9 1983 Weir Associates served on Mr Stroud a notice of increase of rent from January 1 1984 to £705.12 per annum, or £58.56 per month. Mr Stroud objected to that by a letter of December 6 1983 in which he said:

I refer to your statement dated August 8 1983 as to the proposed pitch fee for the above payable with effect from January 1 1984. I do not agree with the pitch fee and consider it excessive.

Following past practice, ie, using the Retail Price Index of June 1983 of 3.7%, the pitch fee would be £48.74 per month. However, as a gesture of goodwill and to avoid possible costly and lengthy court action, I would agree, in this instance, to a pitch fee increase of double the June 1983 figure, that is 7.5% in which case the monthly pitch fee would be £50.53. I am, of course, prepared to discuss the matter with you.

In the event of our failing to reach agreement, the County Court is empowered to consider your written statement to me dealing with such matters as the pitch fee.

Mr Gordon for Weir Associates sought to argue that in that open letter Mr Stroud was in some way committing himself to paying more than the figures for the increase in the Retail Price Index would dictate. I do not take that view: albeit this was an open letter, it was clearly an open letter seeking to compromise the dispute in order to try to avoid the litigation which led the parties to this court. The offer was not accepted, and that really is the end of the matter.

In reply to that letter Mr Weir, on January 14 1984, wrote a general letter to all the residents (it was addressed to "Dear Resident") on the site. He set out the reasons which he was suggesting justified the increased pitch fee for which he was asking. In the second paragraph of that letter he said:

The first point I think to make clear is that by virtue of the provisions in the Mobile Home Act 1983 reducing the commission payable upon the sale of Mobile Homes the site is no longer economically viable at the rent for 1983. Naturally, as a businessman I seek to run the site at a viable profit and to assist me in ensuring that I achieve this, I naturally take specialist advice from accountants.

Then he went on to say that he was in fact asking for less than the accountants had advised.

In the meantime on December 30 1983 Mr Stroud had applied to the county court by originating application "for an Order in such terms as are just and equitable varying or deleting the following express terms of an agreement offered to the Applicant". The first of the matters dealt with is that the "pitch fee of £705.12 payable from the 1st day of January 1984 as demanded by the Respondent in the letter is excessive". The application further asked that para 7 of the 1983 agreement should itself be altered so as to make provision for determination of the review of the pitch fee, and one other matter is referred to. So far as that is concerned the learned judge found against the applicant and, there being no cross-appeal on that, no issue now arises about it. The sole issue before this court was as to the judge's determination in relation to the rent.

The judge determined, at the conclusion of a lengthy hearing (he, as he said, was starting his judgment at 7 o'clock on the evening of the second day of the hearing), that the rent should only be increased to take account of the increases in the Retail Price Index. The evidence before him is that that increase had been some 3.7 per cent, so he simply added 3.7 per cent to the 1983 rental in order to arrive at a figure of £584.88. It is against that decision that Weir Associates now appeal.

In his judgment the judge set out the three factors which under clause 7 had to be taken into account. He then looked at them and commented as follows: "*Item (1)*. The RPI increase . . . went up by 3.7%. So this rent must go up by at least that amount." Mr Gordon submits that in so saying the learned judge was wrong, and strictly I think that comment is justified. I think that the rent is not required to go up by the increase in the Retail Price Index, however large or small that may be: it is a factor to be taken into account. But if the judge is to be understood as saying that with a relatively small rate of inflation in the relevant year it is right that the rent should go up by at least that amount, that remark would be understandable and I would agree with it.

The second item related to sums expended by the owner for the benefit of the occupiers of the park. The judge found against Weir Associates on that issue and nothing now turns on it.

One therefore comes to the third factor, "any other relevant factors including the effect of legislation applicable to the operation of the park". The first matter to be decided was, "What is the proper interpretation of that clause?" and, second, having decided that, "What falls under it?"

The judge construed clause 7 as meaning "any other relevant factors applicable to the operation of the park, including the effect of legislation applicable to the operation of the park". In other words, he expressly applied the phrase "applicable to the operation of the park" to the whole of the clause. He said:

I think that the words "Any other relevant factors" must be read in the context of the whole of clause 7(a). I think that these words mean that I must take into account any other relevant factors applicable to the operation of the park, including the effect of legislation applicable to the operation of the park. In addition I think that the draftsman must be deemed to be intending to limit those factors by reference to the context of the whole clause. The principle of *noscitur a sociis* applies. Factors which are not applicable to the operation of the park should not be taken into consideration. If there is any ambiguity about this clause, it should be construed *contra proferentem*: and that means that it should be construed in favour of the applicant.

Before I come to express my own view about that, I think it right to say a word or two about this whole exercise of reviewing a pitch fee and how a dispute as to the review of pitch fee is to be dealt with where there is no arbitration clause in the agreement. The 1983 Act provides in effect that an agreement may contain an arbitration clause but if it does not, in England and Wales, the issue shall be determined by the county court. It is to my mind clear that when an agreement, as this one does, talks about "determining" the pitch fee, the determination is not a unilateral process. Mr Gordon I think was inclined to argue that the word "determination" means simply "determination by the site owner". I believe that to be wrong. It is clear to my mind that determination involves the agreement of the parties or an order of the court. Clause 7(c) reads:

A note of the reviewed pitch fee shall be endorsed hereon in the form set out in the Second Schedule.

As my Lords pointed out during the course of argument, if one looks at the Second Schedule the form in which the reviewed pitch fee is to be recorded provides:

On

such and such a date

the annual pitch fee payable hereunder was reviewed in accordance with express term 7 hereof and it was agreed by the owner and the occupier that the annual pitch fee payable during the review period would be. . . .

So the agreement itself envisages, as one would expect, that the pitch fee should be agreed between the parties.

If there is no agreement the disagreement may be as to either or both of two matters. The dispute may be as to whether the site owner has taken into account factors which fall within clause 7 of the agreement, or it may be as to whether, taking into account those factors, he has arrived at fair and appropriate figures. Whichever it is, such a dispute is in my view "any question arising under any agreement" to which the 1983 Act applies and it is therefore a dispute within section 4 of the Act.

Mr Gordon argues that the task of the court when faced with such a dispute is merely to decide whether the fee fixed by the site owner is or is not in accordance with the terms of the agreement: in other words, he argues that if it becomes apparent that the site owner has taken appropriate factors into account the court is not concerned to go into the actual figures. However, if the site owner had not taken into account appropriate factors or had taken into account a factor which did not fall into the agreement, the figure would have to be altered and the court would have to decide what then was an appropriate figure. In my view, Mr Gordon's argument on this point is wrong. There is nothing, so far as I can see, in either the agreement or the Act itself which trammels the court once it has before it a dispute as to the review of a rent or pitch fee. In my view, the court cannot merely decide whether relevant factors have been taken into account but it can also decide the figures themselves: in other words, in this respect the court is acting as an arbitrator would do if there were an arbitration clause in the agreement.

I come back to the major issue between the parties, that is to say, "What is the proper interpretation of clause 7(a)(iii) of the 1983 agreement?" As I have already said, the change from the 1978 agreement is that instead of the phrase, "Any other factors which he", ie the arbitrator, "shall consider relevant", the clause now reads, "Any other relevant factors including the effect of legislation applicable to the operation of the park."

The judge interpreted "applicable to the operation of the park" as applying to the whole phrase. For myself I think he was right to do so. Grammatically there is no break, no comma or any other indication to show that the phrase "applicable to the operation of the park" is intended only to include the effect of legislation. The words make sense read as a whole, and for myself I would so interpret them. Whether it is necessary to look at neighbouring words in order to arrive at that conclusion I rather doubt, but if it were I apprehend that what the judge had in mind was that the immediately preceding subclause relating to sums expended by the owner for the benefit of occupiers related to this park and this park only, and he was suggesting other relevant factors which were to be taken into account. For myself I simply base this interpretation on what I believe to be the clear meaning of the phrase itself.

How does that support the judge's decision? Mr Taylor for Mr Stroud reminded us that the evidence which really is in issue here, that is to say the evidence put before the judge as to rents or pitch fees at various other mobile home sites, was not at first apparently the basis upon which Mr Weir sought to support his increased pitch fees. In the letter of January 14 to which I have referred he relied upon his loss of commission as being a major factor, together with subsequent loss of viability of the site. The judge held (in my view he was right to do so) that the loss of commission was a relevant factor applicable to the operation of the site, but he also held that it was in effect counterbalanced by the greater sale value of the homes: in other words, although Mr Weir would get commission at a lesser percentage the actual amount would not be appreciably affected. The judge of course did not have any precise figures in front of him, but he said that one balanced out the other so there was no increase to take account of that factor. There is no appeal so far as that part of his decision is concerned.

So far as improvements made to the site are concerned, that was examined and it was found that the evidence did not justify the suggestion. Reference to what has been called "fair market rent", Mr Taylor submits, was really something of an afterthought, though it became a main issue when the matter came to court. It is not entirely clear from the documents before us when this issue was first raised, but it appears to be the case that it must have been raised some time before July 1984 and certainly not at the initial stages of the dispute between the parties.

The judge, having reached the conclusion which I have already recited as to the meaning of clause 7(a)(iii) of the agreement, ruled that evidence as to pitch fees on other sites was not a relevant factor applicable to the operation of this park. Again I agree with him. However, he went on to say what he might have done had he come to a different conclusion on the interpretation of clause 7(a)(iii). He said this:

The difficulty which I find, in trying to assess a fair market rent, is that Mr [Peter D] King

a chartered surveyor who gave evidence on behalf of Weir Associates Ltd

has isolated 18 characteristics which he says are important on a site: and, according to his schedule, this site is much better than most of the others on the basis of these 18 characteristics. No doubt this site is a very attractive site. I have considered everything that Mr King and Mr Stroud have told me; and, from what they have told me, I find it just about impossible to determine what a fair market rent would be. The position is wholly different from the problems which arise in assessing a fair market rent for, for example, shops in a parade. There is not that sort of similarity or comparability.

A little further on he said:

Looking at the schedule, I would be inclined to take the view that £705 is on the high side, whereas £564 is on the low side. It looks as though a fair market rent would be somewhere between these figures: but the sites are not truly comparable with one another, so that the use of comparables is very difficult.

The judge was asked by the site operators to say that taking into account a fair market rent was a relevant factor, and for the reasons I have already explained he decided that it was not. The phrase "a fair market rent" is not to be found in the agreement or the legislation. It is of course a phrase borrowed from the Rent Acts. In my view, the real question the judge was determining was not whether "a fair market rent" or indeed a market rent was a relevant consideration or not: it was whether evidence of the pitch fees or rents charged at other sites was a relevant factor applicable to the operation of the park within clause 7(a)(iii). I believe that that is the real question the judge was deciding, and he has in effect answered that question, "No, it is not a relevant factor". I think that he was right so to do. The parties say that this case is in effect a test case to give guidance to courts in the future who are confronted with this sort of problem. If that be right, the guidance I would give would be that the exercise of putting before the court detailed evidence of rents on allegedly comparable sites does not provide, in relation to agreements in similar terms to this (of course it all depends on the terms of the agreement), evidence of relevant factors applicable to the operation of the park which the court can properly take into account.

That is really the end of the matter. We did have some discussion about whether evidence as to the rent charged on a new letting of a pitch on the same site would be relevant. That question does not arise in relation to this appeal and I am not to be thought therefore to be deciding it, because that question may very well arise at some time in the future and no doubt it will then be fully argued. For myself I would merely say that I can see that within the interpretation which the judge has given, and which I would give, to clause 7(a)(iii) of this form of agreement, such evidence could be considered to be relevant. But it would not, of course, present anything like the difficulties which are obvious in the sort of evidence with which this learned judge was confronted.

There is one other point which I should make. It may well be that if some evidence is given as to a relevant factor which is applicable to the operation of the park, that evidence may conflict with the increase in the Retail Price Index. If that be the case, the judge will simply have to decide what weight to give to each factor. Whether it will happen in practice remains to be seen. So far as the question which this appeal raises is concerned, my judgment is that the learned judge was right in determining that evidence of rents or pitch fees at other mobile home sites were not relevant factors within clause 7(a)(iii) of the agreement. He was therefore right to disregard that evidence. Having dealt with all the other matters, the only factor that did provide scope for an increase was the Retail Price Index, and he applied that. I would therefore uphold his judgment and dismiss the appeal.

Agreeing, LLOYD LJ said: I add only a word on *Beer* v *Bowden* [1981] 1 WLR 522, the case on which Mr Gordon relied. That was an ordinary rent review case. The lease provided for a rent of £1,250 per annum until March 24 1973, and thereafter "such rent as shall . . . be agreed between the landlords and the tenant". The court held that you must imply the word "fair" between the words "such" and "rent". Mr Gordon argued that by the same token you should imply some reference to fair market rent in clause 7 of the 1983 agreement. I do not agree. In *Beer* v *Bowden* there was no basis for arriving at the new rent should the parties fail to agree. There was a blank — a hiatus. Some implication, therefore, had to be made in that case. There were only two alternatives: one was that the rent should continue at £1,250; the other was to imply a fair market rent. This court very naturally preferred the second of those two alternatives.

Buckley LJ, in a very short judgment, put the point as follows:

It appears to me that the introduction by implication of a single word in the clause in the lease relating to the rent to be payable solves the problem of this case; that is, the insertion of the word "fair" between the words "such" and "rent". If some such implication is not made, it seems to me that this would be a completely inoperative rent review provision, because it is not to be expected that the tenant would agree to an increase in the rent if the rent to be agreed was absolutely at large. Clearly the parties contemplated that at the end of five years some adjustment might be necessary to make the position with regard to the rent a fair one, and the rent review provision with which we are concerned was inserted in the lease to enable such an adjustment to be made. The suggestion that upon the true construction of the clause it provides that the rent shall continue to be at the rate of £1,250 a year unless the parties otherwise agree would, in my opinion, render the provision entirely inoperative, because, as I say, one could not expect the tenant voluntarily to agree to pay a higher rent.

Here there is no problem such as there was in *Beer* v *Bowden*. There is no need to imply any term: indeed there is no room to imply any term, for clause 7 provides for the factors which are to be taken into account. *Beer* v *Bowden* does not therefore help Mr Gordon. For the reasons which my Lord has given I, too, would dismiss the appeal.

O'CONNOR LJ said: I agree with both judgments that have been delivered. The appeal must be dismissed.

The appeal was dismissed with costs.

Chancery Division

March 10 1987
(Before Mr Justice MILLETT)

RIGNALL DEVELOPMENTS LTD v HALIL

Estates Gazette June 13 1987

282 EG 1414-1424

Vendor and purchaser — Whether vendor had shown a good title — Registered local land charge relating to housing improvement grant — Knowledge of vendor's solicitor treated as knowledge of vendor — Effect of conditions in contract of sale — Eve J's "alternative ground" in *Re Forsey and Hollebone's Contract* discussed and criticised — Plaintiff purchasers, who had not searched local land charges register prior to auction, refused to complete unless safeguarded against risk of claim for recovery of improvement grant — Defendant vendor contended that she had shown a good title, the conditions in the contract providing that the purchaser was deemed to have made local searches and to have knowledge of all matters that would have been disclosed thereby — After the plaintiffs' solicitors obtained the removal of the entry in the local register, the parties took up entrenched positions, the plaintiffs claiming specific performance on payment of balance of purchase price without interest, the defendant claiming to rescind the contract and forfeit the deposit — The issue in the present vendor and purchaser summons was whether the vendor had shown a good title and this depended on whether the plaintiff purchasers were entitled to object to the register entry notwithstanding the conditions in the contract — On this issue the judge ruled that, subject to the consideration of certain defences put forward by the defendant vendor, the plaintiffs were so entitled — It was an established principle, from such authorities as *Nottingham Patent Brick & Tile Co* v *Butler,* that, if there is a defect of title or an incumbrance of which the vendor is aware, the vendor cannot rely on conditions such as those in the present case unless full and frank disclosure is made of its existence — The vendor in the present case was aware, through her solicitor, of the entry — In answer to this principle the vendor relied on several submissions, of which the only one of importance was the alleged effect of section 198(1) of the Law of Property Act 1925 as amended by section 24 of the Law of Property Act 1969, considered in conjunction with Eve J's judgment in *Re Forsey and Hollebone's Contract* — Section 198(1) as so amended still applies to local land charges and provides that in relation to them registration is deemed to constitute "actual notice" — In his alternative ground in *Re Forsey* Eve J equated such "actual notice" with the knowledge which prevents a purchaser from objecting to an irremovable incumbrance of which he is aware at the date of the contract — In his comments on this submission by the vendor the judge made the following points — (1) In the present case the incumbrance was not irremovable, but could be removed by payment: Housing Act 1974, section 77 — (2) Eve J's equation of "actual notice" with the state of mind required for terms to be implied into an open contract was "deeply suspect" — (3) The natural reading of section 198 of the Law of Property Act 1925 is that registration constitutes actual notice for the purpose of the enforcement of third parties' rights — The defendant vendor in the present case was asking not merely for the application of *Re Forsey* but for its extension — The reasoning of Eve J was "too unsound to permit of any extension" — However, Millett J was not prepared to condemn Eve J's "alternative ground" on its own facts; "such cases can be dealt with as and when they arise" — Declarations granted in favour of plaintiff purchasers on the basis that the defendant had failed to show a good title

The following cases are referred to in this report.

Ellis v *Rogers* (1885) 29 ChD 661
Faruqi v *English Real Estates Ltd* [1979] 1 WLR 963; (1978) 38 P&CR 318; [1979] EGD 986; 251 EG 1285
Forsey and Hollebone's Contract, Re [1927] 2 Ch 379
Gloag and Miller's Contract, Re (1883) 23 ChD 320
Nottingham Patent Brick & Tile Co v *Butler* (1885) 15 QBD 261

This was a vendor and purchaser summons whereby the plaintiffs, Rignall Developments Ltd, purchasers at auction of a freehold dwelling in Peckham, 218 Bellenden Road, London SE15, sought against the defendant vendor, Gulseren Halil, *inter alia,* a declaration that she had failed to show a good title in accordance with the contract of sale.

Jonathan Ferris (instructed by Armstrong & Co) appeared on behalf of the plaintiffs; Justin Fenwick (instructed by Bazley White & Co) represented the defendant.

Giving judgment, MILLETT J said: For 60 years, ever since the judgment of Eve J in *Re Forsey and Hollebone's Contract* [1927] 2 Ch 379, a prudent purchaser has searched the register of local land charges before contract and has not relied exclusively on making his search in the course of the normal process of investigation of title between contract and completion. Recently, however, the time taken by many local authorities, particularly in London, to reply to inquiries and to deal with applications for official searches of the registers kept by them has become a scandal which threatens to impede the proper working of a free market. Where land is sold by auction, it may be impossible for prospective bidders to obtain official searches in the time available. Where residential property is sold by private treaty, delay can cause havoc with the long chain of transactions which may be involved. There is a growing and useful practice for the vendor's solicitors to obtain and supply the purchaser's solicitor with a copy of the entries on the register at the same time as the draft contract. The present case calls for reconsideration of the consequences of the purchaser's failure to search the register or to obtain copies of the entries thereon before contract, and the correctness of the decision in *Re Forsey and Hollebone's Contract* has been challenged.

The case concerns a freehold dwelling in Peckham known as 218 Bellenden Road ("the property"). At an auction held on December 3 1985 the plaintiffs agreed to purchase the property from the defendant for £16,000, and paid a deposit of £1,600 to the auctioneers as stakeholders. The contract incorporated the National Conditions of Sale — 20th edition — as well as certain general and special conditions. These included the following:

General Condition 11.

The purchaser shall be deemed to have made local searches and inquiries, and to have knowledge of all matters that would be disclosed thereby and shall purchase subject to such matters.

Special Condition 5.

The property is also sold subject to any matters which might be disclosed by a search and/or inquiries of the relevant local authority, either at the date of sale or at the date of completion, and whether or not he has carried out any such search and/or inquiries, the purchaser shall be deemed to buy with full notice and knowledge of all such matters, and shall not raise any objection thereon or requisition relating thereto.

The date fixed for completion was December 31 1985. The plaintiffs did not complete on that date because of a subsisting entry on the register of local land charges to which they took objection. The defendant, relying on the conditions which I have read, denied that this was an objection which the plaintiffs were entitled to raise and served notice to complete.

The plaintiffs' solicitors eventually obtained the removal of the entry, but not until April 14 1986. The plaintiffs then sought to complete, but the defendant refused to complete unless the plaintiffs paid interest on the balance of the purchase price since December 31 1985, which the plaintiffs declined to do. The amount involved was small, and the sensible course would have been for the parties to complete and leave the question of interest to be resolved later. Instead, both sides took up entrenched positions, and the sale has still not been completed. The plaintiffs, having elected not to pursue an alternative claim to rescind the contract and recover the deposit, now seek specific performance on payment of the balance of the purchase price, but without interest; and the defendant seeks to rescind the contract and forfeit the deposit. The case turns on whether the defendant had by December 31 1985 shown a good title to the property in accordance with the contract, and this depends on whether the plaintiffs were entitled to object to the entries on the register notwithstanding the conditions of the contract which I have read.

At all material times the property was let to a protected tenant, and it was sold to the plaintiffs subject to the tenancy. In 1978 the freehold owner had applied to the local authority for an improvement grant under the Housing Act 1974. The application was accompanied by a certificate of availability for letting dated November 10 1978. It was approved on April 4 1979 and registered in the register of local land charges on April 6 1979. The date which the local authority in due course certified as the date on which the property first became fit for occupation after the completion of the relevant works ("the certified date") was June 2 1980. The grant was paid on February 15 1982.

The plaintiffs had not searched the register of local land charges prior to the auction and were unaware of the entries it contained. They were not disclosed by the defendant. On December 11 1985 the plaintiffs' solicitors applied to the local authority for an official search of the register and on the same day they submitted requisitions on title. One of these asked whether any improvement grant had been obtained in respect of the property. The defendant's solicitors replied, disputing the plaintiffs' right to make the inquiry and claiming that it was barred by the contract. Without prejudice to that contention, however, they disclosed that there was with the deeds a search dated October 1984 which revealed that a certificate of availability for letting had been issued on November 10 1978, and an improvement grant had been paid on February 15 1982. Eventually the plaintiffs' solicitors received the official certificate of search, which confirmed the existence of the relevant entries. They disclosed the dates of the certificate of availability for letting and of the approval and payment of the grant, but not the certified date.

The proprietorship register at the Land Registry shows that the defendant was registered as proprietor of the property with title absolute on November 1 1984. It does not disclose whether she was a purchaser for value; but a donee does not normally investigate his donor's title, and from the proximity of the dates I infer that the search of the local land charges register, made in October 1984, was made on her behalf at the time of her acquisition of the property, and that she was.

The significance of the entries on the register cannot be understood without reference to the relevant provisions of the Housing Act 1974. With immaterial passages omitted, they are as follows:

Section 73(1): Where an application for an improvement grant has been approved by a local authority, the provisions of this section shall apply with respect to the occupation, during the period of 5 years beginning with the certified date (in this section referred to as "the initial period") of the dwellings to which the grant relates.

Subsection (3): For the purposes of this section, the following are "qualifying persons" in relation to a dwelling, namely, —

(a) the applicant for the grant and any person who derives title to the dwelling, through or under the applicant, otherwise than by a conveyance for value.

Subsection (4): In any case where the application for the grant was accompanied by a certificate of availability for letting with respect to the dwelling, it shall be a condition of the grant that throughout the initial period:

(a) the dwelling will be let or available for letting as a residence, and not for a holiday, by a qualifying person.

Section 75(1): The provisions of this section shall apply in any case where, under or by virtue of any provision of this Part of this Act, a condition (in this section referred to as a "grant condition") is imposed as a condition of grant.

Subsection (2): If and so long as a grant condition remains in force —

(a) it shall be binding on any person who is for the time being the owner of the dwelling to which the grant relates.

Subsection (3): A grant condition shall be enforced throughout the period of 5 years beginning on the certified date.

Subsection (5): A grant condition shall be treated as not being registrable by virtue of section 15 of the Land Charges Act 1925 but, as soon as may be after an application for a grant has been approved, any condition of that grant shall be registered in the register of local land charges.

Subsection (6): In this Part of this Act "the certified date", in relation to a dwelling in respect of which an application for a grant has been approved, means the date certified by the local authority by whom the application was approved as the date on which the dwelling first becomes fit for occupation after completion of the relevant works to the satisfaction of the local authority.

Section 76(1): The provisions of this section shall have effect in the event of a breach of a condition of grant (in this section referred to as "the relevant grant") at a time when the condition is binding on the owner of the dwelling concerned by virtue of section 75(2) above.

Subsection (2): Where the relevant grant related to a single dwelling, an amount equal to the amount of the relevant grant, together with compound interest thereon as from the certified date, calculated at the appropriate rate . . . shall, on being granted by the local authority, forthwith become payable to the authority by the owner for the time being of the dwelling.

Subsection (4): Nothing in subsection (2), or as the case may be, subsection (3) above shall prevent a local authority from determining not to demand any such amount as is referred to in that subsection or from demanding an amount less than that which they are entitled to demand under that subsection.

Subsection (5): Upon satisfaction of the liability of the owner of a dwelling to make a payment under this section to a local authority in respect of a breach of a condition of a grant, the condition shall cease to be enforced with respect to that dwelling.

Section 77(1): If, at any time while a condition of grant remains in force, the owner of the dwelling to which the condition relates pays to the local authority by whom the grant was made the like amount as would (on a demand by the local authority) become payable under section 76 above in the event of a breach of that condition, all conditions of the grant shall cease to be enforced with respect to that dwelling.

The definition of "qualifying person" in section 73(3)(a) has the result that a grant is repayable on any sale of the property during the initial period, even though the property continues to be occupied by a protected tenant. The object evidently is to prevent short-term speculators from buying properties, improving them at the expense of the ratepayers, and then selling them at a profit which reflects the expenditure of public money. The sale to the defendant in 1984, if, as I infer, she had bought the property, and the present sale to the plaintiffs, if within the initial period, would both constitute breaches of a condition of the grant and make the grant repayable. The certified date did not appear on the register, but from the information available to the plaintiffs' solicitors in December 1985 it must have seemed probable that there had been a breach of condition in 1984 and possible that another would occur if completion took place on December 31. In those circumstances, the plaintiffs refused to complete in the absence of confirmation from the local authority that it would not seek to recover the amount of the grant from the plaintiffs, or from the defendant that, if required, she would repay it out of the proceeds of sale.

On April 14 1986 the local authority's legal department, whose dilatoriness had largely caused the problem, finally notified the plaintiffs' solicitors that the initial period had expired in June 1985, that there would not be a breach of grant condition if the plaintiffs completed their purchase, and that it had given instructions for the entries on the register of local land charges to be removed. The plaintiffs then offered to complete, and the parties adopted the positions I have described.

The defendant relied upon the express terms of the contract. The property, it was submitted, was not sold free from incumbrances, but subject to the entries on the register of local land charges; and the plaintiffs had no right to object to them. The defendant had, therefore, shown a good title to that which she had agreed to sell and was entitled to serve notice to complete. Moreover, by General Condition 11 the plaintiffs were deemed to have searched the register and to have knowledge of the entries thereon.

It is, however, a well-established rule of equity that, if there is a defect in title or incumbrance of which the vendor is aware, the vendor cannot rely upon conditions such as those in the present case unless full and frank disclosure is made of its existence. The leading authority is *Nottingham Patent Brick & Tile Co* v *Butler* (1885) 15 QBD 261. Wills J said at p 271:

The fourth condition provides that the property is sold subject to any matter or thing affecting the same, whether disclosed at the time of the sale or not. Such a condition, however, does not relieve the vendor from the necessity of disclosing any encumbrance or liability of which he is aware, but simply protects him if it should afterwards turn out that the property is subject to some burden or right in favour of a third person of which he is unaware. . . . It would be nothing short of a direct encouragement to fraud if a vendor were at liberty by a condition of this kind to sell to a purchaser as an absolute and unburdened freehold a property which he knows to be subject to liabilities which would materially reduce its market value. . . . In honesty and in law alike he was bound to give the purchaser full and fair information what it was that he had for sale, and was inviting him to buy, and having failed to do so, he cannot insist upon the bargain procured by the suppression of material facts affecting the nature of the subject of sale. I entirely acquit the defendant of anything like intentional misconduct, but in the preparation of the particulars of sale he unfortunately relied upon his solicitor, who, as I cannot help believing, was under the mistaken impression that he could better the position of the vendor by abstaining from making himself acquainted with the contents of the earlier deeds in his possession, and open to his perusal.

As that case shows, the knowledge of the vendor's solicitor is treated as that of the vendor, and it is no answer for him to say that he has not read the contents of his own conveyancing file. In the present case, therefore, the defendant must be taken to have known of the entries on the register, since a search had been made on her behalf at the time of her purchase and a copy of the entries was with the deeds in her solicitor's possession. To entitle her to rely on the relevant conditions of the contract in these circumstances, it was incumbent on her to disclose the existence and nature of the entries to the plaintiffs before contract. Had the information disclosed in the answers to requisitions been included in the particulars of sale, there could have been no objection to conditions precluding all further inquiry and making the sale subject to the entries in question. In the absence of such disclosure the conditions cannot be relied on. It is hardly necessary to add that the equitable principle cannot be circumvented by the inclusion in the contract of a condition deeming the purchaser to have searched the register and to know of its contents. The purchaser's acceptance of such a condition is on the basis that the vendor has made the disclosure required of him.

In answer to this, it was first submitted on behalf of the defendant that the conditions of grant did not create an incumbrance or burden on the property, but only a personal liability upon the owner. But the grant is repayable on demand by the owner for the time being of the property, so that the potential liability binds successive owners of the property affected — which is why it is required to be registered — and in my judgment that is enough. Then it was submitted that the entries on the register were only "bare" entries, without any reality behind them. Unknown to the parties, it was said, the initial period had expired, so that the grant was not repayable and the entries were obsolete. In fact that appears to be incorrect if, as I infer, there had been a breach of grant condition in 1984. As I read the statutory provisions, once there has been a breach of grant condition during the initial period, repayment may be demanded, even after the expiration of that period, from the owner of the property at the date of the demand. But in any case, neither the entries on the register nor the defendant's answers to requisitions disclosed the certified date or showed that the initial period had expired. It was for the defendant to show a good title to the property free from the risk that repayment of the grant might be demanded from the plaintiffs, and she failed to do so.

Next it was submitted that the plaintiffs could have inspected the register before the auction, that a prudent purchaser would have done so, and that there was no reason for equity to come to the assistance of the imprudent. I cannot accept that submission. In *Faruqi* v *English Real Estates Ltd* [1979] 1 WLR 963, the equitable principle was applied in a case where the relevant documents were made available for inspection and the vendors had given the purchasers a fair and proper opportunity of seeing what they were buying. But, as Walton J pointed out, any purchaser reading the conditions of sale would be entitled to assume that, while there were no doubt entries on the register, they were only the usual sort of entries which would not adversely affect the value of the property. That observation applies with equal force to the present case.

As to the plaintiffs' alleged imprudence, the modern practice of making pre-contract searches dates only from 1927 and is a result of the decision in *Re Forsey and Hollebone's Contract*. But even if that is the only safe and prudent course, why should the purchasers' imprudence relieve the vendor of the obligation of candour?

Finally, and most formidably, reliance was placed on section 198(1) of the Law of Property Act 1925 as amended by the Local Land Charges Act 1975. That deems the registration of any instrument or matter in any local land charges register to constitute actual notice of such instrument or matter, and of the fact of such registration, to all persons and for all purposes connected with the land affected.

Equity, it was submitted — and I agree — does not insist on the performance of idle rituals and does not require a vendor to disclose to a purchaser matters already known to him. For this submission to help the defendant, however, the actual notice, of which section 198 speaks, must be equated with knowledge. That is the crucial equation which was made in *Re Forsey and Hollebone's Contract*.

In that case, land was sold free from incumbrances. In accordance with normal conveyancing practice, the purchaser's solicitor did not search the local land charges register until after exchange of contracts and shortly before the date fixed for completion. He then discovered that the local authority had resolved to prepare a town planning scheme. The resolution had been registered as a local land charge, but neither the vendor nor the purchaser was aware of its existence at the date of the contract.

The purchasers applied for a declaration that the vendor had not shown a good title to the property sold in accordance with the contract and for repayment of the deposit. Her application was dismissed by Eve J and the Court of Appeal on the ground that a resolution to prepare a town planning scheme did not operate to impose a subsisting incumbrance on the land affected. At first instance, however, Eve J gave a second ground for his decision — upon which the Court of Appeal expressed no opinion — that even if the registered resolution was an incumbrance, it was an incumbrance of which the purchasers must, under section 198, be deemed to have contracted with actual notice, and she was therefore precluded from refusing to complete the contract. In his words at p 387:

The vendor

has only to point to the section . . . to show that when the contract was entered into vendor and purchaser must alike be deemed to have known of the existence of the incumbrance which the purchaser insists on as a good ground for avoiding the contract.

Eve J thus equated the actual notice attributed to the purchaser by section 198 with knowledge for the purpose of the well-known rule that, in the case of an open contract for the sale of land, a purchaser cannot object to an irremovable incumbrance of which he was aware at the date of the contract. An incumbrance is irremovable if the owner of the land is not entitled as of right to procure its discharge by the payment of money. The rule rests upon implication. A person who knows that a property has some incurable defect or is subject to some irremovable incumbrance, and yet contracts to buy it, must impliedly be taken to have agreed to accept the vendor's title despite the existence of that defect or incumbrance. To this extent the implication in an open contract that the property is sold free from incumbrances is negated. (Normally, the purchaser's knowledge of the state of the title cannot deprive him of the benefit of an express term that the land is sold free from incumbrances. Eve J attributed this to the parol evidence rule, but it would, I think, be more accurate to attribute it to the obvious impossibility, even if the evidence were received, of implying a term inconsistent with an express term of the contract.) The rule has no application to an incumbrance like that in the present case, which can be removed on payment: see section 77 of the Housing Act 1974. The purchaser's knowledge of such an incumbrance is not inconsistent with the vendor's obligation to make the payment necessary to obtain its discharge on completion: see *Re Gloag and Miller's Contract* (1883) 23 Ch D 320. Hence the crucial importance of the contractual provisions on which the defendant relies.

This part of Eve J's judgment represents an alternative ground for his decision and cannot be dismissed as merely obiter. It has stood for nearly 60 years, but not without challenge. It was greeted at the time by conveyancers with consternation and incredulity. Its interpretation of section 198 was described as "startling" and its effect as "revolutionary". It has since been subjected to severe criticism by the editors of *Emitt on Title* and other eminent conveyancers. It has led to a change in conveyancing practice, which is both inconvenient and time-consuming, and which is unlikely to be adopted by those who buy at auction or who contract before consulting solicitors. For such purchasers, the decision constitutes a potential trap. Even in ordinary sales by private treaty, the need to search the local land charges register before contract rather than at the same time as the other registers, that is to say shortly before completion, is inconvenient and entails a delay which can put the contract at risk. Pre-contract searches of the land charges register — to which section 198 also applied — are, of course, impractical, since registration in that register — unlike the local land charges register — is effected against the name of the estate owner for the time being, and until contract it is not known against what names the search should be made.

These problems, aggravated by the reduction in the statutory period of title from 30 to 20 years, led to the reversal of the relevant part of Eve J's decision by section 24 of the Law of Property Act 1969, following a report of the Law Commission (No 18). That section, however, does not apply to local land charges, in respect of which pre-contract searches are feasible and had by 1969 become normal practice. In relation to local land charges, therefore, the decision in *Re Forsey and Hollebone's Contract* has not been reversed by statute, but neither has it been confirmed thereby.

In my judgment, the equation of the actual notice referred to in section 198 with the state of mind required for terms to be implied into an open contract is deeply suspect. I find it impossible to reconcile with principle or authority. If a purchaser knows, or even mistakenly believes, that he cannot expect to obtain a title free from incumbrances, and yet enters into a contract of purchase on that basis, the inference is obvious. But the inference depends upon his state of mind, which may be affected by error, or ignorance, or forgetfulness. Notice — even actual notice — however, has nothing to do with the person's state of mind and is not affected by such matters. In the absence of knowledge, notice cannot support the necessary inference.

This is neatly demonstrated by the decision of the Court of Appeal in *Ellis* v *Rogers* (1885) 29 Ch D 661. In that case, the purchaser knew before he entered into the contract that the land was subject to restrictive covenants, but he wrongly believed that they had been extinguished when the land had been compulsorily acquired by a railway company. He later discovered that the covenants were still extant and would bind him if he completed. When he refused to complete, the vendor contended that, having known of the existence of the covenants from the outset, the purchaser must be taken to have agreed to accept the title subject to them. The Court of Appeal rejected this contention. As Cotton LJ put it at p 671:

The vendor knew nothing of the covenants. The purchaser knew of them, but thought they had been discharged, so that both parties were contracting on the footing that a good title was to be made, and as a good title cannot be made, the purchaser is not bound.

This shows that notice and knowledge are not synonymous. The purchaser had actual notice of the existence of the covenants. Had he completed before discovering his error, he would unquestionably have been bound by them as a purchaser with notice. His ignorance of their continuing subsistence, while negativing any inference which might otherwise have been drawn from his *knowledge* of them, would not avail him against the covenantee, for it would not affect his *notice* of them.

There are two further grounds for suspecting Eve J's reasoning. First, in *Ellis* v *Rogers* Kay J pointed out at p 666 that to force the title on the purchaser it is essential that he should have knowledge not only of the existence of the incumbrance but of the vendor's inability to remove it. Section 198 cannot help with this. Unless consciously aware of the existence of an incumbrance, a purchaser cannot form any useful opinion on the vendor's ability to remove it.

Second, it is unlikely that the notice attributed to the purchaser by section 198 was intended to have any greater effect than actual notice would have had before 1926; and notice of equities before 1926 had nothing to do with the relationship of vendor and purchaser or with the interpretation and effect of their contract of sale. It was concerned exclusively with the enforcement of third parties' rights. The fundamental rule of equity is that an equitable interest is binding on everyone except a bona fide purchaser for value without notice. The Land Charges Act 1925 substituted registration for notice and was likewise concerned exclusively with the protection of third parties' rights. Section 198 of the Law of Property Act 1925 forms an integral part of the machinery of registration. In this context, the natural reading of section 198 is that registration constitutes actual notice to all persons and for all purposes *for which such notice is material,* that is to say for the purpose of the enforcement of third parties' rights against the land affected.

In the present case, the defendant asks me not merely to apply the decision in *Re Forsey and Hollebone's Contract* but to extend it by applying it in a different though closely related context. There, both vendor and purchaser were equally ignorant of the existence of the local land charge; it constituted an irremovable incumbrance; and the question was whether section 198 had the effect of modifying the vendor's contractual obligation to make a good title. Here, the defendant, through her solicitor, knew of the existence of the entry on the register; she could have procured its removal by repaying the grant if necessary; and the question is whether her failure to disclose what she knew prevents her from relying on the express terms of the contract. I cannot think that a vendor who knew of the existence of a registered charge, and who deliberately deceived the purchasers by telling them that there was no such charge, or that it was not registered, could escape liability for fraud by claiming that by virtue of section 198 the purchasers must be taken to have had actual notice of the truth. Similarly, I am not prepared to hold that a vendor who knows of a registered charge and who wishes to make the sale subject to it is exonerated by section 198 from his obligation to make full and frank disclosure of its existence before he can take advantage of an appropriate condition of sale.

It is therefore not strictly necessary to reach any conclusion whether the decision in *Re Forsey and Hollebone's Contract* can be supported where the existence of a registered local land charge is unknown to the vendor at the date of contract — so that it is not unconscionable for him to rely upon a special condition of sale without disclosing it — or where it represents an irremovable defect of title — so that he does not need to rely on any special condition. Such cases can be dealt with as and when they arise. In my judgment the reasoning of the decision is too unsound to permit of any extension, however logical, to a situation not directly covered by it.

I shall grant a declaration that the defendant failed to show a good title in accordance with the contract before April 14 1986; and that the defendant is not entitled to interest on the balance of the purchase money; and I shall grant a decree of specific performance of the contract upon payment by the plaintiffs of the balance of the purchase price, less the costs to the plaintiffs of obtaining the removal of the relevant entries on the local land charges register.

Declarations and decree of specific performance granted as mentioned at the end of the judgment. Leave to appeal, if necessary, granted.

TOWN AND COUNTRY PLANNING

Queen's Bench Division
October 14 1986
(Before Mr Justice MANN)

R v SECRETARY OF STATE FOR THE ENVIRONMENT AND OTHERS, EX PARTE BOURNEMOUTH BOROUGH COUNCIL

Estates Gazette February 7 1987

281 EG 539-544

Town and Country Planning Act 1971, section 205 — Blight notice — "Appropriate authority" — Application by borough council for judicial review, seeking orders of certiorari and mandamus to quash decision of Secretary of State and to direct him to exercise his powers under section 205(2) by determining which of two local authorities, the borough council or a county council, was the appropriate authority for the purpose of the blight provisions — A blight notice under section 193 of the Act was served on the borough council by a lessee of land affected by highway proposals, one of the respondents to the present proceedings, and a similar notice was served on the county council — A reference of the objection was made to the Lands Tribunal, but proceedings in the tribunal were stayed pending the resolution of the judicial review application — This application followed requests by the local authorities for a determination under section 205(2) and a decision letter from the Secretary of State — The decision letter stated that the Secretary of State was not empowered to choose between the two authorities, each of whom was entitled to acquire the subject land but in relation to different sets of circumstances — The Secretary of State's power under section 205(2)(c) to decide which of two local authorities was the appropriate authority was limited to cases where the land in question fell within the same specified class — Where, as here, the authority liable to acquire the land in relation to one class was different from the authority in relation to the other, the Secretary of State could not make a selection — The upshot was that a blight notice could be served on each authority — It was submitted before Mann J on behalf of the Secretary of State that this was the correct interpretation — Where there were two sets of circumstances there would be two appropriate authorities — This would enable a claimant to obtain a disclaimer of intent to acquire from both, whereas the selection of one would leave the blight imposed by the other unresolved — Held that this submission was correct, although there was no satisfactory answer to the problem — Application for judicial review dismissed

The following case is referred to in this report.

Bolton Corporation v *Owen* [1962] 1 QB 470; [1962] 2 WLR 307; [1962] 1 All ER 101; (1962) 61 LGR 7, CA

This was an application by Bournemouth Borough Council for judicial review of a decision by the Secretary of State for the Environment given in the exercise of his powers under section 205(2) of the Town and Country Planning Act 1971. The first respondent was the Secretary of State, the second and third respondents being Dorset County Council and Stephen Harry Saunders, the lessee of premises affected by highway proposals included in the Boscombe local plan.

Duncan Ouseley (instructed by Sharpe Pritchard & Co, agents for S J C Chappell, solicitors' department, Bournemouth Corporation) appeared on behalf of the applicant authority; Michael Rich QC and D Holgate (instructed by the Treasury Solicitor) represented the Secretary of State; the second and third respondents were not represented and took no part in these proceedings.

Giving judgment, MANN J said: There is before the court an application for judicial review. Leave to move was given by Woolf J (as he then was) on June 21 1985. The applicant is the Bournemouth Borough Council. The respondent is the Secretary of State for the Environment. There are two other respondents. The second is the Dorset County Council and the third is a Mr S H Saunders. They neither appeared nor were represented during this hearing.

The decision impugned is a decision of the respondent dated March 18 1985 given in the exercise of his powers under section 205(2) of the Town and Country Planning Act 1971. The relief sought is, first, an order of certiorari to quash the decision and, second, an order of mandamus directing the respondent to exercise his powers under section 205(2) according to law.

Section 205 of the Act of 1971 is within a fasciculus of sections headed "Interests of owner-occupiers affected by planning proposals." The object of the provisions is to enable owners of property affected by planning blight either to sell their properties to a public authority at the price which it would have obtained in the open market had there been no planning proposal or to obtain a disclaimer from the public authority concerned that it has any intention of purchasing the property. The sections had their genesis in Part IV of the Town and Country Planning Act 1959.

The sections concerned are sections 192 to 207. I at once refer to such of them as are presently material.

Section 192 deals with the scope of the provisions.

Subsection (1) provides:

The provisions of sections 193 to 207 of this Act shall have effect in relation to land which (a) is land indicated in a structure plan in force for the district in which it is situated either as land which may be required for the purposes of any of the following functions, that is to say, those of a government department, local authority or statutory undertakers, or of the National Coal Board or the establishment or running by a public telecommunications operator of a telecommunications system, or as land which may be included in an action area; or (b) is land allocated for the purposes of any such functions by a local plan in force for the district or is land defined in such a plan as the site of proposed development for the purposes of any such functions; or

and I omit (bb) and (bc) and pass to (c)

is land indicated in a development plan (otherwise than by being dealt with in a manner mentioned in the preceding paragraphs) as land on which a highway is proposed to be constructed or land to be included in a highway as proposed to be improved or altered . . .

The descriptions in paras (a), (b) and (c) are among what is subsequently termed "the specified descriptions": see subsection (6) of section 192. The word "functions" in paras (a) and (b) include both powers and duties: see section 290(1).

Section 193 deals with the power to serve a blight notice. I should read a part of subsection (1):

Where the whole or part of a hereditament . . . is comprised in land of any of the specified descriptions, and a person claims that (a) he is entitled to an interest in that hereditament . . . and (b) the interest is one which qualifies for protection under these provisions; and (c) he has made reasonable endeavours to sell that interest; and (d) in consequence of the fact that the hereditament . . . or a part of it was, or was likely to be, comprised in land of any of the specified descriptions, he has been unable to sell that interest except at a price substantially lower than that for which it might reasonably have been expected to sell if no part of the hereditament . . . were, or were likely to be, comprised in such land, he may serve on the appropriate authority a notice in the prescribed form requiring that authority to purchase that interest to the extent specified in, and otherwise in accordance with, these provisions.

The reference to "the appropriate authority" should be noted. Section 194 deals with objection to a blight notice. I should read some of it. Subsection (1):

Where a blight notice has been served in respect of a hereditament . . . the appropriate authority, at any time before the end of the period of two months beginning with the date of service of that notice, may serve on the claimant a counternotice in the prescribed form objecting to the notice.

(2) Subject to the following provisions of this section, the grounds on which objection may be made in a counternotice to a notice served under section 193 of this Act are (a) that no part of the hereditament . . . to which the notice relates is comprised in land of any of the specified descriptions; (b) that the appropriate authority (unless compelled to do so by virtue of these provisions) do not propose to acquire any part of the hereditament . . . in the exercise of any relevant powers; . . . (g) that the conditions specified in paragraph (c) and (d) of section 193(1) of this Act are not fulfilled.

Again the reference to the "appropriate authority" should be noted.

Section 195 deals with the reference of an objection to a blight notice to the Lands Tribunal. It is sufficient to read subsection (1):

Where a counternotice has been served under section 194 of this Act objecting to a blight notice, the claimant, at any time before the end of the period of two months beginning with the date of service of the counternotice, may require the objection to be referred to the Lands Tribunal.

The tribunal may uphold or not uphold the objection. Where no objection is made, or where an objection is referred and not upheld, then the appropriate authority is deemed to be authorised to acquire the claimant's interest compulsorily and to have served a notice to treat: see section 196.

It is to be remarked that if an objection on ground (b) of section 194(2) is either not referred or is upheld, then in practice the blighting effect of a planning proposal is removed.

I can pass through the intervening sections direct to section 205. It has the shoulder note "'Appropriate authority' for purposes of these provisions". Subsection (1) reads:

Subject to the following provisions of this section, in these provisions "the appropriate authority", in relation to any land, means the government department, local authority or other body or person by whom, in accordance with the circumstances by virtue of which the land falls within any of the specified descriptions, the land is liable to be acquired or is indicated as being proposed to be acquired or, as the case may be, any right over the land is proposed to be acquired.

(2) If any question arises — (a) whether the appropriate authority in relation to any land for the purpose of these provisions is the Secretary of State or a local highway authority; or (b) which of two or more local highway authorities is the appropriate authority in relation to any land for those purposes; or (c) which of two or more local authorities is the appropriate authority in relation to any land for those purposes, that question shall be referred to the Secretary of State, whose decision shall be final.

(3) If any question arises which authority is the appropriate authority for the purposes of these provisions — (a) section 194(1) of this Act shall have effect as if the reference to the date of service of the blight notice were a reference to that date or the date on which that question is determined, whichever is the later.

I do not pause upon paras (b) and (c).

It is the provisions of subsections (1) and (2) which agitate in the present case. It is conceded and agreed that the reference to the Secretary of State's decision being final does not preclude a decision being subject to judicial review. I agree with that concession and agreement.

The facts of the instant case are as follows. On November 21 1983 the Bournemouth Borough Council adopted the Boscombe Local Plan. That plan contained a number of proposals. Among them are proposals in regard to an area identified on Inset Plan No 2 and referred to as land at Haviland Road. The total area concerned is 6.3 hectares.

The proposals are to be found at s4(ii) and (v) and T2.

s4(ii). A new section of road from Palmerston Road to Ashley Road, as identified on Inset Plan No 2, is a pre-requisite of any redevelopment to achieve satisfactory vehicular access to car parks, bus station and servicing areas. Any planning application submitted for the redevelopment should, therefore, include the new road, together with improvements to the existing Ashley Road and Palmerston Road and improvements to their junctions with Christchurch Road. A substantial contribution towards the provision of this road and its associated improvements will be required from the developer.

S4(v). Appropriate uses upon redevelopment on that part of the site south of the new road include: (a) retail floorspace (b) sports facilities (c) a limited number of small suites of offices above ground floor (d) residential units.

T2. A two-way single carriageway relief road for that part of Christchurch Road identified as the main shopping centre (Palmerston Road to Ashley Road) shall be built in association with redevelopment in the Haviland Road West area. This road shall use the southern ends of Palmerston Road and Ashley Road suitably widened, improved and linked to Christchurch Road and there shall be roundabouts at the junction of the new road with Palmerston Road and with Ashley Road (Inset Plan 2).

Implementation of these proposals is dealt with in chapter 5 of the plan where, at para 5.4.1, it is said:

It is expected that the majority of land required for development proposals within the Local Plan area will be assembled through normal negotiating procedures between developers and the respective existing owners of land. The Council has agreed to enter into discussions as landowner in respect of its future interest in land now owned by the Authority in accordance with normal procedures following approval of the Local Plan. Only in exceptional circumstances will the Local Planning Authority consider utilising its Compulsory Purchase Order powers as Planning Authority to obtain sites required for development. If, however, the implementation within the Plan period of comprehensive development proposals identified below are prejudiced because a certain piece of land is not available or cannot be negotiated between the owners and potential developers, the Local Planning Authority will consider Compulsory Purchase of the site.

Among the "proposals identified below" is proposal s4, the Haviland Road area. The reference to compulsory purchase powers is a reference to the familiar powers contained in section 112 of the Act of 1971.

The paragraph continues:

It must be made clear that this option will only be pursued in circumstances in which the objectives of a proposal are likely to founder in the absence of a particular part of the development site. Accordingly, it is anticipated that the use of Compulsory Purchase powers by the Local Planning Authority will occur in only a few instances where the problems of site assembly are complex and cannot be resolved by negotiation. It is anticipated that the Local Planning Authority will invite developers to submit their schemes in order to determine their acceptability, prior to discussions in respect of any use of Compulsory Purchase powers.

Implementation of certain road proposals is dealt with in para 5.4.2:

The Local Highway Authority may use compulsory purchase powers under the appropriate Highways Acts to achieve implementation of the schemes identified in the following policies and proposals, where the land cannot be acquired by negotiation: (i) That part of proposal T2 which requires alterations to the junctions of Christchurch Road with Palmerston Road and Ashley Road, and of Ashley Road and Palmerston Road between Christchurch Road and Haviland Road.

The local highway authority is the Dorset County Council, in which capacity they are the second respondent to these proceedings.

A Mr S H Saunders, who is the third respondent to these proceedings, is the lessee of premises at 8-10 Palmerston Road. The premises are wholly within the area subject to proposal s4 and is in part affected by the proposal T2.

On February 20 1984 he served a notice under section 193 of the Act of 1971 asserting that his hereditament was land falling within para (b) of section 192(1). The notice was served upon the Bournemouth Borough Council.

On April 6 the Bournemouth Borough Council served a notice of objection on the grounds that the conditions specified in paras (c) and (d) of section 193(1) were not fulfilled.

It is to be observed that no objection was based on grounds (a) and (b) in section 194(2). The notice of objection was referred, on a date unknown to me, by Mr Saunders to the Lands Tribunal. Proceedings in the Lands Tribunal have been stayed pending the resolution of this application.

On August 24 1984 Mr Saunders served a second blight notice, this time upon the Dorset County Council. Again his assertion was that his hereditament was land falling within para (b) of section 192(1). I believe, but do not know, that the county council served a notice of objection.

On September 26 1984 the borough council wrote to the Secretary of State a letter which contained the following passage:

I would be grateful if this letter could be taken as a formal request pursuant to section 205(2) of the 1971 Act for the Secretary of State to determine which of this Council and the Dorset County Council is the appropriate authority to receive a blight notice in this matter.

On October 17 the Dorset County Council made a similar application. The two applications were determined by the Secretary of State on March 18 1985. On that day he wrote letters in identical terms, save as to addressee. It is the decision sent to Bournemouth Borough Council which is challenged upon this application. The Dorset County Council has not, I am told, challenged the decision upon its application.

I should read the decision letter so far as material.

I am directed by the Secretary of State for the Environment to refer to applications from Bournemouth Borough Council and Dorset County Council for him to determine under section 205(2) of the Town and Country Planning Act 1971 the appropriate authority for the purpose of service under section 193 of that Act of a blight notice in relation to property at 8/10 Palmerston Road, Boscombe.

There is then a reference to the correspondence.

Para 2 refers to the definition in section 205(1) and I pass to para 3.

In the present case it is noted that the Boscombe Local Plan indicates that part of the property will be required for a proposed new road which, if built, will be the responsibility of the Dorset County Council as the Highway Authority. At the same time the whole of the property also falls within the area bounded by Palmerston Road, Ashley Road, Gladstone Road and Haviland Road which the Plan indicates is to be redeveloped for commercial purposes. The Bournemouth Borough Council have said that if necessary they will use their compulsory purchase powers to acquire land to implement these proposals. On these facts it is concluded that the property falls within both paragraph (b) and paragraph (c) of section 192(1) of the 1971 Act; and that "the appropriate authority" is Bournemouth Borough Council in so far as the property is within paragraph (b) and Dorset County Council in so far as the property is within paragraph (c).

4. The representations made to the Secretary of State appear to have been based on the assumption that he would decide which of the local authorities concerned should be the recipient of a blight notice in this case; and this is presumably because section 205(2) provides that where any question arises, the Secretary of State is to determine "which of two or more local authorities is the appropriate authority in relation to any land". That provision has to be read, however, in the context of the definition of "the appropriate authority" in subsection (1) of section 205, under which "the appropriate authority" is identified as the authority by whom, "in accordance with the circumstances by virtue of which the land falls within any of the specified descriptions", the land is liable to be acquired. The Secretary of State is advised that the power which is conferred on him by the section is to determine, in respect of any class into which the land falls, which authority is liable to acquire the land: and that where the land falls within more than one specified class and the authority whom he determines to be liable to acquire the land in relation to one class is different from the authority liable to acquire the land in relation to the other or others, the section does not provide for the Secretary of State to choose between them and determine that only one of those authorities can be served with a notice under section 193. In such a situation, it would in fact appear to be open to a person entitled to an interest in that land to serve a blight notice on any of the authorities which is an "appropriate authority".

5. Accordingly the Secretary of State hereby determines under section 205(2) of the Town and Country Planning Act 1971 that the appropriate authority for the service of a blight notice under section 193 of the 1971 Act in respect of the property known as 8/10 Palmerston Road, Boscombe is: (a) in relation to the whole of the property, the Bournemouth Borough Council (in the light of the allocation of the land in the local plan for redevelopment purposes); and (b) in relation to that part which is defined in the local plan as the site of a proposed highway, the Dorset County Council (as local highway authority).

Mr Rich for the Secretary of State concedes that the reference to para (c) of section 192(1) in regard to the Dorset County Council is wrong. It is wrong because para (c) applies only where the land is not dealt with in a preceding para of subsection (1), and this land, so far as the county council was concerned, was within para (b) as being land defined for a highway function. The reference accordingly, in regard to Dorset, should also have been to para (b), but Dorset County Council do not complain.

Mr Ouseley for the applicant seeks to flaw the Secretary of State's decision on two grounds: (1) under section 205(2) the Secretary of State must select one and only one authority in regard to one blight notice in regard to the same land; (2) Bournemouth Borough Council could not be an appropriate authority because the land does not fall within para (b) of section 192(1).

In regard to the first ground of challenge, Mr Ouseley draws attention to the language of section 205(1) and (2) and to the structure and purpose of the group of sections of which it is a part.

Section 205(1) refers to "the appropriate authority" and is drawn in the singular. Section 205(2) postulates the selection of one from two or more. The use of the definite article in both subsections echoes the use of that article in sections 193 to 196. It would, said Mr Ouseley, be surprising if there could be more than one deemed notice to treat in regard to the same land.

However, Mr Rich for the Secretary of State contemplates a plurality of notices to treat with equanimity, while agreeing that there can be but one appropriate authority in relation to any set of circumstances by virtue of which land falls within any of the specified descriptions.

If, he said, there be two sets of circumstances then there will be two appropriate authorities. Such an approach, says Mr Rich, would enable a claimant to obtain a disclaimer of an intent to acquire from both, while the selection of one would leave the blight imposed by the other unresolved.

I find the problem posed in this case perplexing in that there is no satisfactory answer. Neither counsel suggested that their argument achieved a satisfactory solution. However, and not without hesitation, I conclude that Mr Rich is correct in his submissions. If there are two sets of circumstances by virtue of which land apparently falls within any of the specified descriptions, then there may be two appropriate authorities. There could not be three or four, but two. Such an approach does enable a claimant to obtain a disclaimer from both and avoids the question of how the Secretary of State would decide between the two. There would seem to be no rational basis upon which he could.

The case where there is a single set of circumstances by virtue of which land falls within any specified description but where there is dubiety in those circumstances as to which one authority is to acquire is of course different and is a more obvious case for resolution under section 205(2)(c).

I have used the phrase "by virtue of which land apparently falls". In so doing it may be objected that I am putting a gloss on the language of section 205(1). I think that I am, but I do so in order to prevent the Secretary of State adjudicating upon whether land falls within any of the specified descriptions. That is a matter which Parliament has entrusted to the Lands Tribunal.

If the Secretary of State did so adjudicate, he could decide that land did not fall within any specified descriptions and that accordingly there was no appropriate authority at all, with the consequence that pursuit of the claimant's notice becomes impossible in practice.

I cannot think that Parliament could have intended an elaborately constituted procedure to be avoided by an adjudication under section 205(2). In my judgment the Secretary of State must proceed on the basis that what is asserted by the claimant is the case.

I am conscious that my answer to the question posed leaves open the possibility of two deemed notices to treat in relation to the same land. However, I do not think that this possibility gives rise to any problem in point of law. The claimant at his option could proceed with either and any difficulties are essentially of a pragmatic nature.

In my judgment Mr Ouseley's second argument, that is to say that in regard to the Bournemouth Borough Council this land is not within section 192(2)(b), does not arise. The Secretary of State cannot determine that matter. I have sought to explain why. It will have to be determined by the Lands Tribunal in consequence of the notice of objection served by the Bournemouth Borough Council on May 14 1985, in consequence of the liberty afforded by section 205(3)(a). That notice goes beyond the original objection.

In considering that notice of objection the tribunal will no doubt have regard to the decision of the Court of Appeal in *Bolton Corporation* v *Owen* [1962] 1 QB 470. I have been asked to analyse that decision for the benefit of the tribunal. I do not regard it as any part of the function of this court, tempting as it may be, to offer advice upon questions which are not before it and which will not be before it.

For the reasons I have endeavoured to give, this application will be dismissed. I add that the Secretary of State's decision letter has a conceded imperfection. As I have said, it affects Dorset County Council, who do not complain of it. I do not regard it as affecting Bournemouth Borough Council or as giving them any entitlement to relief.

The application was dismissed with costs.

LANDS TRIBUNAL
RATING

TRUSTEE SAVINGS BANK ENGLAND AND WALES AND OTHERS v SAUNDERS (VO)
(LVC/999-1011/1984)
June 5 1986
(C R MALLETT FRICS)

Estates Gazette January 24 1987
281 EG 314-320

Shops in Killingworth township shopping centre (the Citadel) — S20 assessments — Common ground that (1) s19 assessments would have been higher than assessments in valuation list, (2) the township development had been a disappointment and (3) rental values of shops in Killingworth centre have lagged behind those in other shopping centres in valuation area, particularly nearby Longbenton (Arndale Development) — Shops in the two centres assessed on similar basis, and not disputed that in s19 terms Longbenton shops more valuable than those in Killingworth — Ratepayers contended for similar distinction in s20 valuations while VO maintained that all matters giving rise to difference were outside s20 — Finding that Citadel consists of limited number of shops supplying limited day-to-day requirements of people in immediate neighbourhood and that large number of residents shop elsewhere — Tribunal considers s20 requires valuation to be made having regard *inter alia* to occupation and use of shops and houses in Killingworth and neighbouring areas — Not convinced, as VO contended, that unemployment in Killingworth (above regional average) had caused residents to shop elsewhere for essential day-to-day needs, that the architectural "merits" of the Citadel were uppermost in their minds in considering where to shop and that the "best gets better" factor (referring to Newcastle shopping centres) was relevant — Parties relied on agreed 1976-77 assessments at £15 per m² for two large store units in centre, one twice the size of the other with similar frontage but considerably greater depth — Tribunal comments that some of the confusion which arose between the parties when analysing rents was due to rigid application of zoning system to two essentially different properties — Ratepayers' valuer's application of £15 per m² to the small appeal shops closer to reality than VO's £22 per m² (confirmed by LVC), but tribunal not satisfied it did not reflect some aspect of quantum — Determination of £18 per m² zone A

The following case is referred to in this report.

Summers Ltd v Procter (VO) [1985] 2 EGLR 215; (1985) 275 EG 381, LT

David Trotter (instructed by Linsley & Mortimer, of Newcastle upon Tyne) appeared for the nine appellant ratepayers; the Solicitor of Inland Revenue for the respondent valuation officer.

Giving his decision, MR MALLETT said: These appeals, which were heard together, are concerned with the provisions of section 20 of the General Rate Act 1967 in valuing shops in the Killingworth Township shopping centre [2, 4, 5, 7, 8, 11, 13, 15, 17 and 18 Citadel East and 22 and 24 Citadel West]. At the outset the appeals in respect of 5, 13 and 18 Citadel East were withdrawn as the appellants have now vacated their premises.

The remaining appeals stem from proposals made by Storey Sons & Parker, as agents acting for the various shops tenants, seeking substantial reductions in the assessments. The valuation officer objected to all the proposals and when the matter came before the Tyne and Wear Local Valuation Court all the assessments in the valuation list were confirmed but no reasons were recorded.

The ratepayers appeal to this tribunal seeking reductions in all the assessments. The respondent valuation officer seeks to maintain the assessments in the valuation list as confirmed by the local valuation court with the exception of 7 Citadel East where, on correction of an error made in the measurements of the premises, he maintains that the assessments should be slightly reduced to gross value £1,235. The relevant figures are set out in Appendix I.

APPENDIX I

Address	Assessments confirmed by local valuation court		Appellant's contended figures		Respondent's contended figures	
	GV £	RV £	GV £	RV £	GV £	RV £
2 Citadel East	1,300	1,055	900	722	1,300	1,055
4 Citadel East	1,260	1,022	865	693	1,260	1,022
5 Citadel East	1,220	988	Withdrawn			
7 Citadel East	1,300	1,055	830	663	1,235	1,001
8 Citadel East	1,240	1,005	835	668	1,240	1,005
11 Citadel East	1,260	1,022	845	676	1,260	1,022
13 Citadel East	1,265	1,026	Withdrawn			
15 Citadel East	1,230	997	830	663	1,230	997
17 Citadel East	1,220	988	820	665	1,220	988
18 Citadel East	1,920	1,572	Withdrawn			
22 Citadel West	1,550	1,263	1,100	888	1,550	1,263
24 Citadel West	1,450	1,180	1,035	834	1,450	1,180

Mr Trotter appeared on behalf of the ratepayers and called Mr B N Furniss FRICS, a partner in the firm of Storey Sons & Parker. Mr Sainer, for Solicitor of Inland Revenue, appeared on behalf of the valuation officer, and called Mr D M Saunders ARICS, a deputy valuation officer in the Tyneside Valuation Office.

It is common ground that at the material date the assessments of the shops under section 19 of the General Rate Act 1967 would have been in excess of the assessments in the valuation list. It is also common ground that for one reason or another the development of Killingworth Township has been a disappointment and the rental values of shops in the Killingworth centre have lagged behind those in other shopping centres in the valuation area, particularly those at Longbenton, which is close by.

Shops in Killingworth and Longbenton shopping centres are currently assessed at a similar basis of assessment and it is not disputed that in terms of section 19 of the General Rate Act 1967 shops in Longbenton are more valuable than those in Killingworth. The ratepayers contend that a similar distinction should be made in valuations under section 20 of the 1967 Act and the valuation officer maintains that all the matters which give rise to a difference in value are outside the terms of section 20.

Killingworth is a new township development in association with the Northumberland County Council and the Longbenton Rural District Council to act as an overspill area for the city of Newcastle upon Tyne.

The township is about five and a quarter miles to the north of the centre of Newcastle and about three miles to the south of Cramlington (another new township) and just over two miles from the Longbenton shopping centre.

Work on Killingworth and Cramlington was started in the early 1960s. It was estimated that the population of 645 in Killingworth in 1965-66 would increase to 20,000 by 1975.

The township was designed around the "Central Citadel", which was intended to contain offices, shops, schools and recreation facilities with no home more than 15 minutes' walking distance from the Citadel. In order to separate pedestrians from vehicular traffic the shops were built at first-floor level and connected to the residential area by a system of elevated walkways and underpasses. Killingworth is close to the A1 and the A188 and is therefore connected to major traffic routes north and south of the Tyne.

At an early stage of the development a hypermarket (Woolco) with an area of 100,000 sq ft and extensive car-parking facilities was incorporated in the shopping centre and it was hoped that Killingworth would develop into a regional shopping area.

By the time of the revaluation in 1973 the Woolco hypermarket was open and the first stage of the central area was completed. This consisted of 24 small shop units and three large units (units A, B, C), a recreation centre, a swimming pool and a multi-storey car park. These shops were clustered on each side of a walkway at first-floor level, known as the Citadel East and West, situated at the rear of the Woolco hypermarket.

At the 1973 revaluation the five shops and units A and C in Citadel West were let and occupied and assessed on the basis of £30 per m^2 for zone A. Unit B was empty and remains so to this day. In Citadel East nine of the shops were let and occupied and assessed on the basis of £22 per m^2 for zone A.

Mr Furniss, acting for the occupiers of units A and C, eventually agreed reductions in the assessments on these large units on the basis of £17.50 per m^2 for zone A. The agreements were reached in 1976. The reductions took effect back to April 1 1973.

The tenants of unit A vacated in 1975 and by January 1977 a new tenant had not been found. Furthermore, of the shop units which were vacant at April 1 1973 only one had been let (unit 5 on March 26 1976). Mr Furniss made further proposals to reduce the assessments on units A and C and in May 1977 reached agreement with the valuation officer that the two assessments should be on the basis of £15 per m^2 for zone A. It is this figure which Mr Furniss now seeks to apply to the assessments of all the small shops in Citadel East and West.

Subsequently, the valuation officer on his own initiative took steps to reduce the assessments on the shops in Citadel West based upon a reduction in the zone A rate from £30 to £22 per m^2, the same figure as in Citadel East.

Tribunal's findings

I turn now to consider the facts as of March 1982. There is no agreed statement of facts and I find it hard to believe that some agreement was not possible which would have effected a considerable saving in time at the hearing.

From the evidence I find:
1. The proposed increase in population at Killingworth of 20,000 in 1976 had never been reached and by 1982 was in the region of 10,000.
2. Problems had arisen with the high-rise housing units linked to the shopping centre by elevated walkways. One hundred and fifty flats were vacant and the population of the tower blocks thereby reduced from about 3,000 to between 2,300 and 2,500. With hindsight it can now be said that the situation has worsened and the decision has been taken to demolish all these tower blocks and replace them with 300 new two-storey houses. In 1973 only 10 of the high-rise units were vacant.
3. By the material date it was evident that a high proportion of residents in Killingworth worked, and possibly shopped, elsewhere, so that the shopping centre did not have the monopoly of the day-to-day shopping requirements of the Killingworth Township.
4. In August 1980 the metro rail system was opened, providing a rapid and cheap access to the centre of Newcastle and the new covered shopping centre at Eldon Square. The nearest station is two miles from Killingworth at Four Lane Ends, to which there is a frequent bus service, and there is a car park at the station.
5. Other shopping centres on the periphery of Newcastle have been created or extended, particularly at Cramlington, North Shields and Gosforth.
6. There was no sign at the material date that the increase in car ownership over the years had resulted in extending the shopping catchment area of Killingworth. In this respect the Woolco hypermarket could be regarded as having a different type of trade and a different catchment area.
7. The second stage of the shopping centre had not been developed and therefore the centre might more correctly be described as a parade of shops providing for the day-to-day needs of local residents.
8. Unit B still remained vacant. Unit A was vacant from 1975 to 1978 and again from 1979 to 1983. Three shop units were vacant and, it is now known, remained so for two to three years. There was a record of other shops which had been vacant for a year or more. There were other cases where the lease of occupied shops had been offered on the market without being sold. Several shops were empty at the material date and one is now occupied as a solicitors' office and another as a Citizens Advice Bureau.
9. All the rents in the early lettings had been increased on the first seven years' rent review. The original lettings were on the basis of a full repairing lease with the landlord providing the shell of the shop and the finishings being carried out by the tenant to the approval of the landlord. There is a dispute between the parties as to whether the leases provided that the rent review after the first seven years should be on the basis of the rental value of the shell only or on the completed shop. Without deciding the legal interpretation of the rent review clauses in the lease, and the clauses in the agreement for the lease, as to the fitting out of the shops, I accept Mr Furniss' evidence that his firm were acting as letting agents for the owners and they assumed that the rent reviews were on the basis that the rents were for the completed shops. There is no evidence as to the tenants' views.
10. The most comparable shops are in the Arndale Development at Longbenton. This is a parade of 18 small shops and two supermarkets developed in the early 1970s. There is no evidence of any vacant shops. The 1973 assessments were agreed on the basis of £24 per m^2 for zone A for shops facing on to the main thoroughfare and £20 per m^2 for zone A shops on a return frontage. No allowance for quantum was agreed in respect of the supermarkets.
11. Forest Hill Shopping Centre is in Station Road about halfway between the centres at Longbenton and Killingworth. The principal shops are on the north side of the railway line, with supermarkets and a number of national and regional retailers as well as local traders. There is no evidence of any vacant shops. The evidence of three agreed assessments on the north side confirms the basis of assessment in that section at £22 per m^2 for zone A.
12. In 1971 the population in Killingworth was at 4,813, with an unemployment rate of about 5%. By 1981 the population had risen to 10,429, with an unemployment rate of about 13.6%, this percentage being above the regional average. It would appear that between 1971 and 1981 there was a slight decline in the population of Longbenton.

Parties' contentions

Mr Furniss looks at the period between 1973 and 1986 and compares the two shopping centres in Killingworth and Longbenton. In his view, what he terms the non-section-20 matters are common to both areas. These, he submits, include national inflation, regional recession, and local unemployment. Therefore, in his view any change in the relativity of rental values in Killingworth and Longbenton will tend to relate to matters which are within section 20 including the occupation and use of houses and shops in the neighbourhood and the transport and other facilities in the neighbourhood.

He compares an analysis of 54 rents agreed between 1970 and 1986 in respect of shops at Killingworth with 34 rents agreed between 1971 and 1986 in respect of shops at Longbenton. The years of abundant rental evidence do not coincide in the two centres, but by interpolation it is possible to see that rental values in Longbenton have always exceeded those in Killingworth by a clear margin and that the margin has noticeably increased since 1973. This is during a period when Killingworth had the advantage of an expanding population with a hypermarket in the centre serving a large catchment area and when Longbenton had a declining population and was better served by the metro railway system into the centre of Newcastle.

Mr Furniss is of the view that there is a relative decline in Killingworth and that this is attributable to matters that clearly come within section 20(1)(b) of the 1967 Act.

Mr Saunders relies on much the same rental evidence. He admits that there is a change in the relative values of the two areas between 1973 and 1982, but he assumes that the rent reviews at Longbenton are in respect of finished shops and those at Killingworth are respect of the shells only. Therefore, in his eyes, this reduces the margin of difference between the two shopping areas.

He refers to those matters that could be considered to be within section 20(1)(b). In his view, there is no evidence to show that the improved transport services into the centre of Newcastle had any effect on the shop values in Killingworth and Longbenton, which have continued to show a steady increase in rental value. The declining population of the tower blocks had, in his view, been offset by the considerable increase in the number of new houses built in Killingworth.

In his view, the sluggish rental growth of shops in Killingworth could be attributed to the following factors:
(a) unemployment: rate of unemployment in Killingworth was above the regional average.
(b) the best-gets-better factor: a concept referred to in the Lands Tribunal decision in *Plummers Ltd v Procter (VO)* [1985] 2 EGLR 215; (1985) 275 EG 381 (one of my decisions).
(c) the change in public taste which now finds the Citadel development less attractive than more conventional shopping areas.

Mr Saunders had not been personally involved in the previous agreements on shop assessments in Killingworth. From his office records, which were not produced, he assumed that the agreed assessments of units A and C were made on the basis of a zone A rate of £22 per m^2 less 32.6% to allow for quantum. This considerable allowance is, he maintains, justified by comparing the rent for unit A with the average of the rents obtained for the small shops. He admits that no such allowance is justified by the rent for the unit C.

In his view, the basis of £22 per m^2 for zone A had now been established in the Citadels East and West and there were no grounds within section 20(1)(b) for amending the basis. Therefore, he maintained, the assessments in the valuation lists should be upheld with the exception of no 7, where the correction of an error resulted in a slight reduction in the assessment.

I have viewed internally three typical small shops in Killingworth and viewed externally the remaining comparables in Killingworth and Longbenton and Forest Hill. I have also visited other shopping centres in the region and have travelled between Eldon Square and the Citadel using the public transport system.

Decision

From the evidence, there can be little doubt that at the time of the 1973 revaluation the rents of shops in Killingworth were agreed in the hope that there would be rapid development of housing in the area which would result in an increase in the population to 20,000 by 1975, before the next rent review, and that the shopping area would be developed to meet the increased demand.

By 1982 the reality was that the population had reached only 10,000 or thereabouts; the second-stage development of the shopping area had not started; the development of houses was continuing but at a slower pace and a number of the high-rise units had become empty; it was evident that a high proportion of the residents in Killingworth worked elsewhere and had the opportunity to shop elsewhere; the improved transport facilities from Killingworth to other places did not discourage this trend; there was a record of shops that had been empty and available for letting and the limit of the shopping centre facilities provided by the occupied shops was plain to see.

It seems to me that the original concept of Killingworth has not materialised and the Citadel now consists of a limited number of shops supplying the limited day-to-day requirements of those persons in the immediate neighbourhood. The evidence supports the view that a large number do their shopping elsewhere.

A detailed comparison of the rents passing at the two shopping centres at Killingworth and Longbenton clearly shows that rents at Killingworth have always been below those at Longbenton and the increase in rental values has been at a slower rate. Broadly speaking, it may be said that "non section 20" factors are common to both shopping areas and therefore tend to cancel out in any relative comparison. However, the valuation officer has referred to the increase in unemployment in Killingworth and the changing shopping habits of its residents as being two factors which do not come within section 20. I was not referred to any authority for this view and I am not aware of any. The relevant parts of section 20 read as follows:

(1) For the purposes of any alteration of a valuation list to be made under Part V of this Act in respect of a hereditament in pursuance of a proposal, the value or altered value to be ascribed to the hereditament under section 19 of this Act shall not exceed the value which would have been ascribed thereto in that list if the hereditament had been subsisting throughout the relevant year, on the assumptions that at the time by reference to which that value would have been ascertained — (a) the hereditament was in the same state as at the time of valuation and any relevant factors (as defined by subsection (2) of this section) were those subsisting at the last-mentioned time; and (b) the locality in which the hereditament is situated was in the same state, so far as concerns the other premises situated in that locality and the occupation and use of those premises, the transport services and other facilities available in the locality, and other matters affecting the amenities of the locality, as at the time of valuation.

Section 20(2) defines "relevant factors", which are not the subject of the dispute in these appeals.

It seems to me that section 20 requires the valuation to be made having regard, *inter alia*, to the occupation and use of the premises situated within the locality; in these appeals, the occupation and use of shops and houses in Killingworth and neighbouring areas. Statute does not require an examination of the reasons which give rise to the occupation and use.

That said, I am not convinced from the evidence that unemployment in Killingworth caused residents to shop elsewhere for their essential day-to-day needs or that the architectural merits of the Citadel were uppermost in their minds when considering where to shop. As to the "best-gets-better" factor, as I understand the expression, for one reason or another the values of central shopping areas are said generally to appreciate more rapidly than those of secondary shopping areas.

If that be the case, and if the Newcastle shopping centres have appreciated in value at the expense of the outlying shopping areas, then the effects on Killingworth and Longbenton are likely to be similar and to be cancelled out in any comparison of the relative values of those two shopping areas. Therefore it is unnecessary for me to decide whether this is a matter that can be taken into account under section 20.

At the end of the day, in arriving at the appropriate basis of assessment for the appeal premises, the parties relied on the agreed assessments on units A and B which took effect in 1976-77. Interpretation of these agreed assessments is a matter of opinion and judgment. I feel that some of the confusion which arises between the parties when analysing the rents and assessments is due to the rigid application of a zoning system to two properties which are essentially different. Unit A in effect is equivalent to rather more than five small shops rolled into one. Unit C is about twice the size of unit A, with a similar frontage but a considerably greater depth.

It seems to me that Mr Furniss' interpretation of the agreements is closer to reality, but I am not satisfied that a zone A rate of £15 per m^2 does not reflect some aspect of quantum. Looking at all the rental evidence in both Killingworth and Longbenton, in my view a basis of £18 per m^2 for zone A would be appropriate for small shops in Citadel East and West.

There are very minor differences between the parties as to the treatment of mezzanine and basement floors and I have accepted Mr Furniss' figures which reflect an increase in value where the standard of finish is slightly superior.

Therefore the assessments in the valuation lists are reduced to the following figures:

Descriptions	Address	Gross value £	Rateable value £
Bank and premises	2 Citadel East, Killingworth	1,040	838
Shop and premises	4 Citadel East	1,015	817
Shop and premises	7 Citadel East	1,070	863
Shop and premises	8 Citadel East	970	780
Shop and premises	11 Citadel East	995	801
Shop and premises	15 Citadel East	975	784
Shop and premises	17 Citadel East	960	772
Shop and premises	22 Citadel West	1,260	1,022
Shop and premises	24 Citadel West	1,200	972

To that extent the appeals succeed.

I heard submissions as to costs. The valuation officer is to pay the appellants' costs, in the event of disagreement to be taxed by the Registrar of the Lands Tribunal on the High Court Scale of Costs.

For further cases on this subject see pp 164 and 248 Note: The next case appears on p 209 in the green section

SUPPLEMENT
CASES NOT REPORTED IN "ESTATES GAZETTE"

LANDLORD AND TENANT
GENERAL

Chancery Division

March 28 1979

(Before Mr Justice SLADE)

BRADDON TOWERS LTD v INTERNATIONAL STORES LTD

Landlord and tenant — Whether court was precluded in the present state of the law from granting a mandatory injunction compelling someone to carry on a business — Defendant tenants had covenanted to keep premises intended to be used as a supermarket open at all normal times as a first-class shop, with a display window suitably and attractively placed and to use their utmost endeavours to develop, improve and extend the business and not to do or suffer anything to injure the goodwill — The venture proved to be unprofitable and the defendants decided, in deliberate breach of their covenant, and without prior consultation with the plaintiffs, to close the supermarket — The present proceedings by the landlords sought interlocutory relief by notice of motion, asking for an injunction until judgment in the action, which had been commenced by writ claiming remedies of a permanent nature — The notice of motion sought an injunction restraining the defendants from ceasing to operate the premises as a supermarket and from opening another supermarket in the area; also requiring the defendants to keep open the premises as a first-class supermarket during business hours until the trial of the action or further order (This last relief was the one sought at the hearing) — Slade J said that so far as merits were concerned there was no criticism of the plaintiff landlords, but the defendants' attitude was unattractive — The question, however, was whether, having regard to the present state of the authorities, the court could grant the plaintiffs the interlocutory relief they sought — The weight of authority was that an injunction would not be granted to compel someone to carry on a business and a reason frequently given was the impossibility for the courts to supervise the performance of the order — In some recent judgments there were indications of a change of attitude, but there was as yet no decision where the court had granted such an injunction — For many years practitioners had advised their clients that it was the settled and invariable practice of the court never to grant mandatory injunctions requiring persons to carry on a business — There was no real prospect of the trial judge in the present case being persuaded to depart from the practice — That was the decisive reason why, "reluctantly", an injunction could not be granted on the present interlocutory application

The following cases are referred to in this report.

Attorney-General v *Colchester Corporation* [1955] 2 QB 207; [1955] 2 WLR 913; [1955] 2 All ER 124; (1955) 53 LGR 415
Attorney-General v *Staffordshire County Council* [1905] 1 Ch 336
Blackett v *Bates* (1865) 1 Ch App 117
Dowty Boulton Paul Ltd v *Wolverhampton Corporation* [1971] 1 WLR 204; [1971] 2 All ER 277; (1971) 69 LGR 192
Giles (CH) & Co Ltd v *Morris* [1972] 1 WLR 307; [1972] 1 All ER 960
Gravesham Borough Council v *British Railways Board* [1978] Ch 379; [1978] 3 WLR 494; [1978] 3 All ER 853; (1977) 76 LGR 202

Greene v *West Cheshire Railway Co* (1871) LR 13 Eq 44
Hooper v *Brodrick* (1840) 11 Sim 47
Jeune v *Queens Cross Properties Ltd* [1974] 1 Ch 97; [1973] 3 WLR 378; [1973] 3 All ER 97; (1973) 26 P&CR 98; [1973] EGD 976; 228 EG 143
Powell Duffryn Steam Coal Co v *Taff Vale Railway Co* (1874) LR 9 Ch 331
Ryan v *Mutual Tontine Westminster Chambers Association* [1893] 1 Ch 116
Shepherd Homes Ltd v *Sandham* [1971] Ch 340; [1970] 3 WLR 348; [1970] 3 All ER 402; (1970) 21 P&CR 863; [1970] EGD 583; 215 EG 580
Shiloh Spinners Ltd v *Harding* [1973] AC 691; [1973] 2 WLR 28; [1973] 1 All ER 90; (1973) 25 P&CR 48, HL

Editor's note: It will be noted that the judgment of Slade J (as he then was) in this case was given on March 28 1979. The case has hitherto been unreported. The decision to report it, despite the lapse of time, was influenced by the reference to it in F W Woolworth plc v Charlwood Alliance Properties Ltd, *a decision of Judge Finlay QC sitting as a High Court judge, reported at [1987] 1 EGLR 53; (1987) 282 EG 585. Judge Finlay pointed out that he was dealing with a covenant of substantially the same character as the covenant in the* Braddon Towers *case to keep the demised premises open as a department store or shop — Judge Finlay, like Slade J, decided reluctantly that, in the present state of the authorities, he could not grant an injunction by way of, in effect, an order of specific performance to carry on a business. The relevant authorities are considered in some detail in Slade J's judgment.*

This was a notice of motion by Braddon Towers Ltd, the landlords, seeking interlocutory relief by way of injunction, in an action against tenants, International Stores Ltd, in relation to premises in the Vincent Park Shopping Centre, Peel Drive, Vincent Park Estate, Sittingbourne, Kent. The centre consisted of a supermarket, the subject of the present dispute, and three smaller retail units, all built by the plaintiffs in 1969 and 1970. The action and the interlocutory motion arose out of the closing of the supermarket in January 1979.

Ronald Bernstein QC and Jonathan Gaunt (instructed by Berwin Leighton) appeared on behalf of the plaintiffs; Gerald Godfrey QC and T Lloyd (instructed by Kenneth Brown Baker Baker) represented the defendants.

Giving judgment, SLADE J said: This is an interlocutory motion by which extraordinary relief is sought in somewhat unusual circumstances. The plaintiffs in the action, the applicants in the motion, are a company called Braddon Towers Ltd, which is a subsidiary company of Allied London Properties Ltd. The defendants in the action and respondents to the motion are a company, International Stores Ltd, which possesses very large financial resources. According to the defendants' unchallenged evidence, they have a paid-up capital of £15,758,318 and an annual turnover of about £500 million. Another company within the International Stores Group, called Pricerite Ltd, will feature in the history of this matter, though it is not a party to the proceedings.

The relief sought by the notice of motion, according to its terms, is an order that the defendants be restrained by injunction until judgment in the action or further order from

(a) ceasing to operate the premises identified in paragraph 1 of the Statement of Claim herein as a first class supermarket; and (b) opening in Sittingbourne another supermarket or other retail shop selling goods of the same kinds as have been sold in the said premises; and for a further Order that the Defendants do (if necessary) reopen and keep open the said premises as a first-class supermarket during the normal business hours of the locality until the trial of this action or further Order.

Mr Bernstein on behalf of the plaintiffs, however, has indicated that in the events which have happened he does not seek relief in the terms of paras (a) and (b) of the notice of motion. He asks for relief substantially in the terms of the final limb of the notice of motion, though he recognises that the court, if it were to grant relief, might think it appropriate or necessary to depart from the precise wording of the injunction sought.

While quite a large amount of evidence has been filed on both sides in the application, there is remarkably little dispute as to fact. The premises referred to in the notice of motion are comprised in an area known as the Vincent Park Shopping Centre, Peel Drive, the Vincent Park Estate, Sittingbourne, Kent. The centre consists of a supermarket, which is the property there referred to, and three smaller retail units. It was built by the plaintiffs in 1969 and 1970 to serve the Vincent Park Estate, which is a residential area on the outskirts of Sittingbourne, where 400 new homes have recently been erected by an associated company of the plaintiffs, Sterling Homes Ltd.

The shopping area in question is somewhat isolated, in the sense that it is not situated in a shopping street, and, apart from the three shops which I have mentioned, there are no other shops in the immediate vicinity.

The evidence of the plaintiffs' managing director, Mr Leigh, is that since the viability of the whole centre, if and when built, was going to depend on obtaining a tenant for the proposed supermarket, they obtained a tenant before they began the building operation. This tenant was called Granville Supermarkets Ltd (which I will call "Granville"). Its decision to proceed with the building, according to Mr Leigh's evidence, was substantially influenced by the readiness of Granville to take a 21-year lease of the supermarket and to give positive covenants to keep the supermarket open as a supermarket and develop its trade.

The lease of the supermarket was executed by the plaintiffs in favour of Granville on August 10 1970. It was for a term of 21 years and provided for an annual rental of £2,500. There was, however, provision for rent reviews after seven and 14 years and the rent has, I understand, now been increased.

Clause 2(3) of the lease imposed wide repairing obligations on the lessee. Clause 2(8) and 2(9) contained covenants by the lessee which are particularly relevant for present purposes. They read:

(8) Not to use or permit or suffer to be used the demised premises or any part thereof for any illegal or immoral purposes nor for the carrying on of any offensive noisy or dangerous trade business manufacture or occupation . . . and not without the written consent of the Lessor (such consent not to be unreasonably withheld) to use the building of the demised premises otherwise than as a retail shop for the carrying on the trade of a supermarket with liberty to sell intoxicating liquor for consumption off the premises.

(9) At all times of the year during the normal business hours of the locality except when the demised premises shall be closed for alterations or upon a change of occupier to keep the shop open as a first-class shop and at all times to keep the shop or display window suitably and attractively dressed and to use the utmost endeavours to develop improve and extend the said business and not to do or permit or suffer to be done anything which may injure the goodwill of the said business and or any other businesses carried on in the Centre.

Clause 2(20)(b) contained a covenant on the part of the tenant:

Not to assign the whole or charge underlet or part with possession or occupation or share the occupation of the whole or any part of the demised premises without the prior consent in writing of the Lessor.

There followed three provisos, one of which was that, on the granting by the lessor of such consent, every assignment should contain a covenant by the assignee directly with the lessor to observe and perform the covenants and conditions contained in the lease, including the provisions of subclause (20). Clause 4(1) of the lease contained a provision for re-entry on breach, in common form.

In April 1972 Granville applied for consent to assign the property to the defendants. It was one item in a package deal, by which the defendants acquired a number of supermarkets in Granville. The defendants suggested that they should be permitted to occupy this particular supermarket for a trial period of 18 months, pending completion of the assignment, upon terms that they should be responsible for the tenant's obligations only during that limited period in the event of the assignment not proceeding. This trial period, it would appear, had been suggested on behalf of the defendants, because they had doubts as to the profitability of carrying on a supermarket business on these premises, having regard to their previous trading record.

The plaintiffs, however, would not agree to this suggestion of a trial period, and insisted, as they were entitled to insist, under clause 2(20) of the lease, on the defendants' entering into a direct covenant with them to observe and perform the tenant's covenants for the whole of the residue of the term.

The defendants in due course accepted this requirement, and a deed of licence was executed on November 12 1973, by clause 2 of which the defendants entered into such a covenant with the plaintiffs in consideration of the licence being granted for the assignment to them of the term. The assignment was not completed until May 2 1974. It appears from their own evidence that the defendants occupied and traded from the premises before the assignment was completed, but I do not think anything turns on this fact.

After taking possession, the defendants ran the premises for a number of years as a supermarket. The three shops in the centre are let respectively, as to shop 1, to a Mr R N Kelly, a baker; as to shop 2, to a Mr and Mrs Mountford, as newsagents; and as to shop 3, to a Mr and Mrs Bridges, as greengrocers. The plaintiffs are still the landlords of all these various premises. Since the centre is not in a shopping street or shopping area, the fortunes of the three smaller shops naturally depend substantially on the continued presence in the centre of a supermarket.

Mrs Bridges, Mr Mountford and Mr Kelly's wife have all sworn affidavits deposing to the fact that the decision to take a lease of the respective shops was influenced by, among other things, the combination of shops in a block which included a well-known store supermarket, which it was thought would attract shoppers to this rather isolated shopping centre.

For some years the defendants ran the supermarket as an International Stores shop. In an affidavit sworn on their behalf by a Mr Evans, a retail executive director of Pricerite Ltd, he said that the shop showed a profit in the year to September 1977 of £11,000, but in the following year it made a loss of £9,000. This trading loss, he said, was "unacceptable", and the defendants were considering closing the shop in the summer of 1978 at a time when its gross receipts were £2,200 to £2,300 per week. It was decided, however, that the shop be changed in character and run by Pricerite Ltd on a different basis, in the hope of increasing its trade. Whereas International Stores shop stock a large number of different lines, Pricerite shops, according to Mr Evans' evidence, are designed to increase profit margins by stocking a more limited range of goods.

The change to a Pricerite shop in the summer of 1978, however, did not bring about the hoped for increase in receipts, which rose only to a weekly average of £2,500. By the end of November a forecast was made of the store's likely trading results, which showed a prospective loss of £4,000 for the year to September 1979; and it was thought that thereafter things would get worse. In the light of these figures (I quote from Mr Evans' affidavit) "Pricerite Ltd took the decision that the shop should be closed, and that the date for closure should be shortly after Christmas."

To anyone who was aware of the terms of the lease, it must have been obvious that such a closure, if and when it occurred, would involve a serious breach of clause 2(9) of the lease. The defendants' evidence on this motion gives no further particulars of the precise date on which or the manner in which this decision to close the supermarket was taken. It has not, however, been suggested in their evidence that the persons who took the decision on their behalf were ignorant of their obligations to the plaintiffs under the lease; and indeed it would be surprising if an important decision of this kind had been taken by a large commercial company entirely without reference to the terms of the lease affecting the relevant premises.

It must, furthermore, have been obvious to anyone who knew of the locality that closure of the supermarket, besides constituting a breach of the terms of the lease, would be likely to cause serious potential financial loss and anxiety both to the plaintiffs themselves and to the tenants of the adjacent shops. The defendants, however, took no steps to notify either the plaintiffs or these other tenants of the impending closure. However, on the evidence before me, until January 9 1979 they took no steps whatever to attempt to mitigate the loss to the other interested parties by finding a suitable assignee or sublessee of the premises, who would be prepared to carry on a supermarket business and to move in more or less as soon as they themselves vacated the premises. All they did was simply to make their own arrangements for closure, for example by giving notice to the staff employed there. No public announcement of the impending closure was made until a notice was put up in the shop on December 28 1978 announcing the impending closure on Saturday, January 1 1979.

On January 3 1979 the plaintiffs learned for the first time of the notice and of the fact that the shop was nearly devoid of goods and staffed by only three persons. Mr Walker, the plaintiff company's secretary, immediately telephoned Miss Edwards, in the defendant's legal department. On January 4 the plaintiffs' solicitors caused a letter to be delivered by hand to Miss Edwards, referring to the notice and to the fact that any closure would be a breach of the terms of the lease. They asked for confirmation that the premises would not be closed and that the defendants would cease to exhibit the notice.

Even if, therefore, for any reason as yet unexplained, the relevant provisions of the lease had not before January 4 1979 been drawn to the attention of the persons principally concerned with the matter in the defendants' offices, there seems no doubt that they would have known of them shortly after that date. Nevertheless, it would appear, the deliberate decision was taken to go ahead with the closure, in breach of the terms of the lease.

On January 8 Mr Leigh spoke on the telephone to a Mr Deakin, a director of the defendants, to whom he had been referred by another director, Mr Muir. Mr Leigh's account of this conversation, as given in an affidavit, is as follows:

Mr Deakin said that he knew of the decision to close the supermarket but also claimed to be unaware of the covenant requiring the supermarket to be kept open. He promised to ring me back that afternoon after consulting his legal department, but did not do so. I told him that if the defendants did close down the premises on January 13 that it would be the plaintiffs' intention to take proceedings for an injunction, to which he replied, "It seems to my devious mind that it might be possible to leave a couple of girls in the shop just selling a few packets of cigarettes a day." Upon my pointing out how damaging that would be both to the supermarket business and to the adjoining traders, he said that he quite understood.

In an affidavit in answer sworn by Mr Deakin, he says that the substance of what Mr Leigh says about this telephone conversation is correct. He adds:

I was in fact unaware of the covenant in the lease requiring the supermarket to be kept open but I was well aware of the losses being incurred. I pointed out to Mr Leigh that we were trading at a loss and could not be expected to carry on doing so.

This comment was made, as I understand the evidence, after — not before — the existence of the covenants had been brought to Mr Deakin's attention.

On January 9 a further letter was sent by the plaintiffs' solicitors to the defendants' solicitors saying that, since no reply to their letter of January 4 had been received, they were instructed to issue proceedings. Also on January 9, no doubt stimulated by the mounting pressure from the plaintiffs, the defendants, apparently for the first time, put the premises in the hands of estate agents, Ward & Partners, of Chatham. These instructions were to offer the premises on the market on the basis of a sale of the leasehold interest with fixtures and fittings.

Looking ahead for one moment, Ward & Partners in due course did find a potential purchaser for the premises, a Mr Patel, but he withdrew in February. The defendants then verbally instructed other agents, J Trevor & Sons, in addition to Ward & Partners, to find a purchaser. At the moment, however, so far as the evidence shows, there is no other potential purchaser or tenant in sight.

On January 11 letters were delivered to the joint managing directors of the defendants from Mr Leigh, enclosing copies of the plaintiffs' solicitors' letter of January 9. They did not reply to Mr Leigh. On Saturday, January 13, the defendants opened the supermarket for the last time. They did not reopen it on Monday 15 and have never reopened it since. It has remained vacant and unoccupied since that time. Meantime, a Pricerite shop remains open in Sittingbourne High Street which, under present circumstances, is being spared the measure of competition that it would presumably face if a supermarket under another management were reopened at the Centre, though it is fair to say that Mr Godfrey, on behalf of the defendants, tells me on instructions that the two premises are situated about two miles away from one another.

From January 15 onwards, it appears to me on the evidence clear that the defendants were in deliberate and conscious breach of each and every limb of clause 2(9) of the lease. They were not keeping the shop open as a shop during the normal business hours of the locality; they were not keeping the shop or display window suitably and attractively dressed; they were not using their utmost endeavours, or any endeavours at all, to develop, improve and extend the business; they were doing something which might injure both the goodwill of the business and of the other businesses carried on in the centre.

Against this background, the writ in the present action was issued promptly on January 15 itself, seeking relief of a permanent nature, substantially in terms similar to those of the notice of motion, which I have already read. The plaintiffs do not seek forfeiture of the lease, and there is no reason why they should be expected to do so.

On the same day, January 15 1979, the notice of motion was issued. Though he has reserved his position at the trial, Mr Godfrey on behalf of the defendants has not sought to argue before me that they have not been continuously in breach of clause 2(9) of the lease since January 15. Nor has he sought to suggest that the covenants contained in clauses 2(8) and 2(9) of the lease are anything other than valid covenants.

At least for the purposes of the present motion, therefore, the plaintiffs have established not merely a triable issue on the question of liability; at present they have established what seems to be clear liability on the part of the defendants. The only question is what remedy, if any, should be afforded to the plaintiffs pending the trial of the action. While the plaintiffs would, I understand, be content that this motion should be treated as the trial of the action, the defendants are not so content.

The reasons why the plaintiffs seek interlocutory relief of the nature mentioned in the notice of motion are readily understandable. First, the closure of the supermarket is likely to cause a substantial diminution in the value of the plaintiffs' interests both in the supermarket premises themselves and in the reversions of the adjacent shops which are comprised in the centre. It is also likely to cause them embarrassment and injury to their reputation, in a manner which could not be compensated by an award of damages at the trial. The longer the period continues during which the supermarket is left empty and is not used for its intended purpose, the more difficult it will probably be to relet the premises. Furthermore, any substantial period of vacancy might seriously impair the trade of the plaintiffs' other three tenants and in the end force them out of business and leave their shops unlettable. This could not only render the plaintiffs' reversionary interests in the supermarket and shops very difficult to market but it could seriously damage their reputation as developers and landlords in the eyes of potential tenants.

The plaintiffs and their sister company have developed many other residential areas involving shopping centres in the Home Counties over the last 25 years and intend to continue doing so. A disintegration such as is now in the process of occurring in the Sittingbourne Centre can only do harm to their reputation while it continues. Though it is suggested in the defendants' evidence that these anxieties of the plaintiffs are somewhat exaggerated, they are to a considerable extent borne out not only by evidence from Mr Leigh but also by evidence in the affidavits sworn by Mrs Bridges, Mr Mountford and Mrs Kelly. The first two say that they will be reducing their hours of opening, since the expense of opening at certain stated supermarket hours will henceforth exceed any compensating benefit. All state their belief that, as a result of the closure of the supermarket, trade at the centre will decline and it will become increasingly difficult to make a living comparable with that enjoyed over the past years of business.

The plaintiffs have indeed now received a written request from Mr Bridges that he be released from his tenancy, due to financial losses following the closure of the supermarket. They have felt obliged to refuse this request, but the position for them is clearly an embarrassing one.

On this present application they may fairly be said to be fighting a battle not only on their own behalf but on behalf of the tenants of the three shops and the many members of the public who wish to see the supermarket kept open. The plaintiffs, at about the end of January, received a petition headed "Petition for Grocery Shop" signed by a large number of residents in the shopping centre area, I think about 200. If the plaintiffs' battle is unsuccessful there must be a substantial risk of damage to their reputation in the eyes of tenants and local inhabitants, who may not realise how hard they have fought it. For all these reasons, I accept Mr Bernstein's submission that, if interlocutory relief is refused at the present stage, the plaintiffs are likely to suffer irreversible damage, some of which will not be capable of being adequately compensated by a monetary award at the trial.

Mr Bernstein has made it clear that his clients' object in bringing the action, in launching the present motion, is not necessarily to compel the defendants themselves to continue trading in the supermarket for the whole of the residue of the term of the lease, which expires in 1991. All they do wish to do is to ensure that a supermarket business continues to be conducted on the premises by respectable and responsible persons, be they the defendants or other persons, for the residue of that term. If the interim injunction sought were now granted against the defendants, and the defendants were to make really strenuous efforts to find a new tenant or assignee for the supermarket who was prepared to carry on the supermarket business, it might well be that such a person would emerge before the trial and thus render the grant of a more permanent injunction then unnecessary.

Even though one particular potential tenant, Mr Patel, has dropped out of the running, there is as yet no evidence that it will never be possible to find a tenant who will be prepared to take on the obligation of the lease, including clauses 2 (8) and (9), either as assignee or sublessee. Though there is evidence that the defendants believe that they themselves cannot fulfil these obligations except at a financial loss, there is no evidence that shows that other persons, arranging their staffing and so forth in a different manner, might not be able to make a profit out of the shop. Such evidence, of course, may emerge at the trial and, if it did, would doubtless weigh with the court, in so far as it regarded itself as having discretion to grant or withhold the relief by way of injunction.

In the meantime, in Mr Bernstein's submission, the balance of convenience heavily favours an order compelling the defendants to fulfil their obligations under clauses 2 (8) and 2 (9).

So far as merits have anything whatever to do with the present matter, I know of no possible criticism that can justifiably be made of the plaintiffs. On the other hand, I confess that I find the defendants' attitude an unattractive one. They undertook their obligations under the lease with their eyes open and, after negotiation, expressly contracted with the plaintiffs to observe them. They now commit a serious continuing breach of the covenants under clause 2 (9) without any apparent consideration for the plaintiffs or the other occupants of the shopping centre. So complete has been their disregard of the plaintiffs that they failed both to give them any advance notice of their intentions and to make any response whatever to their letters of protest before action.

Furthermore, if they had chosen, they could at least have made attempts to mitigate the loss to the plaintiffs, likely to be caused by their contemplated breach of covenant, by making active attempts to find another tenant to take their own place immediately they themselves vacated the premises. So far as the evidence shows, they did nothing whatever towards this end until January 9, four days before the intended closure, by which time they had received the plaintiffs' letter before action.

In effect, they now say they propose to disregard their obligations under clause 2 (9) of the lease, not because they cannot perform them, nor because they cannot afford to perform them, but simply because they do not choose to do so. So far as the evidence shows, they have offered not one word of excuse, apology or regret to the plaintiffs for their deliberate flouting of these obligations. Certainly not one word of excuse, apology or regret has been expressed by their counsel. The only explanation offered on their behalf has been that given by Mr Deakin, a director of the defendants, in a conversation which I have already quoted: "I pointed out to Mr Leigh that we were trading at a loss and could not be expected to carry on doing so." This explanation was echoed by Mr Godfrey on behalf of the defendants, who submitted that his clients made what he called a correct commercial decision. This so-called correct commercial decision was no doubt made in the expectation that the court would not grant relief by way of injunction to the plaintiffs to compel the performance of the defendants' contractual obligations under clauses 2 (8) and (9) of the lease and under the licence, and after comparison of the likely cost to the defendants of performing the covenants with the probable award of damages to be made against them in the event of their breach. If this is the standard by which the defendants measure the correctness of all their commercial decisions, I have some sympathy for the persons who find themselves doing business with them.

The principal question for this court, however, is not whether the defendants have behaved in a shabby manner. It is whether Mr Godfrey is correct in submitting that, on the present state of the law, the court is precluded from granting a mandatory injunction compelling someone to carry on a business. I have reluctantly been driven by him to the conclusion that, at least so far as this court of first instance is concerned, this submission is correct.

In *Hooper* v *Brodrick* (1840) 11 Sim 47 a lessee had entered into a positive covenant to use and keep open the demised premises during the term as an inn. The inn proved to be a losing concern and the defendant threatened to break the covenant. The landlord obtained *ex parte* an injunction restraining him *inter alia* from discontinuing during the term and to use and keep open the demised premises as an inn. Sir Lancelot Shadwell V-C dissolved the injunction, saying (at p 49 of the report):

The court ought not to have restrained the defendant from discontinuing to use and keep open the demised premises as an inn, which is the same in effect as ordering him to carry on the business of an innkeeper.

In *Attorney-General* v *Colchester Corporation* [1955] 2 QB 207, an attempt was made to obtain an injunction against a ferry owner compelling him to continue to run a ferry which had become a losing concern. Lord Goddard CJ rejected this claim, saying (at p 217):

The duty to maintain and work the ferry is the same whatever the circumstances of the ferry owner. No authority has been quoted to show that an injunction will be granted enjoining a person to carry on a business, nor can I think that one ever would be, certainly not where the business is a losing concern.

Finally, in *Dowty Boulton Paul Ltd* v *Wolverhampton Corporation* [1971] 1 WLR 204, the plaintiff sought an injunction compelling the maintenance of an airfield as a going concern. Pennycuick J, in rejecting this claim, said (at p 211):

I turn now to consider the remedy available to the company should the corporation persevere in its intention to appropriate this land for housing. It seems to me that the remedy of the company must lie in damages only and that the company is not now entitled, and will not be entitled at the hearing of the action, if it is then otherwise successful, to any relief by way of injunction or mandatory order. The right vested in the company necessarily involves the maintenance of the airfield as a going concern. That involves continuing acts of management, including the upkeep of runways and buildings, the employment of staff, compliance with the Civil Aviation Act 1949 and so forth, that is to say, in effect the carrying on of a business. That is nonetheless so by reason that so far the corporation has elected to engage Don Everall Aviation Ltd to manage the airfield on its behalf. It is very well established that the court will not order specific performance of an obligation to carry on a business or, indeed, any comparable series of activities.

One could not have a much clearer statement of the law than that.

The weight of authority against the plaintiffs is thus a formidable one. Not one example has been cited to me of a case where the court has granted an injunction compelling someone to carry on a business, but as I have shown, there are now at least three decisions of first instance in which, effectively, it has been clearly stated that such an injunction will never be granted.

It is, I think, fair to say that none of these three decisions explored in detail the rationale that lay behind the refusal of the courts in the past to make orders compelling the carrying on of a business. Mr Godfrey suggested that it stemmed from two principles, namely, that (1) the court cannot specifically enforce a contract to do continuous successive acts, since the execution of such a contract requires watching over and supervision by the court, which it is not prepared to undertake: (2) it is a necessary requisite of every mandatory injunction that it should be certain and definite in its terms.

In support of his first principal proposition Mr Godfrey cited *Ryan* v *Mutual Tontine Westminster Chambers Association* [1893] 1 Ch 116 and *Blackett* v *Bates* (1865) 1 Ch App 117 (see at p 124, *per* Lord Cranworth). In support of his second proposition he cited a passage from Joyce J's judgment in *Attorney-General* v *Staffordshire County Council* [1905] 1 Ch 336 (at p 342). As regards the latter decision, I entirely accept that, as Joyce J said, in relation to every injunction or mandatory order

it must or ought to be quite clear what the person against whom the injunction or order is made is required to do or to refrain from doing

I am not, however, wholly persuaded that an order to carry on a business would in every case, and by its very nature, be of so uncertain a character that it could not make it plain what the person concerned was required to do. In particular, for all Mr Godfrey's colourful examples of hypothetical borderline cases, I am not convinced that the defendants in the present case would have any real, practical difficulty in knowing what they had to do or to refrain from doing to comply with an order of the court effectively compelling them to observe the provisions of clauses 2 (8) and 2 (9) of the lease. It is not suggested that these provisions are void for uncertainty. As Mr Bernstein pointed out, if contempt proceedings were to be brought against the defendants alleging that they were in breach of such an order, the burden or proof would lie fairly and squarely on the plaintiffs, and the benefit of any doubt would be given to the defendants. I am not persuaded that, if an order of the court were made in terms substantially obliging the defendants to observe the provisions of clauses 2 (8) and 2 (9) of the lease, and if they themselves thereafter acted in good faith, they would have any real difficulty in knowing how to comply with it, without risk of successful contempt proceedings. If, on the other hand, such an order were made and they were, for example, to adopt Mr Deakin's idea of leaving a couple of girls in the shop just selling a few packets of cigarettes a day, the court would no doubt be well capable of dealing with the matter.

As regards the asserted principle that the court will not specifically enforce a contract to do continuous and successive acts, this is supported by a number of earlier decisions, such as the *Ryan* case. Furthermore, James LJ in *Powell Duffryn Steam Coal Co* v *Taff Vale Railway Co* (1874) 9 Ch App 331, said in a judgment, with which Mellish LJ agreed (at p 335):

Where what is required is not merely to restrain a party from doing an act of wrong, but to oblige him to do some continuous act involving labour and care, the court has never found its way to do this by injunction.

A little later he said:

The plaintiffs fail only because of the difficulty in the way of this court's enforcing such a right — a difficulty which to my mind is insuperable.

It appears, however, from recent authorities that the court may to some extent be abandoning this rigid differentiation between mandatory and prohibitory injunctions of a continuing nature. Lord Wilberforce in *Shiloh Spinners Ltd* v *Harding* [1973] AC 691, in a speech with which Viscount Dilhorne, Lord Pearson and Lord Kilbrandon wholly agreed, said:

Where it is necessary, and, in my opinion, right, to move away from some 19th century authorities, is to reject as a reason against granting relief, the impossibility for the courts to supervise the doing of work.

Megarry J in *C H Giles & Co* v *Morris* [1972] 1 WLR 307 (at p 318) said that the so-called rule that contracts involving the continuous performance of services would not be specifically enforced was plainly not absolute and without exception and that it could not be based on any narrow considerations such as difficulties of constant supervision by the court. The logical anomaly involved in a complete differentiation between mandatory and prohibitory injunctions of a continuing nature appears from the following passage in his judgment (at p 318):

Mandatory injunctions are by no means unknown, and there is normally no question of the court having to send its officers to supervise the performance of the order of the court. Prohibitory injunctions are common, and again there is no direct supervision by the court. Performance of each type of injunction is normally secured by the realisation of the person enjoined that he is liable to be punished for contempt if evidence of his disobedience to the order is put before the court.

In a recent decision of my own, *Gravesham Borough Council* v *British Railways Board* [1978] Ch 379, after a brief reference to authority, I expressed the view *obiter* that it cannot now be regarded as an absolute and inflexible rule that the court will never grant an injunction requiring a person to do a series of acts requiring the continuous employment of persons over a number of years, though the jurisdiction was one that would be exercised only in exceptional circumstances.

The willingness of the courts to grant mandatory orders in an appropriate case, even where there is no precedent for such an order, is well illustrated by the decision in *Jeune* v *Queens Cross Properties Ltd* [1974] 1 Ch 97, in which Pennycuick J made an order for specific performance against a landlord in a case where there was a plain breach of covenant to repair. He said (at p 99):

In common sense and justice, it seems perfectly clear that this is the appropriate relief.

This and other recent authorities indicate that the attitude of the court to the making of mandatory orders has to some extent been developing in recent years and that it may nowadays be prepared to consider granting relief of this nature in circumstances in which in the 19th century such relief would have been regarded as unthinkable.

While none of these recent authorities has related to the carrying on of a business, Mr Bernstein submitted that common sense and justice required the grant of a mandatory injunction in the present case and invited me to create a precedent by ordering the defendants to carry on the business of a supermarket accordingly. If I felt free to do so, I would be inclined to accept this invitation, even though this would involve the granting of mandatory relief on an interlocutory application.

This is a case where the defendants, after express notice, have committed and are committing a clear violation of an express contract and can, I think, expect little forbearance from the court, even on an interlocutory application (compare *Shepherd Homes Ltd* v *Sandham* [1971] 1 Ch 340 at p 347). The liability, as I have said, does not appear to be in doubt; only the remedy is in question. Provided that I thought that there was a good prospect of the court continuing mandatory relief at the trial, so far as appropriate in the then circumstances, the balance of convenience would seem to me to favour the grant rather than the withholding of an injunction at the present stage. If it is withheld, the plaintiffs, for reasons already explained, are in my view likely to suffer damage, some of which cannot be compensated in terms of money at the trial. If it were granted, so that the defendants were obliged to reconstitute the supermarket business at the centre, and then, contrary to expectation, they succeeded in resisting liability at the trial, they would, I think, have an adequate remedy in damages, which could be quite easily quantified. The plaintiffs have offered an undertaking in damages, supported by their parent company, Allied London Properties Ltd, and the defendants' counsel did not suggest that such an undertaking would not have sufficient financial backing. A mandatory injunction, if granted at the present stage, would operate only until trial or further order and could be discharged if a subsequent change of circumstances made it appropriate or just to do so. It might well serve the additional useful purpose of persuading the defendants to bestir themselves further in finding new tenants for the premises who were prepared to carry on a supermarket.

In all the circumstances, without the fetter of authority, I would be inclined, on the special facts of this case, to grant an injunction broadly of the nature sought. In the absence of other authority I would approach the case much as Bacon V-C approached the case of *Greene* v *West Cheshire Railway Co* (1870) 13 Eq 44 when he ordered specific performance of an agreement to construct and maintain a siding alongside a railway line upon land belonging to the plaintiff. As he said (at p 50):

A more direct, wilful, and determined violation of a plain contract cannot be suggested. No excuse is offered for it — no suggestion that it is impracticable, or even that it is inconvenient, for the company to perform their part of the contract of which the plaintiff has performed his; but what they say is, that the plaintiff may, by an action at law, recover against them in money such amount of damages as a jury may think he has sustained by their wilful branch of their contract; and that, therefore, a court of equity will not entertain the complaint. I do not understand that the law, as administered in this court, countenances any such defence. If that were the law, the great majority of the cases in which this court has exercised its authority for the purpose of compelling specific performance of contracts might be readily disposed of, because in the great majority of cases a payment in money might satisfy the wrong which the breach of such contracts inflicts. But it would be a total departure from all principles by which the administration of this branch of the law has hitherto been guided, to hold that it is at the option of a man who has persuaded another to part with his rights upon a specific condition to say: "I can, but I will not perform the obligation I have entered into; and instead of keeping faith and honestly fulfilling my promise, I will leave you to take the chances of an action for damages, and reserve to myself the power of endeavouring to defeat your claim; and, instead of acknowledging your just rights, will compel you to receive, instead of them, such a sum as I may be able to persuade a jury will compensate you for the loss and injury and disappointment which my wilful wrongdoing may have occasioned to you".

Many of Vice-Chancellor Bacon's observations appear to me to fit the present facts. In the *Greene* case, however, he was not dealing with a covenant to carry on a business; nor was he faced with at least three reported cases in which courts of first instance had clearly held that the court will never grant a mandatory injunction to compel a person to perform a covenant of the particular nature under consideration.

Whether or not this may be properly described as a rule of law, I do not doubt that for many years practitioners have advised their clients that it is the settled and invariable practice of this court never to grant mandatory injunctions requiring persons to carry on business. In my judgment, there is no real prospect of the trial judge in this case being persuaded to depart from the practice, which has been endorsed and acted upon by at least three different judges sitting at first instance; still less do I think it would be right for a court of first instance to depart from it on an interlocutory application. The rationale which lies behind the rule or practice may perhaps need rethinking, at least in relation to those cases where it would be possible to define with sufficient certainty the obligations of the person enjoined to carry on a business. This process, however, is not an appropriate function for this court on this present interlocutory motion.

Accordingly, I must reluctantly decline to grant any injunction at the plaintiffs' instance on the present application. If they ask me to do so, however, I will give directions for a speedy trial, since it seems to me that this is a case where it is plainly desirable that there should be a quick resolution of the dispute between the parties in one way or the other.

As regards the costs of the motion, subject to any further submission of counsel, I propose simply to direct that the costs be costs in the cause.

After discussion the judge ordered that the costs of the motion be reserved to the judge at the trial.

Court of Appeal
March 19 1987

(Before Lord Justice KERR, Lord Justice NICHOLLS and Lord Justice BINGHAM)

BASS HOLDINGS LTD v MORTON MUSIC LTD

Landlord and tenant — Option for tenants to obtain a further lease — Condition precedent to the exercise of the option that tenants should have paid the rent and performed and observed all the stipulations contained in the lease — Appeals from decision of Scott J on preliminary questions concerning the validity of the exercise of the option by the tenants — Main issue turned on whether the tenants were disqualified from exercising the option by reason of past breaches of covenants, positive and negative, the effects of which were spent before the exercise — Scott J had upheld the landlords' case in respect of one of the negative covenants, while rejecting their conditions in respect of other covenants — Tenants appealed in regard to the one and landlords cross-appealed in regard to the others — Scott J had decided, in accordance with an accepted body of authority, that a condition precedent of the kind found in the present case applied in general only to subsisting breaches of covenant at the material date, not to spent breaches — He considered, however, that spent breaches of negative covenants were an exception to this general rule — In the present case there had been a breach by the tenants of a negative covenant not to apply for planning permission in respect of the demised premises without the landlords' written permission — Although these applications had been rejected and had caused no loss or damage to the landlords, Scott J held that the breach had been irremediable — While the tenants' spent breaches of the covenant to pay rent did not disqualify them, their breach of this negative covenant did have that effect — The Court of Appeal, after an extremely detailed review of the authorities, held that Scott J was in error in treating negative covenants as an exception to the general rule in favour of a "no subsisting breach", as opposed to a "never any breach", construction of the kind of condition precedent found in the present case — The court also decided in favour of the tenants a minor issue as to the effectiveness of the wording of the tenants' letter purporting to exercise the option — Tenants' appeal allowed — Landlords' cross-appeal dismissed

The following cases are referred to in this report.

Bassett v *Whiteley* (1983) 45 P&CR 87
Bastin v *Bidwell* (1881) 18 ChD 238
Expert Clothing Service & Sales Ltd v *Hillgate House Ltd* [1986] Ch 340; [1985] 3 WLR 359; [1985] 2 All ER 998; [1985] 2 EGLR 85; (1985) 275 EG 1011 & 1129, CA
Finch v *Underwood* (1876) 2 ChD 310
Germax Securities Ltd v *Spiegal* (1978) 37 P&CR 204; 250 EG 449, CA
Grey v *Friar* (1854) 4 HLC 565
Porter v *Shephard* (1796) 6 Term Rep 665

Rugby School (Governors) v *Tannahill* [1934] 1 KB 695; [1935] 1 KB 87, CA
Scala House & District Property Co Ltd v *Forbes* [1974] QB 575; [1973] 3 WLR 14; [1973] 3 All ER 308; (1973) 227 EG 1161, CA
Simons v *Associated Furnishers Ltd* [1931] 1 Ch 379

This was an appeal by tenants, Morton Music Ltd, with a cross-appeal by landlords, Bass Holdings Ltd, from a decision of Scott J (reported at [1986] 2 EGLR 50, (1986) 280 EG 1435). The appellant tenants appealed against the judge's decision that the breach of a negative covenant not to apply for planning permission without the landlords' consent amounted to a failure to fulfil a condition precedent to the exercise of an option for a further lease. The landlords' cross-appeal challenged the judge's rejection of other alleged failures to fulfil the condition. The case concerned a lease of the Queen's Hotel, Hamlet Court Road, Westcliff-on-Sea, Essex, premises which consisted of a tied public house and hotel.

Robert Pryor QC and Nicholas Dowding (instructed by Nutt & Oliver) appeared on behalf of the appellant tenants; Paul Morgan (instructed by Nabarro Nathanson) represented the respondent landlords.

Giving judgment, KERR LJ said: This is an appeal from a judgment of Scott J given on July 30 1986 and reported in [1987] 2 WLR 397.* It arises from the trial of a number of preliminary issues concerning the validity of the exercise by the defendant tenants of an option for a further lease from the plaintiff landlords. The main controversy turns on the materiality, if any, of past breaches of covenants by the tenants, both positive and negative, whose effect was spent before the exercise of the option. This raises a point of considerable general importance. There is also a relatively minor issue on the validity of the wording of the letter which purported to exercise the option. The landlords issued an originating summons claiming declarations that on all these grounds there had been no valid exercise of the option. The learned judge agreed on the ground of the tenants' past breaches of one of the negative covenants in the lease, but he rejected the landlords' other contentions. The tenants now appeal to reverse this decision, and the landlords cross-appeal to uphold it on all the grounds raised by their summons.

The leased premises are the Queen's Hotel, Hamlet Court Road, Westcliff-on-Sea, Essex. The plaintiff freeholders form part of the well-known brewery group and the premises include a tied public house as well as an hotel. By a lease dated September 20 1982 the plaintiffs let the premises to the defendants for "15 years computed from 1st April 1982" at a yearly rent of £15,000 "for the first three years of the said term". Thereafter the rent was to be subject to upward adjustments in accordance with a rent review procedure. The lease contained the usual types of covenants to be found in lettings of commercial premises, as well as a series of further covenants specifically designed for tied public houses. The covenants were both positive and negative without any purported distinction between them. Some contained both positive and negative obligations in the same provision and the effect of others was put both positively and negatively in different provisions, eg a covenant to deal exclusively with the landlords and not to purchase any supplies from anyone else, or not to do anything which might cause the licence for the premises to lapse on the one hand and to use best endeavours to ensure that it was regularly renewed on the other. There was also a provision for re-entry for any breach of covenant in the normal form.

Clause 9 contained the option which has given rise to the dispute.

If the tenant shall be desirous of taking a further lease of the demised premises for a further term of 125 years from the date of the term hereby granted and shall not later than 29 September 1985 give to the lessors notice in writing of such its desire and if it shall have paid the rent hereby reserved and shall have performed and observed the several stipulations on its part herein contained and on its part to be performed and observed up to the date thereof then the lessors will on payment to them by the tenant of the sum of £300,000 let the demised premises to the tenant for a further term of 125 years from the date of the term hereby granted at a rent of one peppercorn per annum (if demanded) subject in all other respects to the same stipulations as are herein contained except this clause for renewal and save for the alterations referred to in Part III of the schedule hereto.

By a letter of September 19 1985 from the tenants' then solicitors, which the landlords received on the following day, the tenants purported to exercise this option in the following terms:

We act for Morton Music Ltd the tenant of the above premises under a lease granted by you on 20 September 1982. We hereby give you notice of our clients

---
*Editor's note: See also [1986] 2 EGLR 50; (1986) 280 EG 1435.

desire under clause 9 of the lease to take a further term of the demised premises for 125 years from 1 April 1982 and otherwise upon the terms referred to in the lease. We are sending a copy of this letter to Messrs Nabarro Nathanson your solicitors and we look forward to hearing from them in connection with the new lease accordingly.

The plaintiffs deny that this letter constituted an effective exercise of the option. First, they challenge the wording of the letter because it referred to the commencement of the new term as from April 1 1982 whereas they contend that it should have been either from September 20 1982, the date of the lease, or from March 31 1997, the date of the expiry of the 15-year term. Second, and mainly, the plaintiffs rely on a number of breaches of covenants on the ground that these preclude the right to exercise the option in any event. One of these is disputed and its determination remains in issue, viz whether the premises complied with the repairing and decorating covenants throughout the period from September 20 1982 to September 20 1985 inclusive. In that respect the tenants are said to have been in breach of covenant at the time when they purported to exercise the option. This is disputed by the tenants, but it is common ground that if the tenants were in breach at the time when they purported to exercise the option, then it must necessarily follow that it could not have been exercised validly. However, in order to avoid possibly needless proceedings to decide this issue of disputed fact, it was agreed that there should be a trial of the other two issues which I have mentioned, the form of the exercise of the option and the materiality, if any, of past breaches of covenant. I will deal first with the latter and main issue, which raises questions of principle.

The tenants' past breaches of covenant fall into two groups, one positive and one negative, and all of them are undisputed. The course of events was briefly as follows. By September 1984 rent due under the lease was in arrears to the extent of about £18,000 and the tenants were also in default in payment of water rates. These were breaches of positive covenants. On September 11 1984 the plaintiffs re-entered the premises and forfeited the lease on account of the rent arrears. On October 9 1984 the defendants issued a summons seeking relief from forfeiture. In reply to this application the plaintiffs relied on two breaches of a further negative covenant that:

The tenant will not apply under the Planning Law for planning permission in respect of the demised premises without the Lessors' prior written consent (such consent not to be unreasonably withheld).

In that connection it is agreed that in March and again in October or November 1984 respectively two applications for outline planning permission relating to the premises were made on behalf of the tenants without the landlords' consent. Both were rejected by the local planning authority and it is common ground that neither has caused any quantifiable loss or damage to the plaintiffs. When the plaintiffs relied upon these breaches of covenant as an additional ground for forfeiting the lease, the defendants applied for relief in relation to these breaches as well. Their application was granted by Master Dyson by an order made on March 12 and slightly varied on April 1 1985. This recited an undertaking by the defendants at all times thereafter to comply with the covenant not to apply for planning permission without the plaintiffs' written consent. It also granted the defendants relief from forfeiture both in respect of the arrears of rent and water rates and in respect of these breaches of covenant, but subject to certain conditions. These required the defendants to pay the arrears of rent and water rates together with interest and the plaintiffs' taxed costs. Each of these conditions was complied with, so that the defendants' lease had been reinstated unconditionally before the purported exercise of the option.

The relief from forfeiture and reinstatement of the lease could obviously not undo any of the breaches of covenant in the sense that it could not be said that they had never taken place. On the other hand, the breaches lay wholly in the past and their effect was spent. It was not contended that any cause of action based upon them subsisted at the time of the purported exercise of the option other than a possible theoretical claim for nominal damages on the ground that the planning applications had been made without the landlords' consent. But this is a submission without any commercial or other sensible reality, and the judge rightly did not base himself upon it. Moreover, these admitted breaches of covenant, like the others, became subject to the proceedings between the parties and were dealt with in the resulting order. It follows that their effect was then spent.

This brings me to the classification of breaches of covenants by reference to the time of their occurrence and subsistence, on which the present issues turn. As explained below, there is clearly a long-standing conveyancing practice, going back more than two centuries, whereby tenants' options in leases are made subject to provisos dealing with the observance of the tenants' covenants, similar to the option in the present case, whether the option be for a new lease as here, or for premature termination of an existing lease (a "break" option), or for the purchase of the freehold. In all such cases the provisos link the required observance of the covenants to a point of time in the nature of a *terminus ad quem*. This may be either the date of the exercise of the option as here, or the date of the expiry of the option, or — in cases of options for renewals or for the purchase of the freehold — the date of the expiry of the lease. For present purposes the particular date referred to in the proviso ("the operative date") does not matter. What matters is whether the breach (or breaches) of covenant on which the landlord relies as precluding the exercise of the option has occurred only in the past, so that its effect is spent by the operative date in question, or whether there is still a breach — or at any rate a cause of action based upon a breach, whether for forfeiture or damages or both — which subsists on the operative date. I will refer to the breaches of covenant in these two situations respectively as "spent breaches" and "subsisting breaches". The present appeal is concerned exclusively with spent breaches.

In my view the position on the authorities and on the issues raised by the present case can be summarised as follows:

(1) The first question is whether, on the true construction of the proviso in question, the absence of any material breaches of covenant by the tenant is a condition precedent to the exercise of the option as well as the giving of the requisite notice purporting to exercise the option.

Generally, and admittedly in the present case, the provision contains a double condition precedent, viz (i) the absence of any material breaches of covenants and (ii) compliance with the requirement as to notice.

(2) That, however, leaves the crucial question whether the condition precedent (i), that there must be no material breaches of covenant by the tenants, applies to spent as well as to subsisting breaches. This question is covered by dicta in numerous cases, going back in particular to *Grey v Friar* (1854) 4 HL Cas 565, and by the decision of Clauson J, in *Simons v Associated Furnishers Ltd* [1931] 1 Ch 397. The upshot of these authorities is that spent breaches will not destroy the tenant's right to exercise the option, but subsisting breaches will. As shown by the passages to which I refer below, the reasoning is in effect as follows. First, it must be accepted that absolute and precise compliance by the tenant with every single covenant throughout the period of the lease prior to the operative date is virtually impossible of attainment. If this were required as a condition precedent, then the option would in practice be worthless or merely at the mercy of the landlord. Therefore the parties cannot have intended that the absence of spent breaches should be a condition precedent. Second, however, it is natural and sensible that the landlord should require the tenant not to be in breach of any covenant on the operative date and that all outstanding claims for breach of covenant should have been previously satisfied, so that the lease is then effectively clear. The proviso is therefore to be construed as intended to apply to subsisting breaches, with the result that the relevant condition precedent is the absence of any subsisting breach.

(3) The only suggestion that spent breaches might be able to preclude the exercise of the option is to be found in the judgment of Griffiths LJ (as he then was) in *Bassett v Whiteley* (1982) 45 P & CR 87. But this lies in a different line of authority, as explained below, and cannot properly be applied to the present case.

(4) Subject to the question whether the wording of any particular proviso imposes a condition precedent to the exercise of the option, which was the issue in *Grey v Friar*, the precise words used in such provisos in relation to the observance of the covenants have played no part in the conclusions summarised in (2) above. While it would of course be possible to formulate a proviso which is sufficiently explicit to cause spent breaches to preclude the exercise of the option, there appears to be no reported case in which this was so; and the wording of the proviso in the present case is in a form similar to, and effectively indistinguishable from, the formulations adopted in all the cases subsequent to *Grey v Friar*. Accordingly, in mentioning some of these cases

below, it would serve no purpose to set out the precise terms of the particular provisos.

(5) The reasoning summarised in (2) above, which has led to the generally accepted conclusion that the condition precedent imposed by provisos like the present was intended to apply only to subsisting breaches, is of course particularly cogent in relation to "break" options. In such cases it will obviously be of great importance to the landlord that the demised land or premises should be surrendered to him free from any subsisting breaches of covenant. In cases of options for renewals it may also be of some importance to the landlord that the slate should be clean before the new lease takes effect. In cases of options to buy the freehold, the absence of subsisting breaches may be less important to the landlord. However, the authorities suggest no distinction between these types of option in relation to the conclusions summarised in (2) above. Admittedly, *Grey* v *Friar* in 1854, and *Porter* v *Shephard* (1796) 6 TR 665, which was followed in *Grey* v *Friar,* both involved "break" options. But their reasoning has been followed and applied in relation to tenants' options generally. Accordingly, there is nowadays no reason for imposing any different or special construction on the familiar forms of provisos governing any particular type of tenants' option, such as the option for a further lease in the present case.

(6) Strong arguments can admittedly be raised in extreme cases against the construction that the effect of the condition precedent is to be limited to the absence of subsisting breaches. For instance, what about a tenant who has persistently broken his covenants and only puts matters right just before the operative date? As mentioned below, this troubled Griffiths L J in *Bassett* v *Whiteley (supra),* albeit in a different context. But it must also be borne in mind that the condition requiring the absence of any subsisting breach on the operative date involves not only that any breach should have ceased but also that there should be no subsisting cause of action in respect of any breach. Thus, a landlord may perhaps take advantage of the latter condition, in order to defeat a tenant's option, by postponing any claim for forfeiture or damages in respect of an earlier breach until after the operative date. Anomalies on both sides are inevitable on any construction. But the consensus to be found in the authorities is that the anomaly which must be rejected as too great to be acceptable is that any spent breach should disqualify.

(7) The upshot, in cases such as the present, is that we are nowadays dealing with what has become a standard conveyancing formula imposing a condition precedent concerning the absence of (material) breaches of covenant in relation to the exercise of virtually any tenants' option in leases. The unanimous consensus on the authorities in relation to a proviso such as the present is as summarised in (2) above, namely, that the condition applies to subsisting breaches and not to spent breaches. There is no reported case in which any tenant's option has been defeated by a spent breach, and virtually none in which this has even been suggested. In these circumstances it is, in my view, of great importance that this accepted long-standing interpretation of provisos such as the present should be respected and upheld, since it must have been relied upon and applied in countless transactions.

(8) Up to this point my conclusions accord entirely with those of the learned judge. But I regret that at this juncture there comes a parting of the ways between our views. He points out, entirely correctly, that the jurisprudence leading to the accepted interpretation which I have summarised above, that spent breaches are irrelevant to the exercise of tenants' options, has been concerned throughout with spent breaches of positive covenants, such as failures to pay rent, to repair, etc. The issue has never arisen in the context of a spent breach of a negative covenant. Admittedly, none of the authorities contains any suggestion that there is any difference between breaches of positive and negative covenants for present purposes, whether the breaches be spent or subsisting. But the judge nevertheless concluded that there was a material difference between them. He appears to have based this on the consideration that a breach of a positive covenant can be remedied in the sense, or to the extent, of belated performance. In effect, the tenant can take some positive action to put things right. But a breach of a negative covenant, albeit "spent" in the sense in which I have used this term, is irremediable. It cannot be undone, any more than Omar Khayyám's "moving finger . . . having writ", or an indelible stain on the tenant's escutcheon. The judge accordingly concluded that while the tenants' failure to pay the rent and water rates did not disqualify them from exercising the option, their applications for planning permission without the landlords' consent did have this effect. The reason was that the former were breaches of positive covenants whereas the latter were breaches of a negative covenant. The judge therefore granted the landlords' application for a declaration that the tenants' option had not been validly exercised.

(9) With all due respect, I cannot agree with this analysis or with the conclusion to which it leads. It is not based on any commercial or other practical consideration. Its effect would be entirely fortuitous. Thus, as mentioned at the beginning of this judgment, many of the covenants in this lease in common form contain both positive and negative obligations in one provision, and other obligations to the same or similar effect have been expressed in the form of both positive and negative covenants. I do not see on what realistic grounds it could be said that the intention of the parties was that spent breaches of positive covenants should be irrelevant, but that any breach of any negative covenant, albeit equally "spent" in its practical effect, should disqualify.

This is a summary of the reasons why, while agreeing with the judge on the main part of this analysis of the authorities, I respectfully differ from his ultimate conclusion.

I must now briefly review the authorities. Although I agree with the judge about their effect, I must do so because Mr Morgan challenged the judge's analysis as a whole. He did not abandon reliance upon the distinction which he drew between the different effect of spent positive and negative covenants. But his primary submission was that any breach of any covenant during the currency of the lease prior to the operative date, whether spent or subsisting, disqualified the tenant from exercising the option. In relation to the present case he therefore placed equal reliance on the tenants' failures to pay the rent and water rates in due time as on their unsuccessful planning applications without the landlords' consent.

In reviewing the authorities one must necessarily start with *Grey* v *Friar (supra).* Although of great importance for present purposes, it was a remarkably unattractive case, which occupied the courts repeatedly between July 1848 and August 1853. *Bleak House* was published in instalments between March 1852 and September 1853, and *Jarndyce* v *Jarndyce* must have appeared familiar to the parties. There were two successive actions by the landlords for breaches of covenants contained in a mining lease for 42 years with a "break" option in favour of the tenants after eight years, and every third year thereafter, subject to 18 months' prior notice and the following proviso:

Then and in such case (all arrears of rent being paid, and all and singular the covenants and agreements on the part of the said lessees having been duly observed and performed), this lease, and every clause and thing therein contained, shall, at the expiration of the first eighth year, and thereafter at the expiration of any such third year, cease, determine, and be utterly void. But nevertheless, without prejudice to any claim or remedy which any of the parties hereto may then be entitled to for breach of any of the covenants or agreements hereinbefore contained.

As pointed out in the headnote to the first action reported in 15 QB 81, "the covenants on the part of the lessees . . . were very numerous . . . and some . . . were negative covenants". The landlords sued for arrears of rent due in the ninth year and alleged breaches of covenant. The tenants pleaded that they had validly exercised the "break" option so as to determine the lease at the end of the eighth year and that there were then no subsisting breaches. The landlords replied to the effect that they put the tenants to proof of due compliance with all covenants at the end of the eighth year. This was held to be a good replication by the Court of Exchequer but bad by the Exchequer Chamber on a writ of error, holding that the landlords had to limit their joinder of issue to particularised breaches. In addition, the Exchequer Chamber held, obiter, that in view of the last sentence of the proviso, performance of the covenants "could never have been meant to be a condition precedent to the power to terminate the lease".

But the landlords were undeterred and brought a fresh action. On this occasion their replication relied, *inter alia,* on a breach of a particular covenant which gave a right of re-entry "if anything should be done or neglected to be done whereby the colliery might be drowned with water or otherwise damnified". It is interesting to note that this covenant was both positive and negative in its terms. There was then a demurrer joining issue on the question whether due performance of all the covenants throughout the period of the lease was a condition precedent to the tenants' right to exercise the option.

That issue does not directly affect the subsequent cases, including the present, since it turned largely on the last sentence of the proviso which thereafter fell out of use. But the division of opinion which followed was supported on both sides by comments which are of great relevance, and the history of the action was remarkable. The Court of Exchequer held that the proviso was not a condition precedent to the exercise of the option. This was unanimously reversed by the judges of the Exchequer Chamber, having evidently changed their minds about the obiter passage in the first action mentioned above. The case then went to the House of Lords, who called on the judges for assistance. There were 11 reported judgments, eight in favour of the view that the proviso contained a condition precedent and three against. The Judicial Committee consisted only of the Lord Chancellor and Lord Brougham, but the Lord Chancellor gave the only judgment, since Lord Brougham was absent due to illness. The Lord Chancellor had been a party to the decision of the Court of Exchequer (as he pointed out) and was in favour of allowing the appeal. But he said that Lord Brougham had told him that he took the opposite view and agreed with the Exchequer Chamber. The Lord Chancellor added that he derived some comfort (but evidently little conviction) from the fact that there was such a large majority of the judges who took the same view. Accordingly, since the House of Lords was equally divided, the decision of the Exchequer Chamber stood.

For present purposes, it is not the outcome but the reasoning which is important. None of the judgments supported the conclusion that spent breaches barred the exercise of the option. There was general consensus that the parties could not have intended to make it a condition precedent that none of the covenants should ever have been broken in the slightest degree during the currency of the lease. Both sides proceeded on the basis that the absence of any breach to which I have referred as a "spent" breach could not have been intended as a condition precedent. The division of opinion between them was accordingly on the following lines. The minority thought that, in view of this, the proviso could not have been intended as a condition precedent, particularly having regard to the concluding sentence. The majority, on the other hand, did take the view that the proviso as a whole was in the nature of a condition precedent, but then concluded that spent breaches did not bar the exercise of the option, provided that there were no subsisting breaches at the operative date, the end of the first eight years of the term. That was the outcome. But since all the proceedings were by way of demurrer, without any investigation of the facts, history does not relate whether Mr Grey or Mr Friar was ultimately successful.

Many passages could be quoted from the judgments on both sides in support of the view summarised in para (2) above, which Scott J also accepted. But the clearest are four passages cited by him from those who were in the majority, as set out below:

Talfourd J said at p 592:

the truth probably is, that in the framing of the proviso in question, the parties did not intend to use the words "duly observed and performed" in their technical sense, as importing that no covenant during the eight years or longer period had ever been broken; in which sense they are certainly unreasonable; but in a sense in which they import a condition perfectly natural and just, namely, that before the expiration of the notice, the objects of the covenants should be attained, that is, that the works should be put into repair, the water pumped out of the mine, and everything done which the lessees were bound to do in order that they might deliver up the premises in a proper condition to their landlord.

At p 595 Alderson B said:

But I think that the condition precedent, even taking the words of it, may really mean that covenants broken, if the breach shall be compensated for before the expiration of notice, shall be considered as covenants duly performed within this proviso. For as rent in arrear, if paid before the expiration of the notice, clearly is within it, so the performance of the other covenants being found in conjunction with it, may bear the like interpretation.

Erle J said at p 599:

It is said that there would be inconvenience in restricting the power of determining it to the event of all the covenants having been performed, which would be almost an impossibility. To this one answer is, that if the parties agree so to stipulate, the law must give effect to the stipulation. It may also be answered, that the stipulation does not mean that there should not have been any breach of covenant during the term, but that when the notice expires there should not exist any cause of action in respect of performance of covenants. The stipulation for arrears of rent being paid, refers to a covenant which had been broken; but all cause of action for the breach having been satisfied by subsequent accord, and the covenant for rent would, within the meaning of this clause be observed and performed, if all arrears of rent were paid before the expiration of the notice. So the covenant for repair, though broken during the term, would be observed if all repairs were at last completed. So in respect of other breaches; if the damage had been settled by arbitration and the amount paid, or if an action had been brought and the judgment satisfied, the legal duty of the covenantor, by reason of his covenant, would have been so far observed and performed, that all liability in respect thereof would be at an end. In this sense, the stipulation would be free from any hardship towards the lessee, as he might obtain the privilege if he did his duty. This construction does not depend upon giving a peculiar effect to the words of this instrument, for it seems to me that the same principle is applicable to all contracts. The legal effect of the promise in every contract at common law is alternative, either to do the thing promised or make compensation instead. In some contracts the alternative is expressed when liquidated damages are stipulated for, in others the liability arises by implication of law, either to do or to compensate for not doing, according as may be settled by accord, or arbitration, or judgment. In all contracts the legal duty thereunder has been performed, and so the contract may be said in one sense to be performed, when either the thing contracted for has been done, or compensation instead thereof has been made.

Finally, Coleridge J said at p 608:

The condition, thus expressed, I think it reasonable to understand as requiring that the account between the parties must, both as to rent and covenants, be clear; the rent need not have been always paid on the day; but all arrears, if any, must have been paid up; the covenants must have been strictly kept, or, if broken, must have been satisfied for. So understood, the words import a condition precedent neither impossible nor unreasonable; and where that is clearly the case, the mere difficulty of performance, from the number or nature of the covenants to be performed — a fact which must have been perfectly within the knowledge of the party contracting — seems to me a very unsatisfactory reason for holding it to be otherwise.

This is the reasoning which has in effect been adopted in the line of cases which followed. The learned judge referred to *Finch* v *Underwood* (1876) 2 Ch D 310, *Bastin* v *Bidwell* (1881) 18 Ch D 238 and the decision of Clauson J in *Simons* v *Associated Furnishers Ltd* [1931] 1 Ch 379. In the first of these an obiter passage in the judgment of Mellish LJ expresses the same view. In the second, Kay J did not have to decide the issue, but there is nothing in his judgment suggesting the contrary. But in the latter case Clauson J decided it in the same sense, with references to *Grey* v *Friar* in the following terms at p 386:

The next question is, what does the clause mean? Upon a possible construction, it may make it essential that the tenant should comply with all the covenants throughout the whole period of the five years or, in other words, that the tenant must be able to say that in no single instance during that period has rent been in arrear or a covenant broken. On that question there has been from time to time a certain amount of difference of judicial opinion. If the condition imports that it is unfulfilled if there has been any breach of covenant, even if it has been remedied, the condition may be a very hard one and such as can scarcely be supposed that parties would enter into; but here I am bound by a very heavy weight of judicial opinion to hold that the true meaning of that clause is this, that it will have been complied with, if at the end of the five years "there should not exist any cause of action in respect of performance of covenants": or, I may put it this way, the condition must be understood as "requiring that the account between the parties must, both as to rent and covenants, be clear; the rent need not have been always paid on the day; but all arrears, if any, must have been paid up; the covenants must have been strictly kept, or, if broken, must have been satisfied." In the language I have used I have ventured to quote the language in the first case of Erle J and in the second case the language of Coleridge J in advising the House of Lords in the case to which I have already referred.

The only expression of judicial opinion the other way, but in a different context, is to be found in the judgment of Griffiths LJ in this court in *Bassett* v *Whiteley* in 1982 (*supra*). A lease for eight years contained an option for renewal for a further term of the same length by giving to the landlord a written notice not more than 12 nor less than six months before the expiration of the term, provided that the tenants:

shall have paid the rent hereby reserved and shall have reasonably performed and observed the several stipulations herein contained and on their part to be performed and observed up to the termination of the tenancy hereby created . . .

It will therefore be seen that the wording is virtually identical with the proviso in the present case, but subject to the addition of the important word "reasonably"; and whereas the "operative date" in the present case was the exercise of the option, in that case it was the termination of the original term. The tenants gave notice of their desire to exercise the option and this was accepted. Subsequently, however, they were twice late with their payments of rent, but by the operative date all the rent had been paid and there was no subsisting breach of covenant. The issue before the court, consisting of Waller and Griffiths LJJ, was whether the condition precedent of the proviso had been satisfied on the ground that the tenants "shall have paid the rent hereby reserved and shall have reasonably performed and observed (the covenants up to the operative date)". The court unanimously upheld the exercise of the option. The ratio can be seen from the following passage towards the end of the judgment of Griffiths LJ at p 93:

I agree with the approach of Waller LJ. I think this clause has to be read as a whole. It means that the whole of the rent must have been paid by the time that the eight-year term ended, and it further means that there must have been a reasonable performance of all the stipulations. For the reasons that he has given I think there was a reasonable performance of the stipulations, and in particular the one concerning the payment of rent.

But Mr Morgan relied strongly on the following other passages in the judgment of Griffiths LJ with reference to the citation by Mr Maddocks for the tenants of *Finch* v *Underwood* and *Bastin* v *Bidwell* (*supra*). He said in that regard at the beginning of his judgment:

Mr Maddocks submitted that in construing an option of this kind the critical date at which to look was the termination of the tenancy, and he submitted that if, at that date, it could be seen, first, that all the rent had now been paid and, second, that any previous breaches of covenant, however troublesome, had now been remedied, then the tenants were, on a true construction of the clause, entitled to exercise their option.

Now I emphatically disagree with that approach for it would mean this, that if throughout the whole of the eight-year term the tenants had continuously been woefully in arrears with their rent and had constantly had to be chased for it and were obviously late and poor payers, the landlord could, nevertheless, be saddled with them as tenants for another term if they paid up the arrears at the eleventh hour.

This highlights one of the anomalies to which I referred in paragraph (6) above flowing from the construction that spent breaches do not bar the exercise of an option. Nevertheless, and in agreement with the judge, I think that the current of judicial authority the other way is by now far too strong to be overcome by this short passage, and I cannot think that on a full consideration of the extracts from the judgments which I have cited Griffiths LJ would have taken a different view. His remarks were obiter and concerned with a different point. My impression is that he was rejecting the argument of Mr Maddocks in the context of the word "reasonably." He meant that unreasonable conduct throughout the term could not have been cured by late performance just before the operative date, since the stipulations in the proviso would in that event not have been "reasonably performed and observed".

Apart from placing great reliance upon that decision, Mr Morgan also sought to rely on the unusual terms of the proviso in the present case. He pointed out that the option was exercisable at the beginning of the term, at any time within the first three years of a 15-year term, and he submitted that in these circumstances it was no hardship to require the tenants to perform all their covenants meticulously. He said that the purpose of the proviso was to enable the landlords to decide whether the tenants were satisfactory. But, in the same way as the judge, I cannot accept that this proviso, which is otherwise in common form and covered by all the authorities to which I have referred, should for those reasons receive some different and special construction in the present case.

I then come to the point where I differ from the judge. While agreeing that spent breaches of positive covenants did not bar the option, he took a different view about spent breaches of negative covenants. I have already explained in paras (8) and (9) above why I cannot accept this distinction. I can see no basis for it from the point of view of the sensible commercial construction of the clause. Nor is there any support for it to be found in the authorities. On the contrary, it seems to me that the policy line, if one may so call it, which has prevailed ever since *Grey* v *Friar*, must logically apply in precisely the same way both to positive and negative covenants. Indeed, it is very often pure chance how a particular obligation is worded.

The judge appears to have been influenced by some of the cases dealing with relief from forfeiture under section 146 of the Law of Property Act 1925. He referred to passages in *Rugby School Governors* v *Tannahill* [1934] 1 KB 695, *Scala House & District Property Co Ltd* v *Forbes* [1974] QB 575 and *Expert Clothing Service & Sales Ltd* v *Hillgate House Ltd* [1986] Ch 340.* These cases contain discussions about the irremediability of negative as opposed to positive covenants. But this issue does not go to the substance of the rights and obligations of the parties; but merely to the form of the notices which must be given pursuant to the section. As illustrated by the present case, in appropriate circumstances a tenant will obtain relief from forfeiture following breaches of negative covenants just as readily as for non-compliance with positive covenants. In both cases the effect of the breach becomes spent in the same way. Accordingly I cannot accept the judge's conclusion that the breaches of covenant in relation to rent and rates did not preclude the tenants from exercising the option to renew, but that the applications for planning permission without the landlords' consent had this effect.

This only leaves the landlords' cross-appeal against the judge's conclusion that the form of wording in the letter of September 19 1985 from the tenants' then solicitors complied with the formal requirements for the exercise of the option. It will be remembered that the lease was for "15 years computed from 1st April 1982", that the option was "for a further term of 125 years from the date of the term hereby granted", and that the letter referred to a further term "for 125 years from 1st April 1982". I entirely agree with the judge's conclusion on this part of the case. As regards the landlords' alternative date of March 31 1997, the judge pointed out the improbability that the parties could have intended that the new term should run from then, when the option had to be exercised more than 12 years earlier and the condition precedent requiring compliance with the covenants ceased to have effect thereafter. Moreover, March 31 1997 was not "the date of the term hereby granted" but the date on which it expired. As regards the other alternative date, September 20 1982, when the lease was executed, this cannot have been of any particular materiality to the parties and may well have been unpredictable at the time when the proviso was negotiated. Moreover, as the judge pointed out, if this had been intended, then the proviso would no doubt have said simply "from the date hereof" or words to that effect. In my view, the words "from the date of the term hereby granted" are perfectly apt to refer to April 1 1982, and this is the most likely date which the parties had in mind. On behalf of the tenants Mr Pryor QC additionally submitted that in any event the terms of the letter were sufficiently clear to constitute a valid exercise of the option even if there had been a mistake in regard to the date, and he relied in particular on *Germax Securities Ltd* v *Spiegal* (1978) 37 P & CR 204. If it had been necessary to decide this point I would have been inclined to follow that decision in the present case, but in the circumstances I express no concluded opinion about it.

It follows that I would allow this appeal and refuse the landlords' application for declarations that — on the basis of the preliminary issues before the court — the tenants' exercise of the option to renew was invalid.

Agreeing, NICHOLLS LJ said:

The tenant's appeal: the condition precedent

Leases frequently contain a clause entitling the tenant at his option to determine the lease prematurely or, conversely, to extend his interest in the demised property by obtaining a further lease. Such "break" or "renewal" clauses often, although by no means always, include a provision to the effect that the exercise by the tenant of his option is conditional upon his having paid his rent and performed and observed his covenants and agreements under the lease up to a specified date. Typically the specified date is either the date upon which the tenant gives notice of his intention to exercise his option or the date from which the notice would take effect: for example, in the case of a break clause, the date on which, if the option is exercised, the lease will determine.

Clauses in this form have been widely used in leases for a long time. Speaking in 1930, Clauson J said that this "exceedingly familiar form" of clause had been in common use "for more than a century past" (*Simons* v *Associated Furnishers Ltd* [1931] 1 Ch 379, 384). Shortly stated, the question, of general importance, which arises on this appeal is whether in these clauses a condition to the effect I have mentioned requires for its fulfilment that throughout the whole term of the lease up to the specified date there shall have been no breach of any of the tenant's covenants and agreements or whether the

---

* Editor's note: Reported also at [1985] 2 EGLR 85; (1985) 275 EG 1011.

condition is fulfilled if at the specified date there is no subsisting breach of any of these covenants or agreements. By the shorthand expression "no subsisting breach" I mean that in respect of the rent, covenants and agreements there is at the specified date no outstanding cause of action. I shall call the first of these two alternatives the "never any breach" construction and the other alternative the "no subsisting breach" construction.

I preface my observations by noting that, although each lease falls to be construed having regard to its own particular language and terms, the degree of similarity of language in break clauses and renewal clauses in common use is sufficiently marked in crucial respects for it to be possible and sensible to consider the matter, initially, in fairly general terms.

The two alternative constructions have only to be stated for it to be apparent that the "never any breach" construction would mean that in practice the condition would be impossible of fulfilment in almost all cases of leases of buildings containing a full range of repairing and other covenants by a tenant. However diligent or even punctilious a tenant may be in carrying out his obligations under his lease, in such cases there will in practice inevitably be occasions when there will be outstanding some dilapidations which would, strictly, constitute breaches of the repairing or redecorating covenants. Thus the practical consequence of the "never any breach" construction in such cases would be that the break or renewal option would seldom, if ever, be exercisable by a tenant.

Even in the case of other leases, where the tenant's covenants might be less far-reaching, this construction would lead to much uncertainty for tenants and their assigns. Break options and renewal options may be valuable but, on this construction, after a few years and particularly if there have been assignments or sublettings, the current tenant or a would-be assignee of the lease would be unable in many cases to discover whether or not a break option or a renewal option had already lapsed by reason of a breach of covenant. Indeed, short of a positive answer from a co-operative landlord, it is difficult to see how in this type of situation a tenant or would-be assignee could ever be sure that there had not been a breach, maybe trifling, of one covenant or another in the history of the lease.

It is considerations of this nature that have led the court, for well over a century, to reject the "never any breach" construction. When parties to a lease include an option for the tenant to determine or renew the lease, their intention must be, adopting the words of Martin B in his advice to the House of Lords in *Grey* v *Friar* (1854) 4 HL Cas 565, 582, that "this was meant to be a real power and not merely a delusive one." On this question of construction there was a divergence of view in 1854, but thereafter there have been judicial dicta, of long standing, preferring the "no subsisting breach" construction. For many years that has been the conventional construction. The underlying rationale is that, while it is open to parties to agree that the exercise of a break option or a renewal option shall be subject to a condition as onerous and difficult to fulfil as a "never any breach" condition, where the language used is capable of another, more sensible, meaning that meaning is to be preferred.

I turn to the authorities. In *Grey* v *Friar* the House of Lords considered a break clause in a 42-year lease of a colliery and a farm. The material part of the break clause provided that upon the lessees' giving 18 months' notice expiring on certain dates,

. . . then in such case (all arrears of rent being paid and all and singular the covenants and agreements on the part of the lessees having been duly observed and performed), this lease . . . shall . . . cease; But nevertheless without prejudice to any claim or remedy which any of the parties hereto . . . may then be entitled to for breach of any of the covenants or agreements hereinbefore contained.

The issue before the House of Lords was whether the words in parentheses constituted a condition precedent to the exercise of the right of termination. The Court of Exchequer held not. The Court of Exchequer Chamber reversed that decision. The House of Lords was equally divided, and so the decision of the Court of Exchequer Chamber stood. An important feature in that case, not found in the present case, was the "without prejudice" proviso at the end of the clause. On the one hand, it could be said, the terms of this proviso were inconsistent with the right of termination being exercisable only if the covenants and agreements had all been duly observed. On the other hand, if the words in parentheses did not constitute a condition precedent, they were wholly inoperative.

Of the 11 judges who attended the House of Lords to advise their lordships, three were of the view that the words in parentheses did not create a condition precedent and eight were of the contrary view. In reaching their conclusions each of the three judges in the minority placed some reliance on the view that to construe the words as creating a condition precedent would be likely to defeat the purpose of the clause, because (as Martin B stated, at p 582) "the slightest deviation from any covenant would put an end to the power". Crompton J made the same point (at pp 585-6), as did Parke B (at pp 612-3). Thus these judges supported the literal, "never any breach", construction.

In coming to the contrary conclusion, that the words created a condition precedent, some of the eight judges favoured the "no subsisting breach" construction of these words, a construction assisted in that case by the reference to "arrears of rent". This phrase assumed that the rent had not always been paid punctually and this, in turn, assisted in giving a liberal meaning to the words "duly observed and performed" in relation to the lessees' covenants. Nevertheless, some of the observations made are of relevance. In particular, Talfourd J (at p 592) thought that "probably" the words "duly observed and performed" were not intended to be used in their technical sense, "as importing that no covenant during the eight years or longer period had ever been broken", but in the sense that before the expiration of the notice

the objects of the covenants should be attained, that is . . . everything done which the lessees were bound to do in order that they might deliver up the premises in a proper condition to their landlord.

Erle J (at p 599) considered the argument that the words in parentheses did not mean that there should not have been any breach during the term but that when the notice expired "there should not exist any cause of action in respect of performance of covenants." He observed that the requirements of arrears of rent having been paid referred to a covenant which had been broken but in respect of which all cause of action for the breach had been satisfied by subsequent accord. From this starting point he was attracted by the view that, likewise, if all repairs were at last carried out, and damages paid or any judgment satisfied in respect of breaches of other covenants, the legal duty of the covenantor would have been observed and performed to the extent that "all liability in respect thereof would be at an end."

In this sense the stipulation would be free from any hardship towards the lessee, as he might obtain the privilege if he did his duty.

He regarded this construction as not depending upon giving a peculiar effect to the words of that lease, but as an application of a general principle applicable to all contracts whereby

in all contracts the legal duty thereunder has been performed, and so the contract may be said in one sense to be performed, when either the thing contracted for has been done, or compensation instead thereof has been made.

Likewise Coleridge J (at p 608) construed the words in parentheses as requiring that at the relevant time

the account between the parties must, both as to rent and covenants, be clear: the rent need not have always been paid on the day; but all arrears, if any, must have been paid up; the covenants must have been strictly kept, or, if broken must have been satisfied for.

That case concerned a break clause. With such a clause the commercial purpose achieved by a condition construed as meaning "no subsisting breach" is readily apparent: before the lease can be ended prematurely all the rent due must have been paid, the property must have been put into a proper state of repair, and the other covenants must have been observed and performed in the sense that all liability in respect of any previous breaches must be at an end. What commercial purpose, in such a case, would be served by the "never any breach" construction of the condition precedent is not so readily apparent. In this respect an option to renew may not stand on precisely the same footing. Where a lease is being renewed a landlord may well be concerned not merely with the state of the account between the parties when the option is exercised but with the tenant's conduct, at least in the recent past. Furthermore, as already noted, *Grey* v *Friar* was a case in which the "no subsisting breach" construction was aided by the presence of the reference to arrears of rent.

However, 20 years later the Court of Appeal had to consider a clause which did not contain any mention of arrears of rent and which was concerned with renewal. In *Finch* v *Underwood* (1876) 2 Ch D 310, a seven-year lease contained a covenant for renewal by the landlord at the expiration of the term thereby granted "(in case the covenants and agreements on the said tenants' part shall have been duly observed and performed)''. Notice exercising the option was given shortly before the end of the term, and at that time and also when the lease expired the interior of the property was in need of repairs. The repairs were not major, for a builder was ready to undertake the necessary work for £13.10s. The court construed the material words as creating a condition precedent to entitlement to a new lease. Hence the tenants' claim to be entitled to a new lease failed. However, both James LJ and Mellish LJ firmly adopted the "no subsisting breach" construction of the condition. James LJ (at p 315), having observed that no doubt every property must at times be somewhat out of repair, said:

... but where it is required as a condition precedent to the granting a new lease that the lessee's covenants shall have been performed, the lessee who comes to claim the new lease must shew that *at that time* the property is in such a state as the covenants require it to be.

(Emphasis added.)

Mellish LJ (at p 315) added, even more explicitly:

Under the terms of the covenant in the present case the lease is to be granted only in case the covenants and agreements on the part of the tenants shall have been duly observed and performed. What does that mean? I think it does not mean that the tenants must have strictly observed and performed the covenants all through the term. For the expression is, "shall have been duly observed and performed;" and I think that this is satisfied if they have been so observed and performed that *there is no existing right of action under them at the time when the lease is applied for*.

(Emphasis added.)

Those observations were *obiter dicta,* but they are clear and strong statements, made in a case in which there had been repeated failures to pay rent. On four occasions distress had been levied, and on two other occasions a bailiff had gone to distrain but the rent had been paid before the levy had actually been made. Despite this unattractive record the court was firm in its adoption of the "no subsisting breach" construction.

In 1930 Clauson J [in *Simons* v *Associated Furnishers Ltd* [1931] Ch 379] had to consider the effect of a break clause in a 17-year repairing lease under which the tenant company had the right to determine the lease after five or 10 years if it should give six months' notice and if it

shall up to the time of such determination pay the rent and perform and observe the covenants and conditions on their part hereinbefore contained.

In that case dilapidations existed when the notice was given but they had been remedied by the time the notice expired. Clauson J decided that this fulfilled the requirements of the condition.

In that case the issue before the court concerned the particular date as at which there must be performance of the tenant's covenants: was it the date of the giving of the notice? Or was it the date of expiry of the notice? It is not without significance that no argument was advanced in favour of a "never any breach" construction. Nevertheless, in deciding in favour of the latter date, and having been referred to the authorities, Clauson J (at p 386) did consider that construction as "a possible construction", but he concluded that he was bound by "a very heavy weight of judicial opinion" to hold that the true construction was what I have called the "no subsisting breach" construction.

I come to the most recent authority, much relied upon by Mr Morgan. In *Bassett* v *Whiteley* (1982) 45 P&CR 87 this court considered a renewal clause in an eight-year lease, where the condition was expressed in these terms:

If (the tenants) shall have paid the rent hereby reserved and shall have reasonably performed and observed the several stipulations herein contained and on their part to be performed and observed up to the termination of the tenancy hereby created ...

There, on two occasions after service of a notice exercising the option, the tenants withheld rent for short periods in an attempt to put pressure on the landlord to repair a leaking roof. The rent was fully paid up when the original tenancy expired at the end of 1981. Both Waller LJ (at p 91) and Griffiths LJ (at p 93) adopted the "no subsisting breach" construction of the first limb of the condition regarding payment of rent and rejected the argument to the contrary advanced for the landlord.

On the second limb of the condition, regarding performance and observance of the stipulations, they took the view that since one of the "stipulations herein contained" was for payment of rent, this limb imposed an additional requirement on the tenant in the case of rent. Under that requirement it was

permissible to look at the conduct of the tenants throughout the term to determine whether or not they have reasonably performed and observed the several stipulations:

per Griffiths LJ (at p 92).

I do not read those judgments as casting any doubt on the established, conventional approach to the construction of conditions in break and renewal clauses. A notable feature of that case was the presence of the word "reasonably" in the second limb of the condition. As pointed out by Griffiths LJ at p 93, the word "reasonably" was introduced to mitigate the great hardship that would flow from the possibility that a very trivial breach of the stipulations might result in the loss of the option, but it did so by giving to the court a discretion. I can well understand, if I may respectfully say so, the disinclination of Waller LJ and Griffiths LJ to construe this clause in such a way that, in exercising this discretion, (namely, in determining whether the tenants' stipulations had reasonably been performed), the court should not be able to have regard to the conduct of the tenants.

Those are the principal authorities. In attacking the "no subsisting breach" construction, Mr Morgan submitted that to treat a covenant as performed and observed if at a particular date there is no outstanding cause of action in respect of a past breach, either for forfeiture or for damages (other than nominal damages), is an unworkable and impractical test. Applying this test, compliance would be a matter outside the tenant's control, because in the event of a dispute the tenant could not ensure that the dispute would be determined by the court, and any appeal finally disposed of, before the relevant date. I do not find this persuasive. Difficulties may arise in particular cases, but this test was spelled out by Mellish LJ in 1876, following the even earlier observations of Erle J and Coleridge J, and the paucity of reported decisions on this point since then would suggest that fears in this direction may be exaggerated.

Mr Morgan further submitted that, in any event, the option clause in the present case was significantly different from the conventional renewal clause. Typically a renewal clause provides for the grant of a new lease of the same length as the original term and on the same terms (save as to the amount of rent), the option being exercisable towards the end of the original term, and the new term being consecutive to the original one. Here the initial lease was for 15 years and the new term was for 125 years; the new term would, in part, be concurrent with the original term (if the judge was right in his construction of the phrase "the date of the term hereby granted"); the transaction changed from being a rental transaction to one of a capital nature (a premium of £300,000 was payable for the new lease, under which the rent was to be a peppercorn); the tenant's covenants in the new lease were modified so as to be appropriate for a long lease; and the option was exercisable at the beginning of the original term, not at the end. I agree that there are unusual features about this renewal clause. But I do not see how, considered separately or together, they lead to the conclusion, or even suggest, that in the present case the "never any breach" construction, and not the conventional "no subsisting breach" construction, is the true meaning of the relevant words in clause 9 (which, for convenience, I reproduce):

If the tenant shall be desirous of taking a further lease of the demised premises for a further term of 125 years from the date of the term hereby granted and shall not later than 29 September 1985 give to the lessors notice in writing of such its desire and if it shall have paid the rent hereby reserved and shall have performed and observed the several stipulations on its part herein contained and on its part to be performed and observed up to the date thereof then the lessors will ...

In my view, the condition in this clause is in a well-established conventional form. It is indistinguishable in its essentials from the form considered in *Finch* v *Underwood* and regarding which Mellish LJ made the observations already quoted, over a century ago. One can but wonder how many leases have been entered into and implemented since then on the footing that this form of condition in a break clause or a renewal clause bears the "no subsisting breach" meaning and not the, even stricter, "never any breach" construction. Of course, questions of construction depend upon the particular language of the particular instrument, but this is a field in which the court should be slow to find that small, inexplicit differences in language lead to a clause being construed, contrary to the norm, as imposing a "never any breach" requirement.

Accordingly, and in entire agreement with the judge thus far, I would reject Mr Morgan's submission that this is a "never any breach" condition.

Positive and negative covenants

Mr Morgan's alternative submission was that not all covenants in leases are to be regarded as observed and performed once there is no longer any existing cause of action in respect of a past breach. Repairing covenants may be performed late, by the work being done. But, in ordinary speech, a positive covenant such as a covenant to carry on a particular business if broken over a lengthy period, would not be regarded as "performed and observed" by reason only of the tenant's once more carrying on the business and paying damages for the past breaches. Again, with negative covenants: in ordinary speech, cessation of a user carried on in breach of a user covenant would be regarded as having the consequence that the covenant was currently being observed, but not that it had been observed.

Accordingly, submitted Mr Morgan, the correct test is to ask in relation to each covenant "Has the covenant been performed?", and to treat this not as a question of law but as a question of fact to be answered having regard to the terms of the particular covenant and to what has occurred.

Mr Morgan's further alternative submission was to draw a distinction, accepted by the judge, between negative and positive covenants: once a negative covenant has been broken, a condition that requires performance and observation of all covenants, cannot be fulfilled. Although, hitherto, no suggestion has been made that, in applying the "no subsisting breach" construction, different covenants in a lease might fall to be treated in different ways, no case has previously arisen where the covenants which had been breached were negative.

I am unable to accept either of these submissions, and it is at this point that I must, with respect, part company from the judge, for this reason. The question under consideration is the true construction of a condition, viz a condition requiring for its fulfilment the performance and observance of the tenant's covenants. In adopting the "no subsisting breach" construction the court has given to the requirement that all the tenant's covenants must have been performed and observed a secondary meaning in preference to the literal meaning: strict observance and performance is not necessary, all that is needed is that at the material date any past breaches have been made good in the way which the law recognises as a discharge of the tenant's obligation under the covenant in question. "Either the thing contracted for has been done, or compensation instead thereof has been made", in the words of Erle J in *Grey v Friar*. Depending upon the particular covenant and the particular breach, this secondary meaning will be more, or less, attractive. But if in accordance with what I have referred to as the conventional approach, the requirement of "performance and observance" of the tenant's covenants as a condition in a break clause or a renewal clause is given that secondary meaning, that secondary meaning must, in my view, then be applied to all the tenant's covenants. I can see no basis for construing words such as "perform and observe", appearing as a condition in a break clause or a renewal clause and expressly related without distinction to all the tenant's covenants, as bearing the secondary meaning in relation to some covenants but as bearing their literal, strict meaning in relation to other covenants.

For the same reason I do not think that the distinction, made in section 146(1) of the Law of Property Act 1925, between a breach of a covenant or condition which is "capable of remedy" and one which is not, is relevant in the present case.

For these reasons, and it being common ground that the landlord suffered no loss by reason of the tenant's applications for planning permission having been made in breach of covenant, I, too, would allow the tenant's appeal and answer "no" to the second of the two questions directed to be tried as preliminary issues.

The landlord's appeal: the date of commencement of the 125-year term

I am in complete agreement with the judge's conclusion and reasoning regarding the date on which, on the true construction of clause 9, the new 125-year term would commence. I add a few comments only out of deference to the arguments of counsel addressed to us. Of the three dates competing for the honour of being "the date of the term hereby granted" in clause 9, the date of expiry of the initial 15-year term (March 31 1997) is supported by the references in clause 9 to a "further" lease and a "further term". These suggest that the new term would be consecutive and not (and, indeed, this would be very unusual with a renewal clause) a concurrent term. But in this case the new lease is to be granted on payment of £300,000 and for a peppercorn rent. The existing lease provides for a rental of £15,000 per annum until April 1 1985, and a market rental thereafter. The option was exercisable until September 29 1985. For the tenant to exercise his option and pay £300,000 in or before 1985, and then pay a market rental until 1997, and then enjoy a peppercorn rent for the next 125 years does seem to be an improbable arrangement. When one couples with this the consideration that "the date of the term hereby granted" is not, in my view, a natural way to refer to the date of expiry of the original term, I am left in no doubt that in this case the new lease was not intended to run from March 31 1997, even though the consequence of this construction is that the new term will run from a date earlier than the expiry of the original term.

This leaves two dates: April 1 1982, which was the date from which the initial term of 15 years was measured, and September 20 1982, which was the date borne by the lease. Two features support the first of these two dates. First, in the lease, including in the rent review provisions in clause 3, the term thereby granted was treated as having begun on April 1 1982. Second, it seems unlikely that, in a lease that granted a 15-year term from a fixed, past date (April 1 1982), the parties would have intended that the new 125-year term should be calculated not from that date but from whatever might be the date on which, as events happened to turn out, the initial lease should actually be executed.

I, too, would give an affirmative answer to the first question raised as a preliminary issue and dismiss the landlord's appeal.

Also agreeing, BINGHAM LJ said: Under clause 9 of the lease dated September 20 1982, the tenant's right to require the grant of a new lease for 125 years was subject to fulfilment of two conditions precedent.

The first of these conditions related to the giving of notice not later than September 29 1985. I agree with what the learned judge and my lords have said about this. There is nothing I can usefully add.

The second condition was:

and if it [the tenant] shall have paid the rent hereby reserved and shall have performed and observed the several stipulations on its part herein contained and on its part to be performed and observed up to the date thereof

(ie up to the date of the notice).

Two principal questions arise in this appeal on the construction of this condition. The first is whether it requires the tenant, as a condition of its exercising the right to call for a new lease, to perform and observe all the covenants in the existing lease, whether positive or negative up to the time of giving notice. The lessors' primary submission is that it does, but that submission was not accepted by the judge. The second is whether it requires the tenant, as a condition of its exercising the right to call for a new lease, to perform and observe all negative covenants in the existing lease of which breaches, once committed, cannot be remedied, and to remedy all breaches of positive covenants before giving notice. This is the lessors' alternative submission and it formed the basis of the judge's decision.

The basic 1982 lease here was of a usual enough kind. It was for a 15-year term at a rack-rent with three-yearly rent reviews. It contained covenants of the kind one would expect in a lease between a brewery company and a tenant of hotel or public house premises. But Mr Morgan for the lessors suggests, in my view rightly, that clause 9 makes this a very unusual transaction. On exercise of the option the tenant can acquire a lease some eight times longer than the original term. On payment of a premium the tenant acquires this long lease free from any rental obligation. And the option is one that can, apparently, be exercised at any time after the execution of the lease in September 1982 provided it is exercised before the specified terminal date of September 29 1985. Mr Morgan submits, again rightly in my view, that these novel features require the court to look at the language of this clause in this lease with care and, if necessary, a fresh eye.

The judge summarised Mr Morgan's main submissions in his judgment at pp 403G-404B:

Mr Morgan has made three submissions. First he has submitted that the condition required the rent due under the lease to be paid at the times fixed by the lease for payment. If rent were paid late, then, he submitted, the covenant for the payment of rent would not have been performed and observed. Late payment is made in order to remedy a breach of covenant. It is not a performance of the covenant.

Secondly, he has submitted that the condition required all positive covenants in the lease to be performed in the manner and at the times specified in the lease. This submission is only different from the first in that the first relates specifically to rent and this relates to positive covenants generally. Thus, Mr Morgan submitted, there would be a failure to perform and observe a covenant to keep demised property in repair if the demised property were at any time during the term to be in disrepair. Subsequent repair would remedy the past breach of covenant and would also, perhaps, discharge a continuing obligation under the covenant, but it would not remove, and nothing could remove, the historical fact that the tenant had been in breach of the covenant to repair.

Thirdly, Mr Morgan has submitted that the condition required that no breach should have occurred of any negative covenant. In short, Mr Morgan's case is that any breach, however slight, of any covenant, whether positive or negative, prevents the defendant from claiming to have performed and observed the covenants and bars exercise of the clause 9 option.

Not surprisingly, Mr Morgan makes very much the same submissions to us. The judge observed (at p 404C) "If the condition is to be construed literally according to its strict language, Mr Morgan's submissions are, in my view, correct." I agree. The words used, together with the future perfect tense employed, read quite literally, suggest to me that exact compliance with the terms of the lease up to the time of exercising the option is required as a condition of its exercise.

Mr Morgan urges that we should, in accordance with ordinary canons of construction, give effect to the plain meaning of what the parties have agreed. If the tenant doubts its ability to comply exactly with its covenants, it can give notice under clause 9 very shortly after executing the 1982 lease. If it chooses, or is for any reason obliged, to delay in giving notice, it must pay the price of exact compliance meanwhile on pain of losing the right to exercise this potentially valuable option. Having chosen to give a series of detailed covenants the tenant cannot be heard to say that exact compliance with the covenants is impossible or unreasonable or oppressive.

It is of course true that the law has intervened to mitigate the rigours of the landlord-tenant relationship, for example in providing for relief against forfeiture and restricting the damages recoverable for breach of covenant to repair. But an option of the present kind has little to do with the ordinary relationship of landlord and tenant, and if the matter were free from authority I should see very great force in the argument that the effect of this option should be determined on ordinary contractual principles. Unfortunately for the lessors, however, questions closely related to the present have been intermittently litigated for 200 years and the current of authority has been against the lessors' contention.

Of the leading cases, the first two concerned a tenant's right to determine a lease before expiry of the contractual term. This is of significance, because the context coloured the judicial approach to the question of construction. Where a tenant wished to take advantage of a break clause, the landlord was not greatly concerned with the history of the tenant's performance before the break. The worse the tenant's performance, the readier the landlord might reasonably be to get rid of him. But whatever the tenant's defaults in the past, the landlord would be very much concerned that at the time of the break the rent should be fully paid (because he could no longer distrain) and the covenants fully observed (so that the property could be relet or sold without delay or additional expenditure).

In *Porter* v *Shephard* (1796) 6 TR 665, the lease was for seven years with a right for the tenant to determine after three or five years subject to giving notice and

payment of all rents and arrears of rent and duties on the tenant's part to be paid, and performance of the covenants contained on the part of the lessee until the expiration of the said first three or five years.

The case largely turned on whether this provision was a condition precedent, as it was held to be. But the trend of the court's thinking is reflected in the judgment of Lawrence J (at p 670):

It seems to me that it would be very unjust as against the landlord, if this were not considered to be a condition precedent; for if it were not, the lessee, after neglecting to pay the duties which are a charge on the estate, after suffering the house to be out of repair, and the rest of the premises to lie waste, might give them up to the landlord in that state, and the landlord might be left without any remedy.

Thus attention was firmly concentrated on the state of affairs at the time of the break.

I need not rehearse the curious and tortured history of *Grey* v *Friar* (1850) 5 Exch 584, (1854) 4 HL Cas 565. Most of the difficulty in that case arose from the apparently contradictory terms of the break clause in question. The issue was again whether the clause contained a condition precedent. In the result it was held to do so. There was no ruling on the question in issue here. Among the judges summoned to advise the House of Lords, however, a large majority supported the Court of Exchequer Chamber's ruling that the clause did contain a condition precedent, and in some of their opinions there are clear and influential expressions of view on the present point. The most telling passages are in the opinions of Talfourd J, Alderson B, Erle J and Coleridge J. The learned judge has cited these passages in his judgment at pp 404G-406B and I will not repeat them. All very plainly make the point (undoubtedly influenced by the form of the clause in question) that past breaches do not matter provided that these have been fully remedied or compensated for when the break clause is exercised.

Finch v *Underwood* (1876) 2 Ch D 310 is a significant case for several reasons, First, it concerned a renewal clause, which is in substance closer to the present clause than is a break clause. Second, the wording of the clause was not dissimilar from the present:

(in case the covenants and agreements on the said tenants' part shall have been duly observed and performed).

Third, the tenants' performance in paying rent had been highly unsatisfactory. During a seven-year term the landlords had levied a distress on four occasions, and on two others the bailiff had gone to distrain but the rent had been paid before the levy had actually been made. The case, however, turned on the fact that when the tenant sought to exercise his right to renew there was an extant breach of a repairing covenant, the estimated cost of remedying which was some £13. Malins V-C at first instance held that the landlord was bound by the terms of the covenant to renew and that as the want of repair was trifling it furnished no ground of defence. He appears to have attached no significance to the earlier breaches of the covenant to pay rent. The Court of Appeal took a different view of the failure to repair. In the passage which the learned judge more extensively quoted (at p 406E of his judgment) James LJ said that:

the lessee who comes to claim the new lease must show that *at that time* the property is in such a state as the covenants require it to be.

(My emphasis.)

Mellish LJ (as quoted by the judge at p 406A) was even more specific

Under the terms of the covenant in the present case the lease is to be granted only in case the covenants and agreements on the part of the tenants shall have been duly observed and performed. What does that mean? I think it does not mean that the tenants must have strictly observed and performed the covenants all through the term, for the expression is "shall have been duly observed and performed"; and I think that this is satisfied if they have been so observed and performed that there is no existing right of action under them at the time when the lease is applied for.

No member of the Court of Appeal referred to the earlier and (as one would have thought) serious breaches of the rental covenant, clearly regarding these as irrelevant.

Bastin v *Bidwell* (1881) 18 Ch D 238 also concerned an option to renew, conditional "upon the lessees' paying the rent and performing and observing the covenants of this present lease" According to Kay J (at p 250):

It is obvious the meaning was this: the landlord must have intended by covenant worded like this to say: "I shall have the term in which to see whether you are such a tenant as I shall think it right and expedient to grant a new lease to, and the test I propose in words is this, whether when you come for your new lease, or whether at some time or other (I will not at the present moment say what time) you have paid your rent and performed your covenants."

There was argument in that case (which the judge did not have to resolve) whether the relevant time for testing performance was the date of the notice or the date of its expiry, but consistently with earlier cases it was not suggested that reliance could be placed on breaches existing before either of those dates.

In *Simons* v *Associated Furnishers Ltd* [1931] 1 Ch 379, Clauson construed a break clause of which the condition was that the tenant

shall up to the time of such determination pay the rent and perform and observe the covenants and conditions on their part hereinbefore contained

He held himself (at p 386)

bound by a very heavy weight of judicial opinion to hold that the true meaning of that clause is this, that it will have been complied with, if at the end of the five years "there should not exist any cause of action in respect of performance of covenants": or, I may put it this way, the condition must be understood as "requiring that the account between the parties must, both as to rents and covenants, be clear; the rent need not have been always paid on the day; but all arrears, if any, must have been paid up; the covenants must have been strictly kept, or, if broken, must have been satisfied''.

These were quotations from Erle J and Coleridge J in *Grey* v *Friar*. Although Clauson J made his decision at first instance, it has stood unquestioned (so far as I know) for over half a century.

The only decision which runs against the current of authority I have mentioned is *Bassett* v *Whiteley* (1982) 45 P & CR 87, a decision of this court. The tenants in that case were entitled to renew the lease

If they shall have paid the rent hereby reserved and shall have reasonably performed and observed the several stipulations herein contained and on their part to be performed and observed up to the termination of the tenancy hereby created.

Save for the important word "reasonably" this covenant is not very different from that in *Simons*. The facts were that rent was fully paid both when notice was given and when the original lease expired, but between those dates the tenants delayed payment of rent to encourage the landlords to mend a badly leaking roof. Waller LJ held that in the circumstances the tenants had reasonably performed and observed their covenants. Griffiths LJ (as he then was) agreed with the approach of Waller LJ. It is plain that Griffiths LJ did not think it was correct simply to look at the position at the date of determination of the original lease and I think that Waller LJ, although less explicitly, took the same view. But the crux of the decision was the word "reasonably". It was plain that the tenants had acted altogether reasonably, whichever period or point of time was looked at. This made it unnecessary for the court to decide on the precise time to which the obligation related and I do not think the decision is binding authority on that point. The lessors in this appeal understandably rely heavily on this case, but I do not think it can of itself stem or divert the flow of previous authority.

The learned judge said (at pp 408G to 409B of his judgment):

I would respectfully accept that, in the end, the questions raised in the present case turn on the construction of clause 9 of the lease. None the less, it does seem to me that the line of authorities to which I have referred do establish that a condition precedent which requires that there shall have been performance and observance of a tenant's covenants does not fail simply on account of there having been a past, but remedied, breach of covenant. It would be possible to formulate a condition precedent which did require that there should not at any time have been any breach of a tenant's covenants. But the fairly common form of condition precedent that is to be found in clause 9, not materially different from the corresponding conditions precedent contained in the respective leases in *Grey* v *Friar* (1854) 4 HL Cas 565; *Finch* v *Underwood* (1876) 2 Ch D 310; *Bastin* v *Bidwell* (1881) 18 Ch D 238 and *Simons* v *Associated Furnishers Ltd* [1931] 1 Ch 379, does not, in my judgment, fail on account of a past breach of covenant provided that the breach has been remedied.*

I agree. Whether this is the rule which, starting from scratch, one would devise, I do not know but rather doubt. It does, however, appear that the rule stated by the judge is one which has over a very long period been understood by practitioners as representing the law on this subject. It would in my opinion be a source of mischief if this court were to disturb an assumption, reasonable on the authorities, on which very many landlords and tenants may have based their conduct and their agreements. I am, up to this point, very much in agreement with the judge.

* Editor's note: see [1986] 2 EGLR 50 at p 53M.

But the judge distinguished between positive and negative covenants. A covenant to do something can be substantially performed, even if late. A covenant not to do something, once broken, is broken for ever. As Lady Macbeth, referring to her breach of the sixth (negative) commandment, observed: "what's done is done." The learned judge held that the rule established by the authorities should not be extended to cover negative covenants. In reaching that conclusion he derived help from the authorities on section 146 of the Law of Property Act 1925.

I part company with the judge at this point, and for several reasons. First, the distinction is not one which in this context emerges from the authorities on break and renewal clauses which I have mentioned. They were, it is true, concerned with breaches of positive covenants to pay rent and to repair, but the cases contain much reasoning of a general nature in which this distinction is never mentioned. Second, in a detailed lease such as the present there are numerous covenants, some positive and some negative, and it would in my view be unwieldy and unnecessary to differentiate between them for this purpose. Many of these covenants could be expressed in either a negative or a positive way and even if regard is had to the substance rather than the words used there is sometimes a positive and a negative element in the covenants. Third and most important, I consider that this distinction in the present context distracts attention from the commercial core of the contract and directs it towards what may often be matters of form. One can take for examples the covenants not to sublet the premises or permit them to be used for an unlawful or immoral purpose. Once a breach of such a covenant has occurred, nothing can in a literal sense wipe the slate clean; and, on principles already discussed, it is plain that a tenant who seeks to exercise the option during the currency of a breach or without making proper compensation for a previous breach will be rightly refused. Suppose, however, that an unlawful subletting or an improper user has been brought to an end, and suitable undertakings proffered and appropriate compensation paid, in circumstances such that the landlord could show no continuing damage. It would seem to me highly and undesirably formalistic to hold that the tenant was forever debarred by those breaches from exercising the option, even though breaches of positive covenants at an earlier time could have been committed and made good. The present facts vividly illustrate the point. The tenant should not have applied for planning permission on two occasions without consent. These were plain, and perhaps indefensible, breaches of covenant. But they have caused the lessors no harm of any kind and would not (I think) have done so even if the applications had not been refused. This whole matter was investigated on the tenant's application for relief from forfeiture, and undertakings were given. There was no reason to apprehend a further breach. When the option was exercised, the account between the parties in this respect was (in the words of Coleridge J in *Grey* v *Friar*) clear; a covenant had been broken, but it had been satisfied for. In my judgment these breaches do not have the effect, simply because the covenant was negative, of preventing the tenant from exercising the option.

On these grounds I would allow the appeal.

The appeal was allowed with costs; the form of order to be agreed between counsel.

LANDLORD AND TENANT
RENT ACTS

Court of Appeal

September 25 1986

(Before Lord Justice RALPH GIBSON and Lord Justice WOOLF)

ROBERTS AND ANOTHER v MACILWRAITH-CHRISTIE AND ANOTHER

Rent Act 1977 — Appeal against possession order — Owners of large early-19th-century house, valued at £850,000 to £1 million with vacant possession, wished to sell the property for medical and financial reasons — The basement was occupied by an elderly lady who was a statutory tenant and by another lady whom the county court judge and the Court of Appeal found to be a subtenant — The landlords sought possession and the judge decided in their favour both under Case 1 in Schedule 15 to the 1977 Act (breach of covenant against subletting without consent) and on the ground that suitable alternative accommodation had been offered — Defendants appealed, attacking both grounds for the decision — Under the heading of Case 1 it was suggested that there was no term against subletting, that in any case any breach of the term had been waived, and that the second lady was a lodger and not a subtenant — These criticisms were rejected by the Court of Appeal, but more substantial issues arose in regard to the question of alternative accommodation, where the judge had not separated clearly matters relating to the suitability of the accommodation from the general question whether it was reasonable to make an order for possession — The alternative accommodation offered was in a modernised ground-floor flat, smaller but in better condition than the basement which the tenant had occupied, on a very low rent which the landlords had bound themselves and their successors in title not to increase during the tenancy — The disadvantages were exchanging access to the public garden in Kensington Square for access to Shepherd's Bush Green, lack of room for the large amount of furniture which the tenant possessed, and lack of accommodation for the lady who had been her subtenant — The county court judge, in a "highly compressed judgment", had not made it clear whether he had taken into account the issue of reasonableness as well as the issues of the extent and character of the alternative accommodation — A more serious criticism, and one which the Court of Appeal held to indicate an error of law, was in regard to the judge's expressed opinion, for which there was no basis in the evidence, that the presence of a sitting tenant in the basement of the property would make a very big difference to the selling price "and might approach 50%" — Despite this error the court decided to uphold the judge's order for possession and to determine the issue of reasonableness themselves based on the judge's findings of primary fact — Held that in all the circumstances it was reasonable to make an order for possession in the terms of the judge's order.

The following cases are referred to in this report.

Battlespring Ltd v *Gates* [1983] EGD 573; (1983) 268 EG 355

Cresswell v *Hodgson* [1951] 2 KB 92; [1951] 1 TLR 414; [1951] 1 All ER 710, CA

Hill v *Rochard* [1983] 1 WLR 478; [1983] 2 All ER 21; (1983) 46 P&CR 194; 266 EG 628, CA

Metropolitan Properties Co Ltd v *Cordery* (1979) 39 P&CR 10; (1979) 251 EG 567, CA

Mykolyshyn v *Noah* [1970] 1 WLR 1271; [1971] 1 All ER 49; (1970) 21 P&CR 679, CA

Street v *Mountford* [1985] AC 809; [1985] 2 WLR 877; [1985] 2 All ER 289; [1985] 1 EGLR 128; (1985) 274 EG 821, HL

This was an appeal by the defendants, Winifred Mary Tennyson Macilwraith-Christie and Margaret Maguire, from an order for possession made by Judge Stucley in the West London County Court in an action by the plaintiffs, the present respondents, Brigadier John Mark Herbert Roberts and Nicola Helen Lechmere Roberts, his wife, in respect of a basement flat at 27 Kensington Square, London W8.

K S Munro (instructed by Coles & Stevenson) appeared on behalf of the appellants; H J Barnes (instructed by May May & Merrimans) represented the respondents.

Giving judgment, RALPH GIBSON LJ said: On June 10 1986 in the West London County Court an order for possession was made by His Honour Judge Stucley against the two defendants, Mrs Macilwraith-Christie and Miss Maguire, requiring that possession be given of the basement flat at 27 Kensington Square, London W8, on September 10 1986 subject to compliance by the plaintiffs with certain conditions. The plaintiffs are Brigadier and Mrs Roberts. The conditions, in brief, require the plaintiffs to provide alternative accommodation for the first defendant in a ground-floor flat at 30 Rockley Road, Shepherd's Bush, London W14. I refer to the two flats hereafter as "the basement flat" and the "ground-floor flat".

Both defendants appealed to this court claiming that the order for possession should be set aside. The appeal was heard on September 11 and 12 1986. At the conclusion of the hearing we announced that the appeal was dismissed for reasons to be given at a later date. For my part the reasons which led me to that decision are as follows.

The first defendant became tenant of the flat in September 1958 under an agreement in writing made between her and the late Mr N L Macaskie QC, who held the property on Crown lease. The house at 27 Kensington Square is a handsome early-19th-century house on four floors above the basement. There are four reception rooms and 10 bedrooms. It is in need of modernisation according to the sale particulars to which we were referred. The second plaintiff, Mrs Roberts, is the daughter of Mr Macaskie and the house was her family home. Her father died in 1967. By family arrangement the house came into the ownership of her husband, Brigadier Roberts, the first plaintiff. The plaintiffs, who have 10 children, lived there. The freehold was purchased from the Crown. The plaintiffs decided that they had to move out: the first plaintiff has had severe trouble with a hip and wanted to live in a less vertical house and they considered that they could no longer afford to live in and to maintain the house. Having obtained advice as to the sale value of the property with full vacant possession, a figure of £850,000 to £1 million, they looked for another house for their family which, so far as concerns those still at home, consists of three children still at school, one at university and four who have finished their education and are now at work. They found a house at Twickenham and have contracted to buy it for £400,000. They have paid £200,000 and have spent £100,000 on repairs including eradication of dry rot.

There had been previous requests to the first defendant since about 1972 that she move out of the basement flat. Neither Mr Macaskie nor Mrs Roberts took the steps open to them to require the first defendant to pay any increase in the rent of £8 per week agreed in

1958 or to pass on any part of the increases in the amount of rates payable. By 1986 a fair rent for the basement flat likely to be set by the rent officer was about £40 per week exclusive of rates. It consists of two bedrooms, each about 15 ft by 10 ft, a living room 20 ft by 16 ft, with a kitchen and bathroom. It is reached down 10 steps from pavement level and then along a covered alley at the side of the house. The basement is dark, damp and unmodernised and was said to require a new kitchen and bathroom, rewiring and plumbing and the installation of central heating.

The plaintiffs decided to seek an order for possession against the first defendant, who was entitled to the protection of the Rent Act 1977, by offering what was said to be suitable alternative accommodation. The plaintiffs arranged to buy the long lease of the recently modernised ground-floor flat at Shepherd's Bush, which was the home of their daughter and son-in-law, for £50,000 so as to be able to offer to the first defendant a tenancy of that flat, which would be a protected tenancy at the same low rent which she had been paying for the basement flat in Kensington Square and on terms that the rent should not be increased during her tenancy and that all rates and other outgoings on the flat be paid by the plaintiffs or their successors in title. The choice of the home of their daughter and son-in-law for this purpose was no doubt prompted by their ability to make with them an effective conditional contract of purchase.

The first defendant is now 79 years old. She is in good physical state. Her only disability is a cataract. She stands well and moves as a fully sighted person. No medical evidence relating to her physical condition was put forward.

In about 1968 Miss Maguire, the second defendant, moved into the basement flat. She is now aged 56 years. She has always had one room of her own, a bedroom, and she has shared the use of the other rooms with the first defendant. When she first came she paid £5 per week. By 1986 she was paying £25 per week to the first defendant. At the time of the hearing the second defendant was away from the flat: she had become ill from a form of agitated depression and was in hospital in Lancashire. We were told that she had recovered and was back at the flat. She has always been in work and her present employment is with a firm of solicitors in Marylebone. The relationship between the first and second defendant is not as close as their living circumstances and the long time that they have shared the flat together might suggest. They do not cook for each other and in that sense are not one household. They help each other in some ways: the second defendant does some shopping for the first defendant but not a lot. They take or collect laundry to or from the launderette for each other. The judge found that the relationship between the first and second defendants had been and still was "purely a matter of financial accommodation".

The plaintiffs offered the ground-floor flat, together with new carpets and curtains, removal expenses etc, by letter of December 19 1985 which described the ground-floor flat as having one bedroom. They said they wanted to sell 27 Kensington Square with vacant possession. The first defendant's reply was sent on January 10 1986 by her solicitors: she was 79 years old; would find difficulty in adapting to new surroundings; had friends and neighbours to assist her where she is; was not afraid to return home after dark in the well-lighted area of Kensington High Street; and had been in her present home for some 30 years and expected to die there. There was, she said, no point in a discussion. There was no reference to the second defendant.

The plaintiffs started proceedings on January 28 1986 claiming possession on the ground that "suitable alternative accommodation" would be available for the first defendant when the order took effect: section 98(1)(a) of the Rent Act 1977. As to the second defendant, the plaintiffs' claim said that she was in the premises by some agreement with the first defendant and was not entitled to the protection of the Rent Act 1977. As to the first defendant, the plaintiffs alleged that the flat was let to her by an agreement in writing dated September 29 1958 for the term of three years and that since September 1961 the first defendant had occupied the flat as a statutory tenant at the rent of £8 per week.

Separate defences were delivered by the two defendants. The first defendant admitted that she held under the agreement in writing of September 29 1958 pleaded by the plaintiffs. She contended that the accommodation offered was not reasonably suitable to the means and to the needs of the first defendant as regards extent and character. The second defendant alleged that she had entered into exclusive occupation of one bedroom in 1968, sharing the use of the other rooms with the first defendant and paying £100 per month. She disputed the plaintiffs' claim that she was not entitled to the protection of the Rent Act. The plaintiffs accordingly, in the apparent and reasonable belief that the second defendant was claiming to be a subtenant, amended their claim to assert in the alternative that the first defendant had, in breach of the terms of the tenancy agreement, sublet part of the flat without the consent of the plaintiffs and to claim possession on the ground of that breach under Case 1 of Schedule 15 to the 1977 Act.

At the hearing there was no issue between the two defendants and they were both represented by Mr Munro who appeared for them in this court. The first defendant gave evidence. The second defendant did not. The case made for them was that the second defendant was a lodger and not a tenant. If the second defendant was a tenant, there was no breach because there was no term of the tenancy which restricted the first defendant's right to sublet. If there was a breach it had been waived by acceptance of rent with knowledge in the plaintiffs of the presence of the second defendant. The alternative accommodation was not reasonably suitable and in any event it was unreasonable to make an order.

The judge heard the evidence on May 29 and the submissions of counsel on June 6. Judgment was delivered in writing and dated June 10 1986. The main findings of the judge were that the document produced by the plaintiffs as the tenancy agreement which was headed "Draft Agreement", dated September 29 1958 and signed by Mr Macaskie and the first defendant, was evidence of the agreement between the parties. The document in clause 11(9) provided that the tenant was not to sublet the flat or any part thereof without the consent in writing of the landlord. Next he found that the second defendant was the tenant of the first defendant and that the breach had not been waived by the plaintiffs. The plaintiffs were, accordingly, entitled to an order for possession on that ground under Case 1 of Schedule 15 to the 1977 Act but any order for possession based thereon would be suspended for a term to enable the first defendant to persuade the second defendant to leave. Nothing further was said of that possible ground for an order because the judge proceeded next to consider the plaintiffs' case on alternative accommodation and concluded that the recently modernised ground-floor flat was suitable alternative accommodation for the first defendant and that it would be reasonable to make an order for possession of the basement flat against her on the terms already stated.

The defence of the two defendants was tenaciously and skilfully conducted by Mr Munro at the trial and it was presented by him to this court with sustained vigour. On the first part of the case I will take the points made by him in the order in which he presented them. The first is that there was no sufficient evidence to sustain the judge's finding that there was any term against subletting. He pointed to the heading of the document as a "Draft Agreement", to the various ink amendments, and to the incomplete renumbering of the paragraphs. There was no evidence as to whether the amendments were made before or after signature. Part of clause 11(9) was struck through in ink.

This point is in my judgment of no force. Mr Macaskie was dead. The first defendant had formally admitted that she became tenant under an agreement in writing dated September 29 1958 for a term of three years. The document produced was such an agreement and signed by her. Mr Macaskie was a lawyer of much practical experience. It is highly likely that he would wish to include such a term in the tenancy agreement for the basement of his house. The parties moreover acted as if such a term existed: in September 1960 the first defendant sublet her flat to a Mr Osmond for three months and, when she did so, she told Mr Macaskie, who assisted her with reference to the form of the agreement. The judge found that the first defendant knew well her obligation to seek leave before subletting. Mr Munro submitted that there was no evidence to support that finding, but in my judgment there plainly was. The first defendant said in evidence that before Miss Maguire came she had had "several to stay with her — Mr Macaskie always allowed it but I don't think I asked when Maguire came". I doubt if the first defendant was aware of the difference between subletting part, which she was not free to do without consent, which consent could not unreasonably be withheld, and the taking in of a lodger, which she was free to do without consent. It is clear to me, however, that when she spoke of Mr Macaskie "allowing it" she showed that she knew of the existence of a term in the agreement under which Mr Macaskie might claim not to "allow it". The first defendant gave no evidence whatever with reference to the tenancy agreement. The judge was entitled, as I think, to reach the conclusion which he did and I would have been surprised if he had reached any other.

Next it was submitted that there was no evidence of subletting and that the judge should have found that the second defendant was a lodger only. Reliance was placed upon the decision of the House of Lords in *Street* v *Mountford* [1985] AC 809 and the principles stated in the speech of Lord Templeman. The first defendant said that she shared the kitchen, bathroom and living room with the second defendant. She said nothing to show that the second defendant did not have exclusive occupation of her bedroom. She did not claim that she provided attendance or services of any sort to the second defendant which required the first defendant to have access to the second defendant's room: the only mutual service mentioned was, as I have said, the taking and collection of washing to and from the launderette. Mr Munro referred to the evidence of the second defendant to the effect that the room of the second defendant had been flooded by water from a washing machine in the plaintiffs' kitchen, and that work to repair the ceiling would take three weeks. It was argued that since nothing was said as to who would be doing the work it must be assumed that it would be done by the first defendant and that this assumption of responsibility was a clear indication of unlimited access by the first defendant to the second defendant's room. I see no force in this submission. If the first defendant was assuming responsibility for repairing the ceiling it is at least as consistent with a willingness to get the room repaired, with the consent of the second defendant, while she was away and ill, as with a pre-existing right of access to the room for all or any purposes. The first defendant had given no other evidence suggesting the existence of or the exercise of any right of access to the second defendant's room. Again, in my judgment, this conclusion of the judge that the second defendant was a tenant and not a lodger was one which was open to him on the evidence. I reject the further submission by Mr Munro to the effect that the judge must be treated as having misdirected himself as to the test to be applied in law for this purpose because he failed to refer to *Street* v *Mountford*. The case was cited to him. He recorded reference to the case in his notes. The judgment, which was reserved, is brief and very compressed. The reasons for that were, I am sure, that the learned judge is under considerable pressure of work and inclines to brevity from preference. The judgment — which I shall have to criticise later on some important matters — dealt effectively with all the main points made in a case of some complication and shows that the judge had attended with care both to the evidence and to the arguments addressed to him. I am unable to accept that we could sensibly treat the absence of express reference to the case of *Street* v *Mountford* as sufficient indication that the learned judge had misdirected himself in some unspecified way as to the proper approach in law for determining whether an occupier of residential accommodation is a tenant or a lodger, an issue of law with which he is no doubt long familiar.

The next point is waiver. The second defendant did not claim to have said anything of relevance to either plaintiff or to anyone else as to the status of the second defendant. The first plaintiff did not give evidence. The second plaintiff had known for many years of the presence of the second defendant in the flat, but there was no evidence to suggest that the plaintiffs knew that the second defendant was a tenant and not a lodger. The submission was that, since the plaintiffs knew that the second defendant was there and made no inquiry, they must be treated as having waived the breach. Reliance was placed upon the decision of this court in *Metropolitan Properties Co Ltd* v *Cordery* (1979) 39 P&CR 10. The judge held that *Cordery*'s case was distinguishable and said: "I find no duty on any landlord to interrogate persons who may be guests or lodgers while a lawful tenant continues in occupation."

In *Cordery*'s case the flat was in a block of flats where the landlords employed porters whose duties included that of informing the landlords of any observed changes in occupation. The relevant covenant required the tenant not to assign or underlet or part with possession or share the possession or occupation of the flat without consent. The porters were for a period of time aware of the presence in the flat of the subtenant in circumstances indicating that she was at least sharing occupation with the tenant. The landlords were held to have waived the breach, which was in fact of subletting the whole without consent, despite the fact that the precise nature of the breach was not known to them. In this case, however, the presence of the second defendant did not point to a breach of the clause of the tenancy agreement: it was consistent with the second defendant being there as a lodger and her presence as such would not have constituted breach. I see in the evidence nothing to show that the plaintiffs were obliged to inquire or, if they did not, to be treated as having waived the unsuspected subletting. In circumstances which were left unexplained in evidence, in 1975 the question of the terms of the first defendant's tenancy was raised and the solicitors for the first defendant, by letter of February 25 1975, wrote:

we have no information regarding any restriction on our client's right of subletting but we would confirm the premises are not sublet at present.

At that time the first and second defendants were both living in the flat and the plaintiffs knew they were there. It is true that the statement in the letter did not say that "neither the premises nor any part" were sublet, but it would be surprising if the plaintiffs had not supposed — if they thought about it — that the statement confirmed that the second defendant was a guest or a lodger. The judge was, in my view, entitled to reach the conclusion that waiver was not established and there is nothing to show that he misdirected himself in reaching that conclusion.

I should add that, as admitted in the pleadings between the plaintiffs and the first defendant, the first defendant was, at all times material to this issue of waiver, a statutory tenant. The submissions for both sides proceeded before us on the assumption that the common law rules as to the waiver in relation to breach of a term of a contractual tenancy applied in their full force to breach of a term by a statutory tenant. It has not been necessary for us to consider whether that assumption was right in law.

The next point arises out of the judge's observation that any order based upon the breach by unlawful subletting would be suspended for a term to enable the first defendant to "persuade the second defendant to leave". It became common ground in the argument before this court that, if the order for possession against the first defendant on the ground of alternative accommodation was properly made, it made no difference to either defendant whether the second defendant was held to be a subtenant or a lodger. If she was a lodger, the plaintiffs are entitled to possession against her if they can lawfully require the first defendant to go. If the second defendant had been held to be a lawful subtenant of the first defendant, nevertheless Mr Munro conceded that the second defendant could not on the facts of this case have claimed to remain upon the making of an order against the first defendant. Mr Munro argued, however, that the judge misdirected himself on one aspect of the relationship between the first and second defendants and that this misdirection continued into the judge's assessment of the factual issues relevant to the suitability of the alternative accommodation and to the reasonableness of making an order on that ground.

The argument went thus. If the second defendant was an unlawful subtenant, as the judge held, the first defendant was nevertheless entitled to give her notice to quit and to force her to go: the second defendant was, said Mr Munro, not entitled to the protection of the Rent Acts as against the first defendant. It was possible therefore for the first defendant to change the status of the second defendant from unlawful subtenant to lawful lodger without the second defendant having to leave. Mr Barnes accepted that this point was argued before the judge and he did not seek to challenge the validity of the argument thus far. Mr Munro then contended that the judge's reference to "persuading the second defendant to leave" demonstrated that he had misapprehended the position of the first defendant and had approached the remaining issues in the case, namely those concerned with the suitability of the alternative accommodation and the reasonableness of making an order, in the belief that the first defendant, if she remained in the basement flat, would not have been free to keep the second defendant with her. Therefore, said Mr Munro, the judge put the second defendant, and the importance to the first defendant of her presence, "out of the picture entirely" and thereby misdirected himself. It will be necessary to return later in this judgment to the matter of the importance to the first defendant of the ability to have the second defendant or some other person to live with her and to the judge's treatment of that matter. I am now dealing with the contention that the judge's reference to "persuading the second defendant to leave" demonstrated that the judge was in error and thereby misled himself on this point. I rejected this contention. I saw no reason to suppose that the judge misled himself in this way. The whole basis of the defence of the two defendants was that the second defendant was a lodger and that, if she was, the plaintiffs could not object to her presence. The plaintiffs never argued, and could not have argued, that the plaintiffs could object to the presence in the flat as a lodger of the second defendant or of anyone else. It is impossible to suppose that the judge could have thought otherwise. The judge, I am sure, knew that, if the defendant retained possession of the flat as a statutory tenant, she would, upon bringing the unlawful subtenancy to an end, be free to have whom she pleased as a lodger. The two defendants at the trial were making common cause. No one could have supposed that the second defendant, if she wished to remain in the flat, would refuse to surrender her subtenancy so as to continue as a lodger. The judge's reference to "persuading the second defendant to leave" was no more than a reference to the termination of the unlawful subtenancy. It would have precluded this argument, unfounded as I think it is, if he had used other and more precise words and I am confident that he did not because, on his view of the case as a whole, there was no question of an order being made on the grounds of the unlawful subletting.

I come now to the second part of the case, which is concerned with the judge's approach to and handling of the issues of the suitability of the alternative accommodation and of "reasonableness". On this part of the case Mr Munro's main submissions contain a great deal more substance. Before describing the submissions it is necessary to state the judge's conclusions and his reasons in more detail.

The plaintiffs obtained and put before the judge certificates from the Royal Borough of Kensington and Chelsea and from the London Borough of Hammersmith and Fulham. The first stated that borough would provide accommodation consisting of a self-contained, single-person, one-bedroomed flat for a single elderly female housing applicant. The second certificate stated that

the extent of the accommodation which would be afforded by this Authority to meet the needs of a single person would be a one-bedroomed flat consisting of separate bedroom, living room, kitchen, bath and wc.

I have omitted references to rent because nothing turns on it.

The relevance of these certificates arose thus. By section 98(1) of the 1977 Act "a court shall not make an order for possession of a dwelling-house which is . . . subject to a statutory tenancy unless the court considers it reasonable to make such an order and", and so far as relevant to this case, "the court is satisfied that suitable alternative accommodation is available for the tenant or will be available for him when the order in question takes effect". Then subsection (4) provides that Part IV of Schedule 15 shall have effect for determining whether "suitable alternative accommodation is or will be available for a tenant". So far as is relevant to this case para 4 in Part IV of Schedule 15 provides that "accommodation shall be deemed to be suitable for the purposes" of section 98(1)(a) of this Act if the security of tenure is as required and "in the opinion of the court, the accommodation fulfils the relevant conditions as defined in paragraph 5 below". Nothing in this case now turns upon the security of tenure of the alternative accommodation. Para 5(1) provides that

. . . the relevant conditions are that the accommodation is reasonably suitable to the needs of the tenant and his family as regards proximity to place of work, and either — (a) similar as regards rental and extent to the accommodation afforded by dwelling-houses in the neighbourhood by any housing authority for persons whose needs as regards extent are, in the opinion of the court, similar to those of the tenant and his family; or (b) reasonably suitable to the means of the tenant and to the needs of the tenant and his family as regards extent and character.

The first certificate produced related to "a single elderly female" and the second to a "single person". No point was taken before us as to whether either or both of these two boroughs provide dwelling-houses "in the neighbourhood". It was agreed that the extent of the accommodation afforded by the ground-floor flat is similar to that afforded by the accommodation provided by Hammersmith and Fulham to meet the needs of a single person or by Kensington and Chelsea to meet the needs of a single elderly female. If the "needs as regards extent" of the persons described in the certificates — they are, I think, categories of persons — were in the opinion of the court similar to the needs of the first defendant as regards "extent" then the ground-floor flat must be deemed to be suitable for the purposes of section 98(1)(a). If the needs of the first defendant as regards "extent" were in the opinion of the court not similar to the needs of the categories of persons mentioned in the certificates then the plaintiffs had to show that the flat was (ignoring rent and means on which nothing turns) "suitable to the needs of the tenant and his family as regards extent and character". It is not clear why "character" appears only in para 5(1)(b).

In a case where a certificate is proffered it seems to me that in the ordinary course the judge will be right first to decide whether in his opinion the certificate is conclusive of the issue of suitability under para 4 and para 5. If it is, he will say so and then deal with reasonableness. If it is not, he will state the reasons for his opinion and then deal with para 5(1)(b) and, if it arises, reasonableness. In this case the judge approached this issue as follows. After dealing with the issue of unlawful subletting he turned to consider the main plank of the plaintiffs' claim based on alternative accommodation. He continued:

To this the first defendant says that no 30 is not suitable, even if it measures up to local authority standards which regards a one-bedroomed flat as suitable for a single person. There were certificates to that effect from two local authorities. The first defendant says that it would not be reasonable to require her to move to the "ground-floor flat" from the "basement flat" for a number of reasons which I will consider below.

The judge then dealt under five numbered headings with matters which were directed both to suitability and to reasonableness, namely: (i) the defendant's age, health and desire not to have to move; (ii) the lack of sufficient room in the ground-floor flat to house her furniture; (iii) the lack of room for the second defendant; (iv) loss of access to the public garden in Kensington Square; and (v) the unreasonableness of forcing a tenant of long standing to move. The judge then considered the contrasting benefits and disadvantages of making or of not making the order for possession to the plaintiffs on the one hand and the first defendant on the other; he listed and considered five other objections to the ground-floor flat advanced by Mr Munro; and he concluded, as stated above, both that no 30 was suitable alternative accommodation and that it would be reasonable to make an order for possession.

Mr Munro submitted that it was not clear what view the judge formed on the question of the certificates or whether he was deciding the case on para 5(1)(a) or 5(1)(b). I agree that the judgment did not make this clear and it would be better if it had. For my part I had no doubt from the totality of the judge's judgment that he found primary facts which showed that he rejected the first defendant's contention that "the ground-floor flat was not suitable even if it measured up to local authority standards" and, after considering all the matters raised, intended to hold that the ground-floor flat was "similar as regards . . . extent to the accommodation afforded by dwelling-houses provided in the neighbourhood by any housing authority for persons [ie a single elderly female or a single person] whose needs as regards extent" were similar in his opinion to those of the first defendant. In the passage of his judgment to which I have referred he was, in my view, dealing comprehensively with the defendant's contentions both on her needs as regards extent and as to reasonableness. It is clear that the Act makes two separate requirements. The jurisdiction to make an order arises on proof of the availability of suitable alternative accommodation, but, given that proof, the court is still forbidden to make an order unless the court considers it reasonable to do so. It seems to me to be a better course for the court to deal separately and distinctly with the two issues.

It was necessary to deal at some length with these matters because in the course of argument the question was raised by this court of the extent to which the judge at trial may properly have regard on the issue of reasonableness to matters raised on the issue of suitability and held not to demonstrate the unsuitability of the alternative accommodation. Thus, in this case, if the correct view of the judge's judgment was that he had found the relevant condition satisfied as set out in para 5(1)(a) notwithstanding the contention of the first defendant that her needs as regards extent of accommodation included space for the second defendant or another companion, so that the ground-floor flat was "deemed to be suitable for the purposes of section 98(1)(a)", was it permissible for the court in considering reasonableness to have regard to this absence of space for the second defendant? Mr Munro and Mr Barnes united in submitting that the court, in considering reasonableness, was required to consider all relevant matters; and in assessing the consequences to the plaintiffs and to the first defendant of refusing to grant or granting the order for possession, for the purposes of deciding the issue of reasonableness, the court could not disregard a matter, such as the loss by the first defendant of the ability to have a lodger in her home, merely because the court had determined that, despite the lack of space for a lodger, the accommodation was reasonably suitable. I agree with their submissions on this point. Mr Barnes submitted that the court should consider the judgment on the basis that the judge considered that the plaintiffs had to satisfy the requirements of para 5(1)(b) and then held them to have been satisfied.

On this part of the case Mr Munro advanced five main grounds for his contention that the judge's conclusion was based upon error of law and that this court should either set aside the order for possession and dismiss the plaintiffs' claim or remit the claim for retrial. I will deal first with the three of those five grounds which, as it seemed to me, were of no real substance and then with the two grounds on which error by the judge was, in my judgment, demonstrated. Lastly I will set out my reasons for reaching the conclusion that, notwithstanding those errors, the order for possession should be upheld and the appeal dismissed.

The first point was directed to the judge's treatment of the consequences for the first defendant, on being required to move to the ground-floor flat, of losing access to the public garden in Kensington Square, and in getting in exchange no more than access to Shepherd's Bush Green: the judge said:

The lack of the public garden in Kensington Square, which is undoubtedly more attractive than the Green at Shepherd's Bush, seems covered by the dicta in Hill v Rochard [1983] 1WLR 478 where a lack of stables at an alternative house was found to be no bar to the premises being suitable.

It was argued that the judge was wrong to consider himself bound by dicta in that case. I was unable to find any substance in this submission. It seemed to me that the judge directed himself in accordance with the principles stated in Hill v Rochard where (at p 484G) Dunn LJ, after citing para 5 of Part IV of Schedule 15, continued:

Then subparagraph (b) provides: "reasonably suitable to the means of the tenant and to the needs of the tenant and his family as regards extent and character;" . . . It was (b) which was the material provision in this case.

In my judgment the word "needs" means "needs for housing", and the question is whether the accommodation offered is reasonably suitable for the tenant's housing needs as regards extent and character . . . .

The argument in this court has revolved around the word "character." The subparagraph does not provide, and it is not necessary, that the character of the alternative accommodation should be similar to that of the existing premises . . . the question is whether

the alternative accommodation

is reasonably suitable to the tenants' housing needs as regards its character. In considering those needs the cases to which I have referred show that it is permissible for the court to look at the environment to which the tenants have become accustomed in their present accommodation, and to see how far the new environment differs from that . . . .

In my view the judge was right to say that these tenants were not ordinary tenants. By that I take her to mean that, accommodation aside, the present tenancy enabled them to enjoy the use of certain amenities, including the paddock and out-buildings, so that they could keep their animals. Even on a liberal construction of the statutory provisions I do not think that the Rent Acts were intended to protect incidental advantages of that kind. The Rent Acts are concerned with the provision of housing and accommodation.

Eveleigh LJ agreed with the judgment of Dunn LJ. In this case the judge was entitled to conclude, as I think he did in the terse passage upon which this submission is founded, that on the facts of this case the loss of access to the public garden in Kensington Square was of the nature of the loss by the defendants in Hill v Rochard of the use of stables and that the ground-floor flat was not thereby rendered unsuitable to the needs of the first defendant as regards habitation. The judge did not separately mention the needs of the first defendant as regards the character of her housing including environment, but I find it impossible to accept that he did not have it in mind having regard both to the submissions which had been made to him and to his express reference to the less attractive nature of Shepherd's Bush Green.

Mr Munro's next submission was concerned with the first defendant's inability to get all her furniture into the ground-floor flat. She has a lot of furniture: the living room in the basement flat is "chock a block" in the words of her surveyor. The bedroom in the ground-floor flat is about the same size as her present bedroom, but she would, of course, have only one bedroom, and the living room at no 30 is less than half the size of the living room that she has now: 10 ft 6 in by 12 ft as against 20 ft by 15 ft 9 in. The judge referred to Mykolyshyn v Noah [1970] 1 WLR 1271, which had been cited to him, and said that there appeared to be no requirement for a landlord to provide space for storing furniture and that it would be wrong to reject alternative accommodation merely on the ground that it would not accommodate all the tenant's furniture without considering how much of that furniture the tenant reasonably required. Basing himself on the plan of the ground-floor flat and upon photographs of its interior he held that it would accommodate all the furniture a single person would reasonably require. Mr Munro's submission, if I understood it correctly, was that the judge had misunderstood and misapplied the decision in Mykolyshyn's case in that the court there had approved the disregarding of furniture which the defendant was shown not to use whereas, on the evidence in this case, the first defendant made use of all the large amount of furniture which she had. He submitted that the judge should have had regard not only to what he found to be the reasonable requirements of the first defendant as to furniture but also to her wishes and expectations. I was unable to accept this submission. The judge did not, as I think, misunderstand or misapply the principles stated in Mykolyshyn v Noah. In that case, in which the defendant was ordered in effect to give up one room on the basis that her flat without that room was suitable alternative accommodation, Widgery LJ (as he then was) said at p 1278:

In deciding whether the accommodation offered was reasonably suitable to the tenant's needs, the judge had to consider, amongst other things, whether it would take her furniture, so far as that furniture was required to enable her to live in reasonable comfort. He would, I think, have been right to conclude that the premises were not rendered unsuitable merely because there was no accommodation for additional furniture for which the tenant had no foreseeable need. This is in effect what the judge has done, because he has adopted the defendant's own conduct as the best test of what furniture she really required and has come to the conclusion that the furniture in the sitting-room is surplus to those requirements and, therefore, the alternative offered is not unsuitable merely because it cannot provide a storage place for that furniture.

In this case, the fact that the first defendant was using all of the large amount of furniture in the basement flat does not mean that any alternative accommodation which is unable to hold all or most of that furniture must therefore be unsuitable. On that point I see nothing wrong in the judge's approach.

Upon these last two points, loss of access to Kensington Square and lack of space for furniture, Mr Munro submitted that even if the judge had not gone wrong with reference to either matter on the issue of reasonable suitability of the ground-floor flat yet, from the way he dealt with those points in his judgment, he was shown to have taken no account of them on the separate issue of reasonableness and thereby to have misdirected himself. He makes the same complaint with reference to the next matter, lack of space in the ground-floor flat for the second defendant. He supported that submission by pointing to the words of the judge under the third of the five numbered headings which Mr Munro says was the third objection put forward on the ground of reasonableness. The judge said: "Thirdly there would be no room for the second defendant; but there is no obligation on the plaintiffs to house the second defendant." Earlier the judge had said: "I agree with the plaintiffs that they have no responsibility to rehouse the second defendant if an order for possession should be made against her or against the first defendant."

Mr Munro acknowledged that the judge was correct in both those statements but argued that the absence of obligation in the plaintiffs to rehouse the second defendant did not entitle the judge to dismiss as irrelevant to the issue of reasonableness the detriment to the first defendant of being required to move to accommodation in which she could not receive the second defendant, or another person, as a lodger. Mr Barnes submitted that the cited words of the judge in this highly compressed judgment are an insufficient basis for concluding that the judge had misdirected himself by excluding on the issue of reasonableness the loss of space for the second defendant or the other matters.

I reached the conclusion that the judgment, read fairly and as a whole, did show that the judge appears to have thought that absence of space for the second defendant was made irrelevant for all issues by reason of the absence of obligation in the plaintiffs to the second defendant. The way in which he dealt in the words of his judgment with lack of space for furniture and loss of access to Kensington Square also provides, I think, some basis for Mr Munro's submission, although I regard it as highly unlikely that this experienced judge would direct himself that either of those matters was irrelevant on the issue of reasonableness. If he so directed himself, he was, I think, wrong in law for the reasons which have been given. Whether the basement flat in the judge's opinion fulfilled the conditions set out in para 5(1)(a) of Part IV of Schedule 15, or the conditions in para 5(1)(b), any significant detriment to the first defendant caused by being required to move from the basement flat to the ground-floor flat is, in my view, to be considered and assessed by the judge in considering the issue of reasonableness. The weight to be given to such a detriment is of course for him to assess. If the alleged detriment is primarily a ground for contending that the offered accommodation is not reasonably suitable, and that accommodation is either to be deemed to be suitable under para 5(1)(a) or is held to be suitable under para 5(1)(b), it seems to me likely in the ordinary case that no great weight will properly be given to the point on reasonableness, but a matter of no great weight may prevail when all that is proved on the other side weighs even less.

I turn now to the point which Mr Munro described as the "fatal flaw" in the judge's reasoning. The judge considered the position of the plaintiffs as follows:

I find the first plaintiff to be a retired Brigadier and he and his wife have bought themselves a new home at Twickenham. In anticipation of selling 27 Kensington Square they have agreed to pay £400,000, have paid half, and been allowed to spend a further £100,000 making good damage by dry rot. Although the second plaintiff has an imperfect knowledge of her husband's financial affairs I accept that it would be a great relief to the plaintiffs if they were able to sell their freehold in Kensington Square for £850,000 or £1m as is expected with vacant possession. I have no evidence of the effect on such figures that a sitting tenant in the basement would have. I did not accept Mr French's evidence that it would make little difference. I am of the opinion that it would make a very big difference indeed and might approach 50 per cent.

Mr Munro complains that there was no evidence which justified the rejection of Mr French's evidence that the presence in the basement flat of this house of a protected tenant would make "little difference" or which justified the judge's finding that it would make a very big difference indeed and might approach 50 per cent. The course of the trial as regards this matter must be noted: the plaintiffs called Mr Steele, an estate agent, who dealt with descriptions of the two flats and other matters but said nothing of the effect upon the price realisable on the sale of 27 Kensington Square of a protected tenant in the basement flat. It seemed to me a surprising omission which may have resulted from a belief that the presence of a protected tenant obviously would have a large effect. The defendants called an estate agent, Mr French. In evidence-in-chief he also said nothing on the point. Counsel for the plaintiffs (not Mr Barnes) chose to ask a question of Mr French to which the judge recorded the answer: "I don't agree that vacant possession of the basement flat would make a great deal of difference to a purchaser having regard to Mrs Christie's age". It is not clear whether Mr French also said that vacant possession would make little difference or whether the judge regarded that as equivalent to what he did say. It does not matter. Mr Munro was, in my view, right in his submission that there was no basis in the evidence of the judge's assertion that "this difference might approach 50 per cent" and that that was a misdirection. I did not, however, accept that the judge was wrong in not accepting evidence from Mr French that vacant possession would make "little difference". Upon the evidence before him as to the nature and condition of the house and of the basement flat and of the probable sale price with full vacant possession the judge was entitled to reject the opinion that the presence on a protected tenancy of a healthy 79-year-old woman would make "little difference" to the sale price to be realised solely because of her advanced age. The position therefore is that on this matter the judge made an error of law. He directed himself that, if this defendant was left in occupation, the plaintiffs might lose as much as £425,000 or £500,000 if forced to sell. He had regard to that in assessing the position of the plaintiffs and of the defendants on the issue of reasonableness. The error was, in my judgment, potentially of substantial effect. Therefore the judge's conclusion upon the issue of reasonableness could not stand and this court was required to decide the issue upon the material before us, if it was clear that it was safe and just so to do, or to remit the plaintiffs' claim for retrial.

There was one more point made by Mr Munro. It was directed to what the judge called "uncommon generosity" on the part of the plaintiffs. The judge in an earlier passage of his judgment referred to the fact that the tenant was still paying the 1958 rent of £400 per annum inclusive of rates and that the fair rent in 1986 would be set at about £2,000 per annum exclusive of rates. Later the judge continued:

What seems beyond doubt is that the plaintiffs have behaved with uncommon generosity towards their tenant by allowing her to continue at what has become a notional rent and paying her rates for her. She, conversely, says that were she to move to alternative accommodation provided by the plaintiffs, she suspects the plaintiffs would dishonour their undertakings and that in particular she does not trust Mrs Roberts the second plaintiff. This lack of appreciation of past favours is evidenced by the first defendant standing on her undoubted rights to deny her landlords a site in the area for a central heating oil tank or parking space for their children's bicycles.

The ground of appeal put forward was to the effect that the judge wrongly allowed himself to be influenced by what he called the plaintiffs' past "uncommon generosity" and the first defendant's "lack of appreciation of past favours" and in so doing ignored the evidence: (i) that the plaintiffs had never carried out any work on the first defendant's flat, externally or internally, save for limited damp-proofing to the structural walls and (ii) that the first defendant had made good at her own expense on four occasions damage to ceilings, plaster and decorations and furnishings caused by water escaping from the plaintiffs' kitchen. Mr Munro submitted that the judge had misapprehended the evidence. I was unable to accept this submission. The tenancy agreement of 1958 provided that the first defendant should take the flat "in its present condition" and she agreed to repair the flat during the term. The correspondence before the court shows that in 1980 the first defendant had relied upon the plaintiffs' responsibility under section 32 of the Housing Act 1961 to deduct from the rent a sum of money for renewal of the gas supply. There was no evidence of the plaintiffs being asked to do any other work to the basement flat or of failing to do what it was their obligation to do. The comment of the judge seems to me to be fair and supported by the evidence.

It remains for me to give my reasons for upholding the order for possession. Despite what I regarded as a substantial error on the part of the judge in permitting himself to proceed upon his own view as to the likely extent in percentage terms of the effect upon the sale price of the property of the presence of this protected tenant in the basement flat, and the fact that he appears to have misdirected himself as to the relevance on the issue of reasonableness of the loss by the first defendant of some of the advantages enjoyed by the first defendant in her occupation of the basement flat, it was clear to me that the judge had attended with care to all the main issues of fact in this case and that it was safe and right for this court to act upon his findings of primary fact.

As to the matter of the second defendant, I have already set out the judge's findings, which were clearly supported by the evidence of the first defendant herself. No reliance was placed by Mr Munro on the loss of the financial assistance from payments to be made by a lodger — that was expressly disclaimed. There was no question of close personal friendship between the first and second defendants or of the breaking up of a long-standing shared home or household. The presence of the second defendant was from start to finish a matter of financial accommodation, upon the loss of which, as I have said, no reliance is placed. The second defendant did not act as a companion and there is no evidence that the first defendant needs a companion. To shrink from two bedrooms to one, and from a large to a modest living-room, is a hardship, but having regard to the other matters fully noted and considered by the judge and on all the facts of this case it is not a hardship of any large weight on the issue of reasonableness. I take the same view of the loss of access to Kensington Square and of the need to dispose of some furniture.

As to the effect on the sale value of 27 Kensington Square of the presence of a protected tenant in the basement flat, I have referred already to many of the judge's findings on the facts which explain the decision of the plaintiffs to sell and to try to sell with vacant possession. The plaintiffs wish to sell their property, which has been the family home of Mrs Roberts for very many years. The basement flat, of course, was not part of that home, but it had been let to the first defendant in 1958 and for 27 years Mrs Roberts and her father had been content to leave the first defendant in occupation at what became over the years a very low rent indeed. The plaintiffs' decision to sell was based on their proved need to move out both to get a home more convenient for the first plaintiff's disability and because of their difficulty in meeting the expense of living there. They sought vacant possession of the basement flat in order to be able to obtain the best price possible for their property. The offer of alternative accommodation made by them seems to me to have been wholly fair in the circumstances in that not only is the ground-floor flat reasonably suitable to the needs of the first defendant as regards extent and character but it is, as the judge found, in many respects better than the accommodation provided in the basement. In addition the plaintiffs have committed themselves, and are bound by the judge's order, to provide to the first defendant a tenancy of the ground-floor flat at the very low rent of £33.33 per calendar month upon terms, binding upon themselves and their successors in title, that that rent is never to be increased during the first defendant's tenancy and that all rates, water rates, maintenance and service charges be paid by the plaintiffs. The plaintiffs thus, out of the price which they hope to get for the sale of their property, are bound by the judge's order to use some part of that price to provide for the first defendant, while she remains in the ground-floor flat, security of tenure of a pleasant well-maintained flat at a very low rent.

I considered again every one of the matters put forward on behalf of the first defendant in support of her contention that this court should conclude that it would not be reasonable to uphold the order for possession. I considered also the case of *Battlespring Ltd v Gates* (1983) 268 EG 355, a decision of the Court of Appeal (Watkins and May LJJ) upon which Mr Munro relied but in which I find nothing which suggests that on the facts of this case it should be held not to be reasonable to make such an order. Watkins LJ there referred to the judgment of Somervell LJ in *Cresswell v Hodgson* [1951] 2 KB 92 where he said:

I think the words of the section themselves indicate that the county court judge must look at the effect of the order on each party to it. . . I do not think we should say anything which restricts the circumstances which the county court judge should take into consideration. I think he is entitled to take into consideration that this is a case where the landlord is making a pecuniary gain. That might in other cases be a fact in the landlord's favour, and it might be thought reasonable that he should be given the chance of making pecuniary gain.

It seems plain to me that, on the facts found by the judge, including the evidence of Mr French as recorded by him, the plaintiffs acted in the belief that the continued presence of the second defendant in the basement of the property would make a significant difference to the price which they would obtain on sale and that belief was reasonable and well founded. The property has apparently become worth what to most of us is an astonishingly large sum of money. It is, I think, obvious that to the learned judge this was a very clear case for the making of an order for possession. Despite the defects in his judgment to which I have referred, I agree with him. On all the facts in this case it was in my judgment reasonable for an order for possession to be made on the terms imposed by the judge's order.

WOOLF LJ agreed and did not add anything.

The appeal was dismissed. There was no order for costs except for legal aid taxation. Counsel were asked to agree a schedule of dates.

For the other cases on this subject see p 99

NEGLIGENCE

Queen's Bench Division

November 24 1986

(Before Mr Justice SCHIEMANN)

HARRIS AND ANOTHER v WYRE FOREST DISTRICT COUNCIL AND ANOTHER

Negligence — Houseowners' action against local authority in their capacity as mortgagees and their staff valuer — Two-storey terraced house with brick and slate extension at the rear — House had suffered from extensive settlement and there was a tie bar round it — Valuer noticed tie bar and signs of settlement but concluded that any movement was a thing of the past — He valued the house at £9,450 (the asking price) and recommended a 90% loan — Local authority made the loan to the plaintiffs and they purchased the house — Three years later the plaintiffs wished to sell the house but were put off when the local authority, acting on a report by the same staff valuer, drew attention to the tie bar and settlement and proposed to retain the whole amount of a projected loan until a structural survey had been carried out and any recommended works completed — Plaintiffs then obtained a report from a structural engineer who commented adversely on the condition of the house, which was subsequently taken off the market — In the action against the mortgagees and the valuer the structural engineer gave evidence that the house lacked stability and that there was "a fairly large risk of a fairly substantial disaster" — The judge found that a speculative builder might have paid £3,500 for the house in its unstable condition — Held that the valuer was negligent in his inspection and report and that the local authority were negligent in offering a mortgage to an applicant when, having regard to the condition of the house, they were not empowered to do so under the relevant legislation and would have known this fact if their valuer had done his job properly — The local authority were both vicariously and primarily liable — An exemption clause was held to provide no immunity from a negligence claim — Judgment in favour of plaintiffs, who were still in the house, for £12,000 — *Per* Schiemann J, "It is erroneous to think of there being several values attributable to a house . . . In principle a valuer's valuation should be the same whomever he is acting for."

The following cases are referred to in this report.

Anns v *Merton London Borough Council* [1978] AC 728; [1977] 2 WLR 1024; [1977] 2 All ER 492; [1977] EGD 604; (1977) 243 EG 523 & 591, HL
Curran v *Northern Ireland Housing Executive* [1985] 8 NIJB 22
Donoghue v *Stevenson* [1932] AC 562, HL
Hedley Byrne & Co Ltd v *Heller & Partners Ltd* [1964] AC 465; [1963] 3 WLR 101; [1963] 2 All ER 575; [1963] 1 Lloyd's Rep 485, HL
Leigh & Sillavan Ltd v *Aliakmon Shipping Co* [1985] QB 350; [1985] 2 WLR 289; [1985] 2 All ER 44; [1985] 1 Lloyd's Rep 199, CA; [1986] 2 WLR 902; [1986] 2 145, HL
Peabody Donation Fund (Governors of) v *Sir Lindsay Parkinson & Co Ltd* [1985] AC 210; [1984] 3 WLR 953; [1984] 3 All ER 529, HL
Ward v *McMaster, Louth CC and Nicholas Hardy & Co Ltd* [1986] I LRM 43
Yianni v *Edwin Evans & Sons* [1982] QB 438; [1981] 3 WLR 843; [1981] 3 All ER 592; [1981] EGD 803; (1981) 259 EG 969

This was an action by Adam Charles Harris and his wife, Kim Harris, against Wyre Forest District Council and Trevor James Lee, a valuer on the council's staff, alleging negligence in relation to a mortgage offered to the plaintiffs for the purchase of a house at 74 George Street, Kidderminster.

Malcolm Stitcher (instructed by Thursfield & Adams with Westons, of Kidderminster) appeared on behalf of the plaintiffs; Nicholas Worsley (instructed by Rowleys & Blewitts, of Birmingham) represented the defendants.

Giving judgment, SCHIEMANN J said: In 1978 the plaintiffs were thinking of getting married and were looking in Kidderminster for a house to buy. They were aged 22 and 19, had left school after obtaining some "O" levels and this was their first house-buying transaction. They only had a few hundred pounds saved up and their joint earnings were less than £5,000 per annum. They were first-time buyers, looking for a property costing less than £10,000, with a mortgage for nearly all of the purchase price. Their position was typical of many others. They were looking at Victorian terraced houses. I find them to be honest witnesses.

In the course of their inquiries they came to the conclusion that the most suitable source of mortgage finance was the defendant local authority. At that time and for many years previously, banks and building societies were in general unwilling to lend on pre-1919 small properties. Successive governments have taken the view that local authorities should supplement the activities of the main mortgage institutions and be prepared to assist "applicants who are wishing to buy older and smaller property unlikely to attract a commercial mortgage advance" — (see Circular 22/71 and Circular 42/54, which stresses that "It is important that everything which can be done without undue risk to Public Funds should be done to help everyone willing to buy himself a house").

The defendant council therefore, like innumerable others, had a policy of making loans to intending purchasers of property unlikely to attract a commercial mortgage advance.

The details of the scheme operated by the council at the time are set out in a document, which appears on pp 93-100 in bundle A of the documents that were used at the trial. The document was available to interested parties, but I find as a fact that the plaintiffs never saw it and that no steps were taken by the defendant council to bring it to the attention of potential mortgagors. Mr Lee, the valuer on the defendant council's staff, knew the substance of its contents and in particular knew what was said in the second para on p 97:

> The granting of a loan by the Council, whether or not subject to any requirement that work of repair, etc, is to be carried out, is not to be considered as any guarantee or warranty that the property is sound and free from any structural or other defect. The report of the Council's Valuer is intended for the information of the Council only and the contents thereof will not be disclosed to an applicant.

I find as a fact that Mr Lee thought of himself as having responsibility to the defendant council but did not think of himself as having any responsibilities to potential mortgagors.

The plaintiffs, in their search for a house, came across one which seemed suitable — 12 East Street. If, like the plaintiffs, one wished to apply for a local authority mortgage, there was a standard application form to fill in. This form was readily available to estate agents. It seems probable that the plaintiffs filled in such a form in relation to 12 East Street, but that purchase went off.

Then the plaintiffs saw that 74 George Street was for sale and that a Mr Young was the estate agent. His particulars, after describing the property as "The extremely well situated and utterly deceiving fully modernised Victorian freehold Terraced Residence", set out an asking price of £9,450 and stated, under the heading "Mortgage":

subject to satisfactory status the Agent is in a position to arrange an advance of up to 95% for any purchaser — minimum deposit only £475 — and consequently he will be pleased to discuss such arrangements with all interested parties.

The plaintiffs saw the property and went back to Mr Young and asked about the mortgage. He pulled out a mortgage application form, filled in the particulars with their help, indicated on the form that they were prepared to contribute £500 and showed them where to sign. They signed then and there on August 23 1978. I find as a fact that neither of them read the words that appeared immediately above their signatures although they did enclose the £22 referred to. Those words were:

TO BE READ CAREFULLY AND SIGNED PERSONALLY BY ALL APPLICANTS. I/WE enclose herewith the Valuation Fee & Administration Fee £22.00. I/WE understand that this fee is not returnable even if the Council do not eventually make an advance and that the Valuation is confidential and is intended solely for the information of Wyre Forest District Council in determining what advance, if any, may be made on the security and that no responsibility whatsoever is implied or accepted by the Council for the value or condition of the property by reason of such inspection and report. (You are advised for your own protection to instruct your own Surveyor/Architect to inspect the property). I/WE agree that the Valuation Report is the property of the Council and that I/WE cannot require its production.

The plaintiffs told me that neither Mr Young nor their own solicitor nor indeed their parents (who saw the house with them) suggested that they should have their own survey done and they themselves did not think of it, although they knew that such a thing could be arranged at a price. I interpose to say that the going price at that time for a structural survey was somewhere between £75 and £100. The plaintiffs themselves did not make any inspection of the outside of the house, being more concerned with the colour of the wallpaper, where to put the settee and so on.

When the mortgage application form reached the council, they arranged for Mr Lee, the second defendant, to value the property. He gave evidence before me and I accept him as a truthful and reliable witness. He told me that he must have seen Mr Young's particulars which set out the asking price of £9,450 and included the following:

The residence is constructed in a sound traditional Victorian manner incorporating 9 in solid brickwork which has recently been entirely pointed-up under a pitched slate roof that is sealed to give additional water proof and maintenance free security.

He went to the house on September 8 conscious of the fact that he was dealing with the poorer end of the property market. He told me that when he inspected he looked out of the bedroom window. He noticed the slope in the bedroom; he noticed a tie bar round the house and noticed that settlement had taken place. He then looked more closely and checked to see what movement had taken place. So far as he could see, cracks had been filled in with mortar some considerable time ago. He examined the mortar joints in great detail and he could see no further cracks that had taken place since the mortar had been applied. He looked for signs of continuing movement but did not find them. He was there for about 30 minutes. He thought about calling an engineer the moment he saw the tie bar, but decided against it. He said that if he had been asked to give a guarantee of no further movement he would have said: "I am certain there won't be any significant further movement in the next 25 years". He noticed bulges in the walls, but he put them down to past movement. The basis of his valuation was the tone of the valuation of other properties in the area. He said that if he had thought that the house would fall down he would have told the housing officer that the loan was not to be advanced. He said the council expected him to get the standard and condition of the property right and if he was not sure to say so, and that he did on occasions recommend further investigation by an engineer, but that this had to be paid for by the potential mortgagor. He admitted in cross-examination that he could not tell whether the cracks that he observed had appeared after the tie had been inserted or before that insertion. And he accepted that, if there had been any further movement after the insertion of the tie bar, that would be significant because it would show that the tie bar was not wholly effective. He saw that the walls were considerably out of plumb, but he did not anticipate differential settlement.

Having done his inspection, he went back to the office and had his report typed up on September 13. He valued the house at £9,450 — the asking price — and recommended a loan of 90% of the valuation for a maximum period of 25 years. Under the heading "Essential Repairs" he wrote:

Obtain report for District Council from MEB

that is the Midland Electricity Board

regarding electrics and carry out any recommendations. Make good mortar fillets to extension.

It is implicit in that report that structural repairs were not essential in his view. Although his handwritten note made on site referred to tie bars on the outrigger (what I would call the back addition), the typed-up version does not do so. Mr Lee knew that his report would not be forwarded to the intending mortgagors, but he also knew that the intending mortgagors would be told the amount of any loan and of any repairs the valuer regarded as essential and that the typical applicant was a first-time buyer of modest means.

By September 25 Mr Young wrote to the plaintiffs' solicitors that the vendors had agreed to a price reduction to £9,000. Two days later the council wrote to the plaintiffs that they were prepared to make an advance of £8,505 over a period of 25 years, subject to a number of conditions, one of which was that the plaintiffs sign an undertaking on completion "to carry out within twelve months therefrom the work detailed overleaf". Overleaf there was to be found the following:

ESSENTIAL REPAIRS. 1. Obtain report for District Council from Midland Electricity Board regarding electrics and carry out any recommendations. 2. Make good mortar fillets to extension.

It seems probable that the council had rung Mr Young to tell him the amount of the loan before September 25 and that it was this which caused the vendor to reduce his price.

Completion took place on November 23 1978 and the plaintiffs duly signed an undertaking to carry out the works which had been specified.

So far, so good. The problems which have resulted in this litigation arose when, three years later, the plaintiffs tried to sell. Before turning to those problems I ought to describe the house. There is nothing before me to suggest that anything has changed in the house itself between 1978 and 1986. The property comprises a small, two-storey, terraced house built in brick with a slate roof dating from about 1860, with a two-storey brick and slate extension at the rear. It is difficult to say when this extension was added, but it was probably during the period of 1890 to 1930 and, from a noticeable difference in the colour of the bricks above the first-floor joist level, it seems probable that the extension originally comprised a single-storey wash-house built across the rear of nos 73 and 74, with a chimney stack from the copper boilers later extended upwards to provide bedroom and/or bathroom accommodation above.

It is obvious from looking at the rear extension to the building that considerable settlement and movement has occurred and remedial action has been taken in the past. There are tie rods at the first-floor joist level around the three sides of the rear extension of nos 73 and 74, about 9 ft above ground level. A large metal plate with tie rods going back into the building above and to the left of the bathroom window about 1 ft below eaves level and further S-plates with tie rods about 2 ft below the external tie rods near to the corner of no 73.

There are extensive signs of settlement and movement in the brickwork on the side and rear elevations of the extension and in the entry passage between nos 73 and 74. It is apparent that the rear and side walls of the extension have bulges and that the rear walls of the house and extension are leaning over. Measurements taken reveal that the rear wall of the house is about 4 in out of plumb and that the rear wall of the extension is about 6 in out of plumb.

Internally there are signs of settlement in the brickwork to the cellar. The staircase is not level and slopes towards the back of the house. The door frames of the front bedroom and corridor to the bathroom are distorted and the doors have been cut down accordingly. The front bedroom door swings back against the wall on its hinges, showing that the frame is out of plumb, and the floors of the rear bedroom and bathroom slope down towards the rear of the house.

It appears that the original extension could have been a single-storey structure built on made ground owing to the slope of the hill. As the original rear wall was only 4½ in thick with support from the chimney breast it is possible that the person building the extension did not consider a substantial foundation to be necessary. It then seems probable that the rear extension was added to at first-floor level without having proper regard to the adequacy of the foundations.

I now move back to describe what has happened since 1978. In 1981 Park J heard a case called *Yianni* v *Edwin Evans & Sons* [1982] QB

438. He gave judgment in favour of the plaintiff purchaser against an independent surveyor, engaged by a building society, who had negligently overvalued a property, on the strength of which valuation the building society lent and the purchaser bought and borrowed. That decision alerted surveyors to a potential liability. The Royal Institution of Chartered Surveyors about this time issued a valuation standard VS2A, intended to cover original instructions for valuations carried out for concerns which normally use their own standard report forms, such as building societies, banks, insurance companies and local authorities. That standard stated:

It is commonplace for some prospective mortgagees to disclose the contents of the valuer's report to the applicant or to provide the applicant with a copy. If any disclosure is to be made, it is essential that the applicant receives a full copy of the valuer's report which should include a disclaimer of liability to third parties, except in those cases where the mortgagee has made separate and special arrangements to indemnify the valuer against such liability. Some prospective mortgagees provide forms containing a disclaimer clause, but if no such clause is provided (and there is no specific arrangement with the prospective mortgagee) appropriate clauses should be added as referred to in Section A of Annexe A.

and when one looks at that, it recommends as follows:

The report and valuation for mortgage purposes should contain the following: TO THE MORTGAGE APPLICANTS. IMPORTANT. This report has been obtained solely for the lending institution. IT IS NOT AND SHOULD NOT BE TAKEN AS A DETAILED REPORT ON THE CONDITION OF THE PROPERTY. THE VALUATION FIGURE GIVEN DOES NOT NECESSARILY REPRESENT THE VALUE OF THE PROPERTY TO YOU AS PURCHASER.

Then para 3 says:

This is a report to the lending institutions by its valuers and neither the lending institution nor the valuer gives any warranty, representation or assurance to you that statements, conclusions and opinions expressed or implied in this document are accurate or valid.

Then para 4 says:

The valuer has made this report without any acceptance of responsibility on his or their part to you.

About the same time as this valuation standard was being issued, a new district estates officer had become Mr Lee's superior and had instituted a new system whereby the valuer's report was made available to the intending mortgagor.

In February 1981 75 George Street was offered for sale, but failed to sell because of a problem with settlement at the rear wall and a rear extension similar to that at the rear of no 74.

In November 1981 the Harrises wanted to sell and move upmarket. They contacted Mr Young, the estate agent through whom they had bought, and he found some potential buyers at £14,000. Once more a mortgage with the council was envisaged. However, these two buyers backed off for a reason unknown to the parties.

In April 1982 two more buyers came on the scene and applied for a 20-year loan from the council. Mr Lee inspected once more. He told me, and I believe him, he could see no dramatic difference in the condition of the house he had inspected in 1978. But because of what had happened to no 75, which he thought might have an adverse knock-on effect on no 74, and because of a new style of reporting introduced by his new superior, his report included the following:

VALUATION: £13,750.
RETENTION: Whole amount pending (1) below.
REMARKS: Some settlement of the property has taken place and tie bars and straps are in place around the outrigger, furthermore the rear bedroom floor slopes to the rear. Some pointing of settlement cracks has taken place on both the rear elevation of outrigger and entry wall but it does not appear recent.

Then under the heading "RETENTION WORKS":

(1) Obtain a structural survey report on the stability of the property by an independent architect or structural engineer and carry out any recommended works.

The council informed the would-be purchasers that they were willing to make an advance of £13,050 but that the full amount would be withheld pending a structural survey report and the carrying out of any recommended works. It subsequently became clear that the amount of the cost of the recommended works was many thousands of pounds. The effect of this was that, in substance, the council was refusing to lend and that in consequence there would be no buyers. (I record at this stage that Mr Lee accepted that, had he in 1978 made a report in identical terms to the one he made in 1982, then the council would have acted in the way they did act in 1982 and a potential mortgagor would in all probability have refused to go through with a mortgage and purchase.)

After receipt of this news from the council the would-be purchasers backed off and the plaintiffs themselves instructed Mr C A Moulder, who is a chartered structural engineer and a member of the Institution of Structural Engineers and is also an Incorporated and Corporate Building and Civil Engineering Surveyor, being a Fellow of the Faculty of Architects and Surveyors, the Incorporated Association of Architects and Surveyors, and the Construction Surveyors' Institute. He visited the property in June 1982 and made a report, from which I shall quote at length later in this judgment. In substance, he said the building was unstable and required considerable underpinning etc.

Mrs Harris told me that when she saw Mr Moulder's report she cried for hours. She could not believe that the house was as bad as he said it was. She said that she had never really got over it since and that she does not want to live in that house for the rest of her life. Mr Harris was very annoyed, but, being more phlegmatic, decided to get on with life. In any event he went to his present solicitor, who sent the Moulder report to the council. It seems that the possibility of a grant for works of repair was discussed, and on August 2 1982 the council prepared "a specification of repair works required in conjunction with the proposed repair grant works", which included but went beyond Mr Moulder's suggestions. One firm approached was not prepared to tender, since they regarded the proposed works as impractical and unsafe. Another quoted £13,048 to carry out those works. In the event, a repair grant was applied for, but the council had no surplus funds. The house had been taken off the market upon receipt of Mr Moulder's report in July 1982 and has never been on the market since then because Mr Young advised that this action should be concluded first. The plaintiffs are still there.

The council's insurers' reaction has throughout been to rely on the fact that there has been no recent movement and on the exclusion of liability contained immediately above the plaintiffs' signatures on the mortgage application form. On September 16 1983 the present proceedings were launched by the issue of a writ. In December 1983 Mr Moulder once more received instructions, and in March 1984 he produced a schedule of work and specification for remedial works. The cost of doing those works was about £13,000.

In October 1984 the defence was served. Since about that time, local authorities by reason of financial guidelines from government have ceased to lend on mortgage.

Statutory background

I now look at the statutory background. The relevant statutes have been repealed and replaced by the Housing Act 1985. Although the current Act is in many respects similar to the legislation which it replaced, I shall refer only to the legislation current at the relevant time. The power to lend was contained in the Housing (Financial Provisions) Act 1958. This provided, in section 43(1):

A local authority . . . may . . . advance money, subject to the provisions hereinafter contained, to any persons for the purpose of . . . acquiring houses . . . (2) Before advancing money under this section . . . the local authority . . . shall satisfy themselves that the house . . . to be acquired is . . . or will be made, in all respects fit for human habitation. . . .

The expression "fit for human habitation" is a term of art. By virtue of section 58 of the 1958 Act, section 4 of the Housing Act 1957 is to apply as if references therein to that Act included references to the 1958 Act. Section 4 provides:

In determining for any of the purposes of this Act whether a house is unfit for human habitation, regard shall be had to its condition in respect of the following matters, that is to say — (a) repair; (b) stability; (c) freedom from damp; (d) natural lighting; (e) ventilation; (f) water supply; (g) drainage and sanitary conveniences; (h) facilities for storage, preparation and cooking of food and for the disposal of waste water; and the house shall be deemed to be unfit for human habitation if and only if it is so far defective in one or more of the said matters that it is not reasonably suitable for occupation in that condition.

Returning to section 43 of the 1958 Act, one sees that subsection (3) provided as originally enacted:

The following provisions shall have effect with respect to an advance under this section: — (a) the advance, together with interest thereon, shall be secured by a mortgage of lands the subject of the carrying out of the purpose for which the advance is made; (b) the amount of the principal of the advance shall not exceed . . . ninety per cent of the value of the mortgaged security . . . (e) the advance shall not be made except after a valuation duly made on behalf of the local authority or county council.

Subsection (4) as originally enacted provided:

An advance under this section shall not be made if the estimated value of the fee simple in possession free from incumbrances of the house in respect of which assistance is to be given exceeds five thousand pounds. . . .

It is in my judgment clear from the foregoing that the authority were not empowered to make advances unless and until: 1 They had satisfied themselves that the house in question was or would be made in all respects fit for human habitation. This will normally involve the making of an inspection by someone qualified to form a view as to fitness or as to the need of specialist advice. 2 A valuation had been duly made on behalf of the local authority and that valuation had revealed: (a) that the estimated value of the fee simple did not exceed £5,000, and (b) that the amount of the principal of the advance did not exceed 90% of the value of the mortgage security. I note that by virtue of the House Purchase and Housing Act 1959 the reference to 90% in subsection (3)(b) and the reference to £5,000 in subsection (4) of the 1958 Act were repealed. The consequence of that is that in 1978 the authority could as a matter of statutory *vires* lend even on more expensive houses and could lend up to 100% of the value of the mortgaged security. However, I do not consider that these amendments affected the meaning of the word "value" in the 1958 Act.

The point is of some significance and of general importance, because Mr Worsley, on behalf of the authority, argued that because of the heavy housing responsibilities of local authorities a house might be of greater value to them than it was to the outside world and that the value referred to is the value to the housing authority. It was, he submitted, legitimate for a valuer valuing for the purpose of the 1958 Act to bear in mind that, if the proposed purchaser did not have access to mortgage funds from a council he and his family might well apply to the council for a place in accommodation provided by the council. He drew my attention in this context to appendix 1 of Circular 42/54, issued by the Ministry of Housing and Local Government in connection with one of the forerunners of the 1958 Act. This provides in para 1:

As a result of increases in building costs since the war the amount of the deposit payable by the intending purchaser is often more than he is able or willing to find. On the other hand, if his need of accommodation has to be met by the Local Authority, an addition to the heavy and increasing burden of subsidy on the exchequer and on the local rates is inevitable. It is therefore in the public interest that everything possible without undue risk of loss to Public Funds should be done to help everyone wishing to provide himself with a home.

He also drew my attention to the evidence before me, which I accept established that, as at 1978, for this type of house, there were in practice only two types of buyers — speculative builders and private buyers very heavily reliant on local authority mortgages. At that time other lending institutions were simply not interested in lending moneys secured on this type of Victorian terraced property.

In my judgment it is erroneous to think of there being several values attributable to a house — a value to a millionaire with a sentimental attachment, a value to a building society, a value to a local authority, and so on. The proper approach is to say that the house has a value which is arrived at by looking at the market of potential buyers of that house, which market in turn will be influenced by the availability of mortgage facilities. Now it may be that because of their heavy housing responsibilities local authorities may be generous with their mortgages and this may in turn affect value, but that is not the same thing as saying that a local authority valuer is, in his valuation, allowed to take into account the length of the housing list, the possible application of the Housing (Homeless Persons) Act or its successor, and so on. He is not. In principle a valuer's valuation should be the same whomever he is acting for. In the context of the 1958 Act as originally enacted, when a valuer was asked "what is the value of the mortgage security?", he would arrive at that value by the same process as he adopted when asking himself "does the value of the fee simple in possession free from incumbrances of the house in respect of which assistance is to be given exceed £5,000?". It is right to record that Mr Lee himself claimed to have done exactly that which in my view he ought to have done, namely to have valued on the tone of the valuation of other properties.

Was no 74 fit for human habitation?

The only respect in which it is alleged that the house was not fit for human habitation is lack of stability. Mr Worsley rightly pointed out that there was no express requirement in the 1958 Act that the authority should be satisfied that the house would *remain* fit, as opposed to being fit at the time of the loan. However, the concept of fitness for human habitation takes into account stability, which in the context of this case one can define with the *Oxford English Dictionary* as "freedom from liability to fall or be overthrown." In my judgment the mere fact that there is *a* risk of further movement at some time in the future does not render a house so far defective in stability that it is not reasonably suitable for occupation in that condition. I am conscious of the fact that innumerable consequences are attached by statute to a finding that a house is unfit, many of which are not necessarily welcomed by an owner-occupier of that house. I am also conscious of Circular 69/67 (which was not cited to me and which is of course not legally definitive), which includes among other matters a note on section 4, suggested by the report of the Standards of Fitness Sub-Committee of the Central Housing Advisory Committee, which states, "evidence of instability is only significant if it indicates the probability of further movement which would constitute a threat to occupants of the house".

In my judgment, however, if there is a significant risk of damage to a house by reason of further movement occurring at any time during the next few years which damage would render the house not reasonably suitable for occupation in that condition, then the house is not fit for human habitation during the time that that risk is extant.

The evidence which I have relevant to this point is essentially that of Mr Moulder. His report, made in 1982, is not challenged as such. It ends as follows:

In our opinion the situation regarding this house has been caused by a classic example of differential settlement coupled with bad design in several aspects of its construction. To begin with the front part of the house is well founded and stable because it is on a deep cellar foundation perhaps down to rock level. On the other hand the back part of the original house built in the nineteenth century was taken down on foundations which are in loose soft ground without any spread or consolidation. We could also not discount the action of water in the subsoil due to leaking drains at some time. Thus the structure has settled towards the back only and rotated to cause the bulging.

The later construction of the wing at the back comprising bathroom and kitchen perhaps around the earlier part of the century was built in an even worse fashion with foundations only marginally better and upon even more suspect filled ground. This later structure was not even properly bonded to the original and was at some stage weakened by the removal of the chimney breast in the kitchen and this could have contributed to the rotation of the walls resulting in such serious bulging. Perhaps around this time the tie bars and plates were put in to try to contain the situation and prevent further movement. The settlement has probably continued over many years. There is often a substantial initial settlement due to differential movement but after a while it stabilises to a large degree and then there is only a tiny "creep" which often goes on over a long period of time. In the last few years this movement has probably been imperceptible but we could not guarantee that it would not continue and we think that if those tie rods were taken out there would be serious danger of collapse. Even with the tie rods as they are there is a strain on the original structure which is reflected throughout the house. This house has been weakened in other directions by the removal of the chimney breast and the inadequate roof and foundation design and certain as yet unknown factors relative to upper tie rod positions and the bathroom wall over the entry. However, the main problem is the stability of the rear wall which we consider must be dealt with. It is an unstable building which could one day move again and maybe even collapse, particularly now that there is increased vibration from heavy traffic movement on the nearby ring road.

In his statement prepared for trial in May 1986, which was taken as having been given in evidence before me, he said in relation to the ties:

I have not previously come across this particular way of tying up a structure. It must be regarded as extremely unusual. Unfortunately the girdle is anchored only onto the rear wall of the main part of the house rather than to the front part of the house which is quite well founded onto the cellar walls going down to a considerable depth. Normal tie rods anchor a wall face well back to a secure point and an "S" shaped or flat plate grips the brickwork. By way of contrast this girdle of bars only holds the corners of the projecting rear part of the house and the stability rests on the end anchorage. Thus this arrangement depends on anchorage for effective restraint and the system does not work unless it is fixed back properly and this is not the case as the rear wall itself is not well restrained and leans out 3 inches in places.

The statement ends:

In my opinion the house is potentially dangerous and could disintegrate if, for example, the girder tie bars were taken out during works to this or the adjoining property. A serious knock on the tie bars would dislodge them and could cause collapse. Careless erection of scaffolding or removal of walls or putting in extra windows to this or the adjoining property is a danger. In my view because of the structural instability it is not fit for human habitation unless modified in line with my specification and drawings. It is an old property, the original part carried out in lime mortar and these sort of mortar joints are plastic in essence and easily suddenly give way if overstrained.

In cross-examination, he asserted there was "a fairly large risk of a fairly substantial disaster".

I am conscious of the fact that the house has not moved perceptibly during the last eight years and that this is a relevant fact in considering whether Mr Moulder's evidence is reliable. It is, however, clearly not conclusive. My position is made more difficult by the fact that I did not find Mr Moulder a very impressive witness. However, he *is* qualified and experienced, and I am satisfied that he genuinely held the opinions he expressed. No engineering evidence was called on the other side, and on balance I am satisfied that there was in 1978, in Mr Moulder's words, "a fairly large risk of a fairly substantial disaster". In those circumstances I find that the house was unfit. I record that I am not satisfied that there was a *probability* in 1978 that during the next few years there would be further movement which would constitute a threat to the occupants of the house.

What was the value of the house in 1978?

I find, in so far as it is not agreed, that:
1 A speculative builder would have paid £3,500.
2 The only other potential buyers would have required a substantial mortgage.
3 The only available source of mortgage for this type of property was the local authority.
4 In mortgageable condition the value of the house was £9,500.
5 If the authority had made the mortgage subject to a condition that an engineer's report be obtained and the works recommended in it be done, then a potential occupier would either have backed off straightaway or would have done so once he had seen the engineer's report and learned the anticipated cost of works.

I find that an engineer's report would probably have been broadly the same as Mr Moulder's and that the works would have cost many thousands of pounds. In those circumstances I find the value of the house to have been £3,500.

What was Mr Lee's duty to the authority in 1978 and was he in breach of it?

Mr Lee had, in my judgment, a duty to inspect in order to advise the authority both on the matter of fitness and on valuation. It is clear that the value of this house depended in part on its fitness. I accept Mr Lee as honest and I accept he considered the question of structural stability and genuinely came to the conclusion that there would not be any significant further movement during the next 25 years. Notwithstanding my finding as to risk, I accept that it may be that he was right in that conclusion.

However, he was not a building surveyor, still less an engineer, and I am satisfied by the evidence of Mr Kenchington — Fellow of the Royal Institution of Chartered Surveyors and a partner in the leading Midland firm of Colliers, Bigwood & Bewlay — that a surveyor, seeing what Mr Lee saw, should have advised his authority to obtain a structural survey. In my opinion, he did not have, and should have known that he did not have, sufficient expertise to form a firm view as to fitness or as to value. I interpose to say he was not a building surveyor but a valuation surveyor. He should have done in 1978 what he did in fact do in 1982 when advising on the application by Mr Small and Miss Derbyshire for a loan on no 74, namely, indicate that they should obtain a structural survey report on the stability of the property by a structural engineer and carry out any recommended work.

I find that he was negligent both in advising the authority that the house was fit and in advising it that its value was £9,450. However, this is not an action by the council against Mr Lee. Indeed, the Harrises having so far not defaulted on their mortgage obligations, the council has so far not suffered any cash loss from Mr Lee's negligence. I now go on to consider whether either or both of the defendants are in breach of any duty owed to the plaintiffs.

Are either or both of the defendants in breach of any duty owed to the plaintiffs?

In my judgment, Mr Lee was negligent in his inspection and report, and the council was negligent in offering a mortgage to an applicant when it was not empowered so to do under the relevant legislation and would have known this fact if its surveyor had done his job properly. Further, I consider that a council offering a mortgage to an applicant represents that it is empowered to do so under the relevant legislation. That representation in the circumstances of the present case was made negligently. There was no evidence before me to suggest that the Harrises were familiar with the statutory provisions under which the council was empowered to lend. Their position was that they were attracted by the prospect of a 95% mortgage, that if the council was willing to lend them the money the house must be worth something of that order, and that they thought that the essential repairs specified in the council's mortgage offer were the only essential repairs or that in any event there were no substantially worse defects. They assumed what as a matter of law should have been true, namely that the council would not offer to lend a sum in excess of the value of the house. I am satisfied that the plaintiffs' purchase of this house was the foreseeable consequence of the council's offer of a mortgage and that the council's offer of a mortgage and the plaintiffs' purchase of the house were both foreseeable consequences of Mr Lee's report to the council. This finding of foreseeability, although a prerequisite of, is not conclusive of the liability of the defendants. There are many cases where the courts have held that, although the damage was foreseeable, no liability attaches to the defendant notwithstanding recent developments in the law of negligence — see the judgments in *Leigh and Sillavan Ltd* v *Aliakmon Shipping Co Ltd* [1985] QB 350 in the Court of Appeal and [1986] 2 WLR 902 in the House of Lords, in which judgments there is a lengthy appraisal of those developments.

It is clear that those developments are not to be taken as having upset by a sidewind established rules of law. There is, however, no established rule of law which prevents recovery in the circumstances of the present case. I was not referred to any English case where liability attached to a statutory body on the basis that it had negligently acted *ultra vires* or had negligently impliedly represented that it was acting *intra vires*.

However, I bear in mind that

the position has now been reached that in order to establish that a duty of care arises in a particular situation, it is not necessary to bring the facts of that situation within those of previous situations in which a duty of care has been held to exist

per Lord Wilberforce in *Anns* v *Merton London Borough Council* [1978] AC 728 at p 751.

In *Peabody Donation Fund (Governors of)* v *Sir Lindsay Parkinson & Co Ltd* [1985] AC 210, Lord Keith put the matter thus, at p 240:

The true question in each case is whether the particular defendant owed to the particular plaintiff a duty of care having the scope which is contended for, and whether he was in breach of that duty with consequent loss to the plaintiff. A relationship of proximity in Lord Atkin's sense must exist before any duty of care can arise, but the scope of the duty must depend on all the circumstances of the case ... So in determining whether or not a duty of care of particular scope was incumbent upon a defendant it is material to take into consideration whether it is just and reasonable that it should be so.

In that case the plaintiffs were themselves in breach of their statutory duty not to build save in accordance with approved plans, and their lordships held that it would be neither reasonable nor just to allow such plaintiffs to recover against a local authority whose inspector had negligently permitted a departure from the approved plans. However, although it is submitted in the present case that a more cautious family than the Harrises would have instructed their own surveyor and relied upon him or her, it cannot be, and is not, submitted that the plaintiffs were under a statutory duty so to do.

The defendants relied upon the exemption clause, which I have set out towards the beginning of this judgment, to negative their liability. I accept that the plaintiffs did not read the disclaimer and I accept that it is common, even, I suspect, among lawyers, for people not to read documents which they sign. However, in circumstances such as the present, if the Harrises choose to sign immediately below a perfectly visible disclaimer without reading it then they run the risk that the courts will treat them as if they had read it. I shall so treat them.

However, I do not read the exemption clause as relieving the surveyor from any liability arising from a negligent valuation of the property and from negligently assuring the council that the property was fit for human habitation with the foreseeable result that the council makes a loan which it is not empowered to make by statute. Nor do I read it as exempting the council from any liability arising from a decision to lend on an unfit house more than the market value of the house or from the representations contained in the mortgage offer. In those circumstances I propose to ignore the exemption clause when considering whether or no liability attaches to either defendant.

When considering the potential liability of Mr Lee, it is conceded by Mr Worsley that, if the defects were such that one could not say that the house was fit for human habitation, then Mr Lee was under a duty to the council to draw that to the council's attention. It is also conceded by Mr Worsley that, if Mr Lee had been a building society surveyor, he would have been in breach of his duty to that society in ascribing a value to the house without benefit of a fuller survey. I have already held that Mr Lee was negligent in the performance of his duties on this occasion. His position in carrying out a valuation is the same as that of a valuer acting for a building society who ascribes too high a value to a property. Such a valuer has been held to be liable to the mortgagor in the *Yianni* case and I see nothing on the grounds of policy or in the subsequent case law which should prevent me from following that decision. Mr Worsley did not seek to argue that if Mr Lee was liable none the less the council was not vicariously liable.

But *Yianni* trod new ground and may be overruled, and so I go on to consider the primary liability of the council. Apart from his submissions going to lack of negligence, the stability of the property and the disclaimer, the main submission made by Mr Worsley in favour of negating the council's liability was that the provisions of section 43 of the 1958 Act were inserted for the protection of ratepayers and not for the protection of potential purchasers. He referred me to *Curran* v *Northern Ireland Housing Executive* (unreported), a recent decision of the Court of Appeal in Northern Ireland, in which a preliminary point was taken on pleadings which seem to have been in a rather muddled form. It was a case in which the plaintiff alleged negligence against the Executive in two respects: (i) the making of a mortgage advance which he used in order to purchase his house; (ii) the granting of an improvement grant to a predecessor in title of the plaintiff for the purpose of improving the house which the plaintiff ultimately purchased*. For the purposes of the case based on the mortgage advance, the facts assumed against the Executive were that its officer was negligent in his inspection of the house and that the house suffered from substantial defects which should have become apparent on an inspection procured by the Executive. The mortgage was made under the provisions of the Housing Executive Act (Northern Ireland) 1971, which, so far as appears from the report and so far as my own researches go, did not include provisions similar to section 43(2) and (3) of the Housing (Financial Provisions) Act 1958, which, it will be recalled, provide that the amount of any advance is not to exceed the value of the house and that before making a loan the authority shall satisfy themselves that the house is fit for human habitation.

After lengthy citation from *Hedley Byrne & Co Ltd* v *Heller & Partners Ltd* [1964] AC 465, Gibson LJ, in a judgment with which the remaining members of the court concurred, said:

It will be seen that the criteria indicated by their lordships by which one must judge whether a duty of care exists is not fulfilled in the present case. The valuation was obtained exclusively for the use and benefit of the Executive. The plaintiffs never saw or were made aware of the contents of the valuation of the property procured by the Executive. Nor was it ever intended or contemplated by anyone that they should do so. All that they knew was that the Executive did not advance money on mortgage without the support of a valuation to justify it. The plaintiffs assumed, and were probably entitled to assume that the valuation was at least as much as the figure which the Executive was willing to advance on mortgage, and one must assume that the Executive knew that the plaintiffs would probably rely on that figure to the extent that they would without further evidence treat it as warranting the price at which they agreed to buy the house. One must further assume that the Executive was lacking in proper care in selecting its valuer and that the valuer was negligent in failing to note or take into account defects in the premises which a competent valuer would have done; and one must further assume that because of this the valuation was excessive and the sum advanced on mortgage greater than the value of the house, and that in consequence the plaintiffs suffered loss. But whichever of the criteria propounded by their Lordships in the *Hedley Byrne* case is adopted, it will be seen that the facts of the present case do not meet the requirements. The Executive made no representation to the plaintiffs on the matter, and there was no conduct by it which could be regarded as bringing it virtually into the position of a party contracting with the plaintiffs, for there was no agreement or anything approaching an agreement that the Executive would exercise care in choosing its valuer. Nor were there any facts from which one could probably infer an implied undertaking on the part of the Executive to assume responsibility for any loss resulting from reliance by the plaintiffs on the propriety of the valuation underlying the amount of the mortgage.

It is true that the Executive could when it opened discussions with the plaintiffs have warned them that it would accept no responsibility for any loss consequent upon the plaintiffs' reliance on the proposed mortgage sum offered; but failure to do so cannot, in my opinion, create an implied undertaking to accept responsibility for any damage suffered by the plaintiffs in the event of their assumption that the offer is based upon a valuation competently given for the Executive's sole use. One person cannot impose upon another by whose words he may be affected a duty to exercise care towards him merely because he informs the other or lets it be known that he is aware of the other's course of business and that he intends to act as though the other does owe him that duty and that he will hold the other responsible if as a result of any lack of care in making the statement he should suffer loss. As I understand it, either there must be an express agreement to that effect or the special nature of the relationship must by implication give rise to the duty. That is to say, to adopt Lord Devlin's summary in the *Hedley Byrne* case at p 529, there must be an assumption of responsibility in circumstances in which, but for the absence of consideration, there would be a contract. Responsibility can only attach if the defendant's act implied a voluntary undertaking to assume responsibility. Were it otherwise a person who offered to an expert any object for sale, making it clear that he was unaware of its value and that he was relying on the other to pay a proper price, could sue the other should he later discover that he had not received the full value even though the purchaser had made no representation that he was doing any more than look after his own interests. Nor can any class of persons who to the knowledge of another habitually fail to take precautions for their own protection in a business relationship cast upon another without his consent an obligation to exercise care for their protection in such a transaction so as to protect them from their own lack of ordinary business prudence. Generally a mortgage contract in itself imports no obligation on the part of a mortgagee to use care in protecting the interests of a mortgagor.

It will be seen that there was not in that case an argument such as there is in the present case to the effect that the lender in making the loan is representing that the value of the house is not less than the loan and that it is satisfied that it is fit for human habitation. In those circumstances, the claim against the Executive based on the making of the loan was dismissed.

It was otherwise so far as the case was based on the making of the improvement grant. As to that, the learned lord justice said:

Article 47(1) of the Housing (Northern Ireland) Order 1976 provides that the Executive shall pay an improvement grant if an application for the grant has been approved by it and the conditions for payment of the grant are fulfilled. But by paragraph (2) the Executive is forbidden to approve an application unless it is satisfied that, on completion of the works, the house will attain the required standard. That is to say, the Executive must be satisfied that the house will, *inter alia*, be in good repair, conform to the Department's standards of construction, and be likely to provide satisfactory housing accommodation for 15 years. All these conditions must be fulfilled to the satisfaction of the Executive at the time when it approves the application; that is to say, before any work is done. . . . But article 60(5) of the 1976 Order makes the payment of a grant by the Executive conditional upon the works being executed to its satisfaction. It is only payable if the Executive is satisfied as to the standard of the work both when it approves the application and after the completion of the work.

A little later on in the judgment he says:

I do not understand that the Executive has any power to control the work of construction which is carried out by the building owner. All it can do is to withhold payment of the grant if the work is not satisfactorily executed. Has it then a duty to inspect or supervise the work as it progresses so as to be in a position properly to judge whether the work does conform to the requisite standard so that it may with proper knowledge decide whether to pay the grant? The Order does not in terms impose such a duty. But the Executive does have a duty to be satisfied that the work is up to standard. It is left to the Executive as to what steps it should take in order to be so satisfied. Provided that it had adopted a procedure which was directed towards indicating whether the work was satisfactory there could be no challenge to the sufficiency of that course because that would have been a policy decision, and this would be so even though in the present case it failed to bring to light the fact that the foundations were defective. But it is perfectly possible — and indeed not unlikely — that the Executive did in fact adopt a procedure of supervision or inspection or both. If that were so, then the fact that the work was not done to the requisite standard would be regarded as indicative of negligence by the appropriate official of the Executive which would have been a failure within the operative sphere. The Statement of Claim does not, therefore, preclude the possibility of negligence by the Executive. But the further question which must be answered in favour of the plaintiffs is whether a duty of care is owed by the Executive to the plaintiffs. Carswell J was of the opinion that it is not one of the functions of the Executive to protect against loss the successors in title of the recipients of an improvement grant when it is exercising its statutory duty of paying the grant. So one has to ask, what is the legislative purpose in directing the payment of such grants; and is there a duty of care imposed on the Executive in making the payments; and, if so, to whom is that duty owed?

There is then a consideration in the judgment of *Anns* case and the *Peabody* case, and the learned judge says later:

---

*Editor's note: The decision in respect of (ii) was reversed by the House of Lords on April 8 1987 on an appeal on this point *sub nom Curran* v *Northern Ireland Co-ownership Housing Asssociation Ltd* [1987] 2 WLR 1043.

Assuming, therefore, that the Executive did exercise the function of supervising or inspecting the work in progress prior to paying the grant as a means of satisfying itself that the work was up to the requisite standard, did its negligence in so doing amount to a breach of a duty of care to the then owner of the house? The purpose which Parliament had in mind in directing the payments to be made was to help the owner occupier of a dwelling house not only financially but also by ensuring in some measure that his occupation would be more beneficial to him and his family. If by reason of any neglect by the Executive that object was frustrated or he otherwise benefited less than was intended, I consider that a duty of care owed to him was breached and that he would have been entitled to recover damages . . . .

A similar problem arose in *Ward* v *McMaster, Louth County Council and Nicholas Hardy & Co Ltd,* a decision of Costello J sitting in the High Court of the Republic of Ireland. It is reported at [1986] ILRM 43. The headnote so far as presently relevant reads:

The plaintiff, the purchaser of a seriously defective bungalow . . . sought to recover damages against its amateur builder, the housing authority which had advanced money to finance the purchase, and a firm of auctioneers employed by the housing authority to inspect the property prior to purchase.

The claim against the housing authority was that the valuation which it commissioned, pursuant to regulation 12(b) of the Housing Authorities (Loans for Acquisition or Construction of Houses) Regulations 1972, should have revealed the defects. It was held by Costello J that the principle in *Donoghue* v *Stevenson* [1932] AC 562 applies when statutory functions are being performed, provided there is a relationship of proximity between the parties (such to be determined in the light of all the circumstances, and especially of the relevant statutory provisions) to the extent that the court finds it just and reasonable.

The learned judge after reviewing the authorities said this:

Whilst not attempting in any way to summarise all the conclusions which are to be derived from the authorities which I have just quoted, it seems to me that for the purposes of this case I can apply the following principles: (a) When deciding whether a local authority exercising statutory functions is under a common law duty of care, the court must firstly ascertain whether a relationship of proximity existed between the parties such that, in the reasonable contemplation of the authority, carelessness on their part might cause loss. But all the circumstances of the case must in addition be considered, including the statutory provisions under which the authority is acting. Of particular significance in this connection is the purpose for which the statutory powers were conferred and whether or not the plaintiff is in the class of persons which the statute was designed to assist. (b) It is material in all cases for the court in reaching its decision on the existence and scope of the alleged duty to consider whether it is just and reasonable that a common law duty of care as alleged should in all the circumstances exist.

The factual relationship in this case should be ascertained first by considering the statutory framework in which the local authority was operating. Section 39 of the Housing Act 1966 provides that a housing authority may, subject to regulations made by the Minister for the Environment, lend money to a person for the purpose of acquiring or constructing a house. Regulations made from time to time imposed a limit on the amount of each loan and at the time of the plaintiff's application this stood at £12,000. The amount of the loan was based on the market value of the house and paragraph 12(b) of the . . . Regulations . . . expressly imposed an obligation on every housing authority to satisfy itself that the value of the ownership of the house was sufficient to provide adequate security for the loan.

A little later on the learned judge said:

Although the plaintiff did not expressly inform any member of the staff of the council that he was relying on their valuation, and although the council carried it out for their own purposes and to comply with obligations imposed on them by the statutory regulations to which I have referred, I am satisfied that they ought to have been aware that it was probable that the plaintiff would not have gone to the expense of having the house examined by a professionally qualified person and that he would have relied on the inspection which their Scheme indicated would be carried out.

At p 62 he said:

But before concluding that a common law duty of care existed it is necessary to consider all the circumstances of the case and, in particular, the statutory framework in which the relationship between the parties existed. The scheme which the council had adopted for giving effect to the powers conferred on it by section 39 of the 1966 Act was one designed to help persons of limited means to buy their own houses, and in order to permit this to be done a valuation of the house had to be carried out. As the purpose for which the statutory powers were being exercised was to help persons like the plaintiff it seems to me to be consistent with the council's public law powers that they should be accompanied by a private law duty of care in his favour.

For similar reasons it seems to me just and reasonable that the court should hold that a duty of care arose in this case. The plaintiff was relying on the council's valuation and they should have been aware that he was doing so. In such circumstances it would not be just to hold that no duty of care was imposed on the council and it seems to me to be perfectly reasonable that it should be.

As to the scope of the duty of care, the test is again one of foreseeability and reasonableness. The council had a duty to see that the valuation was carried out with reasonable care, and that implied not only that the person who carried out the inspection would not act carelessly, but that the council would ensure that the person carrying out the valuation would be competent to discover reasonably ascertainable defects which would materially affect its market value.

The judgment in *Curran's* case in so far as it related to the improvement grant and the judgment in *Ward* v *McMaster* indicates a similar approach to the serious policy question which the present case raises to the one I have formed.

I see no policy reason why a potential mortgagor should not recover in the circumstances of the present case. It seems just and reasonable to me that an individual who has been foreseeably exposed to a disastrous situation by reason of the actions of the local authority, which actions would not have taken place had the authority and its staff acted non-negligently, should be able to recover damages from the authority.

Those damages are agreed in the instant case as £12,000 to include interest. That agreement is subject to any finding I might make to the effect that the plaintiffs have contributed to their own misfortune by failing to instruct their own surveyor, as indeed they were advised to do on the mortgage application form. I reject the submission that I should make any deduction from the award for contributory negligence. In my judgment it would not be right to hold, in effect, that the plaintiffs should have known that local authorities might lend on unfit properties amounts substantially in excess of the value of those properties. No doubt that occasionally happens, just as occasionally doctors, solicitors, engineers and others are negligent. But knowledge of the foregoing should not, in my judgment, give rise to a deduction for contributory negligence for a plaintiff who failed to get a second opinion.

I therefore give judgment for the plaintiffs in the sum of £12,000. Subject to any submissions by counsel, I am minded to order that the first defendant pay the plaintiffs their costs.

Judgment was given in favour of the plaintiffs for £12,000 with costs against the first defendants and legal aid taxation of the plaintiffs' costs. A stay of execution was granted for 42 days if a notice of appeal was not filed; if such a notice was filed, a stay until hearing or withdrawal of appeal.

House of Lords
January 22 1987
(Before Lord KEITH OF KINKEL, Lord BRANDON OF OAKBROOK, Lord TEMPLEMAN, Lord GRIFFITHS and Lord GOFF OF CHIEVELEY)

KETTEMAN AND OTHERS v HANSEL PROPERTIES LTD AND OTHERS

Negligence — Houses with defective foundations — Claims by houseowners against builders, local authority and architects — Liability — Limitation Act — Unsatisfactory result of prolonged litigation — House of Lords divided on procedural point, with serious consequences for architects — In the end House of Lords had to deal only with appeal by architects without hearing any argument from other parties — Builders had not challenged in Court of Appeal the judgment against them by Judge Hayman at the trial and the local authority did not pursue in the House their challenge to the Court of Appeal's decision against them — Architects' appeal was dismissed by the majority in the House (Lords Templeman, Griffiths and Goff of Chieveley), the minority who would have allowed the appeal being Lords Keith of Kinkel and Brandon of Oakbrook — The majority decided that the House should not interfere with the discretion of the Court of Appeal in disallowing an amendment of the defence, submitted very late at the trial, which would, *inter alia*, have allowed the architects to plead that their liability was statute-barred — If the architects had been allowed to rely on the plea

of limitation they would have succeeded — There were certain matters on which the House as a whole was agreed — The architects were validly joined as defendants when they lodged their defence, thereby waiving the necessity for the service on them of the amended writ (a step which had been overlooked) — The decision in *Seabridge* v *H Cox & Sons (Plant Hire) Ltd* was overruled — The doctrine of "relation back", questioned in *Liff* v *Peasley*, was finally discredited — The "doomed from the start" dictum of Lord Fraser in *Pirelli* v *Oscar Faber & Partners* was still further explained away — Appeal dismissed

The following cases are referred to in this report.

Anns v *Merton London Borough Council* [1978] AC 728; [1977] 2 WLR 1024; [1977] 2 All ER 492; [1977] EGD 604; (1977) 243 EG 523 & 591, HL
Byron v *Cooper* (1844) 11 Cl&Fin 556
Clarapede & Co v *Commercial Union Association* (1883) 32 WR 262
Junior Books Ltd v *Veitchi Co Ltd* [1983] 1 AC 520; [1982] 3 WLR 477; [1982] 3 All ER 201, HL
Liff v *Peasley* [1980] 1 WLR 781; [1980] 1 All ER 623, CA
Pirelli General Cable Works Ltd v *Oscar Faber & Partners* [1983] 2 AC 1; [1983] 2 WLR 6; [1983] 1 All ER 65; [1983] EGD 889; (1982) 265 EG 979, HL
Seabridge v *H Cox & Sons (Plant Hire) Ltd* [1968] 2 QB 46; [1968] 2 WLR 629; [1968] 1 All ER 570, CA
Sparham-Souter v *Town & Country Developments (Essex) Ltd* [1976] QB 858; [1976] 2 WLR 493; [1976] 2 All ER 65, CA

This was an appeal by the architects, Jamieson Green Associates, against a decision of the Court of Appeal (reported at (1984) 271 EG 1099) against them. For reasons explained in the speech of Lord Keith the houseowners and the local authority were represented before the House on the question of costs only. The result was that on the merits of the appeal the House heard only the arguments for the appellants.

Michael Ogden QC and Mark Smith (instructed by Hewitt Woolacott & Chown) appeared on behalf of the appellants; Christopher Symons (instructed by Herbert Smith & Co) represented the plaintiff houseowners on the question of costs only; A E M Cooper (instructed by Barlow Lyde & Gilbert) represented the Mid-Sussex District Council on the question of costs only.

In his speech, LORD KEITH OF KINKEL (who would have allowed the appeal) said: The proceedings giving rise to this appeal have been characterised by an unusual number of unfortunate vicissitudes, not the least unsatisfactory of which has been its arrival in this House with the appearance of only one of the two appellants to whom the Court of Appeal granted leave to bring it here and with respondents who had no interest in presenting a contradiction of the appellants' argument, and consequently did not do so.

On May 27 1980 the respondents (whom I shall call "the houseowners") commenced proceedings against Hansel Properties Ltd ("Hansel"), the builders of five houses in Burgess Hill, West Sussex, which the houseowners had respectively purchased from them in 1975. In the summer of 1976 all five houses began to show signs of structural damage in the form of internal and external cracks. The houseowners claimed damages from Hansel on grounds of breach of contract and in tort. The claim was based on the inadequacy of the foundations of the houses. On April 28 1981 Hansel issued a third party notice to Mid-Sussex District Council ("Mid-Sussex"), claiming indemnity or contribution upon allegations of negligence by the latter in approving the plans for the foundations and in inspecting them during the course of construction. Mid-Sussex denied the allegations against them and on August 18 1981 issued a fourth party notice to Jamieson Green Associates ("the architects") claiming indemnity or contribution from them on the ground of negligence in the design of the foundations. The architects lodged a denial. The houseowners at first took no steps to have Mid-Sussex and the architects joined as parties to the action, because their solicitors believed that Hansel were sound financially. Later they began to have doubts about this (which have since proved only too well-founded), and in June 1982 they issued a summons for leave to join Mid-Sussex and the architects as defendants to the action. The summons was heard by Judge Sir William Stabb QC on June 25 1982, and on that day, no opposition being offered by counsel for Mid-Sussex and the architects, he made an order in these terms:

It is ordered that: (1) The first third party and the fourth party be joined as defendants to the action. (2) A statement of claim to be served on the first third party and the fourth party within 21 days. (3) Defences to be served 14 days thereafter . . . (7) The trial date for July 12 to be vacated and that the date for trial to be fixed for November 22 1982 with an estimated length of ten days.

What happened next was thus described by Lawton LJ in the course of his judgment in the Court of Appeal [1984] 1 WLR 1274 at p1282:

This order did not state explicitly that the specially endorsed writ should be amended. Counsel for the plaintiffs was instructed to settle an amended statement of claim. This he did. On or about July 26 1982 a court clerk employed by the plaintiffs' solicitors went to the central office to get the amended specially endorsed writ stamped. A clerk there refused to apply the stamp on the ground that the order of June 25 1982 did not provide for the writ endorsed with the statement of claim to be amended. It did so by implication because it provided that the third and fourth parties should be joined as defendants to the action. It is to be regretted that the clerk decided as he did. By letters dated July 30 1982 the plaintiffs' solicitors sent the third and fourth parties an amended specially endorsed writ together with a draft consent order to put right that which the clerk in the central office had said was wrong. The fourth parties' solicitors returned the draft consent order duly endorsed on August 4 1982. The third parties did the same on August 9. The plaintiffs' solicitors returned to court on September 8 1982. Judge Newey QC then made an order in these terms: "Upon reading the parties' agreed terms it is ordered that: The plaintiffs have leave to amend the statement of claim in the form annexed hereto."

On September 9 1982 the writ was reissued. On September 17 1982 the plaintiffs' solicitors sent the third and fourth parties a copy of the order dated September 8, but they did not serve them with a copy of the reissued amended writ, as they should have done. Thereafter, the third and fourth parties behaved as if they were the second and third defendants.

It is to be added that copies of an amended statement of claim were also sent to the solicitors for Mid-Sussex and for the architects.

The result is that technically Mid-Sussex and the architects have never been properly joined as defendants. But having entered defences and since participated in the proceedings as defendants they could not now be heard to take that point.

The trial of the action began before Judge Hayman, sitting as deputy Official Referee, on November 23 1982 and continued for some time. On December 10 1982, after counsel for the architects had concluded his submissions on the evidence and counsel for the plaintiffs was in the course of making his, counsel became aware of the decision of this House in *Pirelli General Cable Works Ltd* v *Oscar Faber & Partners* [1983] 2 AC 1, judgment in which had been delivered on the previous day. In that case the House, disapproving of certain dicta in the Court of Appeal in *Sparham-Souter* v *Town and Country Developments (Essex) Ltd* [1976] QB 858, held that the date of accrual of a cause of action in tort for damage caused by the negligent design or construction of a building was the date when the damage came into existence, and not the date when the damage was discovered or could with reasonable diligence have been discovered. The leading speech, concurred in by the others of their lordships who heard the appeal, was that of Lord Fraser of Tullybelton, who said at p16:

The plaintiff's cause of action will not accrue until *damage* occurs, which will commonly consist of cracks coming into existence as a result of the defect even though the cracks or the defect may be undiscovered and undiscoverable. There may perhaps be cases where the defect is so gross that the building is doomed from the start, and where the owner's cause of action will accrue as soon as it is built, but it seems unlikely that such a defect would not be discovered within the limitation period. Such cases, if they exist, would be exceptional.

Counsel for Mid-Sussex and for the architects, reading the report of the decision in *The Times* newspaper, perceived that they might have available to them a defence of limitation, as the decision indicated that in a case of this kind the limitation period began to run from an earlier date than had previously been thought to apply. They accordingly applied for leave to amend their defences, the proposed plea in the case of the architects being in these terms:

If (which is denied) the third defendants were guilty of negligence or breach of statutory or other duty whether as alleged or at all, time for the purposes of the Limitation Acts began to run on the occurrence of one of the following events: (a) the submissions to and/or approval by the local authority for building regulations purposes of the plans drawn by the third defendant and/or the approval by such authority of the excavations and/or foundations of the plaintiffs' premises allegedly constructed in reliance upon the said plans, whereby the homes or back of them were defectively designed and/or built as alleged by the plaintiffs; (b) the defective construction and/or completion of the houses and/or of their foundations, as alleged by the plaintiffs; (c) the purchase by the plaintiffs and each of them of their respective houses,

designed and/or constructed defectively as alleged by the plaintiffs; (d) the settlement of the foundations wholly or in part or other movement or damage at the said houses caused or contributed to by the defective design and/or construction as alleged by the plaintiffs.
The third defendants will contend that each of the events above mentioned occurred more than six years prior to the joinder of the third defendants.

At this stage neither counsel appears to have appreciated that, due to the events described earlier, their clients had not been joined as defendants until at the earliest September 9 1982. It seems to have been taken for granted that they had been joined very soon after Judge Stabb's order of June 25 1982. The evidence led in the action had been to the effect that cracks in the five houses had been discovered in the course of August and early September 1976. So counsel had in mind to argue primarily that the houses were "doomed from the start" within the meaning of Lord Fraser of Tullybelton's dictum, and paras (a), (b) and (c) of the proposed amendments were directed to this aspect. Para (d) was directed to a date such as was actually decided in the *Pirelli* case to have been that upon which the limitation period had started to run. They did not believe that they might have had a limitation defence under the law as it was understood to have been before the *Pirelli* decision, and it seems that they so informed Judge Hayman when moving the proposed amendments. Judge Hayman allowed the amendments upon terms as to costs and offered all parties an opportunity of leading further evidence if so advised, an opportunity of which none of them took advantage. December 10 was a Friday, and it seems that in the course of the ensuing weekend counsel discovered the true state of affairs regarding the joinder as defendants of Mid-Sussex and the architects, and the judge was told about it on the following Monday, December 13. The case was adjourned until January 18 1983, when Judge Hayman, on the plaintiffs' application, made an order amending that of Judge Sir William Stabb QC dated June 25 1982 to the effect that time for joinder of Mid-Sussex and the architects was extended to September 9 1982 and that the joinder was to be effective as from July 30 1982.

On February 14 1983 Judge Hayman delivered a judgment in favour of three of the plaintiff houseowners against Hansel and the architects and in favour of two of them against Hansel only. Judgment was entered for Mid-Sussex against all the plaintiffs. He found negligence to have been proved against all of Hansel, Mid-Sussex and the architects, but that Mid-Sussex was not liable because it had not been established that the condition of the houses presented any present or imminent danger to the health or safety of the occupants — which in the light of *Anns v Merton London Borough Council* [1978] AC 728 he held to be a necessary condition of liability. In the case of the two plaintiffs who failed against the architects he held that the latter succeeded in their limitation defence in respect that it had not been established on the evidence that physical damage to the two houses concerned had first occurred on or after July 30 1976, viz less than six years before the date of joinder of the architects, which he took as being July 30 1982.

The houseowners and the architects appealed to the Court of Appeal. All the houseowners appealed against that part of the judgment which dismissed their claims against Mid-Sussex. The two whose claims against the architects had been dismissed appealed against that part of the judgment. The architects appealed against the part of the judgment which decided against them in favour of the other three houseowners. All the houseowners also appealed out of time, with leave, against the order of Judge Hayman, on December 10 1982, allowing amendment of the defences of Mid-Sussex and the architects.

By a judgment given on July 25 1984 the Court of Appeal [1984] 1 WLR 1274 (Lawton, Stephen Brown and Parker LJJ) allowed the houseowners' appeal against Mid-Sussex and also that of two of them against the architects. They dismissed the architects' appeal.

The appeal against Mid-Sussex was decided primarily on the issue whether there was proof of imminent danger to the health or safety of the occupants of the houses, as to which the Court of Appeal reached a different conclusion from that of the trial judge. There has been no appeal to this House on that issue and no more need be said about it. The Court of Appeal also, however, in disposing of the appeals, decided in favour of the houseowners a number of points on the issue of limitation, and it is their decisions on these points which have given rise to the instant appeal to this House by the architects. The Court of Appeal granted leave to appeal here both to the architects and to Mid-Sussex, but the latter decided not to prosecute their appeal and were

represented before the House on the question of costs only at the closing stages of the argument. The houseowners, being holders of a judgment against Hansel, the architects and Mid-Sussex, jointly and severally, appreciated that even if the architects succeeded in their appeal they would still be in a position to recover the whole of their damages from Mid-Sussex, and consequently they were not represented for the purpose of argument on the merits of the architects' appeal but lodged a case and appeared by counsel only in relation to costs. The result is that upon the merits of the appeal, which involve issues by no means easy of resolution, your lordships have had the benefit of argument on one side only.

The issues which arise in connection with limitation are four in number; first, what was the date of joinder as defendants of the architects; second, whether the joinder of the architects as defendants is for limitation purposes to be related back to the date when the writ was originally issued against Hansel; third, whether leave to amend so as to raise the defence of limitation should have been granted as respects the whole of the proposed amendments or only as respects paras (a), (b) and (c) of those amendments; and fourth, whether or not, assuming that leave should have been granted only as respects paras (a), (b) and (c), ie to the effect of allowing a limitation defence to be pleaded solely on the "doomed from the start" basis, that defence succeeds.

I will consider the four issues in that order.

(1) *Date of joinder*
The houseowners obtained leave to join the architects as additional defendants on June 25 1982, but an effectively amended writ was not issued until September 9 1982, and that writ was never served on the architects. The architects were, however, sent copies of an amended statement of claim and they served defences to it dated October 6 1982. There can be no doubt that by thus serving defences they waived the necessity of serving the amended writ on them. Assuming that the defences were served on the date which they bore, the architects became parties to the action at the latest on October 6 1982. The question is whether by virtue of RSC Ord 15, r8(4) they must be taken to have become parties on September 9 1982, the date when the effectively amended writ was issued. The relevance of the question lies in the finding of the trial judge that the damage to the fifth plaintiff's house occurred on September 9 1976, precisely six years earlier.

It is to be observed at this stage that the Court of Appeal held that the trial judge had no power to declare, as he purported to do by his order of January 18 1983, that Mid-Sussex and the architects were defendants from July 30 1982. The houseowners did not appear before your lordships to argue the contrary, and the decision was clearly correct.

RSC Ord 15, r8(4) provides:

Where by an order under rule 6 or 7 a person is to be added as a party . . . that person shall not become a party until –
(a) where the order is made under rule 6, the writ has been amended in relation to him under this rule and (if he is a defendant) has been served on him . . .

In the present case the order for joinder was made under rule 6.

The natural meaning of RSC Ord 15, r8(4)(a), according to the ordinary use of language, would appear to be that a person added as a defendant does not become a party until not only has the writ been amended but also the amended writ has been served upon him. But in *Seabridge v H Cox & Sons (Plant Hire) Ltd* [1968] 2 QB 46, upon the construction of an earlier version of the rule in the same terms, the Court of Appeal took a different view. The plaintiffs in an action for damages for personal injuries sought to add an additional defendant and obtained leave accordingly under RSC Ord 15, r6. Amended writs were filed and stamped at the central office under rule 8 on the third anniversary of the accident, but were not served on the new defendant (India Tyres Ltd) until a few days later. The Court of Appeal held that the amended writs took effect, for limitation purposes, on the date of stamping, not that of service. Lord Denning MR said, at pp 51-52:

The old rule, Order 16, r11, said that when a party is added, the proceedings against such party should "be deemed to have begun only on the service of such writ." If that rule had been the rule in existence today, I feel that the words would have compelled the court to hold that, as against India Tyres Ltd, the proceedings only began on September 21 1967, that is, more than three years after the date of the accident and they would be entitled to the benefit of the Statutes of Limitation. But that rule has been altered. In 1962 the wording of it was changed. Again in 1964 it was changed once more. We have to construe the new rule, Order 15, r8(4). That says: "Where by an order under rule 6

that is, the one we are concerned with

. . . a person is to be added as a party . . . that person shall not become a party until — (a) . . . the writ has been amended in relation to him under this rule and (if he is a defendant) has been served on him . . ."

Mr Kidwell argued that this new rule has the same effect as the old rule. It means, he said, that the proceedings are not brought against the added defendant until he is served. That argument convinced the master and the judge. They set aside the service. But I am unable to accept the argument. We must read the new rules as a whole. It seems to me that when the amendment is made in the prescribed manner, namely, by the amendment being taken to the Central Office and filed and the amended writ stamped, then at that moment the amended writ takes effect as against the added defendant. That procedure is equivalent to the issue of a writ against an original defendant. Once the amendment is made, the rules as to service apply as against the added defendant, just as they do to an original defendant on the issue of a writ: see Order 15, r8(2). The result is that the plaintiff has to serve the added defendant within 12 months from the date when the amendment is made.

I am prepared to hold, therefore, that when an amendment is made adding a defendant, the amendment takes effect when it is stamped in the Central Office. It takes effect at that moment against that defendant equivalent to the issue of the writ against him. If the amendment is so made within the three years, it is in time, even though it is served later. In this particular case the amendment was just in time. It was on the very last day of the three years. This still leaves ample scope for Order 15, r8(4) when it says that "that person shall not become a party until [he] has been served." That only means that he is not affected by notices, and he cannot enter an appearance until he is served. We were referred to the county court rule, which is the same as the old Order 16, r11. That has not been amended. In the county court, therefore, the old position may prevail. But in the High Court I think a better rule now prevails. The amendment takes effect when it is stamped in the Central Office.

Mr Kidwell was forced to concede that, if the plaintiffs had issued another writ on September 18 1967, on India Tyres Ltd, it would undoubtedly be in time, even though it was served later. I think it should be the same when they add them as defendants by amendment on that day. In good sense there ought to be no difference. On a broad and liberal interpretation of the rules, there is no difference.

Diplock LJ agreed and so did Salmon LJ, albeit with some hesitation. That was a very powerful court, but with the greatest respect to it, I have been unable to accept the reasoning of Lord Denning as convincing. In my opinion, the plain language of the rule must prevail and *Seabridge* should be overruled as wrongly decided.

It follows that the architects were not validly joined as defendants until they lodged defences on October 6 1982, more than six years after damage to the fifth plaintiff's house occurred on September 9 1976.

(2) *Relation back*

It has long been a rule of practice that amendment should not be allowed for the joinder of an additional defendant in a situation where a relevant period of limitation has already expired in relation to the cause of action against him. The reason for that rule of practice is, however, debatable. One theory is that an additional defendant, joined unconditionally, becomes a party to the action as from the date of issue of the writ against the original defendant, that is to say, that joinder is related back to that date. That being so, it would be unjust to join him as a defendant at a time when limitation has run in his favour, because to do so would have the effect of depriving him of a valid defence. This theory has the support of a long line of authority in the Court of Appeal, which was examined by Stephenson and Brandon LJJ in *Liff* v *Peasley* [1980] 1 WLR 781. In that case the correctness of the theory was challenged by the citation of an old case in this House, *Byron* v *Cooper* (1844) 11 Cl & Fin 556. Brandon LJ expressed the opinion, obiter, that the relation back theory was incorrect. He considered the true basis of the rule of practice to be that no useful purpose would be served by joining an additional defendant at a time when limitation had run in his favour, because he was not to be deemed to have become a party at any earlier date than the actual date of joinder and therefore would have an unanswerable defence.

In the present case, if the relation back theory is correct, the result is that the joinder of the architects would date back to May 27 1980, when the writ was originally issued against Hansel. Limitation would therefore not be a good defence to any cause of action that arose after May 27 1974, and since all the plaintiffs purchased from Hansel in 1975 the defence raised by paras (c) and (d) of the proposed amendments would be bound to fail. The Court of Appeal held that they were bound by authority to accept the relation back theory and accordingly that amendment in terms of these paragraphs should not have been allowed.

The Limitation Act 1939, which was applicable at the material time, provides by section 2(1):

The following actions shall not be brought after the expiration of six years from the date on which the cause of action accrued, that is to say:
(a) actions founded on simple contract or on tort . . .

A cause of action is necessarily a cause of action against a particular defendant, and the bringing of the action which is referred to must be the bringing of the action against that defendant in respect of that cause of action. The causes of action here against Mid-Sussex and the architects were separate and distinct from the cause of action against Hansel. In my opinion, there are no good grounds in principle or in reason for the view that an action is brought against an additional defendant at any earlier time than the date upon which that defendant is joined as a party in accordance with the Rules of Court. Further, I consider *Byron* v *Cooper* to constitute clear authority in this House against the relation back theory. That was a case under the Tithes Act 1832, which by section 1 provided for shortening of the time required for establishment of a claim to *inter alia* exemption from tithes. Section 3 provided:

That this Act shall not be prejudicial or available to or for any plaintiff or defendant in any suit or action relative to any of the matters before mentioned, now commenced, or which may be hereafter commenced, during the present session of Parliament, or within one year from the end thereof.

The rector of a parish filed a bill for an account of tithes against five defendants before the expiry of the time so limited, and after the expiration of that time the bill was amended by order of the court so as to add a further four defendants. The House held that the latter were not deprived of the protection of section 1 of the Act. Lord Brougham said at p579:

The first miscarriage in the court below, however, was to consider the whole defendants to the suit, the whole nine appellants, as excluded from the operation of the Act. The ground of this opinion was that the bill being originally filed before August 16 1833, and the four last-named appellants being, under an order of the Court of Exchequer, made defendants to that same bill, were as much excluded by the third section of the Act as if they had been made originally defendants to the bill filed on August 5 1833. This is as great and as manifest an error as could well be committed.

And later, at pp 579-580:

The parson is permitted to add new defendants to his amended bill, in order to save delay and expense; but each defendant so added is to be considered as sued by the proceeding which makes him a defendant, and the date of his being added is the date of the suit's commencement quoad him; consequently the four last-named and last-added defendants in this case were only sued in November 1834, and quoad them the bill and the suit bear the date of November 1834. They do not fall, therefore, within the description of the third section of the statute. They are not defendants, to use the words of that statute, "in a suit or action commenced within one year" after August 16 1832, being the last day of the session in which the Act passed.

Although the wording of section 2(1) of the Limitation Act 1939 naturally differs from that of section 3 of the Tithes Act 1832, that difference is not such as to demonstrate an intention that a defendant brought into an action by amendment, in respect of a cause of action separate from that against an existing defendant, should be treated as having had the action brought against him at the same time as it was brought against the latter.

Accordingly, I find myself in respectful agreement with the provisional view expressed by Brandon LJ in *Liff* v *Peasley* [1980] 1 WLR 781 at p 804 that the "no useful purpose" theory, and not the "relation back" theory, is the true basis of rule of practice to which I have referred. It follows that, contrary to the view taken in the Court of Appeal, the date of joinder of the architects does not fall to be related back to May 27 1980 and that the Court of Appeal was wrong to disallow paras (c) and (d) of the proposed amendments on the ground of such relation back.

(3) *Amendment*

Having regard to the conclusions I have reached upon the date of joinder issue and the relation back issue, and irrespective of the dates upon which the houseowners' causes of action against the architects are taken to have arisen, there can be no doubt that on October 6 1982, when the architects effectively became defendants to the action, they had available to them a cast-iron defence of limitation. Whatever view be taken as to the dates when the causes of action arose, these dates must all have fallen more than six years before October 6 1982. The latest possible dates were in late August and early September 1976, when damage to the houses first occurred. These were the dates picked on by para (d) of the proposed amendments.

The Court of Appeal decided that irrespective of the relation back aspect para (d) of the proposed amendments should not have been allowed. Lawton LJ said at [1984] 1 WLR 1274 at p 1285:

It would be convenient, in my opinion, to start the unravelling of this surprising mish-mash of legal issues by considering the procedural ones. The first in point of time, and importance, is that raised by the amendments which were allowed on December 10 1982. On that date the local authority and the architects learned that they might have a limitation defence which they had not known about before, that is to say the houses might have been "doomed from the start". This is what they then wanted to plead. Nothing more. They told the judge that they did not want to plead that the claims had been statute-barred when they were joined as defendants. They could have done so and, as I have already commented, if they had, on the evidence, they could have argued that the foundations had settled when or shortly before the cracks appeared. Paragraph (d) of the amendments which was allowed was in substance nothing more than what could have been put into the defences served but which was not. In my judgment, an amendment in terms of paragraph (d) should not have been allowed. The amendments allowed in paragraphs (a), (b) and (c) were of a different kind. The local authority and the architects and their advisers had had no reason to think before December 10 1982 that a cause of action could accrue if a house were erected which, due to negligent design or construction was "doomed from the start". They wanted that issue tried and, in my judgment, they were entitled to have it tried, provided that the plaintiffs were not prejudiced by a late amendment. They were not because they were given an opportunity of adducing more evidence. They decided not to take advantage of that opportunity.

Stephen Brown LJ agreed with him. Parker LJ said at pp 1292-1293:

If . . . the correct date for the purposes of limitation is September 9 1982, I would allow the appeal as to sub-paragraph (d) of the amendment. It was clearly granted by the judge upon a mistaken view of the situation and the fact that leave to amend is discretionary is therefore of no significance. In the circumstances of the case, no amendment should be allowed.

Whether or not a proposed amendment should be allowed is a matter within the discretion of the judge dealing with the application, but the discretion is one that falls to be exercised in accordance with well-settled principles. In his interlocutory judgment of December 10 1982, allowing the proposed amendments, Judge Hayman set out and quoted at some length from the classical authorities on this topic. The rule is that amendment should be allowed if necessary to enable the true issues in controversy between the parties to be resolved, and if allowance would not result in injustice to the other party not capable of being compensated by an award of costs. In *Clarapede & Co v Commercial Union Association* (1883) 32 WR 262 at p 263, Brett MR said:

The rule of conduct of the court in such a case is that, however negligent or careless may have been the first omission, and, however late the proposed amendment, the amendment should be allowed if it can be made without injustice to the other side. There is no injustice if the other side can be compensated by costs; but, if the amendment will put them into such a position that they must be injured, it ought not to be made.

The sort of injury which is here in contemplation is something which places the other party in a worse position from the point of view of presentation of his case than he would have been in if his opponent had pleaded the subject-matter of the proposed amendment at the proper time. If he would suffer no prejudice from that point of view, then an award of costs is sufficient to prevent him from suffering injury and the amendment should be allowed. It is not a relevant type of prejudice that allowance of the amendment will or may deprive him of a success which he would achieve if the amendment were not to be allowed. In my opinion, no sensible distinction is to be drawn for this purpose between an amendment seeking to plead limitation and any other sort of amendment. I am not aware of any authority for drawing such a distinction, nor have I experience of any practice to that effect.

In the present case Judge Hayman allowed the proposed amendments believing, as did at the time counsel who proffered the amendments, that the decision of this House in the *Pirelli* case [1983] 2 AC 1 offered some prospect of a limitation defence which, owing to the time element, would not have been available under the law as it was understood to be prior to that decision. The effect of the decision was to set back, in some cases at least, the date at which the limitation period started to run. It is to be noted that in their original pleadings the houseowners had made averments as to the date of the first defect in the houses "becoming apparent." By amendment in answer to those allowed by the order of January 10 1983 they altered that to "occurring." So the ground upon which Judge Hayman allowed the amendments was that it had come to counsel's notice because of *Pirelli* that they might have a limitation defence which it had not previously occurred to them that they might have. At that time counsel appear to have been unaware of the procedural blunders which had led to their clients not having been joined as defendants until a date which, upon the law as it then stood under the *Seabridge* case [1968] 2 QB 46, was capable of being identified as September 9 1982. There does not seem to be any question of an earlier deliberate decision, taken in knowledge of these blunders, not to plead limitation. In his interlocutory judgment of January 18 1983 Judge Hayman stated that at the hearing of the application to amend all counsel apparently took the date of joinder to have been June 25 1982. It may be inferred that if they had known of the blunders counsel would have pleaded limitation from the outset.

One of the reasons given by Judge Hayman, in that interlocutory judgment, for backdating the joinder of Mid-Sussex and the architects, was that counsel for these parties had disclaimed on December 10 1982 any intention of arguing that the houseowners' claims would be time-barred apart from *Pirelli*, and that in these circumstances it would be unjust to the plaintiffs to hold that these parties should be discharged at least as regards the first four plaintiffs (the evidence being that the damage to the fifth plaintiff's home occurred on September 9 1982), because these parties had in some way waived any right they might have had to be discharged, through not having applied to have their joinder set aside. It is not altogether easy to follow the reasoning of the judge. The fact that counsel have overlooked the availability of some defence does not mean that they have elected not to pursue it. Mid-Sussex and the architects had the opportunity to lodge defences only some three months before the trial started. By lodging defences they must be taken to have waived the procedural irregularities to the effect of being properly joined when they did so, which in the case of the architects was October 6 1982. But if, having discovered the true position, they had proposed to amend in order to plead limitation at any time up to the start of the trial, it is clear that they could not reasonably have been refused leave to do so. When the amendments were actually proffered, during closing speeches towards the end of the trial, it was a matter for very careful consideration whether allowance of them would result in any prejudice to the plaintiffs beyond what could be compensated by an award of expenses. That would be so whether the proffer of the amendments was prompted by *Pirelli* or by the late discovery of an extraordinary and hitherto unsuspected series of procedural blunders. The effect on the plaintiffs by way of prejudice or lack of it was exactly the same in either case.

In the event, Judge Hayman was told about the discovery of the blunders and that counsel for the second and third defendants now wished to rely not only on the "doomed from the start" argument but also upon a limitation defence based on a period starting when the damage to the houses actually occurred, on December 13 1982. He then neither recalled his order of December 10 nor altered it so as to exclude para (d) of the proposed amendments, though at that stage and in those circumstances it would clearly have been competent for him to have done so. It does not seem to have occurred to counsel for the plaintiffs to invite him to take that course. If he had been so invited, he might in my opinion, in the light of all the circumstances, have reasonably and properly exercised his discretion against excluding para (d). I do not consider that the interlocutory judgment of January 18 1983 can be read as indicating that he would not have done so.

The judgments in the Court of Appeal on this issue do not canvass the principles to be applied in the exercise of the discretion to allow or disallow late amendments. They do not allude to the short period of time which elapsed between the second and third defendants being given the opportunity to lodge defences and the commencement of the trial, nor to their counsel having informed Judge Hayman on December 13 1982 of their altered position in the light of their discovery of the procedural blunders. They do not allude to any practice of disallowing late amendments designed to raise a defence of limitation. Upon the whole matter I am of the opinion that the very brief reasons given by the Court of Appeal for their decision do not indicate that they had due regard to the applicable principles and to all relevant circumstances. I would, therefore, hold that, in so far as they purported to exercise their discretion afresh, their exercise of it was erroneous and capable of being corrected. I understand, however, that a majority of your Lordships have reached a different conclusion.

(4) The "doomed from the start" argument
The appellants' presentation of this argument involved two aspects. In the first place it was maintained that the houseowners' respective causes of action accrued not when the physical damage to their houses occurred but when they became the owners of houses with defective foundations. It was argued that they then suffered economic loss because the houses were less valuable than they would have been if the foundations had been sound. The proposition that a cause of action in tort accrued out of negligence resulting in pure economic loss was sought to be vouched by reference to *Junior Books Ltd* v *Veitchi Co Ltd* [1983] 1 AC 520. That case was also cited in *Pirelli* v *Oscar Faber & Partners* [1983] 2 AC 1, in support of the argument that, since in that case there was economic loss when the chimney was built, the cause of action arose then. The argument was clearly rejected in the speech of Lord Fraser of Tullybelton, concurred in by all the others of their lordships who participated in the decision. At p 16 he expressed the opinion that a latent defect in a building does not give rise to a cause of action until damage occurs. In the present case there can be no doubt that the defects in the houses were latent. No one knew of their existence until damage occurred in the summer of 1976. This branch of the argument for the architects is, in my opinion, inconsistent with the decision in *Pirelli* and must be rejected.

In the second branch of the argument it was maintained that a distinction fell to be drawn between the case where the defect in a building was such that damage must inevitably eventuate at some time and the case of a defect such that damage might or might not eventuate. The former case was that of a building "doomed from the start" such as was in the contemplation of Lord Fraser of Tullybelton when he made reference to that concept in his dicta in *Pirelli*. In the present case the houses were doomed from the start because the event showed that damage was bound to occur eventually. My Lords, whatever Lord Fraser may have had in mind in uttering the dicta in question, it cannot, in my opinion, have been a building with a latent defect which must inevitably result in damage at some stage. That is precisely the kind of building that *Pirelli* was concerned with and in relation to which it was held that the cause of action accrued when the damage occurred. This case is indistinguishable from *Pirelli* and must be decided similarly. The second branch of the architects' argument fails. I understand that all your lordships agree.

My Lords, in view of the decision of the majority on the amendment issue, the appeal must be dismissed. The appellants will pay the costs of the respondents and of Mid-Sussex.

In his speech, LORD BRANDON OF OAKBROOK (who would also have allowed the appeal) said: The five plaintiffs in the action in which this appeal arises were the purchasers in 1975 of houses which had been built on defective foundations. By reason of those defective foundations the houses later developed serious structural damage. The plaintiffs might have incurred very substantial costs in making good the defects and repairing the damage, but they chose instead to sell their houses at prices much reduced by the defects and damage concerned. Subsequently, the five plaintiffs began an action to recover the losses which they had suffered by reason of these events. At first the plaintiffs sued the builders as sole defendants. Later they joined the local authority and the architects who designed the houses as second and third defendants respectively.

The action was tried by Judge Hayman, sitting as a deputy Official Referee. By an order made on February 14 1983 he decided: (1) that all five plaintiffs succeeded against the first defendants; (2) that all five plaintiffs failed against the second defendants; (3) that the first, second and fifth plaintiffs succeeded against the third defendants; (4) that the third and fourth plaintiffs failed against the third defendants; and (5) that, as between the first and third defendants where both were liable, liability should be apportioned 60% to the first defendant and 40% to the third defendants. The judge further stated in his judgment that, if he had found the second defendants also liable to the plaintiffs he would, as between all three defendants, have apportioned liability 50% to the first defendants, 15% to the second defendants and 35% to the third defendants.

The third defendants appealed and the plaintiffs cross-appealed to the Court of Appeal [1984] 1 WLR 1274 (Lawton, Stephen Brown and Parker LJJ). By the order of that court dated July 25 1984: (1) the third defendants' appeal against so much of the judge's order as made them liable to the first, second and fifth plaintiffs was dismissed; (2) the cross-appeal of all five plaintiffs against so much of the judge's order as dismissed their claims against the second defendants was allowed; (3) the cross-appeal of the third and fourth plaintiffs against so much of the judge's order as dismissed their claim against the third defendants was also allowed; (4) liability as between the three defendants was apportioned 50% to the first defendants, 15% to the second defendants and 35% to the third defendants; and (5) leave was given to both the second and third defendants to appeal to your Lordships' House.

Appeals to your Lordships' House were initiated by both the second and the third defendants, but the former later withdrew their appeal. The result of such withdrawal was that the plaintiffs, who by the order of the Court of Appeal had a judgment in their favour against the second defendants, ceased to have any interest in the outcome of the third defendants' appeal. The consequence of that was that, so far as the substance of the third defendants' appeal is concerned, your lordships heard only counsel for them, with no opposition on behalf of either the plaintiffs or the second defendants. Counsel for these two latter parties were, however, heard on the subject of costs.

My Lords, four questions arise on this appeal. The first question is when did the plaintiffs' causes of action against the third defendants accrue. The second question is when was action brought by the plaintiffs against the third defendants. The third question is whether, having regard to the answers to the first and second questions, the plaintiffs' claims against the third defendants were time-barred. The fourth question is whether, the trial judge having allowed a very late application by the third defendants for leave to amend their defence by adding a plea that the claims against them were time-barred, the Court of Appeal was right to disallow a crucial part of that amendment. I shall call these four questions the accrual issue, the joinder issue, the time-bar issue and the amendment issue respectively, and I shall consider them in the order in which I have stated them. Before doing so, however, it will be convenient to draw attention to a decision of your Lordships' House which was given at a time when the trial of the present action had reached a very late stage and had an important bearing on the course which the trial then took. That decision is *Pirelli General Cable Works Ltd* v *Oscar Faber & Partners* [1983] 2 AC 1, which was given on December 9 1982.

The decision in the Pirelli case
It was held by the Court of Appeal in *Sparham-Souter* v *Town and Country Developments (Essex) Ltd* [1976] QB 858 that the date of accrual of a cause of action in tort for damage caused by the negligent design or construction of a building is the date when the damage is discovered or could with reasonable diligence have been discovered. In the *Pirelli* case your Lordships' House held that this decision of the Court of Appeal was wrong and that the date of accrual of the cause of action was the date when the damage resulting from the negligent design or construction first came into existence. Lord Fraser of Tullybelton, who delivered the leading speech in that case, while holding that the general rule was that which I have just stated, referred in two passages in his speech to the possible existence of special cases in which a building might be regarded as so defective as to be "doomed from the start", in which event the date of accrual of the cause of action might be the date when the building was constructed. These passages, however, were not necessary to the decision of the appeal and were therefore no more than obiter dicta.

The accrual issue
The Court of Appeal, applying the ground of decision in the *Pirelli* case, held that the plaintiffs' causes of action against the third defendants accrued at the various times when the structural damage to their houses, consequential on their originally defective foundations, first came into existence. The court rejected the contention put forward for the third defendants that the houses were "doomed from the start," and that, on the basis of the observations of Lord Fraser of Tullybelton in the *Pirelli* case to which I referred earlier, the plaintiffs' causes of action accrued when the houses were built. This contention was renewed by the third defendants before your Lordships' House. The argument of counsel, as I understand it, proceeded as follows. Where a house was built on defective foundations, a buyer of it might suffer two kinds of damage. The first kind of damage was physical, in the form of consequential structural failure or damage. The second kind of damage was economic loss, in the form of diminution in market value. In the case of the first kind of

damage, the buyer's cause of action against any party for negligence in respect of the defective foundations accrued when the consequential structural failure or damage occurred. But, in the case of the second kind of damage, the diminution of market value was present from the time of the original construction, and it was at that earlier time that the buyer's cause of action in respect of such diminution accrued. The plaintiffs in the present case had sued for the second kind of damage, namely, diminution of market value. Their causes of action had, therefore, accrued at the date when the houses were built.

In my opinion, this contention cannot be supported. I do not know what special cases Lord Fraser of Tullybelton had in mind when he referred in his speech in the *Pirelli* case to buildings "doomed from the start". It may be that he was only keeping open the possibility of the existence of such special cases out of major caution. Be that as it may, however, I am quite sure that he was not seeking to differentiate between causes of action in respect of making good defects or damage on the one hand and causes of action in respect of diminution in market value on the other. In any case, on the facts of the present case it seems that the plaintiffs, in reselling their houses at a loss, were acting reasonably in mitigation of their damage, so that the distinction between the two kinds of damage relied on is one of form rather than substance.

In my view, there is nothing in the facts of the present case which would take it out of the general principle laid down in the *Pirelli* case and put it into some special class of case, if there be one, of buildings "doomed from the start".

It follows that I would answer the first question by saying that the plaintiffs' causes of action against the third defendants accrued at the dates on which the consequential structural damage to their houses first came into existence.

The time of joinder issue

My Lords, the unhappy course of the plaintiffs' action, so far as the joinder of the second and third plaintiffs is concerned, is set out fully and clearly in the judgment of Lawton LJ in the Court of Appeal [1984] 1 WLR 1274 at pp 1281-1284, and I do not think that it is necessary for me to recount it in detail again here. The outcome of it all was that, although the plaintiffs obtained leave to join the local authority and the architects as additional defendants on June 25 1982, no effectively amended writ was issued until September 9 1982, and even then it was never served on the new parties. The latter, however, were sent copies of the amended statement of claim and later served defences to it without waiting for any amended writ to be served on them. The defence of the second defendants is undated; that of the third defendants was served on October 6 1982. By serving defences in this way the second and third defendants waived the service on them of the amended writ, and in the case of the third defendants they must be taken to have done so on the date of service of their defence, that is to say on October 6 1982.

Leave to join the local authority and the architects as additional defendants had been given under RSC Ord 15, r6. Rule 8(4) of Order 15 provides:

Where by an order under rule 6 or 7 a person is to be added as a party . . ., that person shall not become a party until —
(a) Where the order is made under rule 6, the writ has been amended in relation to him under this rule and (if he is a defendant) has been served on him . . .

My Lords, in my opinion the plain effect of this rule is that defendants added under RSC Ord 15, r6, become parties to the action in which they are joined when, and only when, an amended writ is served on them. In *Seabridge* v *H Cox & Sons (Plant Hire) Ltd* [1968] 2 QB 46 the Court of Appeal (Lord Denning MR and Diplock and Salmon LJJ) held that, on the true construction of an earlier rule in the same terms, a writ amended to join an additional defendant took effect from the time when it was stamped in the central office, and that the provision that an added defendant shall not become a party until the amended writ has been served on him only meant that he was not affected by notices and could not enter an appearance until he has been served. With great respect to the exceptionally distinguished judges who constituted the Court of Appeal on that occasion, I cannot agree with their interpretation of the rule, which appears to me to be contrary to the plain language of it. I would, therefore, hold that the *Seabridge* case was wrongly decided and should be overruled.

For the reasons which I have given, I am of opinion that the third defendants did not become parties to the action, and were not therefore effectively joined, until October 6 1982.

The time-bar issue

My Lords, the Court of Appeal held that the joinder of the second and third defendants, which they regarded as having occurred on September 9 1982, the date of issue of the amended writ, once it had been made related back to the date of the original writ. They so held because they regarded themselves as bound by previous decisions of the Court of Appeal to do so.

It was common ground that, despite the passing of the Limitation Amendment Act 1980 (c24) and the Limitation Act 1980 (c58), the Limitation Act 1939 applied to the action in this case. The reason for this is that the action was begun on May 27 1980 before either of the Acts of 1980 had come into force. Section 2(1) of the Limitation Act 1939 provides:

The following actions shall not be brought after the expiration of six years from the date on which the cause of action accrued, that is to say:-
(a) actions founded on simple contract or on tort . . .

The question, therefore, is when was action brought by the plaintiffs against the third defendants within the meaning of section 2(1) above. In my opinion, there can be only one answer to that question: it was when the third defendants, by waiving service on them of the amended writ, first became parties to the action, that is to say on October 6 1982.

The theory that, at common law, and apart from anything in either of the Acts of 1980, the joinder of a defendant to an existing action relates back to the date of the original writ was examined in depth by a two-judge Court of Appeal in *Liff* v *Peasley* [1980] 1 WLR 781. Stephenson LJ was one of the two judges and I was the other. That case proceeded on the basis, which is not open to question, that there was an established rule of practice prohibiting the joinder by amendment of an additional defendant after the expiration of a relevant period of limitation in his favour. The question discussed in the judgments, however, was whether that rule of practice was founded on the concept of relation back (the relation back theory), in that to allow such joinder would deprive the added defendant of an accrued defence, or on the concept that to allow such joinder would serve no useful purpose (the no useful purpose theory), in that the added defendant would, if joined, have an unanswerable defence of time-bar. It was not necessary for the disposal of that case to decide finally which of these two concepts was correct. However, in the judgment which I gave I expressed a strong provisional view that the rule of practice was founded on the no useful purpose theory and not on the relation back theory, and I gave what appeared to me to be convincing reasons for that view. While it is more appropriate for the rest of your lordships than for me to decide whether the provisional view which I expressed in *Liff* v *Peasley* is correct or not, I stand by that view and the reasons which I gave for holding it. It follows that I consider that the Court of Appeal was wrong to accept and apply the relation back theory in this case.

With the relation back theory out of the way, it is clear on the findings of fact of the judge that the structural damage to the houses complained of occurred more than six years before October 6 1982, which is the date on which I have held that action was first brought by the plaintiffs against the third defendants. It follows that the plaintiffs' claims were by then time-barred.

The amendment issue

My Lords, the decision in the *Pirelli* case was given, as I indicated earlier, on December 9 1982. By then the trial was nearing its end with counsel making their final speeches. On December 10 1982 counsel for the third defendants, having become aware of the decision in the *Pirelli* case, applied to the judge for leave to amend the third defendants' defence by raising a plea of limitation. He subsequently put before the judge a draft of the proposed amendment, which involved adding to the existing defence of the third defendants a new para 3A in these terms:

Further or alternatively the Plaintiffs' claims are statute-barred by reason of the following facts or matters:

PARTICULARS

If (which is denied) the third defendant was guilty of negligence or breach of statutory or other duty whether as alleged or at all, time for the purposes of the Limitation Acts began to run on the occurrence of one of the following events:
(a) the submission to and/or approval by the local authority for building regulations purposes of the plans drawn by the third defendants and/or the approval by such authority of the excavations and/or foundations at the

plaintiffs' premises allegedly constructed in reliance upon the said plans, whereby the homes or back of them were defectively designed and/or built as alleged by the plaintiffs
(b) The defective construction and/or completion of the houses and/or of their foundations, as alleged by the plaintiffs.
(c) The purchase by the plaintiffs and each of them of their respective houses, designed and/or constructed defectively as alleged by the plaintiffs.
(d) The settlement of the foundations wholly or in part or other movement or damage at the said houses caused or contributed to by the defective design and/or construction as alleged by the plaintiffs.
The third defendants will contend that each of the events above mentioned occurred more than six years prior to the joinder of the third defendants.

Counsel for the second defendants also applied for leave to make a similar amendment at the same time.

It will be observed that there is a fundamental distinction between paras (a), (b) and (c) of the above particulars on the one hand and para (d) on the other. Paras (a), (b) and (c) were designed to enable the third defendants to contend that the houses, the subject-matter of the action, came within the special category of buildings "doomed from the start", as contemplated by Lord Fraser of Tullybelton in the observations made by him in the *Pirelli* case to which I referred earlier. On that basis the third defendants wished to contend that the plaintiffs' cause of action against them accrued when the houses were built or even earlier, with the result that, even if the doctrine of relation back applied, such cause of action was already time-barred when the original writ was issued. By contrast, para (d) was designed to enable the third defendants to contend, in reliance on the ground of decision in the *Pirelli* case, that the plaintiffs' cause of action against them accrued when the structural damage to the houses first came into existence. On that basis they might also succeed in a defence of limitation, but only if they could persuade the judge that, having regard to *Liff* v *Peasley,* the doctrine of relation back did not apply. It is material to add that, on the facts as found by the judge, para (d) would have served the third defendants just as well for this purpose if the plaintiffs' cause of action against them were treated as having accrued at the time laid down in *Sparham-Souter* v *Town and Country Developments (Essex) Ltd* [1976] QB 858, namely, the time when the structural damage was discovered or could with reasonable diligence have been discovered. The significance of this is that a plea analogous to that made in para (d), but based on the *Sparham-Souter* case instead of the *Pirelli* case, could readily have been included in the third defendants' defence at a much earlier stage of the proceedings and certainly well before trial.

The second and third defendants' applications to amend their defence were, as might have been expected, strongly opposed by the plaintiffs. The judge, however, in a reasoned judgment of considerable length, decided to allow them on terms that the second and third defendants should pay the plaintiffs' costs of the amendment and of the adjournment which he directed should follow it in any event. In his judgment the judge said:

I now have before me a draft amended defence from Mr Cordara on behalf of the third defendants. Of the relevant sub-para (a), (b) and (c) relate to the "doomed from the start" point. Sub-para (d) is based on the ratio decidendi of the *Pirelli* case.

That statement was, so far as it went, entirely correct.

With regard to the principles on which his discretion to allow or refuse the applications to amend should be exercised, the judge referred to the notes to RSC Ord 20, r5, in the *Supreme Court Practice* 100th ed (1982) and to the authorities there cited. The effect of these authorities can, I think, be summarised in the following four propositions. First, all such amendements should be made as are necessary to enable the real questions in controversy between the parties to be decided. Second, amendments should not be refused solely because they have been made necessary by the honest fault or mistake of the party applying for leave to make them: it is not the function of the court to punish parties for mistakes which they have made in the conduct of their cases by deciding otherwise than in accordance with their rights. Third, however blameworthy (short of bad faith) may have been a party's failure to plead the subject-matter of a proposed amendment earlier, and however late the application for leave to make such amendment may have been, the application should, in general, be allowed, provided that allowing it will not prejudice the other party. Fourth, there is no injustice to the other party if he can be compensated by appropriate orders as to costs.

Guiding himself by the authorities, the effect of which I have summarised above, the judge decided, not without some doubt and hesitation, to allow the proposed amendments. The further hearing of the case was adjourned to the beginning of the following term.

On January 18 1983 the judge, on the application of the plaintiffs, made an order that the joinder of the second and third defendants should take effect from July 30 1982. The Court of Appeal rightly held that the judge had no jurisdiction to make such an order and the making of it can therefore be disregarded. It is, however, material to refer to certain passages in the judgment, again of considerable length, which the judge gave when making the order. In paras 2 to 5 of his judgment the judge said:

. . . On Thursday December 9 1982 the House of Lords delivered judgment in *Pirelli General Cable Works Ltd* v *Oscar Faber & Partners* (now reported: [1983] 1 All ER 65). That judgment caused a stir, to say the least, in the present case. On Friday December 10 application was made by the second and third defendants to amend their defences in order to plead limitation in the light of the *Pirelli* judgment.

At the time of the application, both Mr Cordara and Mr Brunner [counsel for the third and second defendants respectively] made it plain that neither of them was seeking to say that the claims were statute-barred . . . under the law as it was understood to be before the *Pirelli* judgment, the law as laid down in *Sparham-Souter* v *Town and Country Developments (Essex) Ltd* [1976] QB 858 CA.

The matter was fully argued on that basis, and in the event I gave leave to these two defendants to amend their defences.

Some reference was made at the time to the date of the order giving leave to join the second and third defendants which was June 25 1982. It would seem that the matter was further investigated over that weekend, because by Monday December 13 Mr Cordara and Mr Brunner had apparently changed their minds. (I imagine that over the weekend it was discovered, or appreciated for the first time, that the actual joinder was not effected until September 9 1982.) Mr Symons [counsel for the plaintiffs] informed me on that Monday morning that Mr Cordara and Mr Brunner wished to argue that the claims were statute-barred in any event as the cracks had appeared over six years before the date of joinder.

My Lords, three matters appear to me to emerge from this part of the judge's judgment. First, the judge gave leave to reamend on December 10 1982 after being told by counsel for the second and third defendants that they were not seeking to say that the claims against them were statute-barred in any event, that is to say even on the basis of the law as it was thought to be before the *Pirelli* case. Second, on December 13 the judge was told by counsel for the plaintiffs that counsel for the second and third defendants had resiled from what they had said to the judge in this respect on Friday December 10. Third, despite counsel for the second and third defendants having resiled in this manner, the judge did not consider that their doing so gave him reason to disallow para (d) of the amendment, which he could well, if he had thought it right, have done.

The Court of Appeal took a different view about para (d) of the amendment. Lawton LJ, after explaining the various issues in the case, said [1984] 1 WLR 1274 at p 1285:

It would be convenient, in my opinion, to start the unravelling of this surprising mish-mash of legal issues by considering the procedural ones. The first in point of time, and importance, is that raised by the amendments which were allowed on December 10 1982. On that date the local authority and the architects learned that they might have a limitation defence which they had not known about before, that is to say the houses might have been "doomed from the start". This is what they then wanted to plead. Nothing more. They told the judge that they did not want to plead that the claims had been statute-barred when they were joined as defendants. They could have done so and, as I have already commented, if they had, on the evidence, they could have argued that the foundations had settled when or shortly before the cracks appeared. Paragraph (d) of the amendments which were allowed was in substance nothing more than what could have been put into the defences served but which was not. In my judgment, an amendment in terms of paragraph (d) should not have been allowed. The amendments in paragraphs (a), (b) and (c) were of a different kind. The local authority and the architects and their advisers, had no reason to think before December 10 1982 that a cause of action could accrue if a house were erected which, due to negligent design or construction, was "doomed from the start". They wanted that issue tried and, in my judgment, they were entitled to have it tried, provided that the plaintiffs were not prejudiced by a late amendment. They were not because they were given an opportunity of adducing more evidence. They decided not to take advantage of that opportunity.

Stephen Brown LJ said [1984] 1 WLR 1274 at p 1291 that he agreed with the judgment of Lawton LJ.

Parker LJ said [1984] 1 WLR 1274 at p 1292:

If, upon joinder, there is a relation back as, in common with the judgment given by Lawton LJ and for his reasons, I hold that there is, any amendment seeking to raise limitation in respect of any cause of action accruing after May 26 1974 ought to be and to have been refused, because any such amendment would be purposeless. It would inevitably fail. Since none of the plaintiffs purchased until 1975, it follows that sub-paragraphs (c) and (d) of the amendment should not, in any event, have been allowed.

The situation, therefore, is that three separate reasons were given in the Court of Appeal for disallowing para (d) of the amendments. The first reason, relied on by Lawton and Stephen Brown LJJ, was that, when the application for leave to amend was made on December 10 1982 counsel for the second and third defendants had told the judge that they did not wish to plead that the claims against them had been statute-barred in any event, that is to say even on the basis of the law as it was understood to be before the *Pirelli* case. The second reason, also relied on by Lawton and Stephen Brown LJJ, was that the plea raised by para (d) was in substance a plea which could have been raised, on the basis of the law as it was understood to be before the *Pirelli* case, when the defences were first served, but had not been so raised. The third reason, relied on by Parker LJ, was that, because of the relation back theory, para (d) of the amendment was bound to fail and was therefore purposeless.

With regard to the first reason it is true that, when counsel for the second and third defendants applied for leave to amend on Friday December 10 1982, they told the judge that they did not wish to argue that the claims against them were statute-barred even on the basis of the law as it was thought to be before the *Pirelli* case. However, as I pointed out earlier, by reference to what the judge said later on the occasion of the plaintiffs' misconceived application for an order back-dating the joinder of the second and third defendants, the judge was informed by counsel for the plaintiffs on Monday December 13 1982 that counsel for the second and third defendants had changed their minds in this respect and now did wish to argue that the claims against them were statute-barred even on the basis of the law as it was understood to be before the *Pirelli* case. I pointed out further that, despite the judge having been informed of this change of intention, the judge did not see fit to revoke his previous allowance of para (d) of the amendments, as he would certainly have been entitled to do. In the result, while it may be right to say that the judge allowed para (d) of the amendments originally on a basis which subsequently proved to be incorrect, he nevertheless adhered to his original decision, instead of revoking it, after being informed of the true position. In so far, therefore, as the first reason relied on in the Court of Appeal is that, when the judge allowed para (d) of the amendments on December 10 1982, he was misled as to its purpose, it seems to me clear that, after all misapprehension in that respect had been removed by what he was told on December 13 1982, he nevertheless stood by his previous decision. In these circumstances I do not consider that there is any real substance in the first reason relied on in the Court of Appeal.

With regard to the second reason, it amounts to no more than this, that the second and third defendants made a serious mistake in the conduct of their cases by not raising a plea of limitation in their original defences. On the authorities, to the effect of which I referred earlier, this circumstance ought not to be regarded as of itself precluding the giving of leave to amend, in the absence of any prejudice to the plaintiffs which could not be compensated for by appropriate orders as to costs. I cannot see that there was any such prejudice.

With regard to the third reason, I have already expressed my opinion that the relation back theory is not well-founded, and, if I am right about that, it follows that the third reason is not well-founded either.

The opinion which I have formed on this part of the case is that, having regard to the authorities to which the judge referred in his judgment, he had a discretion, when he was properly informed about the purpose of para (d) of the amendments on December 13 1982, to let stand the order relating to it which he had made, on a partly mistaken basis, on December 10 1982. I stress that I think that he had a discretion to do this: I do not go so far as to say that he was bound to do so.

On the footing that the judge had a discretion to allow para (d) of the amendment to stand, I do not consider, for the reasons which I have given, that the Court of Appeal were justified, on any of the three grounds relied on by them, to interfere with the exercise of that discretion.

Conclusion

The result of the view which I have expressed on the four issues which I set out earlier is that the third defendants were entitled to rely on limitation as a defence to the plaintiffs' claims and that they were entitled to succeed in that defence. It follows that I would myself allow the third defendants' appeal.

I understand, however, that a majority of your lordships have reached a conclusion different from mine on what I have called the amendment issue, being of the opinion that the Court of Appeal were right to disallow para (d) of the particulars under para 3A of the third defendants' amended defence. Having regard to the views which I have expressed on what I have called the accrual issue, the third defendants' defence of limitation can only succeed if they are allowed to rely on para (d). It follows that, if as a result of the majority view of your lordships para (d) is disallowed, the third defendants' appeal, instead of succeeding as I think that it should, must fail.

LORD TEMPLEMAN (in favour of dismissing the appeal) said:

The houses were negligently designed and sited by the fourth party architects. The houses were negligently constructed by the defendant builders. The foundations of the houses were negligently inspected by the third party local authority. The houses were completed, purchased and occupied by the end of 1975. As a result of the negligence of the architects, the builders and the local authority, damage was sustained by the houses. According to the trial judge, the damage to no 34 occurred about August 4 1976 and was discovered on August 11 1976. The damage to no 32 occurred about August 7 1976 and was discovered on August 14 1976. The damage to no 42 occurred about September 2 1976 and was discovered on September 9 1976. The damage to no 36 was discovered on August 11 1976. The damage to no 38 was discovered on September 3 1976.

Under the law as understood in *Sparham-Souter* [1976] QB 858 the claims against the architects and the local authority were time-barred six years after the date when the damage was discovered, ie not later than September 9 1982. Under the law laid down in *Pirelli* [1983] 2 AC 1 the claims against the architects and the local authority were time-barred six years after the date when the damage occurred, ie not later than September 9 1982 unless the houses were "doomed" in which case time may have expired before July 30 1982. The trial judge granted leave to plead the Limitation Act but he also ordered that the joinder of the third and fourth parties to the action was to take effect from July 30 1982.

The architects and the local authority now successfully contend that they were not joined in the action until October 6 1982. But in my opinion they cannot have the benefit of that success and at the same time ignore the conditions which the trial judge imposed upon them. The architects and the local authority cannot have it both ways. If they were not joined in the action until October 6 1982 then, when they were joined, time had expired under the law as expounded in *Sparham-Souter* or as corrected in *Pirelli*. The architects and the local authority delivered their defences in October 1982. They did not plead the Limitation Act. They did not apply to amend their defences until December 10 1982 after the trial had been completed save for closing speeches and judgment. The architects and the local authority now rely on the leave granted to them by the trial judge to rely on the Limitation Act, but they can only rely on his order to the extent to which it was intended. By back-dating, ineffectually as it turned out, the joinder of the third and fourth defendants to July 30 1982 the trial judge clearly intended to prevent any reliance on the Limitation Act save on the footing that the houses were "doomed".

I agree with my noble and learned friend, Lord Keith of Kinkel, about the date of joinder, the inappropriateness of relation back, and the "doomed from the start" arguments. But for the reasons I have indicated, and in agreement with the views expressed by my noble and learned friend, Lord Griffiths, about the late amendment in order to plead the Limitation Act, I would dismiss this appeal.

LORD GRIFFITHS (also in favour of dismissing the appeal) said:

The amendment of the defences

I am of opinion that there are no grounds that would justify interfering with the discretion exercised by the Court of Appeal to disallow para 3A(d) of the amendment of the third and fourth defendants' defences made on December 10 1982. In order to explain my reasons it is necessary to set out the history of the litigation before the amendment.

By a writ dated May 27 1980 the plaintiffs claimed damages against the first defendants, the builders, for breach of contract and/or negligence in the construction of houses which they had sold to the plaintiffs. By a third party notice dated April 28 1981 issued against the local authority, the Mid-Sussex District Council, the first defendants claimed to be indemnified against the plaintiffs' claim on the ground that the third party had negligently approved the foundations of the houses and granted building regulation consents in breach of their statutory duties. By a fourth party notice dated August 18 1981 the third party claimed an indemnity against the fourth party, the architects of the houses, alleging breach of duty to the plaintiffs.

The plaintiffs did not immediately join the third and fourth parties because they did not realise that the builders might be insolvent. By the summer of 1982, however, the plaintiffs had anxieties about the solvency of the builders and so on June 25 1982 they obtained an order that the third and fourth parties be joined as defendants to the action and that a statement of claim be served on them within 21 days. This order was unopposed by the solicitors for the third and fourth parties because as the law was understood at that time pursuant to the decision of the Court of Appeal in *Sparham-Souter* v *Town & Country Developments (Essex) Ltd* [1976] QB 858, the plaintiffs' causes of action were believed to arise when they first observed the damage to their houses. In the statement of claim it had been pleaded that the dates upon which the defects first became apparent fell between August 11 1976 and September 9 1976 and accordingly they all fell within the limitation period of six years and the third and fourth parties raised no objection to the order.

On July 26 1982 the plaintiffs' solicitors went to the Central Office to have the amended specially endorsed writ stamped. The clerk in the Central Office refused to apply the stamp because the order of June 25 1982 did not explicitly provide for the writ endorsed with the statement of claim to be amended. To overcome this difficulty the plaintiffs' solicitors wrote a letter dated July 30 1982 to the third and fourth parties' solicitors enclosing an amended specially endorsed writ together with a draft consent order. Both solicitors returned the draft consent order duly endorsed. On September 8 1982 the plaintiffs' solicitors obtained an order from Judge Newey in the following terms:

Upon reading the parties' agreed terms it is ordered that: the plaintiffs have leave to amend the statement of claim in the form annexed hereto.

On September 9 1982 the writ was reissued. On September 17 1982 the plaintiffs' solicitors sent the third and fourth parties a copy of the order dated September 8 1982. The plaintiffs' solicitors did not serve a copy of the reissued amended writ on the third and fourth parties as they should have done, but nevertheless the third and fourth parties both delivered defences.

For the reasons given by Lord Keith of Kinkel, a defendant does not become a party until not only has the writ been amended but also the amended writ has been served on him. When the third and fourth parties delivered their defences they thereby waived the necessity for service of the amended writ and accordingly became parties to the action from the date of delivery of defence, which in the case of the fourth party, the appellants, was October 6 1982.

It will be observed from the chronology that at the time the fourth party delivered their defence they both knew that they had not been joined as defendants until October 6 because they had been asked to consent to the amendment of the writ on July 30 and on September 17 they had received a copy of the order dated September 8 which in fact gave leave to amend the writ. By the time they were joined, namely on October 6 1982, over six years had elapsed since the date upon which it was pleaded that the plaintiffs first discovered the defects in their houses. But the third and fourth defendants did not plead by their defences that the claims were statute-barred. Whether this was deliberate or inadvertent, I do not know. The weight of authority at that time was in favour of the view that the joinder would relate back to the date of the original writ for the purpose of limitation and on that view of the law there would be no purpose in pleading the statute of limitations. On the other hand, if they had wished to pursue the defence that the claims were time-barred because time ran until the date of joinder then they had all the information necessary to enable them to do so. I cannot accept that they were in any way misled by the plaintiffs' failure to act as quickly as they might have done after they had obtained the order to join them as defendants on June 25.

The trial of the action started on November 23 1982 and lasted some two weeks during which the second and third defendants fought the claim on its merits. During the course of the plaintiffs' final speech an application was made on behalf of the second and third defendants to amend their defences and raise a plea that the claims were time-barred pursuant to the statute of limitations. This application was occasioned by the publication of the decision of your Lordships' House in *Pirelli General Cable Works Ltd* v *Oscar Faber & Partners* [1983] 2 AC 1 which had been delivered on December 9 1982.

I will pause here to consider what would have been the fate of this application if the *Pirelli* decision had not been published during the course of the hearing. I have never in my experience at the Bar or on the Bench heard of an application to amend to plead a limitation defence during the course of the final speeches. Such an application would, in my view, inevitably have been rejected as far too late. A defence of limitation permits a defendant to raise a procedural bar which prevents the plaintiff from pursuing the action against him. It has nothing to do with the merits of the claim which may all lie with the plaintiff; but as a matter of public policy Parliament has provided that a defendant should have the opportunity to avoid meeting a stale claim. The choice lies with the defendant and if he wishes to avail himself of the statutory defence it must be pleaded. A defendant does not invariably wish to rely on a defence of limitation and may prefer to contest the issue on the merits. If, therefore, no plea of limitation is raised in the defence the plaintiff is entitled to assume that the defendant does not wish to rely upon a time-bar but prefers the court to adjudicate on the issues raised in the dispute between the parties. If both parties on this assumption prepare their cases to contest the factual and legal issues arising in the dispute and they are litigated to the point of judgment, the issues will by this time have been fully investigated and a plea of limitation no longer serves its purpose as a procedural bar.

If a defendant decides not to plead a limitation defence and to fight the case on the merits he should not be permitted to fall back upon a plea of limitation as a second line of defence at the end of the trial when it is apparent that he is likely to lose on the merits. Equally, in my view, if a defence of limitation is not pleaded because the defendant's lawyers have overlooked the defence, the defendant should ordinarily expect to bear the consequences of that carelessness and look to his lawyers for compensation if he is so minded.

Mr Ogden submitted that the authorities obliged a judge to allow an amendment no matter how late it was made nor for what reason, provided the other party could be properly compensated by an award of costs. He relied upon the authorities set out in the *Annual Practice* and in particular the decision of Brett MR in *Clarapede & Co* v *Commercial Union Association* (1883) 32 WR 262 at p 263:

The rule of conduct of the court in such a case is that, however negligent or careless may have been the first omission, and, however late the proposed amendment, the amendment should be allowed if it can be made without injustice to the other side. There is no injustice if the other side can be compensated by costs . . .

This was not a case in which an application had been made to amend during the final speeches and the court was not considering the special nature of a limitation defence. Furthermore, whatever may have been the rule of conduct a hundred years ago, today it is not the practice invariably to allow a defence which is wholly different from that pleaded to be raised by amendment at the end of the trial even on terms that an adjournment is granted and that the defendant pays all the costs thrown away. There is a clear difference between allowing amendments to clarify the issues in dispute and those that permit a distinct defence to be raised for the first time.

Whether an amendment should be granted is a matter for the discretion of the trial judge and he should be guided in the exercise of the discretion by his assessment of where justice lies. Many and diverse factors will bear upon the exercise of this discretion. I do not think it possible to enumerate them all or wise to attempt to do so. But justice cannot always be measured in terms of money and in my view a judge is entitled to weigh in the balance the strain the litigation imposes on litigants, particularly if they are personal litigants rather than business corporations, the anxieties occasioned by facing new issues, the raising of false hopes, and the legitimate expectation that the trial will determine the issues one way or the other. Furthermore to allow an amendment before a trial begins is quite different from allowing it at the end of the trial to give an apparently unsuccessful defendant an opportunity to renew the fight on an entirely different defence.

Another factor that a judge must weigh in the balance is the pressure on the courts caused by the great increase in litigation and the consequent necessity that, in the interests of the whole community, legal business should be conducted efficiently. We can no longer afford to show the same indulgence towards the negligent conduct of litigation as was perhaps possible in a more leisured age

There will be cases in which justice will be better served by allowing the consequences of the negligence of the lawyers to fall upon their own heads rather than by allowing an amendment at a very late stage of the proceedings.

I now turn back to the present case. The *Pirelli* decision showed that the plaintiffs' causes of action arose when the damage to their houses occurred and not when they first discovered the damage. The decision also raised the possibility that a cause of action might arise at a much earlier date if a building was "doomed from the start". If a plea of limitation based upon an assumption that the cause of action arises at the date of discovery of the damage bars the claim it must follow that the plea will also bar the claim if it arises at the date of damage, because the date of damage must be earlier in time than the discovery of the damage. Therefore, if the third and fourth defendants had raised the plea of limitation based on the date of the discovery of the damage in their defences it would have been available to defeat the claim which arose on the date of the damage occurring. As I have already pointed out, whether or not the plea succeeded depended upon whether the *terminus ad quem* for the purposes of limitation was October 1982, the date of joinder, or May 27 1980, the date of the original writ. As in the circumstances of this case the damage, namely the cracks, must have occurred relatively shortly before they were discovered, the claim of limitation would succeed if October 1982 was the crucial date but fail if it was May 1980. The decision in *Pirelli*, therefore, in so far as it held that the cause of action arose at the date of the damage had no effect on any decision made by the defendants as to whether or not they should plead limitation nor on their failure to do so.

In these circumstances it is hardly surprising that when on December 10 1982 the third and fourth defendants applied for leave to amend to raise the plea of limitation, they told the judge that they were not seeking leave to amend to allege that the claims were statute-barred in any event.

The possibility of arguing that these houses were "doomed from the start" did, however, introduce a new factor into the defence of limitation. The design and construction of the houses had begun over six years before the date of the original writ so that even if the date of the original writ was the *terminus ad quem* the claims might still be statute-barred. Taking the law as it stood before the decision in *Pirelli*, this possible defence of limitation could not reasonably have been expected to be in the minds of the defendants' legal advisers at the time the defences were settled. The first three amendments contained in sub-paragraphs (a), (b) and (c) all related to a defence of limitation based upon an allegation that the houses were "doomed from the start". The final amendment (d) was, however, in wider terms and permitted the defendants on the face of its wording to argue that even if the houses were not "doomed from the start" and the cause of action arose at the date of the damage they were all statute-barred by October 1982. It is, however, quite apparent that neither the judge at the time he allowed the amendment nor the defendants when they applied for it had any intention that such an argument should be received by the court. The defendants had said that they did not intend to argue that the claims were statute-barred in any event.

December 10 was a Friday, and over the weekend I suspect that the third and fourth defendants' advisers awoke for the first time to the possibility that these claims might have been statute-barred at the time they were joined as defendants even under the law as it was previously understood. I say suspect because it may be that a conscious decision had been taken to accept that their joinder would relate back to the date of the original writ and it was, therefore, not worth pleading a time-bar. However that may be, the judge was informed on December 13 1982 that the defendants now wished to argue that on the basis of amendment (d) the claims were time-barred in any event.

In order to give the parties an opportunity to consider the position the judge adjourned the case until the following term. The plaintiffs' counsel, appreciating the danger that the claims might be statute-barred if it was held that the *terminus ad quem* was the date of joinder, took out a summons seeking a declaration that the proceedings against the third and fourth defendants had commenced or were deemed to commence on or before July 30 1982. The judge after hearing argument made an order on January 18 1983 that the joinder of the third and fourth defendants was to take effect from July 30 1982. The judge was not asked to reconsider the terms of amendment (d) or to alter it. But the effect of his order radically altered the effect of amendment (d) by providing that the *terminus ad quem* for limitation purposes moved back from October 1982 to July 30 1982. It necessarily follows that the judge would not have allowed amendment (d) on December 10 1982 if he had appreciated that its effect was to provide that the claims were statute-barred by the time that the defendants were joined in October 1982. If that had been his intention, there would have been no point in making the order he did on January 18 1983. When the judge exercised his discretion to allow the amendment on December 10 1982 he clearly did not appreciate that it permitted or was intended to permit the third and fourth defendants to argue that the claims were statute-barred in any event by October 1982.

In these circumstances the Court of Appeal having held that the judge had no power to rectify the effect of the amendment by backdating the joinder to July 30 1982 were free to substitute their own discretion for that of the judge, which had clearly been based on a misapprehension. The appellants in their written case concede that the Court of Appeal were free to exercise their own discretion but contend that in the circumstances of this case they were bound as a matter of law to exercise it in favour of allowing the amendment.

My Lords, assuming that the third and fourth defendants knew, or ought to have known, that they were not joined until October 1982, ample grounds exist to justify the refusal of the Court of Appeal to allow amendment (d). The only extra argument that the *Pirelli* decision made available for limitation purposes was that these claims were statute-barred if the *terminus ad quem* was the date of the original writ and that argument depends upon the view that the houses were "doomed from the start". This argument is available under amendments (a), (b) and (c). If the defendants had wished to plead that these claims were time-barred in any event by October 1982 they should have pleaded it in their defence. It would have been a truly technical defence as they could hardly have claimed to be taken by surprise and therefore at a disadvantage as they had been parties to the proceedings in their capacity as third and fourth parties since April and August 1981 respectively.

In considering how they should exercise their discretion, the Court of Appeal were entitled to take into account the following among other factors, that these were personal plaintiffs, that the trial was nearly ended and the merits of the case had been fully investigated, that the defendants had in fact been parties to the litigation since August 1981, and that they had no one but themselves to blame for not pleading limitation in the original defence if they had wished to do so. Here were ample grounds upon which to exercise the discretion in favour of refusing amendment (d) and I agree with the decision of the Court of Appeal.

I have had the advantage of reading the speech of my noble and learned friend, Lord Keith of Kinkel, on the remaining issues in this appeal, namely "relation back," "date of joinder," and "doomed from the start". I agree with his reasoning and conclusions on each of these issues. However, as I agree with the Court of Appeal that amendment (d) should not have been allowed by the trial judge, it follows that I would dismiss this appeal.

LORD GOFF OF CHIEVELEY (also in favour of dismissing the appeal) said:

I have had the opportunity of reading in draft the speeches of my noble and learned friends Lord Keith of Kinkel and Lord Griffiths. Like Lord Griffiths, I am in agreement with the opinion expressed by Lord Keith of Kinkel on the issues concerned with "relation back," "date of joinder," and "doomed from the start". I also agree with Lord Griffiths that, for the reasons given by him, there were ample grounds upon which the Court of Appeal could exercise their own discretion in favour of refusing amendment *(d)*. I would therefore dismiss the appeal.

The appeal by the architects was dismissed, the appellants to pay the costs of the houseowners and the local authority in the House of Lords.

RATING

House of Lords
December 4 1986
(Before Lord KEITH of KINKEL, Lord TEMPLEMAN, Lord GRIFFITHS, Lord MACKAY of CLASHFERN and Lord ACKNER)

DEBENHAMS PLC v WESTMINSTER CITY COUNCIL

Rating of unoccupied property — Exemption for listed buildings under Town and Country Planning Act 1971 — House of Lords reverses decision of Court of Appeal who had held, affirming decision of Hodgson J, that the whole of the hereditament constituting the old Hamleys toy shop in Regent Street, including part on the east side of Kingly Street, was exempt from the unoccupied property rate as a listed building — The part of the Hamleys toy shop on the east side of Kingly Street was formerly connected to the Regent Street shop by a tunnel at basement level and a footbridge at second-floor level — Hodgson J and Court of Appeal held that it was, within the meaning of section 54(9) of the 1971 Act, a "structure fixed" to the main building and was also "comprised within the curtilage" of the latter, with the consequence that the whole hereditament was exempt from the rate — Held by the majority of their lordships (Lord Ackner dissenting) that this construction of section 54 (9) produced unreasonable results and must be regarded as incorrect — "Structure" in this context is intended to convey a limitation to such structures as are ancillary to the listed building itself, eg the stable block of a mansion house or the steading of a farmhouse — The building on the east side of Kingly Street was an independent building — The hereditament comprised one building which was listed, namely the main Hamleys building in Regent Street with its rear on the west side of Kingly Street, and another, the building on the east side, which was not — It was not the intention of Parliament to grant exemption from the rate to a hereditament of which a part only was listed — Lord Ackner in a dissenting speech considered that "structure" in section 54(9) covered any building, and he would have dismissed the appeal — Appeal allowed; magistrate's decision and issue of distress warrant restored

The following cases are referred to in this report.

Attorney-General, ex rel Sutcliffe v *Calderdale Borough Council* (1982) 46 P&CR 399, CA
Corthorn Land & Timber Co Ltd v *Minister of Housing and Local Government* (1965) 17 P&CR 210; 63 LGR 490
Providence Properties Ltd v *Liverpool City Council* [1980] RA 189, DC

This was an appeal by Westminster City Council, the rating authority, from a decision of the Court of Appeal (reported at [1986] 1 EGLR 189; (1986) 278 EG 974) in favour of a claim by Debenhams plc for exemption from unoccupied property rates in respect of the premises in Regent Street and Kingly Street, London W1, formerly Hamleys toy shop.

Graham Eyre QC and Richard Hone (instructed by the City Solicitor, Westminster City Council) appeared on behalf of the appellants; Matthew Horton and Michael Humphries (instructed by Forsyte Kerman) represented the respondents, Debenhams plc.

In his speech, LORD KEITH OF KINKEL said: This appeal raises difficult questions as to the proper construction of certain of the unoccupied rates provisions of the General Rate Act 1967 (as amended) in their application to buildings listed by the Secretary of State, under section 54 of the Town and Country Planning Act 1971, as being of special architectural or historic interest. The proper construction of certain provisions of section 54 is also in issue.

The appellants are the rating authority for the City of Westminster, and the respondents were at the material time owners of a hereditament described in the valuation list which came into force on April 1 1973 as "200/202 Regent Street (and 50/52 Kingly Street)". This hereditament was formerly Hamleys toy shop. It comprised premises fronting on to Regent Street and running back to the west side of Kingly Street at the rear and also (notwithstanding the description in the list) further premises on the east side of Kingly Street and known as 27/28 Kingly Street. The back part of the former premises (50/52 Kingly Street) was formerly connected to the latter premises by a footbridge passing over Kingly Street at second-floor level and by a tunnel passing underneath it. The tunnel was filled in by operations which concluded in January 1983 and the footbridge was removed in March 1983. These works were done to enable 27/28 Kingly Street to be sold separately. No physical demarcation existed between 200/202 Regent Street and 50/52 Kingly Street.

In 1973 the Secretary of State for the Environment compiled, under section 54 of the Act of 1971, a list of buildings of special architectural or historic interest which included a number of properties in Regent Street. Under the heading "Regent Street, W1 (East Side)" there appeared *inter alia* "Nos 172 to 206 (even)."

The respondents occupied the hereditament and carried on Hamleys toy shop there until October 31 1981, when they vacated it, and it remained unoccupied when, on July 22 1982, the appellants made a complaint against the respondents for non-payment of rates amounting to £68,696.91 upon the hereditament in respect of the period from February 1 1982 to March 31 1983. At the same time they issued a summons applying for a distress warrant, which was heard by Mr Campbell, a metropolitan stipendiary magistrate, on March 29 1983. The respondents claimed exemption from unoccupied rates under Schedule 1 para 2(c) to the Act of 1967 as amended, but the magistrate rejected the claim and issued a distress warrant, holding that the exemption for listed buildings there provided for was not available when, as he found to be the case, part only of the hereditament was listed. He found that 200/202 Regent Street and 50/52 Kingly Street were listed but that 27/28 Kingly Street was not. At the request of the respondents the magistrate stated a case for the opinion of the High Court, in which he made findings of fact upon which the foregoing account is based and posed the following questions of law:

(i) Did I err in law in holding that only part of the hereditament was listed?
(ii) Did I err in law in holding that the listed part was 200/202 Regent Street and 50/52 Kingly Street?

The respondents' appeal by stated case was heard by Hodgson J, who allowed it and answered the questions of law in the affirmative. On appeal by the rating authority, the Court of Appeal (Fox, Neill and Ralph Gibson LJJ) affirmed that decision. The authority now appeals to your Lordships' House.

Section 17 of the Act of 1967 provides that a rating authority may resolve that the provisions of Schedule 1 to the Act shall apply to their area, and the appellants have done so. Para 1(1) of Schedule 1 provides that in these circumstances where any relevant hereditament in the area is unoccupied for a continuous period exceeding three months "the owner shall, subject to the provisions of this Schedule, be rated in respect of that hereditament for any relevant period of vacancy; and the provisions of this Act shall apply accordingly as if the hereditament were occupied during that relevant period of vacancy by the owner."

By para 15:

"relevant hereditament" means any hereditament consisting of, or of part of, a house, shop, office, factory, mill or other building whatsoever, together with any garden, yard, court or other land ordinarily used or intended for use for the purposes of the building or part;

Para 2, as amended by section 291 of and Schedule 23 to the Town and Country Planning Act 1971, provides:

2. No rates shall be payable under paragraph 1 of this Schedule in respect of a hereditament for, or for any part of the three months beginning with the day following the end of, any period during which —
(a) the owner is prohibited by law from occupying the hereditament or allowing it to be occupied;
(b) the hereditament is kept vacant by reason of action taken by or on behalf of the Crown or any local or public authority with a view to prohibiting the occupation of the hereditament or to acquiring it;
(c) the hereditament is the subject of a building preservation notice as defined by section 58 of the Town and Country Planning Act, 1971 or is included in a list compiled or approved under section 54 of that Act, or is notified to the rating authority by the Minister as a building of architectural or historic interest;
(d) the hereditament is the subject of a preservation order or an interim preservation notice under the Ancient Monuments Acts 1913 to 1953, or is included in a list published by the Minister of Public Building and Works under those Acts;
(e) an agreement is in force with respect to the hereditament under section 56(1)(a) of this Act; or
(f) the hereditament is held for the purpose of being available for occupation by a minister of religion as a residence from which to perform the duties of his office.

It is in subpara (c) of that paragraph which is directly in point here. If it be a correct conclusion that the whole of the hereditament is to be regarded as included in the list compiled by the Secretary of State in 1973, notwithstanding that only part of it is specifically mentioned in that list, then exemption from unoccupied rates will be available. Counsel for the respondents argued that, on a proper construction and application of the relevant provisions of the Act of 1971 as regards listing, this was indeed the position.

Section 54(1), (2) and (9) of that is in these terms:

(1) For the purposes of this Act and with a view to the guidance of local planning authorities in the performance of their functions under this Act in relation to buildings of special architectural or historic interest, the Secretary of State shall compile lists of such buildings, or approve, with or without modifications, such lists compiled by other persons or bodies of persons, and may amend any list so compiled or approved.

(2) In considering whether to include a building in a list compiled or approved under this section, the Secretary of State may take into account not only the building itself but also —
(a) any respect in which its exterior contributes to the architectural or historic interest of any group of buildings of which it forms part; and
(b) the desirability of preserving, on the ground of its architectural or historic interest, any feature of the building consisting of a man-made object or structure fixed to the building or forming part of the land and comprised within the curtilage of the building.

(9) in this Act "listed building" means a building which is for the time being included in a list compiled or approved by the Secretary of State under this section; and, for the purposes of the provisions of this Act relating to listed buildings and building preservation notices, any object or structure fixed to a building, or forming part of the land and comprised within the curtilage of a building, shall be treated as part of the building.

The argument for the respondents, which was accepted by Hodgson J and the Court of Appeal, was that the building 27/28 Kingly Street, not mentioned in the list compiled by the Secretary of State, was, within the meaning of section 54(9) a "structure" which by the footbridge and the tunnel was fixed to the building 200/202 Regent Street and 50/52 Kingly Street, which was mentioned in that list, or alternatively formed part of the land and was within the curtilage of the latter building. Accordingly it fell to be treated as part of that listed building. The argument was supported by reliance on *Attorney-General ex rel Sutcliffe* v *Calderdale Borough Council* (1982) 46 P&CR 399. That case concerned a disused mill and a terrace of cottages with a bridge linking the two, the cottages having been formerly owned by the millowners and occupied by their workers, though they had later come to be in separate ownership. The mill was listed but the cottages were not. The Court of Appeal held that, within the meaning of section 54(9), the terrace of cottages was a structure fixed to the mill and further was one which formed part of the land and was comprised within the curtilage of the mill. The cottages could not, therefore, be demolished without the consent of the Secretary of State.

In my opinion, the success or failure of the argument must turn on the meaning to be attributed to the word "structure" in section 54(9). In its ordinary significance the word certainly embraces anything built or constructed and so would cover any building. The question is whether its context here requires a narrower meaning to be attributed to it. The wider meaning could lead to some strange results. For example, if one house in an architecturally undistinguished terrace were listed as having once been the birthplace of an historically famous personage, it appears that all the houses in the terrace, being fixed to the listed building either directly or through each other, would require to be treated as part of it, as indeed might many other terraces connected to that one. Many other such examples may be figured. Notice of listing, under section 54(7), is required to be given only to the owner and occupier of the listed building itself, so the owner of some quite remote building might unwittingly undertake its demolition and become liable to penalties under section 55(1). The incongruous results which might follow from the decision in the *Calderdale* case were recognised by Stephenson LJ, giving the leading judgment in the Court of Appeal, 46 P&CR 399, 405, but he took the view that the argument from incongruity was met by the fact that the listing building code of control did not prevent demolition or alteration but merely required consent to it. It is to be observed that the words in section 54(9) "any object or structure fixed to a building, or forming part of the land and comprised within the curtilage of a building" echo similar words in section 54(2), where, however, the words are prefaced by "man-made", and the relevant object or structure must be a feature of the building. It is, I think, clear that in the context of subsection (2) the word "structure" is not intended to embrace some other complete building in its own right. This indicates that the draftsman of the relevant part of the Act has thought it appropriate to use the word in a narrow sense the first time that he introduced the quoted phrase, and it is a reasonable inference that he intended to use it in the same sense the second time. At all events, the result is to introduce an ambiguity into subsection (9) or, perhaps more accurately, to deepen the ambiguity which is there already. In resolving a statutory ambiguity, that meaning which produces an unreasonable result is to be rejected in favour of that which does not, it being presumed that Parliament did not intend to produce such a result. In my opinion, to construe the word "structure" here as embracing a complete building not subordinate to the building of which it is to be treated as forming part would, in the light of the considerations I have mentioned, indeed produce an unreasonable result. Stephenson LJ in the *Calderdale* case considered that objection to be offset by what he regarded as part of the purpose of the listing provisions, namely that of protecting the setting of an architecturally or historically important building. But if that was part of the purpose, it would have been to be expected that Parliament would not have stopped at other buildings fixed to or within the curtilage of such a building, but would have subjected to control also buildings immediately adjoining but not fixed to the listed building or on the opposite side of the street. All these considerations and the general tenor of the second sentence of subsection (9) satisfy me that the word "structure" is intended to convey a limitation to such structures as are ancillary to the listed building itself, for example the stable block of a mansion house, or the steading of a farmhouse, either fixed to the main building or within its curtilage. In my opinion the concept envisaged is that of principal and accessory. It does not follow that I would overrule the decision in the *Calderdale* case, though I would not accept the width of the reasoning of Stephenson LJ. There was, in my opinion, room for the view that the terrace of cottages was ancillary to the mill.

The question thus comes to be whether the building 27/28 Kingly Street was at the material time ancillary to the building 200/202 Regent Street and 50/52 Kingly Street. The former was not of its nature ancillary to the latter, in the sort of sense that a steading is ancillary to a farmhouse. It was historically an independent building. It is true that for a very considerable period of time both buildings were occupied and used together for the purposes of Hamleys toy shop, but throughout the rating year 1982-83 neither of them was being used for any purpose whatsoever, and indeed it must have been in contemplation that there would be no resumption of joint use, as is evidenced by the circumstance that in October 1982 steps began to be taken to sever the links between the two buildings with a view to 27/28

Kingly Street being sold off separately. These considerations tend to show that on a broad perspective 27/28 Kingly Street was not ancillary to the Regent Street building. The matter of listing or not listing cannot turn upon the business purposes or manner of use of adjoining properties of a particular occupier. Fox LJ, giving the leading judgment in the Court of Appeal in this case, said that 27/28 Kingly Street was really an annexe fixed to the rest of the hereditament. From the point of view of the occupier that may have been so, but the subordination of one building to another for the particular purposes of someone who happens for the time being to occupy both does not mean that, objectively speaking and for the purposes of the listing legislation, one of the buildings is ancillary to the other. In my opinion, 27/28 Kingly Street was an independent building and does not fall within section 54(9).

A large part of the argument for the appellants was directed to the proposition that the words in section 54(9) "for the purposes of the provisions of this Act relating to listed buildings and building preservation notices" had the effect that the enactment which followed them was not to be taken into account for the purposes of Schedule 1 to the General Rate Act 1967. In my opinion that proposition is ill-founded. The quoted words have the effect, for the purposes of the listed building provisions of the Act, of widening the definition of "building" in section 290(1) of the Act of 1971. No other effect can properly be attributed to them. It would be an absurd result, such as cannot have been intended by Parliament, if a structure subjected to listed building control by the Act of 1971 were to be treated as not so subjected for the purpose of some other Act dealing with the consequences of listing.

Having reached the conclusion that only the building 200/202 Regent Street and 50/52 Kingly Street is listed, and not the building 27/28 Kingly Street, it is necessary to consider whether or not para 2(c) of Schedule 1 to the Act of 1967 as amended applies to that situation. The construction of para 2(c) presents difficulty owing to the draftsman, as it would appear, not having kept in view the distinction between a hereditament and a building. It is buildings, not hereditaments, which may be the subject of building preservation notices (under section 58 of the Act of 1971) and which are included in lists compiled under section 54. Although a hereditament may consist in a building and no more, there are a great many hereditaments which comprise a building and also something more, even if only a small garden or yard. Some hereditaments may comprise more than one independent building, as is the position here. "Hereditament" throughout para 2 of Schedule 1 to the Act of 1967 must, in my opinion, be read as "relevant hereditament" as defined in para 15. The Schedule is, after all, dealing only with relevant hereditaments. So it is clearly in contemplation that a hereditament which attracts the exemption from rates afforded by para 2 may be not only one which is a building and no more but also one which is a building with a garden, yard, court or other land ordinarily used for the purposes of the building. It follows that the presence of such garden etc would not deprive the hereditament of the exemption, notwithstanding that it is only the building, and not the whole hereditament, which, for example, is included in a list compiled under section 54 of the Act of 1971. Likewise, the presence of some ancillary structure such as a garage or outhouse, either fixed to the main building or within its curtilage, would not affect the exemption, since by virtue of section 54(9) such ancillary structure would fall to be treated as part of the building.

The position in the present case is that the hereditament comprises two independent buildings, one of which is listed and the other of which is not. In the event that one of the buildings, but not the other, were the subject of a building preservation notice made under section 58 of the Act of 1971, it could be said, without any undue straining of language, that the hereditament as such was the subject of the notice, even though the notice applied to part only of it. If one only of two buildings on the hereditament were included in a list compiled under section 54, it could surely not be said that the hereditament as such was included in the list. On the other hand, it is not likely that Parliament would have intended to treat the two cases differently. Para 2 should be construed so as to accord the same treatment to both. In making the choice between the stricter and the more liberal constructions some assistance can, in my opinion, be derived from para 3 of Schedule 1, which provides:

The Minister may by regulations provide that rates shall not be payable under paragraph 1 of this Schedule in respect of hereditaments of such descriptions as may be prescribed by the regulations or in such circumstances as may be so prescribed and the regulations may make different provision for hereditaments of different descriptions and of different circumstances.

This provision enables the minister to enlarge the classes of hereditaments in respect of which the exemption is afforded. It does not enable him to restrict it. There would seem to be nothing to prevent the minister, if so advised, from prescribing by regulations hereditaments comprising both a listed and an unlisted building in relation to which the value of the listed building amounted to more than some specified proportion of the value of the whole hereditament. It might seem unfair that exemption should be denied where the value of the listed building accounted for a very large proportion of the value of the whole hereditament. Yet from the other point of view it might seem unreasonable that exemption should be afforded where the value of the listed building formed a very small proportion of the total value. The minister has been given power which would enable him to alleviate the former anomaly but not the latter. Accordingly, para 3 is an indication in favour of the view that Parliament intended the stricter construction of para 2(c) of Schedule 1 to the Act of 1967.

In *Providence Properties Ltd* v *Liverpool City Council* [1980] RA 189, a Divisional Court consisting of Lord Lane CJ and Boreham J had occasion to consider the scope of the para 2(c) exemption in relation to a hereditament which comprised three warehouses, one of which was listed and the other two of which were not. It was decided that the exemption was not available to a hereditament part only of which was listed. The reasoning was that if Parliament had intended to afford the exemption to such a hereditament it would have done so in express terms. There is much force in that view of the matter, and taken in conjunction with the other considerations set out above it must, in my opinion, determine the issue in favour of the appellants.

My Lords, for these reasons I would allow the appeal and restore the adjudication of the stipendiary magistrate and the distress warrant. The questions posed in the case stated do not deal exhaustively with the issues raised in the appeal. They should be answered in the affirmative, but in addition it should be found that on a true construction of para 2 of Schedule 1 to the Act of 1967 as amended exemption from rates is not available to the respondents' hereditament in respect of the rating year 1982-83.

LORD TEMPLEMAN agreed that the appeal should be allowed for the reasons given in the speech of Lord Keith of Kinkel.

LORD GRIFFITHS agreed that the appeal should be allowed for the reasons given in the speeches of Lord Keith of Kinkel and Lord Mackay of Clashfern.

In a speech concurring that the appeal should be allowed, LORD MACKAY OF CLASHFERN said: I have the advantage of reading in draft the speeches prepared by my noble and learned friends Lord Keith of Kinkel and Lord Ackner.

I agree with both that the principal argument relied upon by the appellants in the present case is ill-founded. I agree with my noble and learned friend Lord Keith of Kinkel that this appeal should be allowed for the reasons which he has given. Since I differ from my noble and learned friend Lord Ackner on this aspect of the appeal and also from the unanimous judgment of the Court of Appeal I shall add some observations.

Although the question in this appeal arises in the context of relief from rates on unoccupied property, the point that has divided us is of considerable importance in the administration of the system of listed building control now governed by the provisions of the Town and Country Planning Act 1971.

By the first part of section 54(9) for the purposes of the Act of 1971 "listed building" means a building which is, for the time being, included in a list compiled or approved by the Secretary of State under the section. The list so compiled or approved is to be a list of buildings of special architectural or historic interest. Since it is obviously necessary that the list should identify the buildings contained in it, the question whether a particular physical entity is listed or not listed depends whether on reading the list and taking account of the statutory provisions that entity is to be regarded as a building or part of a building included in the list. In the present case the contention for the respondents is that the entry in the list under the heading "Regent Street, W1 (East Side) Nos 172 to 206 (even)" meant not only that the building which has the address 200/202 Regent Street and, since it physically carries through to Kingly Street, also has the address 50/52 Kingly Street and which I shall refer to as "the Regent Street building," was included in the list but also that a building on the opposite side of Kingly Street, namely 27/28 Kingly Street which I shall refer to as "the Kingly Street building," was in the list. If this contention is correct, inevitably a considerable number of other buildings in Kingly Street, in Foubert's Place, in Carnaby Street and in Fuch's Place, which were included in the block of which the Kingly Street building formed part, were also included in the list. If the intention of the Secretary of State in compiling or approving this list was to include all these buildings in it one would have expected the entry to have clearly included them. If the effect of the action taken by the Secretary of State was that all of these buildings should be included, in my opinion the entry in the list is positively misleading.

The magistrate found that the entry in the list referred, and referred only, to the Regent Street building and did not extend to the Kingly Street building. He has asked, for the opinion of the court, whether he erred in law in so holding. The respondents urge that he was wrong, and the reason for this submission is the second part of section 54(9), which provides that:

for the purposes of the provisions of this Act relating to listed buildings and building preservation notices, any object or structure fixed to a building, or forming part of the land and comprised within the curtilage of a building, shall be treated as part of the building.

Before considering this submission further I think it is necessary to refer to the definitions of "land" and "building" provided in the interpretation section, section 290(1) of the Act of 1971:

"land" means any corporeal hereditament, including a building, and, in relation to the acquisition of land under Part VI of this Act, includes any interest in or right over land

and

"building" (except in sections 73 to 86 of this Act and Schedule 12 thereto) includes any structure or erection, and any part of a building, as so defined, but does not include plant or machinery comprised in a building.

Cases under the Income and Corporation Taxes Act 1970 demonstrate that the word "plant" is a word of very extensive import and it is obvious that plant or machinery could be fixed to a building and might include structures so fixed.

The statutory provision which is now the latter part of section 54(9) first appeared in the Town and Country Planning Act 1968. As an illustration of a question that had arisen prior to that statutory provision which might throw light on the reason for its insertion in the legislation in 1968, your lordships were referred to the decision of Russell LJ sitting as an additional judge in the Queen's Bench Division in *Corthorn Land & Timber Co Ltd* v *Minister of Housing and Local Government* (1965) 17 P&CR 210. In that case a building preservation order had been made in respect of a mansion of outstanding architectural merit. The building preservation order provided *inter alia* that the mansion should not, without the consent of the planning authority, be demolished, altered or extended and that the following items *inter alia* should not be altered or removed:

1   27 portrait panels in the King's Room, being 19th-century copies of Tudor and Stuart Kings and Queens.
2   Carved oak panels in the wall of the Oak Room dating from the 15th to mid-17th centuries.
3   A large wood carving in the Great Hall.
4   Large wooden medieval equestrian figures on the main landing.
5   A pair of painted wooden panels depicting the Hall in the ornate mantelpiece in one of the drawing rooms.

The owner applied to quash the building preservation order on the ground that the above-mentioned items were not properly included in it. It was held that any chattel which was affixed definitely to a building became part of the building; that there was no doubt that the items in dispute were all fixed and annexed in their places as part of an overall and permanent architectural scheme and were intended in every sense to be annexed to the freehold, and were accordingly part of the building; and that, in these circumstances, the restriction on their removal was properly made. Russell LJ, after saying that he did not propose to detail the effect of the evidence laid before him as to the methods of fixing employed in relation to the various items in dispute, said at 17 P&CR 210 at p 213:

It suffices to say that all the items would properly be described as fixtures as that phrase is commonly applied in law.

Russell LJ went on to quote from a number of authorities which can be summarised by saying that the ancient rule of the common law was that whatever is planted or built in the soil or freehold becomes part of the freehold or inheritance; thus a house becomes part of the land on which it stands and anything annexed or affixed to any building (not merely laid upon or brought into contact with the building) was treated as an addition to the property of the owner of the inheritance in the soil and was termed a "fixture." This rule of the common law was relaxed in favour of trade to enable tenants to affix their machinery or plant to a building or to the land and not thus make a present of it to the landlord, so that machinery or plant fixed to the inheritance for the purposes of trade may be removed by the tenant during the tenancy under certain conditions.

In the Act of 1971 the general definition of "building" excludes from its scope plant and machinery. Certain items that otherwise would be "fixtures" and form part of the building are therefore excluded. Against this background it appears to me that the word "fixed" is intended in section 54(9) to have the same connotation as in the law of fixtures and that what is achieved by the latter part of section 54(9) is that the ordinary rule of the common law is applied so that any object or structure fixed to a building should be treated as part of it. The provision is dealing with the question whether certain things, namely objects or structures, are to be treated as part of a building, not whether what is undoubtedly a building or part of a building is to be regarded as part of another building. The use of the indefinite article in describing the subject-matter of the provision tends to suggest this, in my opinion. The result would be to put beyond question the matter that was decided by Russell LJ in the case to which I have referred and I consider that in its context this is the natural interpretation of the provision. I think it is not a natural use of language to describe two adjoining houses in a terrace by saying that one is an object or structure fixed to the other. It would, I think, be a perfectly appropriate provision in a contract for the sale of a house that there was included in the sale any object or structure fixed to the house, but I think it highly unlikely that the purchaser would expect under the terms of such a contract to become the owner of the house next door, with which it shared a mutual wall.

The respondents' contention involves reading the word "structure" in its context as including a completely distinct building which is connected structurally to the first building. This reading seriously restricts the power of the Secretary of State in relation to listed buildings, since on this view he could not select one out of a terrace of houses nor could he select a part of a building to be listed. Part of a building necessarily is fixed in the sense contended for to the rest of the building. It is suggested that having regard to the purpose of the Act of 1971 no harm is done by forcing the Secretary of State if he wishes to list a building in a terrace to list the whole terrace, since in respect of the buildings in the terrace not of architectural or historic interest, permission for alteration or demolition could readily be given. However, section 54(2)(a) gives ample power to the Secretary of State, if he chooses, to list the whole terrace in respect of architectural or historic interest possessed by the whole; and the suggestion involves a compulsory notification to the owners of the whole terrace that their houses have been included in the list in view of the terms of section 54(7). I consider the respondents' construction is hard to reconcile with the provisions of section 190 of the Act of 1971 dealing with the service of a purchase notice when listed building consent has been refused, conditionally granted, or modified. In certain circumstances the owner may require the listed building to be purchased. Section 190(3) provides:

In this section and in Schedule 19 to this Act, "the land" means the building in respect of which listed building consent has been refused, or granted subject to conditions, or modified by the imposition of conditions, and in respect of which its owner serves a notice under this section, together with any land comprising the building, or contiguous or adjacent to it, and owned with it, being land as to which the owner claims that its use is substantially inseparable from that of the building and that it ought to be treated, together with the building, as a single holding.

This seems to envisage that a listed building will normally constitute a single holding.

In my opinion it is inconsistent with this provision to interpret section 54(9) as having the effect that if the Secretary of State lists one building in a terrace the consequence is that all the other buildings in the terrace which are distinct from the listed building and are owned separately from it and from one another and whose uses are completely independent from that of the listed building are to be treated as part of the listed building.

I see no practical difficulty in the operation of the Act of 1971 on the construction which appears to me to be correct. If a listed building is extended the extension will form part of the listed building without any need to rely on section 54(9). So far as buildings are concerned which exist at the date of the listing the Secretary of State will have the right to include or exclude without being constrained to include by reason of physical connection between a building he wishes to include and one he wishes to exclude.

Applying these considerations to the present case leads me to the conclusion that the magistrate decided it correctly. He concluded that what is found in the list was intended to refer only to the Regent Street building and was not intended to include the Kingly Street building. I consider that he was not bound to hold that the Kingly Street building was a structure fixed to the Regent Street building. It was a completely distinct building which at the end of the period in respect of which exemption was in question was completely separate from the Regent Street building and even when connected to it by footbridge and tunnel it was not a structure fixed to it within the meaning of section 54(9). The magistrate was not, in my opinion, in any way bound to hold that the Kingly Street building was within the curtilage of the Regent Street building as it was separated from that at ground level by a public street, although, for rating purposes, when the buildings were in common occupation they were treated as a single hereditament. The effect of the respondents' contention, in my opinion, is to say that the list compiled or approved by the Secretary of State included not only the Regent Street building but also the Kingly Street building. The list, so long as it remained unaltered, therefore included the Kingly Street building even after the footbridge was demolished and the tunnel was closed and the two buildings were completely distinct and separated by a public road. If the Secretary of State should now decide that the Kingly Street building should be deleted from the list it is not entirely clear to me what action he could take to achieve this purpose. The same consideration applies to all the other buildings in the block, of which the Kingly Street building forms part. According to the respondents' contention, as I have already said, these were all listed when 200/202 Regent Street was inserted in the list and notice should have been served on the owners of all these properties in terms of section 54(7) that they were so included. If this were truly the position the only proper course, in my opinion, would have been to include their addresses in the list along with 200/202 Regent Street. The concept of a building impliedly in the list when it consists of premises distinct from those whose address is given in the list seems to me calculated to lead only to confusion in a case where the list is a document which requires to be registered in the register of local land charges and which should, consequently, have the precision necessary to enable a person inspecting that register to appreciate all the subjects to which it relates.

In my opinion, *Attorney-General ex rel Sutcliffe* v *Calderdale Borough Council* (1982) 46 P&CR 399 is a very special case on its facts and I believe that it was possible to treat the terrace and the mill, having regard to the history of the properties, as a single unit. At the time the listing was made the whole property was in one ownership and therefore when the mill was included a notice to that effect was served on the only person who was interested as owner in the terrace. For the reasons which I have already given, I cannot regard, with respect, the reasoning by which the Court of Appeal in that case reached its conclusion as according with the true construction of section 54(9) of the Act of 1971.

In a dissenting speech, LORD ACKNER said: When Mr Graham Eyre QC opened this appeal, he submitted to your lordships, consistent with the appellants' written case, that the appeal raised only one question and that this question, although wrongly answered by the Court of Appeal, had been properly formulated by Fox LJ in the course of his judgment, with which Neill LJ and Ralph Gibson LJ concurred. The question was:

Can a building which is treated as part of the listed building by the provisions of section 54(9) be properly regarded as "included in a list compiled or approved under section 54 of the Town and Country Planning Act 1971"?

Neither in the Court of Appeal nor in his case before your Lordships' House did Mr Eyre seek to suggest that if that question, contrary to his submissions, was answered in the affirmative the appeal could succeed on any other ground or that it gave rise to any other issue. Until invited by your lordships to consider the matter, he did not suggest that the Court of Appeal were in error in their following conclusions:

(a) that 27/28 Kingly Street was a "structure fixed to" 200/202 Regent Street, the building expressly included in the list; and;

(b) that 27/28 Kingly Street was a "structure forming part of the land and comprised within the curtilage" of that listed building.

And accordingly, by reason of the provision of section 54(9) of the Act of 1971, 27/28 Kingly Street was to be treated as part of 200/202 Regent Street.

Hodgson J and the Court of Appeal, in reaching their decisions on the above two matters, derived assistance from *Attorney-General ex rel Sutcliffe* v *Calderdale Borough Council* (1982) 46 P&CR 399 in which the leading judgment was given by Stephenson LJ and with which Sir Sebag Shaw and I concurred. Mr Eyre, neither in his case nor in your Lordships' House, until invited to do so, sought to criticise that decision. He submitted that the essential issue in the *Calderdale* case was whether the terrace in question was subject to the statutory control of works of demolition under section 55 of the Act of 1971. He submitted that the Court of Appeal in that case did not have any reason to consider the provisions of Schedule 1 to the General Rate Act 1967 relating to the exemption from rates in respect of an unoccupied hereditament included in a list compiled or approved under section 54 of the Act of 1971. He maintained that there was no conflict between that case and *Providence Properties Ltd* v *Liverpool City Council* [1980] RA 189. Mr Eyre contended that the question as to what is or is not a "listed building" or what is or is not to be treated as part of such a building for the purpose of the provisions of the Act of 1971 is wholly irrelevant to the question whether a hereditament is exempt, by virtue of para 2(c) of Schedule 1 to the General Rate Act 1967, as being "included in a list compiled or approved" by the Secretary of State under [section 54 of the Act of 1971].

Mr Eyre's contention was that the words in section 54(9) "for the purposes of the provisions of this Act relating to listed buildings" confined the effect of what is provided in the subsection — namely the widening of the definition of "listed building" — to the provisions of the Act of 1971. It could not be taken into account for the purposes of Schedule 1 to the General Rate Act 1967. I entirely agree with the view expressed by my noble and learned friend Lord Keith of Kinkel, whose speech in draft I have had the privilege of reading, that this proposition is ill-founded. If a structure is by virtue of the Act of 1971 to be treated as a part of a building which has been expressly included in the list, it cannot cease to be so treated for the purposes of some other Act which itself makes special provision (exemption from rates in certain circumstances) for buildings which are included in the list.

The fact that the appellants were willing to accept the Court of Appeal's construction of the word "structure" as used in section 54(9) of the Act of 1971 and their almost total absence of enthusiasm in espousing the critical comments of that decision made by your lordships during the course of argument is not, and cannot in any sense be, decisive of the point. If the Court of Appeal in the *Calderdale* case and in the instant appeal has misinterpreted those words, then your lordships must so declare. However, the course which the appellants' argument took, particularly when they stressed at the outset of the appeal the importance of their success in the appeal, is not lightly to be dismissed. It certainly suggests that if the Court of Appeal was in error, in the *Calderdale* case and/or in this case that error is not easily discernible. For the reasons which I now set out I am unable to discern the error and accordingly I would have dismissed this appeal.

(1) *The literal interpretation of the words "structure fixed to . . . "*

It has at no time been disputed that "structure" in its ordinary everyday sense includes a building. Section 290 of the Act of 1971 is the definition section. It does not define "structure." It provides, however, that, except so far as the context otherwise requires, building "includes any structure or erection and any part of a building, as so defined, but does not include plant or machinery comprised in a building." Thus, the power given by section 54 of the Act of 1971 to the Secretary of State to compile lists of buildings of special architectural or historic interest includes the power to list a part only of a building. Thus, it is accepted that if the Secretary of State should include in the list only the facade of a building, as indeed we are told he does from time to time, then by virtue of section 54(9) the whole building, that is the structure of which the facade is but a part, falls to be treated as part of that which is expressly included in the list, ie the facade.

But in the example given above, what is the building? The facade, the "listed building", may only extend across part of the face of the original building, let alone the original building as subsequently extended. Clearly on a literal interpretation of the words of subsection (9) every part of the building, original or extended, is "a structure fixed to" the listed building, the facade, and is by virtue of the subsection to be treated as part of the facade. Thus far, I believe, there is no dispute. Yet to treat such a structure as ancillary or subordinate or as a feature of the facade would, to my mind, be quite unrealistic.

Approaching the matter from another angle, one can well envisage a listed Georgian mansion, far too large to provide a convenient private residence, whose optimum, or most profitable, use, bearing in mind its situation and the size of its grounds, is that of a high-class hotel. However, for such a development considerable extensions and alterations are necessary to provide more bedroom accommodation, conference halls and other facilities. Clearly such alterations and extensions would affect the character of the Georgian mansion and accordingly listed building consent would be required (see section 55). If the consent were to be given on terms, *inter alia*, that the extensions should be achieved by building two extensive wings on to the Georgian mansion, again quite clearly on a literal interpretation of the words of the subsection the wings, when built, would fall to be treated as part of the Georgian mansion.

However, assume that the extra accommodation which is contemplated as being necessary to make the development a success would so dwarf the listed building, if it were to consist of two wings built on to the listed building, that listed building consent is only given on terms that two large buildings were erected on either side of the listed building but each connected thereto with a bridge. Again, applying the ordinary meaning to the word "structure," each of those new buildings would by reason of the terms of section 54(9) be treated as part of the listed building.

(2) *The purposive "approach"*

The purpose of "listing" buildings is to ensure the protection and enhancement of the local heritage of buildings. This is achieved by making it an offence for a person to execute "any works for the demolition of a listed building or for its alteration or extension in any manner which would affect its character as a building of special architectural or historic interest" (section 55(1)). To confine this control to the building which is expressly included in the list, because of its special architectural or historic interest, may be often quite insufficient — the example of the Georgian facade referred to above is but one obvious example. Hence the extended definition of "listed building" contained in section 54(9). If the two additional buildings connected to the Georgian mansion in the example given above are not, when built, to be treated as part of the mansion, the purpose of the Act would be frustrated. Listed building consent would not have been necessary in the first instance and demolition of *one* of the new buildings, thereby destroying the whole harmony of the development, could take place at the whim of the owner. It is common ground that no planning permission would be required for such demolition and that the building by-law control is designed only to ensure that the physical stability of the remaining building is not affected, that control not being concerned with aesthetics.

Thus, both the ordinary meaning of the words used in section 54(9) and the very purpose of the legislation strongly supports the proposition that the word "structure" covers any building and therefore includes 27/28 Kingly Street. It is not disputed that it was "fixed to," that is joined on to, 200/202 Regent Street both by the underground tunnel and by the footbridge. Moreover, it formed part of the land and was comprised within the curtilage of 200/202 Regent Street.

(3) *Does the context require a different meaning from that normally associated with the ordinary use of the word?*

In the *Calderdale* case there was a terrace of cottages, a mill and a bridge which linked the two. The mill was expressly included in the list but the terrace was not. The issue was — was listed building consent necessary for the demolition of the terrace? Skinner J, dealing with the first limb of section 54(9), held that the terrace was not a "structure fixed to" the mill on the ground, not argued by counsel or put to them by the judge, that the terrace could not be both fixed to the mill and comprised in the curtilage of the mill and that the two alternative limbs of section 54(9) were mutually exclusive. He then proceeded to give his reasons for holding that the terrace was within the curtilage of the mill. The Court of Appeal rejected this proposition, concluding that a structure can be both fixed to a listed building and comprised within its curtilage, as indeed had been common ground before the judge. No reliance was placed in your Lordships' House on this aspect of Skinner J's decision.

Skinner J's interpretation of the first limb of section 54(9) was designed to avoid what he considered was the incongruity of deciding, in accordance with the ordinary sense of the words of the subsection, not only that the first of the cottages was fixed to the mill but so was the whole terrace.

Stephenson LJ in the *Calderdale* case expressly accepted that the literal construction could give rise to incongruity. He said at 46 P&CR 399 at p 405:

a multiple store adjacent to the birthplace of a statesman might have to be treated as part of the birthplace because it was a structure fixed to it. A block of flats replacing the stables of the mansion house might have to be treated as part of the mansion because they are within the curtilage of the mansion.

However, he concluded, rightly in my judgment, that the theoretical absurdities are fairly met by the nature of the control imposed on listed buildings and all their parts, actual and deemed. The code of listed building control does not *prevent* demolition or alteration or extension. It merely requires *consent* to such works. As Mr Eyre emphasised, there is likely to be far less difficulty in obtaining permission to demolish, alter or extend a structure fixed to a building, which has not been expressly referred to in the list, since *ex-hypothesi* it is not of itself of sufficient architectural or historic interest to merit specific mention.

In the *Calderdale* case a subsidiary argument, not taken before the judge, was raised in the notice of appeal. This argument, which was rejected by the Court of Appeal and not in terms adopted by Mr Eyre in your Lordships' House, has apparently found some favour with your lordships. As I understand it, it proceeds as follows.

Section 54(2) of the Act of 1971 provides:

In considering whether to include a building in a list compiled or approved under this section, the Secretary of State may take into account not only the building itself but also —

. . .

(b) the desirability of preserving, on the ground of its architectural or historic interest, *any feature of the building consisting of* a made-made object or structure fixed to the building or forming part of the land and comprised within the curtilage of the building [emphasis added].

The submission in the *Calderdale* case was that the words which I have emphasised were intended by the draftsman to be included in section 54(9) immediately prior to the words "any object or structure fixed to a building", thereby qualifying and very substantially limiting the ordinary meaning, *inter alia*, of the word "structure" to a mere feature or characteristic of the building which has been expressly listed. Thus the words in section 54(9) with which your lordships are concerned are said to be mere shorthand for the words in section 54(2)(b).

I cannot attribute to the draftsman of section 54 some invincible repugnance to repeat in subsection (9) some eight words which he had used in an earlier subsection of the same section. The words in section 54(2)(b) are used in a quite different context, namely in the context of the factors, other than the building, which the Secretary of State may take into account when considering whether or not to include a building in the list. Section 54(2)(b) empowers the Secretary of State to list a building which may have little or no architectural or historic interest if it includes a special feature, eg a staircase, a painted ceiling or a 17th-century folly within its grounds. That subsection is not concerned with extending the definition of "listed building." If such a narrow construction is acceptable then it really defeats the purpose of the Act of 1971.

However, I do not understand your lordships to limit "structures" to mere features or characteristics of the building expressly included in the list. Your lordships are prepared to accept that "structure" has a wider meaning and includes a separate building provided it is ancillary or subordinate or an accessory to the building expressly included in the list. Your lordships would not, therefore, overrule the decision in the *Calderdale* case, that the terrace of cottages were a "structure fixed to" the mill, but would not accept the width of the reasoning of Stephenson LJ.

The only support for this approach appears to me to be a combination of the language of section 54(2)(b) and the incongruity argument. With all proper respect, I cannot accept that this provides sufficient justification for departing so radically from the ordinary meaning of the words in section 54(9) and the very purpose for which they were used.

However, to my mind, even on that interpretation, 27/28 Kingly Street qualified to be treated as part of 200/202 Regent Street. It was attached to the main shop in order to extend its shopping facilities. Understandably, listed building consent was thought to be necessary for the removal of the connecting footbridge and this was accordingly applied for and granted prior to its demolition in March 1983. Nos 27/28 Kingly Street was used as a subordinate part of the main shop and an ancillary thereto. It was, as Fox LJ aptly described it, an annexe. It is thus in no way surprising that it, together with the main shop 200/202 Regent Street, formed a single hereditament for rating purposes.

The appeal was allowed.

For the other cases on this subject see pp 164 and 201